ANALYTICAL MECHANICS

BOSTON STUDIES IN THE PHILOSOPHY OF SCIENCE

VOLUME 191

J. L. LAGRANGE

ANALYTICAL MECHANICS

Translated from the *Mécanique analytique*, *novelle édition* of 1811

Translated and edited by

AUGUSTE BOISSONNADE
Lawrence Livermore National Laboratory, Livermore, CA, U.S.A.

and

VICTOR N. VAGLIENTE
Department of Civil Engineering, San Jose State University, San Jose, CA, U.S.A.

Springer-Science+Business Media, B.V.

A C.I.P. Catalogue record for this book is available from the Library of Congress

DOI 10.1007/978-94-015-8903-1

Printed on acid-free paper

TABLE OF CONTENTS

VOLUME II: DYNAMICS

PREFACE

to the English translation of Lagrange's Mécanique Analytique

Lagrange's *Mécanique Analytique* appeared early in 1788 almost exactly one century after the publication of Newton's *Principia Mathematica.* It marked the culmination of a line of research devoted to recasting Newton's synthetic, geometric methods in the analytic style of the Leibnizian calculus. Its sources extended well beyond the physics of central forces set forth in the *Principia.* Continental authors such as Jakob Bernoulli, Daniel Bernoulli, Leonhard Euler, Alexis Clairaut and Jean d'Alembert had developed new concepts and methods to investigate problems in constrained interaction, fluid flow, elasticity, strength of materials and the operation of machines. The *Mécanique Analytique* was a remarkable work of compilation that became a fundamental reference for subsequent research in exact science.

During the eighteenth century there was a considerable emphasis on extending the domain of analysis and algorithmic calculation, on reducing the dependence of advanced mathematics on geometrical intuition and diagrammatic aids. The analytical style that characterizes the *Mécanique Analytique* was evident in Lagrange's original derivation in 1755 of the δ-algorithm in the calculus of variations. It was expressed in his consistent attempts during the 1770s to prove theorems of mathematics and mechanics that had previously been obtained synthetically. The scope and distinctiveness of his 1788 treatise are evident if one compares it with an earlier work of similar outlook, Euler's *Mechanica sive Motus Scientia Analytice Exposita* of 1736.[1] Euler was largely concerned with deriving the differential equations in polar coordinates for an isolated particle moving freely and in a resisting medium. Both the goal of his investigation and the methods employed were defined by the established programme of research in Continental analytical dynamics. The key to Lagrange's approach by contrast was contained in a new and rapidly developing branch of mathematics, the calculus of variations. In applying this subject to mechanics he developed during the period 1755–1780 the concept of a generalized coordinate, the use of single scalar variables (action, work function), and standard equational forms (Lagrangian equations) to describe the static equilibrium and dynamical motion of an arbitrary physical system. The fundamental axiom of his treatise, a generalization of the principle of virtual work, provided a unified point of view for investigating the many and diverse problems that had been considered by his predecessors.

In what was somewhat unusual for a scientific treatise, then or now, Lagrange preceded each part with an historical overview of the development of the subject. His study was motivated not simply by considerations of priority but also by a genuine interest in the genesis of scientific ideas. In a book on the calculus published several years later he commented on his interest in past mathematics.

He suggested that although discussions of forgotten methods may seem of little value, they allow one "to follow step by step the progress of analysis, and to see how simple and general methods are born from complicated and indirect procedures."[2]

Lagrange's central technical achievement in the *Mécanique Analytique* was to derive the invariant-form of the differential equations of motion

$$\frac{\partial T}{\partial q_i} - \frac{\mathrm{d}}{\mathrm{d}t}\frac{\partial T}{\partial \dot{q}_i} = \frac{\partial V}{\partial q_i},$$

for a system with n degrees of freedom and generalized coordinates q_i $(i = 1, \ldots, n)$. The quantities T and V are scalar functions denoting what in later physics would be called the kinetic and potential energies of the system. The advantages of these equations are well known: their applicability to a wide range of physical systems; the freedom to choose whatever coordinates are suitable to describe the system; the elimination of forces of constraint; and their simplicity and elegance.

The flexibility to choose coordinates is illustrated in the simplest case by a calculation of the inertial reactions for a single mass m moving freely in the plane under the action of a force. It is convenient here to use polar quantities r and θ to analyze the motion. We have $x = r\cos\theta$ and $y = r\sin\theta$, where x and y are the Cartesian coordinates of m. The function T becomes

$$T = \frac{m}{2}(\dot{x}^2 + \dot{y}^2) = \frac{m}{2}(\dot{r}^2 + r^2\dot{\theta}^2).$$

Hence

$$\frac{\partial T}{\partial r} - \frac{\mathrm{d}}{\mathrm{d}t}\left(\frac{\partial T}{\partial \dot{r}}\right) = m(r\dot{\theta}^2 - \ddot{r}),$$

$$\frac{\partial T}{\partial \theta} - \frac{\mathrm{d}}{\mathrm{d}t}\left(\frac{\partial T}{\partial \dot{\theta}}\right) = -\frac{\mathrm{d}}{\mathrm{d}t}(mr^2\dot{\theta}).$$

By equating these expressions to $\partial V/\partial r$ and $\partial V/\partial \theta$ we obtain the equations of motion in polar coordinates. If the force is central, $V = V(r)$, this procedure leads to the standard form

$$m(r\dot{\theta}^2 - \ddot{r}) = V'(r),$$

$$mr^2\dot{\theta} = \text{constant}.$$

Lagrange derived his general equations from a fundamental relation that originated with the principle of virtual work in statics. The latter was a well-established rule to describe the operation of such simple machines as the lever, the pulley and the inclined plane. The essential idea in dynamics — due to d'Alembert — was to suppose that the actual forces and the inertial reactions form a system in equilibrium or balance; the application of the static principle leads within a variational framework to the desired general axiom. Historian Norton Wise has called attention to the pervasiveness of the image of the balance in Enlightenment scientific thought.[3] Condillac's conception of algebraic analysis emphasized the balancing of terms on each side of an equation. The high-precision balance was a central laboratory instrument in the chemical revolution of Priestley, Black and Lavoisier. A great achievement of eighteenth-century astronomy, Lagrange and Laplace's theory of planetary perturbations, consisted in establishing the stability of the various three-body systems within the solar system. The *Mécanique Analytique* may be viewed as the product of a larger scientific mentality characterized by a neo-classical sense of order and, for all its intellectual vigour, a restricted consciousness of temporality.

A comparison of Lagrange's general equations with the various laws and special relations that had appeared in earlier treatises indicates the degree of formal sophistication mechanics had reached by the end of the century. The *Mécanique Analytique* contained as well many other significant innovations. Notable here were the use of multipliers in statics and dynamics to calculate the forces of constraint; the method of variation of arbitrary constants to analyze perturbations arising in celestial dynamics (added in the second edition of 1811); an analysis of the motion of a rigid body; detailed techniques to study the small vibrations of a connected system; and the Lagrangian description of the flow of fluids.

In addition to presenting powerful new methods of mechanical investigation Lagrange also provided a discussion of the different principles of the subject. The *Mécanique Analytique* would be a major source of inspiration for such nineteenth-century researchers as William Rowan Hamilton and Carl Gustav Jacobi.[4] The seminal character of Lagrange's theory is evident in the way in which they were able to use it to derive new ideas for organizing and extending the subject. Combining results from analytical dynamics, the calculus of variations and the study of ordinary and partial differential equations Hamilton and Jacobi constructed on Lagrange's variational framework a mathematical-physical theory of great depth and generality. Within the calculus of variations itself the Hamilton-Jacobi theory would become a source for Weierstrassian field theory at the end of the century; within physics it took on new importance with the advent of quantum mechanics in the 1920s.

Beyond its historical and scientific interest the *Mécanique Analytique* is a work of considerable significance in the philosophy of science. It embodies a type of empirical investigation which emphasizes the abstract power of mathematics to link and to coordinate observational variables. The concepts of an idealized constraint, a generalized coordinate and a scalar functional allow one to describe the system without detailed hypotheses concerning its internal physical structure and working.[5] In the third part of his *Treatise on Electricity and Magnetism* James Clerk Maxwell (1892) stressed this aspect of Lagrange's theory as he used it to create a "dynamical" theory of electromagnetism.[6] Beginning with Auguste Comte and continuing with such later figures as Ernst Mach and Pierre Duhem, Lagrange's analytical mechanics has attracted the attention of leading positivist philosophers of physics.[7] In 1883 Mach praised Lagrange for having brought the subject to its "highest degree of perfection" through his introduction of "very simple, highly symmetrical and perspicuous schema."[8]

Lagrange's book remains valuable today as an exposition of subjects of ongoing utility to engineering physics and applied mathematics. Its value to the historian of mechanics, its intrinsic interest to the practising scientist and its contribution to the philosophy of physics ensure its place as an enduring classic of exact science.

CRAIG G. FRASER
Victoria College, University of Toronto,
Ontario, Canada

[1] Euler's work was published as volumes 1 and 2 of series 2 of his *Opera Omnia* (Leipzig and Berlin: Teubner, 1912)

[2] These remarks appear in the section on calculus of variations in Lagrange's *Leçons du calcul des fonctions* (1806), p. 315 of volume 10 of his *Oeuvres* (1884).

[3] M. Norton Wise and Crosbie Smith, "Work and Waste: Political Economy and Natural Philosophy in Nineteenth Century Britain", *History of Science* 27 (1989), pp. 263-301. Wise contrasts earlier scientific thought with the emerging consciousness of temporality (change, evolution, dissipation) that took place in British natural philosophy in the 1840s.

[4] Hamilton, William Rowan, "On a general method employed in Dynamics, by which the study of the motions of all free Systems of attracting or repelling points is reduced to the search and differentiation of one central solution or characteristic function", *Philosophical Transactions of the Royal Society of London* 124, (1834) 247-308; and "Second essay on a general method in Dynamics", *Philosophical Transactions of the Royal Society of London* 125 (1835), 95-144. Carl Gustav Jacobi, "Über die Reduction der Integration der partiellen Differentialgleichungen erster Ordnung zwischen irgend einer Zahl Variabeln auf die Integration eines einzigen Systemes gewohnlicher Differentialgleichungen", *Journal für die reine und angewandte Mathematik* 17 (1838), 97-162.

[5] See Mario Bunge, "Lagrangian Formulation and Mechanical Interpretation", *American Journal of Physics* 25 (1957), pp. 211-218.

[6] J. Clerk Maxwell, *A Treatise on Electricity and Magnetism*, 3rd edition (Oxford, 1892).

[7] Auguste Comte, *Cours de Philosophie Positive*, Volume 1 (1830). Ernst Mach, *The Science of Mechanics A Critical and Historical Account of its Development* (Open Court, 1960). The English translation appeared in 1893. The German first edition was published in 1883 as *Die Mechanik in Ihrer Entwicklung Historisch-Kritisch Dargestellt.* Pierre Duhem, *The Aim and Structure of Physical Theory* (New York: Athenum, 1974). Duhem's book appeared originally in French in 1906 as *La Théorie physique, son object et son structure.* The English translation is of the second 1914 edition.

[8] Ibid Mach, pp. 561-2.

TRANSLATOR'S INTRODUCTION

Let the Translation Give the Thought of the Author in the Idiom of the Reader

More than two hundred years have passed since Lagrange published the *Méchanique analitique*[1] in 1788. During this period four editions followed and were each reviewed and annotated. The second edition was prepared almost entirely by Lagrange toward the end of his life. It is a greatly expanded version of the first edition and was published in two volumes rather than only one as in the case of the first edition. The first volume of the second edition appeared in 1811 and the second volume had just reached the printer when Lagrange died. It is this edition which is translated here. This edition was completed by de Prony and Garnier and it appeared in 1815. A third edition was prepared by Bertrand in 1853. The third edition includes mathematical corrections by Bertrand, mathematical notes by various scientists of the day and finally, three memoirs by Lagrange. The *Oeuvres*[2] of Lagrange contain a fourth edition of this work which was edited by Darboux. It reproduces nearly all of the third edition. The added memoirs of Lagrange have been deleted because they appear in other volumes of the *Oeuvres*. Two mathematical notes by Darboux have been added to the text. This edition appeared in 1888 to mark the hundredth anniversary of the first edition and comprises volumes XI and XII of the *Oeuvres*. Finally, a fifth edition of this work appeared in 1965 and incorporated the text and notes of the third edition with the notes of the fourth edition.

There have been three translations of this work — one into German, a second into Portuguese and a third into Russian.[3] All three translations were of the first edition. The German translation appeared in 1797 with a second printing in 1887, the Portuguese translation in 1798 and the Russian translation in 1938 with a second edition in 1950. It is not surprising that there have been only three attempts at translation. During the two hundred years that this book has been in existence, the French language is understood worldwide and consequently, there is no need for a translation. However, this explanation overlooks a second reason for the lack of an English translation; the inherent difficulty of translating such a philosophically and mathematically sophisticated work as the *Mécanique analytique*. The broad mathematical and language skills required of a would-be translator makes a translation of this work a formidable undertaking. The work has never been translated into English. We thought it time to offer an English translation especially since the prominent place of the French language in the world has been taken by English.

LIFE OF LAGRANGE

Whenever an individual attains universal recognition by a monumental achievement, curiosity concerning the life of its creator is only natural. We wonder what Lagrange may have said or thought about more mundane topics or how he may have earned his livelihood. It is a trait of human nature. After all, in our case, the *Mécanique analytique* received an acclaim which put the reputation of its creator in an exalted circle. In its own right, it is as great a book as Newton's *Principia*. While the *Principia* created and organized the science of mechanics, Lagrange's effort was to bring a large portion of what was known about mechanics in his day under one principle — the Principle of Virtual Work. In the course of this undertaking, Lagrange contributed a great deal to the further organization of mechanics. In addition, he displayed a depth and breadth of abstract analysis in the *Mécanique* which puts him far beyond his contemporaries. In this regard, the *Mécanique analytique* displays the elegance and simplicity which is characteristic of all of his works.

Newton's scientific work, in addition, had a great impact on philosophy. The empirical nature of his science was carried over into the creation of philosophical systems. For example, the British empirical school of philosophy developed from Newton's scientific achievements. In a somewhat different fashion, Lagrange's work is the realization of a philosophical program which formed a significant part of an 18th century movement embracing all of human knowledge. This broad movement came to be known as the Enlightenment in English-speaking countries. In order to understand the claim we have made, it is necessary to describe the thought of a leading representative of this movement-Jean le Rond d'Alembert (1717–1783). D'Alembert held that a science should be deduced from clear and distinct mathematically formulated concepts of natural phenomena. He also claimed that a science is fully-developed when its principles are reduced to the least possible number and its methodology becomes more abstract and general. In the *Mécanique*, Lagrange shows that this program is attainable. He reduces statics to a single formula and he does the same for dynamics. Both sciences rest on a single principle-the Principle of Virtual Work. Lagrange further emphasizes the laws connecting phenomena and he makes no attempt to search for a final cause of the phenomena.

This last statement is important for two reasons. Prior to Lagrange a great deal of effort is spent uselessly by scientists in the search for final causes. For example, the proponents of the Principle of Least Action had at the basis of their arguments a theological interest. Lagrange would have none of this speculation. Secondly, Lagrange's achievement became a standard for 18th century scientists. In keeping with his characterization of himself as a subtle philosopher, his work became an important basis for Auguste Comte's philosophy of positivism.

F.C. Seaman
Drug Dynamics Institute, College of Pharmacy, The University of Texas at Austin, Austin, TX 78712-1074, USA

P.M. StHilaire
Carlsberg Laboratory, Department of Chemistry, Gamle Carlsberg Vej 10, 2500 Valby, Denmark

J.-S. Sun
Biophysics Laboratory, National Natural History Museum, INSERM U 201, CNRS URA 481, 43, rue Cuvier, 75231 Paris Cédex 05, France

B.M. Trost
Department of Chemistry, Stanford University, Stanford, CA 94305-5080, USA

could offer him much more than Italy. During his entire life, Lagrange is definitely a Francophile. He sets his sights on Paris in his youth but his first step is to Berlin. Once he left Turin, he never returned.

Whatever his reasons for remaining in Berlin and later, in Paris, he nevertheless referred to Piedmont as his country and he expressed on occasion the kinship he felt for the inhabitants of Piedmont and for Italy. In 1781, when he was asked to support the creation of the **Società italiana delle scienze**, he replied that he was anxious to merit that honour and to prove himself a good fellow countryman.[8] In addition, when speaking of France to French mathematicians, Lagrange always spoke of "your country". He also retained throughout his life the habit of rolling his r's when speaking French which is characteristic of Italian. It may have been his ambition which kept him in Berlin. It was during this period that he contributed some of his best memoirs and wrote the *Méchanique analitique*. These efforts required concentration and almost total preoccupation with the task he had set for himself. It could be that he was content with his memories of Piedmont.

However, it seems strange that during the twenty-one years he spent in Berlin and the twenty-six years he spent in Paris he never found time to visit his family, relatives or colleagues in Turin. Certainly, he had the financial resources to do so and we can speculate that over that long period of time, he should have been able to find an opportunity to visit Turin. To be fair to Lagrange, there were constant threats of war and frequent wars in this period. These factors made travel hazardous. This situation was especially true during his stay in Paris. But the distance from Paris to Turin is not great. For a brief period after the French Revolution, the Kingdom of Piedmont-Savoy was at war with France. Later, when Piedmont-Savoy was part of the coalition which had been defeated and it was annexed by France in 1802, Lagrange became a French citizen during the Consulate and accepted the awards offered by the French government. His loyalty to France may have upset his family and former colleagues and thereby, he may have been uneasy in their presence. But from this period on, he would have to be considered a Frenchman.

The benign attitude on the part of the Duke of Savoy towards the great-grandfather extended to Lagrange's grandfather. An office was created specifically for him, Treasurer of the Office of Public Works and Fortifications in Turin. Lagrange's father and later, a brother held this office until it was abolished in 1800 for administrative reasons.[9] Lagrange's grandfather was married to a Countess Bormiolo di Vercelli who was also descended from a prominent Italian family.

Lagrange's father — Giuseppe Francesco Lodovico Lagrangia — married a Teresa Grosso or as her last name is sometimes spelled, Gros. She was the only daughter of a wealthy physician from Cambiano, a small town near Turin. From this union, there were eleven children of whom only two survived to become adults. The eldest was Lagrange who was christened Giuseppe Lodovico Lagrangia. The spelling of the last name was an attempt to Italianize an originally French name. Lagrange himself wrote his name in various ways. At times, he used Ludovico de la Grange or Luigi de Lagrange. The use of the particle which suggested nobility appears to be common practice in his time in France, where it was sometimes adopted illegitimately. (It appears that Lagrange's great-grandfather — Louis de Lagrange — used the particle legitimately. Consequently, there is some basis for the use of the particle by Lagrange.) Moreover, in imitation of the French, it was rather common to use the particle in Piedmont-Savoy. However, during the period of the French Revolution, when Lagrange lived in Paris, the use of the particle was more of a burden than a distinction. Consequently, he later favored the spelling La Grange and finally, Lagrange, which is the spelling we use throughout this work and which is how he is usually known. In Lagrange's day, the spelling of last names was not as fixed as it is today. Modern living requires more precise identification and today, the arbitrary change in spelling of a last name would cause more than a little distress.

During Lagrange's boyhood, his father had speculated financially without success and had nearly lost the entire family fortune. In the light of this event, it was decided by the family that Lagrange should pursue a career with the promise of a reasonable income. The law profession was chosen. Consequently, at the age of fourteen years, he was enrolled in the University of Turin to begin his studies of law.

It is during this period that Lagrange discovered his mathematical talent. His instructors in mathematics and physics at the University were Filippo Antonio Revelli (1716–1801) and Giovanni Battista Beccaria (1716–1781), respectively. Both men had established reputations in their fields. It was perhaps their capable instruction which awakened in Lagrange his native mathematical ability. In fact, Beccaria was noted for his teaching ability and his ability to motivate his students. He is also given credit for introducing the physics of Newton into Italy and thereby, replacing the physics of Descartes.

At the outset, Lagrange studied geometry but his attention quickly turned to the rapidly developing field at that time of mathematical analysis. The leaders in this area were the Bernoulli's, especially Daniel, Maclaurin, d'Alembert and Euler. Lagrange read their works as is evidenced by his references to them in his early memoirs. It can be assumed that he also read the work *Istituzione analitiche ad uso della gioventù italiana*[10] of a very remarkable woman Maria Gaetana Agnesi (1718–1799). This work provided a good foundation in analytical methods and prepared Lagrange to master swiftly the works of the major mathematicians of his day. It had been translated into the other major European languages and was used as a text.

A testimony to Lagrange's recognized and promised ability as a mathematician came very early in his life. As a youth of nineteen years, he was appointed substitute professor by a Royal Decree dated September 28, 1755, at the Royal Artillery School in Turin with an annual salary of 250 crowns. At this time, the technical and mathematical standards in a European artillery school surpassed those of a university. His duties consisted of teaching mathematics and mechanics in the Italian language. It is not known for certain whether or not Lagrange was an able teacher. However, from what we know of people like Lagrange, we might venture to say that he probably found teaching burdensome and would have preferred to substitute the time required for his teaching duties for research.

Before Lagrange left Turin in 1766, he and two colleagues — the chemist, Count Saluzzo di Monesiglio (1734–1810) and the anatomist, Giovanni Cigna (1734–1790) — had founded a private scientific society which would later become the Royal Turin Academy of Sciences. One of the goals of this society was to publish in Latin and French a miscellany which is usually known by its Latin title Miscellanea Taurinensia but at times by its French title Mélanges de Turin.[11] The society was founded in 1757 and issued its first volume of memoirs in 1759.

The use of the French language in the Turin Academy is an indication of the French influence in the Kingdom of Piedmont-Savoy but for that matter French was used by the educated throughout Europe. Even in the Academy of Sciences of Berlin the language used was French.

During this period, Lagrange was demonstrating his immense talent for mathematics. The first volume of the Miscellanea Taurinensia contained three papers by Lagrange. In the first paper[12] , Lagrange extended a procedure first worked out by Maclaurin[13] for the determination of an ordinary maximum or minimum of a function. This paper is an indication of his interest in this field of mathematics. Later, in the second volume published by this society,[14] Lagrange wrote on what would become known as the Calculus of Variations, but which he called the Method of Variations.[15] He developed a purely analytical procedure to find the function which would extremize an integral, that is, the function which would render the value of the integral a maximum or minimum. In this paper, Lagrange's famous δ operator is defined for the first time and a complete description of the methodology is provided.

Lagrange did not invent the Calculus of Variations since this branch of mathematics has a long history. However, what leads to his great contribution is that he developed a totally analytical and thus tractable methodology. It avoided the partly geometrical, partly analytical approach of Euler and John Bernoulli. Furthermore, it clearly distinguished between a differential and a variation which is not done in Euler's work.

This development will be very important since it is Lagrange's somewhat later recognition that the variation of an integral is analogous to making a virtual displacement in mechanics. These two concepts taken together will form a major part of the foundation of the *Mécanique analytique*.

Lagrange was very much aware of his achievement. Later, in a letter dated November 20, 1769[16] written to d'Alembert he said that "I have always regarded that method as the best I have ever done in geometry." Considering the contribution that the Calculus of Variations has made to physics and engineering, it is clear that Lagrange's claim was justified.

Lagrange grew restless in Turin. It is important for a developing mathematician or for that matter, anyone who has selected learning as their vocation to be in an environment where learning is respected and the calibre of one's associates are of a high order. Turin could not provide that type of environment. Naturally, a man like Lagrange would look towards obtaining a position in Paris or Berlin where this type of environment existed.

Lagrange's desire to leave Turin was especially annoying to the Duke of Savoy, Charles Emmanuel III. The teaching position at the school of artillery had been given to Lagrange in order to keep him in Piedmont-Savoy. Although a university position would have been more suitable to Lagrange, there was no instruction in the infinitesimal calculus at the University of Turin at this time. Lagrange was paid only a modest salary at the school of artillery but Charles Emmanuel III was known to be very frugal when it came to granting government controlled salaries. Many important functionaries in his government were paid at the same level as Lagrange. But what is more, Lagrange's desire to leave was interpreted by Charles Emmanuel as an ungrateful act on his part in view of what had been done for him.

Furthermore, it appears that Lagrange was ill-suited to teaching the practical mathematics required of an instructor of artillery officers. An evaluation of Lagrange's teaching ability by Alessandro d'Antoni (1714–1786), the Director of the school, reads

> The sublime talent of the substitute professor La Grangia, which placed him rightfully among the outstanding Academicians in Europe, did not make it possible for him to reduce the fundamentals necessary to us to their elements. Thus the individuals, who in the earlier course criticized his treatises as too advanced, metaphysical, diffused throughout with extraneous material and lacking applications to the professions of artillery and fortification engineering, had judged correctly. However, if someone made this identical pronouncement on the contents of the most recent course, he would display his ignorance of the subject matter.[17]

It is clear from the lecture notes left by Lagrange that d'Antoni was accurate in his evaluation. Lagrange was not contributing to the mission of the school. His lecture notes reveal his desire to advance the calculus without regard to the needs of his students and to instruct in topics which were of little use to these same students. It was probably the fact that he was appointed by Royal Decree and that he had the attention of Charles Emmanuel that he was able to retain his position. From the last part of d'Antoni's evaluation, it is clear that with Lagrange's departure the status of the classes which had been taught by Lagrange were now more in keeping with the needs of the student body. It is probably also true that Lagrange resented the restrictions placed on his desire to win recognition through his mathematical ability. His dissatisfaction with his position is readily understood in these circumstances.

Euler and d'Alembert were both impressed by Lagrange's ability. At this time, Euler is in Berlin where he is Director of the Mathematical Section of the Berlin Academy. Maupertuis (1698–1759) had died in 1759 leaving vacant the office of President of the Berlin Academy and had not been replaced. Euler was discharging the duties of President but his standing and relationship with the reigning monarch-Frederick the Great, King of Prussia-were not good. Consequently, he was not offered the position of permanent President. D'Alembert was offered the position by Frederick but since he was quite happy in Paris, he declined. Euler grew increasingly dissatisfied with the situation in Berlin and resolved to leave. Both men recommended Lagrange as the Director of the Mathematical Section to Frederick. Since d'Alembert enjoyed the respect and confidence of Frederick, his recommendation was especially important. Frederick then accepted the yet unknown Lagrange as Director of the Mathematical Section. Lagrange was not entirely unknown to the members of the Berlin Academy since Euler had arranged for Lagrange to become an associate foreign member on September 2, 1756.

Thus early in his life, Lagrange was on intimate terms with two of the outstanding figures of European science. Euler and d'Alembert, though older and established as mathematicians, corresponded with him as an equal. Although he never personally met Euler, he had befriended both d'Alembert and Euler.[18] For their part, they had recognized Lagrange's ability and had sought to help his career. For his part, he appreciated their efforts in his behalf and derived inspiration from both of them.

Lagrange arrived in Berlin on October 27, 1766, after a lengthy negotiation with the Duke of Savoy who very reluctantly permitted him to leave. He was formally installed as Director of the Mathematical Section on November 6th replacing Euler who had left earlier for St. Petersburg. The next 21 years of his life were passed in Berlin. They represent for Lagrange a very fruitful and rewarding period of his life. But it appears that he always saw himself a foreigner in Berlin. There is no evidence that he ever travelled in Prussia or even in the city of Berlin. Perhaps the cultural differences between Prussia and his own country were much too great to overcome.

Within a year of his arrival in Berlin, Lagrange married for the first time. According to Lagrange's own testimony, the marriage was one of convenience. He married a cousin — Victoria Conti (1747–1783) — in 1767, whom he had known while a student in Turin. The actual date of the marriage is unknown. There were no children from this marriage since it appears that Lagrange wanted none as he did not want any distraction from mathematics. His first wife died in 1783 after a lengthy illness. Lagrange tenderly saw to her needs during this period and with her death he felt her loss deeply. The marriage that began as one of convenience had become something quite different. Nearly ten years will pass before Lagrange marries for a second time.

It was stated earlier that Lagrange's stay in Berlin was a very productive period. During this time, he wrote on all branches of mathematics and intensively studied mechanics. The majority of his memoirs during this period deal with celestial mechanics. It is also during this period that the foundations for the *Méchanique analitique* were laid. More will be said about this claim when we discuss the origin of this work.

Lagrange's life was intimately bound to the academies of which he was a member. In his day, these academies served as research institutions unlike the more recent time where research is carried on by professors in universities and by commercial laboratories or engineering companies. The primary responsibility of the various academies was to promote learning.

Probably the single greatest influence on Lagrange's life was his association with the Academies of Turin, Berlin and Paris. Although scientists at these academies did not form a well-defined community in the 18th century, they did provide a forum for individuals who shared common values of reason, observation and experimentation to meet and present their research.[19]

The Turin academy was modelled after the Berlin academy and the Berlin academy was modelled on the Parisian Académie des Sciences. This is understandable since French science in the 18th century outshone that of any other country. France had professionalized science under the **Ancien Régime** and this professionalization carried through the French Revolution into the early decades of the 19th century. The success of French science was due to a large degree to government support. For example, the French government supported the Académie des Sciences, the Collège Royal, the Observatoire de Paris, the École militaire du génie at Mézières and the Jardin du Roi. The members of these academies and institutions were paid a modest pension which generally had to be supplemented by additional income from other activities. The prestige associated with the membership in one of these organizations made it somewhat easier to obtain a teaching position which was the usual activity of a member or act as a consultant to a government agency or commercial organization. The furtherance of a rational understanding of nature did not always coincide with the goals and needs of society which the French government sought to nurture. Therefore, many academicians had to perform duties which were mundane in order to have some time for scientific research. It should be understood that membership in an organization such as the Académie des Sciences was not awarded for past accomplishment as it would be today. At that time, a member was usually selected at a much younger age and mainly for the promise of future accomplishment.

Members of the Académie des Sciences attended bi-weekly meetings whose duration was on the order of two hours. At these meetings, members discussed technical questions on various topics and with regular attendance the academicians became well-informed on a large number of topics, including those outside of their particular area.

Membership in the various Académies was very limited. For example, the Académie des Sciences in Paris was composed of less than fifty members. Each member negotiated their own pension directly with the government. And as we noted earlier, membership in the Académie seems to have brought with it more prestige than income.

The responsibility of the Académie des Sciences to promote learning was partly addressed in the 18th century by sponsoring competitions for the best memoir on a topic which was a current and significant problem. For example, in this period, improved astronomical methods showed that celestial motions were not accurately and completely described by the mathematical equations which were in use for this purpose. There seems to have been two reasons for this interest: an intellectual interest which implied that Newton's Theory of Universal Gravitation was only approximate and a commercial interest which sought to find a ship's longitude at sea from the position of the planets such as Jupiter and Saturn relative to the Earth or by the location of the Earth's moon. In addition, the exact shape of the Earth had been brought into question. Geophysical measurements in the early 18th century had indicated that Newton's claim and demonstration that the Earth is flattened at the poles may not be correct.

Thus three of the major scientific problems of the 18th century were in the areas of astronomy and geophysics. This state accounts for the fact that most of Lagrange's memoirs during this period were in these areas. The problems were defined as

1. To describe mathematically the precise motion of the Moon.
2. To account for the apparently secular (non-periodic) inequality in the motions of Jupiter and Saturn.
3. To determine the precise shape of the Earth.

The first problem required that the motion of the Moon around the Earth be described accurately. Apart from its primary "monthly" orbit about the Earth, there are perturbations to this primary motion which are caused by the attraction of other celestial bodies. These secondary motions make a complete description of its absolute motion extraordinarily complex. A great deal of effort was put into the solution of this problem since it would aid navigation by contributing to the determination of longitude at sea.

The second problem was primarily theoretical, although it was believed for a time that its solution might have commercial applications. During the 18th century, astronomical observations indicated that the solar system might be unstable. The calculation of the mean velocity of Jupiter and Saturn had indicated that for the former it was increasing and for the latter it was decreasing. It was believed that this phenomenon was due to the mutual attraction of the two planets. Hence, the problem consisted of accounting for this observation. This problem, like the problem of the motion of the Moon, reduced to the consideration of three bodies in mutual attraction. Since the Sun's attractive force is primary its effect had to be included. By solving this problem the Académie hoped to show that Newton's Theory of Universal Gravitation was accurate and consequently, the stability of the solar system would be restored.

The third problem dealt with the shape of the Earth. Newton demonstrated from his theory that the Earth is an oblate spheroid. However, astronomical observations in the 18th century implied that the Earth was a prolate spheroid or a sphere flattened at the equator rather than at the poles as Newton's theory implied. Most scientists believed Newton to be correct since other observational data was available which corroborated his theory. One example was the change of the period of a pendulum at different degrees of latitude. This fact correlated well with Newton's theory.

The competitions held by the Académie des Sciences which dealt with these questions were all in the area of celestial mechanics. Besides the prestige associated with winning the competition there was also a monetary award.

Lagrange won the competitions held by the Académie des Sciences five times.[20] His prize-winning memoirs include:

1. Recherches sur la libration de la Lune, 1764. *Oeuvres de Lagrange* VI, pp. 5-61.
2. Recherches sur la inegalités de satellites de Jupiter causées par leur attraction mutuelle, 1766. *Oeuvres de Lagrange* VI, pp. 65-225.
3. Essai d'une nouvelle méthode pour résoudre le problème des trois corps, 1772. *Oeuvres de Lagrange* VI, pp. 229-324. (There is a note from the editor on pp. 324-331.)
4. Sur l'equation séculaire de la lune, 1774. *Oeuvres de Lagrange* VI, pp. 335-399.
5. Recherches sur la théorie des perturbations que les comètes peuvent éprouver par l'action des planètes, 1780. *Oeuvres de Lagrange* VI, pp. 403-503.

There is one interesting note which we might add to our discussion. The second paper which was submitted for the prize of 1766 dealt with the problem which we discussed earlier; namely, the apparent change in average velocity of the planets Jupiter and Saturn. Lagrange thought that the deviations were secular and due to the mutual gravitational attraction between the various planets. Later, in 1787 Laplace showed that the inequalities were in fact periodic with very long periods. This problem was solved by Laplace using a statistical approach.

Although Lagrange was anxious to retain his position in Berlin, with the death of Frederick the Great on August 17, 1786, he became very dissatisfied with his situation. The new king, Frederick Wilhelm II, was from a very different mold than his illustrious uncle. His administration was reactionary and provincial. The cosmopolitan atmosphere which his uncle had tried to foster quickly disappeared. Also, Lagrange foresaw the Prussian government's declining interest and support of the Berlin Academy and he resolved to leave.

Word reached representatives of the French government that Lagrange was not satisfied with his situation in Berlin and that he could probably be persuaded to accept a position in Paris. French science at this time enjoyed the respect of scientists all over the world and this fact made it difficult for Lagrange to reject any reasonable offer.

After a lengthy negotiation, Lagrange agreed to the offer of the French government. He left Berlin on May 18, 1787 and journeyed to Paris where on July 29, 1787, he became pensionaire vétéran of the Académie des Sciences. This title was created especially for him and it replaced the title of associé étranger which he had held since May 22, 1772.

It can be assumed that Lagrange expected to continue his life in Paris as he had in Berlin. But Paris at this time is vastly different from Berlin. Political unrest had grown steadily in France since the early 1700's. It was heightened in the larger cities such as Paris. By the late 1780's the country was ready to explode. If Lagrange had foreseen this upheaval, he probably would not have accepted the French offer. His attitude seems to have been a-political and his personal tranquility of foremost importance. Thus it is very likely in this case that he would have accepted one of the Italian offers and returned to Italy.

Upon his arrival in Paris, he was lodged in the Louvre at the invitation of the Royal family. He resided in the Hotel de la Briffe, Quai des Théatins near the rooms set aside in the Louvre for the meetings of the Académie des Sciences. He attended his first meeting on June 13, 1787. He remained in the Louvre only a short time. At the beginning of the year 1788 he left to live in a private residence.[21]

Paris presented a tremendous contrast to Berlin, while the latter possessed a tranquil and somber city atmosphere, the former was turbulent and on the eve of the Revolution seethed with excitement. Lavoisier's discoveries in chemistry were the topics of discussion in the city's fashionable Salons. For the quiet and pensive Lagrange, Paris was not what he needed. Nevertheless, he had brought along the *Méchanique analitique*. An excerpt from this work was presented to the Académie des Sciences by Lagrange on April 5, 1788. A committee was selected from the academy to review the work and to decide whether it deserved the approbation of the Académie. Its response was positive and the first edition was published in 1788.

The very next year was to have a tremendous effect on Lagrange's life in Paris. In order to understand this effect, we will outline a portion of the political and economic effects brought about during the period of time known as the French Revolution. Our main interest is to show how it affected Lagrange.

The royal government of Louis XVI had remained solvent during the 1780's by borrowing large sums of money within France and from foreign banks. By the end of the decade, the ability of the King to obtain revenue or credit to finance his government had been exhausted. To solve his government's fiscal insolvency, Louis XVI called the Estates General into session. With this action the King unwittingly initiated a series of events which would lead to the destruction of the Bourbon monarchy.

The Estates-General met on May 5, 1789. The three estates composing this body were

First Estate	Clergy
Second Estate	Nobility
Third Estate	Commoners

At the outset, the nobility claimed an ancient right which exempted them from taxation. After the preliminary discussion, it appeared that the Third Estate would bear the burden of new taxation. Their representatives tried to obtain some compensating privilege and they claimed political rights. In addition, arguments ensued over strictly procedural questions in the conduct and voting rights of the three estates. The result of these discussions was an impasse. Finally, the Third Estate refused to recognize the traditional distinctions between the orders and claimed that it had the sole power to represent the French people. The Third Estate declared itself the National or Constituent Assembly on June 17, 1789. From then on, a sequence of events occurred which culminated in the storming of an ancient royal prison-the Bastille-on July 14, 1789. This act marks the beginning of what is known as the French Revolution.

Louis XVI was unable to provide the requisite leadership to bring about the need for change. Consequently, the Constituent Assembly undertook to reorganize the country. The unusual part of this effort was that scientists were to be used for these national political purposes.

As a member of the Académie des Sciences, Lagrange was expected to participate and contribute expertise to the solution of national problems. On May 8, 1790, the Constituent Assembly ordered the standardization of weights and measures throughout France. The Académie des Sciences was asked to found the system on a fixed basis with the intent that it would be nationally adopted. On October 27, 1790, the Académie appointed the five members of the commission who were Lagrange, Borda, Condorcet, Lavoisier and Tillet. This commission was retained in its capacity even after the academies were closed. The metric system was eventually sanctioned on Dec. 10, 1799.

The Royal family appeared to acquiesce to these goings on but secretly opposed them. The Constituent Assembly was moving towards a parliamentary system which limited the prerogatives of the monarchy. Consequently, on June 21, 1791, the Royal family fled Paris only to be captured and returned to the capital before realizing their object which was to organize resistance to the democratic trends underway.

In the midst of this turmoil, Lagrange married for a second time. The marriage took place on May 31, 1792, nearly ten years after the death of his first wife. His second wife was Reneé-Françoise Adélaide Le Monnier (1767–1833).[22] She was the daughter of a fellow academician, the astronomer Pierre-Charles Le Monnier (1715–1799). Lagrange was fifty-six and she was twenty-five years old. She was young enough to bear children and probably capable of doing so, but again, Lagrange had no desire for children. He had said to Delambre (1749–1822), "I had no children from my first marriage. I do not know whether I shall have any from my second, but I do not wish for any."[23] Lagrange's second marriage was like the first, a very happy one. His wife provided the companionship and the domestic tranquility that he seemed to need.

Shortly after Lagrange married, economic and political chaos in France increased rapidly. Inflation quickly made Lagrange's income insufficient to meet his needs. Furthermore, the convocation of the National Convention on September 20, 1792 made the political situation uncertain for Lagrange.

The Constituent Assembly had confirmed and continued to pay the pension that was due to Lagrange because of his contract. However, the National Convention viewed the academies as elitist and therefore, relics of the **Ancien Régime**. On August 8, 1793, the academies were closed. All pension salaries were reduced by one-half but in the case of Lagrange an exception was made. Lagrange kept his entire pension because his agreement was considered a foreign contract.

By the year 1793, he was in danger of expulsion or arrest as an enemy alien. During September of 1793, a decree of the National Convention ordered that all foreigners born in countries at war with France should be arrested. The chemist Lavoisier intervened on behalf of Lagrange and obtained an exemption for him.

It has often been asked why Lagrange chose to remain in Paris at the outbreak of the French Revolution. It appears that a threat to his life existed-albeit a small one. It also appears to us that a natural act would have been to flee Paris. However, the effect of flight in the eyes of the French authorities would have been difficult to judge. An attempt to flee may have led to his arrest. At this time, he still carried a passport issued by the Kingdom of Piedmont-Savoy. Although he had resided in France for many years, he had not applied for French citizenship. In addition, the countries to which he may have been able to flee were limited. An attempt to return to Piedmont-Savoy may not have been all that welcome. Although Lagrange's reputation as a mathematician would have helped, there were instances of the arrest of returning emigrés to Piedmont-Savoy since it appears that the government of that country believed them to harbor Jacobin ideals. He probably could have returned to Prussia, but this abode would not have suited Lagrange at all. The Prussian government was reactionary and its support of scholarly activities had all but disappeared. Generally, the monarchies of Europe were trying to contain the Jacobin ideals of the French Revolution and anyone coming from such an environment was suspect. Thus in reality, Lagrange had no place to go and had to remain in Paris.

In place of the academies, the National Convention created the **Institut de France** or **Institut National**. The body consisted of three classes

1. Sciences Physiques et Mathématiques
2. Sciences Morales et Politiques
3. Littérature et Beaux-Arts

It was founded on August 22, 1795. The Académie des Sciences became the first class of the **Institut**. This class contained 66 of the 144 members of the **Institut**. However, the **Institut** was not basically a scientific institution but a cultural institution as its full title suggests, a "National Institute of Arts and Sciences". The legislators hoped that this institution would provide an enduring cultural and scientific foundation for the nation.

Lagrange was a member of the First Class of the **Institut**. He attended the first meeting of the **Institut** in the Salle des Cariatides in the Louvre on April 4, 1796.

The revolutionary government continued its efforts to stimulate the sciences. The Bureau des Longitudes was created on June 25, 1795 to administer the nation's observatories and to plan a program of national astronomical research. Lagrange and Laplace were appointed to the Bureau. Much later, a new responsibility was added to the Bureau's charter. Napoleon decreed on September 24, 1803 that the metric system's standards would be preserved by the Bureau des Longitudes.

One of the goals of the French Revolutionaries was to eliminate the elitism which resulted from the status of one's birth and which pervaded most levels of French society during the **Ancien Régime**. This elitism existed in the schools of the country which were for the most part controlled by the Roman Catholic Church. Consequently, it was in this period that the church supported primary and secondary schools lost the pre-eminence they enjoyed before the revolution. The church was compromised in the eyes of the French revolutionaries by its association with the monarchy in France. Thus the French government undertook to secularize the nation's school system.

To this end, the National Convention in 1794 ordered the creation of the Ecole Normale. Its purpose was to train teachers for the nation's school system. The most learned individuals in the nation were appointed as instructors. Lagrange, and Laplace as his assistant, were to lecture on elementary mathematics. The Ecole Normale began instruction on January 20, 1795 but closed after slightly more than three months of operation. As part of his duties as a professor at the Ecole Normale, Lagrange wrote a series of lectures on elementary mathematics. The lectures dealt with algebra in general and contained a discussion of the theory of equations and the solution of algebraic equations. The elegance and lucidity of these lectures are a model to be emulated and show that they originated in the mind of a first-rate algebraist. These lectures were later collected and published in volume seven of the *Oeuvres de Lagrange* with the title *Leçons elementaires sur les mathématiques*.[24] Lagrange also began teaching mathematics along with de Prony at the Ecole Centrale des Travaux Publics which later became the École Polytechnique. This school was founded on March 11, 1794 and began instruction on December 21, 1794. It is still in existence today.

During the Consulate and the Empire, Lagrange received more honours. On May 18, 1802, the government created the Legion of Honour to reward military valor and service, and also to recognize outstanding contributions to the nation by intellectual attainment and civil service. On October 2, 1803, Lagrange along with Berthollet, Laplace and Monge were made members of the Legion of Honour. Later, Lagrange was further elevated when he was made a Grand Officer of the Legion of Honour on July 14, 1804.

As a member of the Legion of Honour, Lagrange received an annual pension of 5000 francs. In order to evaluate the magnitude of this pension it can be compared with the annual salary of a university professor at this time which amounted to about 1500 francs. Lagrange in a very generous act gave all of this pension to his younger brother Michele in Turin. The latter had a large family and probably found this money to be very useful.

Lagrange became a member of the Sénat-Conservateur on December 24, 1799. This election was not merely to a ceremonial body. The Sénat-Conservateur was instituted during the Consulate by the Constitution of December 24, 1799 and it was charged with insuring that the laws of the nation were constitutional and with amending the Constitution by Senate deliberations and a majority vote. This body also had the right to nominate individuals to various high government offices and also, the right to nominate judges to the high courts of appeal.

Unlike Laplace who became secretary and later president of the Sénat-Conservateur, Lagrange held no office in the body. However, he did participate in various commissions. For example, he was chairman of the commission charged with restoring the Gregorian calendar to France. The Revolutionary calendar, adopted by the National Convention in 1793, had made it more difficult for France in its intercourse with other European nations who were using the Gregorian calendar. The Gregorian calendar was put into effect again in France during the Napoleonic era when on January 1, 1806, it was restored by Napoleon. As a member of the Sénat-Conservateur, Lagrange's social status rose significantly. In addition, membership in this body brought with it an annual pension of 25,000 francs. The significance of this pension is clearly evident if it is compared, as we did earlier, to a university professor's annual salary of 1500 francs.

The title page of the second edition of the *Mécanique analytique* lists all of Lagrange's titles and honours bestowed by the government.

> *Mécanique analytique* par J. L. Lagrange de l'Institut des Sciences, Lettres et Arts, du Bureau des Longitudes: Membre du Sénat-Conservateur, Grand-Officier de la Legion d'Honneur et Comte de l'Empire.

Lagrange insisted on the inclusion of these titles on the title page. He was proud of what he had achieved and proud of the honour he had brought to his profession of mathematician.

Lagrange died in his home on April 10, 1813[25] in Paris. His death was marked by two religious ceremonies even though it appears that Lagrange personally was not very religious. Two religious ceremonies may have appeared appropriate for such a distinguished member of the French government because of the government's relations with the Roman Catholic Church. Napoleon had promulgated a Concordat with the Church on July 15, 1801. It would not have looked well if an individual to whom the government had given so many honours was buried in a simple civil ceremony.

The first ceremony was held in his Parish of St. Philippe du Roule and the second a few days later at the Panthéon[26] (Parish of Sainte-Geneviève) in Paris where Lagrange was laid to rest. Lacépède (1756–1825) delivered a eulogy in his role as the principal officer of the Legion of Honour and Laplace delivered a second eulogy as the representative of the Sénat-Conservateur.[27]

Thus the life of a man who had lived only for mathematics and who had been recognized as a leading mathematician in Europe for nearly sixty years had ended. He was close to a pure mathematician and only in so far as mathematics could be applied to mechanics, could he be called a physicist. His interest in subjects outside of mathematics was never deep enough to induce him to publish on those subjects and consequently, as we have pointed out, we know so very little about the other dimensions which make a man.

Lagrange was a man of average height, relatively thin and with a delicate constitution. He seems to have habitually worried about his health and this was perhaps due also to his view that the sedentary life of a man of letters was an unhealthy one. Based on the testimony of his colleagues in the academies of which he was a member, he was generally well-liked, very reluctant to take offense and careful not to give offense. His modesty and a calm nature, appears to have won the friendship of nearly all who knew him.

Lagrange generally spoke and wrote in French to the exclusion of other languages. His speech showed a slight tendency to roll the letter 'r'. The letters written to Euler at the beginning of his career were written in Latin but nearly all his correspondence afterward, including his letters to his family in Turin, were written in French.

Lagrange appears to have been a very kind man. However, he lived through the French Revolution and none of its excesses, with one exception, moved him to action or to comment. The exception was the execution of the chemist, Lavoisier, on May 8, 1794. After the execution, Lagrange remarked to Delambre, "It has taken them only a moment to cause that head to fall, and a hundred years may not suffice to produce a like one."[28]

The question of what were Lagrange's religious beliefs is difficult to answer. Authors who have investigated this aspect of Lagrange's personality have not been able to reach a conclusion. We believe that there are a number of aspects of this problem which haven't been given significant attention. The first is the time in which Lagrange lived. During this period, generally referred to as the Enlightenment, organized religion, namely, the Christian religion, was the object of severe criticism. Many of Lagrange's colleagues were not members of the Christian church and therefore, Lagrange, as prudent as he was in the conduct of his personal affairs, kept his religious beliefs to himself. He may have viewed religious belief as he did metaphysical arguments, that is, with the attitude that they would not produce any tangible results. Secondly, his private library contained a number of books on religion — not all on the Christian religion but bibles and works on various world religions. These books indicate an interest in religion on the part of Lagrange.

On the other hand, in his letters to his family, where it is not necessary to be prudent, there are no references to religion. Even when a death in the near family seemed to require words of consolation such as the death of his mother, brother and father, there are none. He generally sent gifts to his family in Turin — mainly, his brother's family — on religious holidays such as Christmas but again there are no references to religion or any expression of religious feelings.

In addition, during the period when he lived in Turin, Lagrange was accused of being an atheist. But there was never any indication from Lagrange that he was an atheist. In fact, he attended Mass regularly and during his sojourn in Paris during the Empire, there is a record of his many chance meetings with Napoleon at Mass.

Lagrange's name is found in the *Dictionnaire des athées anciens et modernes*[29] by Sylvain Maréchal along with the names of Napoleon, Frederick the Great, Laplace, Monge, de Prony, Lalande, Peyrard, Fourcroy, etc. Maréchal noted that these men held that it was impossible to prove the existence of God. He referred to them as atheists but it would have been more accurate for Maréchal to point out that a declaration of agnosticism rather than atheism was really in question.

Lagrange carried on an extensive correspondence with d'Alembert and Condorcet. In this correspondence, he often showed his hostility towards the Jesuit Order of the Catholic Church. However, such a demonstration was not unusual for the time. In fact, d'Alembert had written a book[30] advocating the suppression of the Jesuit Order in France. Lagrange had praised the work of d'Alembert and had said of a well-known scientist in the Jesuit Order, Ruggiero Boscovich (1711–1787), that he was a "... moine et jesuite à bruler."[31] Even Euler, who issued from a staunchly Lutheran family, was judged harshly in the correspondence between Lagrange and d'Alembert over his religious views. Euler had written a work — Lettres à une princesse d'Allemagne[32] — in which he attempted to reconcile science and religion. Lagrange held that this effort was inappropriate for a man of science.

Finally, there is an anecdote about a meeting between Napoleon and Laplace in which the former thanked the latter for dedicating the fourth volume of the *Mécanique celeste* to him. Napoleon remarked that Laplace had not mentioned God in his work and to which Laplace replied "Sire, I had no need of the hypothesis." Lagrange upon hearing Laplace's reply observed "But it is a beautiful hypothesis with which many things can be explained away."

ORIGINS OF THE *Mécanique analytique*

Lagrange wrote the *Mécanique analytique* during an epoch which is usually referred to as the Age of Reason or the Enlightenment. It is a period of time extending from the time of Newton and John Locke to about the end of Lagrange's life. Roughly speaking, it is the interval of time between 1675 and 1805. This is an extraordinary period in history. It cannot be understood easily for it encompasses the interrelated concepts of God, nature, reason and mankind. The outlook during this period held that Nature is inherently simple. Intellectuals searched for a single principle which would unify thought. During this period, the solution of a problem is not complete unless it could be demonstrated from obvious and certain principles. It was recognized that various approaches utilizing different principles would provide a solution. Clearly, the utilization of different principles to solve the same problem must give the same result. Thus the various principles of science had to be related in such a way that if one were true the rest of them could be derived by mathematical deduction. Consequently, there is an interest in this period to show how the principles are related and to reduce the various principles to a single fundamental principle. At this point, if it could be shown that the fundamental principle is independent of experience then a deeper understanding of natural phenomena could be found. A simple example of this view is the effort by Daniel Bernoulli to show that what was then taken for the fundamental principle of statics — the Composition of Forces — is a geometric truth independent of experience. Investigators were interested in a rational understanding of science not solely the enumeration of experimental facts. Today, we believe that the root of a scientific principle lies in experience.

This preceding discussion explains to a degree the esteem accorded Lagrange's effort. He had derived all of what was known in mechanics from two principles — the Principle of Virtual Work and d'Alembert's Principle. The solution of a problem of mechanics was thereby reduced to the application of a mathematical formula. In addition, the mathematical framework which he developed around the Principle of Virtual Work and d'Alembert's Principle in order to link the various areas of mechanics contributed an aesthetic quality to the entire treatise. This effort began what today is called analytical mechanics.

The intellectual development of Lagrange can be followed in his published memoirs. Another source is his correspondence. Indeed, for his early development, his correspondence with Euler is particularly revealing because of the strong influence exerted by Euler on Lagrange.

A study of Lagrange's memoirs while he was in Turin and Berlin shows that the *Mécanique analytique* is the culmination of the mathematical discoveries that he made during this period. His discoveries were a result of his research and his teaching duties. The development will be summarized briefly here since it probably is the best approach to understanding the foundation of Lagrange's work and accomplishment.

In a letter[33] dated July 4, 1754, Lagrange wrote to Euler from Turin to tell him that he has been studying Euler's memoirs on variational methods and that he himself has made some observations about the occurrence of maxima and minima in the actions of nature. In another letter[34] dated August 12, 1755, Lagrange again wrote to Euler from Turin to inform him that he has developed a new methodology for the solution of isoperimetrical problems, that is, problems of maxima and minima. In the following pages, Lagrange presents his method using the familiar operator δ. Euler replied on September 6, 1755,[35] and greeted the young Piedmontese's discovery with genuine enthusiasm and admiration. Euler's letter was particularly flattering to Lagrange. Euler was nearly thirty years older than Lagrange and with an established reputation. Coming from such an illustrious and established mathematician, Euler's praise must have meant a great deal to Lagrange. In fact, Euler's enthusiasm was genuine. He later withheld publication of his most recent research in the variational calculus in order that Lagrange would receive undisputed credit for his discovery.

The Principle of Least Action had a stimulating effect on Lagrange. He never imbibed the metaphysics associated with the principle but he recognized that his new method for treating problems of maxima and minima conjointly with the Principle of Least Action could be used advantageously in problems of mechanics. In fact, his method is ideally suited for this application.

Lagrange wrote to Euler on May 19, 1756, from Turin[36] to discuss his mediations on the Principle of Least Action. He envisions the application of this Principle to the whole of dynamics. This inclination is repeated in his first paper published three years later in 1759[37] and in his discussion of the maxima and minima of functions he adds

> I reserve the right to treat this subject, which indeed, I believe to be entirely new, in a particular work which I am preparing on this subject and in which after having presented a general analytical method to resolve all problems related to maxima and minima, I will deduce the mechanics of solid and fluid bodies completely using the Principle of Least Action.

At this point in his career, Lagrange has seen how he can systematize and structure all of mechanics including statics. But his all-encompassing principle is at this point the Principle of Least Action. It is clear that he considers this principle to be the basis of mechanics.

Lagrange continued his research while teaching in Turin. In a letter[38] of November 15, 1759, he wrote to Daniel Bernoulli about his research, to inquire about the death of Maupertuis and to ask for news about Euler. In a postscript to this letter, he advised Bernoulli of his continuing research into problems of maxima and minima.

> Since I am actually working on a treatise whose object is to deduce in a simple and general fashion the solution of the most complicated problems of equilibrium or of the motion of bodies and fluids using only the principle of the least quantity of action, I would like very much to know all that you have learned about elastic curves by means of the formula $\int ds/r^2$ which Euler demonstrated represents least action in this case. I have added a great deal to what Euler wrote on this subject in his book[39] on isoperimetrical curves and in the memoirs of Berlin with the aid of a new method, totally analytical, for dealing with these types of problems and much more general than the one used by this author.... This work should have appeared two or three years ago, because the principles in it were communicated to Euler in 1756. But because of a lack of leisure time, I have not been able to complete it. Now I intend to complete it as soon as possible.

Lagrange is referring to his work with regard to mechanics and mathematics which he was writing for his students at the Royal Artillery School of Turin. Note that the Principle of Least Action is foremost in his approach to mechanics and that he is interested in the connection between this dynamical principle and statics. Thus it is evident that his intent is to find a principle which will encompass statics and dynamics.

In a letter[40] written to Euler on November 24, 1759, Lagrange comments on the research he has done in connection with his teaching duties at the Royal Artillery School of Turin.

> I have written down the elements of mechanics and the differential and integral calculus for the use of my students and I believe that I have developed as far as is possible the true metaphysics of their principles.

Unfortunately, the book on mechanics is lost. It can be assumed that in this work it is unlikely that we would find any background to his later use of the Principle of Virtual Work in conjunction with d'Alembert's Principle.

In particular, this formulation appears in Lagrange's prize memoir on the theory of the libration of the Moon in 1764. However, some investigators have suggested that Lagrange's use of the Principle of Virtual Work would naturally occur in this work. They point out that it would have been natural for Lagrange to begin with statics and to follow this subject with dynamics in his class presentations. This is the usual order for these two subjects and it would have been most natural to use the Principle of Virtual Work in statics. Then this application of the Principle of Virtual Work would naturally carry over to dynamics. However, this explanation overlooks the fact that at this time the Principle of Virtual Work was not the basis for statics. Statics was based on the Composition of Forces. Lagrange appears to have found the Principle of Virtual Work by reading the *Nouvelle Mécanique* of Varignon where it appears in a letter to Varignon from John Bernoulli. It appears more likely that Lagrange would have used Maupertuis' Law of Rest as a basis for statics at this early period if he had wanted an analytical function.

The second work[41] is available and is entitled *Principj di Analisi sublime dettati da La Grange alle Reggie Scuole di Artiglieria.*[42] This work consists of two parts — the algebraic theory of curves and the calculus of differentiation and integration. The first part deals with commonly encountered geometric figures such as the conic sections. There are numerous figures in this treatise which is unusual in Lagrange's finished work. The second part discusses the basics of the calculus and it is in this part that Lagrange discusses the "metaphysics" of the calculus which he referred to in his letter to Euler. He begins by defining the concept of function.

> Then, in general, we call functions of one or more variables an algebraic expression composed in any fashion of these variables and containing any number of additional constants.

The discussion continues with an examination of the differential calculus. Lagrange follows a Newtonian formulation but with Leibnitzian notation. The emphasis is on algebra and it is certainly clear that this discussion is a harbinger to his later work in this area, to wit, the *Théorie des fonctions analytiques*. The work includes figures which is unusual in Lagrange's finished work and it was supposed to be written from the viewpoint of the army engineer or artilleryman. However, in reality, it is much too difficult for these practical minded individuals.

It is clear at this point that Lagrange's main interest was to synthesize mechanics. The next step in the development of his thought is a memoir which appeared in the proceedings of the Turin Academy for the years 1760–61.[43] This memoir generalizes the Principle of Least Action following the lead of Euler.

> General Principle - Let there be an arbitrary number of bodies denoted by M, M', M'' ... which interact in an arbitrary fashion and which, if it is wished, are acted upon by central forces proportional to arbitrary functions of the distance between them. Let s, s', s'' ... denote the displacements made by these bodies in the time t and let u, u', u'' ... denote their velocities at the end of the time interval t, then the equation
>
> $$M\int u\,\mathrm{d}s + M'\int u'\,\mathrm{d}s' + \dots$$
>
> is always a maximum or a minimum.

Lagrange develops this principle further to obtain the development and methodology given in the *Mécanique analytique*.

He begins by applying the Calculus of Variations to the integral

$$\sum_{i=1}^{n} m_i \int u_i\,\mathrm{d}s_i.$$

The first variation of this integral produces the following equation

$$\sum_{i=1}^{n} m_i \int (\delta u_i\,\mathrm{d}s_i + u_i\delta\mathrm{d}s_i) = 0.$$

The equation of force vive

$$m\frac{u^2}{2} + U = \text{ constant}$$

permits the elimination of the variation of the velocity u

$$mu\,\delta u + \delta U = 0.$$

After some algebraic operations, Lagrange arrives at the following result

$$\mathrm{S}\Big(P\,\delta p + Q\,\delta q + R\,\delta r + \cdots + \mathrm{d}\left(\frac{\mathrm{d}x}{\mathrm{d}t^2}\right)\delta x + \mathrm{d}\left(\frac{\mathrm{d}y}{\mathrm{d}t^2}\right)\delta y + \mathrm{d}\left(\frac{\mathrm{d}z}{\mathrm{d}t^2}\right)\delta z\Big)m = 0.$$

He observes that this latter equation is a statement of d'Alembert's Principle in conjunction with the Principle of Virtual Work. Consequently, in his prize essay of 1764 on the theory of the libration of the Moon, he abandons the Principle of Least Action since he now views it as a result of the laws of mechanics. He substitutes d'Alembert's Principle in conjunction with the Principle of Virtual Work.

The memoir of 1764 on the theory of the libration of the Moon[44] marks the change in this point of view. It is also in this memoir that Lagrange mentions the letter from John Bernoulli to Varignon on the Principle of Virtual Work.

A second memoir on the libration of the Moon appeared in 1780.[45] It is clear in this memoir that Lagrange has fully accepted his earlier view that d'Alembert's principle in conjunction with the Principle of Virtual Work is the basic approach to dynamics. It is only a simple step now to apply the Principle of Virtual Work to problems in statics.

Lagrange began writing a treatise on mechanics while in Berlin. This treatise is very likely the *Mécanique analytique*. A letter[46] written to Laplace by Lagrange on September 15, 1782 includes this statement:

> I have almost completed a treatise on analytical mechanics founded solely on the principle or formula which I gave in the first section of the above memoir.[47] But since I still do not know when and where I will be able to have it published, I am not in any hurry to finish the treatise.

The principle or formula to which Lagrange referred is the Principle of Virtual Work. Note also that Lagrange includes statics in the treatise.

THE FOUNDATION OF THE *Mécanique analytique*

The foundation of the *Mécanique analytique* is the Principle of Virtual Work. Credit for resurrecting the Principle of Virtual Work is generally given to Lagrange.[48] Prior to Lagrange's use of this Principle, statics had come to be based on the Law of the Composition of Forces and this law in conjunction with Newton's Second Law was the basis for dynamics. It is no wonder that Lagrange devotes a great deal of discussion to demonstrating the Principle of Virtual Work in PART I, SECTION II. Lagrange retains the old terminology which referred to this principle as the Principle of Virtual Velocities. We know today that this designation is a misnomer. In the 1840's, James Prescott Joule (1818–1889) performed some exhaustive experiments to show that it is force times incremental distance which is the significant quantity and not the rate at which the force moves. Lagrange passively accepted Jean Bernoulli's terminology but in his demonstrations, it is clear that he recognizes work, i.e. force multiplied by displacement, as the significant quantity.

Lagrange's work brought the Principle of Virtual Work to a level on par with the Law of the Composition of Forces in statics. But he must have been troubled by this result because he makes a significant effort to show that the Principle of Virtual Work is on a sound foundation. During his time, the Principle of Virtual Work is somewhat suspect since it appears to be founded on Aristotelian dynamics, that is, the law which claimed that force is proportional to velocity.

To remove any doubt as to the veracity of the Principle of Virtual Work, he returns to a consideration of simple machines, in particular, pulley systems. Lagrange invents a new principle which he calls the Principle of Pulleys. This principle has been known for centuries and has come down to us as the statement that what is gained in force is lost in velocity. But it is clear especially from the research of Descartes that it is displacement and not velocity that is the crucial element. Lagrange demonstrates that the quantity which must be considered is work as it is defined today: to wit, force multiplied by displacement. Then he demonstrates that the principle is applicable to an arbitrary number of forces. Thus he has demonstrated that the Principle of Virtual Work is a very general principle and he concludes his discussion with a statement of the principle. It says very basically that once a system has reached equilibrium, the further expenditure of work is impossible. Consequently, if the configuration of the system is varied near the equilibrium configuration, which is tantamount to assuming that the system is given a virtual displacement, the increment in the quantity of work must be zero.

The status of the Principle of Virtual Work in 18th century mechanics can be clarified by considering its use by two of the most illustrious of the many investigators of this period-Leonhard Euler and Jean le Rond d'Alembert.

Euler, as a member of the Académie de Berlin, was involved in the controversy over the Loi du repos and the Loi de moindre action of Maupertuis, the president of the academy. He sought to defend the Loi du repos in his memoir entitled: Harmonie entre les principes géneraux de repos et de mouvement.[49] Towards the end of this memoir, he shows that the basic laws of statics can be derived from the Loi du repos. He begins by considering the equilibrium of a body on an inclined plane. He defines the "effort" discussed in the Loi du repos as

$$A \cdot x + B \cdot y = 0,$$

where A represents the vertical force due to a body on an inclined plane and B is the applied force which holds it in equilibrium on the plane. The variables x and y

are the displacements in the directions of the forces A and B, respectively. The Loi du repos requires that this quantity be an extremum for equilibrium. Consequently, after differentiation, there results

$$A \cdot \mathrm{d}x + B \cdot \mathrm{d}y = 0.$$

Euler applies the Principle of Virtual Work, although he never states what principle he is using, in order to show that the principle leads to the summation of forces parallel to the plane that is, $-A \cdot \sin\gamma + B \cdot \cos\delta = 0$ is the equilibrium equation parallel to the plane. His analysis is given below.

> Let the body O make an infinitesimal displacement on the inclined plane so that it traverses the length Oo = ds. From point o draw the perpendicular oa to the vertical line OA and from O draw the perpendicular Ob to Bo. After the displacement has been made, it is clear that Oa $= -\mathrm{d}x$ and ob $= \mathrm{d}y$. Since the angle Ooa is equal to γ, we will have that Oa $= \sin\gamma\, \mathrm{d}s$ and since the angle Oob = EOB is equal to δ, we will also have ob $= \cos\delta \cdot \mathrm{d}s$. Therefore, $\mathrm{d}x = -\sin\gamma\, \mathrm{d}s$ and $\mathrm{d}y = -\cos\delta\, \mathrm{d}s$. Then in the state of equilibrium, it is necessary that $-A \cdot \sin\gamma \cdot \mathrm{d}s + B \cdot \cos\delta \cdot \mathrm{d}s = 0$ or $A \cdot \sin\gamma = B \cdot \cos\delta$. Thus the force OB is to the weight of the body O as the sine of the elevation of the inclined plane is to the cosine of the angle EOB which defines the direction of the force OB with the inclined plane. This same ratio could be found using the ordinary principles of statics.

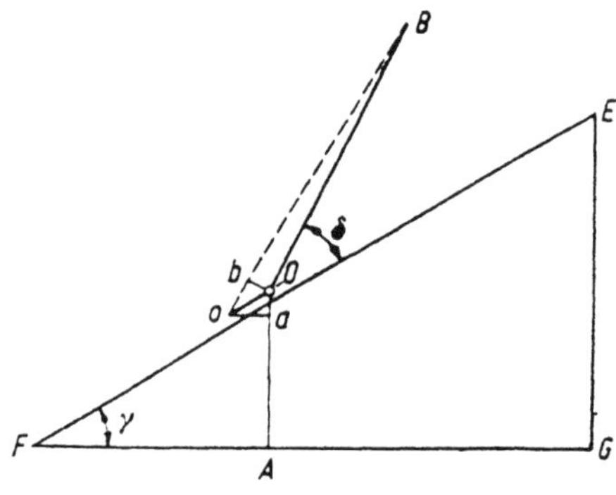

Fig. I EULER'S Fig. 10

Euler goes on to say that, in the same fashion, all of the problems of statics can be solved.

It is noteworthy to mention at this point that the Principle of Virtual Work had been replaced in statics by the Composition of Forces. Consequently, he does not give the former principle any special mention. But it is significant that he has reversed the order of the application of the Principle of Virtual Work and the Potential Function. He begins with the potential function and derives the equation which represents the Principle of Virtual Work. It is this operation which Lagrange will reverse and use as the basis for his mechanics.

D'Alembert also recognized the broad applicability of the Principle of Virtual Work. However, in his works, d'Alembert very seldom uses this principle. He viewed this principle as based on the concept of energy and since he knew that energy was lost in inelastic collisions, it held no primary interest for him. Moreover, he was attempting to banish the concept of force from mechanics and of course, the Principle of Virtual Work depends on the definition of force. His attitude made his approach to mechanics very narrow and in fact, it was very difficult for him to treat problems dealing with static equilibrium. In his *Traité de Dynamique*, d'Alembert attempted to demonstrate the law of the Conservation of Vis Viva.[50]

> It follows from all of our preceding discussion that in general the conservation of **vis viva** depends on this principle, that when the powers [puissances] are in equilibrium, the velocities of the points where they are applied in the directions of these powers are in inverse ratio to these same powers. This principle has long been recognized by geometers as the fundamental principle of equilibrium; but no one that I know has yet demonstrated the principle or shown that the conservation of **vis viva** necessarily results from it.

Many years will pass before Lagrange leaves Turin and departs for Berlin. However, the influence of Euler on his thinking even while he is in Turin is manifest throughout his work. Maupertuis and Euler had been involved for a lengthy period in a controversy over a principle which stated that nature minimizes "action" in all its phenomena. The quantity which expresses "action" was not clearly stated by Maupertius. However, Euler saw how to make the principle precise. He stated that under the action of central forces the following integral

$$\int M u \, \mathrm{d}s$$

where M is the mass of the body, u is the velocity and s is its displacement must be a minimum.

The controversy over the Law of Rest and the Principle of Least Action is very important because in the hands of Lagrange they would lead to the creation of analytical mechanics. The significant results were obtained by Euler whose interest in this controversy was fueled by his interest in metaphysics.[51]

Euler's singular result is the derivation of the Principle of Least Action from the Law of Rest. The significance of the derivation is that it provides a link between statics and dynamics and therefore, it reduces mechanics to a single principle. It must have been clear to Lagrange from Euler's demonstration that the Principle of Virtual Work provided the link between statics and dynamics.

A short summary of Euler's work will be presented here because of its relation to Lagrange's work. Euler began the derivation by considering central forces which he denoted by the letter V_i. The central force V_i is assumed to be acting on a body of mass M_i. Further, let v_i be the distance from the mass M_i to an arbitrary point lying on the line of action of the central force V_i. Then the integral

$$\int V_i \, \mathrm{d}v_i$$

expresses what Euler called the effort of the force V_i and later, Lagrange would call this quantity the potential. This integral can also be recognized as the work of the force V_i or the statement of the Law of Rest for this single force. Since there can be more than one central force, the Law of Rest states that an integral of this form exists for every force V_i and that the sum of all the integrals represents the total effort of all the forces acting on the masses of the system and that this sum is a maximum or a minimum. To put these statement in modern terms, Euler is asserting the following: when the particle displaces such that a component of the gravitational field causes the displacement, the potential decreases. The decrease in potential is equal to the work done by the gravitational force.

Euler continues the development by stating that

> ... what is more natural than to maintain that the principle of equilibrium [i.e. the Law of Rest] should also hold for the motion of bodies acted upon by similar forces. If the intent of nature is to economize the total effort as much as possible, this intent must extend to motion provided that the effort is considered for the duration of the motion and not only for an instant of time. If the effort or the sum of the efforts at any instant during the motion is denoted by Φ and if dt denotes the element of time, the integral $\int \Phi \, dt$ must be a minimum. Thus if in the state of equilibrium, the quantity Φ is a minimum, the same laws of nature seem to require that for motion, the integral $\int \Phi \, dt$ should also be a minimum.

It should be noted that Euler was by no means required to derive the Principle of Least Action from the Law of Rest. He could have derived the Law of Rest from the Principle of Least Action. Apparently, he considered this route for he remarks that it is much easier and more convincing to use the former approach. Since the two principles are equivalent demonstrating either principle and deriving the remaining principle from it is sufficient.

The second part of the derivation treats the actual equality between the Law of Rest and the Principle of Least Action. Euler says the following:

> Now it is precisely in this formula that Maupertuis' other formula concerning motion is contained, however different the two principles may appear on the surface. In order to demonstrate how well they accord, I need only remark that when a body moves under the action of the forces V_i the effort Φ to which the body is subject expresses simultaneously the **vis viva** of the body, i.e. the product of the mass M of the body and the square of its velocity.

If the quantity u represents the velocity of a body and if the quantity s represents the path of the body then $u \, \mathrm{d}t = \mathrm{d}s$. Consequently, the integral of the **vis viva** becomes $\int Mu^2\,\mathrm{d}t = \int Mu\,\mathrm{d}s$. Recall that the **vis viva** did not include the factor 1/2 which is included today because of the principle that the work done equal the change in kinetic energy and this quantity is called the kinetic energy.

Of course, it is clear from the last equation that the second integral represents the Principle of Least Action. It remains to demonstrate that the first integral which is the **vis viva** leads to the Law of Rest. Euler continues

> Then, since $\int V_1\,\mathrm{d}v_1 + \int V_2\,\mathrm{d}v_2 + \cdots + \int V_i\,\mathrm{d}v_i$ expresses the effort of the forces on the body M which I said earlier is equal to Φ, it is clear that $Mu^2 =$ constant $- \Phi$. It is easily seen that the constant does not disturb the harmony which I have sought to establish between the effort Φ and the vis viva Mu^2 of the body because if $\int \Phi\,dt$ is a maximum or minimum the formula $\int Mu^2\,\mathrm{d}t$ or $\int Mu\,\mathrm{d}s$ will also be a maximum or minimum since the term $\int$ constant $\cdot$ dt $=$ constant $\cdot\, t$ does not enter into the consideration of the maximum or minimum.
> Thus this term has no effect on the existence of an extremum. And besides, since the effort Φ is expressed by integral formulas, it already has an arbitrary constant so that I could neglect this constant entirely and simply write $Mu^2 = \Phi$ which makes the equivalence more obvious.

The mathematical formulation of the Principle of Least Action by Euler was to make a strong impression on Lagrange as we showed when we presented a broad outline of Lagrange's development which we took from his correspondence. This principle inspired his research because of the new methodology he developed for the Calculus of Variations.[52] It appears that he had no interest in Maupertuis' attempt to harmonize the Principle of Least Action with the existence of a deity. In his view, it was simply a mathematical proposition. However, once he recognized that a mathematical variation and a virtual displacement were analogous operations, he had come upon the basic foundation of mechanics and from this point, he could construct analytical mechanics.

Lagrange is not alone in his view that the principles of mechanics are simply propositions, but such a view was clearly the exception among scientists of his time. However, during this period of the Enlightenment, the Deist view, which held that the universe evolved without God's intervention, was adhered to by many intellectuals of the period. This view is, of course, contrary to Christian teachings. The endless discussions over this question, which appeared to lead nowhere, appeared meaningless to Lagrange. Hence, he undertook to describe the mechanical behavior of material systems without any metaphysical commentary at all.

Lagrange was careful to derive his analytical formulas from simple propositions whose truth was undeniable. Thus the *Mécanique analytique* continues a development which began with Newton where the analytical formulas are carefully derived from demonstrated empirical results. Consequently, Lagrange avoids a fallacy which many investigators saw in mathematical reasoning and which d'Alembert was to describe. D'Alembert observed that mathematics had destroyed the search for metaphysical systems to explain the natural world but that it was not, in itself, sufficient. Mathematical deduction had limitations which required empirical data if the resulting physical system was to be substantial.

The *Mécanique analytique* embodies this observation succinctly described by d'Alembert. The treatise is divided into two parts: Part I deals with statics and Part II with dynamics. Each part is divided into sections: Part I has eight sections while Part II has twelve sections. The entire treatise is based on the Principle of Virtual Work. For problems of statics and hydrostatics, Lagrange introduces the Lagrangian multiplier in order to treat constraints. In statics, the multiplier is the force or moment which enforces the rigidity of the structure and in hydrostatics, it represents the pressure in the fluid. The treatment of problems in dynamics requires that the Principle of Virtual Work be applied in conjunction with d'Alembert's Principle in order to account for the forces of inertia. The resulting equations, after some algebraic treatment, leads to the Lagrangian form of the equations of motion for any system.

Lagrange began the sections on statics, hydrostatics, dynamics and hydrodynamics with a short summary of the fundamental principles which serve as their basis.[53] These summaries were not intended in any sense to serve as histories of the subject, although some investigators have viewed them in this fashion. They simply fulfilled the Enlightenment ideal that a system be demonstrably derived from indubitably understood and known premises. There is one bias that was common to Enlightenment intellectuals which is also reflected in the historical sketches of Lagrange. The science and technology of the Middle Ages was completely ignored. It is probably true that not much was known about this subject during the Enlightenment. However, the attitude of intellectuals was that the Middle Ages represented a period of time in which religion dominated all human activities and since a basic tenet of the Enlightenment was that religion was hindering human progress, they refused to recognize any contribution in this period to intellectual growth.

Lagrange makes a point of stating in the Preface to the *Mécanique* that "No figures will be found in this work." No other comprehensive work of the time completely excluded figures from its pages. Lagrange had used figures in his notes for his students on mechanics and mathematics at the Royal School of Artillery and he certainly must have used figures as he composed his memoirs on mathematics even though they weren't included in the finished work.

The lack of figures in Lagrange's work is due to developments in mathematics which began in the 16th century. During the 16th century, primarily due to the efforts of the Italian mathematicians such as Pacioli, Cardan, Tartaglia and others, the methodology of algebra had expanded enormously. Algebra had developed to a point that in this period it is on an equal footing with geometry and henceforth, it will now surpass it in scope and ease of application. It is true that during this period, geometric proofs of the rules of algebra are still given. However, this fact will soon change. By the time of Lagrange, algebra stands supreme. Lagrange uses no figures in his work since he is fully aware that algebra has overcome geometry and figures are no longer necessary.

ABOUT THE TRANSLATION

The object of every translator is to remain true to the original text, yet make it comprehensible to an audience speaking a different language in perhaps a different period of time. Thus this object has a two-fold manifestation. The first is to translate accurately the meaning of the original text. However, languages evolve in time, and expressions and terminologies change with time. Our use of modern terminology may bring with it much more than Lagrange intended. Consequently, we have been very careful with our translation of technical terms and we have tried to keep the language simple and direct. The second object is to capture the style of the text and render it into the language of translation. This second object presents the greatest difficulty. Lagrange's style of writing is very austere and blunt. It is almost as if he wrote in this fashion so as not to detract from the mathematics. He is seemingly not conscious of the repetition of basic phrases which make the work somewhat tiresome to read. We have tried to convey the austere and blunt style of Lagrange while eliminating his use of the indefinite pronoun in the active voice by replacing it with a phrase usually in the passive voice. In so doing, we have tried to retain the essence of Lagrange's style of writing without improving on the text.

The third edition of the *Mécanique analytique* was edited by Joseph Bertrand. He added numerous footnotes in order to clarify the text. However, we believe that most of his footnotes were unnecessary since the text should be clear to the modern reader without them. But we have retained a few of Bertrand's footnotes and indicated that they are by him if we thought that they contributed to an understanding of the text.

We have added footnotes wherever we believed that they would be useful to a modern audience. This led us to consider very carefully the composition of our audience. We anticipate our audience to be composed of individuals from four main groups — historians of science, physicists, mathematicians and engineers. Each group has its own interests which made it difficult to put our footnotes in a form pleasing to all. Moreover, the footnotes are intended to complement the text where necessary and are not intended to be complete in any sense.

Excerpt from the Registers of l'Académie Royal des Sciences[1]

Messrs. de La Place, Cousin, LeGendre and I have thoroughly reviewed the work entitled: *Méchanique analitique* by J.L. de La Grange. The Academy has judged this work worthy of its approbation and of publication under its privilege.

I certify that this excerpt conforms to the registers of
the Academy. At Paris, February 27, 1788.

The Marquis de Condorcet

ACKNOWLEDGEMENT

The translators wish to thank Ms. Barbara Hines Buccuzzo for her time and effort in preparing this translation of Lagrange's *Mécanique analytique*.

MÉCANIQUE ANALYTIQUE

Par J. L. LAGRANGE, *de l'Institut des Sciences, Lettres et Arts, du Bureau des Longitudes; Membre du Sénat Conservateur, Grand-Officier de la Légion d'Honneur, et Comte de l'Empire.*

NOUVELLE ÉDITION,

REVUE ET AUGMENTÉE PAR L'AUTEUR.

TOME PREMIER.

PARIS,

Mme Ve COURCIER, IMPRIMEUR-LIBRAIRE POUR LES MATHÉMATIQUES.

1811.

LAGRANGE

HEIM PINXT — ACH MARTINET SCULPT

TABLE OF CONTENTS

PART II. DYNAMICS

PREFACE
to the First Edition

There already exist several treatises on mechanics, but the purpose of this one is entirely new. I propose to condense the theory of this science and the method of solving the related problems to general formulas whose simple application produces all the necessary equations for the solution of each problem. I hope that my presentation achieves this purpose and leaves nothing lacking.

In addition, this work will have another use. The various principles presently available will be assembled and presented from a single point of view in order to facilitate the solution of the problems of mechanics. Moreover, it will also show their interdependence and mutual dependence and will permit the evaluation of their validity and scope.

I have divided this work into two parts: Statics or the Theory of Equilibrium, and Dynamics or the Theory of Motion. In each part, I treat solid bodies and fluids separately.

No figures will be found in this work. The methods I present require neither constructions nor geometrical or mechanical arguments, but solely algebraic operations subject to a regular and uniform procedure. Those who appreciate mathematical analysis will see with pleasure mechanics becoming a new branch of it and hence, will recognize that I have enlarged its domain.

PREFACE
to the Second Edition

There already exist several treatises on mechanics, but the purpose of this one is entirely new. I propose to condense the theory of this science and the method of solving the related problems to general formulas whose simple application produces all the necessary equations for the solution of each problem.

In addition, this work will have another use. The various principles presently available will be assembled and presented from a single point of view in order to facilitate the solution of the problems of mechanics. Moreover, it will also show their interdependence and mutual dependence, and will permit the evaluation of their validity and scope.

I have divided this work into two parts: Statics or the Theory of Equilibrium, and Dynamics or the Theory of Motion. In each part, I treat solid bodies and fluids separately.

No figures will be found in this work. The methods I present require neither constructions nor geometrical or mechanical arguments, but solely algebraic operations subject to a regular and uniform procedure. Those who appreciate mathematical analysis will see with pleasure mechanics becoming a new branch of it and hence, will recognize that I have enlarged its domain.

This is the purpose which I tried to fulfill in the first edition of this work published in 1788. The present edition is in many respects a new work based on the same outline but augmented. I have further developed the principles and general formulas and I have introduced numerous additional applications in which the solutions to the major problems in the domain of mechanics will be found.

I have kept the ordinary notation of the differential calculus because it fits the system of infinitesimals adopted in this treatise. Once the spirit of this system has been grasped well and the accuracy of its results established by either geometrical methods[1] or by the analytical method of derived functions,[2] the infinitesimal calculus can then be applied as a certain and manageable tool to shorten and simplify the demonstrations. It is in this way by using the method of indivisibles that the demonstrations of the Ancients are shortened.

We are now going to point out the principle additions which distinguish this edition from the First Edition.

SECTION I of Part I contains a more complete analysis of three principles of statics with new remarks on the nature and relation of these principles. The section ends with a direct demonstration of the Principle of Virtual Velocities which is completely independent of the two other principles.

In SECTION II, it is demonstrated in a more rigorous manner that the Principle of Virtual Velocities for an arbitrary number of forces can be deduced from the case where there are no more than two forces, which leads this principle back directly to the principle of the lever. The equations which result from this principle are then reduced to a more general form and the necessary conditions are given for a system of forces to be equivalent to and able to replace another system.

In SECTION III, the formulas for instantaneous rotational motion and for the composition of these types of motion are established in a more straight-forward manner. In addition, the theory of moments and their composition is deduced from the preceding development. Also, a little known property of the center of gravity is presented and a new demonstration is given for maxima and minima in the state of equilibrium.

SECTION IV contains more general and simpler formulas for the solution of problems which depend on the calculus of variations and from the comparison of these formulas with those for the equilibrium of bodies of variable shape, it is shown how the problems relative to their equilibrium belong to the category of those which are known under the title of General Problems of Isoperimetrics and which are solved in the same manner.

SECTION V presents some new problems and some important comments on some of the solutions already given in the first edition.

In SECTION VI, some details are added to the historical analysis of the principles of hydrostatics.

In SECTION VII, the calculation of the variations associated with the molecules of a fluid have been treated more rigorously and with more generality. The analysis of the terms which refer to the limits of the fluid mass have been greatly simplified. From these terms, the theory of the action of fluids on the solids which they cover or on the walls of vessels which contain them is deduced and a direct demonstration is given of the following theorem: In the case of equilibrium between a solid and a fluid, the forces which act on the solid are the same as if the fluid and solid formed a solid mass. Much more has been added to this section and to the following section which treats the equilibrium of elastic fluids and presents some applications of the general formulas of the equilibrium of fluids.

Part II, which treats dynamics, has also been considerably augmented.

In SECTION I, a more complete and more accurate analysis is given of some topics in the history of dynamics.

There is an important addition in SECTION II. It is shown for which cases the general formula of dynamics and consequently, the equations which result for the motion of a system of bodies, is independent of the position of the coordinate axes in space. This demonstration gives a means of completing a solution by the introduction of three new arbitrary constants where some constants would have otherwise been assumed to be equal to zero.

In SECTION III, more development is given to the properties relative to the motion of the center of gravity and to the areas described by a system of bodies. There are additions to the theory of principal axes or of uniform rotation, deduced from the consideration of the instantaneous motion of rotation by an analysis which is different from the one used earlier.

Also, some new theorems are demonstrated on the rotation of a solid body or a system of bodies, when they depend on an initial impulse.

There is very little difference between SECTION IV of the first edition and this edition.

SECTION V is entirely new. It contains the theory of the Method of the Variation of Arbitrary Constants[3] which is the subject of three memoirs printed among those of the First Class of the Institute in 1808. It is presented here in a much simpler manner and as a general method of approximation for all the problems of mechanics, where there are perturbing forces which are small compared to the principle forces.

In order to extend this theory as far as possible, the function V, which depends on forces principally, can only be an exact function of the independent variables ξ, ψ, φ, etc., and of the time t, but it is not necessary that the function denoted by Ω and which depends on perturbing forces, also possess the same nature. Whatever the forces, if they are resolved for each body m of the system into three components X, Y, Z in the positive directions of the coordinates x, y, z and with the tendency to increase them, there remains only to reduce the coordinates to functions of the independent variables ξ, ψ, φ, etc. and to substitute in place of the partial derivatives $\mathrm{d}\Omega/\mathrm{d}\xi$, $\mathrm{d}\Omega/\mathrm{d}\psi$, etc. their respective sums

$$Sm\left(X\frac{\mathrm{d}x}{\mathrm{d}\xi}+Y\frac{\mathrm{d}y}{\mathrm{d}\xi}+Z\frac{\mathrm{d}z}{\mathrm{d}\xi}\right), \qquad Sm\left(X\frac{\mathrm{d}x}{\mathrm{d}\psi}+Y\frac{\mathrm{d}y}{\mathrm{d}\psi}+Z\frac{\mathrm{d}z}{\mathrm{d}\psi}\right), \qquad \ldots$$

and as a consequence, the quantity

$$Sm(X\Delta x+Y\Delta y+Z\Delta z)$$

will be obtained in place of $\Delta\Omega$, where the operator Δ refers to the arbitrary constants in such a way that the derivative $\mathrm{d}\Omega/\mathrm{d}\alpha$ can be changed to

$$Sm\left(X\frac{\mathrm{d}x}{\mathrm{d}\alpha}+Y\frac{\mathrm{d}y}{\mathrm{d}\alpha}+Z\frac{\mathrm{d}z}{\mathrm{d}\alpha}\right),$$

and so on for the other partial derivatives of Ω. In this fashion, the method is applicable to perturbing forces represented by arbitrary variables.

Finally, SECTION VI, which is the last section of this volume and which corresponds to the first paragraph of SECTION V of the first edition, is augmented by various remarks and above all, by the solution of some problems on the small vibrations of bodies. It ends with the theory of vibrating strings which I presented earlier in the first volume of the Mémoires of the Académie de Turin[4] and which is presented here in a very simple manner and free of the objections which d'Alembert made against this theory, in the first volume of his *Opuscules*.[5]

PART I

STATICS

SECTION I
THE VARIOUS PRINCIPLES OF STATICS

Statics is the science of the equilibrium of forces. In general, **force** or **power** is the cause, whatever it may be, which induces or tends to impart motion to the body to which it is applied. The force or power must be measured by the quantity of motion produced or to be produced. In the state of equilibrium, the force has no apparent action. It produces only a tendency for motion in the body it is applied to. But it must be measured by the effect it would produce if it were not impeded. By taking any force or its effect as unity, the relation of every other force is only a ratio, a mathematical quantity, which can be represented by some numbers or lines. It is in this fashion that forces must be treated in mechanics.

Equilibrium results from the equilibration of several forces which oppose and negate the actions which they exert on each other. The goal of statics is to formulate the laws under which this equilibration takes place. These laws are founded on some general principles which can be reduced to three: the **lever**, the **composition of forces** and **virtual velocities**.[1]

1. Among the Ancients, Archimedes is the sole author who left us a theory of equilibrium in his treatise of two books *de Aequiponderantibus* or *de Planorum aequilibriis*.[2] He is also the author of the principle of the lever, which consists, as every mechanician knows, of the following: If a straight lever is loaded by two weights placed on each side of the fulcrum at distances inversely proportional to these same weights, the lever will be in equilibrium. Moreover, its point of support would be loaded by the sum of the two weights. Archimedes considers this principle applied to two equal weights placed at equal distances from the fulcrum as an axiom of mechanics obvious by itself or at least as a principle derived from experience. He uses this simple and fundamental case to analyze the one of unequal weights by imagining the weights, when they are commensurable, decomposed into several equal and smaller weights and by assuming that these small weights are distributed at equal intervals over both arms of the same lever such that the lever is loaded by several small equal weights placed at equal distances on each side of the fulcrum. Then he proves the same theorem for incommensurable weights with the aid of the **Method of Exhaustion**.[3] He shows that the weights could not be in equilibrium unless they were placed inversely proportional to their distances from the fulcrum.

Some modern authors, such as Stevin in his statics and Galileo in his dialogues on motion, have simplified Archimedes' demonstration by assuming that the weights attached to the lever are two horizontal parallelepipeds suspended from their midpoints. The parallelepipeds are assumed to have equal widths and heights but with lengths double the lever arms which inversely correspond to them. Thus, in this fashion, the two parallelepipeds are in an inverse relation to their lever arms, and at the same time, they are positioned end to end, so that they form an integral whole for which the midpoint corresponds precisely to the point of support of the lever. Archimedes had already employed a similar construction to determine the center of gravity of a plane figure composed of two parabolic segments in the First Proposition of the Second Book of the *Equilibrium of Planes*.

On the other hand, some authors believing that they had found some deficiencies in Archimedes' demonstration, have reworked it differently to make it more rigorous. However, it is clear that by altering the simplicity of this demonstration, they have added almost nothing in terms of cogency.

Among those who have tried to supplement Archimedes' demonstration on the equilibrium of the lever must be mentioned Huygens, who has written a short work entitled *Demonstratio aequilibrii bilancis*.[4] This work was published in 1693 in the Recueil des anciens Mémoires de l'Académie des Sciences.[5]

Huygens[6] observes that Archimedes assumed implicitly that if a single weight is replaced by a number of equal weights, which are distributed over a horizontal lever at equal intervals, they will exert the same moment to incline the lever. Either they are all on the same side of the fulcrum, or some are on one side with the rest on the other side. To avoid this precarious assumption, instead of distributing two commensurate weights on the same lever, as Archimedes had done, Huygens distributes them on two other horizontal levers placed perpendicular to the ends of the first lever. In this way, one has a horizontal plane loaded with several equal weights which is evidently in equilibrium with respect to the first lever, because the weights are equally and symmetrically distributed from the two arms of this lever. But Huygens demonstrates that this plane is also in equilibrium with respect to an axis inclined to the former and passing through the point which divides the first lever in segments inversely proportional to the weights which load it. Since the small weights are also placed at equal distances on each side of the same straight line, he concludes that this configuration and consequently, the lever under consideration, must be in equilibrium with respect to the same point. This demonstration is ingenious, but it does not replace entirely what can be found in the one given by Archimedes.[7]

2. The equilibrium of a straight and horizontal lever with its extremities loaded by equal weights and with the fulcrum located at its midpoint is a self-evident truth because there is no reason that either of the weights should move since they are both equal to one another and located at equal distances from each side of the fulcrum. This observation is not equivalent to the assumption that the reaction at the fulcrum is equal to the sum of the two weights. It appears that all mechanicians have accepted as a fact derived from daily experience that the weight of a body depends only on its total mass, and not on its shape.[8] Nevertheless, this truth can be deduced from the first one by considering, as Huygens did, the equilibrium of a plane about a line.

In order to carry out this demonstration, it is only necessary to imagine a triangular plane loaded by two equal weights at the two corners of its base and by one of double weight at its top. This plane, which is supported on a straight line or fixed axis passing through the midpoints of the two sides of the triangle, will obviously be in equilibrium since each of the sides can be visualized as a lever loaded at each end by an equal weight and with its fulcrum on the axis passing through its midpoint. This state of equilibrium can be conceived in yet another fashion, by viewing the base of the triangle as a lever whose ends are loaded by two equal weights and by imagining a transverse lever connecting the top of the triangle to the midpoint of its base. If the transverse lever which is loaded at its top by a weight double the one at each corner of the base is used as a fulcrum to the lever which forms the base, it is obvious that the base lever is in equilibrium with respect to the transverse lever which it carries at its midpoint. Also, the transverse lever will be in equilibrium with respect to the axis to which the plane is already in equilibrium. And because this axis passes through the midpoint of the sides of the triangle, it will necessarily pass through the midpoint of the transverse lever. Hence, the transverse lever will have its fulcrum at its midpoint and consequently, it will have to be loaded equally at the two ends. Furthermore, the load which is supported by the fulcrum of the base lever will be equal to the weight at the top of the transverse lever and consequently, it will be equal to the sum of the two weights of the base lever.

If, instead of a triangle, a trapezoid loaded at its four corners by four equal weights is considered, it would be found, in the same fashion, that the two levers of unequal lengths, constituting the parallel sides of the trapezoid, exert equal forces on their fulcrums.

3. Once this proposition is established, it is clear that a weight in equilibrium on a lever can be replaced by two weights, each equal to half of this weight and placed at equal distances on each side of the point where the weight was originally suspended, as Archimedes did. The action of this weight on the point to which it is attached is the same as the action of a lever suspended by this point which is also its midpoint and loaded at its two extremities by two weights each equal to half of this weight. Moreover, it is obvious that nothing prevents us from approaching this latter lever from the standpoint of the former. In short, that it belongs to or which is perhaps more rigorous, this latter lever can be considered as held in equilibrium by a force applied at its midpoint directed vertically upward and equal to the weight of the two halves which are assumed applied at its ends. Then, total equilibrium will always be present if this lever which is in equilibrium is superimposed on the first lever which is also assumed to be in equilibrium with respect to its support point. And, if the application is done such that the midpoint of the second lever coincides with the extremity of one of the arms of the first lever, the force which supports the second lever will be canceled by the weight applied to the extremity of the first lever. Therefore, the first lever will be in equilibrium. Finally, the weight applied to the extremity of the first lever will then be replaced by two weights each equal to half of it and placed at the extremities of the second lever which is now part of an extended first lever. The superposition of equilibria in mechanics is a principle as fecund as the superposition of figures in geometry.

4. Thus the equilibrium of a straight and horizontal lever loaded with two weights inversely proportional to their distances from the fulcrum can be viewed as a rigorously demonstrated truth. Using the principle of superposition, it is easy to extend it to any bent lever with the fulcrum at the apex of the angle and with the arms pulled in opposite directions by forces perpendicular to the directions of the arms. Indeed, it is obvious that a bent lever with equal arms and able to rotate about the apex of the angle will be held in equilibrium by two equal forces acting in opposite directions and applied perpendicularly to the ends of the two arms. Therefore, if there is a straight lever in equilibrium with one arm of the same length as one arm of the bent lever and if this straight lever is loaded at one of its ends by a weight equal to each of the forces applied to the bent lever with the other arm loaded such that equilibrium holds and if these two levers are superimposed in such a way that the apex of the bent lever coincides with the fulcrum of the straight lever, and if the equal arms of both levers coincide and form one lever arm, then the force applied to the arm of the bent lever will equilibrate the weight hung from the arm with the same length as the straight lever such that both arms can be removed from the system which we have defined. The state of equilibrium will still exist between the two remaining arms which constitute a bent lever pulled at its extremities by perpendicular forces inversely proportional to the lengths of the arms, as in the case of the straight lever.

But a force can be considered applied to any point along its direction. Therefore, two forces applied at arbitrary points on a plane, which is fixed at one point, and directed arbitrarily in this plane are in equilibrium when they are inversely proportional to perpendiculars drawn from this fixed point to their respective directions. Indeed, these perpendiculars can be considered part of a bent lever for which the fulcrum is the fixed point in the plane. This result is called today the **Principle of Moments**, where moment is defined as the product of a force by the [length of the] arm of the lever on which it acts.

This general principle suffices to solve all the problems of statics. The initial investigations following Archimedes in the study of the pulley in the theory of simple machines, led to a recognition of this principle. The results of these investigations can be seen in the treatise of Guido Ubaldo[9] entitled *Mechanicorum liber*[10] which was published at Pesaro in 1577. However, this author did not know how to apply this principle to the inclined plane or to other machines such as the wedge and the screw where it is also applicable.[11] For the wedge and screw, he presented only an approximate theory.

5. The ratio of the component of a weight parallel to an inclined plane to the total weight of the body was a problem for modern mechanicians for a long time. Stevin is the first to solve it but his solution is founded on a proposition which is indirect and independent of the theory of the lever.[12]

Stevin considers a solid triangle placed on a horizontal plane so that its two sides form two inclined planes. He imagines further that a wreath made of several equal weights, stringed at equal intervals or rather a closed chain of constant thickness is placed on the two sides of this triangle in such a manner that the entire upper part is applied to the two sides of the triangle and the lower part hangs freely beneath the triangle as if it were attached to the two ends of the base.[13]

Now Stevin remarks that even if the wreath or chain were assumed capable of slipping freely on the triangle, it must nevertheless remain at rest. Indeed, if it started to slip by itself in either direction under its own weight, it would continue to slip, because the cause of the motion remains. The chain, because of the uniformity of its segments, is always in the same configuration on the triangle. Thus a perpetual motion would result, which is absurd.

Therefore, equilibrium necessarily exists between all the elements of the chain and the portion which hangs beneath the triangle is in equilibrium under its own weight. Thus it must be that the action of all the weight resting on one of the inclined sides counterbalances the action of all the weight resting on the other side. However, the ratio of the total weight on one of the inclined sides to the total weight on the other side is equal to the ratio of the lengths of the sides on which they lie. Hence, the same force will always be required to maintain one or more weights at rest on inclined planes when the total weight is proportional to the lengths of the planes assuming that the height of these planes is the same. When the plane is vertical, the force is equal to the weight and therefore, for any inclined plane, the ratio of the force to the weight is equal to the ratio of the height of the plane to its length.

I mention Stevin's demonstration because it is very ingenious and besides little is known about it. Also, Stevin deduces from this theory the equilibrium between three forces which act at the same point, and he finds that equilibrium exists when the forces are parallel to and proportional to the three sides of any plane triangle. Refer to the **De staticae elementis** and to the **Additamentum staticae** by this author in the *Hypomnemata Mathematica*[14] printed at Leyden in 1605 and in the works of Stevin translated into French and printed in 1634 by the Elzevirs.[15] But it should be recognized that this fundamental theorem of statics, although it is commonly attributed to Stevin, has been demonstrated by this author only for the case where the directions of the two forces form a right angle.

Stevin remarks with reason that the problem of a weight placed on an inclined plane and maintained by a force parallel to the plane, is the same as if this weight were held by two strings-one perpendicular and the other parallel to the plane. Then applying his theory of the inclined plane, he finds that the ratio of the weight to the force parallel to the plane is equal to the ratio of the hypotenuse to the base of a right triangle constructed on the inclined plane by two lines, one vertical and the other perpendicular to the plane. Stevin limits himself thereafter to applying this ratio to the case where the string which holds the weight on the inclined plane would also be inclined to this plane by constructing an analogous triangle with the same lines-one vertical and the other perpendicular to the plane-and by taking the base in the direction of the string. But for this case, it should be demonstrated that the same ratio holds for the equilibrium of a weight held on an inclined plane by a force oblique to the plane, which cannot be deduced from the consideration of the wreath of spheres visualized by Stevin.

6. In the *Le Meccaniche*[16] of Galileo, first published in French by Father Mersenne in 1634, equilibrium on an inclined plane is reduced to the case of a bent lever with two equal arms. One arm is assumed perpendicular to the plane and loaded by the weight resting on the plane while the other arm is horizontal and loaded with a weight equivalent to the force required to maintain the weight on the plane. This equilibrium is then reduced to the case of a straight and horizontal lever by considering the weight attached to the arm perpendicular to the plane as suspended from a horizontal arm which, when combined with the horizontal arm of the bent lever produces a straight lever. Hence, the weight is to the force which holds it on the inclined plane in inverse proportion to the two arms of the straight lever and it is easy to prove that these two arms are to one another as the height of the plane is to its length.

It can be said that this presentation is the first direct demonstration of the equilibrium of a body on an inclined plane. Galileo used it later to rigorously demonstrate the equality of the velocities acquired by heavy bodies descending from the same height but on planes with different inclinations: an equality that he only assumed in the first edition of his *Dialogues.*[17]

It would have also been easy for Galileo to solve the problem where the force which maintains the weight is inclined to the plane. But this additional step was taken by Roberval only a short time later in a treatise on mechanics published as part of Mersenne's *Harmonie universelle* in 1636.[18]

7. Roberval also treats the equilibrium of a body placed on an inclined plane as if it were attached to the arm of a lever perpendicular to the plane and he considers the force applied to the same arm in a given direction. He has, therefore, a lever with only one arm for which one of the ends is fixed while the other end is acted upon by two forces-the weight and the force which holds the weight. He later substitutes for this lever a bent lever with two arms perpendicular to the directions of these two forces and having the same fixed point as the point of support. He assumes that the two forces are applied to the arms of the lever so that at equilibrium the ratio of the weight to the force is inversely proportional to the ratio of the two arms of the bent lever, that is, the perpendiculars drawn from the fixed point to the directions of the weight and force.

From the prior result, Roberval deduces the equilibrium condition for a weight held by two strings having an arbitrary angle between them. This is achieved by substituting for the lever perpendicular to the plane a rope which is attached to the point of support of the lever and by substituting for the force another rope pulling in the direction of this force. And through diverse constructions and analogies which are a bit complicated, he arrives at this conclusion: If from any point lying on the vertical passing through the weight, a parallel line to one of these ropes is drawn to intersect the other rope, the triangle thus formed will have its sides proportional to the weight and to the forces which act in the direction of the same sides. It is obvious that this is the theorem given by Stevin.

I believe that I had to mention this demonstration of Roberval, not only because it is the first rigorous demonstration that has been given for Stevin's theorem, but also because it remained forgotten in a treatise on harmony[19] rare enough nowadays, where no one would think to look for it. Moreover, I presented these details regarding the theory of the lever only to please those who like to follow a train of thought in science and who like to follow the paths which seminal thinkers have followed as well as the more direct paths that they could have followed.

8. The treatises on statics which appeared after Roberval's, up to the time of the discovery of the **Composition of Forces**, have added nothing to this branch of mechanics. Only the well-known properties of the lever and the inclined plane and their applications to the other simple machines can be found. Moreover, there exist treatises which contain theories which are only approximate such as the one by Lami[20] on the equilibrium of solids, in which he gives an erroneous ratio for the weight of a body to the force which holds it on an inclined plane. Descartes, Torricelli and Wallis will not be mentioned here because they adopted for the definition of the state of equilibrium a principle which is related to the **Principle of Virtual Velocities**, and for which they did not have a demonstration.

9. The second fundamental principle of statics is that of the **Composition of Forces**. It is founded on this supposition: If two forces from different directions act simultaneously on a body, the forces are equivalent to a unique force capable of imparting to this body the same velocity that the two forces acting individually would have imparted to it. Indeed, a body which is moved uniformly in two different directions simultaneously must necessarily traverse the diagonal of the parallelogram whose sides it would have followed separately in virtue of each of the two velocities. From this fact, it can be concluded that two arbitrary forces which act simultaneously on the same body are equivalent to a single force represented in magnitude and direction by the diagonal of the parallelogram for which the sides represent the magnitudes and directions of the two given forces. This result is the principle which is called the **Composition of Forces**.

This principle alone suffices to determine the laws of equilibrium in all cases because by taking all the forces successively two at a time, an equivalent, unique force would be obtained. In the case of equilibrium, the magnitude of this force must be zero if there is no fixed point in the system. But if there is a fixed point, this unique force would have to pass through the fixed point. This is what is found in all books on statics and in particular, in the *Nouvelle Mécanique* of Varignon[21] where the theory of machines is deduced solely from this principle.

It is obvious that Stevin's theorem on the equilibrium of three forces parallel to and proportional to the three sides of a triangle is an immediate and necessary consequence of the principle of the **Composition of Forces** or rather, it is this same principle formulated differently. But the former has the advantage of being founded on simple and natural concepts, unlike Stevin's theorem which is founded only on indirect considerations.

10. The Ancients knew the **Composition of Velocities** as can be seen from some passages in Aristotle's Μηχανικά προβλήματα.[22] Geometers have used it principally for the description of curves, for example, Archimedes for the spiral, Nicomedes for the conoid, etc. Recently, Roberval deduced from it an ingenious method to draw tangents to curves which can be described by two motions for which the equations are given. But Galileo is the first to have used the concept of composed motion in mechanics to determine the trajectory of a heavy body under the action of gravity and the force of projection.

In the second proposition of the Fourth Day of his *Dialogues*[23] , Galileo demonstrates that a body moving with two uniform velocities, one horizontal and the other vertical, must have a total velocity represented by the hypotenuse of a triangle where the sides represent the two component velocities. At the same time, it seems that Galileo did not recognize the importance of all aspects of this proposition to the theory of equilibrium. In the third dialogue, where he treats the motion of heavy bodies on inclined planes, instead of applying the principle of the **Composition of Velocities** to determine directly the positional gravity of a body on an inclined plane, he makes this evaluation from the theory of equilibrium applied to inclined planes deduced from what he had previously established in his treatise *Della Scienza Meccanica.*[24] In the latter treatise, he considers everything from the inclined plane to the lever.

Later, the theory of composed velocities is found in the works of Descartes, Roberval, Mersenne, Wallis, etc. But up to the year 1687, when Newton's *Principia* and Varignon's *Projet de la nouvelle Mécanique* were published, no one thought to substitute force for the velocity that it would produce in the **Composition of Velocities** and to determine the resultant force composed of two given forces as one determines the resultant velocity of two given rectilinear and uniform velocities.[25]

In the second corollary to the Third Law of Motion, Newton describes in a few words how the laws of equilibrium are easily deduced from the composition and resolution of forces, by taking for the resultant force the diagonal of a parallelogram for which its sides represent the component forces. But this subject is treated in more detail in the works of Varignon and the *Nouvelle Mécanique*, which was published after his death in 1725, contains a complete theory on the equilibrium of forces in various machines deduced solely from the consideration of the composition or resolution of forces.

11. The principle of the **Composition of Forces** gives immediately the conditions of equilibrium between three forces which act at a point. These conditions could only be deduced by a chain of thought beginning with the equilibrium of the lever. But, on the other hand, when the conditions of equilibrium between two parallel forces applied to the extremities of a straight lever are to be found using this principle, one is obliged to use an indirect approach. For example, by substituting a bent lever for the straight lever, as Newton and d'Alembert have done or by adding two external forces which cancel each other, but which after being combined with the applied forces makes their resultant intersect on the same line or finally by imagining that the directions of the forces extend to infinity and by proving that the resultant force must pass through the point of support.[26] This last

approach is taken by Varignon in his mechanics. Hence, although the **Principles of the Lever** and of the **Composition of Forces** always lead to the same results, it is remarkable that the most simple case for one of these principles becomes the most complicated case for the other.

12. But an immediate relation can be established between these two principles, by using the theorem that Varignon gave in his *Nouvelle Mécanique* (SECTION I, Lemma XVI). This relation consists of the following: If from an arbitrary point in the plane of a parallelogram, perpendiculars are drawn to the two sides and the diagonal of the parallelogram, the product of the diagonal with its perpendicular is either equal to the sum or the difference of the products of its two sides with their respective perpendiculars depending on whether or not the point lies outside of or inside of the parallelogram, respectively.[27] Varignon demonstrates simply that by constructing triangles which have the diagonal and the two sides of the parallelogram for bases and for sides the lines drawn from the extremities of the component parts of the parallelogram to the arbitrary point, the triangle[28] formed with the diagonal is, in the first case, equal to the sum and in the second case, to the difference of the two triangles formed with the sides. This is by itself a very nice theorem of geometry, independent of its application to mechanics.

This theorem and its demonstration would be equally valid if arbitrary segments equal to the diagonal and the sides were taken on the prolongations of these straight lines. Indeed, since every force can be assumed applied at an arbitrary point along its direction, it can generally be concluded that two forces represented in magnitude and direction by two lines lying in a plane, have a composed line or a resultant represented in magnitude and direction by a line lying in the same plane which, when prolonged, passes through the point of intersection of these two lines and such that having taken an arbitrary point in this plane and drawing from this point perpendiculars to the prolongation of these three lines, if necessary, the product of the resultant with its perpendicular is equal to the sum or the difference of the products of the two composed forces with their respective perpendiculars, depending on whether the point at which the three perpendiculars meet, is outside or inside the area defined by the lines representing the composed forces.[29]

When this point is assumed to lie on the line of direction of the resultant, this force does not contribute to the equation of equilibrium. Therefore, equality exists between the two products of composed forces with their perpendiculars. This is true for any straight or bent lever for which the point of support is the same as the point referred to earlier, because the action of the resultant is equilibrated by the resistance of the support.

The theorem attributed to Varignon, is the foundation of almost all of modern statics in which it constitutes a general concept called the **Principle of Moments**. Its great advantage is that the composition and resolution of forces are reduced to additions and subtractions so that whatever the number of forces to be combined, it is easy to find the resultant force which must be zero in the case of equilibrium.

13. I mentioned earlier that Varignon's discovery occurred at the time of the publication of his *Nouvelle Mécanique*, although in the preface of the *Nouvelle Mécanique*, it is claimed that it was two years earlier. In the **Histoire de la République des Lettres**, there is a memoir on the block and tackle in which he claims to use the **Composition of Velocities** to derive all that applies to this device. But I must say that this article "lacks in accuracy".[30] This memoir appears only in the **Nouvelles de la République des Lettres** for the month of May 1687 under the title: **Nouvelle Démonstration générale de l'usage des Poulies à Moufle.**[31] The author analyzes the equilibrium of a weight supported by a rope which is wrapped around a pulley and which extends out obliquely. He does not use or even mention the principle of the **Composition of Forces**, but he uses known theorems on the problem of weights suspended by ropes and cites the statics of Pardies[32] and Dechales.[33] In a second demonstration, he reduces the problem of the pulley to that of the lever by comparing the line which joins the two points where the rope leaves the pulley to a lever loaded with the weight applied to the shaft of the pulley and whose extremities are pulled by the two segments of the rope which supports the pulley.

With regard to the history of the discovery of the **Composition of Forces**, I must say a word about a short work published by Lami in 1687 entitled: **Nouvelle manière de démontrer les principaux Théorèmes des élémens des Mécaniques**[34] so that nothing is omitted. The author asserts that if a body is moved by two forces acting in two different directions, it will necessarily follow a middle path. But if the path in this direction is blocked, it would remain at rest and the two forces would be in equilibrium. He determines the middle direction from the composition of the two velocities that the body would follow due to each of the two forces as if they were acting separately. This gives the diagonal of a parallelogram having for sides the distances covered in the same time by the action of the two forces. Therefore, the length of the sides is proportional to these forces. From this result he deduces immediately the theorem that two forces are themselves proportional to the sines of the angles that their two directions make with the middle direction that the body would take if it were not impeded. Subsequently, he applies it to the inclined plane and to the lever when its extremities are acted on by nonparallel forces. But in the case where they are parallel, he uses a vague and rather inconclusive reasoning.

The similarity between the principle used by Lami and by Varignon led the author of the **Histoire des Ouvrages des Savans**[35] in April, 1688 to say that it appears that the former was indebted to the latter for the discovery of this principle. Lami defended himself in a letter published in the **Journal des Savans** on September 13, 1688. The author replied in December of the same year. But this controversy, in which Varignon did not take part, did not go any further and Lami's letters have been forgotten.

In fact, the simplicity of the principle of the **Composition of Forces** and the ease with which it is applied to all problems of equilibrium led to its adoption by mechanicians as soon as it was discovered. And it can be said that it is used as a basis in almost all the treatises on statics which have been published since then.

14. However, it is easy to recognize that the **Principle of the Lever** is the only principle which has the advantage of being founded on the essence of equilibrium itself, independent of motion. Also, there is an essential difference in the way the forces which are in equilibrium are calculated in these two principles. If it were impossible to correlate them through their results, one could reasonably doubt that the principle of the lever could be substituted for the principle which results from the consideration of composed velocities.

In fact, in the equilibrium of the lever, the forces are the applied weights or can be considered as such and a force can not be considered as double or triple another unless it is formed of two or three equal forces. But the potential for motion is assumed to be the same for each force whatever its magnitude. On the other hand, in the case of the principle of the **Composition of Forces**, the magnitudes of the forces are calculated from the magnitudes of the velocities which they would impart to the body to which they are applied if each were free to act independently. And it is perhaps this difference in the way of conceiving forces which has prevented mechanicians for a long time from using known laws of the **Composition of Velocities** in the theory of equilibrium for which the simplest case is the equilibrium of heavy bodies.[36]

15. Since then, one has tried to formulate the principle of the **Composition of Forces** independently of the consideration of motion and to establish it solely from arguments obvious by themselves. Daniel Bernoulli was the first to present in the **Novi Commentarii Academiae Petropolitaneae**,[37] Volume I, a very ingenious but lengthy and complicated demonstration of the parallelogram of forces.[38] Later, d'Alembert simplified this demonstration in the first volume of his *Opuscules*.

This demonstration is based on the following two principles:

1. If two equal forces act in different directions at the same point, their resultant is a unique force which bisects the angle formed by them. This resultant is equal to their difference when the angle is equal to 180 degrees.
2. Multiples of the same forces or arbitrary forces which are proportional, have a resultant which is also a multiple of their resultant or proportional to this resultant if the angles are the same.

This second principle is obvious if forces are viewed as quantities which can be added or subtracted.

The first principle is demonstrated by considering the direction of the motion of a body moved by two forces not in equilibrium. Since this motion is necessarily unique, it can be attributed to an individual force acting on the body in the direction of its motion. Therefore, it can be said that this principle is not entirely exempt from the consideration of motion.

With regard to the direction of the resultant in the case of two equal forces, it is obvious that there is no reason that it should be nearer to one than the other of these two forces. Consequently, it must bisect the angle formed by these two forces.

Later, the essence of this demonstration was formulated analytically in more or less simple fashion by considering the resultant as a function of the forces and the angle between them. Refer to the second volume of the **Miscellanea taurinensia**,[39] the **Mémoires de l'Académie des Sciences** of 1769, the Sixth Volume of the *Opuscules* of d'Alembert, etc. But it must be recognized that by isolating in this manner the principle of the **Composition of Forces** from the principle of the **Composition of Velocities**, its main advantages which are its clarity and simplicity are lost. It is reduced to a result of geometrical or analytical constructions.

16. Finally, I come to the third principle, the one of **virtual velocities**.[40] One must understand by the term **virtual velocity**, the velocity which a body in equilibrium would take if the state of equilibrium ceased to exist, that is, the velocity that the body would have in the first instant of its motion. The principle requires the forces to be in a state of equilibrium if they are to be inversely proportional to their virtual velocities taken in the direction of these forces. For one who examines the conditions of equilibrium of the lever and other devices, it is easy to recognize the law that the acting forces and resisting weights are always inversely proportional to the distances which they traverse in the same interval of time. Yet, it appears that the Ancients never knew this law. Guido Ubaldo was perhaps the first to perceive it in the lever and in the block and tackle.[41] Galileo recognized it later in the inclined plane and in machines which depend on the properties of the inclined plane. He considers it as a general property of the equilibrium of machines. Refer to his treatise on *Mechanics* and to the scholium to the second proposition of the Third Dialogue in the Bologna edition of 1655.[42]

Galileo means by the **moment** of a weight or any acting force applied to a mechanical device, the effort, action, energy, or **impetus** of the force which puts the machine in motion so that there is equilibrium between two forces when their **moments** during the motion of the machine are equal and opposite. He shows that **moment** is always proportional to the force multiplied by the virtual velocity, depending on the manner in which the force acts.[43]

This notion of moment was also adopted by Wallis in his *Mechanics* published in 1669 in which the author expounds the principle of the equality of moments as a foundation of statics.[44] From this principle, he deduces the theory of equilibrium for common types of mechanical devices.

At this time, **moment** is more commonly understood as the product of a force and the distance from its line of direction to a point, line or plane, that is, by the lever arm over which it acts. But it seems to me that the notion of moment given by Galileo and Wallis is much more natural and general, and I do not see why it has been abandoned and another substituted which expresses only the value of the **moment** in certain cases, such as in the lever, etc.[45]

Descartes has similarly reduced all statics to a unique principle which is in reality similar to Galileo's principle but Descartes presented it in a less general manner than Galileo. This principle states that the same amount of force[46] is required to raise a heavier weight to a correspondingly smaller elevation or a less heavy weight to a correspondingly greater elevation. Refer to the letter numbered 73[47] of the first volume published in 1657 and the *Traité de Mécanique*[48] published in his posthumous works. The result to be derived from this principle is that equilibrium holds between two bodies when they are positioned such that the vertical paths which they traverse simultaneously are inversely proportional to their weights. But in the application of this principle to various machines, it is only necessary to consider the displacements made in the first instant of motion which are proportional to their virtual velocities. Otherwise, one would not have the true laws of equilibrium.

Also, whether the **Principle of Virtual Velocities** is viewed as a general property of equilibrium, in the fashion of Galileo, or is taken for the true cause of equilibrium, as did Descartes and Wallis, it must be said that it has all the simplicity that can be desired in a fundamental principle. We will see later how often this principle is used because of its generality.

Torricelli, the famous disciple of Galileo, is the author of another principle which also depends on the **Principle of Virtual Velocities**. That is, when two weights are joined together and placed such that their center of gravity can not descend further, they are in equilibrium. Although Torricelli only applies this principle to the inclined plane, it is easy to see that it is also applicable to other machines. Refer to his treatise entitled: *De motu gravium naturaliter descendentium*[49] which was published in 1644.

The **Principle of Torricelli** led to another principle which some authors have used to resolve several problems of statics more easily. This principle is: For a system of heavy bodies in equilibrium, the center of gravity must be at the lowest possible position. Indeed, it is known from the theory of maxima and minima that the center of gravity is in its lowest position when the differential of its descent is zero or, which is the same, when the center of gravity does not rise or descend should the system displace by an infinitesimal amount.

17. The **Principle of Virtual Velocities** can be expressed in a more general manner: If an arbitrary system of any number of bodies or mass points, each acted upon by arbitrary forces, is in equilibrium and if an infinitesimal displacement is given to this system, in which each mass point traverses an infinitesimal distance which expresses its virtual velocity, then the sum of the forces, each multiplied by the distance that the individual mass point traverses in the direction of this force, will always be equal to zero. Furthermore, the small distances traversed in the direction of the forces are considered positive and the distances traversed in the opposite direction are considered negative.

John Bernoulli is the first, as far as I know, to have recognized in the **Principle of Virtual Velocities** its great generality and its relevancy to solving the problems of statics. This can be seen from one of the letters he wrote to Varignon in 1717, which Varignon[50] put at the beginning of the ninth section of his new mechanics, a section devoted exclusively to demonstrating by different applications the verity and use of this principle.

This same principle is the basis for the one that Maupertuis[51] proposed in the **Mémoires de l'Académie des Sciences** of Paris for the year 1740 under the name of the **Loi du Repos**.

Later, Euler developed it further in the **Mémoires de l'Académie de Berlin** for the year 1751. Finally, this is again the same principle which is used as a basis for the one that Courtivron[52] gave in the **Mémories de l'Académie des Sciences** for the years 1748 and 1749. In general, I believe that I can say that all general principles that might possibly be discovered in the science of equilibrium will be tantamount to the **Principle of Virtual Velocities**, but viewed and expressed in a different fashion.

But this principle is not only very simple and very general, it has also the value and unique advantage of expression in a general formula which covers all the problems that might arise on the equilibrium of bodies. We will develop this formula completely and we will even try to present it in a more general fashion than has been done previously and to use it in new applications.

18. With respect to the concept of the **Principle of Virtual Velocities**, it must be said that it is not sufficiently obvious by itself to be erected as a founding principle. But it can be viewed as a general expression of the laws of equilibrium, deduced from the two principles which we have just discussed. Therefore, in the known demonstrations of this principle, it is always made to depend on these other two principles in a more or less direct fashion. But there exists another general principle in statics which is independent of the **Principle of the Lever** and the **Composition of Forces**. Although mechanicians commonly refer this principle to the **Principle of the Lever** and the **Composition of Forces**, it seems to be the natural foundation of the **Principle of Virtual Velocities**. This principle can be called the **Principle of Pulleys.**[53]

If several pulleys are mounted together on the same block, the assembly is called a **polispaste** or **pulley block** and the combination of two pulley blocks, one fixed and the other mobile, joined by the same rope for which one end is fixed and the other is free to be pulled by a force, constitutes a machine for which the ratio of the force to the weight carried by the mobile block is proportional to the number of ropes which converge at this block, assuming they are all parallel and neglecting friction and the stiffness of the rope. It is obvious that because of the constant tension along the length of the rope that the weight is carried by a number of forces equal to the number of ropes supporting the pulley block and each equal to the force applied to the free end of the rope. Also, since these ropes are parallel, they can be considered a single rope, if desired, by imagining the diameters of the pulleys reduced to an infinitely small dimension.

In the same fashion, the fixed and mobile pulleys wrapped with the same rope can be replicated by means of various fixed and counter pulleys and thus the same force applied to the free end of the rope can support as many weights as there are mobile blocks, for which the contribution of each mobile block to this force is proportional to the number of ropes which holds this block to the unit.

For greater simplicity, let us substitute a weight for the force, after having wrapped the last rope over a fixed pulley which holds this weight which we will take as unity. Imagine that the various mobile blocks, instead of supporting weights, are attached to bodies viewed as mass points and arranged such that they can model any given arbitrary system. In this fashion, the same weight will produce, by means of the rope which wraps around all the pulleys, different forces which act on the various points of the system along the direction of the ropes which converge at the pulley attached to these points. The ratio of the weight to one of these forces is equal to the number of ropes so that these forces will be represented by the number of ropes which contribute to their creation by their tension.

In order for this system to remain in equilibrium, it is obvious that the weight must not descend when portions of the system are subjected to any arbitrary and infinitesimal displacement. Indeed, since the weight always has an inclination to descend, it will descend if there is a displacement of the system which allows it to descend. The weight will then necessarily descend and it will produce a displacement in the system.

Let us denote by **p**, **q**, **r**, etc. the infinitesimal displacements that this motion will impart to the various parts of the system in the direction of the forces which are applied to them and by P, **Q**, **R**, etc. the number of ropes of the pulley applied to these parts to produce these same forces. Then it is obvious that the displacements p, q, r, etc. will also be the magnitude of the displacements by which the mobile pulleys will approach the corresponding fixed pulleys. Since the pulleys draw nearer to each other, the length of the rope which is wound about the pulleys would be reduced by the quantities Pp, Qq, Rr, etc. Therefore, because of the invariable length of the rope, the weight would descend by the following amount Pp+ Qq+ Rr+ $\cdots$. Consequently, the forces represented by the letters P, Q, R, etc. will be in equilibrium if the following equation holds

$$Pp + Qq + Rr + \cdots = 0,$$

which is the analytical expression for the general **Principle of Virtual Velocities**.

19. If the quantity Pp+ Qq+ Rr+ $\cdots$ were negative instead of equal to zero, it seems that this condition would suffice to establish equilibrium because it is impossible for the weight to ascend by itself. But it must be recognized that whatever the connection between the parts which form the given system, the relations which can result between the infinitesimal quantities p, q, r, etc. can only be expressed by differential equations. Consequently, the relations between these quantities are linear and one or several among them will necessarily be indeterminate and could be taken as positive or negative. Therefore, the values of all those quantities are such that they could change sign. Hence, if in a given displacement of the system, the value of the quantity Pp+ Qq+ Rr+ $\cdots$ is negative, it can become positive by taking the quantities p, q, r, etc. with opposite signs. Furthermore, since the reverse displacement is equally possible, the weight can descend and upset the state of equilibrium.

20. Conversely, it can be proved that if the equation

$$Pp + Qq + Rr + \cdots = 0$$

is valid for all possible infinitesimal displacements of the system, the system will necessarily be in equilibrium and since the weight is immobile during these displacements, the forces which act on the system remain in the same position. There is no reason that they should produce one or any other displacement in which the quantities p, q, r, etc. have opposite signs. This is the reason that the balance remains in equilibrium because there is no reason that it should incline to one side rather than the other.

Thus the **Principle of Virtual Velocities** is demonstrated for commensurate forces and will also be valid for incommensurate arbitrary forces because it is known that every proposition demonstrated for commensurate quantities can also be demonstrated for incommensurate quantities by a **reductio ad absurdam.**

SECTION II
A GENERAL FORMULA OF STATICS AND ITS APPLICATION TO THE EQUILIBRIUM OF AN ARBITRARY SYSTEM OF FORCES

1. The general law of equilibrium for machines is that the forces or powers are inversely proportional to the velocities of the mass points[54] to which they are applied, taken in the direction of these forces.

There exists in this law what is commonly called the **Principle of Virtual Velocities.**[55] This principle has been recognized for a long time as the fundamental principle of equilibrium as we have shown in the preceding section. Consequently, it can be considered as a kind of axiom of mechanics.

In order to reduce this principle to a formula, let us assume that the forces P, Q, R, etc. acting in given directions are in equilibrium. And let us draw from the points where these forces are applied straight lines along the line of application of these forces equal to p, q, r, etc. Let us denote by $\mathrm{d}p$, $\mathrm{d}q$, $\mathrm{d}r$, etc., the variations or increments of these lines due to an infinitesimal change in the positions of the different bodies or mass points of the system.

It is evident that these increments will represent the distance traversed during the same instant of time by the forces P, Q, R, etc. along their own lines of application. Thus, assuming that these forces extend the lines p, q, r, etc., the increments $\mathrm{d}p$, $\mathrm{d}q$, $\mathrm{d}r$, etc. will be proportional to the virtual velocities of the forces P, Q, R, etc. and can, for more simplicity, be substituted for these velocities.

Now having stated these facts, let us first consider two forces P and Q in equilibrium. From the law of equilibrium between two forces, the two quantities P and Q must be inversely proportional to the increments $\mathrm{d}p$ and $\mathrm{d}q$. But it is easy to see that there would be no equilibrium between these two forces unless they were oriented such that when one of them is displaced along its line of application, the other is constrained to move in the opposite direction along its line of application. From this result, it can be seen that the values of the increments $\mathrm{d}p$ and $\mathrm{d}q$ must be of opposite signs. Therefore, if the forces P and Q are both assumed to be positive, the equilibrium condition will be expressed by

$P/Q = -\mathrm{d}q/\mathrm{d}p$ or $P\,\mathrm{d}p + Q\,\mathrm{d}q = 0$. This is the general formula for the equilibrium of two forces.

Let us now consider the equilibrium of three forces P, Q, R with the virtual velocities represented by $\mathrm{d}p$, $\mathrm{d}q$, $\mathrm{d}r$. Let $Q = Q' + Q''$ and assume, which is permissible, that the fraction Q' of the total force Q is such that $P\,\mathrm{d}p + Q'\,\mathrm{d}q = 0$. This force will then be in equilibrium with the force P. In order to obtain overall equilibrium, the remaining fraction Q'' of the force Q must be in equilibrium with the force R which will give the equation $Q''\,\mathrm{d}q + R\,\mathrm{d}r = 0$. When the latter expression is added to the former and recalling that $Q' + Q'' = Q$, the following equation is obtained

$$P\,\mathrm{d}p + Q\,\mathrm{d}q + R\,\mathrm{d}r = 0.$$

If there is a fourth force S for which the virtual velocity is represented by the increment $\mathrm{d}s$ and furthermore, if one postulates $Q = Q' + Q''$ and $P\,\mathrm{d}p + Q'\,\mathrm{d}q = 0$, then $R = R' + R''$ and $Q''\,\mathrm{d}q + R'\,\mathrm{d}r = 0$. Then Q' will be in equilibrium with P, R' will be in equilibrium with Q'' and to obtain overall equilibrium, R'' must be in equilibrium with S so that one must have $R''\,\mathrm{d}r + S\,\mathrm{d}s = 0$. These three equations can then be combined to obtain

$$P\,\mathrm{d}p + Q\,\mathrm{d}q + R\,\mathrm{d}r + S\,\mathrm{d}s = 0.$$

In the same fashion, this formulation can be further extended to an arbitrary number of forces.

2. Therefore, in general, for the equilibrium of an arbitrary number of forces P, Q, R, etc. acting in the directions p, q, r, etc. and applied to any number of bodies or mass points arranged in an arbitrary fashion, an equation such as the following will be obtained

$$P\,\mathrm{d}p + Q\,\mathrm{d}q + R\,\mathrm{d}r + \cdots = 0.$$

This is the general formula of statics for the equilibrium of an arbitrary system of forces.

We will call each term of this formula, such as $P\,\mathrm{d}p$, the moment of the force P by defining the meaning of the term in the same fashion as Galileo,[56] that is, the product of the force by its virtual velocity. Then, the general formula of statics is that the sum of the moments of all the forces is equal to zero. The only difficulty in applying this formula will be to determine, depending on the nature of the system, the values of the differentials $\mathrm{d}p$, $\mathrm{d}q$, $\mathrm{d}r$, etc.

Therefore, the system will be considered in two configurations which differ infinitesimally and the most general expressions for the differentials will be sought by introducing in these expressions as many unknown quantities as there are arbitrary coordinates which define the change in position of the system. Then the expressions for $\mathrm{d}p$, $\mathrm{d}q$, $\mathrm{d}r$, etc. will be substituted in the proposed equation and the equation set equal to zero, independent of all unknowns such that global equilibrium of the system exists. Consequently, the sum of the terms related to each unknown will be individually set equal to zero and by this means as many independent equations as there are unknowns will be obtained. But it is not difficult to recognize that this number must always be equal to the number of unknown quantities which define the configuration of the system. Thus as many equations will be obtained by this method as will be required to determine the state of equilibrium of the system.[57]

All the authors who previously applied the **Principle of Virtual Velocities** to solve the problems of statics have applied it in this fashion. However, using this principle in this fashion often requires constructions and geometrical considerations which produce solutions as lengthy as if they were derived by the ordinary principles of statics. This is perhaps the reason which prevented its application as often as it should be applied in view of its simplicity and generality.

3. The purpose of this work is to reduce mechanics to purely algebraic operations and the formula which we have just developed is fully adequate for this purpose. It is only necessary to express algebraically, in the most general fashion, the values of p, q, r, etc. taken in the directions of the forces P, Q, R, etc. and the values of the virtual velocities $\mathrm{d}p$, $\mathrm{d}q$, $\mathrm{d}r$, etc. will be obtained by simple differentiation.

Attention must be given solely to the fact that when several quantities vary simultaneously in the differential calculus they can all be assumed to change simultaneously with their differentials. And if, depending on the nature of the problem, some differentials decrease while others increase, then a negative sign must be given to the differentials which decrease.

The differentials $\mathrm{d}p$, $\mathrm{d}q$, $\mathrm{d}r$, etc. which represent the virtual velocities of the forces, P, Q, R, etc. must be taken positive or negative depending on whether the forces will lengthen or shorten the lines p, q, r, etc. which are their respective lines of action. But since the general formula of equilibrium does not change when all the signs of its terms are changed, it is permissible to retain as positive the differentials of the displacements which together increase or decrease and as negative the differentials of the displacements of a contrary direction. Thus by considering all forces as positive, their moments $P\,\mathrm{d}p$, $Q\,\mathrm{d}q$, $R\,\mathrm{d}r$, etc. will be positive or negative if the virtual velocities $\mathrm{d}p$, $\mathrm{d}q$, $\mathrm{d}r$, etc. are positive or negative. And if the forces act in opposite directions, it is only necessary to give a negative sign to the quantities which represent these forces or change the signs of their moments.

The general property of equilibrium derives from this result, which is, that an arbitrary system of forces in the state of equilibrium remains in equilibrium when each of the forces acts in the opposite direction as long as the configuration of the system does not change with a change in direction of all the forces.

4. Whatever the forces which act on a given system of bodies or mass points, they can always be considered to act toward points located on their lines of action. We will call these points the "force centers". It is possible to define the lines p, q, r, etc. as the respective distances from these centers to the points of the system at which the forces P, Q, R, etc. are applied. In this case, it is clear that these forces will have a tendency to shorten the

lines p, q, r, etc. Consequently, a negative sign must be given to their differentials. But if all signs are changed, the general formula will remain

$$P\,\mathrm{d}p + Q\,\mathrm{d}q + R\,\mathrm{d}r + \cdots = 0.$$

The force centers can be either outside or inside the system and still be part of it. Thus a distinction can be made between external and internal forces.

In the first case, it is obvious that the increments $\mathrm{d}p$, $\mathrm{d}q$, $\mathrm{d}r$, etc. express total variations of the lines p, q, r, etc. resulting in a change of configuration for the system. They are consequently total differentials of the quantities p, q, r, etc. when all quantities related to the configuration of the system are treated as variables and when those related to the location of the different force centers are viewed as constants.

In the second case, some of the bodies of the system will be the force centers which act on other bodies of the same system and because of the equality between action and reaction, these latter bodies will be at the same time the force centers which act on the former.

Let us consider two bodies which act on one another with an arbitrary force P. This force can result either from attraction or repulsion between the bodies, from a spring placed between them or from some other cause. Let p be the distance between these two bodies and $\mathrm{d}p'$ the variation of this distance. Depending on the change of position of one body with respect to the other, it is clear that the virtual moment of the force P with respect to this body is $P\,\mathrm{d}p'$. Similarly, if the variation of this same distance p resulting from a change in position of the other body is designated by $\mathrm{d}p''$, one will have with respect to this second body the moment $P\,\mathrm{d}p''$ of the same force P. Therefore, the total moment of this force will be expressed as $P(\mathrm{d}p' + \mathrm{d}p'')$, but it is obvious that $\mathrm{d}p' + \mathrm{d}p''$ is the total differential of p which we will call $\mathrm{d}p$ because the distance p can only change with the displacement of the two bodies. Therefore, the moment will be expressed simply by $P\,\mathrm{d}p$. This reasoning can be generalized to apply to as many bodies as desired.

5. Consequently, in order to obtain the sum of the moments of all the forces in a given system, whether they are external or internal forces, it is only necessary to consider, in particular, each of the forces acting on different bodies or points of the system and to take the sum of the products of these forces by the differential of the distance between the point on which it acts and the point to which it is moving. While carrying this out, the quantities in the differentials are considered to be either variable or constant. The quantities which are related to the configuration of the system are considered variables and the quantities which are related to the external centers or points are constants. That is, these points are considered fixed while the configuration of the system is varied. When this sum is set equal to zero, it will give the general formula of statics.

6. In order to present all the generality as well as the simplicity that the analytical expression of this formula is capable of, the position of all bodies or mass points within a given system and the position of the force centers will be expressed with respect to three orthogonal axes fixed in space.

In general, x, y, z will designate the coordinates of the points where the forces are applied and subsequently, we will distinguish between the different points of the system by one or more primes. We will also designate by a, b, c the coordinates of the force centers.

It is clear that the distances p, q, r, etc. between the points of application of the forces and the force centers can be expressed in general by a formula of the form

$$\sqrt{(x-a)^2+(y-b)^2+(z-c)^2},$$

in which the quantities a, b, c will be constants or at least, should be considered as such, while x, y, z vary in the case where they are related to points placed outside of the system and where the forces are external. But in the case where the forces are internal and originate from some of the bodies within the system itself, the quantities a, b, c will become

$$x^{m\ etc.}, \qquad y^{m\ etc.}, \qquad z^{m\ etc.}$$

and consequently, they will be variables.

Thus having the expressions of the finite quantities p, q, r, etc. as a known function of the coordinates of the various bodies of the system, it is only necessary to differentiate considering these coordinates as the sole variables in order to obtain the required expressions for the increments dp, dq, dr, etc., which are in the general formula of equilibrium.

7. Although the forces P, Q, R, etc. can always be considered directed to given centers, the consideration of these centers is not part of the problem in which only the magnitude and direction of each force is considered as data. The differentials dp, dq, dr, etc. are presented here in a more general fashion.

At the outset, let us assume, which is always permissible, that the force P is directed towards a fixed center. Then one has

$$p=\sqrt{(x-a)^2+(y-b)^2+(z-c)^2}$$

and from there, by differentiating, holding a, b, c constant and for an external force P, the following equation will be obtained

$$\mathrm{d}p=\frac{x-a}{p}\,\mathrm{d}x+\frac{y-b}{p}\,\mathrm{d}y+\frac{z-c}{p}\,\mathrm{d}z.$$

It is easy to see that

$$\frac{x-a}{p}, \qquad \frac{y-b}{p}, \qquad \frac{z-c}{p}$$

are the direction cosines of the angles which the line p makes with the lines $x-a$, $y-b$ and $z-c$. Therefore, in general, if we call α, β, γ the angles that the direction of the force

P makes with the coordinate axes x, y, z or with lines parallel to these axes, the direction cosines will be

$$\frac{x-a}{p} = \cos\alpha, \qquad \frac{y-b}{p} = \cos\beta, \qquad \frac{z-c}{p} = \cos\gamma.$$

Consequently, there results

$$\mathrm{d}p = \cos\alpha\,\mathrm{d}x + \cos\beta\,\mathrm{d}y + \cos\gamma\,\mathrm{d}z$$

and in the same fashion, the expressions for the other increments $\mathrm{d}q$, $\mathrm{d}r$, etc. are obtained.

But if the force P is an internal force and acts on two points with coordinates x, y, z and x', y', z' where these points can be near or far from one another, the expression for p will have $a = x'$, $b = y'$, $c = z'$. The result will be the following equation

$$\mathrm{d}p = \cos\alpha(\mathrm{d}x - \mathrm{d}x') + \cos\beta(\mathrm{d}y - \mathrm{d}y') + \cos\gamma(\mathrm{d}z - \mathrm{d}z').$$

It will be noted at the outset with respect to the angles α, β, γ that $\cos^2\alpha + \cos^2\beta + \cos^2\gamma = 1$, which is obvious because of the preceding formulas. Secondly, if the angle of projection of the line p on the xy-plane with the x-axis is designated by ϵ, one will have

$$\frac{x-a}{\pi} = \cos\epsilon, \qquad \frac{y-b}{\pi} = \sin\epsilon$$

assuming that

$$\pi = \sqrt{(x-a)^2 + (y-b)^2}.$$

Then after replacing $x - a$ and $y - b$ by the expressions $p\cos\alpha$, $p\cos\beta$, the equation for π becomes

$$\pi = p\sqrt{(\cos^2\alpha + \cos^2\beta)} = p\sqrt{(1 - \cos^2\gamma)} = p\sin\gamma$$

and

$$\frac{x-a}{p} = \sin\gamma\cos\epsilon, \qquad \frac{y-b}{p} = \sin\gamma\sin\epsilon.$$

Consequently, there results, $\cos\alpha = \sin\gamma\cos\epsilon$, $\cos\beta = \sin\gamma\sin\epsilon$.

8. Since $\mathrm{d}p$ represents the small interval that the body or point to which the force P is applied traverses in the direction of this force, this point can only move in a direction perpendicular to the direction of this force if $\mathrm{d}p$ is set equal to zero. Hence, $\mathrm{d}p = 0$ will be

the differential equation of a surface to which the direction of the force P is perpendicular. This surface will be a sphere if the quantities a, b, c are constants, but it is an arbitrary surface when these quantities are variables.

Let us now assume in general that the force P acts perpendicular to a surface given by the equation

$$A\,\mathrm{d}x + B\,\mathrm{d}y + C\,\mathrm{d}z = 0.$$

In order to make this equation compatible with the following equation

$$(x - a)\,\mathrm{d}x + (y - b)\,\mathrm{d}y + (z - c)\,\mathrm{d}z = 0,$$

which results from the assumption that $\mathrm{d}p = 0$, it is necessary to set

$$\frac{A}{C} = \frac{x-a}{z-c}, \qquad \frac{B}{C} = \frac{y-b}{z-c},$$

for which

$$x - a = \frac{A}{C}(z - c), \qquad y - b = \frac{B}{C}(z - c)$$

and after substituting these quantities in the expression for $\mathrm{d}p$, the following equation is obtained

$$\mathrm{d}p = \frac{A\,\mathrm{d}x + B\,\mathrm{d}y + C\,\mathrm{d}z}{\sqrt{(A^2 + B^2 + C^2)}}.$$

Therefore, when the differential equation of the surface to which the force P is perpendicular is known, the expression for its virtual velocity $\mathrm{d}p$ will also be known.

It can be assumed that

$$A\,\mathrm{d}x + B\,\mathrm{d}y + C\,\mathrm{d}z = \mathrm{d}u,$$

where u is a function of x, y, z, because it is known that a differential equation of the first order with three variables can not represent a surface unless it can be integrated or it can be made integrable by a multiplier. Also, by the algorithm of partial differences,

$$A = \frac{\mathrm{d}u}{\mathrm{d}x}, \qquad B = \frac{\mathrm{d}u}{\mathrm{d}y}, \qquad C = \frac{\mathrm{d}u}{\mathrm{d}z},$$

so that the expression for $\mathrm{d}p$ will become

$$\mathrm{d}p = \frac{\mathrm{d}u}{\sqrt{\left[\left(\frac{\mathrm{d}u}{\mathrm{d}x}\right)^2 + \left(\frac{\mathrm{d}u}{\mathrm{d}y}\right)^2 + \left(\frac{\mathrm{d}u}{\mathrm{d}z}\right)^2\right]}}.$$

Thus the moment of a force P perpendicular to a surface given by the equation $\mathrm{d}u = 0$ will be

$$\frac{P\,\mathrm{d}u}{\sqrt{\left[\left(\frac{\mathrm{d}u}{\mathrm{d}x}\right)^2 + \left(\frac{\mathrm{d}u}{\mathrm{d}y}\right)^2 + \left(\frac{\mathrm{d}u}{\mathrm{d}z}\right)^2\right]}}.$$

In the same fashion, the values of the other differentials $\mathrm{d}q$, $\mathrm{d}r$, etc. will be determined from the differential equations of surfaces to which the directions of the forces Q, R, etc. are perpendicular.

9. But without considering the surface to which a force is perpendicular, in the same fashion as an arbitrary quantity is represented by a line, the variable p can be considered as an arbitrary function of the coordinates and the force P as the force which changes the value of p. Then $P\,\mathrm{d}p$ will also be the virtual moment of the force P and similarly, $Q\,\mathrm{d}q$, $R\,\mathrm{d}r$, etc. will be the virtual moments of the forces Q, R, etc. respectively. In the same fashion, these forces are considered to vary the quantities q, r, etc. which are assumed to be arbitrary functions of the same coordinates. Considering the moments in this fashion gives a much larger domain of application to the general formula of equilibrium.

10. Since the expressions for the differentials $\mathrm{d}p$, $\mathrm{d}q$, $\mathrm{d}r$, etc. are known as differential functions of the coordinates of different bodies of the system, it is only necessary to substitute them in the general formula

$$P\,\mathrm{d}p + Q\,\mathrm{d}q + R\,\mathrm{d}r + \cdots = 0$$

and then to arrange this equation in a manner independent of the differentials that it contains.

Thus if the system is completely free such that there is neither a given relationship between the coordinates of the different bodies nor consequently, between their differentials, the preceding equation must be satisfied independent of these differentials. Therefore, the sum of all the terms multiplied by each of these differentials must be equal to zero individually. This will give as many equations as there are unknowns and consequently, as many equations as are needed to determine all the variables. Thus the position of all the parts of the system in the state of equilibrium will be known.

But if the nature of the system is such that the motions of the bodies are constrained in some fashion, these particular conditions should be expressed at the outset by algebraic equations which we will call "Equations of Condition". This is always easy. For example, if some of the bodies were constrained to move on given lines or surfaces, their coordinates would be expressed by the equations of these lines or surfaces. Also, if two bodies were always at a constant distance k apart, the following equation would always hold

$$k^2 = (x' - x'')^2 + (y' - y'')^2 + (z' - z'')^2,$$

and so on.

After the equations of condition are found, they must be used to eliminate as many differentials as possible in the expressions for $\mathrm{d}p$, $\mathrm{d}q$, $\mathrm{d}r$, etc. so that the remaining differentials are absolutely independent of one another and express only what is arbitrary in the change of configuration of the system. Then, since the general formula of statics must be satisfied, whatever the change in configuration of the system might be, the sum of all terms which are affected by each of the indeterminate differentials must be equal to zero. Thus there will be as many particular equations as there are differentials. These equations added to the given equations of condition will contain all the necessary conditions for the determination of the state of equilibrium of the system, because it is easy to see that the total number of these equations will always be the same as the number of different variables which are used as coordinates for all the bodies of the system. Therefore, these equations will suffice to determine these variables uniquely.

11. Also, if we have always defined the locations of bodies with rectangular coordinates, it is because this approach has the advantage of simplicity and ease of calculation. But this does not prevent us from using other systems of coordinates in the methodology since it is clear that nothing prevents us from using a system of coordinates other than rectangular coordinates.[58] For example, instead of using the two coordinates x and y, one could use, when the circumstances appear favorable, the radius vector $\rho = \sqrt{(x^2 + y^2)}$ and an angle φ, for which the tangent is x/y, which will give $x = \rho \cos \varphi$ and $y = \rho \sin \varphi$, keeping the third coordinate z as it is. In a similar fashion, a radius vector $\rho = \sqrt{(x^2 + y^2 + z^2)}$ with two angles φ and ψ could be used such that

$$\tan \varphi = \frac{y}{x}, \qquad \tan \psi = \frac{z}{\sqrt{x^2 + y^2}},$$

which will give $x = \rho \cos \psi \cos \varphi$, $y = \rho \cos \psi \sin \varphi$ and $z = \rho \sin \psi$.[59] Finally, it should be clear that any other system of coordinates could be used.

Let us also note that since the methodology requires only the differentials $\mathrm{d}x$, $\mathrm{d}y$, $\mathrm{d}z$, the origin of the coordinates can be located freely which can help simplify the expressions for these differentials.

For example, if $\rho \cos \varphi$ and $\rho \sin \varphi$, are substituted for x and y, one would have $\mathrm{d}x = \mathrm{d}\rho \cos \varphi - \rho \sin \varphi \, \mathrm{d}\varphi$, $\mathrm{d}y = \mathrm{d}\rho \sin \varphi + \rho \cos \varphi \, \mathrm{d}\varphi$ but by setting $\varphi = 0$, which is the same as placing the origin of the angle on the radius, one will have more simply $\mathrm{d}x = \mathrm{d}\rho$ and $\mathrm{d}y = \rho\,\mathrm{d}\varphi$. Similar transformations hold for the other cases.

12. In general, whatever the system of forces for which the state of equilibrium is sought and whatever the manner in which the points to which the forces are applied are related to one another, the variables which determine the position of these points in space can always be reduced to a smaller number of independent variables by eliminating as many variables as there are equations of condition: that is, by expressing all the variables, three for each point, by a smaller number of them or by other arbitrary variables which, because they are not constrained by any condition, will be independent and indeterminate. Equilibrium must then exist for each of these independent variables because they can describe individually various changes of configuration for the system.

13. Indeed, if we denote the independent variables by ξ, ψ, φ, etc. and consider the values of p, q, r, etc. as functions of these variables, the following equations will result

$$\begin{aligned}
\mathrm{d}p &= \frac{\mathrm{d}p}{\mathrm{d}\xi}\,\mathrm{d}\xi + \frac{\mathrm{d}p}{\mathrm{d}\psi}\,\mathrm{d}\psi + \frac{\mathrm{d}p}{\mathrm{d}\varphi}\,\mathrm{d}\varphi + \cdots \\
\mathrm{d}q &= \frac{\mathrm{d}q}{\mathrm{d}\xi}\,\mathrm{d}\xi + \frac{\mathrm{d}q}{\mathrm{d}\psi}\,\mathrm{d}\psi + \frac{\mathrm{d}q}{\mathrm{d}\varphi}\,\mathrm{d}\varphi + \cdots \\
\mathrm{d}r &= \frac{\mathrm{d}r}{\mathrm{d}\xi}\,\mathrm{d}\xi + \frac{\mathrm{d}r}{\mathrm{d}\psi}\,\mathrm{d}\psi + \frac{\mathrm{d}r}{\mathrm{d}\varphi}\,\mathrm{d}\varphi + \cdots \\
&\ \vdots
\end{aligned}$$

and the equation of equilibrium $P\,\mathrm{d}p + Q\,\mathrm{d}q + R\,\mathrm{d}r + \cdots = 0$ will become

$$\begin{aligned}
&\left(P\frac{\mathrm{d}p}{\mathrm{d}\xi} + Q\frac{\mathrm{d}q}{\mathrm{d}\xi} + R\frac{\mathrm{d}r}{\mathrm{d}\xi} + \cdots\right)\mathrm{d}\xi \\
&+ \left(P\frac{\mathrm{d}p}{\mathrm{d}\psi} + Q\frac{\mathrm{d}q}{\mathrm{d}\psi} + R\frac{\mathrm{d}r}{\mathrm{d}\psi} + \cdots\right)\mathrm{d}\psi \\
&+ \left(P\frac{\mathrm{d}p}{\mathrm{d}\varphi} + Q\frac{\mathrm{d}q}{\mathrm{d}\varphi} + R\frac{\mathrm{d}r}{\mathrm{d}\varphi} + \cdots\right)\mathrm{d}\varphi \\
&+ \cdots = 0,
\end{aligned}$$

in which owing to the indeterminate nature of $\mathrm{d}\xi$, $\mathrm{d}\psi$, $\mathrm{d}\varphi$, etc., the following distinct equations will be obtained

$$\begin{aligned}
&P\frac{\mathrm{d}p}{\mathrm{d}\xi} + Q\frac{\mathrm{d}q}{\mathrm{d}\xi} + R\frac{\mathrm{d}r}{\mathrm{d}\xi} + \cdots = 0 \\
&P\frac{\mathrm{d}p}{\mathrm{d}\psi} + Q\frac{\mathrm{d}q}{\mathrm{d}\psi} + R\frac{\mathrm{d}r}{\mathrm{d}\psi} + \cdots = 0 \\
&P\frac{\mathrm{d}p}{\mathrm{d}\varphi} + Q\frac{\mathrm{d}q}{\mathrm{d}\varphi} + R\frac{\mathrm{d}r}{\mathrm{d}\varphi} + \cdots = 0 \\
&\quad\vdots
\end{aligned}$$

whose number will be equal to the number of variables ξ, ψ, φ, etc. These equations will be used to determine the variables.

It is obvious that each of these equations represent a particular equilibrium condition with a known ratio between the virtual velocities. And it is from the combination of all these partial equilibriums that the general equilibrium of the system is composed.

It should also be noted that the reasoning of Article 1 of this section applies without exception to these partial and determinate equilibrium states. And since, as in the case of two forces, their state of equilibrium can always be reduced to the equilibrium of a straight lever for which the arms are proportional to their virtual velocities, the general **Principle of Virtual Velocities** can be made to depend by this means solely on the **Principle of the Lever**.

14. When the quantity $P\,\mathrm{d}p + Q\,\mathrm{d}q + R\,\mathrm{d}r + \cdots$ is not equal to zero with respect to all independent variables, the forces P, Q, R, etc. will not be in equilibrium and therefore, the bodies will move under the action of these forces according to these forces and their reciprocal action.

Let us assume that other forces represented by P', Q', R', etc., with lines of action p', q', r', etc., are acting on the bodies of the system and therefore, are the cause of the motions. These forces would be equivalent to the former and in all cases, they could replace them because their effect is seen to be exactly the same. And if the forces P', Q', R', etc., retain their magnitudes but reverse their directions, it is obvious that they will induce in the same bodies equal motions, but in opposite directions. Consequently, if in this new state they acted on the bodies of the same system simultaneously with the forces P, Q, R, etc., the bodies would remain at rest. The motion induced in one direction is cancelled by an equal motion in the opposite direction. Therefore, by necessity, there would be equilibrium between all the forces which results in the following equation (Article 2)

$$P\,\mathrm{d}p + Q\,\mathrm{d}q + R\,\mathrm{d}r + \cdots - P'\,\mathrm{d}p' - Q'\,\mathrm{d}q' - R\,\mathrm{d}r' - \cdots = 0$$

from which the following equation derives

$$P\,\mathrm{d}p + Q\,\mathrm{d}q + R\,\mathrm{d}r + \cdots = P'\,\mathrm{d}p' + Q'\,\mathrm{d}q' + R'\,\mathrm{d}r' + \cdots .$$

This is the necessary condition for the forces P', Q', R', etc. acting along the lines p', q', r', etc. to be equivalent to the forces P, Q, R, etc. acting along the lines p, q, r, etc. Since two systems of forces can only be equivalent in one way (the motion of a body is always unique and determinate), it follows that the two systems of forces P, Q, R, etc. and P', Q', R', etc. are such that generally one has, considering all independent variables, the following equation

$$P\,\mathrm{d}p + Q\,\mathrm{d}q + R\,\mathrm{d}r + \cdots = P'\,\mathrm{d}p' + Q'\,\mathrm{d}q' + R'\,\mathrm{d}r' + \cdots,$$

then these two systems will be equivalent and in all cases, they could be substituted for one another.[60]

15. The following important theorem of statics follows from that result: Two systems of forces are equivalent and can be substituted for one another in the same system of bodies interacting in an arbitrary fashion if the sum of the moments of the forces are always equal in both systems. Inversely, when the sum of the moments of the forces of a system is always equal to the sum of the moments of the forces of another system, these two systems of forces are equivalent and can be substituted for one another in the same system of bodies.

If the lengths p, q, r, etc. are made dependent on the lengths ξ, ψ, φ, etc., the formula $P\,\mathrm{d}p + Q\,\mathrm{d}q + R\,\mathrm{d}r + \cdots$ becomes, as in Article 13, $\Xi\,\mathrm{d}\xi + \Psi\,\mathrm{d}\psi + \Phi\,\mathrm{d}\varphi + \cdots$ in which

$$\Xi = P\frac{\mathrm{d}p}{\mathrm{d}\xi} + Q\frac{\mathrm{d}q}{\mathrm{d}\xi} + R\frac{\mathrm{d}r}{\mathrm{d}\xi} + \cdots$$

$$\Psi = P\frac{\mathrm{d}p}{\mathrm{d}\psi} + Q\frac{\mathrm{d}q}{\mathrm{d}\psi} + R\frac{\mathrm{d}r}{\mathrm{d}\psi} + \cdots$$

$$\Phi = P\frac{\mathrm{d}p}{\mathrm{d}\varphi} + Q\frac{\mathrm{d}q}{\mathrm{d}\varphi} + R\frac{\mathrm{d}r}{\mathrm{d}\varphi} + \cdots$$

$$\vdots$$

Therefore, the following general equation results

$$P\,\mathrm{d}p + Q\,\mathrm{d}q + R\,\mathrm{d}r + \cdots = \Xi\,\mathrm{d}\xi + \Psi\,\mathrm{d}\psi + \Phi\,\mathrm{d}\varphi + \cdots .$$

Thus the system of forces P, Q, R, etc. directed along the lines of action p, q, r, etc. is equivalent to the system of forces Ξ, Ψ, Φ, etc. directed along the lines of action ξ, ψ, φ, etc. The latter forces can replace the former in the same system of bodies to which the former system of forces are applied.[61]

SECTION III
THE GENERAL PROPERTIES OF EQUILIBRIUM OF A SYSTEM OF BODIES DEDUCED FROM THE PRECEDING FORMULA

1. Let us consider an arbitrary system or assemblage of bodies or mass points mutually in equilibrium to which are applied various forces. If, for an instant, the action of these forces ceased to be mutually equilibrated, the system would begin to move and whatever its motion, it could always be considered as composed of

1) A translational motion common to all bodies.
2) A rotational motion about an arbitrary point.
3) A relative motion of the bodies expressing their change of position and their distance from one another.

But, if they are to be in equilibrium, the bodies cannot have any of the motions cited above. However, it is obvious that the relative motions depend on the manner in which the bodies are arranged with respect to one another. Consequently, the conditions required to preclude these motions must be specifically fit to each system. Furthermore, the motions of translation and rotation can be independent of the configuration of the system and they can take place without changing the relative position of the bodies composing the system. Thus the consideration of these two types of motion must furnish the general conditions or properties for equilibrium. These conditions and general properties are what we shall investigate.

Subsection I
Properties of the Equilibrium of a Free System Relative to the Motion of Translation

2. Let there be an arbitrary number of bodies viewed as mass points and situated or connected in an arbitrary fashion. These bodies or mass points are acted upon by the forces P, P', P'', etc. along the direction of the lines p, p', p'', etc. From the preceding section, the general formula for the equilibrium of these bodies is

$$P\,\mathrm{d}p + P'\,\mathrm{d}p' + P''\,\mathrm{d}p'' + \cdots = 0.$$

By expressing in rectangular coordinates the coordinates of the different points upon which the forces P, P', etc. act as well as the coordinates of the force centers as in Article 6 of the preceding section, the following expressions for the external displacements will be obtained

$$\begin{aligned} p &= \sqrt{(x-a)^2 + (y-b)^2 + (z-c)^2} \\ p' &= \sqrt{(x'-a')^2 + (y'-b')^2 + (z'-c')^2} \\ &\vdots \end{aligned}$$

But if the bodies with coordinates x, y, z and $\bar{x}$, $\bar{y}$, $\bar{z}$, for example, act on one another through a common force which we will designate by $\bar{P}$, then the rectilinear distance between these two bodies, denoted by $\bar{p}$, will be equal to

$$\bar{p} = \sqrt{(x-\bar{x})^2 + (y-\bar{y})^2 + (z-\bar{z})^2}$$

and the term $\bar{P}\,\mathrm{d}\bar{p}$, resulting from the internal force $\bar{P}$ should be added to the general formula. If several forces acted on the same bodies, a similar procedure should be applied to each force.

3. Let us make the following definitions, which is perfectly acceptable

$$\begin{array}{lll} x' = x + \xi, & y' = y + \eta, & z' = z + \zeta \\ x'' = x + \xi', & y'' = y + \eta', & z'' = z + \zeta' \\ \vdots & & \\ \bar{x} = x + \bar{\xi}, & \bar{y} = y + \bar{\eta}, & \bar{z} = z + \bar{\zeta} \\ \vdots & & \end{array}$$

and let us further assume that these expressions have been substituted in the preceding formula.

Since x, y, z are the global coordinates[62] of the body loaded by the force P, it is clear that ξ, η, ζ, ξ', η', ζ', etc., can only be the relative coordinates of the other bodies with respect to it which is assumed to locate the common origin so that the relative position of the bodies will only depend on the latter coordinates and not on the former. Therefore, if the system is assumed to be entirely free, that is, the bodies are connected in an arbitrary fashion but without being restrained or constrained by fixed supports or arbitrary external impediments, it is easy to recognize that the resulting equations defining the configuration of the system can only be functions of the quantities ξ, η, ζ, ξ', η', ζ', etc. and not of the quantities x, y, z for which, consequently, the differentials will remain independent and indeterminate.

Thus, after making the substitutions referred to earlier, each of the relations affected by $\mathrm{d}x$, $\mathrm{d}y$, $\mathrm{d}z$ should also be equated individually to zero. This will give the following three equations (Article 2)

$$P\frac{\mathrm{d}p}{\mathrm{d}x} + P'\frac{\mathrm{d}p'}{\mathrm{d}x} + P''\frac{\mathrm{d}p''}{\mathrm{d}x} + \cdots + \bar{P}\frac{\mathrm{d}\bar{p}}{\mathrm{d}x} + \cdots = 0$$
$$P\frac{\mathrm{d}p}{\mathrm{d}y} + P'\frac{\mathrm{d}p'}{\mathrm{d}y} + P''\frac{\mathrm{d}p''}{\mathrm{d}y} + \cdots + \bar{P}\frac{\mathrm{d}\bar{p}}{\mathrm{d}y} + \cdots = 0$$
$$P\frac{\mathrm{d}p}{\mathrm{d}z} + P'\frac{\mathrm{d}p'}{\mathrm{d}z} + P''\frac{\mathrm{d}p''}{\mathrm{d}z} + \cdots + \bar{P}\frac{\mathrm{d}\bar{p}}{\mathrm{d}z} + \cdots = 0$$

At the outset, it is obvious that the variables x, y, z are not included in the expression for $\bar{p}$. Therefore, one will have

$$\frac{\mathrm{d}\bar{p}}{\mathrm{d}x} = 0, \qquad \frac{\mathrm{d}\bar{p}}{\mathrm{d}y} = 0, \qquad \frac{\mathrm{d}\bar{p}}{\mathrm{d}z} = 0, \qquad \ldots,$$

so that the terms which contain the internal forces $\bar{P}$, $\bar{P}'$, etc. will disappear.

It is also clear that the values of

$$\frac{\mathrm{d}p'}{\mathrm{d}x}, \quad \frac{\mathrm{d}p'}{\mathrm{d}y}, \quad \frac{\mathrm{d}p'}{\mathrm{d}z}, \quad \frac{\mathrm{d}p''}{\mathrm{d}x}, \quad \frac{\mathrm{d}p''}{\mathrm{d}y}, \quad \frac{\mathrm{d}p''}{\mathrm{d}z}, \quad \ldots$$

will be the same as those of

$$\frac{dp'}{dx'}, \quad \frac{dp'}{dy'}, \quad \frac{dp'}{dz'}, \quad \frac{dp''}{dx''}, \quad \frac{dp''}{dy''}, \quad \frac{dp''}{dz''}, \quad \dots$$

If α, β, γ represent the angles that the line p makes with the axes x, y, z or with the parallels to these axes and if α', β', γ' are the angles that the line p' makes with the same axes, one will obtain, as has been observed earlier (Article 7 of the preceding section)

$$\frac{dp}{dx} = \cos\alpha, \qquad \frac{dp}{dy} = \cos\beta, \qquad \frac{dp}{dz} = \cos\gamma.$$

Similarly

$$\frac{dp'}{dx'} = \cos\alpha', \qquad \frac{dp'}{dy'} = \cos\beta', \qquad \frac{dp'}{dz'} = \cos\gamma', \qquad \dots.$$

Therefore, the three equations above will become

$$P\cos\alpha + P'\cos\alpha' + P''\cos\alpha'' + \dots = 0$$
$$P\cos\beta + P'\cos\beta' + P''\cos\beta'' + \dots = 0$$
$$P\cos\gamma + P'\cos\gamma' + P''\cos\gamma'' + \dots = 0.$$

These equations must be satisfied for the equilibrium of a free system. They are necessary to prevent translational motion.

4. If the forces P, P', P'', etc. were parallel, one would have $\alpha = \alpha' = \alpha''$, etc., $\beta = \beta' = \beta''$, etc., $\gamma = \gamma' = \gamma''$, etc. and the three equations above would become $P + P' + P'' + \dots = 0$ which demonstrates that the sum of the parallel forces must be zero.

In general, it is easy to recognize that when P represents the total action of the force P in its own direction then $P\cos\alpha$ will represent its component action taken in the direction of the x-axis which makes the angle α with the direction of the force P. Similarly, $P\cos\beta$ and $P\cos\gamma$ will be the relative actions of the same force in the directions of the y- and z-axes. The other forces P', P'', etc. are dealt with in a similar fashion.

The theorem of statics which results from the above development states that "the sum of the forces in the directions of the three orthogonal axes must be equal to zero with respect to each of the three axes for a system to be in equilibrium".

Subsection II
Properties of Equilibrium Relative to Rotational Motion

5. Let us now consider, which is permissible, instead of the coordinates x, y, x', y', x'', y'', etc. and $\bar{x}$, $\bar{y}$, etc., the radius vectors ρ, ρ', ρ'', etc. and $\bar{\rho}$, etc. with the angles φ,

φ', φ'', etc. and $\bar{\varphi}$ etc. which these radii make with the x-axis. It is well known that the following relations hold between the coordinates, $x = \rho\cos\varphi$, $y = \rho\sin\varphi$ and similarly $x' = \rho'\cos\varphi'$, $y' = \rho'\sin\varphi'$, etc., $\bar{x} = \bar{\rho}\cos\bar{\varphi}$, $\bar{y} = \bar{\rho}\sin\bar{\varphi}$, etc.

Let us substitute these expressions in the general formula of Article 2 and assume that $\varphi' = \varphi + \sigma$, $\varphi'' = \varphi + \sigma'$, etc., and $\bar{\varphi} = \varphi + \bar{\sigma}$, etc. It is obvious that σ, σ', etc. and $\bar{\sigma}$, etc. will be the angles that the radii ρ', ρ'', etc. and $\bar{\rho}$, etc. make with the radius ρ. Consequently, the distances of the bodies measured with respect to the xy-plane or with respect to the point which is taken as the origin of coordinates will only depend on the quantities ρ, ρ', ρ'', etc., $\bar{\rho}$, etc., σ, σ', etc., $\bar{\sigma}$, etc., z, z', z'', etc. and $\bar{z}$, etc.

Therefore, if a system is free to rotate about this point parallel to the xy-plane, that is to say, about the z-axis which is perpendicular to this plane, the angle φ will be independent of the conditions of the system and consequently, its differential $\mathrm{d}\varphi$ will remain arbitrary. Thus all the terms multiplied by $\mathrm{d}\varphi$ in the general equation of equilibrium must all be equal to zero.

It is obvious that all these terms will be represented by $N\mathrm{d}\varphi$ by setting

$$N = P\frac{\mathrm{d}p}{\mathrm{d}\varphi} + P'\frac{\mathrm{d}p'}{\mathrm{d}\varphi} + P''\frac{\mathrm{d}p''}{\mathrm{d}\varphi} + \cdots + \bar{P}\frac{\mathrm{d}\bar{p}}{\mathrm{d}\varphi} + \cdots,$$

so that the equation for equilibrium is $N = 0$.

By substituting the values of x, y, x', y', etc. and $\bar{x}$, $\bar{y}$, etc. in the expressions for p, p', etc. and $\bar{p}$, etc. (Article 2) and also by letting $a = R\cos A$, $b = R\sin A$, $a' = R'\cos A'$, $b' = R'\sin A'$, etc., the following equations will be obtained

$$\begin{aligned}
p &= \sqrt{\rho^2 - 2\rho R\cos(\varphi - A) + R^2 + (z-c)^2}\\
p' &= \sqrt{\rho'^2 - 2\rho' R'\cos(\varphi' - A') + R'^2 + (z'-c')^2}\\
&\vdots\\
\bar{p} &= \sqrt{\rho^2 - 2\rho\bar{\rho}R\cos(\varphi - \bar{\varphi}) + \bar{\rho}^2 + (z-\bar{z})^2}\\
&\vdots
\end{aligned}$$

in which it still remains to replace φ', φ'', etc., $\bar{\varphi}$, etc. by $\varphi + \sigma$, $\varphi + \sigma'$, etc., $\varphi + \bar{\sigma}$, etc.

With the latter substitutions, it is clear at the outset that the quantities $\bar{p}$, etc. no longer include the angle φ. Thus we will have

$$\frac{\mathrm{d}\bar{p}}{\mathrm{d}\varphi} = 0$$

and consequently, the internal forces $\bar{P}$, etc. will disappear from the equation and only the external forces P, P', etc. will remain.

Therefore, we will have

$$\frac{\mathrm{d}p}{\mathrm{d}\varphi}=\frac{\rho R\sin(\varphi-A)}{p},\qquad \frac{\mathrm{d}p'}{\mathrm{d}\varphi}=\frac{\rho' R'\sin(\varphi'-A')}{p'},\qquad \ldots$$

and the quantity N will become

$$N=\frac{PR\rho\sin(\varphi-A)}{p}+\frac{P'R'\rho'\sin(\varphi'-A')}{p'}+\cdots.$$

Since the force centers of the forces P, P', etc. can be taken at arbitrary locations along the directions of these forces, the forces can be assumed to be represented by the same lines p, p', etc. which are the rectilinear distances from their points of application to their respective centers. In this fashion, the following simpler equation will be obtained

$$N=R\rho\sin(\varphi-A)+R'\rho'\sin(\varphi'-A')+\cdots.$$

In this formula, the radii R and ρ which emanate from the origin of the coordinate system and which include the angle $(\varphi-A)$, are the sides of a triangle which has for its base the projection of the line p on the xy-plane. Consequently, the quantity $R\rho\sin(\varphi-A)$ is twice the area of this triangle and a similar assertion can be made for the other quantities.

Now after having noted above (Article 3) that γ, γ', etc. are the angles which the directions of the forces P, P', etc. make with the z-axis or with lines which are parallel to this axis, it is clear that the complements of these angles will be the inclinations of the lines p, p', etc. to the xy-plane. Thus $p\sin\gamma$, $p'\sin\gamma'$, etc. will be the projections of these lines. If perpendiculars to these projections are drawn from the origin of the coordinate system which we will call Π, Π', etc., one will have $R\rho\sin(\varphi-A)=\Pi p\sin\gamma$, $R'\rho'\sin(\varphi'-A')=\Pi' p'\sin\gamma'$, etc. and the quantity N will be reduced to the following form

$$N=\Pi P\sin\gamma+\Pi' P'\sin\gamma'+\Pi'' P''\sin\gamma''+\cdots$$

by replacing p, p', p'', etc. with P, P', P'', etc.

6. Thus the equation, $N=0$, will give the following theorem: For the equilibrium of a system which is free to rotate about an axis and which is composed of bodies interacting in an arbitrary fashion and which are at the same time acted upon by external forces, the sum of the products of each force resolved parallel to a plane perpendicular to the axis and multiplied by a line drawn from the axis perpendicular to the direction of the force projected on the same plane must be zero. The forces which would cause the system to rotate in an opposite direction are given opposite signs.

This theorem is ordinarily stated more simply as: The moments of the forces about an axis must equilibrate one another to obtain equilibrium with respect to that axis. The moment of a force with respect to a line is now understood in mechanics as the product of the component of this force taken parallel to a plane which is perpendicular to this line, with its lever arm which is the perpendicular drawn from this line to the direction of the force projected on this same plane. Indeed, the action of the force which causes the system to rotate about the axis depends only on this moment, because if this force is resolved into two components, one parallel to the axis and the other in a plane perpendicular to the axis, it is obvious that only the latter could produce rotation. Consequently, we will give to this moment the specific name of moment about the axis of rotation.

7. The coefficient N in the term $N\,\mathrm{d}\varphi$ (Article 5) represents the sum of the moments of all the forces of the system about the axis of instantaneous rotation $\mathrm{d}\varphi$. Thus in order to find the sum of this moment about an arbitrary axis, the general formula $P\,\mathrm{d}p + P'\,\mathrm{d}p' + P''\,\mathrm{d}p'' +$ etc. which represents the sum of the virtual moments of all the forces will only have to be transformed by replacing one of the independent variables with the angle of rotation about the given axis. The coefficient of the differential of this angle will be the sum of all the moments about this axis. This latter expression is occasionally very useful.

8. When the system can rotate in any direction about a point which we take for the origin of the coordinate system, the instantaneous rotations about the three axes x, y, z must be considered simultaneously. An equation similar to the one we just found, and which contains the property of moments will be obtained with respect to each of these axes. But it will be useful to resolve the same problem by a simpler and more general analysis. To this effect, as in Article 5, let us set

$$x = \rho\cos\varphi, \qquad y = \rho\sin\varphi, \qquad x' = \rho'\cos\varphi', \qquad y', = \rho'\sin\varphi', \qquad \ldots.$$

By permitting the angles φ, φ', etc. to vary by the same amount $\mathrm{d}\varphi$, one will have

$$\mathrm{d}x = -y\,\mathrm{d}\varphi, \qquad \mathrm{d}y = x\,\mathrm{d}\varphi, \qquad \mathrm{d}x' = -y'\,\mathrm{d}\varphi, \qquad \mathrm{d}y' = x'\,\mathrm{d}\varphi, \qquad \ldots.$$

These expressions are the variations of x, y, x', y', etc. resulting from the elementary rotation $\mathrm{d}\varphi$ of the system about the z-axis.

Similarly, the variations of y, z, y', z', etc. resulting from an elementary rotation $\mathrm{d}\psi$ about the x-axis can be obtained by simply replacing x, y, x', y', etc. in the preceding formulas by y, z, y', z', etc. and $\mathrm{d}\varphi$ by $\mathrm{d}\psi$. This will give

$$\mathrm{d}y = -z\,\mathrm{d}\psi, \qquad \mathrm{d}z = y\,\mathrm{d}\psi, \qquad \mathrm{d}y' = -z'\,\mathrm{d}\psi, \qquad \mathrm{d}z' = y'\,\mathrm{d}\psi, \qquad \ldots.$$

By replacing y, z, y', z', etc. in these latter formulas with z, x, z', x', etc. and $\mathrm{d}\psi$ by $\mathrm{d}\omega$ the variations resulting from the elementary rotation $\mathrm{d}\omega$ about the y-axis will be obtained

$$\mathrm{d}z = -x\,\mathrm{d}\omega, \qquad \mathrm{d}x = z\,\mathrm{d}\omega, \qquad \mathrm{d}z' = -x'\,\mathrm{d}\omega, \qquad \mathrm{d}x' = z'\,\mathrm{d}\omega, \qquad \ldots.$$

If the three rotations are assumed to occur simultaneously, the total variations of the coordinates x, y, z, x', y', z', etc. will be, by the principles of the differential calculus, equal to the sums of the partial variations resulting from each of these rotations so that the following complete expressions will result

$$\mathrm{d}x = z\,\mathrm{d}\omega - y\,\mathrm{d}\varphi, \qquad \mathrm{d}y = x\,\mathrm{d}\varphi - z\,\mathrm{d}\psi, \qquad \mathrm{d}z = y\,\mathrm{d}\psi - x\,\mathrm{d}\omega,$$

$$\mathrm{d}x' = z'\,\mathrm{d}\omega - y'\,\mathrm{d}\varphi, \qquad \mathrm{d}y' = x'\,\mathrm{d}\varphi - z'\,\mathrm{d}\psi, \qquad \mathrm{d}z' = y'\,\mathrm{d}\psi - x'\,\mathrm{d}\omega,$$

$$\vdots$$

By substituting these expressions in the general formula of equilibrium (Article 2), one will have the terms resulting only from the rotations $\mathrm{d}\varphi$, $\mathrm{d}\omega$, $\mathrm{d}\psi$ about the z-, y- and x-axes, which should be set equal to zero separately when the system is free to rotate in any direction about a point which is at the origin of the coordinate system.

By differentiation, the following equations will be obtained

$$\mathrm{d}p = \frac{(x-a)\,\mathrm{d}x + (y-b)\,\mathrm{d}y + (z-c)\,\mathrm{d}z}{p}$$

$$\mathrm{d}p' = \frac{(x'-a')\,\mathrm{d}x' + (y'-b')\,\mathrm{d}y' + (z'-c')\,\mathrm{d}z'}{p'}$$

$$\vdots$$

$$\mathrm{d}\bar{p} = \frac{(x-\bar{x})(\mathrm{d}x-\mathrm{d}\bar{x}) + (y-\bar{y})(\mathrm{d}y-\mathrm{d}\bar{y}) + (z-\bar{z})(\mathrm{d}z-\mathrm{d}\bar{z})}{\bar{p}}$$

$$\vdots$$

Thus after substitution, the following equations will result

$$\mathrm{d}p = \frac{(ay-bx)\,\mathrm{d}\varphi + (bz-cy)\,\mathrm{d}\psi + (cx-az)\,\mathrm{d}\omega}{p}$$

$$\mathrm{d}p' = \frac{(a'y'-b'x')\,\mathrm{d}\varphi + (b'z'-c'y')\,\mathrm{d}\psi + (c'x'-a'z')\,\mathrm{d}\omega}{p'}$$

$$\vdots$$

And it will be found that $\mathrm{d}\bar{p} = 0$, $\mathrm{d}\bar{p}' = 0$, etc. by replacing $\mathrm{d}\bar{x}$, $\mathrm{d}\bar{y}$, $\mathrm{d}\bar{z}$, etc. by the analogous values $\bar{z}\,\mathrm{d}\omega - \bar{y}\,\mathrm{d}\varphi$, $\bar{x}\,\mathrm{d}\varphi - \bar{z}\,\mathrm{d}\psi$, $\bar{y}\,\mathrm{d}\psi - \bar{x}\,\mathrm{d}\omega$ etc. It can then be concluded immediately that the terms $\bar{P}\,\mathrm{d}\bar{p}$, $\bar{P}'\,\mathrm{d}\bar{p}'$, etc. of the same equation, which would result from the internal forces of the system, will disappear after substitution.

Also, $\mathrm{d}p = 0$, if one sets $a = 0$, $b = 0$ and $c = 0$, that is, if the force centers for the forces P are at the origin of the coordinate system. This will also result in the elimination of this force.

9. Thus if the internal forces are not considered, should there be any, and in addition, if all forces which are directed towards the origin of the coordinate system are not considered,

one will have in general the following equation for all the forces P, P', etc. with directions along the lines p, p', etc.

$$L\,\mathrm{d}\psi + M\,\mathrm{d}\omega + N\,\mathrm{d}\varphi = 0,$$

where

$$L = \frac{P(bz - cy)}{p} + \frac{P'(b'z' - c'y')}{p'} + \cdots$$
$$M = \frac{P(cx - az)}{p} + \frac{P'(c'x' - a'z')}{p'} + \cdots$$
$$N = \frac{P(ay - bx)}{p} + \frac{P'(a'y' - b'x')}{p'} + \cdots$$

and for any system free to rotate in any direction about the origin of the coordinate system three equations $L = 0$, $M = 0$ and $N = 0$ will result. These equations are related to the equation of Article 5, which is written with respect to three coordinate axes.

Also, if we use the angles α, β, γ, α', etc. instead of the coordinates a, b, c, a', etc. of the force centers that the directions of the forces make with the coordinate axes and if we use the relations of Article 7 of the preceding section

$$a = x - p\cos\alpha, \qquad b = y - p\cos\beta, \qquad c = z - p\cos\gamma$$

and if we do the same for all other similar quantities, there results

$$L = P(y\cos\gamma - z\cos\beta) + P'(y'\cos\gamma' - z'\cos\beta') + \cdots$$
$$M = P(z\cos\alpha - x\cos\gamma) + P'(z'\cos\alpha' - x'\cos\gamma') + \cdots$$
$$N = P(x\cos\beta - y\cos\alpha) + P'(x'\cos\beta' - y'\cos\alpha') + \cdots$$

But since $P\cos\alpha$, $P\cos\beta$, $P\cos\gamma$ are the components of the force P calculated along the directions of the three axes x, y, z, one sees immediately that $xP\cos\beta - yP\cos\alpha$ are the moments relative to the z-axis, in which the term $yP\cos\alpha$ has a negative sign because the force $P\cos\alpha$ tends to rotate the system in the opposite direction of the force $P\cos\beta$. Similarly, $zP\cos\alpha - xP\cos\gamma$ will be the moment relative to the y-axis and $yP\cos\gamma - zP\cos\beta$, the moment relative to the x-axis, and the same should be done for all other analogous expressions. Thus the three equations $L = 0$, $M = 0$, $N = 0$ express the fact that the sum of these moments is zero with respect to each of the three axes.

Also, it is clear that the coefficients L, M, N of the instantaneous rotations $\mathrm{d}\psi$, $\mathrm{d}\omega$, $\mathrm{d}\varphi$ are only the moments relative to the axes of instantaneous rotation (Article 7) $\mathrm{d}\psi$, $\mathrm{d}\omega$, $\mathrm{d}\varphi$.

10. One could doubt that the rotations about the three coordinate axes will suffice to represent all the small motions that a system of material points may have about a fixed point, without changing their relative configuration. In order to dispel this doubt we are going to investigate directly these types of motion.

Let us imagine a straight line passing through the given point which is also used as the origin of coordinates x, y, z and through another point in the system. Now let us also imagine a plane defined by this straight line and by a third point of the system. Furthermore, let us represent with respect to this line and this plane the other points of the system by new rectangular coordinates x', y', z' having the same origin as the former coordinates x, y, z. It is clear that these new coordinates will depend only on the relative configuration of the points of the system and consequently, will be constant when the system moves while the old coordinates would have changed.

The well-known theory of the transformation of coordinates gives directly the relations between the first three and the last three coordinates as

$$\begin{aligned} x &= \alpha x' + \beta y' + \gamma z' \\ y &= \alpha' x' + \beta' y' + \gamma' z' \\ z &= \alpha'' x'' + \beta'' y'' + \gamma'' z'' \end{aligned}$$

The nine coefficients α, β, γ, α', etc. depend only on the relative position of the axes of the two systems of coordinates and must be such that the coordinates x, y, z represent the same points as the coordinates x', y', z', and consequently, the two expressions $x^2 + y^2 + z^2$ and $x'^2 + y'^2 + z'^2$ are identical, which gives these six equations of condition

$$\begin{aligned} &\alpha^2 + \alpha'^2 + \alpha''^2 = 1, && \beta^2 + \beta'^2 + \beta''^2 = 1 \\ &\gamma^2 + \gamma'^2 + \gamma''^2 = 1, && \alpha\beta + \alpha'\beta' + \alpha''\beta'' = 0 \\ &\alpha\gamma + \alpha'\gamma' + \alpha''\gamma'' = 0, && \beta\gamma + \beta'\gamma' + \beta''\gamma'' = 0 \end{aligned}$$

so that among the nine quantities α, β, γ, α', etc. three will remain indeterminate.

When the axes of the coordinates x', y', z' coincide with the axes of the coordinates x, y, z then $x = x'$, $y = y'$ and $z = z'$. Consequently, there results $\alpha = 1$, $\beta = 0$, $\gamma = 0$, $\alpha' = 0$, $\beta' = 1$, $\beta'' = 0$, $\gamma = 0$, $\gamma' = 0$, $\gamma'' = 1$. Thus by differentiating the preceding formulas and by making these substitutions one will have the expression for an arbitrary, infinitesimal displacement of the system about a given point.

By differentiating the expressions of x, y, z under the hypothesis that x', y', z' are constant and then by substituting x, y, z instead of these quantities, the result will be

$$\begin{aligned} \mathrm{d}x &= x\,\mathrm{d}\alpha + y\,\mathrm{d}\beta + z\,\mathrm{d}\gamma \\ \mathrm{d}y &= x\,\mathrm{d}\alpha' + y\,\mathrm{d}\beta' + z\,\mathrm{d}\gamma' \\ \mathrm{d}z &= x\,\mathrm{d}\alpha'' + y\,\mathrm{d}\beta'' + z\,\mathrm{d}\gamma'' \end{aligned}$$

But the six equations of condition after differentiation, give by the substitution of the values $\alpha = 1$, $\beta = 0$, $\gamma = 0$, etc. found above, $\mathrm{d}\alpha = 0$, $\mathrm{d}\beta' = 0$, $\mathrm{d}\gamma' = 0$, $\mathrm{d}\beta + \mathrm{d}\alpha' = 0$, $\mathrm{d}\gamma + \mathrm{d}\alpha'' = 0$, $\mathrm{d}\gamma' + \mathrm{d}\beta'' = 0$, from which $\mathrm{d}\alpha' = -\mathrm{d}\beta$, $\mathrm{d}\alpha'' = -\mathrm{d}\gamma$, $\mathrm{d}\beta'' = -\mathrm{d}\gamma'$.

After these values are substituted in the expressions for $\mathrm{d}x$, $\mathrm{d}y$, $\mathrm{d}z$, one will have the following

$$\mathrm{d}x = -y\,\mathrm{d}\alpha' + z\,\mathrm{d}\gamma, \qquad \mathrm{d}y = x\,\mathrm{d}\alpha' - z\,\mathrm{d}\beta'', \qquad \mathrm{d}z = -x\,\mathrm{d}\gamma + y\,\mathrm{d}\beta'',$$

which are equal to those found in Article 8 after setting $\mathrm{d}\alpha' = \mathrm{d}\varphi$, $\mathrm{d}\gamma = \mathrm{d}\omega$, $\mathrm{d}\beta'' = \mathrm{d}\psi$.

These formulas for the variations of x, y, z have all the generality that the statements of the problem may include. The three equations $L = 0$, $M = 0$, $N = 0$ which result from the elimination of the terms affected by $\mathrm{d}\psi$, $\mathrm{d}\omega$, $\mathrm{d}\varphi$ in the general equation of equilibrium are consequently the only ones necessary to maintain the system in equilibrium about a given point, not considering the relative configuration of the points. So that when this configuration is invariant the equilibrium of the system will depend only on the three cited equations.

In his Recherches sur la Précession des équinoxes,[63] d'Alembert is the first to have found the laws of equilibrium for several forces applied to a system of points with an invariant configuration. He arrived at this result by a very complicated procedure using the composition and resolution of forces. Since then these laws have been demonstrated more simply by different authors. But our formulas have the advantage of leading to the conclusion directly.

Subsection III
The Composition of Rotational Motion And of Moments About Different Axes

11. If a point in the system is selected for which the coordinates x, y, z are proportional to $\mathrm{d}\psi$, $\mathrm{d}\omega$, $\mathrm{d}\varphi$, the corresponding differentials $\mathrm{d}x$, $\mathrm{d}y$, $\mathrm{d}z$ will be zero, as can be seen from the formulas of Article 8. This point and all those which will have the same property will thus be immobile during the instant that the system traverses the three angles $\mathrm{d}\psi$, $\mathrm{d}\omega$, $\mathrm{d}\varphi$, by rotating simultaneously about the x-, y-, z-coordinate axes. And it is easy to see that all these points will be on a straight line passing through the origin of the coordinates and making an angle λ, μ, ν with the x, y, z-axes, respectively, such that

$$\cos\lambda = \frac{\mathrm{d}\psi}{\sqrt{(\mathrm{d}\psi^2 + \mathrm{d}\omega^2 + \mathrm{d}\varphi^2)}}$$
$$\cos\mu = \frac{\mathrm{d}\omega}{\sqrt{(\mathrm{d}\psi^2 + \mathrm{d}\omega^2 + \mathrm{d}\varphi^2)}}$$
$$\cos\nu = \frac{\mathrm{d}\varphi}{\sqrt{(\mathrm{d}\psi^2 + \mathrm{d}\omega^2 + \mathrm{d}\varphi^2)}}.$$

This line will be the instantaneous axis of the composed rotation.

By using the angles λ, μ, ν and taking, in order to shorten the equations

$$d\theta = \sqrt{(d\psi^2 + d\omega^2 + d\varphi^2)},$$

one will have

$$d\psi = d\theta \cos\lambda, \qquad d\omega = d\theta \cos\mu, \qquad d\varphi = d\theta \cos\nu$$

and the general expressions for dx, dy, dz (Article 8) will become

$$\begin{aligned} dx &= (z \cos\mu - y \cos\nu)\, d\theta \\ dy &= (x \cos\nu - z \cos\lambda)\, d\theta \\ dz &= (y \cos\lambda - x \cos\mu)\, d\theta. \end{aligned}$$

Since the square of the small distance traversed by an arbitrary point is $dx^2 + dy^2 + dz^2$, it will be expressed by

$$\begin{aligned} &((z \cos\mu - y \cos\nu)^2 + (x \cos\nu - z \cos\lambda)^2 + (y \cos\lambda - x \cos\mu)^2)\, d\theta^2 \\ &\quad = (x^2 + y^2 + z^2 - (x \cos\lambda + y \cos\mu + z \cos\nu)^2)\, d\theta^2 \end{aligned}$$

because $\cos^2\lambda + \cos^2\mu + \cos^2\nu = 1$.

It is easy to prove that $x \cos\lambda + y \cos\mu + z \cos\nu = 0$ is the equation of the plane passing through the origin of the coordinate system and perpendicular to the line which makes the angles λ, μ, ν with the x, y, z coordinate axes. Thus the small distance traversed by an arbitrary point of this plane will be

$$d\theta \sqrt{(x^2 + y^2 + z^2)}$$

and since the instantaneous axis of rotation is perpendicular to this same plane, it results that $d\theta$ will be the angle of rotation about this axis, composed of three partial rotations $d\psi$, $d\omega$, $d\varphi$ about the three coordinate axes.

12. It follows that arbitrary instantaneous rotations $d\psi$, $d\omega$, $d\varphi$ about three orthogonal axes, intersecting at the same point, are composed of one rotation $d\theta = \sqrt{(d\psi^2 + d\omega^2 + d\varphi^2)}$, about an axis passing through this same point of intersection and making the angles λ, μ, ν, with these three axes such that

$$\cos\lambda = \frac{d\psi}{d\theta}, \qquad \cos\mu = \frac{d\omega}{d\theta}, \qquad \cos\nu = \frac{d\varphi}{d\theta},$$

and conversely, an arbitrary rotation $d\theta$ about a given axis can be resolved into three partial rotations expressed by $\cos\lambda\, d\theta$, $\cos\mu\, d\theta$, $\cos\nu\, d\theta$ about three axes which intersect at right angles on a point of the given axis and which make with this axis the angles λ, μ, ν. This gives a simple way to compose and resolve instantaneous motions or rotational velocities.

Thus if one takes three other orthogonal axes which make with the axis of rotation $\mathrm{d}\psi$ the angles λ', λ'', λ''', with the axis of rotation $\mathrm{d}\omega$ the angles μ', μ'', μ''', and with the axis of rotation $\mathrm{d}\varphi$ the angles ν', ν'', ν''', the rotation $\mathrm{d}\psi$ can be resolved into three rotations $\cos\lambda'\,\mathrm{d}\psi$, $\cos\lambda''\,\mathrm{d}\psi$, $\cos\lambda'''\,\mathrm{d}\psi$ about these new axes. The rotation $\mathrm{d}\omega$ will be resolved similarly into three rotations $\cos\mu'\,\mathrm{d}\omega$, $\cos\mu''\,\mathrm{d}\omega$, $\cos\mu'''\,\mathrm{d}\omega$, and the rotation $\mathrm{d}\varphi$ in three rotations $\cos\nu'\,\mathrm{d}\varphi$, $\cos\nu''\,\mathrm{d}\varphi$, $\cos\nu'''\,\mathrm{d}\varphi$ about the same axes. By adding together the rotations about a given axis and calling $\mathrm{d}\theta'$, $\mathrm{d}\theta''$, $\mathrm{d}\theta'''$ the total rotations about the three new axes, one will have

$$\begin{aligned}
\mathrm{d}\theta' &= \cos\lambda'\,\mathrm{d}\psi + \cos\mu'\,\mathrm{d}\omega + \cos\nu'\,\mathrm{d}\varphi \\
\mathrm{d}\theta'' &= \cos\lambda''\,\mathrm{d}\psi + \cos\mu''\,\mathrm{d}\omega + \cos\nu''\,\mathrm{d}\varphi \\
\mathrm{d}\theta''' &= \cos\lambda'''\,\mathrm{d}\psi + \cos\mu'''\,\mathrm{d}\omega + \cos\nu'''\,\mathrm{d}\varphi.
\end{aligned}$$

13. The rotations $\mathrm{d}\psi$, $\mathrm{d}\omega$, $\mathrm{d}\varphi$ are thus reduced in this fashion to three rotations $\mathrm{d}\theta'$, $\mathrm{d}\theta''$, $\mathrm{d}\theta'''$, about three orthogonal axes which consequently, must give by composition the same rotation $\mathrm{d}\theta$ which results from the rotations $\mathrm{d}\psi$, $\mathrm{d}\omega$, $\mathrm{d}\varphi$ so that one will have (Article 11)

$$\mathrm{d}\theta^2 = \mathrm{d}\theta'^2 + \mathrm{d}\theta''^2 + \mathrm{d}\theta'''^2 = \mathrm{d}\psi^2 + \mathrm{d}\omega^2 + \mathrm{d}\varphi^2$$

and since this last equation must be an identity the following relations are obtained

$$\begin{aligned}
&\cos\lambda'^2 + \cos\lambda''^2 + \cos\lambda'''^2 = 1 \\
&\cos\mu'^2 + \cos\mu''^2 + \cos\mu'''^2 = 1 \\
&\cos\nu'^2 + \cos\nu''^2 + \cos\nu'''^2 = 1 \\
&\cos\lambda'\cos\mu' + \cos\lambda''\cos\mu'' + \cos\lambda'''\cos\mu''' = 0 \\
&\cos\lambda'\cos\nu' + \cos\lambda''\cos\nu'' + \cos\lambda'''\cos\nu''' = 0 \\
&\cos\mu'\cos\nu' + \cos\mu''\cos\nu'' + \cos\mu'''\cos\nu''' = 0,
\end{aligned}$$

which can also be found by geometry.

From these relations the values of $\mathrm{d}\psi$, $\mathrm{d}\omega$, $\mathrm{d}\varphi$ in terms of $\mathrm{d}\theta'$, $\mathrm{d}\theta''$, $\mathrm{d}\theta'''$, can be immediately obtained by adding together the values of $\mathrm{d}\theta'$, $\mathrm{d}\theta''$, $\mathrm{d}\theta'''$, successively multiplied by $\cos\lambda'$, $\cos\lambda''$, $\cos\lambda'''$, $\cos\mu'$, $\cos\mu''$, etc. With this procedure the following equations will be obtained

$$\begin{aligned}
\mathrm{d}\psi &= \cos\lambda'\,\mathrm{d}\theta' + \cos\lambda''\,\mathrm{d}\theta'' + \cos\lambda'''\,\mathrm{d}\theta''' \\
\mathrm{d}\omega &= \cos\mu'\,\mathrm{d}\theta' + \cos\mu''\,\mathrm{d}\theta'' + \cos\mu'''\,\mathrm{d}\theta''' \\
\mathrm{d}\varphi &= \cos\nu'\,\mathrm{d}\theta' + \cos\nu''\,\mathrm{d}\theta'' + \cos\nu'''\,\mathrm{d}\theta'''.
\end{aligned}$$

14. Moreover, if π', π'', π''' designate the angles that the axis of the composed rotation $\mathrm{d}\theta$ makes with the axes of the three partial rotations $\mathrm{d}\theta'$, $\mathrm{d}\theta''$, $\mathrm{d}\theta'''$, one will have as in Article 11

$$\mathrm{d}\theta'\cos\pi'\,\mathrm{d}\theta, \qquad \mathrm{d}\theta''\cos\pi''\,\mathrm{d}\theta, \qquad \mathrm{d}\theta'''\cos\pi'''\,\mathrm{d}\theta,$$

and if in the expressions given above (Article 12) for $d\theta$, $d\theta'$, $d\theta'$, the angles $d\psi$, $d\omega$, $d\varphi$ are replaced by the values expressed in $d\theta$ of Article 11, that is, $\cos\lambda\, d\theta$, $\cos\mu\, d\theta$, $\cos\nu\, d\theta$ the comparison of the different expressions for $d\theta'$, $d\theta''$, $d\theta'''$ will give after division by $d\theta$, the new relations

$$\cos\pi' = \cos\lambda\cos\lambda' + \cos\mu\cos\mu' + \cos\nu\cos\nu'$$
$$\cos\pi'' = \cos\lambda\cos\lambda'' + \cos\mu\cos\mu'' + \cos\nu\cos\nu''$$
$$\cos\pi''' = \cos\lambda\cos\lambda''' + \cos\mu\cos\mu''' + \cos\nu\cos\nu'''$$

which can be verified by geometry.

15. It is clear from this development that the composition and resolution of rotational motions are entirely analogous to rectilinear motions. Indeed, if on the three axes of rotation $d\psi$, $d\omega$, $d\varphi$, one takes from their point of intersection lines proportional respectively to $d\psi$, $d\omega$, $d\varphi$, and if one draws on these three lines a rectangular parallelepiped, it is easy to see that the diagonal of this parallelepiped will be the axis of composed rotation $d\theta$ and will be at the same time proportional to this rotation $d\theta$. From this result and because the rotations about the same axis can be added or subtracted depending on whether they are in the same or opposite directions as the motions which are in the same or opposite directions, in general, one must conclude that the composition and resolution of rotational motions is done in the same manner and by the same laws that the composition or resolution of rectilinear motions, by substituting for rotational motions rectilinear motions along the direction of the axes of rotation.

16. Now if in the formula of Article 9

$$L\, d\psi + M\, d\omega + N\, d\varphi$$

which contains the terms resulting from the rotations $d\psi$, $d\omega$, $d\varphi$ in the general formula $P\, dp + P'\, dp' + P''\, dp''$ + etc., one substitutes for $d\psi$, $d\omega$, $d\varphi$, the expressions found in Article 13, this formula becomes

$$(L\cos\lambda' + M\cos\mu' + N\cos\nu')\, d\theta'$$
$$+(L\cos\lambda'' + M\cos\mu'' + N\cos\nu'')\, d\theta''$$
$$+(L\cos\lambda''' + M\cos\mu''' + N\cos\nu''')\, d\theta'''$$

Thus from Article 7, the coefficients of the elementary angles $d\theta'$, $d\theta''$, $d\theta'''$, will express the sum of the motions relative to the axes of rotation $d\theta'$, $d\theta''$, $d\theta'''$. Therefore, the moments which are equal to L, M, N and relative to three rectangular axes, are expressed by the following three equations

$$L\cos\lambda' + M\cos\mu' + N\cos\nu',$$
$$L\cos\lambda'' + M\cos\mu'' + N\cos\nu'',$$
$$L\cos\lambda''' + M\cos\mu''' + N\cos\nu'''$$

relative to three other rectangular axes which make respectively with the former, the angles $\lambda', \mu', \nu'; \lambda'', \mu'', \nu''; \lambda''', \mu''', \nu'''$.

There is a geometrical demonstration of this theorem in Volume VII of the Nova Acta of the Académie de Petérsbourg.[64]

17. If the rotations $d\psi$, $d\omega$, $d\varphi$ are assumed proportional to L, M, N and if

$$H = \sqrt{L^2 + M^2 + N^2},$$

then by Article 11 the following equations will result

$$L = H\cos\lambda, \qquad M = H\cos\mu, \qquad N = H\cos\nu$$

and the three moments which we have just found will be reduced, using the relations of Article 14, to the following simple form

$$H\cos\pi', \qquad H\cos\pi'', \qquad H\cos\pi'''.$$

But π', π'', π''' are the angles that the axes of rotation $d\theta', d\theta'', d\theta'''$, make with the axis of the composed rotation $d\theta$. Therefore, if the axis of rotation $d\theta'$ is made to coincide with the axis of rotation $d\theta$, one has $\pi' = 0$, and π'', π''' each equal to a right angle. Consequently, the moment about this axis will be simply H and the two other moments about the axes perpendicular to this one will be zero.

Thus it results from this development that moments equal to L, M, N and relative to three rectangular axes, will compose a unique moment H equal to $\sqrt{(L^2 + M^2 + N^2)}$ and relative to an axis which makes the angles λ, μ, ν, with respect to the rectangular axes such that

$$\cos\lambda = \frac{L}{H}, \qquad \cos\mu = \frac{M}{H}, \qquad \cos\nu = \frac{N}{H}.$$

These are the known theorems on the composition of moments. And it is obvious that this composition also follows the same rules as the composition of rectilinear motions. One could have immediately deduced this result from the composition of instantaneous rotations by substituting the moments for the rotations that they produce as Varignon substituted forces for rectilinear motions.

Subsection IV
Properties of Equilibrium Relative to the Center of Gravity

18. If in the formulas of Article 9, all of the forces P, P', P'', etc. are assumed to act along directions mutually parallel, one will have $\alpha = \alpha' = \alpha''$, etc., $\beta = \beta' = \beta''$, etc.,

$\gamma = \gamma' = \gamma''$, etc. Consequently, if the following relations are defined, in order to shorten the formulas

$$X = Px + P'x' + P''x'' + \cdots$$
$$Y = Py + P'y' + P''y'' + \cdots$$
$$Z = Pz + P'z' + P''z'' + \cdots$$

the quantities L, M, N will become

$$L = Y\cos\gamma - Z\cos\beta$$
$$M = Z\cos\alpha - X\cos\gamma$$
$$N = X\cos\beta - Y\cos\alpha$$

and the equations of equilibrium will be $L = 0$, $M = 0$, $N = 0$, of which the third is here a permutation of the first two equations. But since there is also the equation $\cos^2\alpha + \cos^2\beta + \cos^2\gamma = 1$ (SECTION II, Article 7), one will be able to determine with these equations the angles α, β, γ and it will be found that

$$\cos\alpha = \frac{X}{\sqrt{(X^2 + Y^2 + Z^2)}}$$
$$\cos\beta = \frac{Y}{\sqrt{(X^2 + Y^2 + Z^2)}}$$
$$\cos\gamma = \frac{Z}{\sqrt{(X^2 + Y^2 + Z^2)}}.$$

Therefore, if the positions of the bodies are given with respect to three axes, the system must be placed relative to the direction of the forces, such that this direction makes with the three axes the angles α, β, γ that have just been determined, so that all rotational motion of the system is prevented.

19. If the quantities X, Y, Z were equal to zero, the angles α, β, γ will remain indeterminate and the position of the system, relative to the direction of the forces, could be arbitrary. From this result the following theorem is obtained: If the sum of the products of parallel forces with their distances to three orthogonal planes is equal to zero with respect to each of these planes, the ability of the forces to rotate the system about the common point of intersection of these three planes is equal to zero.

It is known that gravity acts vertically and proportional to mass. Thus in a system of heavy bodies, if one seeks a point such that the sum of the masses multiplied by their distances to three orthogonal planes passing through this point is zero, gravity could not act on the system to produce rotational motion about this point. This point is called the center of gravity and it is used often in all branches of mechanics.

In order to determine this point, it is only necessary to find its distance to three given orthogonal planes. But because the sum of the products of the masses with their distances to a plane passing through the center of gravity is zero, the sum of the products of the same masses with their distances to another plane parallel to the former plane will necessarily be equal to the product of the sum of the masses with the distance of the center of gravity to the latter plane. This distance will be obtained by dividing the sum of the products of the masses with their distances by the total sum of the mass. From this result, the well-known formulas for the centers of gravity of lines, surfaces and solids are obtained.

20. But there exists a property of the center of gravity which is not as well-known and which is occasionally useful because it is independent of the planes to which the different bodies of the system are referred and because it is used to determine the center of gravity from the relative position of the bodies. This is what it consists of:

Let A be the sum of the products of the masses taken two at a time with the square of their relative distance divided by the square of the sum of the masses.

Let B be the sum of the products of each mass with the square of its distance to a given arbitrary point divided by the sum of the masses.

Then the quantity $\sqrt{(B - A)}$ is the distance between the center of gravity of all the masses and the point about which the moments of the masses are taken. Since the quantity A is independent of this latter point, the values of B can be determined with respect to three different points taken either in or outside of the system. Hence, the distance of the center of gravity to these three points will be found and consequently, the position of the center of gravity with respect to these points will be known. If the bodies were all in the same plane, it would be sufficient to consider only two points and it will be sufficient to consider only one if all the bodies were on a given straight line.

By requiring the given points to be within the bodies of the system, the position of the center of gravity of the system will be given uniquely by the masses and their respective distances to the point. This is where the main advantage of this approach to determine the center of gravity lies.

To demonstrate it, I begin again with the expressions X, Y, Z of Article 18 and I also take three arbitrary quantities f, g, h and develop these three identities which are easy to verify,

$$\begin{aligned}
&(X - (P + P' + P'' + \cdots)f)^2 \\
&\quad = (P + P' + P'' + \cdots)(P(x - f)^2 + P'(x' - f)^2 + P''(x'' - f)^2 + \cdots) \\
&\qquad - PP'(x - x')^2 - PP''(x - x'')^2 - P'P''(x' - x'')^2 - \cdots \\
&(Y - (P + P' + P'' + \cdots)g)^2 \\
&\quad = (P + P' + P'' + \cdots)(P(y - g)^2 + P'(y' - g)^2 + P''(y'' - g)^2 + \cdots) \\
&\qquad - PP'(y - y')^2 - PP''(y - y'')^2 - P'P''(y' - y'')^2 - \cdots \\
&(Z - (P + P' + P'' + \cdots)h)^2 \\
&\quad = (P + P' + P'' + \cdots)(P(z - h)^2 + P'(z' - h)^2 + P''(z'' - h)^2 + \cdots) \\
&\qquad - PP'(z - z')^2 - PP''(z - z'')^2 - P'P''(z' - z'')^2 - \cdots
\end{aligned}$$

The quantities P, P', P'', etc. represent the weights or the masses of the bodies which are proportional to them and the quantities x, y, z, x', y', z', x'', etc. are the rectangular coordinates of these bodies. But we have seen (Article 19) that when the origin of the coordinates is at the center of gravity, the three quantities X, Y, Z are zero. Therefore, if one sets in the three preceding equations $X = 0$, $Y = 0$ and $Z = 0$, adds them together and assumes in order to shorten the expression, the following relations

$$
\begin{aligned}
&f^2 + g^2 + h^2 = r^2 \\
&(x - f)^2 + (y - g)^2 + (z - h)^2 = (0)^2 \\
&(x' - f)^2 + (y' - g)^2 + (z' - h)^2 = (1)^2 \\
&(x'' - f)^2 + (y'' - g)^2 + (z'' - h)^2 = (2)^2 \\
&\vdots \\
&(x - x')^2 + (y - y')^2 + (z - z')^2 = (0,1)^2 \\
&(x - x'')^2 + (y - y'')^2 + (z - z'')^2 = (0,2)^2 \\
&(x' - x'')^2 + (y' - y'')^2 + (z' - z'')^2 = (1,2)^2 \\
&\vdots
\end{aligned}
$$

After dividing by $(P + P' + P'' + \cdots)^2$, the following expression will be found

$$
\begin{aligned}
r^2 = &\frac{P(0)^2 + P'(1)^2 + P''(2)^2 + \cdots}{P + P' + P'' + \cdots} \\
&- \frac{PP'(0,1)^2 + PP''(0,2)^2 + P'P''(1,2)^2 + \cdots}{(P + P' + P'' + \cdots)^2}.
\end{aligned}
$$

Now if the three quantities f, g, h are taken for the rectangular coordinates of a given point, it is evident that r will be the distance of this point to the center of gravity which is assumed at the origin of the coordinate system and that (0), (1), (2), etc. will be the distances of the weights P, P', P'', etc. to this same point. Also, (0,1), (0,2), (1,2) etc. will be the distances between the bodies or weights P and P', P and P'', P' and P'', etc. Therefore the above equation will become $r^2 = B - A$, from which one has $r = \sqrt{(B - A)}$.

Subsection V
Properties of Equilibrium Relative to Maxima and Minima

21. We are now going to consider the properties of maxima and minima which occur at equilibrium. For this effort, we will consider again the general formula

$$P\,\mathrm{d}p + Q\,\mathrm{d}q + R\,\mathrm{d}r + \cdots = 0$$

of equilibrium between the forces P, Q, R, etc. directed along the lines p, q, r, etc. which pass through the centers of these forces (SECTION II, Article 4).

It can be assumed that these forces are expressed in such a fashion that the quantity $P\,\mathrm{d}p + Q\,\mathrm{d}q + R\,\mathrm{d}r +$ etc. is an exact differential of a function of p, q, r, etc. which is represented by Π, so that the following equation results

$$\mathrm{d}\Pi = P\,\mathrm{d}p + Q\,\mathrm{d}q + R\,\mathrm{d}r + \cdots.$$

Then for equilibrium the equation $\mathrm{d}\Pi = 0$ will hold which indicates that the system must be configured in such a manner that the function Π is, generally speaking, a maximum or a minimum in the state of equilibrium. I say, generally speaking, because it is known from the theory of curves that the equation of a differential set equal to zero does not always represent a maximum or a minimum.

The preceding assumption holds when the forces P, Q, R, etc. are definitely directed toward either fixed points or bodies of the same system, and are proportional to arbitrary functions of distance, which is properly the case in nature. Thus for this type of force, the system will be in equilibrium when the function Π is a maximum or a minimum. This constitutes the principle which Maupertuis called the *Loi du Repos*.[65]

In a system of heavy bodies in equilibrium, the forces P, Q, R, etc. resulting from gravity are, as it is known, proportional to the masses of the bodies and consequently, they are constant. The lines p, q, r, etc. converge at the center of the Earth. Thus one will have in this case that $\Pi = Pp + Qq + Rr + \cdots$.

Consequently, since the lines p, q, r, etc. are assumed to be parallel,[66] the quantity

$$\frac{\Pi}{P + Q + R + \cdots}$$

will represent the distance from the center of gravity of the system to the center of the Earth. This distance will be a minimum or a maximum when the system is in equilibrium. For example, it will be a minimum in the case of the catenary and a maximum in the case of several elements which support one another in the shape of an arch. This principle has been known for a long time.

22. Now if the same system is considered to be in motion and u', u'', u''', etc. designate the velocities and m', m'', m''', etc. designate the respective masses of the various bodies which compose it, the well-known principle of the **Conservation des Forces Vives**[67] for which we will give a direct and general demonstration in PART II of this work, will give the following equation[68]

$$m'u'^2 + m''u''^2 + m'''u'''^2 + \cdots = \text{const.} - 2\Pi.$$

Therefore, since in the state of equilibrium, the quantity Π is a minimum or a maximum, it follows that the quantity $m'u'^2 + m''u''^2 + m'''u'''^2 +$ etc., which expresses the **force vive** of the entire system, will be at the same time a maximum or minimum. This leads to another principle of statics, which is, that in all configurations which the system takes successively, the one where it has the largest or smallest **force vive** is also the one in which it should be placed at the outset so that it remains in equilibrium. In the Mémoires de l'Académie des Sciences of Paris for 1748 and 1749[69] there is a discussion of this principle.

23. We have shown that the function Π is a minimum or a maximum when the configuration of the system is one of equilibrium. We are now going to demonstrate that if this function is a minimum the equilibrium will be stable, so that the system, which is assumed in equilibrium and displaced by a small amount, will return to the configuration of equilibrium by itself while making infinitesimal oscillations. On the contrary, in the case where the same function is a maximum, the equilibrium will not be stable and once disturbed, the system will begin by performing fairly small oscillations but the amplitude of the oscillation will continually grow larger.[70]

In order to demonstrate this proposition in a general fashion, I consider that whatever might be the configuration of the system, the position of the diverse bodies which compose it will always be determined by a given number of variables and that the quantity Π will be a given function of these same variables. Let us assume that in the state of equilibrium the variables in question are equal to a, b, c, etc. and that, in a state very close to the state of equilibrium, they are $a + x$, $b + y$, $c + z$, etc., in which the quantities x, y, z, etc. are very small. Substituting the latter quantities in the function for Π and expanding it in a series according to the order of the very small quantities x, y, z, etc., the function Π will become[71]

$$\begin{aligned}\Pi = A &+ Bx + Cy + Dz + \cdots \\ &+ Fx^2 + Gxy + Hy^2 + Kxz + Lyz + Mz^2 + \cdots,\end{aligned}$$

where the quantities A, B, C, etc. are obtained as functions of a, b, c, etc. But in the state of equilibrium the value of $d\Pi$ must be equal to zero, however the configuration of the system is altered. Therefore, the differential of Π must be zero in general when x, y, z, etc. are equal to zero. Thus it must be that $B = 0$, $C = 0$, $D = 0$, etc.

Thus for an arbitrary state which is very close to equilibrium, the following expression for Π now results

$$\Pi = A + Fx^2 + Gxy + Hy^2 + Kxz + Lyz + Mz^2 + \cdots$$

in which, as long as the variables x, y, z, etc. are very small, it will suffice to consider solely the second order magnitudes of these variables.

24. Now it is clear that when x, y, z, etc. are zero, for the quantity Π to be a minimum the function

$$Fx^2 + Gxy + Hy^2 + Kxz + Lyz + Mz^2 + \cdots,$$

which I will call X, must always be positive whatever the values of the variables x, y, z, etc.

This function can be reduced to the form

$$X = f\xi^2 + g\eta^2 + h\zeta^2 + \cdots$$

by taking

$$\begin{aligned}
f &= F \\
\xi &= x + \frac{Gy}{2f} + \frac{Kz}{2f} + \cdots \\
g &= H - \frac{G^2}{4f} \\
\eta &= y + \left(L - \frac{GK}{2f}\right)\frac{z}{2g} + \cdots \\
h &= M - \frac{K^2}{4f} - \frac{L^2}{4g} \\
\zeta &= z + \cdots \\
&\vdots
\end{aligned}$$

Thus for the function X to be always positive, the coefficients f, g, h, etc. must be positive and at the same time, it can be seen that if these coefficients are positive, the values of X will necessarily be positive because the quantities ξ, η, ζ etc. are real when the variables x, y, z, etc. are also real.

On the contrary, if the quantity Π is always a maximum when x, y, z, etc. are zero, the function X must always be negative. Consequently, the coefficients f, g, h, etc. should be negative. Conversely, if these coefficients were negative, the result would be that the value of X will necessarily be negative.

25. One will have, accounting for only the second order magnitudes of the very small quantities x, y, z, etc. that

$$\Pi = A + f\xi^2 + g\eta^2 + h\zeta^2 + \cdots$$

and the equation for the **Conservation des Forces Vives** (Article 22) will become

$$\begin{aligned}
&M'u'^2 + M''u''^2 + M'''u'''^2 + \cdots \\
&\quad = \text{const.} - 2A - 2f\xi^2 - 2g\eta^2 - 2h\zeta^2 + \cdots.
\end{aligned}$$

But in the state of equilibrium, it has been assumed that $x = 0, y = 0, z = 0$, etc. Thus one also has $\xi = 0, \eta = 0, \zeta = 0$, etc. (Article 19). Therefore, if it is assumed that the system has been disturbed from this state by impressing on the bodies M', M'', M''', etc., the very small velocities V', V'', V''', etc. one should have $u' = V'$, $u'' = V''$, $u''' = V'''$, etc. when $\xi = 0, \eta = 0, \zeta = 0$, etc. Hence, one will have $M'V'^2 + M''V''^2 + M'''V'''^2 +$etc. $=$

const. $-\, 2A$. This equation will serve to determine the arbitrary constant and the preceding equation will become

$$M'u'^2 + M''u''^2 + M'''u'''^2 + \cdots = M'V'^2 + M''V''^2 + M'''V'''^2 + \cdots$$
$$-2f\xi^2 - 2g\eta^2 - 2h\zeta^2 - \cdots$$

from which it is easy to draw these two conclusions:

1. In the case where Π is a minimum for which the coefficients f, g, h, etc. are all positive, the quantity $2f\xi^2 + 2g\eta^2 + 2h\zeta^2 +$ etc. which is always positive must necessarily be less or at least will not be greater than the given quantity $M'V'^2 + M''V''^2 + M'''V'''^2 +$ etc. which is itself very small. Consequently, if this quantity is denoted by the letter T, one will have for each of the variables ξ, η, ζ, etc. the limits

   $$\pm\sqrt{\frac{T}{2f}}, \quad \pm\sqrt{\frac{T}{2g}}, \quad \pm\sqrt{\frac{T}{2h}}, \quad \cdots$$

   by which they will always be bounded. In this case, the result is that the system will be able to depart by only a small amount from its state of equilibrium and will be able to make only very small oscillations of a very finite extent.

2. In the case where Π is a maximum for which the coefficients f, g, h, etc. are all negative, the quantity $-2f\xi^2 - 2g\eta^2 - 2h\zeta^2$, etc. which is always positive could increase to infinity and thus the system could depart further and further from its state of equilibrium. At least the equation above shows that in this case nothing prevents the variables ξ, η, ζ, etc. from constantly increasing. But yet it does not follow that they must actually increase. We will demonstrate this last proposition in SECTION VI of PART II.

If all the coefficients f, g, h, etc. were zero, it is known from the methods of maxima and minima that in order to have a minimum or a maximum that the terms of the third order should vanish and secondly, that the terms of the fourth order must always be either positive or negative. And it is also in this fashion that the stability of the equilibrium given by the vanishing of the terms of the first order can be judged when those terms of the second order also vanish.

26. Incidently, the properties of maxima and minima which are present in the equilibrium of an arbitrary system of forces are only an immediate consequence of the demonstration which we gave for the Principle of Virtual Velocities at the end of SECTION I.

Indeed, let p be the distance between the first two pulleys, one fixed and the other mobile.[72] Furthermore, let the pulleys be connected by P strings which produce a force proportional to P which is represented simply by P. Assume that the weight which pulls on the rope can be taken as unity. Conversely let q be the distance between two pulleys which produce the force Q, r the distance between the pulleys which produce the force R, etc. It is obvious that Pp will be the length of the portion of the rope which wraps around the first two pulleys. Conversely, Qq, Rr, etc. will be the lengths of the portion of the rope which wraps around the other pulleys so that the total length of the rope wrapped around the fixed and mobile pulleys will be $Pp + Qq + Rr + \cdots$.

Let us add to this length, the lengths of the different segments of the rope which are located between fixed pulleys to make the necessary change in direction which we will denote by a. Let us also add the segment of the rope which is located between the last counterpulley and the weight attached to the extremity of the rope which we will denote by u. Finally, let ℓ be the total length of the rope for which one end is attached to a fixed point in space and for which the other end carries the weight. One will obviously have the equation

$$\ell = Pp + Qq + Rr + \cdots + a + u,$$

for which

$$u = \ell - a - Pp - Qq - Rr - \cdots.$$

But assuming that the forces P, Q, R, etc. are constant, that is, independent of p, q, r, etc. which is always possible in an equilibrium state where only infinitesimal displacements are considered, it is obvious that the quantity $Pp + Qq + Rr + \cdots$ will be the same as the one we have designated by Π in Article 21. Thus one will have generally that $u = \ell - a - \Pi$, where ℓ and a are constants.

27. Now, it is clear that since the weight tends to descend as far as possible, equilibrium will generally hold only when the value of u, which expresses the distance traversed by the weight from the fixed pulley, will be a maximum and consequently, the value of Π will be a minimum. It is obvious at the same time that in this case the equilibrium will be stable because a small arbitrary change in the configuration of the system will surely cause the weight to ascend, but the weight will then tend to descend and return the system to a state of equilibrium.

But it has been shown that to have equilibrium, it suffices to have $\mathrm{d}\Pi = 0$ and consequently, $du = 0$ which also holds when the value of u is a minimum, in which case, the weight, will be at its highest position instead of at its lowest position. In this case, it is obvious that a small change in the configuration of the system will only have a tendency to lower the weight, which then will not tend to ascend but to descend further and to move the system further from the initial state of equilibrium. It is clear from this result that this equilibrium state will not be stable and that once disturbed, the system will not return to its initial configuration.

SECTION IV
A MORE GENERAL AND SIMPLER WAY TO USE THE FORMULA OF EQUILIBRIUM PRESENTED IN SECTION II

1. The authors who have written on the Principle of Virtual Velocities in the past have concentrated on proving the veracity of this principle by demonstrating the congruity between solutions obtained using this principle with those obtained from the ordinary principles of statics rather than to demonstrate its application to solve directly the problems of this science. We propose to fill this latter task with all possible generality and to deduce from this principle analytical formulas which contain the solution of all the problems of the equilibrium of bodies, basically in the same fashion that the formulas for sub-tangents, oscillating radii, etc. contain the means of determining these lines in all curves.

The method presented in SECTION II can be used in every case and requires, as has been shown, solely analytical operations. But since the immediate elimination of the variables or their differences by means of the equations of condition, can lead to excessively complicated calculations, we will present the same method in a simpler form, by reducing all cases to the case of an entirely free system.

Subsection I
Method of Multipliers

2. Let $L = 0$, $M = 0$, $N = 0$, etc. be the various equations of condition which are given by the nature of the system. The quantities L, M, N, etc. are finite functions of the variables, x, y, z, x', y', z', etc. By differentiation of these equations, the following expressions result $\mathrm{d}L = 0$, $\mathrm{d}M = 0$, $\mathrm{d}N = 0$, etc. which will give the relation which must exist between the differential of the same variables. In general, we will represent by $\mathrm{d}L = 0$, $\mathrm{d}M = 0$, $\mathrm{d}N = 0$, etc. the equations of condition between these differentials whether or not the equations are exact differentials as long as the differentials are linear.

Now, since these equations are only used to eliminate an equal number of differentials in the general formula of equilibrium, after which the coefficients of the remaining differentials must individually be set equal to zero, it is not difficult to prove by the theory of elimination for linear equations that the same results will be obtained if the different equations of condition $\mathrm{d}L = 0$, $\mathrm{d}M = 0$, $\mathrm{d}N = 0$, etc. each multiplied by an undetermined coefficient are simply added to this formula. Then if the sum of all the terms which are multiplied by the same differential are put equal to zero, as many particular equations as there are differentials will be obtained. Finally, from these latter equations the undetermined coefficients, by which the equations of condition have been multiplied, can be eliminated.[73]

3. An extraordinarily simple rule results from this approach to find the equilibrium conditions of an arbitrary, given system. The sum of the moments of all the forces which must be in equilibrium can be gathered (SECTION II, Article 5) and to which can be added

the various differential functions which must be equal to zero from the conditions of the problem after having multiplied each of these functions by an undetermined coefficient. The whole can be equated to zero and thus a differential equation results which can be treated as an ordinary equation of maxima and minima and from which as many particular finite equations as there are variables can be extracted. These equations from which the undetermined coefficients have been eliminated will give all the necessary conditions for equilibrium.

The differential equation will have the following form

$$P\,\mathrm{d}p + Q\,\mathrm{d}q + R\,\mathrm{d}r + \cdots + \lambda\,\mathrm{d}L + \mu\,\mathrm{d}M + \nu\,\mathrm{d}N + \cdots = 0$$

in which λ, μ, ν, etc. are undetermined quantities. We will call this equation the General Equation of Equilibrium.

This equation will give with respect to each coordinate, such as x, of each body of the system, an equation of the following form so that the number of these equations will be equal to the number of coordinates of the bodies. We will call these equations the Particular Equations of Equilibrium.

$$P\frac{\mathrm{d}p}{\mathrm{d}x} + Q\frac{\mathrm{d}q}{\mathrm{d}x} + R\frac{\mathrm{d}r}{\mathrm{d}x} + \cdots + \lambda\frac{\mathrm{d}L}{\mathrm{d}x} + \mu\frac{\mathrm{d}M}{\mathrm{d}x} + \nu\frac{\mathrm{d}N}{\mathrm{d}x} + \cdots = 0.$$

4. The greatest difficulty will be to eliminate the undetermined coefficients λ, μ, ν, etc. from these latter equations. But this can always be done by known means. It will be appropriate to choose in every case the alternative which can lead to the simplest result. The final equations will contain all the necessary conditions for the proposed equilibrium state. Since the number of these equations will be equal to the total number of coordinates of the bodies of the system minus those of the undetermined coefficients λ, μ, ν, etc. which had to be eliminated and since the number of these same undetermined coefficients is equal to the number of finite equations of condition $L = 0$, $M = 0$, $N = 0$, etc., it follows that these equations added to the last equations will give the same number as the number of coordinates of all the bodies. Consequently, they will be enough to determine these coordinates and to determine the position of each body at equilibrium.

5. I now note that the terms $\lambda\,\mathrm{d}L$, $\mu\,\mathrm{d}M$, etc. of the general equation of equilibrium can also be viewed as representing the moments of the various forces applied to the same system.

Indeed, assuming that $\mathrm{d}L$ is a differential function of the variables x', y', z', x'', y'', etc. which are the coordinates of the various bodies of the system, it will be composed of different parts which will be designated by $\mathrm{d}L'$, $\mathrm{d}L''$, etc. so that $\mathrm{d}L = \mathrm{d}L' + \mathrm{d}L'' + \text{etc.}$, with $\mathrm{d}L'$ having only the terms which depend on $\mathrm{d}x'$, $\mathrm{d}y'$, $\mathrm{d}z'$ and $\mathrm{d}L''$ having only the terms which depend on $\mathrm{d}x''$, $\mathrm{d}y''$, $\mathrm{d}z''$ and so on.

In this fashion, the term $\lambda\,\mathrm{d}L$ of the general equation will be composed of the terms $\lambda\,\mathrm{d}L'$, $\lambda\,\mathrm{d}L''$, etc. But if the following form is given to the term $\lambda\,\mathrm{d}L'$

$$\lambda\sqrt{\left(\frac{\mathrm{d}L'}{\mathrm{d}x'}\right)^2+\left(\frac{\mathrm{d}L'}{\mathrm{d}y'}\right)^2+\left(\frac{\mathrm{d}L'}{\mathrm{d}z'}\right)^2}\times\frac{\mathrm{d}L'}{\sqrt{\left(\frac{\mathrm{d}L'}{\mathrm{d}x'}\right)^2+\left(\frac{\mathrm{d}L'}{\mathrm{d}y'}\right)^2+\left(\frac{\mathrm{d}L'}{\mathrm{d}z'}\right)^2}},$$

it is obvious from what was stated in Article 8 of SECTION II that this quantity can represent the moment of a force equal to

$$\lambda\sqrt{\left(\frac{\mathrm{d}L'}{\mathrm{d}x'}\right)^2+\left(\frac{\mathrm{d}L'}{\mathrm{d}y'}\right)^2+\left(\frac{\mathrm{d}L'}{\mathrm{d}z'}\right)^2}$$

applied to the body with coordinates x', y', z' and directed perpendicular to the surface which has for its equation $\mathrm{d}L'=0$, considering only x', y', z' as variables. Conversely, the term $\lambda\,\mathrm{d}L''$ can represent the moment of a force equal to

$$\lambda\sqrt{\left(\frac{\mathrm{d}L''}{\mathrm{d}x''}\right)^2+\left(\frac{\mathrm{d}L''}{\mathrm{d}y''}\right)^2+\left(\frac{\mathrm{d}L''}{\mathrm{d}z''}\right)^2}$$

applied to the body with coordinates x'', y'', z'' and directed perpendicular to the curved surface with equation $\mathrm{d}L''=0$, which for this case, considers only x'', y'', z'' as variables and so on.

Therefore, in general, the term $\lambda\,\mathrm{d}L$ will be equivalent to the effect of different forces expressed by

$$\lambda\sqrt{\left(\frac{\mathrm{d}L}{\mathrm{d}x'}\right)^2+\left(\frac{\mathrm{d}L}{\mathrm{d}y'}\right)^2+\left(\frac{\mathrm{d}L}{\mathrm{d}z'}\right)^2}$$
$$\lambda\sqrt{\left(\frac{\mathrm{d}L}{\mathrm{d}x''}\right)^2+\left(\frac{\mathrm{d}L}{\mathrm{d}y''}\right)^2+\left(\frac{\mathrm{d}L}{\mathrm{d}z''}\right)^2}$$
$$\vdots$$

and applied respectively to bodies with coordinates x', y', z', x'', y'', z'', etc. along directions perpendicular to different curved surfaces represented by the equation $\mathrm{d}L=0$, in which x', y', z' are varied first, followed by x'', y'', z'' and so on.

6. In general, the term $\lambda\,\mathrm{d}L$ could be interpreted as the moment of a force λ which can change the value of the function L. Since $\mathrm{d}L=\mathrm{d}L'+\mathrm{d}L''+$ etc. the term $\lambda\,\mathrm{d}L$ will also express the moments of several forces equal to λ and all with the tendency to change the function L considering the variability of the various coordinates x', y', z', x'', y'', z'', etc.

A similar development can be made for the terms $\mu\,dM$, $\nu\,dN$, etc. (SECTION II, Article 9).

Furthermore, because of the fact that in the general equation of equilibrium (Article 3) the forces P, Q, R, etc. are assumed directed towards force centers where the lines p, q, r, etc. intersect and with a tendency to shorten these lines, the forces, λ, μ, etc. should be viewed as having a tendency to reduce the magnitudes of the functions L, M, etc.

7. From the preceding development, it is clear that each equation of condition is equivalent to one or more forces applied to the system along given directions or in general, able to change the values of given functions so that the state of equilibrium of the system will remain the same whether these forces or the equations of condition are considered.

Conversely, these forces can take the place of the equations of condition resulting from the nature of the given system so that by using these forces, the bodies can be viewed as entirely free and without any constraints. From this result the metaphysical reason for the introduction of the terms $\lambda\,dL + \mu\,dM +$ etc. in the general equation of equilibrium is obvious. It is that this equation can be analyzed as if all the bodies of the system were entirely free. This is the essence of the method presented in this section.

Strictly speaking, these forces account for the resistance that the bodies must experience through their mutual connection or because of the impediments which, due to the nature of the system, could hinder their motion. Or rather, these forces may only be the forces of resistance themselves, which must be equal and directly opposed to the pressures created by the bodies. Our method provides, as one sees, the means of determining the forces and resistances which is one of the more important advantages of this method.

8. In the case where the forces P, Q, R, etc. are not in equilibrium and where it is required to reduce them to equivalent forces with given directions, it will be enough to add to the sum of the moments of the forces P, Q, R, etc., the moments resulting from the equations of condition $L = 0$, $M = 0$, etc. The sum of the moments of the forces equivalent to the forces P, Q, R, etc. and to the action that these bodies exert on one another will be obtained by virtue of these equations of condition.

Thus by using all the equations of condition given by the nature of the proposed system, the coordinates of every body of the system could be viewed as independent and an expression of the form

$$P\frac{dp}{dx} + Q\frac{dq}{dx} + R\frac{dr}{dx} + \cdots + \lambda\frac{dL}{dx} + \mu\frac{dM}{dx} + \nu\frac{dN}{dx} + \cdots$$

will be obtained, which will express the resulting force acting along the direction of the line x, which must be equal to zero in the case of equilibrium as was seen in Article 3.

Subsection II
Application Of The Same Method To The Formula For The Equilibrium Of Continuous Bodies Where All The Points Are Loaded By Arbitrary Forces

9. We have considered bodies to be material points until now and we have seen how the laws of equilibrium are derived for these points, whatever their number and the forces acting on them. But since a body of arbitrary volume and shape, is only the aggregate of an infinity of parts or material points, it follows that the laws of equilibrium for the bodies of arbitrary shape can also be determined by applying the preceding principles.

Indeed, the ordinary manner of solving the problems of mechanics dealing with bodies of finite mass consists of first considering only a given number of points located at finite distances from one another, then to find the laws of their equilibrium or of their motion, subsequently, to expand this research to any number of points, further, to assume that the number of points becomes infinite and at the same time that the distance between them becomes infinitesimal and ultimately, to apply to the formulas found for a finite number of points the simplifications and modifications which are necessary when extrapolating from the finite to the infinite.

It is clear that this procedure is analogous to the geometric and algebraic methods which preceded the infinitesimal calculus. And if this calculus has the advantage to facilitate and to simplify, in a surprising manner, the solutions of problems dealing with curves, this is because it considers these lines by themselves as curves without having to view them first as polygons and then as curves. Thus nearly the same advantage is obtained by treating the problems of mechanics by direct means and by considering from the beginning, the bodies of finite mass as an aggregate of an infinite quantity of material points or particles, each moved by given forces. And nothing is more easy to modify and simplify by this consideration than the general method which we just gave.

10. But it is necessary to note at the outset that in the application of this method to bodies of finite mass, for which all the points are acted upon by arbitrary forces, two types of differentials are present which must be distinguished. Some are with respect to the various points which compose the body. Others are independent of the relative position of these points and only represent the infinitesimal distances that every point can traverse, assuming that the location of the body changes by an infinitesimal amount. Since heretofore we had only differentials of this type to consider, we designated them by the ordinary symbol 'd'. Now since we must consider the two types simultaneously, it is necessary to introduce a new symbol. It seems judicious to us to use the old symbol 'd' to designate the differences of the first type which are analogous to those which are commonly considered in geometry and to call the differences of the second type, which are peculiar to the matter which we shall treat, by the symbol used in the Calculus of Variations 'δ',[74] which has a necessary and intimate relation with the problem considered here.

For this reason, we will call variations the differences produced by 'δ', and we will retain the name of differentials for those produced by 'd'. Also, the same formulas which define ordinary differentials will give the variations when 'δ' is substituted for 'd'.

11. I now note that instead of considering the given mass as an aggregate of an infinite number of contiguous points, one should, in accordance with the principles of the infinitesimal calculus, consider it rather as composed of infinitesimal elements which are of the same dimensional order as the entire mass. So, in order to have the forces which act upon each of the elements, the forces P, Q, R, etc. should be multiplied by these same elements. These forces are assumed applied at every point of the elements and will be viewed as forces of acceleration similar to those which are generated by the action of gravity.

Thus if the total mass is denoted by m and any one of its elements by $\mathrm{d}m$, one will have $P\,\mathrm{d}m$, $Q\,\mathrm{d}m$, $R\,\mathrm{d}m$, etc. for the forces which act on the element $\mathrm{d}m$ in the directions of the lines p, q, r, etc. Therefore, the moment of this force will be obtained by multiplying these forces by the variations δp, δq, δr, etc., respectively. The sum of the moments for each element $\mathrm{d}m$ will be represented by the formula $(P\,\delta p + Q\,\delta q + R\,\delta r + \text{etc.})\,\mathrm{d}m$. And the sum of the moments of all the forces of the system will be obtained by taking the integral of this formula with respect to the total given mass.

We will denote these total integrations, which are relative to the entire mass by the symbol 'S' retaining the ordinary symbol $\int$ to designate the partial and indefinite integrals.[75]

12. Thus we will have for the sum of the moments of all the forces of the system, the integral formula $\mathrm{S}(P\,\delta p + Q\,\delta q + R\,\delta r + \text{etc.})\,\mathrm{d}m$. In general, this quantity must be equal to zero for the equilibrium state of the system.

Because of the nature of the system, there are necessarily given ratios between the different variations δp, δq, δr, etc. relative to each mass point. They should be reduced to a given number of independent and undetermined variations. And the particular equations for equilibrium will be obtained when the terms multiplied by these latter variations are set equal to zero. But these simplifications can be burdensome and it is then convenient to avoid them by means of the Method of Multipliers which we presented in the preceding subsection.

13. In order to apply this method to the case in question, we will assume that $L = 0$, $M = 0$, etc. are the equations of condition, which must exist with respect to every mass point because of the nature of the problem, and we will call them Undetermined Equations of Condition.

Here, the quantities L, M, etc. will be functions of the finite coordinates x, y, z of each point of the given mass and of their differentials of any order.

After differentiating these equations according to 'δ', the following equations will be obtained $\delta L = 0$, $\delta M = 0$, etc. Then after multiplying the quantities δL, δM, etc. by the undetermined quantities λ, μ, etc., the general equation of equilibrium will be obtained by taking the total integral, which consequently will be represented by the formula $\mathrm{S}(\lambda\,\delta L + \mu\,\delta M + \text{etc.})$ and adding this integral to the one of the preceding article.

We will note that it is not necessary that $\delta L, \delta M$, etc. be exact variations of the functions of x, y, z, $\mathrm{d}x$, $\mathrm{d}y$, etc. but that it is sufficient that $\delta L = 0$, $\delta M = 0$, etc. be the undetermined equations of condition between the variations of x, y, z, $\mathrm{d}x$, $\mathrm{d}y$, etc. (Article 3).

But it must be recognized that besides the forces which in general act on all points of the mass, there may be some which act only on some particular points of this mass. These points are ordinarily at the boundaries of the given mass, that is, at the lower and upper limits of the integral denoted by S.

Conversely, there may be particular equations of condition at these points, which we will call Determined Equations of Condition to distinguish them from those which exist in general for the entire mass. We will express them by $A = 0$, $B = 0$, $C = 0$, etc. or rather by $\delta A = 0$, $\delta B = 0$, $\delta C = 0$, etc.

We will denote by a prime, two primes, three primes, etc. all the quantities which are related to known points of the mass and in particular, those which are at the beginning of the integral denoted by S will be indicated by one prime, and those which are at the end of this integral by two primes and those which are in intermediate positions by three primes. Thus one should add to the integral $\mathrm{S}(P\,\delta p + Q\,\delta q + R\,\delta r + \cdots)\,\mathrm{d}m$, the quantity $P'\,\delta p' + Q'\,\delta q' + R'\,\delta r' + \cdots + P''\,\delta p'' + Q''\,\delta q'' + R''\,\delta r'' + \cdots$ and to the integral $\mathrm{S}(\lambda\,\delta L + \mu\,\delta M + \cdots)$, the quantity $\alpha\,\delta A + \beta\,\delta B + \gamma\,\delta C + \cdots$. Hence, the general equation of equilibrium will be of the form

$$\begin{aligned}
&\mathrm{S}(P\,\delta p + Q\,\delta q + R\,\delta r + \cdots)\,\mathrm{d}m + \mathrm{S}(\lambda\,\delta L + \mu\,\delta M + \cdots)\\
&\quad + P'\,\delta p' + Q'\,\delta q' + R'\,\delta r' + \cdots + P''\,\delta p'' + Q''\,\delta q'' + R''\,\delta r'' + \cdots\\
&\quad + \alpha\,\delta A + \beta\,\delta B + \gamma\,\delta C + \cdots = 0.
\end{aligned}$$

14. Because the functions L, M, etc. not only contain the finite variables x, y, z, but also their differentials, the variations δL, δM, etc. will give the terms multiplied by δx, δy, δz, $\delta\,\mathrm{d}x$, $\delta\,\mathrm{d}y$, etc. And the preceding equation, after one has substituted the values of δp, δq, δr, etc. δL, δM, etc., for δx, δy, δz, $\delta\,\mathrm{d}x$, $\delta\,\mathrm{d}y$, $\delta\,\mathrm{d}z$, etc. as well as those of $\delta p'$, $\delta p''$, etc., $\delta q'$, $\delta q''$, etc., δA, δB, etc. for $\delta x'$, $\delta x''$, etc., $\delta y'$, $\delta y''$, etc., $\delta\,\mathrm{d}x'$, etc. deduced from the particular conditions of each problem, will always have an analogous form to those which the Calculus of Variations gives for the determination of maxima and minima of an indefinite integral. Therefore, the known rules of this calculus must be applied.

Therefore, one will consider that because the symbols 'd' and 'δ' denote two types of differences entirely independent of one another, when these operations are combined, the order in which they are applied does not matter because assuming that a quantity varies in two different manners, the same result will be obtained whatever the order of these operations. Therefore, $\delta\mathrm{d}x$ will be the same as $\mathrm{d}\delta$x and conversely, $\delta\mathrm{d}^2x$ will be the same as $\mathrm{d}^2\delta x$, and so on. Therefore, the order of the operations can always be interchanged at will without changing the values of the differences. And for our purpose, it will be judicious to place the operator 'd' before 'δ', so that the proposed equation contains only the variations of the coordinates and the differentials of these same variations.

Thus, the symbols of integration $\int$ or S, are treated in a fashion similar to the symbol for the variation δ. Consequently the symbols $\delta \int$ or δS can always be substituted for $\int \delta$ or Sδ. This is the first fundamental principle of the Calculus of Variations.

15. But the differentials $\mathrm{d}\delta x$, $\mathrm{d}\delta y$, $\mathrm{d}\delta z$, $\mathrm{d}^2\delta x$, etc. which are under the symbol S, can be eliminated by the well-known operation of integration by parts. Because, in general, $\int \Omega\, \mathrm{d}\delta x = \Omega\, \delta x - \int \delta x\, \mathrm{d}\Omega$, $\int \Omega\, \mathrm{d}^2\, \delta x = \Omega\, \mathrm{d}\delta x - \mathrm{d}\Omega\, \delta x + \int \delta x\, \mathrm{d}^2\Omega$, and similarly for the others, where one must observe that the quantities outside the integral sign $\int$ are naturally evaluated at the boundary points of the integrals. But in order to make these integrals complete, it is necessary to subtract the values of the same quantities outside the symbol which are related to the first points of the integrals so that everything cancels with respect to these points which is obvious from the theory of integration.

Thus denoting by a prime the quantities related to the beginning of the total integrals denoted by S and by two primes those related to the end of these integrals, one will have the following relations

$$\begin{aligned}
&\mathrm{S}\,\Omega\, \mathrm{d}\delta x = \Omega''\, \delta x'' - \Omega'\, \delta x' - \mathrm{S}\ \delta x\, \mathrm{d}\Omega \\
&\mathrm{S}\Omega\, \mathrm{d}^2\, \delta x = \Omega''\, \mathrm{d}\delta x'' - \mathrm{d}\Omega''\, \delta x'' - \Omega'\, \mathrm{d}\,\delta x' + \mathrm{d}\Omega'\, \delta x' + \mathrm{S}\, \delta x\, \mathrm{d}^2\Omega \\
&\vdots
\end{aligned}$$

which will be used to eliminate all the differentials of the variations which could be under the symbol S. These relations are the second fundamental principle of the Calculus of Variations.

16. Thus in this manner the general equation of equilibrium will be reduced to the following form

$$\mathrm{S}(\Xi\, \delta x + \Sigma\, \delta y + \Psi\, \delta z) + \Lambda = 0$$

in which Ξ, Σ, Ψ will be functions of x, y, z and their differentials. The term Λ will contain the expressions affected by the variations $\delta x'$, $\delta y'$, $\delta z'$, $\delta x''$, $\delta y''$, etc. and their differentials.

Therefore, in order that this equation exists independently of the variations of the various coordinates, one should have:

1. The functions Ξ, Σ, Ψ, are equal to zero for the entire domain of the integral S, that is, at every point of the mass.
2. Every term of Λ is also equal to zero.

The undefined equations $\Xi = 0$, $\Sigma = 0$, $\Psi = 0$, will give in general the relation which must exist between the variables x, y, z. But for this purpose, the undetermined variables

λ, μ, etc., whose number is the same as the number of equations of condition $L = 0$, $M = 0$, etc. (Article 13) should be eliminated.

But it can be shown that the number of these equations can not be more than three because being undefined equations between the three variables x, y, z, and their differentials, it is obvious that if there were more than three, there would be more equations than variables so that the fourth one would be a function of the first three and similarly for the others.[76] Therefore, there will never be more than three undetermined variables λ, μ, ν, to eliminate. Consequently, the values of these undetermined variables can always be found as functions of x, y, z. But the equations which were eliminated will be replaced by the equations of condition so that the values of x, y, z can always be found when the system is in the state of equilibrium.

However, the equations of condition $L = 0$, $M = 0$, etc. could also contain other variables u, v, etc. with their differentials which should be eliminated by means of other equations such as $U = 0$, $V = 0$, etc. In this case, these new equations of condition could be treated as those which are given by the nature of the problem. By taking as undetermined coefficients σ, ν, etc. the terms $\sigma\,\delta U + \nu\,\delta V +$ etc. would have to be added to the terms $\lambda\,\delta L + \mu\,\delta M +$ etc., which are under the integral sign in the general equation of Article 13. After having eliminated all the differentials of the variations δx, δy, δz, δu, δv, etc. the final equation of Article 16 will contain under the integral sign the terms affected by the variations δu, δv, etc. which consequently, shall be individually equal to zero. Thus there will be as many new equations as unknowns σ, ν, etc. which should be eliminated. Then, the new variables u, v, etc. will be eliminated by using the given equations $U = 0$, $V = 0$, etc. This method will be very useful when there exist integrable quantities in the functions L, M, etc. Because, by substituting for these quantities new unknowns, all the integration signs can be eliminated which will simplify the computation.

17. With respect to the other equations resulting from the different terms of the quantity Λ which is outside the integral sign, they will only be particular equations which should only exist with respect to determined points of the mass and which will be mainly used to determine the arbitrary constants that the expressions for x, y, z, deduced from the preceding equations, could contain. Thus in order to make use of these equations, the previously determined values of λ, μ, etc. will be substituted and the indeterminates α, β, etc. eliminated, and the equations of condition $A = 0$, $B = 0$, etc. will be added which will be used to replace those which have been eliminated as previously discussed.

18. Moreover, the terms $P\,\delta p$, $Q\,\delta q$, etc. resulting from the forces of acceleration P, Q, etc. do not require any simplification as long as these forces act along the lines p, q, etc. since the quantities p, q, etc. are only functions of the finite variables x, y, z. However, it will be different when forces are included for which the analysis will consist of varying a given function (SECTION II, Article 9). Therefore, if this function contains differentials, the same simplifications used for the terms $\lambda\,\delta L$, etc. should be applied and a final equation of the same form will always be obtained. This case occurs when elastic bodies, whether solids or fluids, are considered.

Subsection III
The Analogy of Problems of this Type with those of Maxima and Minima

19. Not only is the Calculus of Variations applicable in the same fashion to problems of the equilibrium of continuous bodies and to problems of maxima and minima relative to integrable formulas, but it raises a remarkable analogy between the two types of problems which we will consider. We will begin by giving a general formula for the variation of an arbitrary differential function with several variables.

It is known that one differential of the first order can always be taken as constant in functions of several variables and in their differentials of order higher than the first, which simplify the function without reducing its generality. But in differentiation by δ, the variable for which the differential has been assumed to be constant must also be considered as constant. And if variations have to be attributed to all the variables, the variability of the differential which is assumed to be constant should be reestablished.

20. Let U be a function of

$$x, \qquad y, \qquad \frac{\mathrm{d}y}{\mathrm{d}x}, \qquad \frac{\mathrm{d}^2y}{\mathrm{d}x^2}, \qquad \ldots,$$

where $\mathrm{d}x$ is assumed constant. If the following definitions are made, as in the *Théorie des Fonctions*[77]

$$\frac{\mathrm{d}y}{\mathrm{d}x} = y', \qquad \frac{\mathrm{d}y'}{\mathrm{d}x} = y'', \qquad \frac{\mathrm{d}y''}{\mathrm{d}x} = y''', \qquad \ldots,$$

the quantity U will become a function of x, y, y', y'', etc. and the variation δU will be, using the notation of partial differentials,[78] of the form

$$\delta U = \frac{\mathrm{d}U}{\mathrm{d}x}\,\delta x + \frac{\mathrm{d}U}{\mathrm{d}y}\,\delta y + \frac{\mathrm{d}U}{\mathrm{d}y'}\,\delta y' + \frac{\mathrm{d}U}{\mathrm{d}y''}\,\delta y'' + \cdots.$$

Now, by permitting all of the terms to vary, the following equations will be obtained

$$\begin{aligned}
\delta y' &= \delta\frac{\mathrm{d}y}{\mathrm{d}x} = \frac{\delta\,\mathrm{d}y}{\mathrm{d}x} - \frac{\mathrm{d}y}{\mathrm{d}x}\frac{\delta\,\mathrm{d}x}{\mathrm{d}x} = \frac{\mathrm{d}\delta y}{\mathrm{d}x} - y'\frac{\mathrm{d}\delta x}{\mathrm{d}x}\\
&= \frac{\mathrm{d}(\delta y - y'\delta x)}{\mathrm{d}x} + y''\,\delta x\\
\mathrm{d}y'' &= \frac{\mathrm{d}(\delta y' - y''\delta x)}{\mathrm{d}x} + y'''\,\delta x\\
&= \frac{\mathrm{d}^2(\delta y - y'\,\delta x)}{\mathrm{d}x} + y'''\,\delta x\\
\delta y''' &= \frac{\mathrm{d}^3(\delta y - y'\,\delta x)}{\mathrm{d}x^3} + y^{\mathrm{IV}}\,\delta x\\
&\;\;\vdots
\end{aligned}$$

After substituting these expressions and making the following definition in order to shorten the expressions, $(\delta y - y'\,\delta x) = \delta u$, and as a consequence, $\delta y = (\delta u + y'\,\delta x)$ there will result

$$\delta U = \left(\frac{dU}{dx} + \frac{dU}{dy}y' + \frac{dU}{dy'}y'' + \frac{dU}{dy''}y''' + \cdots\right)\delta x$$
$$+\frac{dU}{dy}\,\delta u + \frac{dU}{dy'}\frac{d\delta u}{dx} + \frac{dU}{dy''}\frac{d^2\delta u}{dx^2} + \cdots$$

By differentiating the function U and substituting $y'\,dx$ for dy, $y''dx$ for dy', there will result

$$dU = \left(\frac{dU}{dx} + \frac{dU}{dy}y' + \frac{dU}{dy'}y'' + \frac{dU}{dy''}y''' + \cdots\right)dx$$

from which is derived

$$\frac{dU}{dx} + \frac{dU}{dy}y' + \frac{dU}{dy'}y'' + \cdots = \frac{1}{dx}\,dU$$

Then, finally

$$\delta U = \frac{1}{dx}\,dU\,\delta x + \frac{dU}{dy}\,\delta u + \frac{dU}{dy'}\frac{d\delta u}{dx} + \frac{dU}{dy''}\frac{d^2\delta u}{dx^2} + \cdots$$

If the quantity U contained another variable such as z with its differentials

$$\frac{dz}{dx}, \qquad \frac{d^2z}{dx^2}, \qquad \ldots,$$

then by setting

$$\frac{dz}{dx} = z', \qquad \frac{dz'}{dx} = z'', \qquad \ldots$$

and operating on the equations in the same fashion, the following expression would result

$$\frac{dU}{dz}\,\delta\nu + \frac{dU}{dz'}\frac{d\,\delta\nu}{dx} + \frac{dU}{dz''}\frac{d^2\delta\nu}{dx^2} + \cdots$$

in which $d\nu = (\delta z - z'\,\delta x)$ is to be added to the preceding expression for δU and so on.

21. Therefore, if it is required to make the integral function $\int U\,dx$ a maximum or a minimum, one should proceed as follows using the principles of the Calculus of Variations

$$\delta\int U\,dx = \int \delta(U\,dx) = \int(\delta U\,dx + U\,\delta dx) = 0$$

Substituting the expression for δU, replacing $\delta \mathrm{d}x$ by $\mathrm{d}\delta$x, and eliminating by the integration by parts, the differences δx, δu, $\delta\nu$, there will remain under the integral sign only terms of the form $(\Xi\,\delta x + \Upsilon\,\delta u + \Psi\,\delta\nu)\,\mathrm{d}x$, in which

$$\Xi = \mathrm{d}U - \mathrm{d}U = 0$$
$$\Upsilon = \frac{\mathrm{d}U}{\mathrm{d}y} - \frac{1}{\mathrm{d}x}\mathrm{d}\frac{\mathrm{d}U}{\mathrm{d}y'} + \frac{1}{\mathrm{d}x^2}\mathrm{d}^2\frac{\mathrm{d}U}{\mathrm{d}y''} - \cdots$$
$$\Psi = \frac{\mathrm{d}U}{\mathrm{d}z} - \frac{1}{\mathrm{d}x}\mathrm{d}\frac{\mathrm{d}U}{\mathrm{d}z'} + \frac{1}{\mathrm{d}x^2}\mathrm{d}^2\frac{\mathrm{d}U}{\mathrm{d}z''} - \cdots$$

These terms must be equal to zero whatever the variations δx, δy, δz. By replacing δu and $\delta\nu$ by their expressions $\delta y - y'\,\delta x$, $\delta z - z'\,\delta x$, these terms will become, since $\Xi = 0$, $(\Upsilon\,\delta y + \Psi\,\delta z - (\Upsilon y' + \Psi z')\,\delta x)\,\mathrm{d}x$, from which only the two equations $\Upsilon = 0$, $\Psi = 0$, are obtained. The third equation, dependent on δx, is included in these two.

It is obvious from this result that it is not necessary to vary the quantity x which is assumed constant in the function U because the necessary equations for the solution of the problem result only from the variations of the other variables. This is an observation which was made at the outset in the Calculus of Variations and which is a necessary deduction from this field of mathematics.

However, it can be useful to consider all the variations simultaneously with respect to the integration limits because particular conditions for the points at the boundaries can result as we showed in the last lesson of the *Leçons sur le Calcul des Fonctions*.[79]

22. The integrable function from which the maximum or minimum is derived can also contain other integrals. But whatever it is, it can always be reduced to contain finite variables with their differentials and to depend on one or several equations of condition between these same variables which can always be solved by the Method of Multipliers.

Let us assume, for example, that U is a function of x, y, z and their differentials and that, at the same time, the variable z depends on the equation of condition $L = 0$. After this equation is differentiated with respect to δ the result is $\delta L = 0$. Thus we only need to multiply this equation by an undetermined coefficient λ or by $\lambda\,\mathrm{d}x$ for homogeneity when L is a finite function, add the integrable equation $\int \lambda\delta L\,\mathrm{d}x = 0$ to the equation of maximum or minimum $\delta\int U\,\mathrm{d}x = 0$ and finally, treat the variations δx, δy, δz as if they were independent. But by viewing L as a function of x, y, y', y'', etc., z, z', z'', etc., the following equation is obtained

$$\delta L = \frac{\mathrm{d}L}{\mathrm{d}x}\delta x + \frac{\mathrm{d}L}{\mathrm{d}y}\delta y + \frac{\mathrm{d}L}{\mathrm{d}y'}\delta y'$$
$$+ \frac{\mathrm{d}L}{\mathrm{d}z}\delta z + \frac{\mathrm{d}L}{\mathrm{d}z'}\delta z' + \frac{\mathrm{d}L}{\mathrm{d}z''}\delta z'' + \cdots.$$

Thus, if the same substitutions used above are made for $\delta y'$, $\delta z'$, $\delta y''$, etc. δL will become

$$\delta L = \frac{1}{\mathrm{d}x}\mathrm{d}L\delta x + \frac{\mathrm{d}L}{\mathrm{d}y}\delta u + \frac{\mathrm{d}L}{\mathrm{d}y'}\frac{\mathrm{d}\delta u}{\mathrm{d}x}$$
$$+ \frac{\mathrm{d}L}{\mathrm{d}z}\delta\nu + \frac{\mathrm{d}L}{\mathrm{d}z'}\frac{\mathrm{d}\delta\nu}{\mathrm{d}x} + \cdots$$

and the terms under the integral sign which derive from the equation $\int(\delta U\,\mathrm{d}x+\lambda\delta L\,\mathrm{d}x)=0$ will be of the form $(\Xi\delta x+\Upsilon\delta u+\Psi\delta\nu)\mathrm{d}x$, where

$$\begin{aligned}\Xi &= \lambda\mathrm{d}L\\ \Upsilon &= \left(\frac{\mathrm{d}U}{\mathrm{d}y}+\lambda\frac{\mathrm{d}L}{\mathrm{d}y}-\frac{1}{\mathrm{d}x}\mathrm{d}\left(\frac{\mathrm{d}U}{\mathrm{d}y'}+\lambda\frac{\mathrm{d}L}{\mathrm{d}y'}\right)\right.\\ &\quad\left.+\frac{1}{\mathrm{d}x^2}\mathrm{d}^2\left(\frac{\mathrm{d}U}{\mathrm{d}y''}+\lambda\frac{\mathrm{d}L}{\mathrm{d}y''}\right)-\cdots\right)\mathrm{d}x\\ \Psi &= \left(\frac{\mathrm{d}U}{\mathrm{d}z}+\lambda\frac{\mathrm{d}L}{\mathrm{d}z}-\frac{1}{\mathrm{d}x}\mathrm{d}\left(\frac{\mathrm{d}U}{\mathrm{d}z'}+\lambda\frac{\mathrm{d}L}{\mathrm{d}z'}\right)\right.\\ &\quad\left.+\frac{1}{\mathrm{d}x^2}\mathrm{d}^2\left(\frac{\mathrm{d}U}{\mathrm{d}z''}+\lambda\frac{\mathrm{d}L}{\mathrm{d}z''}\right)-\cdots\right)\mathrm{d}x\end{aligned}$$

But since $L=0$ is the equation of condition, $\mathrm{d}L$ is also equal to zero which, in turn, will give $\Xi=0$. Thus by equating to zero the coefficients of the three variations δx, δy, δz, only the two equations $\Upsilon=0$ and $\Psi=0$, will remain of which one will be used to eliminate the indeterminate λ, so that only one equation in x, y, z remains to be combined with the given equation $L=0$ for the solution of the problem.

23. Since by assuming $\mathrm{d}x$ to be constant, one has

$$\begin{aligned}y' &= \frac{\mathrm{d}y}{\mathrm{d}x}, & y'' &= \frac{\mathrm{d}^2y}{\mathrm{d}x^2}, & &\ldots\\ z' &= \frac{\mathrm{d}z}{\mathrm{d}x}, & z'' &= \frac{\mathrm{d}^2z}{\mathrm{d}x^2}, & &\ldots,\end{aligned}$$

it suffices to replace the variables y, z, etc. by their differentials in the functions U, L, etc. Thus by using the notation of partial differentials with the operator δ the following equation is obtained

$$\begin{aligned}\delta U &= \frac{\delta U}{\delta y}\delta y+\frac{\delta U}{\delta\mathrm{d}y}\mathrm{d}\delta y+\frac{\delta U}{\delta\mathrm{d}^2y}\mathrm{d}^2\delta y+\cdots\\ &\quad+\frac{\delta U}{\delta z}\delta z+\frac{\delta U}{\delta\mathrm{d}z}\mathrm{d}\delta z+\frac{\delta U}{\delta\mathrm{d}^2z}\mathrm{d}^2\delta z+\cdots\end{aligned}$$

and if the variation of U with respect to x is also required simultaneously, the term $(1/\mathrm{d}x)\,\mathrm{d}U\delta x$ must be added to the expression for δU and δy must be replaced by $\delta y-(\mathrm{d}y/\mathrm{d}x)\delta x$ and δz by $\delta z-(\mathrm{d}z/\mathrm{d}x)\,\delta x$, etc.

In this fashion, after reducing the expression, there will result

$$\begin{aligned}\delta\int U\,\mathrm{d}x &= \int(\Upsilon\,\delta y+\Psi\,\delta z+\cdots)\,\mathrm{d}x\\ &\quad+\Upsilon'\,\delta y+\Upsilon''\,\mathrm{d}\delta y+\cdots+\Psi'\delta z+\Psi''\,\mathrm{d}\delta z+\cdots,\end{aligned}$$

where

$$\Upsilon=\frac{\delta U}{\delta y}-\mathrm{d}\frac{\delta U}{\delta\mathrm{d}y}+\mathrm{d}^2\frac{\delta U}{\delta\mathrm{d}^2y}-\cdots$$

$$\Upsilon' = \frac{\delta U}{\delta \mathrm{d}y} - \mathrm{d}\frac{\delta U}{\delta \mathrm{d}^2 y} + \cdots, \qquad \Upsilon'' = \frac{\delta U}{\delta \mathrm{d}^2 y} - \cdots, \qquad \cdots$$

$$\Psi = \frac{\delta U}{\delta z} - \mathrm{d}\frac{\delta U}{\delta \mathrm{d}z} + \mathrm{d}^2\frac{\delta U}{\delta \mathrm{d}^2 z} - \cdots$$

$$\Psi' = \frac{\delta U}{\delta \mathrm{d}z} - d\frac{\delta U}{\mathrm{d}\delta^2 z} + \cdots, \qquad \Psi'' = \frac{\delta U}{\delta \mathrm{d}^2 z} - \cdots, \qquad \cdots$$

and to obtain the expression with respect to x, one will put in all the terms $-(\mathrm{d}y/\mathrm{d}x)\,\delta x$ in place of δy and $-(\mathrm{d}z/\mathrm{d}x)\,\delta x$ in place of δz.

24. This is the general method for the problems of maxima and minima relative to the indefinite integral formulas for which the Calculus of Variations was first formulated. And it is clear that even by varying all the variables, there will nevertheless always be one less equation then there are variables. This is also in conformity with the nature of the problem because it is not the individual value of each of the variables that is sought, as in ordinary problems of maxima and minima, but indefinite relations between these variables by which they become a function of one another and can be represented by curves of single or double curvature.

25. Let us now apply the same method to the problems of mechanics and let us assume for simplicity that the formula $P\,\mathrm{d}p + Q\,\mathrm{d}q + R\mathrm{d}r +$ etc. is integrable and that its integral is Π, as in Article 21 of SECTION III. Thus one will have $P\mathrm{d}p + Q\,\mathrm{d}q + R\,\mathrm{d}r +$etc. $= \delta\Pi$, and the general equation of equilibrium (Article 13) will become $\mathrm{S}(\delta\Pi\,\mathrm{d}m + \lambda\,\delta L + \mu\,\delta M + \text{etc.}) = 0$, as long as no consideration is taken of the equations of condition relative to determined points. Since the mass of every particle dm of the system must not change during the time the position of the system changes, it should be assumed that $\delta\,\mathrm{d}m = 0$ and consequently, $\delta L = \delta\,\mathrm{d}m$.

When the system is linear, one has generally $\mathrm{d}m = U\,\mathrm{d}x$, where U is a function similar to those presented in Article 20. Thus $\delta L = \delta U\,\mathrm{d}x + U\,\delta\mathrm{d}x$ and the formula $\mathrm{S}\,\lambda\,\delta L$ will have the following terms under the integral sign $(\Xi\,\delta x + \Upsilon\,\delta u + \Psi\,\delta\nu)\,\mathrm{d}x$, from which there results (Article 22)

$$\Xi = (\lambda\,\mathrm{d}U - \mathrm{d}(\lambda U))\frac{1}{\mathrm{d}x}$$

$$\Upsilon = \lambda\frac{\mathrm{d}U}{\mathrm{d}y} - \frac{1}{\mathrm{d}x}\,\mathrm{d}(\lambda\frac{\mathrm{d}U}{\mathrm{d}y'}) + \frac{1}{\mathrm{d}x^2}\mathrm{d}^2(\lambda\frac{\mathrm{d}U}{\mathrm{d}y''}) - \cdots$$

$$\Psi = \lambda\frac{\mathrm{d}U}{\mathrm{d}z} - \frac{1}{\mathrm{d}x}\,\mathrm{d}(\lambda\frac{\mathrm{d}U}{\mathrm{d}z'}) + \frac{1}{\mathrm{d}x^2}\,\mathrm{d}^2(\lambda\frac{\mathrm{d}U}{\mathrm{d}z''}) - \cdots.$$

26. Thus if there is no other condition, the equation obtained from the terms under the integral sign S will be

$$\delta\Pi\,\mathrm{d}m + (\Xi\,\delta x + \Upsilon\,\delta u + \Psi\delta v)\,\mathrm{d}x = 0,$$

which should be verified individually with respect to the variables δx, δy, δz.

But since Π is a function of x, y, z, one has

$$\delta\Pi = \frac{\mathrm{d}\Pi}{\mathrm{d}x}\,\delta x + \frac{\mathrm{d}\Pi}{\mathrm{d}y}\,\delta y + \frac{\mathrm{d}\Pi}{\mathrm{d}z}\,\delta z$$

and since

$$\delta u = \delta y - \frac{\mathrm{d}y}{\mathrm{d}x}\,\delta x, \qquad \delta v = \delta z - \frac{\mathrm{d}z}{\mathrm{d}x}\,\delta x$$

the preceding equation becomes

$$\begin{aligned} &(\frac{\mathrm{d}\Pi}{\mathrm{d}x}\,\mathrm{d}m + \Xi\,\mathrm{d}x - \Upsilon\,\mathrm{d}y - \Psi\,\mathrm{d}z)\,\delta x \\ &+(\frac{\mathrm{d}\Pi}{\mathrm{d}y}\,\mathrm{d}m + \Upsilon\,\mathrm{d}x)\,\delta y + (\frac{\mathrm{d}\Pi}{\mathrm{d}z}\,\mathrm{d}m + \Psi\,\mathrm{d}x)\,\delta z = 0 \end{aligned}$$

from which the following three equations are obtained

$$\begin{aligned} &\frac{\mathrm{d}\Pi}{\mathrm{d}x}\,\mathrm{d}m + \Xi\,\mathrm{d}x - \Upsilon\,\mathrm{d}y - \Psi\,\mathrm{d}z = 0 \\ &\frac{\mathrm{d}\Pi}{\mathrm{d}y}\,\mathrm{d}m + \Upsilon\,\mathrm{d}z = 0, \qquad \frac{\mathrm{d}\Pi}{\mathrm{d}z}\,\mathrm{d}m + \Psi\,\mathrm{d}x = 0 \end{aligned}$$

Thus in this case, there are as many equations as variables which seems to be the difference between problems of this type in mechanics and the problems of maxima and minima

27. But I observe at the outset that these three equations can be reduced to two by elimination of the undetermined coefficient λ. And in general, although the equations of condition always replace those which disappear after the elimination of the undetermined coefficient, the condition $\delta \mathrm{d}m = 0$, introduced here, that is, $\mathrm{d}m$ is a constant, cannot furnish a particular equation for the solution of the problem because, according to the logic of the differential calculus, it is always permissible to take an arbitrary element for a constant since it is, properly speaking, only the ratios of differentials and not the differentials themselves which are to be considered in the calculus. Therefore, the three equations will be reduced to two and will be used only to determine the nature of the curve as in problems of maxima and minima.

28. I observe in passing that the problems of statics discussed here can be reduced to simple problems of maxima and minima.

If after multiplying the first equation found above by $\mathrm{d}x$, the second by $\mathrm{d}y$ and the third by $\mathrm{d}z$, the three equations are added together, the equation $\mathrm{d}\Pi\,\mathrm{d}m + \Xi\,\mathrm{d}x^2 = 0$ will result because of the relation

$$\frac{\mathrm{d}\Pi}{\mathrm{d}x}\,\mathrm{d}x + \frac{\mathrm{d}\Pi}{\mathrm{d}y}\,\mathrm{d}y + \frac{\mathrm{d}\Pi}{\mathrm{d}z}\,\mathrm{d}z = \mathrm{d}\Pi$$

But since $\Xi\,dx = \lambda\,dU - d(\lambda U) = -U\,d\lambda$, and $dm = U\,dx$, one will have after dividing by dm, $d\Pi - d\lambda = 0$ from which the equation $\lambda = \Pi + a$ is obtained, where the letter a represents an arbitrary constant.

Therefore, the term $\lambda\,\delta L$ in the equation of Article 25 will become $\Pi\,\delta dm + a\,\delta dm$ since $\delta L = \delta dm$, and since $\delta\Pi\,dm + \Pi\,\delta dm = \delta(\Pi\,dm)$, this equation will become $S\delta(\Pi\,dm) + aS\,\delta dm = 0$, that is

$$\delta(S\,\Pi\,dm) + a\,\delta(S\,dm) = 0$$

This is the necessary equation which makes the integrable formula $S\,\Pi\,dm$ a maximum or a minimum among all those for which the formula $S\,dm$ has the same value.

In this fashion, one of the variables can be considered constant as is done in the problems of maxima and minima relative to the variations by δ, which in turn simplifies the analysis. But the general method has the advantage of providing the value of the coefficient λ, which by the theory developed in the preceding section, will express the force with which the element dm reacts the forces P, Q, R, etc. which act on the system.

29. We have assumed, for greater simplicity, that there were no other equations of condition, but if, in addition, there were the equation $M = 0$, where M is a function of x, y, z, y', y'', etc., z', z'', etc., the term $\mu\delta M$, or rather, for homogeneity, the term $\mu\delta M\,dx$, should be added to the term $\lambda\delta L$ under the integral sign in the equation of equilibrium, which means to add to the values Ξ, Υ, Ψ of Article 25 the following quantities, respectively

$$\frac{1}{dx}\mu\,dM$$
$$\mu\frac{dM}{dy} - \frac{1}{dx}\,d(\mu\frac{dM}{dy'}) + \frac{1}{dx^2}\,d^2(\mu\frac{dM}{dy''}) - \cdots$$
$$\mu\frac{dM}{dz} - \frac{1}{dx}\,d(\mu\frac{dM}{dz'}) + \frac{1}{dx^2}\,d^2(\mu\frac{dM}{dz''}) - \cdots$$

Thus three equations of the same form as those of Article 26 would be obtained which by elimination of the two indeterminates λ and μ would be reduced to only one. But by adding the equation of condition $M = 0$, two equations between the three variables x, y, z would be obtained as before.

These three equations lead to the same equation as in Article 28 which is $\delta\Pi\,dm + \Xi\,dx^2 = 0$. Here we have $\Xi\,dx = -U\,d\lambda + \mu\,dM$. But the equation $M = 0$, gives also $dM = 0$. Therefore, one will have simply, as in the cited article, $\Xi\,dx = -U\,d\lambda$, from which the same result $\delta(S\,\Pi\,dm) + a\,\delta(dm) = 0$ will be obtained.

30. Therefore, in general, the problem of equilibrium of a system of particles dm on which act the forces P, Q, R, etc. along the directions of the lines p, q, r, etc. and with the assumption that $P\,dp + Q\,dq + R\,dr + \text{etc.} = d\Pi$, is simply reduced to making the integrable formula $S\,\Pi\,dm$ a maximum or a minimum with respect to particular conditions of the system, which as one sees, makes all the problems of equilibrium enter the class of problems of maxima and minima, known under the name of "isoperimetrical problems".

In the case of the catenary by taking the ordinates y in the vertical direction one has $\Pi = gy$, where g is the constant force of gravity. Thus the formula $\mathrm{S}y\,\mathrm{d}m$ must be a maximum or a minimum among all those where the value of $\mathrm{S}\,\mathrm{d}m$ is the same. But $\mathrm{S}y\,\mathrm{d}m/\mathrm{S}\,\mathrm{d}m$ is the distance from the center of gravity to the horizontal. So since the entire mass is assumed given, this distance must be the largest or smallest value which is known already.

31. We have only considered functions of variables which are viewed as independent up to this point. But if the variable z were a function of x and y, and one had a function U which contains x, y, z, with the partial differences of z relative to x and y, the variation of δU could be expressed with respect to the simultaneous variations of x, y, z.

For added simplicity, let us define

$$\frac{\mathrm{d}z}{\mathrm{d}x} = z', \qquad \frac{\mathrm{d}z}{\mathrm{d}y} = z_{\prime}, \qquad \frac{\mathrm{d}^2 z}{\mathrm{d}x^2} = z''$$
$$\frac{\mathrm{d}^2 z}{\mathrm{d}x\mathrm{d}y} = z'_{\prime}, \qquad \frac{\mathrm{d}^2 z}{\mathrm{d}y^2} = z_{\prime\prime}, \qquad \frac{\mathrm{d}^3 z}{\mathrm{d}x^3} = z'''$$
$$\frac{\mathrm{d}^3 z}{\mathrm{d}x^2\mathrm{d}y} = z''_{\prime}, \qquad \frac{\mathrm{d}^3 z}{\mathrm{d}x\mathrm{d}y^2} = z'_{\prime\prime}, \qquad \ldots$$

Then the quantity U will be a function of x, y, z, z', $z_{\prime}$, z'', $z'_{\prime}$, $z_{\prime\prime}$ etc., so that the following equation results

$$\delta U = \frac{\mathrm{d}U}{\mathrm{d}x}\delta x + \frac{\mathrm{d}U}{\mathrm{d}y}\delta y + \frac{\mathrm{d}U}{\mathrm{d}z}\delta z$$
$$+\frac{\mathrm{d}U}{\mathrm{d}z'}\delta z' + \frac{\mathrm{d}U}{\mathrm{d}z_{\prime}}\delta z_{\prime} + \frac{\mathrm{d}U}{\mathrm{d}z''}\delta z'' + \frac{\mathrm{d}U}{\mathrm{d}z'_{\prime}}\delta z'_{\prime} + \cdots$$

Thus the problem is reduced to finding the expressions for the variations $\delta z'$, $\delta z_{\prime}$, $\delta z''$, etc. by varying simultaneously the elements $\mathrm{d}x$ and $\mathrm{d}y$ in the partial differences.

We can assume, in order to simplify the calculation, that the variation δx is a function of x independent of y and that the variation δy is a function of y independent of x. We shall see in what follows that this assumption possesses all the generality required to obtain the solution.

32. After having presented this information, differentiation will give the following equation

$$\delta z' = \delta\frac{\mathrm{d}z}{\mathrm{d}x} = \frac{\delta\mathrm{d}z}{\mathrm{d}x} - \frac{\mathrm{d}z}{\mathrm{d}x}\frac{\delta\mathrm{d}x}{\mathrm{d}x}$$

Furthermore, it is clear that $(\delta\mathrm{d}z/\mathrm{d}x) = (\mathrm{d}\delta z/\mathrm{d}x)$ and $(\delta\mathrm{d}x/\mathrm{d}x) = (\mathrm{d}\delta x/\mathrm{d}x)$.

Consequently, the following equation will be obtained

$$\delta z' = \frac{\mathrm{d}\delta z}{\mathrm{d}x} - z'\frac{\mathrm{d}\delta x}{\mathrm{d}x} = \frac{\mathrm{d}(\delta z - z'\delta x)}{\mathrm{d}x} + \frac{\mathrm{d}z'}{\mathrm{d}x}\delta x$$

or

$$\delta z' = \frac{\mathrm{d}(\delta z - z'\delta x - z_{\prime}\delta y)}{\mathrm{d}x} + \frac{\mathrm{d}z'}{\mathrm{d}x}\delta x + \frac{\mathrm{d}z_{\prime}}{\mathrm{d}x}\delta y$$

Similarly

$$\delta z_{\prime} = \frac{\mathrm{d}(\delta z - z'\delta x - z_{\prime}\delta y)}{\mathrm{d}y} + \frac{\mathrm{d}z'}{\mathrm{d}y}\delta x + \frac{\mathrm{d}z_{\prime}}{\mathrm{d}y}\delta y$$

since $\mathrm{d}\delta x/\mathrm{d}y = 0$ and $\mathrm{d}\delta y/\mathrm{d}x = 0$. The following result will then be obtained

$$\delta z'' = \delta\frac{\mathrm{d}z'}{\mathrm{d}x} = \frac{\mathrm{d}\delta z'}{\mathrm{d}x} - \frac{\mathrm{d}z'}{\mathrm{d}x}\frac{\mathrm{d}\delta x}{\mathrm{d}x}$$

Substituting the expression for $\delta z'$, one will have

$$\delta z'' = \frac{\mathrm{d}^2(\delta z - z'\,\delta x - z_{\prime}\,\delta y)}{\mathrm{d}x^2} + \frac{\mathrm{d}^2 z'}{\mathrm{d}x^2}\delta x + \frac{\mathrm{d}^2 z_{\prime}}{\mathrm{d}x^2}\delta y$$

Similarly

$$\delta z'_{\prime} = \delta\frac{\mathrm{d}z'}{\mathrm{d}y} = \frac{\mathrm{d}\delta z'}{\mathrm{d}y} - \frac{\mathrm{d}z'}{\mathrm{d}y}\frac{\mathrm{d}\delta y}{\mathrm{d}y}$$

Substituting also the expression of $\delta z'$, one will have since $\mathrm{d}z_{\prime}/\mathrm{d}x = \mathrm{d}z'/\mathrm{d}y$

$$\delta z'_{\prime} = \frac{\mathrm{d}^2(\delta z - z'\,\delta x - z_{\prime}\,\delta y)}{\mathrm{d}x\mathrm{d}y} + \frac{\mathrm{d}^2 z'}{\mathrm{d}x\mathrm{d}y}\delta x + \frac{\mathrm{d}^2 z_{\prime}}{\mathrm{d}x\mathrm{d}y}\delta y$$

similarly

$$\delta z'_{\prime\prime} = \frac{\mathrm{d}^2(\delta z - z'\,\delta x - z_{\prime}\,\delta y)}{\mathrm{d}y^2} + \frac{\mathrm{d}^2 z'}{\mathrm{d}y^2}\delta x + \frac{\mathrm{d}^2 z_{\prime}}{\mathrm{d}y^2}\delta y$$

and so on.

33. Thus in order to shorten the derivations, if one sets

$$\delta z - \frac{\mathrm{d}z}{\mathrm{d}x}\delta x - \frac{\mathrm{d}z}{\mathrm{d}y}\delta y = \delta u$$

and observes that

$$\frac{\mathrm{d}z_{\prime}}{\mathrm{d}x} = \frac{\mathrm{d}z'}{\mathrm{d}y}, \qquad \frac{\mathrm{d}z'}{\mathrm{d}y} = \frac{\mathrm{d}z_{\prime}}{\mathrm{d}x}, \qquad \frac{\mathrm{d}^2 z'}{\mathrm{d}x^2} = \frac{\mathrm{d}z''}{\mathrm{d}x}, \qquad \frac{\mathrm{d}^2 z}{\mathrm{d}x^2} = \frac{\mathrm{d}z''}{\mathrm{d}y}$$

$$\frac{\mathrm{d}^2 z'}{\mathrm{d}x\mathrm{d}y} = \frac{\mathrm{d}z'_{\prime}}{\mathrm{d}x}, \qquad \frac{\mathrm{d}^2 z_{\prime}}{\mathrm{d}x\mathrm{d}y} = \frac{\mathrm{d}z'_{\prime}}{\mathrm{d}y}, \qquad \frac{\mathrm{d}^2 z'}{\mathrm{d}y^2} = \frac{\mathrm{d}z_{\prime\prime}}{\mathrm{d}x}, \qquad \ldots$$

the following equations result

$$\begin{aligned}
\delta z' &= \frac{\mathrm{d}\delta u}{\mathrm{d}x} + \frac{\mathrm{d}z'}{\mathrm{d}x}\,\delta x + \frac{\mathrm{d}z'}{\mathrm{d}y}\,\delta y \\
\delta z_{\prime} &= \frac{\mathrm{d}\delta u}{\mathrm{d}y} + \frac{\mathrm{d}z_{\prime}}{\mathrm{d}x}\,\delta x + \frac{\mathrm{d}z_{\prime}}{\mathrm{d}y}\,\delta y \\
\delta z'' &= \frac{\mathrm{d}^2\delta u}{\mathrm{d}x^2} + \frac{\mathrm{d}z''}{\mathrm{d}x}\,\delta x + \frac{\mathrm{d}z''}{\mathrm{d}y}\,\delta y \\
\delta z'_{\prime} &= \frac{\mathrm{d}^2\delta u}{\mathrm{d}x\mathrm{d}y} + \frac{\mathrm{d}z'}{\mathrm{d}x}\,\delta x + \frac{\mathrm{d}z'_{\prime}}{\mathrm{d}y}\,\delta y \\
\delta z_{\prime\prime} &= \frac{\mathrm{d}^2\delta u}{\mathrm{d}y^2} + \frac{\mathrm{d}z_{\prime\prime}}{\mathrm{d}x}\,\delta x + \frac{\mathrm{d}z_{\prime\prime}}{\mathrm{d}y}\,\delta y \\
&\vdots
\end{aligned}$$

After making these substitutions in the expression for δU, replacing δz by

$$\delta u + \frac{\mathrm{d}z}{\mathrm{d}x}\,\delta x + \frac{\mathrm{d}z}{\mathrm{d}y}\,\delta y$$

and further ordering the terms with respect to $\delta x, \delta y, \delta u$, the result is the following equation

$$\begin{aligned}
\delta U = &\left(\frac{\mathrm{d}U}{\mathrm{d}x} + \frac{\mathrm{d}U}{\mathrm{d}z}\frac{\mathrm{d}z}{\mathrm{d}x} + \frac{\mathrm{d}U}{\mathrm{d}z'}\frac{\mathrm{d}z'}{\mathrm{d}x} + \frac{\mathrm{d}U}{\mathrm{d}z_{\prime}}\frac{\mathrm{d}z_{\prime}}{\mathrm{d}x}\right. \\
&\quad \left. + \frac{\mathrm{d}U}{\mathrm{d}z''}\frac{\mathrm{d}z''}{\mathrm{d}x} + \frac{\mathrm{d}U}{\mathrm{d}z'_{\prime}}\frac{\mathrm{d}z'_{\prime}}{\mathrm{d}x} + \cdots\right)\delta x \\
&+ \left(\frac{\mathrm{d}U}{\mathrm{d}y} + \frac{\mathrm{d}U}{\mathrm{d}z}\frac{\mathrm{d}z}{\mathrm{d}y} + \frac{\mathrm{d}U}{\mathrm{d}z'}\frac{\mathrm{d}z'}{\mathrm{d}y} + \frac{\mathrm{d}U}{\mathrm{d}z_{\prime}}\frac{\mathrm{d}z_{\prime}}{\mathrm{d}y}\right. \\
&\quad \left. + \frac{\mathrm{d}U}{\mathrm{d}z''}\frac{\mathrm{d}z''}{\mathrm{d}y} + \frac{\mathrm{d}U}{\mathrm{d}z'_{\prime}}\frac{\mathrm{d}z'_{\prime}}{\mathrm{d}y} + \cdots\right)\delta y \\
&+ \frac{\mathrm{d}U}{\mathrm{d}z}\,\delta u + \frac{\mathrm{d}U}{\mathrm{d}z'}\frac{\mathrm{d}\delta u}{\mathrm{d}x} + \frac{\mathrm{d}U}{\mathrm{d}z_{\prime}}\frac{\mathrm{d}\delta u}{\mathrm{d}y} \\
&+ \frac{\mathrm{d}U}{\mathrm{d}z''}\frac{\mathrm{d}^2\delta u}{\mathrm{d}x^2} + \frac{\mathrm{d}U}{\mathrm{d}z'_{\prime}}\frac{\mathrm{d}^2\delta u}{\mathrm{d}x\mathrm{d}y} + \cdots
\end{aligned}$$

Let us designate by $(\mathrm{d}U/\mathrm{d}x)$ and $(\mathrm{d}U/\mathrm{d}y)$ the partial differences[80] of U relative to x and y. Then, by regarding z as a function of these two variables, it is clear that one will have

$$\begin{aligned}
\left(\frac{\mathrm{d}U}{\mathrm{d}x}\right) &= \frac{\mathrm{d}U}{\mathrm{d}x} + \frac{\mathrm{d}U}{\mathrm{d}z}\frac{\mathrm{d}z}{\mathrm{d}x} + \frac{\mathrm{d}U}{\mathrm{d}z'}\frac{\mathrm{d}z'}{\mathrm{d}x} + \frac{\mathrm{d}U}{\mathrm{d}z_{\prime}}\frac{\mathrm{d}z_{\prime}}{\mathrm{d}x} + \cdots \\
\left(\frac{\mathrm{d}U}{\mathrm{d}y}\right) &= \frac{\mathrm{d}U}{\mathrm{d}y} + \frac{\mathrm{d}U}{\mathrm{d}z}\frac{\mathrm{d}z}{\mathrm{d}y} + \frac{\mathrm{d}U}{\mathrm{d}z'}\frac{\mathrm{d}z'}{\mathrm{d}y} + \frac{\mathrm{d}U}{\mathrm{d}z_{\prime}}\frac{\mathrm{d}z_{\prime}}{\mathrm{d}y} + \cdots
\end{aligned}$$

Thus the total variation of U will be reduced to this simple form

$$\begin{aligned}\delta U &= (\frac{\mathrm{d}U}{\mathrm{d}x})\delta x + (\frac{\mathrm{d}U}{\mathrm{d}y})\delta y + \frac{\mathrm{d}U}{\mathrm{d}z}\delta u \\ &+ \frac{\mathrm{d}U}{\mathrm{d}z'}\frac{\mathrm{d}\delta u}{\mathrm{d}z} + \frac{\mathrm{d}U}{\mathrm{d}z_{\prime}}\frac{\mathrm{d}\delta u}{\mathrm{d}y} + \frac{\mathrm{d}U}{\mathrm{d}z''}\frac{\mathrm{d}^2\delta u}{\mathrm{d}x^2} \\ &+ \frac{\mathrm{d}U}{\mathrm{d}z'_{\prime}}\frac{\mathrm{d}^2\delta u}{\mathrm{d}x\,\mathrm{d}y} + \frac{\mathrm{d}U}{\mathrm{d}z_{\prime\prime}}\frac{\mathrm{d}^2\delta u}{\mathrm{d}y^2} + \cdots\end{aligned}$$

34. Therefore, if the double integral SS $U\,\mathrm{d}x\,\mathrm{d}y$ is to be maximized or minimized the following equation will be obtained

$$\delta(\mathrm{SS}\,U\,\mathrm{d}x\,\mathrm{d}y) = \mathrm{SS}\,\delta(U\,\mathrm{d}x\,\mathrm{d}y) = 0$$

But by making all quantities variable, one has $\delta(U\,\mathrm{d}x\,\mathrm{d}y) = \delta U\,\mathrm{d}x\,\mathrm{d}y + U\,\delta(\mathrm{d}x\,\mathrm{d}y)$ where it should be noted that $\mathrm{d}x\,\mathrm{d}y$ represents a rectangle which is the element of the xy-plane. This rectangle will remain a rectangle after assembling the variations δx and δy of the coordinates x and y using the adopted assumption that δx does not depend on y and neither δy on x. The variation $\mathrm{d}x\,\mathrm{d}y$ in this case will be simply $\mathrm{d}y\delta\mathrm{d}x + \mathrm{d}x\,\delta\mathrm{d}y$. Therefore

$$\delta\mathrm{d}x = \mathrm{d}\delta x = \frac{\mathrm{d}\delta x}{\mathrm{d}x}\,\mathrm{d}x, \qquad \delta\mathrm{d}y = \mathrm{d}\delta y = \frac{\mathrm{d}\delta y}{\mathrm{d}y}\,\mathrm{d}y$$

Since δx and δy are assumed to be functions of x and y only, one will have

$$\delta(U\mathrm{d}x\,\mathrm{d}y) = \left(\delta U + U\frac{\mathrm{d}\delta x}{\mathrm{d}x} + U\frac{\mathrm{d}\delta y}{\mathrm{d}y}\right)\mathrm{d}x\,\mathrm{d}y$$

After substituting the expression for δU, and eliminating by partial integration the differentials of the variations δx, δy, δu, the following terms will remain under the double integral sign $(\Xi\delta x + \Upsilon\delta y + \Psi\delta u)\,\mathrm{d}x\mathrm{d}y$ where

$$\begin{aligned}\Xi &= (\frac{\mathrm{d}U}{\mathrm{d}x}) - (\frac{\mathrm{d}U}{\mathrm{d}x}) = 0 \\ \Upsilon &= (\frac{\mathrm{d}U}{\mathrm{d}y}) - (\frac{\mathrm{d}U}{\mathrm{d}y}) = 0 \\ \Psi &= \frac{\mathrm{d}U}{\mathrm{d}z} - (\frac{\mathrm{d}U'}{\mathrm{d}x}) - (\frac{\mathrm{d}U_{\prime}}{\mathrm{d}y}) + (\frac{\mathrm{d}^2U''}{\mathrm{d}x^2}) \\ &+ (\frac{\mathrm{d}^2U'_{\prime}}{\mathrm{d}x\mathrm{d}y}) + (\frac{\mathrm{d}^2U_{\prime\prime}}{\mathrm{d}y^2}) + \cdots\end{aligned}$$

in which the following definitions have been made in order to shorten the expressions

$$\begin{aligned}U' &= \frac{\mathrm{d}U}{\mathrm{d}z'}, \qquad U_{\prime} = \frac{\mathrm{d}U}{\mathrm{d}z_{\prime}}, \qquad U'' = \frac{\mathrm{d}U}{\mathrm{d}z''}, \\ U'_{\prime} &= \frac{\mathrm{d}U}{\mathrm{d}z'_{\prime}}, \qquad U_{\prime\prime} = \frac{\mathrm{d}U}{\mathrm{d}z_{\prime\prime}}, \qquad \cdots\end{aligned}$$

Furthermore, it is assumed that the partial differentials in parentheses represent the total values of these differences by viewing z as a function of x and y.

35. Thus because of the expression

$$\delta u = \delta z - \frac{\mathrm{d}z}{\mathrm{d}x}\delta x - \frac{\mathrm{d}z}{\mathrm{d}y}\delta y$$

the terms under the double integral sign will simply give the equation

$$\Psi\left(\delta z - \frac{\mathrm{d}z}{\mathrm{d}x}\delta x - \frac{\mathrm{d}z}{\mathrm{d}y}\delta y\right) = 0$$

from which, by equating separately to zero the coefficients of δz, δx, δy, the equation for Ψ will be equal to zero, as if only the variable z had been varied.

It is obvious that in problems of maxima and minima relative to double integrals in which one of the three variables is a function of the other two, there is precisely only one equation that can be found directly by varying by δ the variable which is assumed a function of the other two. This is the equation of the surface which satisfies the problem. This is how the equation of partial differences for the least surface is found, by making $U = \sqrt{(1 + (z')^2 + (z_{,})^2)}$. What we have demonstrated proves that this equation fulfills all the conditions of the problem whatever the variations given to the three coordinates of the surface.

36. The formulas of variation which were just found can be applied to the equation of a plane system of particles $\mathrm{d}m$ acted upon by arbitrary forces. In the case where $\mathrm{d}m$ is invariable, the general equation of equilibrium will become, as in Article 25

$$\mathrm{SS}(\delta\Pi\,\mathrm{d}m + \lambda\,\delta\mathrm{d}m) = 0$$

Here the expression for dm will be of the form $U\,\mathrm{d}x\,\mathrm{d}y$ and consequently, the following equation results (Article 34)

$$\delta\mathrm{d}m = \left(\delta U + U\frac{\mathrm{d}\delta x}{\mathrm{d}x} + U\frac{\mathrm{d}\delta y}{\mathrm{d}y}\right)\mathrm{d}x\mathrm{d}y.$$

After substituting this expression, as well as the one for δU from Article 33 in the integrable formula SS $\lambda\,\delta\mathrm{d}m$ and eliminating by the integration by parts the differences of the variations δx, δy, δu, there will remain under the double integral sign only the following terms

$$(\Xi\delta x + \Upsilon\delta y + \Psi\delta u)\,\mathrm{d}x\,\mathrm{d}y,$$

where

$$\begin{aligned}
\Xi &= \lambda(\frac{\mathrm{d}U}{\mathrm{d}x}) - (\frac{\mathrm{d}(\lambda U)}{\mathrm{d}x}) = -U(\frac{\mathrm{d}\lambda}{\mathrm{d}x}) \\
\Upsilon &= \lambda(\frac{\mathrm{d}U}{\mathrm{d}y}) - (\frac{\mathrm{d}(\lambda U)}{\mathrm{d}y}) = -U(\frac{\mathrm{d}\lambda}{\mathrm{d}y}) \\
\Psi &= \frac{\mathrm{d}U}{\mathrm{d}z} - (\frac{\mathrm{d}U'}{\mathrm{d}x}) - (\frac{\mathrm{d}U_{\prime}}{\mathrm{d}y}) + (\frac{\mathrm{d}^2U''}{\mathrm{d}x^2}) \\
&\quad + (\frac{\mathrm{d}^2U'_{\prime}}{\mathrm{d}x\mathrm{d}y}) - (\frac{\mathrm{d}^2U_{\prime\prime}}{\mathrm{d}y^2}) + \cdots
\end{aligned}$$

by retaining the values of U', $U_{\prime}$, U'', $U'_{\prime}$, etc. of Article 34.

Let us add to these terms those which are derived from the integral SS $\delta\Pi\,\mathrm{d}m$, by substituting the values of $\delta\Pi$ and $\mathrm{d}m$

$$\left(\frac{\mathrm{d}\Pi}{\mathrm{d}x}\delta x + \frac{\mathrm{d}\Pi}{\mathrm{d}y}\delta y + \frac{\mathrm{d}\Pi}{\mathrm{d}z}\delta z\right) U\,\mathrm{d}x\,\mathrm{d}y$$

and by replacing δu with the expression (Article 33)

$$\delta z - \frac{\mathrm{d}z}{\mathrm{d}x}\delta x - \frac{\mathrm{d}z}{\mathrm{d}y}\delta y.$$

The general equation of equilibrium will contain under the double integral sign SS, the following terms with respect to the variations δx, δy, δz

$$\left.\begin{aligned}
&((\frac{\mathrm{d}\Pi}{\mathrm{d}x} - (\frac{\mathrm{d}\lambda}{\mathrm{d}x}))U - \Psi\frac{\mathrm{d}z}{\mathrm{d}x})\delta x \\
&+((\frac{\mathrm{d}\Pi}{\mathrm{d}y} - (\frac{\mathrm{d}\lambda}{\mathrm{d}y}))U - \Psi\frac{\mathrm{d}z}{\mathrm{d}y})\delta y \\
&+(\frac{\mathrm{d}\Pi}{\mathrm{d}z}U + \Psi)\delta z
\end{aligned}\right\} \mathrm{d}x\,\mathrm{d}y$$

from which the three following equations are obtained

$$\begin{aligned}
&(\frac{\mathrm{d}\Pi}{\mathrm{d}x} - (\frac{\mathrm{d}\lambda}{\mathrm{d}x}))U - \Psi\frac{\mathrm{d}z}{\mathrm{d}x} = 0 \\
&(\frac{\mathrm{d}\Pi}{\mathrm{d}y} - (\frac{\mathrm{d}\lambda}{\mathrm{d}y}))U - \Psi\frac{\mathrm{d}z}{\mathrm{d}y} = 0 \\
&\frac{\mathrm{d}\Pi}{\mathrm{d}z}U + \Psi = 0
\end{aligned}$$

The last equation gives $\Psi = -U(\mathrm{d}\Pi/\mathrm{d}z)$ and this expression when substituted into the other two and after dividing by U produces

$$\begin{aligned}
&\frac{\mathrm{d}\Pi}{\mathrm{d}x} + \frac{\mathrm{d}\Pi}{\mathrm{d}z}\frac{\mathrm{d}z}{\mathrm{d}y} - \frac{\mathrm{d}\lambda}{\mathrm{d}x} = 0 \\
&\frac{\mathrm{d}\Pi}{\mathrm{d}y} + \frac{\mathrm{d}\Pi}{\mathrm{d}z}\frac{\mathrm{d}z}{\mathrm{d}y} - \frac{\mathrm{d}\lambda}{\mathrm{d}y} = 0
\end{aligned}$$

The first equation gives $\lambda = \Pi + f(y)$ and the second equation gives $\lambda = \Pi + f(x)$. Therefore, one will have $\lambda = \Pi + a$, where a is a constant. After substituting this expression in the general equation of equilibrium, it will become $\mathrm{SS}(\delta(\Pi\, dm) + a\,\delta dm) = 0$, that is, $\delta(\mathrm{SS}\,\Pi\, dm) + a\,\delta(\mathrm{SS}\, dm) = 0$. This is the equation for the maximum or minimum of the integrable formula $\mathrm{SS}\,\Pi\, dm$, among all those in which the value of $\mathrm{SS}\, dm$ is the same.

Thus it is here that the problems of mechanics reduce to a simple equation of maxima and minima, for which the solution depends only on the variation of the coordinate z which is assumed a function of x and y (Article 35). This theory could be expanded to include triple integrable formulas and similar conclusions deduced.

SECTION V
THE SOLUTION OF VARIOUS PROBLEMS OF STATICS

We will now demonstrate the application of our method to various problems in the equilibrium of bodies. It is apparent from the uniform and rapid solution of the problems of statics using our method, how superior it is to those presently used.

Chapter I
THE EQUILIBRIUM OF SEVERAL FORCES APPLIED AT THE SAME POINT AND THE COMPOSITION AND RESOLUTION OF FORCES

1. We propose to find the laws for the equilibrium of an arbitrary number of forces P, Q, R, etc., all applied at the same point and directed towards given points. Denoting by p, q, r, etc. the rectilinear distances between the common point of application of these forces and the respective points to which they are directed, the following formula

$$P\,\mathrm{d}p + Q\,\mathrm{d}q + R\,\mathrm{d}r + \cdots$$

will result for the sum of the moments of all the forces. This equation must be put equal to zero for the state of equilibrium.

Let x, y, z be the three rectangular coordinates of the point at which all the forces are applied and conversely, let a, b, c be the rectangular coordinates of the point towards which the force P is directed; f, g, h those of the point toward which the force Q is directed; ℓ, m, n those of the point toward which the force R is directed and so on for the remaining forces. Since all of these coordinates are considered relative to the same fixed axes in space, the lengths p, q, r can be expressed by the following equations

$$\begin{aligned}
p &= \sqrt{(x-a)^2 + (y-b)^2 + (z-c)^2} \\
q &= \sqrt{(x-f)^2 + (y-g)^2 + (z-h)^2} \\
r &= \sqrt{(x-\ell)^2 + (y-m)^2 + (z-n)^2} \\
&\;\;\vdots
\end{aligned}$$

The quantity $P\,\mathrm{d}p + Q\,\mathrm{d}q + R\,\mathrm{d}r +$ etc. will become $X\,\mathrm{d}x + Y\,\mathrm{d}y + Z\,\mathrm{d}z$, where[81]

$$X = \frac{x-a}{p}P + \frac{x-f}{q}Q + \frac{x-\ell}{r}R + \cdots$$
$$Y = \frac{y-b}{p}P + \frac{y-g}{q}Q + \frac{y-m}{r}R + \cdots$$
$$Z = \frac{z-c}{p}P + \frac{z-h}{q}Q + \frac{z-n}{r}R + \cdots$$

It is worthwhile to note that in these expressions the quantities

$$\frac{x-a}{p}, \qquad \frac{y-b}{p}, \qquad \frac{z-c}{p}$$

are equal to the cosines of the angles which the line p, that is, the direction that the force P, makes with the x, y, z axes. Similarly

$$\frac{x-f}{q}, \qquad \frac{y-g}{q}, \qquad \frac{z-h}{q}$$

are the cosines of the angles that the direction of the force Q makes with the same axes and so on (SECTION II, Article 7).

Subsection I
The Equilibrium of a Body or Material Point Loaded by Several Forces

2. After having stated these facts, let us assume at the outset that the body or point to which the forces P, Q, R, etc., are applied is entirely free. Then there is no equation of condition between the coordinates x, y, z and the quantity $X\,\mathrm{d}x + Y\,\mathrm{d}y + Z\,\mathrm{d}z$ must be equal to zero, independent of the values of $\mathrm{d}x$, $\mathrm{d}y$, $\mathrm{d}z$ (SECTION II, Article 10). This result will immediately give the following three particular equations

$$X = 0, \qquad Y = 0, \qquad Z = 0$$

These equations express the laws of equilibrium for any number of forces applied at the same point.

3. If in the expressions for X, Y, Z, one sets $P = p$, $Q = q$, $R = r$, etc. which is permissible[82] since it is immaterial where the force center is taken along the directions of the forces, the following equations will be obtained

$$x - a + x - f + x - \ell + \cdots = 0$$
$$y - b + y - g + y - m + \cdots = 0$$
$$z - c + z - h + z - n + \cdots = 0$$

and by assuming that the number of forces P, Q, R, etc. is μ, there results

$$x = \frac{a + f + \ell + \cdots}{\mu}, \qquad y = \frac{b + g + m + \cdots}{\mu}, \qquad z = \frac{c + h + n + \cdots}{\mu}$$

These expressions for x, y, z show that the point at which these forces are applied is at the center of the points to which these forces are directed.

The theorem of Leibnitz derives from this development. It states that for any number of forces in equilibrium at a point from which straight lines are drawn to represent the magnitude and the direction of each force, the point under consideration will be the center of gravity of all the points at which the lines terminate.

Thus, if there are only four forces and if a pyramid is imagined for which the four angles are at the extremities of the lines which represent the forces, there will be equilibrium between the four forces when the point on which they act is at the center of gravity of the pyramid, since it is known from geometry that the center of gravity of the pyramid has the same location as the center of gravity of four identical bodies placed at the four corners of the pyramid. This last result is due to Roberval.

4. Let us now assume that the body or mass point on which the forces P, Q, R, etc. act is not entirely free, but it is constrained to move along a surface or on a given line. Then there will be one or two equations of condition between the coordinates x, y, z which will be the equations of the surface or line.

Thus let $L = 0$ be the equation of the surface on which the body can only slide. Then the term $\lambda\, \mathrm{d}L$ (SECTION IV, Article 3) will be added to the sum of the moments of the forces $X\, \mathrm{d}x + Y\, \mathrm{d}y + Z\, \mathrm{d}z$ and the general equation of equilibrium will be $X\, \mathrm{d}x + Y\, \mathrm{d}y + Z\, \mathrm{d}z + \lambda\, \mathrm{d}L = 0$ where λ is an undetermined quantity.

Now L is a known function of x, y, z and after its differentiation, the following equation results

$$\mathrm{d}L = \frac{\mathrm{d}L}{\mathrm{d}x}\,\mathrm{d}x + \frac{\mathrm{d}L}{\mathrm{d}y}\,\mathrm{d}y + \frac{\mathrm{d}L}{\mathrm{d}z}\,\mathrm{d}z$$

Thus by substituting and equating separately to zero the sum of the terms multiplied by each of the differences $\mathrm{d}x$, $\mathrm{d}y$, $\mathrm{d}z$, the following three particular equations of equilibrium will result

$$X + \lambda\frac{\mathrm{d}L}{\mathrm{d}x} = 0, \qquad Y + \lambda\frac{\mathrm{d}L}{\mathrm{d}y} = 0, \qquad Z + \lambda\frac{\mathrm{d}L}{\mathrm{d}z} = 0$$

Then after eliminating the undetermined coefficient λ from these equations, one will obtain

$$Y\frac{\mathrm{d}L}{\mathrm{d}x} - X\frac{\mathrm{d}L}{\mathrm{d}y} = 0, \qquad Z\frac{\mathrm{d}L}{\mathrm{d}x} - X\frac{\mathrm{d}L}{\mathrm{d}z} = 0,$$

which consequently, express the required conditions for the equilibrium of a body on a given surface.

5. If the theory presented in Article 5 of SECTION IV is applied here, the conclusion will be that the surface must support the body with a force equal to

$$\lambda\sqrt{(\frac{\mathrm{d}L}{\mathrm{d}x})^2+(\frac{\mathrm{d}L}{\mathrm{d}y})^2+(\frac{\mathrm{d}L}{\mathrm{d}z})^2}$$

and directed along the perpendicular to the surface which has for its equation $\mathrm{d}L=0$, that is, perpendicular to the surface on which the body is resting. And since

$$\lambda\frac{\mathrm{d}L}{\mathrm{d}x}=-X,\qquad \lambda\frac{\mathrm{d}L}{\mathrm{d}y}=-Y,\qquad \lambda\frac{\mathrm{d}L}{\mathrm{d}z}=-Z,$$

it follows that the force of the body on the surface (a force which must be equal and directly opposed to the vertical force exerted by the surface) will be expressed by $\sqrt{(X^2+Y^2+Z^2)}$ and will act perpendicular to the same surface.[83] The two equations for the equilibrium of a body found above are uniquely reduced to this condition, as can be shown by the composition of forces.

6. Also, for the case of a single body acted on by given forces, the conditions of equilibrium can be found more easily by immediately substituting in the equation $X\,\mathrm{d}x+Y\,\mathrm{d}y+Z\,\mathrm{d}z=0$, the expression

$$-\frac{\frac{\mathrm{d}L}{\mathrm{d}x}\,\mathrm{d}x+\frac{\mathrm{d}L}{\mathrm{d}y}\,\mathrm{d}y}{\frac{\mathrm{d}L}{\mathrm{d}z}}$$

for the differential $\mathrm{d}z$ obtained from the differential equation of the given surface on which the body can slide and by equating separately to zero the coefficients of the differentials $\mathrm{d}x$ and $\mathrm{d}y$, which remain indeterminate, following the general method of Article 10 of SECTION II.

The two equations

$$X-Z\frac{\mathrm{d}L/\mathrm{d}x}{\mathrm{d}L/\mathrm{d}z}=0,\qquad Y-Z\frac{\mathrm{d}L/\mathrm{d}y}{\mathrm{d}L/\mathrm{d}z}=0$$

which are the same as those found above will be obtained immediately.

In the same fashion, if the body were forced to move on a given line defined by the two differential equations $\mathrm{d}y=p\,\mathrm{d}x$ and $\mathrm{d}z=q\,\mathrm{d}x$, it will only be necessary to substitute the values of $\mathrm{d}y$ and $\mathrm{d}z$ in the equation $X\,\mathrm{d}x+Y\,\mathrm{d}y+Z\,\mathrm{d}z=0$ in order to obtain, after dividing by $\mathrm{d}x$, $X+Yp+Zq=0$ for the equilibrium equation.

But in every case where several bodies are in equilibrium, the method of undetermined coefficients presented in the preceding section, will always have the advantage of ease, simplicity and uniformity of calculation.

Subsection II
The Composition and Resolution of Forces

7. The identity $P\,\mathrm{d}p + Q\,\mathrm{d}q + R\,\mathrm{d}r + \cdots = X\,\mathrm{d}x + Y\,\mathrm{d}y + Z\,\mathrm{d}z$ found in Article 1 shows that the system of forces P, Q, R, etc. directed along the lines p, q, r, etc. is equivalent to a system of three forces X, Y, Z directed along the coordinate axes x, y, z (SECTION II, Article 15). Thus the quantities X, Y, Z are the magnitudes of the forces P, Q, R, etc. after these latter forces have been resolved in the direction of the three rectangular coordinates x, y, z, respectively. The forces X, Y, Z are assumed to reduce the magnitudes of these coordinates, in the same fashion that the forces P, Q, R, etc. are assumed to reduce the lengths of p, q, r, etc.

8. In general, if arbitrary forces P, Q, R, etc. directed along the lines p, q, r, etc. act at the same point, these forces can always be reduced to three others directed along the lines ξ, ψ, φ, as long as the three lines are not all in the same plane. Since three lines located in different planes are sufficient to determine the position of an arbitrary point in space, the values of the lines p, q, r, etc. can always be expressed as a function of the three quantities ξ, ψ, φ. And from the theorem of Article 15 of SECTION II, the forces P, Q, R, etc. will be equivalent to three forces Ξ, Ψ, Φ, defined by the following formulas

$$\Xi = P\frac{\mathrm{d}p}{\mathrm{d}\xi} + Q\frac{\mathrm{d}q}{\mathrm{d}\xi} + R\frac{\mathrm{d}r}{\mathrm{d}\xi} + \cdots$$
$$\Psi = P\frac{\mathrm{d}p}{\mathrm{d}\psi} + Q\frac{\mathrm{d}q}{\mathrm{d}\psi} + R\frac{\mathrm{d}r}{\mathrm{d}\psi} + \cdots$$
$$\Phi = P\frac{\mathrm{d}p}{\mathrm{d}\varphi} + Q\frac{\mathrm{d}q}{\mathrm{d}\varphi} + R\frac{\mathrm{d}r}{\mathrm{d}\varphi} + \cdots$$

and directed along the lines ξ, ψ, φ or along the differentials $\mathrm{d}\xi$, $\mathrm{d}\psi$, $\mathrm{d}\varphi$ if some of these lines were curved.

These formulas can be of great utility in many cases such as when it is a question of finding the resultant of an infinity of forces which act at the same point, as in the case of the attraction of a body of arbitrary shape.

9. Let m be the mass of a body for which each of the elements $\mathrm{d}m$ are viewed as the center of a force P proportional to $\mathrm{d}m$ and to a function fp[84] of the distance p. By making $\int fp\,\mathrm{d}p = Fp$, the element $\mathrm{d}m$ will give, in the expression for Ξ, the term $(\mathrm{d}(Fp)/\mathrm{d}\xi)\mathrm{d}m$ for which the integral relative to the total mass m will be reduced to the attraction of this mass. And because this integration is independent of the differentiation relative to ξ, the

preceding integral can be expressed as $\mathrm{d}(\mathrm{S}\, Fp\, \mathrm{d}m)/\mathrm{d}\xi$ so that by defining $\mathrm{S}(Fp\, \mathrm{d}m) = \Sigma$, one has

$$\Xi = \frac{\mathrm{d}\Sigma}{\mathrm{d}\xi}, \qquad \Psi = \frac{\mathrm{d}\Sigma}{\mathrm{d}\psi}, \qquad \Phi = \frac{\mathrm{d}\Sigma}{\mathrm{d}\varphi}$$

and it will only be necessary to substitute in the function Fp, instead of p, its expression in terms of a function of the coordinates which determine the position of each particle dm in space and the coordinates ξ, ψ, φ of the attracted point and then to perform separately the integrations relative to the former and the differentiations relative to the latter. Because of the nature of the problem, one has $fp = 1/p^2$, therefore, $Fp = -1/p$ and consequently, $\Sigma = -\mathrm{S}(\mathrm{d}m/p)$.

Let a, b, c be the coordinates of each particle $\mathrm{d}m$ of the body. By assuming that the density of this particle is expressed by Γ, a function of a, b, c, one will have $\mathrm{d}m = \Gamma\, \mathrm{d}a\, \mathrm{d}b\, \mathrm{d}c$ and thus $\Sigma = -\mathrm{S}(\Gamma\, \mathrm{d}a\, \mathrm{d}b\, \mathrm{d}c/p)$.

But, since x, y, z are the coordinates of the attracted point, one has (Article 1)

$$p = \sqrt{(x-a)^2 + (y-b)^2 + (z-c)^2}$$

Thus

$$\Sigma = -\mathrm{S}\frac{\Gamma\, \mathrm{d}a\, \mathrm{d}b\, \mathrm{d}c}{\sqrt{(x-a)^2 + (y-b)^2 + (z-c)^2}}.$$

10. The simplest case occurs when the attracting body is a sphere. In this case, by defining $\Gamma = 1$ and assuming that the center of the sphere is at the origin of the coordinates x, y, z of the attracted point, one has

$$\Sigma = -\frac{m}{\sqrt{x^2 + y^2 + z^2}},$$

where m is the mass of the sphere which is known to be equal to $(4\pi/3)\alpha^3$, the variable α is the radius of the sphere and π is the ratio of the diameter of the sphere to its circumference. If the function Γ were a variable inside the sphere and if Γ were assumed to be a function of α, one would have

$$m = \frac{4\pi}{3}\mathrm{S}\,\Gamma\, \mathrm{d}(\alpha^3)$$

The value of Σ can also be found when the attracting body is an elliptical spheroid for which the surface is represented by the equation

$$\frac{a^2}{A^2} + \frac{b^2}{B^2} + \frac{c^2}{C^2} = 1$$

where A, B, C are the semi-axes of the three principal planes and a, b, c are the rectangular coordinates of the surface taken along the three axes and having their origin at the intersection of the axes which is the center of the sphere. But the general expression for this value depends on a rather complicated integrable formula by which it is impossible to obtain Σ as a function of x, y, z.

However, if the spheroid is assumed to have nearly the same shape as the sphere or if the distance from the attracted point to the center of the spheroid is very large relative to the axes of the spheroid, the value of Σ can be expressed by a convergent series without integral terms. Laplace has given in his Théorie des attractions des sphéroïdes,[85] a very nice formula by which it is possible to successively calculate all the terms of the series and which shows at the same time that the value of Σ/m, where m is the mass of the spheroid, depends only on the quantities $B^2 - A^2$ and $C^2 - A^2$, which are the squares of the eccentricities of the two sections which go through the same semi-axis A.

I found that by starting from this result and using the theorem that I gave in the Mémoires de Berlin for 1792–3 that this series could be developed at once solely from the radical

$$\frac{1}{\sqrt{x^2 + y^2 + z^2 - 2by - 2cz + b^2 + c^2}}$$

according to the powers of b and c, keeping only the terms which are even powers of b and c and transforming each of these terms such as $Hb^{2m}c^{2n}$, to

$$\frac{(1.3.5...2m-1)(1.3.5...2n-1)H(B^2 - A^2)^m(C^2 - A^2)^n}{5.7.9...2m+2n+3}m$$

where m is the mass of the spheroid which is given by $(4\pi/3)ABC$.

Therefore, in order to obtain immediately the series ordered according to the powers of y and z, make the following definition

$$r = \sqrt{x^2 + y^2 + z^2}$$

and then order the radical

$$(r^2 - 2by - 2cz + b^2 + c^2)^{-1/2}$$

with respect to the powers of y and z, keeping only the terms with even powers. Thus one will have

$$\frac{1}{\sqrt{r^2 + b^2 + c^2}} + \frac{3}{2}\frac{b^2y^2 + c^2z^2}{(r^2 + b^2 + c^2)^{5/2}} + \frac{5.7}{8}\frac{b^4y^4 + 6b^2c^2y^2z^2 + c^4z^4}{(r^2 + b^2 + c^2)^{9/2}} + \cdots$$

Then develop the radical $(r^2 + b^2 + c^2)^{-1/2}$ with respect to the powers of b^2 and c^2 and transform these powers to powers of $B^2 - A^2$ and $C^2 - A^2$ by the formula given above. In this fashion, if one defines for greater simplicity

$$B^2 - A^2 = e^2, \qquad C^2 - A^2 = i^2$$

where e and i are the eccentricities of two ellipsoids formed by the sections which pass through the semi-axes A, B and A, C one will have for Σ a series expression of this form

$$-m(R + Ty^2 + Vz^2 + Xy^4 + Yy^2z^2 + Zz^4 + \cdots)$$

where

$$\begin{aligned}
R &= \frac{1}{r} - \frac{e^2 + i^2}{2.5r^3} + \frac{9(e^4 + i^4) + 6e^2i^2}{8.5.7r^5} + \cdots \\
T &= \frac{3e^2}{2.5r^5} - \frac{9e^4 + 3e^2i^2}{4.7r^7} + \cdots \\
U &= \frac{3i^2}{2.5r^5} - \frac{9i^4 + 3e^2i^2}{4.7r^7} + \cdots \\
X &= \frac{3e^4}{8r^9} + \cdots \\
Y &= \frac{6e^2i^2}{8r^9} + \cdots \\
Z &= \frac{3i^4}{8r^9} + \cdots \\
&\vdots
\end{aligned}$$

The approximation for e and i has only been carried out to the fourth power. But it is easy to expand the series to any number of terms.

If the spheroid were composed of elliptic layers of different densities, then by varying in the expression for Σ the quantities A, B, C and consequently, also e and i, the expression for Σ relative to this spheroid would be $\mathrm{S}\,\Gamma\,\mathrm{d}\Sigma$.

Thus having the value of Σ as a function of the rectangular coordinates x, y, z of the attracted point, the forces

$$\frac{\mathrm{d}\Sigma}{\mathrm{d}x}, \qquad \frac{\mathrm{d}\Sigma}{\mathrm{d}y}, \qquad \frac{\mathrm{d}\Sigma}{\mathrm{d}z}$$

resulting from the total attraction of the spheroid with respect to these coordinates will be immediately obtained by differentiation.

If instead of the coordinates x, y, z, the radius r and the two angles μ and ν are taken such that

$$x = r\cos\mu, \qquad y = r\sin\mu\sin\nu, \qquad z = r\sin\mu\cos\nu$$

the attraction of the spheroid will be resolved in the direction of the radius r which joins the attracted point and the center of the spheroid, perpendicular to this radius in the plane which passes through the semi-axis A and perpendicular to the same radius in a plane parallel to the one which passes through the semi-axes B and C from the three following partial differentials, that is

$$\frac{\mathrm{d}\Sigma}{\mathrm{d}r}, \qquad \frac{\mathrm{d}\Sigma}{r\,\mathrm{d}\mu}, \qquad \frac{\mathrm{d}\Sigma}{r\sin\mu\,\mathrm{d}\nu}$$

These formulas are particularly useful in the theory of the shape of the Earth.

Chapter II
THE EQUILIBRIUM OF SEVERAL FORCES APPLIED TO A SYSTEM OF BODIES TREATED AS POINTS AND JOINED TOGETHER BY STRINGS OR RODS

11. We saw earlier (Article 7) that whatever the forces which act on each body, these forces can always be resolved into three components X, Y, Z directed along the three rectangular coordinates x, y, z of this body and with a tendency to reduce the magnitude of the coordinates.

Thus we will assume for greater simplicity here and in what follows that all the external forces which act at a point are reduced to three: namely, X, Y, Z. Therefore, the sum of the moments of these forces will be expressed in general by the formula $X\,\mathrm{d}x + Y\,\mathrm{d}y + Z\,\mathrm{d}z$. Consequently, the total sum of the moments of all the forces of the system will be expressed by the sum of as many similar formulas as there are bodies or mobile points, denoting by one, two or three primes, etc., the quantities relative to the various bodies which we will call the first, second, third, etc.

In this manner, the sum of the moments of the forces which act on three or more bodies will be expressed by the following equation

$$X'\,\mathrm{d}x' + Y'\,\mathrm{d}y' + Z'\,\mathrm{d}z' + X''\,\mathrm{d}x'' + Y''\,\mathrm{d}y'' + Z''\,\mathrm{d}z''$$
$$+X'''\,\mathrm{d}x''' + Y'''\,\mathrm{d}y''' + Z'''\,\mathrm{d}z''' + \cdots$$

It will only remain to investigate the equations of condition $L = 0$, $M = 0$, $N = 0$, etc. resulting from the nature of the problem.

Thus having L, M, N, etc. or their differentials as a function of x', y', z', x'', etc. and considering the undetermined coefficients λ, μ, ν, etc., the terms $\lambda\,\mathrm{d}L + \mu\,\mathrm{d}M + \nu\,\mathrm{d}N +$ etc. will be added to the preceding equation and then the expressions associated with each of the differences $\mathrm{d}x'$, $\mathrm{d}y'$, $\mathrm{d}z'$, $\mathrm{d}x''$, etc. (SECTION IV, Article 5) should be equated separately to zero.

Subsection I
The Equilibrium of Three or More Bodies Joined by an Inextensible or Extensible and Contractible String

12. Let us consider at the outset three bodies firmly joined by an inextensible string. The constraints of the problem are that the distances between the first and second bodies and between the second and third bodies are invariant. These distances represent the lengths of the segments of the string between the bodies.

Denoting the first of these lengths by the letter f and the second by g, one will have $\mathrm{d}f = 0$, $\mathrm{d}g = 0$ for the equations of condition. Therefore, $\mathrm{d}L = \mathrm{d}f$ and $\mathrm{d}M = \mathrm{d}g$, and the general equation of equilibrium for the three bodies will be

$$\begin{aligned}&X'\,\mathrm{d}x' + Y'\,\mathrm{d}y' + Z'\,\mathrm{d}z' + X''\,\mathrm{d}x'' + Y''\,\mathrm{d}y'' + Z''\,\mathrm{d}z''\\&+X'''\,\mathrm{d}x''' + Y'''\,\mathrm{d}y''' + Z'''\,\mathrm{d}z''' + \lambda\,\mathrm{d}f + \mu\,\mathrm{d}g = 0\end{aligned}$$

But it is obvious that in addition, there are two equations of condition

$$\begin{aligned}f &= \sqrt{(x''-x')^2+(y''-y')^2+(z''-z')^2}\\ g &= \sqrt{(x'''-x'')^2+(y'''-y'')^2+(z'''-z'')^2}\end{aligned}$$

and therefore, after differentiation[86]

$$\begin{aligned}\mathrm{d}f &= \frac{(x''-x')(\mathrm{d}x''-\mathrm{d}x')+(y''-y')(\mathrm{d}y''-\mathrm{d}y')+(z''-z')(\mathrm{d}z''-\mathrm{d}z')}{f}\\ \mathrm{d}g &= \frac{(x'''-x'')(\mathrm{d}x'''-\mathrm{d}x'')+(y'''-y'')(\mathrm{d}y'''-\mathrm{d}y'')+(z'''-z'')(\mathrm{d}z'''-\mathrm{d}z'')}{g}\end{aligned}$$

and after these expressions are substituted in the equilibrium equation, the following nine equations for the equilibrium of the bodies will be obtained

$$\begin{aligned}&X' - \lambda\frac{x''-x'}{f} = 0\\ &Y' - \lambda\frac{y''-y'}{f} = 0\\ &Z' - \lambda\frac{z''-z'}{f} = 0\\ &X'' + \lambda\frac{x''-x'}{f} - \mu\frac{x'''-x''}{g} = 0\\ &Y'' + \lambda\frac{y''-y'}{f} - \mu\frac{y'''-y''}{g} = 0\\ &Z'' + \lambda\frac{z''-z'}{f} - \mu\frac{z'''-z''}{g} = 0\\ &X''' + \mu\frac{x'''-x''}{g} = 0\end{aligned}$$

$$Y''' + \mu\frac{y''' - y''}{g} = 0$$

$$Z''' + \mu\frac{z''' - z''}{g} = 0$$

and it is only necessary to eliminate from these equations the two unknowns λ and μ. This can be accomplished in several ways which will also give different forms of the equations for the equilibrium of the three bodies attached to the string. We will choose the approach which seems the simplest.

It can be seen immediately that if the first three equations are added to the three following and then the resulting equations are added to the last three, respectively, the following three equations, without the unknowns λ and μ, are obtained

$$X' + X'' + X''' = 0$$

$$Y' + Y'' + Y''' = 0$$

$$Z' + Z'' + Z''' = 0$$

which show that the sum of all the forces parallel to each of the three coordinate axes must be equal to zero and are simply a particular case of the general equation found in SECTION III, Subsection I.

Therefore, it remains to find four additional equations. For this purpose, disregard the first three and add the three in the middle to the last three equations, respectively. Thus the following equations, which are free of μ, result

$$X'' + X''' + \frac{\lambda}{f}(x'' - x') = 0$$

$$Y'' + Y''' + \frac{\lambda}{f}(y'' - y') = 0$$

$$Z'' + Z''' + \frac{\lambda}{f}(z'' - z') = 0$$

and which, after the elimination of λ, gives the following two equations

$$Y'' + Y''' - \frac{y'' - y'}{x'' - x'}(X'' + X''') = 0$$

$$Z'' + Z''' - \frac{z'' - z'}{x'' - x'}(X'' + X''') = 0$$

Finally, considering separately the last three equations which contain only μ and after eliminating μ, the following two equations will be obtained

$$Y''' - \frac{y''' - y''}{x''' - x''}X''' = 0, \qquad Z''' - \frac{z''' - z''}{x''' - x''}X''' = 0$$

These seven equations express the necessary conditions for the equilibrium of the three bodies and when they are added to the equations of condition f and g, which are identical to the given quantities, they suffice to determine the location of every body in space.

13. If the string, assumed to be inextensible, were loaded by four bodies, acted upon by the forces $X', Y', Z', X'', Y'', Z'', X'''$, etc. in the direction of the three rectangular coordinate axes, respectively, one would find by similar procedures, which seem unnecessary for me to repeat, the following nine equations for the equilibrium of these four bodies

$$X' + X'' + X''' + X^{\text{IV}} = 0$$
$$Y' + Y'' + Y''' + Y^{\text{IV}} = 0$$
$$Z' + Z'' + Z''' + Z^{\text{IV}} = 0$$
$$Y'' + Y''' + Y^{\text{IV}} - \frac{y'' - y'}{x'' - x'}(X'' + X''' + X^{\text{IV}}) = 0$$
$$Z'' + Z''' + Z^{\text{IV}} - \frac{z'' - z'}{x'' - x'}(X'' + X''' + X^{\text{IV}}) = 0$$
$$Y''' + Y^{\text{IV}} - \frac{y''' - y''}{x''' - x''}(X''' + X^{\text{IV}}) = 0$$
$$Z''' + Z^{\text{IV}} - \frac{z''' - z''}{x''' - x''}(X''' + X^{\text{IV}}) = 0$$
$$Y^{\text{IV}} - \frac{y^{\text{IV}} - y'''}{x^{\text{IV}} - x'''}X^{\text{IV}} = 0$$
$$Z^{\text{IV}} - \frac{z^{\text{IV}} - z'''}{x^{\text{IV}} - x'''}X^{\text{IV}} = 0$$

It is now easy to extend this solution to any number of bodies and even to the case of a funicular or catenary. However, we will treat this case, in particular, by the method presented in Subsection II of the preceding section.

14. In some respects a simpler solution would be obtained if the invariability of the segments f, g, etc. were introduced in the calculation at the outset. Thus by limiting ourselves to the case of three bodies and denoting by ψ, ψ', the angles that the lines f, g make with the xy-plane and by φ, φ', the angles which the projections of these lines on the same plane make with the x-axis, one will have

$$x'' - x' = f\cos\varphi\cos\psi, \qquad y'' - y' = f\sin\varphi\cos\psi$$
$$z'' - z' = f\sin\psi, \qquad x''' - x'' = g\cos\varphi'\cos\psi'$$
$$y''' - y'' = g\sin\varphi'\cos\psi', \qquad z''' - z'' = g\sin\psi'$$

By substituting the values of x'', y'', z'', x''', y''', z''', obtained from these equations in the general formula of equilibrium for three bodies

$$X'\,\mathrm{d}x' + Y'\,\mathrm{d}y' + Z'\,\mathrm{d}z' + X''\,\mathrm{d}x'' + Y''\,\mathrm{d}y'' + Z''\,\mathrm{d}z''$$
$$+X'''\,\mathrm{d}x''' + Y'''\,\mathrm{d}y''' + Z'''\,\mathrm{d}z''' = 0$$

and by simply varying the quantities x', y', z', φ, φ', ψ, ψ', for which the variations remain indeterminate and finally, by equating the quantities multiplied by each of these variations separately to zero, the following seven equations will be obtained

$$
\begin{aligned}
&X' + X'' + X''' = 0\\
&Y' + Y'' + Y''' = 0\\
&Z' + Z'' + Z''' = 0\\
&(X'' + X''')\sin\varphi - (Y'' + Y''')\cos\varphi = 0\\
&X'''\sin\varphi' - Y'''\cos\varphi' = 0\\
&(X'' + X''')\cos\varphi\sin\psi + (Y'' + Y''')\sin\varphi\sin\psi - (Z'' + Z''')\cos\psi = 0\\
&X'''\cos\varphi'\sin\psi' + Y'''\sin\varphi'\sin\psi' - Z'''\cos\psi' = 0
\end{aligned}
$$

where the first five equations coincide exactly with those found in Article 12 after elimination of the undetermined coefficients λ and μ and for which the last two equations are reduced easily by eliminating Y'', Y''' by means of the fourth and fifth equations.

However, if, in this fashion, the final equations are obtained more directly, it is because a preliminary transformation of the variables has been used which contains the equations of condition. Instead of immediately using the equations with undetermined coefficients, as in Article 12, the solution of the problem is reduced to a pure calculation. Moreover, one has from these coefficients, the values of the forces which the strings f and g must support by their resistance to extension, as will be seen later.

15. If the first body were fixed, then the differences $\mathrm{d}x'$, $\mathrm{d}y'$, $\mathrm{d}z'$, would be equal to zero and the terms affected by these differences will drop out of the general equation of equilibrium. Therefore, the three equations of Article 12, that is

$$
X' - \frac{\lambda}{f}(x'' - x') = 0, \qquad Y' - \frac{\lambda}{f}(y'' - y') = 0, \qquad Z' - \frac{\lambda}{f}(z'' - z') = 0
$$

would not exist. Thus, the equations $X' + X'' + X''' + \text{etc.} = 0, Y' + Y'' + Y''' + \text{etc.} = 0$, $Z' + Z'' + Z''' + \text{etc.} = 0$ would not exist either, but all the others would remain the same. It is obvious that this is the case where the string would be firmly secured at one of its ends.

And if the string were fixed at both ends, then not only would one have $\mathrm{d}x' = 0$, $\mathrm{d}y' = 0$, $\mathrm{d}z' = 0$, but also $\mathrm{d}x'''^{\text{etc.}} = 0$, $\mathrm{d}y'''^{\text{etc.}} = 0$, $\mathrm{d}z'''^{\text{etc.}} = 0$, and the terms related to these six differences in the general equation of equilibrium would drop out and consequently, the six particular equations which depend on it will also drop out.

In general, if the two ends of the string were not entirely free, but were attached to mobile points according to a given law, this law, expressed analytically would give one or several equations between the differences $\mathrm{d}x'$, $\mathrm{d}y'$, $\mathrm{d}z'$ which are related to the first body and the differences $\mathrm{d}x'''^{\text{etc.}} = 0$, $\mathrm{d}y'''^{\text{etc.}} = 0$, $\mathrm{d}z'''^{\text{etc.}} = 0$ which are related to the last body. These equations, each multiplied by a new undetermined coefficient, should be added to the general equation of equilibrium found above or it is acceptable to substitute in this general equation the value of one or several of these differences obtained from these equations and then to equate to zero the coefficients of each of those which would remain, as it is carried out in Article 14. Since no difficulty is present here, we will not go further.

16. In order to calculate the forces which are produced by the reaction of the string on the various bodies, the method formulated for this purpose in the preceding section (Article 5) should be used.

Thus it will be observed that, in the present case, one has

$$\begin{aligned}
\mathrm{d}L &= \mathrm{d}f \\
&= \frac{(x''-x')(\mathrm{d}x''-\mathrm{d}x')+(y''-y')(\mathrm{d}y''-\mathrm{d}y')+(z''-z')(\mathrm{d}z''-\mathrm{d}z')}{f} \\
\mathrm{d}M &= \mathrm{d}g \\
&= \frac{(x'''-x'')(\mathrm{d}x'''-\mathrm{d}x'')+(y'''-y'')(\mathrm{d}y'''-\mathrm{d}y'')+(z'''-z'')(\mathrm{d}z'''-\mathrm{d}z'')}{g}
\end{aligned}$$

1. Therefore, the result with respect to the first body with coordinates x', y', z' will be

$$\frac{\mathrm{d}L}{\mathrm{d}x'} = -\frac{x''-x'}{f}, \qquad \frac{\mathrm{d}L}{\mathrm{d}y'} = -\frac{y''-y'}{f}, \qquad \frac{\mathrm{d}L}{\mathrm{d}z'} = -\frac{z''-z'}{f},$$

and

$$\sqrt{(\frac{\mathrm{d}L}{\mathrm{d}x'})^2+(\frac{\mathrm{d}L}{\mathrm{d}y'})^2+(\frac{\mathrm{d}L}{\mathrm{d}z'})^2} = \frac{\sqrt{(x''-x')^2+(y''-y')^2+(z''-z')^2}}{f} = 1$$

Furthermore, the first body will receive from the action of the other bodies a force equal to λ, and for which the direction will be perpendicular to the surface represented by the equation $\mathrm{d}L = \mathrm{d}f = 0$, obtained by simply varying x', y', z'. However, it is clear that this surface is nothing less than a sphere for which the radius is f and for which the center has as coordinates x'', y'', z''. Consequently, the force λ will be directed along the same radius, that is, along the string which links the first and second bodies.

2. Similarly, with respect to the second body whose coordinates are x'', y'', z'', the following equation will be obtained

$$\frac{\mathrm{d}L}{\mathrm{d}x''} = \frac{x''-x'}{f}, \qquad \frac{\mathrm{d}L}{\mathrm{d}y''} = \frac{y''-y'}{f}, \qquad \frac{\mathrm{d}L}{\mathrm{d}z''} = \frac{z''-z'}{f}$$

thus

$$\sqrt{(\frac{\mathrm{d}L}{\mathrm{d}x''})^2+(\frac{\mathrm{d}L}{\mathrm{d}y''})^2+(\frac{\mathrm{d}L}{\mathrm{d}z''})^2} = \frac{\sqrt{(x''-x')^2+(y''-y')^2+(z''-z')^2}}{f} = 1$$

from which it follows that the second body will also have a force λ directed perpendicular to the surface represented by the equation $\mathrm{d}L = \mathrm{d}f = 0$, obtained by varying x'', y'', z''. This surface is also a sphere with radius f but with a center having coordinates x', y', z' which are the coordinates of the first body. Consequently, the force λ which acts on the second body will also be directed along the string f which links this body to the first body.

3. In addition, with respect to the second body, there will be the following equations

$$\frac{\mathrm{d}M}{\mathrm{d}x''} = -\frac{x''' - x''}{g}, \qquad \frac{\mathrm{d}M}{\mathrm{d}y''} = -\frac{y''' - y''}{g}, \qquad \frac{\mathrm{d}M}{\mathrm{d}z''} = -\frac{z''' - z''}{g}$$

thus

$$\sqrt{(\frac{\mathrm{d}M}{\mathrm{d}x''})^2 + (\frac{\mathrm{d}M}{\mathrm{d}y''})^2 + (\frac{\mathrm{d}M}{\mathrm{d}z''})^2} = 1$$

so that the second body will also be acted upon by a force equal to μ for which the direction will be perpendicular to the surface having the equation $\mathrm{d}g = 0$, obtained by varying x'', y'', z''. Since this is the equation of a sphere with radius g, it is true that the direction of the force will act along this radius, that is, along the string which links the second body to the third body. The same reasoning applies to the other bodies and similar conclusions will be drawn.

17. It is evident that the force λ produced in the first body along the direction of the string which links this body to the second and the force equal to λ in magnitude but directed in the opposite direction which acts on the second body in the same direction can only be the forces which result from the reaction of the string on the two bodies, that is, the tension exerted by the segment of the string between the first and second bodies so that the coefficient λ will express the magnitude of this tension. In the same fashion, the coefficient μ expresses the tension in the segment of the string between the second and third bodies and similarly for the other bodies.

Incidently, in the solution of this problem it has been tacitly assumed that each segment of the string was not only inextensible but also rigid [sic] so that it will always keep the same length. Consequently, the forces λ, μ, etc. will define the tensions as long as they are positive and will have a tendency to move the bodies towards one another. But if they were negative and had a tendency to move the bodies apart, they will define rather the resistance that the string exerts on the bodies by means of its rigidity or incontractibility.

18. In order to confirm what we just demonstrated and to present at the same time a new application of our methods, we will assume that the string to which the bodies are attached is elastic along its length and capable of extension and contraction and that F, G, etc. are the forces of contraction in the segments of the string f, g, etc. between the first and second bodies, between the second and third bodies, etc. It is clear from what was said in Article 9 of SECTION II that the forces F, G, etc. will give the moments $F\,\mathrm{d}f + G\,\mathrm{d}g$, etc.

Therefore, these moments should be added to those resulting from the action of the external forces which as we saw above (Article 11) are represented by the formula $X'\,dx' + Y'\,dy' + Z'\,dz' + X''\,dx'' + Y''\,dy'' + Z''\,dz'' + X'''\,dx''' + Y'''\,dy''' + Z'''\,dz'''$ + etc. to obtain the total sum of the moments of the system. Since there is no particular condition to fulfill relative to the configuration of the bodies, the general equation of equilibrium will be obtained by simply equating this sum to zero. Thus, this equation will be

$$X'\,dx' + Y'\,dy' + Z'\,dz' + X''\,dx'' + Y''\,dy'' + Z''\,dz'' + X'''\,dx'''$$
$$+Y'''\,dy''' + Z'''\,dz''' + \cdots + F\,df + G\,dg + \cdots = 0$$

Substituting the values of df, dg, etc. found above (Article 12) and equating to zero the sum of the terms related to each of the differences dx', dy', etc., the following equations are obtained for the equilibrium of the string in this case

$$X' - \frac{F(x'' - x')}{f} = 0$$
$$Y' - \frac{F(y'' - y')}{f} = 0$$
$$Z' - \frac{F(z'' - z')}{f} = 0$$
$$X'' + \frac{F(x'' - x')}{f} - \frac{G(x''' - x'')}{g} = 0$$
$$Y'' + \frac{F(y'' - y')}{f} - \frac{G(y''' - y'')}{g} = 0$$
$$Z'' + \frac{F(z'' - z')}{f} - \frac{G(z''' - z'')}{g} = 0$$
$$X''' + \frac{G(x''' - x'')}{g} = 0$$
$$Y''' + \frac{G(y''' - y'')}{g} = 0$$
$$Z''' + \frac{G(z''' - z'')}{g} = 0$$

which are analogous to those of the same article for the case where the string is inextensible and give by comparison, $\lambda = F$, $\mu = G$, etc.

From this development, it is obvious that the quantities F, G, etc. which express here the forces of the strings which are assumed elastic, are the same as those which we found above (Article 16) to express the forces in these same strings, assuming that they are inextensible.

19. Let us consider further the case of an inextensible string loaded by three bodies, but let us assume at the same time that the body at the center can be moved along the string. In this case, the constraint on the problem will be that the sum of the distances between the first and second bodies and between the second and third bodies is constant. Thus denoting, as we did above, these distances by f and g, we will have $f + g =$ constant and consequently, $df + dg = 0$.

Multiplying the differential quantity $\mathrm{d}f + \mathrm{d}g$ by an undetermined coefficient λ and adding it to the sum of the moments of the different forces which are assumed acting on the bodies, will result in the following general equation of equilibrium

$$\begin{aligned} &X'\,\mathrm{d}x' + Y'\,\mathrm{d}y' + Z'\,\mathrm{d}z' + X''\,\mathrm{d}x'' + Y''\,\mathrm{d}y'' + Z''\,\mathrm{d}z'' + X'''\,\mathrm{d}x''' \\ &\quad + Y'''\,\mathrm{d}y''' + Z'''\,\mathrm{d}z''' + \lambda(\mathrm{d}f + \mathrm{d}g) = 0 \end{aligned}$$

from which (by substituting the values of $\mathrm{d}f$ and $\mathrm{d}g$, and equating to zero the sum of the terms related to each of the differences $\mathrm{d}x'$, $\mathrm{d}y'$, etc.) the following equations for the equilibrium of the string will be obtained

$$\begin{aligned} &X' - \lambda\frac{x'' - x'}{f} = 0 \\ &Y' - \lambda\frac{y'' - y'}{f} = 0 \\ &Z' - \lambda\frac{z'' - z'}{f} = 0 \\ &X'' + \lambda\Big(\frac{x'' - x'}{f} - \frac{x''' - x''}{g}\Big) = 0 \\ &Y'' + \lambda\Big(\frac{y'' - y'}{f} - \frac{y''' - y''}{g}\Big) = 0 \\ &Z'' + \lambda\Big(\frac{z'' - z'}{f} - \frac{z''' - z''}{g}\Big) = 0 \\ &X''' + \lambda\frac{x''' - x''}{g} = 0 \\ &Y''' + \lambda\frac{y''' - y''}{g} = 0 \\ &Z''' + \lambda\frac{z''' - z''}{g} = 0 \end{aligned}$$

in which only the unknown λ will have to be eliminated.

From this analysis, one learns how to proceed when there are a larger number of bodies, some firmly attached to the string and some free to move along the string.

Subsection II
The Equilibrium of Three or More Bodies Attached to an Inflexible and Rigid Rod

20. Let us now assume that the three bodies are linked by an inflexible rod so that they are constrained to remain an equal distance apart. In this case, not only does $\mathrm{d}f = 0$ and $\mathrm{d}g = 0$, but in addition the differential of the distance between the first and third bodies, which we will call h, is also equal to zero.[87] Consequently, by taking three undetermined

coefficients λ, μ, ν, the following general equation of equilibrium will be obtained

$$X'\,\mathrm{d}x' + Y'\,\mathrm{d}y' + Z'\,\mathrm{d}z' + X''\,\mathrm{d}x'' + Y''\,\mathrm{d}y'' + Z''\,\mathrm{d}z'' + X'''\,\mathrm{d}x''' + Y'''\,\mathrm{d}y''' + Z'''\,\mathrm{d}z''' + \lambda \mathrm{d}f + \mu \mathrm{d}g + \nu\,\mathrm{d}h = 0$$

The expressions for $\mathrm{d}f$ and $\mathrm{d}g$ were given earlier. For $\mathrm{d}h$, it is clear that the expressions will be

$$h = \sqrt{(x''' - x')^2 + (y''' - y')^2 + (z''' - z')^2}$$

and as a consequence

$$\mathrm{d}h = \frac{(x''' - x')(\mathrm{d}x''' - \mathrm{d}x') + (y''' - y')(\mathrm{d}y''' - \mathrm{d}y') + (z''' - z')(\mathrm{d}z''' - \mathrm{d}z')}{h}$$

Making these substitutions and equating to zero the sum of all the terms related to each of the differences $\mathrm{d}x'$, $\mathrm{d}y'$, etc. the following nine particular equations will be obtained

$$X' - \lambda \frac{x'' - x'}{f} - \nu \frac{x''' - x'}{h} = 0$$

$$Y' - \lambda \frac{y'' - y'}{f} - \nu \frac{y''' - y'}{h} = 0$$

$$Z' - \lambda \frac{z'' - z'}{f} - \nu \frac{z''' - z'}{h} = 0$$

$$X'' + \lambda \frac{x'' - x'}{f} - \mu \frac{x''' - x''}{g} = 0$$

$$Y'' + \lambda \frac{y'' - y'}{f} - \mu \frac{y''' - y''}{g} = 0$$

$$Z'' + \lambda \frac{z'' - z'}{f} - \mu \frac{z''' - z''}{g} = 0$$

$$X''' + \mu \frac{x''' - x''}{g} + \nu \frac{x''' - x'}{h} = 0$$

$$Y''' + \mu \frac{y''' - y''}{g} + \nu \frac{y''' - y'}{h} = 0$$

$$Z''' + \mu \frac{z''' - z''}{g} + \nu \frac{z''' - z'}{h} = 0$$

from which the three undetermined coefficients λ, μ, ν, should be eliminated so that only six equations remain for the conditions of equilibrium.

21. At the outset, it is clear from the form of these equations that by adding the first three to the following three, respectively and then to the last three, the three equations where the coefficients λ, μ, ν, have been eliminated will immediately be obtained

$$X' + X'' + X''' = 0, \qquad Y' + Y'' + Y''' = 0, \qquad Z' + Z'' + Z''' = 0$$

Nothing is easier than to find three more equations by elimination of λ, μ, ν. But to obtain the equations in a simpler and more general fashion, it is better to begin by deducing from the equations of the preceding article, the following nine transformed equations

$$X'y' - Y'x' - \lambda\frac{y'x'' - x'y''}{f} - \nu\frac{y'x''' - x'y'''}{h} = 0$$
$$X'z' - Z'x' - \lambda\frac{z'x'' - x'z''}{f} - \nu\frac{z'x''' - x'z'''}{h} = 0$$
$$Y'z' - Z'y' - \lambda\frac{z'y'' - y'z''}{f} - \nu\frac{z'y''' - y'z'''}{h} = 0$$
$$X''y'' - Y''x'' + \lambda\frac{y'x'' - x'y''}{f} - \mu\frac{y''x''' - x''y'''}{g} = 0$$
$$X''z'' - Z''x'' + \lambda\frac{z'x'' - x'z''}{f} - \mu\frac{z''x''' - x''z'''}{g} = 0$$
$$Y''z'' - Z''y'' + \lambda\frac{z'y'' - y'z''}{f} - \mu\frac{z''y''' - y''z'''}{g} = 0$$
$$X'''y''' - Y'''x''' + \mu\frac{y''x''' - x''y'''}{g} + \nu\frac{y'x''' - x'y'''}{h} = 0$$
$$X'''z''' - Z'''x''' + \mu\frac{z''x''' - x''z'''}{g} + \nu\frac{z'x''' - x'z'''}{h} = 0$$
$$Y'''z''' - Z'''y''' + \mu\frac{z''y''' - y''z'''}{g} + \nu\frac{z'y''' - y'z'''}{h} = 0$$

which are, as can be seen, analogous to the original equations and will give in the same fashion by simple addition, the three following equations

$$X'y' - Y'x' + X''y'' - Y''x'' + X'''y''' - Y'''x''' = 0$$
$$X'z' - Z'x' + X''z'' - Z''x'' + X'''z''' - Z'''x''' = 0$$
$$Y'z' - Z'y' + Y''z'' - Z''y'' + Y'''z''' - Z'''y''' = 0$$

The three equations found above show that the sum of the forces parallel to each of the three coordinate axes must be zero. The three that we just found contain the known principle of moments (denoting moment here by the product of force by its lever arm), by which the sum of the moments of all the forces, tending to rotate the system about each of the three axes, is also zero. Thus these six equations are only particular cases of the general equations given in SECTION III (Subsections I and II).

22. If the first body were fixed, then the differences $\mathrm{d}x'$, $\mathrm{d}y'$, $\mathrm{d}z'$ would be equal to zero and the first three of the nine equations of Article 20 would not exist. Only six equations will remain which after elimination of the three unknowns λ, μ, ν, will reduce to three.

In order to derive these three equations, a method analogous to the one that was used to find the last three equations of the preceding article can be employed if care is taken that the transformed equations do not contain the indeterminates λ and ν which are present in the first three equations and which now must be eliminated. The following equations are obtained using these combinations

$$X''(y''-y')-Y''(x''-x')-\mu\frac{(y''-y')(x'''-x'')-(x''-x')(y'''-y'')}{g}=0$$
$$X''(z''-z')-Z''(x''-x')-\mu\frac{(z''-z')(x'''-x'')-(x''-x')(z'''-z'')}{g}=0$$
$$Y''(z''-z')-Z''(y''-y')-\mu\frac{(z''-z')(y'''-y'')-(y''-y')(z'''-z'')}{g}=0$$
$$X'''(y'''-y')-Y'''(x'''-x')+\mu\frac{(y'''-y')(x'''-x'')-(x'''-x')(y'''-y'')}{g}=0$$
$$X'''(z'''-z')-Z'''(x'''-x')+\mu\frac{(z'''-z')(x'''-x'')-(x'''-x')(z'''-z'')}{g}=0$$
$$Y'''(z'''-z')-Z'''(y'''-y')+\mu\frac{(z'''-z')(y'''-y'')-(y'''-y')(z'''-z'')}{g}=0$$

and if the first three of these transformed equations were now added to the last three, the following three equations will be immediately obtained

$$X''(y''-y')-Y''(x''-x')+X'''(y'''-y')-Y'''(x'''-x')=0$$
$$X''(z''-z')-Z''(x''-x')+X'''(z'''-z')-Z'''(x'''-x')=0$$
$$Y''(z''-z')-Z''(y''-y')+Y'''(z'''-z')-Z'''(y'''-y')=0$$

which will always exist whatever the state of the first body because they are independent of the equations relative to the position of this body. These equations contain, as can be seen, the same principle of moments but with respect to axes which pass through the first body.

23. Let us assume that a fourth body is attached to the same inflexible rod, with rectangular coordinates x^{IV}, y^{IV}, z^{IV} and whose forces are parallel to these coordinates and given by $X^{\text{IV}}, Y^{\text{IV}}, Z^{\text{IV}}$.

The quantity $X^{\text{IV}}\,\mathrm{d}x^{\text{IV}}+Y^{\text{IV}}\,\mathrm{d}y^{\text{IV}}+Z^{\text{IV}}\,\mathrm{d}z^{\text{IV}}$ should be added to the sum of the moments of the forces, and because the distances between all the bodies must remain constant, one will have not only $\mathrm{d}f=0$, $\mathrm{d}g=0$, $\mathrm{d}h=0$, as in the preceding case, but also $\mathrm{d}\ell=0$, $\mathrm{d}m=0$, $\mathrm{d}n=0$, in which ℓ, m, n are the distances between the fourth body and the other three. Thus in this case the general equation of equilibrium will be

$$\begin{aligned}&X'\,\mathrm{d}x'+Y'\,\mathrm{d}y'+Z'\,\mathrm{d}z'+X''\,\mathrm{d}x''+Y''\,\mathrm{d}y''+Z''\,\mathrm{d}z''\\&\quad+X'''\,\mathrm{d}x'''+Y'''\,\mathrm{d}y'''+Z'''\,\mathrm{d}z'''+X^{\text{IV}}\,\mathrm{d}x^{\text{IV}}+Y^{\text{IV}}\,\mathrm{d}y^{\text{IV}}+Z^{\text{IV}}\,\mathrm{d}z^{\text{IV}}\\&\quad+\lambda\,\mathrm{d}f+\mu\,\mathrm{d}g+\nu\,\mathrm{d}h+\pi\,\mathrm{d}\ell+\rho\,\mathrm{d}m+\sigma\,\mathrm{d}n=0\end{aligned}$$

The expressions for $\mathrm{d}f$, $\mathrm{d}g$, $\mathrm{d}h$ are the same as above. For the values of $\mathrm{d}\ell$, $\mathrm{d}m$, $\mathrm{d}n$, it is obvious that

$$\ell=\sqrt{(x^{\text{IV}}-x')^2+(y^{\text{IV}}-y')^2+(z^{\text{IV}}-z')^2}$$

$$m = \sqrt{(x^{\text{IV}} - x'')^2 + (y^{\text{IV}} - y'')^2 + (z^{\text{IV}} - z'')^2}$$
$$n = \sqrt{(x^{\text{IV}} - x''')^2 + (y^{\text{IV}} - y''')^2 + (z^{\text{IV}} - z''')^2}$$

and consequently

$$d\ell = \frac{(x^{\text{IV}} - x')(dx^{\text{IV}} - dx') + (y^{\text{IV}} - y')(dy^{\text{IV}} - dy') + (z^{\text{IV}} - z')(dz^{\text{IV}} - dz')}{\ell}$$
$$dm = \frac{(x^{\text{IV}} - x'')(dx^{\text{IV}} - dx'') + (y^{\text{IV}} - y'')(dy^{\text{IV}} - dy'') + (z^{\text{IV}} - z'')(dz^{\text{IV}} - dz'')}{m}$$
$$dn = \frac{(x^{\text{IV}} - x''')(dx^{\text{IV}} - dx''') + (y^{\text{IV}} - y''')(dy^{\text{IV}} - dy''') + (z^{\text{IV}} - z''')(dz^{\text{IV}} - dz''')}{n}$$

After making these substitutions and equating to zero the sum of the terms related to each of the differences dx', dy', etc., twelve particular equations will be found of which the first nine will be the same as those of Article 20 after adding respectively to their first members the following quantities

$$-\pi\frac{x^{\text{IV}} - x'}{\ell}, \qquad -\pi\frac{y^{\text{IV}} - y'}{\ell}, \qquad -\pi\frac{z^{\text{IV}} - z'}{\ell},$$
$$-\rho\frac{x^{\text{IV}} - x''}{m}, \qquad -\rho\frac{y^{\text{IV}} - y''}{m}, \qquad -\rho\frac{z^{\text{IV}} - z''}{m},$$
$$-\sigma\frac{x^{\text{IV}} - x'''}{n}, \qquad -\sigma\frac{y^{\text{IV}} - y'''}{n}, \qquad -\sigma\frac{z^{\text{IV}} - z'''}{n}$$

and for which the last three equations will be

$$X^{\text{IV}} + \pi\frac{x^{\text{IV}} - x'}{\ell} + \rho\frac{x^{\text{IV}} - x''}{m} + \sigma\frac{x^{\text{IV}} - x'''}{n} = 0$$
$$Y^{\text{IV}} + \pi\frac{y^{\text{IV}} - y'}{\ell} + \rho\frac{y^{\text{IV}} - y''}{m} + \sigma\frac{y^{\text{IV}} - y'''}{n} = 0$$
$$Z^{\text{IV}} + \pi\frac{z^{\text{IV}} - z'}{\ell} + \rho\frac{z^{\text{IV}} - z''}{m} + \sigma\frac{z^{\text{IV}} - z'''}{n} = 0$$

24. Since there are a total of twelve equations and six indeterminate quantities λ, μ, ν, π, ρ, σ to eliminate, only six final equations will be left as the conditions of equilibrium as in the case of three bodies. And by a method similar to the one of Article 21, the following six equations will be found which are analogous to those of this article

$$X' + X'' + X''' + X^{\text{IV}} = 0$$
$$Y' + Y'' + Y''' + Y^{\text{IV}} = 0$$
$$Z' + Z'' + Z''' + Z^{\text{IV}} = 0$$

$$X'y' - Y'x' + X''y'' - Y''x'' + X'''y''' - Y'''x''' + X^{\mathrm{IV}}y^{\mathrm{IV}} - Y^{\mathrm{IV}}x^{\mathrm{IV}} = 0$$
$$X'z' - Z'x' + X''z'' - Z''x'' + X'''z''' - Z'''x''' + X^{\mathrm{IV}}z^{\mathrm{IV}} - Z^{\mathrm{IV}}x^{\mathrm{IV}} = 0$$
$$Y'z' - Z'y' + Y''z'' - Z''y'' + Y'''z''' - Z'''y''' + Y^{\mathrm{IV}}z^{\mathrm{IV}} - Z^{\mathrm{IV}}y^{\mathrm{IV}} = 0$$

Instead of the last three equations, the following three which are found by the method of Article 22 could also be substituted and since they are independent of the equations relative to the first body, they have the advantage that they will always exist whatever the location of this body

$$\begin{aligned}
&X''(y'' - y') - Y''(x'' - x') + X'''(y''' - y') - Y'''(x''' - x') \\
&\quad + X^{\mathrm{IV}}(y^{\mathrm{IV}} - y') - Y^{\mathrm{IV}}(x^{\mathrm{IV}} - x') = 0 \\
&X''(z'' - z') - Z''(x'' - x') + X'''(z''' - z') - Z'''(x''' - x') \\
&\quad + X^{\mathrm{IV}}(z^{\mathrm{IV}} - z') - Z^{\mathrm{IV}}(x^{\mathrm{IV}} - x') = 0 \\
&Y''(z'' - z') - Z''(y'' - y') + Y'''(z''' - z') - Z'''(y''' - y') \\
&\quad + Y^{\mathrm{IV}}(z^{\mathrm{IV}} - z') - Z^{\mathrm{IV}}(y^{\mathrm{IV}} - y') = 0
\end{aligned}$$

25. Now it is clear how one should proceed in order to find the conditions of equilibrium for an arbitrary number of bodies attached to a rod or rigid lever. In general, it is obvious that for the relative positions of the bodies to remain the same, it suffices that the distances between the first three bodies remain constant and that the distances from each of the other bodies to these three are also constant, because the location of an arbitrary point is always determined by the distances from this point to three given points. For each additional body added to the lever, the same reasoning and operations which were made in Article 23 with respect to the fourth body should be followed. And each of these bodies will produce three new particular equations with three new undetermined coefficients to eliminate so that the final number of equations will always be the same as in the case of three bodies. These equations will have the form which we found in the preceding article.

It is also obvious that these equations are part of the group which we found in general for the equilibrium of an arbitrary free system in Articles 3 and 9 of SECTION III. Indeed, because of the rigidity of the rod, the distances between the bodies remain constant and equilibrium holds if the motions of translation and rotation are nullified. The preceding problem could have been solved with this single consideration using the formulas of the cited articles. But we believe that it is useful to give a direct solution and to obtain the particular conditions of the problem.

Subsection III
The Equilibrium of Three or More Bodies Attached to an Elastic Rod

26. Let us reconsider the case of three bodies linked together by rods. Moreover, let us assume that the rod is elastic at the point where the second body is located so that its

distance from the first and last bodies is constant, but that the angle made by the directions of these distances is variable and that the effect of elasticity[88] consists of increasing this angle and consequently, decreasing the exterior angle made with one of the sides by the prolongation of the other.

Let us denote the forces[89] of elasticity by E and the exterior angle by e. The force E has a tendency to decrease the angle e. The moment of this force will be expressed by Ede (SECTION II, Article 9) so that the sum of the moments of all the forces of the system will be

$$X'\,\mathrm{d}x' + Y'\,\mathrm{d}y' + Z'\,\mathrm{d}z' + X''\,\mathrm{d}x'' + Y''\,\mathrm{d}y'' + Z''\,\mathrm{d}z''$$
$$+X'''\,\mathrm{d}x''' + Y'''\,\mathrm{d}y''' + Z'''\,\mathrm{d}z''' + E\,\mathrm{d}e$$

But the conditions of the problem are the same here as in Article 12, that is, $\mathrm{d}f = 0$ and $\mathrm{d}g = 0$. Thus the general equation of equilibrium will be

$$X'\,\mathrm{d}x' + Y'\,\mathrm{d}y' + Z'\,\mathrm{d}z' + X''\,\mathrm{d}x'' + Y''\,\mathrm{d}y'' + Z''\,\mathrm{d}z''$$
$$+X'''\,\mathrm{d}x''' + Y'''\,\mathrm{d}y''' + Z'''\,\mathrm{d}z''' + E\,\mathrm{d}e + \lambda\,\mathrm{d}f + \mu\,\mathrm{d}g = 0$$

and it will only be necessary to substitute the expressions for de, $\mathrm{d}f$, $\mathrm{d}g$. The values of $\mathrm{d}f$ and $\mathrm{d}g$ are the same as those in the cited article.

In order to find the value of de, denote the rectilinear distance between the first and third bodies by the letter h as in Article 20. In the triangle formed by the three sides f, g, h and for which the angle opposite the side h is $(180° - e)$, the application of a well-known theorem[90] produces the following equation

$$-\cos e = \frac{f^2 + g^2 - h^2}{2fg}$$

from which the expression for de will be obtained by differentiation. From the conditions of the problem, where $\mathrm{d}f = 0$ and $\mathrm{d}g = 0$, it will suffice to vary e and h which will give

$$\mathrm{d}e = -\frac{h\,\mathrm{d}h}{fg\sin e}$$

When this expression is substituted in the preceding equilibrium equation, it is easy to see that it will assume the same form as the general equation of equilibrium for the similar case of Article 20, assuming in this case that

$$\nu = -\frac{Eh}{fg\sin e}$$

Consequently, the particular equations will still be the same in both cases. The only difference is that in the case of the cited article, the quantity ν is indeterminate and consequently, it must be eliminated. On the other hand, in the present case, this quantity is known and there are only the two indeterminates λ and μ to eliminate so that there is one additional equation than in the cited case, that is, seven equations instead of six. But since, whether the quantity ν is known or not, nothing prevents us from eliminating it with the other two quantities λ and μ, it is clear that one will also have in the present case the same equations that were found in Articles 21 and 22. And to find the seventh equation, it is only necessary to eliminate λ in the first three equations or μ in the last three of the nine particular equations of Article 20 and to substitute for ν the following expression

$$-\frac{Eh}{fg\sin e}$$

27. Also, if in the formulation of the expression for de, it is assumed that df and dg are not equal to zero, an expression of the following form would be obtained

$$\mathrm{d}e = -\frac{h\,\mathrm{d}h}{fg\sin e} + A\,\mathrm{d}f + B\,\mathrm{d}g$$

where A and B are functions of f, g, h, $\sin e$. Then the three terms $E\,\mathrm{d}e + \lambda\,\mathrm{d}f + \mu\,\mathrm{d}g$ of the general equation would become

$$\frac{Eh}{fg\sin e}\,\mathrm{d}h + (EA + \lambda)\,\mathrm{d}f + (EB + \mu)\,\mathrm{d}g$$

but since λ and μ are two indeterminate quantities, it is obvious that they could be replaced by $\lambda - EA$, $\mu - EB$, after which the quantity given above would become

$$-\frac{Eh}{fg\sin e}\,\mathrm{d}h + \lambda\,\mathrm{d}f + \mu\,dg$$

as if f and g remained constant when calculating the expression for de.

If several bodies were linked together by elastic rods, the necessary equations for the equilibrium of these bodies would be found in the same manner. In general, our method will always give with the same facility, the conditions of equilibrium for a system of bodies linked together in an arbitrary fashion and loaded by external forces. The procedure is, as one sees, always uniform, which must be accepted as one of the main advantages of this method.

Chapter III
THE EQUILIBRIUM OF A STRING WHERE ALL THE POINTS ARE LOADED BY ARBITRARY FORCES AND WHICH IS ASSUMED FLEXIBLE OR INFLEXIBLE, OR ELASTIC AND AT THE SAME TIME CAN BE EXTENSIBLE OR NOT

28. This is the time to use the method which we developed in Subsection II of SECTION IV. For greater simplicity, we will always assume that all the external forces which act at each point of the string are reduced to three, namely, X, Y, Z, directed along the rectangular coordinate axes x, y, z associated with this point. Thus denoting an element of the string by dm which is proportional to the elemental length $\mathrm{d}s$ along the length of the string, multiplied by the string's thickness, one will have for the sum of the moments of all the forces relative to the total length of the string the following integrable formula (SECTION IV, Article 12)

$$\mathrm{S}(X\,\delta x + Y\,\delta y + Z\,\delta z)\,\mathrm{d}m$$

is obtained since the quantity $X\,\mathrm{d}x + Y\,\mathrm{d}y + Z\,\mathrm{d}z$ is only a transformation of the relation $P\,\mathrm{d}p + Q\,\mathrm{d}q + R\,\mathrm{d}r +$ etc. (Article 1). If the forces P, Q, R, etc. are such that this quantity is integrable and if the function Π is its integral, the following equation results, as in Article 25 of SECTION IV

$$X\,\delta x + Y\,\delta y + Z\,\delta z = \delta\Pi$$

and the sum of the moments will be expressed by $\mathrm{S}\,\delta\Pi\,\mathrm{d}m$.

Subsection I
The Equilibrium of a Flexible and Inextensible String

29. Let us consider at the outset the case of a perfectly flexible and inextensible string. Since the element $\mathrm{d}s$ of the length of this string is expressed by $\sqrt{(\mathrm{d}x^2 + \mathrm{d}y^2 + \mathrm{d}z^2)}$ and since the string is inextensible, $\mathrm{d}s$ must be an invariable quantity and in addition, the indefinite equation of condition $\delta\,\mathrm{d}s = 0$ must hold with respect to each element of the string. After multiplying $\delta\,\mathrm{d}s$ by an undetermined coefficient λ and forming the total integral the quantity $\lambda\,\delta\,\mathrm{d}s$ will be obtained. And if there is no other equation of condition, the general equation of equilibrium will result by equating to zero the sum of the two integrals $\mathrm{S}\,\delta\Pi\,\mathrm{d}m$ and $\mathrm{S}\lambda\,\delta\,\mathrm{d}s$.

Now considering that $\mathrm{d}s = \sqrt{(\mathrm{d}x^2 + \mathrm{d}y^2 + \mathrm{d}z^2)}$, the following equation results after differentiation by δ

$$\delta\,\mathrm{d}s = \frac{\mathrm{d}x\,\delta\,\mathrm{d}x + \mathrm{d}y\,\delta\,\mathrm{d}y + \mathrm{d}z\,\delta\,\mathrm{d}z}{\mathrm{d}s}$$

and thus

$$\mathrm{S}\lambda\,\delta\,\mathrm{d}s = \mathrm{S}\frac{\lambda\,\mathrm{d}x}{\mathrm{d}s}\delta\,\mathrm{d}x + \mathrm{S}\frac{\lambda\,\mathrm{d}y}{\mathrm{d}s}\delta\,\mathrm{d}y + \mathrm{S}\frac{\lambda\,\mathrm{d}z}{\mathrm{d}s}\delta\,\mathrm{d}z$$

After replacing $\delta\mathrm{d}$ by $\mathrm{d}\delta$ and integrating by parts according to the rules given in Article 15 of SECTION IV to eliminate the operator d before the operator δ, the following transformed equations will result

$$\begin{aligned}
\mathrm{S}\frac{\lambda\,\mathrm{d}x}{\mathrm{d}s}\delta\,\mathrm{d}x &= \frac{\lambda''\,\mathrm{d}x''}{\mathrm{d}s''}\delta x'' - \frac{\lambda'\,\mathrm{d}x'}{\mathrm{d}s'}\delta x' - \mathrm{S}\,\mathrm{d}\frac{\lambda\,\mathrm{d}x}{\mathrm{d}s}\delta x\\
\mathrm{S}\frac{\lambda\,\mathrm{d}y}{\mathrm{d}s}\delta\,\mathrm{d}y &= \frac{\lambda''\,\mathrm{d}y''}{\mathrm{d}s''}\delta y'' - \frac{\lambda'\,\mathrm{d}y'}{\mathrm{d}s'}\delta y' - \mathrm{S}\,\mathrm{d}\frac{\lambda\,\mathrm{d}y}{\mathrm{d}s}\delta y\\
\mathrm{S}\frac{\lambda\,\mathrm{d}z}{\mathrm{d}s}\delta\,\mathrm{d}z &= \frac{\lambda''\,\mathrm{d}z''}{\mathrm{d}s''}\delta z'' - \frac{\lambda'\,\mathrm{d}z'}{\mathrm{d}s'}\delta z' - \mathrm{S}\,\mathrm{d}\frac{\lambda\,\mathrm{d}z}{\mathrm{d}s}\delta z
\end{aligned}$$

Therefore, the general equation will become

$$\begin{aligned}
&\mathrm{S}\left(\left(X\,\mathrm{d}m - \mathrm{d}\frac{\lambda\,\mathrm{d}x}{\mathrm{d}s}\right)\delta x + \left(Y\,\mathrm{d}m - \mathrm{d}\frac{\lambda\,\mathrm{d}y}{\mathrm{d}s}\right)\delta y\right.\\
&\left.+\left(Z\,\mathrm{d}m - \mathrm{d}\frac{\lambda\,\mathrm{d}z}{\mathrm{d}s}\right)\delta z\right) + \frac{\lambda''\,\mathrm{d}x''}{\mathrm{d}s''}\delta x'' + \frac{\lambda''\,\mathrm{d}y''}{\mathrm{d}s''}\delta y''\\
&+\frac{\lambda''\,\mathrm{d}z''}{\mathrm{d}s''}\delta z'' - \frac{\lambda'\mathrm{d}x'}{\mathrm{d}s'}\delta x' - \frac{\lambda'\mathrm{d}y'}{\mathrm{d}s'}\delta y' - \frac{\lambda'\,\mathrm{d}z'}{\mathrm{d}s'}\delta z' = 0
\end{aligned}$$

30. The coefficients of δx, δy, δz under the first integral sign must be equated to zero (SECTION IV, Article 16). Then the three following particular and indefinite equations will result

$$X\,\mathrm{d}m - \mathrm{d}\frac{\lambda\,\mathrm{d}x}{\mathrm{d}s} = 0, \qquad Y\,\mathrm{d}m - \mathrm{d}\frac{\lambda\,\mathrm{d}y}{\mathrm{d}s} = 0, \qquad Z\,\mathrm{d}m - \mathrm{d}\frac{\lambda\,\mathrm{d}z}{\mathrm{d}s} = 0,$$

from which after eliminating the multiplier λ, two equations will remain to define the shape of the string.

This elimination is very easy, since it is only necessary to integrate the preceding equations, which results in the following equations

$$\frac{\lambda\,\mathrm{d}x}{\mathrm{d}s} = A + \int X\,\mathrm{d}m, \qquad \frac{\lambda\,\mathrm{d}y}{\mathrm{d}s} = B + \int Y\,\mathrm{d}m, \qquad \frac{\lambda\,\mathrm{d}z}{\mathrm{d}s} = C + \int Z\,\mathrm{d}m$$

where A, B, C are arbitrary constants. Then after eliminating λ, the following equations will be obtained

$$\frac{\mathrm{d}y}{\mathrm{d}x} = \frac{B + \int Y\,\mathrm{d}m}{A + \int X\,\mathrm{d}m}, \qquad \frac{\mathrm{d}z}{\mathrm{d}x} = \frac{C + \int Z\,\mathrm{d}m}{A + \int X\,\mathrm{d}m}$$

which are two equations which are consistent with the formulas for the catenary.

If one wished to obtain exact differential equations directly, without the integral sign, the developed equations could be put in the following form

$$X\,\mathrm{d}m - \lambda\,\mathrm{d}\frac{\mathrm{d}x}{\mathrm{d}s} - \mathrm{d}\lambda\frac{\mathrm{d}x}{\mathrm{d}s} = 0$$

$$Y\,\mathrm{d}m - \lambda\,\mathrm{d}\frac{\mathrm{d}y}{\mathrm{d}s} - \mathrm{d}\lambda\frac{\mathrm{d}y}{\mathrm{d}s} = 0$$

$$Z\,\mathrm{d}m - \lambda\,\mathrm{d}\frac{\mathrm{d}z}{\mathrm{d}s} - \mathrm{d}\lambda\frac{\mathrm{d}z}{\mathrm{d}s} = 0$$

from which, after eliminating $\mathrm{d}\lambda$, the two following equations will be obtained

$$\frac{X\,\mathrm{d}y - Y\,\mathrm{d}x}{\mathrm{d}s}\,\mathrm{d}m = \lambda(\frac{\mathrm{d}y}{\mathrm{d}s}\mathrm{d}\frac{\mathrm{d}x}{\mathrm{d}s} - \frac{\mathrm{d}x}{\mathrm{d}s}\mathrm{d}\frac{\mathrm{d}y}{\mathrm{d}s})$$

$$\frac{X\,\mathrm{d}z - Z\,\mathrm{d}x}{\mathrm{d}s}\,\mathrm{d}m = \lambda(\frac{\mathrm{d}z}{\mathrm{d}s}\mathrm{d}\frac{\mathrm{d}x}{\mathrm{d}s} - \frac{\mathrm{d}x}{\mathrm{d}s}\mathrm{d}\frac{\mathrm{d}z}{\mathrm{d}s})$$

Then, if the same equations are multiplied respectively by

$$\frac{\mathrm{d}x}{\mathrm{d}s}, \qquad \frac{\mathrm{d}y}{\mathrm{d}s}, \qquad \frac{\mathrm{d}z}{\mathrm{d}s}$$

and because of the following relation

$$\frac{\mathrm{d}x}{\mathrm{d}s}\mathrm{d}\frac{\mathrm{d}x}{\mathrm{d}s} + \frac{\mathrm{d}y}{\mathrm{d}s}\mathrm{d}\frac{\mathrm{d}y}{\mathrm{d}s} + \frac{\mathrm{d}z}{\mathrm{d}s}\mathrm{d}\frac{\mathrm{d}z}{\mathrm{d}s} = \frac{1}{2}\mathrm{d}\left(\frac{\mathrm{d}x^2 + \mathrm{d}y^2 + \mathrm{d}z^2}{\mathrm{d}s^2}\right) = 0$$

the following equation will be obtained

$$\frac{X\,\mathrm{d}x + Y\,\mathrm{d}y + Z\,\mathrm{d}z}{\mathrm{d}s}\,\mathrm{d}m = \mathrm{d}\lambda$$

and it will only be necessary to substitute successively in this last equation the values of λ obtained from the two preceding equations.

31. Since the quantity $\lambda\,\delta\mathrm{d}s$ can represent the moment of a force λ having a tendency to shorten the length of the element $\mathrm{d}s$ (SECTION IV, Article 6), the term $\mathrm{S}\lambda\,\delta\mathrm{d}s$ of the general equation of equilibrium of the string (Article 29), will represent the sum of the moments of all the forces λ that can be assumed to be acting on all elements of the string. Indeed, each element resists, by its inextensibility, the action of external forces and generally, this resistance can be viewed as an active force which is called "tension". Thus the coefficient λ will denote the tension in the string.

32. With respect to the condition of inextensibility of the string represented by the invariability of each element $\mathrm{d}s$ of the curve, it cannot be introduced in the equation of this

curve as a replacement for the undetermined coefficient λ as in the case where the string forms a polygon, because by the nature of the differential calculus the absolute value of the elements of the curve and in general, all the infinitesimal elements remain indeterminate. But also, for the same reason, it is not necessary to have as many equations as there are variables and it suffices to have one less equation to determine a curve whether of single or double curvature. Thus the solution which we just found by our method is complete with respect to the differential equations and necessitates only integrations which depend upon the expressions for the forces X, Y, Z.

33. Let us now consider the terms of the general equation of Article 29, which are outside the integral sign S. Furthermore, let us assume at the outset that the string is entirely free. In this case, the variations $\delta x'$, $\delta y'$, $\delta z'$ and $\delta x''$, $\delta y''$, $\delta z''$ which are related to the two end points of the string will all be indeterminate and arbitrary. Consequently, each term affected by these variations must be equal to zero independently. Therefore, one should have $\lambda' = 0$ and $\lambda'' = 0$ that is, the value of λ must be zero at the two extremities of the string. This condition is met by means of constants. Therefore, since the first three integrable equations of Article 30 give, for the first point of the string where the quantities affected by the integral sign $\int$ become zero, and for the last point of the string where $\int$ is changed to S

$$\frac{\lambda'\,\mathrm{d}x'}{\mathrm{d}s'} = A, \qquad \frac{\lambda'\,\mathrm{d}y'}{\mathrm{d}s'} = B, \qquad \frac{\lambda'\,\mathrm{d}z'}{\mathrm{d}s'} = C$$
$$\frac{\lambda''\,\mathrm{d}x''}{\mathrm{d}s''} = A + \mathrm{S}X\,\mathrm{d}m, \qquad \frac{\lambda''\,\mathrm{d}y''}{\mathrm{d}s''} = B + \mathrm{S}Y\,\mathrm{d}m, \qquad \frac{\lambda''\,\mathrm{d}z''}{\mathrm{d}s''} = C + \mathrm{S}Z\,\mathrm{d}m$$

One will have, in this case, $A = 0$, $B = 0$, $C = 0$ and consequently

$$\mathrm{S}X\,\mathrm{d}m = 0, \qquad \mathrm{S}Y\,\mathrm{d}m = 0, \qquad \mathrm{S}Z\,\mathrm{d}m = 0$$

It is obvious that these three equations are related to those of Article 12 of the preceding section.

34. Secondly, let us assume that the string is either fixed at one of its extremities or at both. If it is the first end that is fixed, the variations $\delta x'$, $\delta y'$, $\delta z'$ will be zero and it will suffice to equate to zero the coefficients of $\delta x''$, $\delta y''$, $\delta z''$, that is, it will suffice to make $\lambda'' = 0$.

By the same reasoning, when the other end is fixed, it will suffice to make $\lambda' = 0$. But if the two ends were simultaneously fixed, there will be no particular condition to fulfill because the variations $\delta x'$, $\delta y'$, $\delta z'$, $\delta x''$, $\delta y''$, $\delta z''$ will all be equal to zero.

35. Thirdly, let us assume that the ends of the string are attached to lines or curved surfaces along which they can slide freely. For example, let $\mathrm{d}z' = a'\,\mathrm{d}x' + b'\,\mathrm{d}y'$, $\mathrm{d}z'' = a''\,\mathrm{d}x'' + b''\,\mathrm{d}y''$ represent the differential equations of the surfaces to which the two ends of the string are fixed. In addition, one will have by replacing d with δ, $\delta z' = a'\,\delta x' + b'\,\delta y'$, $\delta z'' = a''\,\delta x'' + b''\,\delta y''$. Then these expressions will be substituted in the terms in question and the coefficients of $\delta x'$, $\delta y'$, $\delta x''$, $\delta y''$ will be equated to zero.

In general, one will treat the expression which is outside the integral sign in the general equation of equilibrium as if it were separate and as if it represented the equation of equilibrium of two distinct bodies located at each end of the string.

36. As an example, let us assume that the string is fixed by its two ends to the extremities of a lever mobile about a fixed point. Let a, b, c be the three rectangular coordinates which determine the position of this fixed point in space, that is, the position of the fulcrum of the lever. Let f be the distance between this support point and the end of the lever to which is fixed the first end of the string, g the distance between the same support point and the other end of the lever to which is fixed the second end of the string, h the distance between the two extremities of the lever and consequently, also between the ends of the string. It is clear that these six quantities, namely, a, b, c, f, g, h derive from the nature of the problem. It is also obvious at the same time that since x', y', z' are the coordinates at the origin of the string and x'', y'', z'' the coordinates at the other end of the string, one will have

$$\begin{aligned} f &= \sqrt{(a-x')^2+(b-y')^2+(c-z')^2} \\ g &= \sqrt{(a-x'')^2+(b-y'')^2+(c-z'')^2} \\ h &= \sqrt{(x''-x')^2+(y''-y')^2+(z''-z')^2} \end{aligned}$$

But since the quantities f, g, h are invariant, by differentiating by δ the following three determinate equations of condition will be obtained

$$\begin{aligned} &(a-x')\delta x'+(b-y')\delta y'+(c-z')\delta z'=0 \\ &(a-x'')\delta x''+(b-y'')\delta y''+(c-z'')\delta z''=0 \\ &(x''-x')(\delta x''-\delta x')+(y''-y')(\delta y''-\delta y')+(z''-z')(\delta z''-\delta z')=0 \end{aligned}$$

which when multiplied respectively by an undetermined coefficient, shall also be added to the general equation of equilibrium. Therefore, taking α, β, γ as the three required coefficients and equating to zero the coefficients of the six variations $\delta x'$, $\delta y'$, $\delta z'$, $\delta x''$, $\delta y''$, $\delta z''$, six particular equations will result

$$\begin{aligned} &\alpha(a-x')-\gamma(x''-x')-\frac{\lambda'\,\mathrm{d}x'}{\mathrm{d}s'}=0 \\ &\alpha(b-y')-\gamma(y''-y')-\frac{\lambda'\,\mathrm{d}y'}{\mathrm{d}s'}=0 \\ &\alpha(c-z')-\gamma(z''-z')-\frac{\lambda'\,\mathrm{d}z'}{\mathrm{d}s'}=0 \\ &\beta(a-x'')+\gamma(x''-x')+\frac{\lambda''\,\mathrm{d}x''}{\mathrm{d}s''}=0 \\ &\beta(b-y'')+\gamma(y''-y')+\frac{\lambda''\,\mathrm{d}y''}{\mathrm{d}s''}=0 \\ &\beta(c-z'')+\gamma(z''-z')+\frac{\lambda''\,\mathrm{d}z''}{\mathrm{d}s''}=0 \end{aligned}$$

which, after elimination of α, β, γ, will be reduced to three. When these equations are combined with the three equations of condition above, they can be used to determine the location of the two ends of the string. This demonstration shows how to proceed in other similar cases.

37. Finally, if besides the forces which act upon each point of the string, there were particular forces applied to the two ends of the string, which are denoted by X', Y', Z' for one end of the string and by X'', Y'', Z'' for the other end, these forces will produce the moments

$$X'\,\delta x' + Y'\,\delta y' + Z'\,\delta z' + X''\,\delta x'' + Y''\,\delta y'' + Z''\,\delta z''$$

and this quantity should be added to the first part of the general equation of equilibrium, that is, to the part which is outside the integral sign, which would then become

$$\begin{aligned}&(X'' + \frac{\lambda''\,\mathrm{d}x''}{\mathrm{d}s''})\delta x'' + (Y'' + \frac{\lambda''\,\mathrm{d}y''}{\mathrm{d}s''})\delta y'' + (Z'' + \frac{\lambda''\,\mathrm{d}z''}{\mathrm{d}s''})\delta z''\\ &+(X' - \frac{\lambda'\,\mathrm{d}x'}{\mathrm{d}s'})\delta x' + (Y' - \frac{\lambda'\,\mathrm{d}y'}{\mathrm{d}s'})\delta y' + (Z' - \frac{\lambda'\,\mathrm{d}z'}{\mathrm{d}s'})\delta z'\end{aligned}$$

and which can be applied to the different cases as was demonstrated in the preceding articles.

38. Let us now assume that the string is loaded at every point by the same forces X, Y, Z, and also pulled at its two ends by the forces X', Y', Z', X'', Y'', Z''. Its configuration is defined by a curved surface for which the equation is $\mathrm{d}z = p\,\mathrm{d}x + q\,\mathrm{d}y$ and that the configuration and location of the string on the same surface is described so that it is in equilibrium.

This problem which might be difficult enough to treat by ordinary principles of mechanics is solved very easily by our method and formulas.[91] Indeed, with the equation of the given surface, one has, after replacing d by δ, $\delta z = p\,\delta x + q\,\delta y$. Therefore, this expression for δz will only have to be substituted in all the terms under the integral sign of the general equation of equilibrium of the string (Article 29) and then to equate separately to zero the quantities affected by δx and δy. By this means the two following indefinite equations will result

$$\begin{aligned}&X\,\mathrm{d}m - \mathrm{d}(\frac{\lambda\,\mathrm{d}x}{\mathrm{d}s}) + p(Z\,\mathrm{d}m - \mathrm{d}(\frac{\lambda\,\mathrm{d}z}{\mathrm{d}s})) = 0\\ &Y\,\mathrm{d}m - \mathrm{d}(\frac{\lambda\,\mathrm{d}y}{\mathrm{d}s}) + p(Z\,\mathrm{d}m - \mathrm{d}(\frac{\lambda\,\mathrm{d}z}{\mathrm{d}s})) = 0\end{aligned}$$

which after combining them with the equation of the surface $\mathrm{d}z = p\,\mathrm{d}x + q\,\mathrm{d}y$ will be used to determine the equation of the string which is now free since the indeterminate λ has been eliminated.

39. Moreover, since the string is assumed to be in the same plane, one will have for its two endpoints $\delta z' = p'\,\delta x' + q'\,\delta y'$ and $\delta z'' = p''\,\delta x'' + q''\,\delta y''$. Thus these substitutions will be made again in the terms outside the integral sign of the general equation or rather in the formula given in Article 37 in which the forces X', Y', etc. are present. Then

the quantities relative to each of the four remaining variations $\delta x'$, $\delta y'$, $\delta x''$, $\delta y''$ will be equated separately to zero. Thus four new equations result

$$X' - \frac{\lambda' \, dx'}{ds'} + p'(Z' - \frac{\lambda' \, dz'}{ds'}) = 0$$
$$Y' - \frac{\lambda' \, dy'}{ds'} + q'(Z' - \frac{\lambda' \, dz'}{ds'}) = 0$$
$$X'' + \frac{\lambda'' \, dx''}{ds''} + p''(Z'' + \frac{\lambda'' \, dz''}{ds''}) = 0$$
$$Y'' + \frac{\lambda'' \, dy''}{ds''} + q''(Z'' + \frac{\lambda'' \, dz''}{ds''}) = 0$$

which should be satisfied by means of constants.

40. But, instead of substituting the expressions for δz in terms of δx and δy obtained from the equation $\delta z - p\,\delta x - q\,\delta y = 0$, as we just did, this same equation could be viewed as a new indeterminate equation of condition. Then, the total integral of this equation multiplied by another undetermined coefficient μ, should be added to the general equation of equilibrium (Article 29). In this manner, the expression under the integral sign would become

$$\mathrm{S}[(X\, dm - d(\frac{\lambda\, dx}{ds}) - \mu p)\delta x + (Y\, dm - d(\frac{\lambda\, dy}{ds}) - \mu q)\delta y$$
$$+(Z\, dm - d(\frac{\lambda\, dz}{ds}) + \mu)\delta z]$$

and these three undefined equations would be obtained immediately

$$X\, dm - d\frac{\lambda\, dx}{ds} - \mu p = 0$$
$$Y\, dm - d\frac{\lambda\, dy}{ds} - \mu q = 0$$
$$Z\, dm - d\frac{\lambda\, dz}{ds} + \mu = 0$$

which, by elimination of μ will give again the same equations already found (Article 38). But these latter equations have, moreover, the advantage that the pressure that each element of the string exerts on the surface is obtained simultaneously from the theory given in Article 5 of SECTION IV.

Indeed, it is easy to deduce from this theory that the terms $\mu(\delta z - p\,\delta x - q\,\delta y)$ coming from the equation of condition $\delta z - p\,\delta x - q\,\delta y = 0$, can represent the effect of a force equal to $\mu\sqrt{(1 + p^2 + q^2)}$ applied to each element ds of the string in a direction perpendicular to the surface which has for its equation $\delta z - p\,\delta x - q\,\delta y = 0$ or $dz - p\, dx - q\, dy = 0$, that is, to the same surface on which the string is assumed resting. This surface, because of its resistance, produces the force $\mu\sqrt{(1 + p^2 + q^2)}$, which consequently, will be equal and directly opposed to the pressure exerted by the string on the same surface (SECTION IV, Article

7). Thus the pressure on each point of the string will be equal to $[\mu\sqrt{(1+p^2+q^2)}]/\mathrm{d}s$ or rather, by substituting the values of μ, μp, μq obtained from the equations above

$$\frac{\sqrt{(X\,dm - \mathrm{d}\frac{\lambda\,\mathrm{d}x}{\mathrm{d}s})^2 + (Y\,dm - \mathrm{d}\frac{\lambda\,\mathrm{d}y}{\mathrm{d}s})^2 + (Z\,dm - \mathrm{d}\frac{\lambda\,\mathrm{d}z}{\mathrm{d}s})^2}}{\mathrm{d}s}$$

Then the same reasoning will be applied to the portion of the general equation which is outside the integral sign S and similar conclusions will be reached.

41. If the string resting on the given surface is under tension as a result of forces applied at its ends only, one would have $X = 0$, $Y = 0$, $Z = 0$, and consequently, $\mathrm{d}\lambda = 0$ (Article 30). Thus the coefficient λ is equal to a constant. Therefore, the tension in the string would be the same everywhere along its length (Article 31) which is consistent with what is already known. In this case, the general formula for equilibrium would be reduced to

$$\lambda\,\mathrm{S}\,\delta\mathrm{d}s + \mathrm{S}\mu(\delta\,z - p\,\delta x - q\,\delta y) = 0$$

of which the first term is the same as $\lambda\,\delta(\mathrm{Sd}s)$ or $\lambda\,\delta s$. Thus this equation expresses the fact that the length of the curve formed by the string on the surface represented by the equation $\mathrm{d}z - p\,\mathrm{d}x - q\,\mathrm{d}y = 0$ must be a maximum or a minimum and that the pressure exerted by the string at each point of this surface will be

$$\frac{\lambda\sqrt{(\mathrm{d}\frac{\mathrm{d}x}{\mathrm{d}s})^2 + (\mathrm{d}\frac{\mathrm{d}y}{\mathrm{d}s})^2 + (\mathrm{d}\frac{\mathrm{d}z}{\mathrm{d}s})^2}}{\mathrm{d}s}$$

But it is known that

$$\sqrt{(\mathrm{d}\frac{\mathrm{d}x}{\mathrm{d}s})^2 + (\mathrm{d}\frac{\mathrm{d}y}{\mathrm{d}s})^2 + (\mathrm{d}\frac{\mathrm{d}z}{\mathrm{d}s})^2}$$

defines the angle of the instantaneous tangent to the curve which is equal to $\mathrm{d}s/\rho$ where ρ is the osculating radius. Therefore, the pressure will be λ/ρ and consequently, it will be inversely proportional to the osculating radius.

Subsection II
The Equilibrium of a Flexible and Simultaneously Extensible and Contractible String or Surface

42. Up to this point, we have assumed that the string was inextensible. Let us treat it now as a spring capable of extension and contraction. And let F be the force with which each element ds of the curve of the string might contract. One will have, as in Article 18

(by replacing f by $\mathrm{d}s$ and by replacing d with δ), $F\,\delta\mathrm{d}s$ for the moment of this force and $\mathrm{S}F\,\delta\mathrm{d}s$ for the sum of the moments of all the forces of compression which act on the length of the string. Thus this integral $\mathrm{S}F\,\delta\mathrm{d}s$ is added to the integral $\mathrm{S}(X\,\delta x + Y\,\delta y + Z\,\delta z)\mathrm{d}m$ which expresses the sum of the moments of all the external forces acting on the string (Article 28) and by equating the expression to zero, the general equation of equilibrium for a deformable string will be obtained.

But it is obvious that this equation will be of the same form as the one in Article 29 for the case of the inextensible string and that by replacing F with λ the two equations will become identical. In the present case, one will have the same particular equations for the equilibrium of a string that were found in Article 30 by replacing F in those equations by λ. And if the quantity F were eliminated, as the quantity λ was eliminated, two equations identical to those which exist for an inextensible string would be obtained to represent the curve formed by an extensible string.

43. With respect to the quantity F which represents the elasticity of the force of contraction of each element $\mathrm{d}s$, it is natural to express it as a function of the extension that this element receives from the action of the forces X, Y, Z. Therefore, assuming that $\mathrm{d}\sigma$ is the initial length of $\mathrm{d}s$, F can be viewed as a given function of $\mathrm{d}s/\mathrm{d}\sigma$. But because of the nature of the differential calculus the absolute value of the element ds remains indeterminate. Thus the value of F will also be indeterminate and can be found only by means of one of the three equations of equilibrium of the string. Therefore, although in the present case our analysis seems to give one more equation than needed, nevertheless, it provides only the necessary equations to determine the curve of the string and the resistance of each of its elements.

Because the quantity λ in the solution of Article 30 corresponds exactly to the quantity F, which is the real force with which each element of the string is pulled by the action of the external forces, it is obvious that this quantity can be viewed as representing the tension in the inextensible string. This is what we have found earlier **a priori** in Article 31.

44. Let us now apply the same principles to the determination of the equilibrium of a surface for which all of its elements dm are extensible and contractible. The element of a surface with coordinates x, y, z, where z is viewed as a function of x and y, is expressed by the formula

$$\mathrm{d}x\,\mathrm{d}y\sqrt{1 + (\frac{\mathrm{d}z}{\mathrm{d}x})^2 + (\frac{\mathrm{d}z}{\mathrm{d}y})^2}$$

Thus, by calling F the force of elasticity with which this element has the tendency to contract, the sum of the moments of all these forces will be expressed by the double integral

$$\mathrm{SS}\,F\,\delta\left(\mathrm{d}x\,\mathrm{d}y\sqrt{1 + (\frac{\mathrm{d}z}{\mathrm{d}x})^2 + (\frac{\mathrm{d}z}{\mathrm{d}y})^2}\right)$$

which, when added to the double integral

$$\mathrm{SS}\,(X\,\delta x + Y\,\delta y + Z\,\delta z)\mathrm{d}m$$

where $\mathrm{d}m$ is the element of the surface, will give the sum of the moments of all the forces, which must be equal to zero at equilibrium.

By making, as in Article 31 of SECTION IV

$$\frac{\mathrm{d}z}{\mathrm{d}x} = z', \qquad \frac{\mathrm{d}z}{\mathrm{d}y} = z_{,}, \qquad \sqrt{1 + z'^2 + z_{,}^2} = U$$

one will have $\mathrm{d}m = U\,\mathrm{d}x\,\mathrm{d}y$ and

$$\frac{\mathrm{d}U}{\mathrm{d}z'} = \frac{z'}{U}, \qquad \frac{\mathrm{d}U}{\mathrm{d}z_{,}} = \frac{z_{,}}{U}$$

then (Articles 33 and 34 of SECTION IV)

$$\delta U = (\frac{\mathrm{d}U}{\mathrm{d}x}\delta x + \frac{\mathrm{d}U}{\mathrm{d}y}\delta y) + \frac{1}{U}(z'\frac{\mathrm{d}\delta u}{\mathrm{d}x} + z_{,}\frac{\mathrm{d}\delta u}{\mathrm{d}y})$$

$$\delta(U\,\mathrm{d}x\,\mathrm{d}y) = (\delta U + U(\frac{\mathrm{d}\delta x}{\mathrm{d}x} + \frac{\mathrm{d}\delta y}{\mathrm{d}y}))\mathrm{d}x\,\mathrm{d}y$$

After substituting these values in the double integral $\mathrm{SS}\,F\,\delta(U\,\mathrm{d}x\,\mathrm{d}y)$ and eliminating through integration by parts the partial differences of the variations indicated by δ, one will have

$$\mathrm{S}(U\,\delta y + \frac{z_{,}}{U}\delta u)F\,\mathrm{d}x + \mathrm{S}(U\,\delta x + \frac{z'}{U}\delta u)F\,\mathrm{d}y$$

$$+\mathrm{SS}\left(\left(\frac{F\,\mathrm{d}U}{\mathrm{d}x} - \frac{\mathrm{d}(FU)}{\mathrm{d}x}\right)\delta x + \left(\frac{F\,\mathrm{d}U}{\mathrm{d}y} - \frac{\mathrm{d}(FU)}{\mathrm{d}y}\right)\delta y - V\,\delta u\right)\mathrm{d}x\,\mathrm{d}y$$

where

$$V = \frac{\mathrm{d}\frac{Fz'}{U}}{\mathrm{d}x} + \frac{\mathrm{d}\frac{Fz_{,}}{U}}{\mathrm{d}y}$$

and $\delta u = \delta z - z'\,\delta x - z_{,}\,\delta y$ (cited articles).

The simple integrals relative to x and y are calculated with respect to limits and vanish identically in the case where the boundaries of the surface are assumed fixed because then the variations δx, δy, δz are equal to zero at all points of the boundaries of the surface.

When the terms under the double integral sign SS are added to those of the double integral $SS(X\,\delta x + Y\,\delta y + Z\,\delta z)U\,\mathrm{d}x\,\mathrm{d}y$, the coefficients of the variations δx, δy, δz will be equated separately to zero to obtain the following three equations

$$XU + \frac{F\,\mathrm{d}U}{\mathrm{d}x} - \frac{\mathrm{d}(UF)}{\mathrm{d}x} + Vz' = 0, \qquad YU + \frac{F\,\mathrm{d}U}{\mathrm{d}y} - \frac{\mathrm{d}(UF)}{\mathrm{d}y} + Vz_{\prime} = 0,$$
$$ZU - V = 0$$

The first two equations will give the expression of the force F that must be substituted in the expression for V of the third so that one will have ultimately only one equation with partial differences to determine the equilibrium surface.

Indeed, although the force F must be assumed a known function of the element dm of the surface in its state of contraction or extension, it is, nevertheless, indeterminate because the absolute value of the elements of the surface cannot enter in the calculation. Therefore, the value of F can only be determined by the conditions of equilibrium. The case considered here is similar to the one of Article 43.

45. In order to eliminate the quantity F, the value of V obtained from the last equation will be substituted in the first two equations. They will become

$$U(X + Z\frac{\mathrm{d}z}{\mathrm{d}x}) + \frac{F\,\mathrm{d}U}{\mathrm{d}x} - \frac{\mathrm{d}(UF)}{\mathrm{d}x} = 0, \qquad U(X + Z\frac{\mathrm{d}z}{\mathrm{d}y}) + \frac{F\,\mathrm{d}U}{\mathrm{d}y} - \frac{\mathrm{d}(UF)}{\mathrm{d}y} = 0$$

Then as in Article 28

$$X\,\mathrm{d}x + Y\,\mathrm{d}y + Z\,\mathrm{d}z = \mathrm{d}\Pi$$

one will have since z is assumed to be a function of x and y

$$\frac{\mathrm{d}\Pi}{\mathrm{d}x} = X + Z\frac{\mathrm{d}z}{\mathrm{d}x}, \qquad \frac{\mathrm{d}\Pi}{\mathrm{d}y} = Y + Z\frac{\mathrm{d}z}{\mathrm{d}y}$$

and the two equations will become after division by U

$$\frac{\mathrm{d}\Pi}{\mathrm{d}x} = \frac{\mathrm{d}F}{\mathrm{d}x}, \qquad \frac{\mathrm{d}\Pi}{\mathrm{d}y} = \frac{\mathrm{d}F}{\mathrm{d}y}$$

which give simply the following equation

$$\mathrm{d}\Pi = \mathrm{d}F$$

from which

$$F = \Pi + a$$

This result is in accordance with Article 36 of Section IV.

Then the third equation will give by viewing Π as a function of x, y, z

$$U\frac{\mathrm{d}\Pi}{\mathrm{d}z} - \frac{\mathrm{d}\frac{Fz'}{U}}{\mathrm{d}x} - \frac{\mathrm{d}\frac{Fz_{\prime}}{U}}{\mathrm{d}y} = 0$$

This last equation will be the equation of the surface.

If the surface departs only slightly from a plane so that the ordinate z is very small, then by neglecting some very small quantities of second order, one would have $U = 1$. Thus $F = \Pi + a$ where a is a constant and the equation of the surface is

$$\frac{\mathrm{d}\Pi}{\mathrm{d}z} + \frac{\mathrm{d}(\Pi + a)\frac{\mathrm{d}z}{\mathrm{d}x}}{\mathrm{d}x} + \frac{\mathrm{d}(\Pi + a)\frac{\mathrm{d}z}{\mathrm{d}y}}{\mathrm{d}y} = 0$$

By assuming that there is no other force than gravity g which acts to increase the ordinate z, one will have $\Pi = -gz$. Consequently, if terms of the second order are always neglected, the following equation results

$$a(\frac{\mathrm{d}^2z}{\mathrm{d}x^2} + \frac{\mathrm{d}^2z}{\mathrm{d}y^2}) = g$$

In general, this is an integrable equation but with imaginary solutions which makes application of the solution difficult.

Subsection III
The Equilibrium of an Elastic String or Strip

46. Let us now reconsider the case of an inextensible string. But instead of assuming in addition that it is perfectly flexible, as was done previously, let us assume that it is elastic so that there is at each point a force which I will call E which opposes the deformation of the string and consequently, which tends to diminish the tangential angle. Calling this angle e, one will have, as in Article 26 (replacing only d by δ), $E\,\delta e$ for the moment of each force E. Thus $\mathrm{S}\,E\,\delta e$ will be the sum of the moments of all the forces of elasticity which act along the full length of the string, which then must be added to the first member of the general equation of equilibrium in the case of an inextensible and perfectly flexible string (Article 29).

All the difficulty rests with trying to reduce the integral $\mathrm{S}\,E\delta e$ to a convenient form. In order to find a convenient form of the integral, the expression for e must first be found. But we have found above (Article 26)

$$-\cos e = \frac{f^2 + g^2 - h^2}{2fg}$$

from which the following relation is obtained

$$\sin^2 e = \frac{4f^2g^2 - (f^2 + g^2 - h^2)^2}{4f^2g^2}$$

In order to apply this formula to the present case, it suffices to remark that the coordinates $x', y', z', x'', y'', z'', x''', y''', z'''$, which we used to define the quantities f, g, h (Articles 12 and 20) are reduced here to $x, y, z; x+\mathrm{d}x, y+\mathrm{d}y, z+\mathrm{d}z; x+2\,\mathrm{d}x+\mathrm{d}^2x, y+2\,\mathrm{d}y+\mathrm{d}^2y, z+2\,\mathrm{d}z+\mathrm{d}^2z$; so that

$$\begin{aligned} f^2 &= \mathrm{d}x^2 + \mathrm{d}y^2 + \mathrm{d}z^2 = \mathrm{d}s^2, \qquad g^2 = (\mathrm{d}x + \mathrm{d}^2x)^2 + (\mathrm{d}y + \mathrm{d}^2y)^2 + (\mathrm{d}z + \mathrm{d}^2z)^2 \\ &= \mathrm{d}x^2 + \mathrm{d}y^2 + \mathrm{d}z^2 + 2(\mathrm{d}x\,\mathrm{d}^2x + \mathrm{d}y\,\mathrm{d}^2y + \mathrm{d}z\,\mathrm{d}^2z) + \mathrm{d}^2x^2 + \mathrm{d}^2y^2 + \mathrm{d}^2z^2 \\ &= \mathrm{d}s^2 + 2\,\mathrm{d}s\,\mathrm{d}^2s + \mathrm{d}^2x^2 + \mathrm{d}^2y^2 + \mathrm{d}^2z^2, \qquad h^2 = (2\,\mathrm{d}x + \mathrm{d}^2x)^2 + (2\,\mathrm{d}y + \mathrm{d}^2y)^2 \\ &\quad +(2\,\mathrm{d}z + \mathrm{d}^2z)^2 = 4\,\mathrm{d}s^2 + 4\,\mathrm{d}s\,\mathrm{d}^2s + \mathrm{d}^2x^2 + \mathrm{d}^2y^2 + \mathrm{d}^2z^2 \end{aligned}$$

then

$$f^2 + g^2 - h^2 = -2\,\mathrm{d}s^2 - 2\,\mathrm{d}s\,\mathrm{d}^2s$$

and

$$\begin{aligned} &4f^2g^2 - (f^2 + g^2 - h^2)^2 - 4\,\mathrm{d}s^4 + 8\,\mathrm{d}s^3\,\mathrm{d}^2s + 4\,\mathrm{d}s^2(\mathrm{d}^2x^2 + \mathrm{d}^2y^2 + \mathrm{d}^2z^2) \\ &-4(\mathrm{d}s^2 + \mathrm{d}s\,\mathrm{d}^2s)^2 = 4\,\mathrm{d}s^2(\mathrm{d}^2x^2 + \mathrm{d}^2y^2 + \mathrm{d}^2z^2 - \mathrm{d}^2s^2) \end{aligned}$$

Finally, neglecting the infinitesimal quantities of third order there results

$$\sin^2 e = \frac{\mathrm{d}^2x^2 + \mathrm{d}^2y^2 + \mathrm{d}^2z^2 - \mathrm{d}^2s^2}{\mathrm{d}s^2}$$

Since this value of $\sin^2 e$ is an infinitesimal quantity of second order, it happens that $\sin^2 e$ and consequently, also the angle e, will be an infinitesimal quantity of the first order. Therefore

$$e = \frac{\sqrt{\mathrm{d}^2x^2 + \mathrm{d}^2y^2 + \mathrm{d}^2z^2 - \mathrm{d}^2s^2}}{\mathrm{d}s}$$

This is the expression for the tangential angle for an arbitrary curve with double curvature. It is similar to the expression of Article 41.

47. Now the value of δe is obtained by differentiating according to δ. Since the string is inextensible, it must be that $\delta\mathrm{d}s = 0$ (Article 29) and also that $\mathrm{d}\delta\mathrm{d}s = \delta d^2 s = 0$. Thus $\mathrm{d}s$ and d^2s could be treated as constants. In this fashion, one will have

$$\delta e = \frac{\mathrm{d}^2x\,\delta\mathrm{d}^2x + \mathrm{d}^2y\,\delta\mathrm{d}^2y + d^2z\,\delta\mathrm{d}^2z}{\mathrm{d}s\sqrt{\mathrm{d}^2x^2 + \mathrm{d}^2y^2 + \mathrm{d}^2z^2 - \mathrm{d}^2s^2}}$$

Substituting in $\mathrm{S}\, E\, \delta e$, and in so doing, the expression is shortened by putting

$$I = \frac{E}{\mathrm{d}s\sqrt{\mathrm{d}^2x^2 + \mathrm{d}^2y^2 + \mathrm{d}^2z^2 - \mathrm{d}^2s^2}}$$

there results

$$\mathrm{S}\, E\, \delta e = \mathrm{S}\, I\, \mathrm{d}^2x\, \delta\mathrm{d}^2x + \mathrm{S}\, I\, \mathrm{d}^2y\, \delta\mathrm{d}^2y + \mathrm{S}\, I\, \mathrm{d}^2z\, \delta\mathrm{d}^2z$$

These expressions are treated according to the rules given in Article 15 of SECTION IV by first replacing $\delta\mathrm{d}$ by $\mathrm{d}\delta$ and integrating by parts to eliminate the d before the δ. The following transformed equations will result

$$\begin{aligned} \mathrm{S}\, I\, \mathrm{d}^2x\, \delta\mathrm{d}^2x = {} & I''\, \mathrm{d}^2x''\, \mathrm{d}\delta x'' - \mathrm{d}(I''\, \mathrm{d}^2x'')\delta x'' - I'\, \mathrm{d}^2x'\, \mathrm{d}\delta x' \\ & + \mathrm{d}(I'\, \mathrm{d}^2x')\delta x' + \mathrm{S}\, \mathrm{d}^2(I\, \mathrm{d}^2x)\delta x, \end{aligned}$$

$$\begin{aligned} \mathrm{S}\, I\, \mathrm{d}^2y\, \delta\mathrm{d}^2y = {} & I''\, \mathrm{d}^2y''\, \mathrm{d}\delta y'' - \mathrm{d}(I''\, \mathrm{d}^2y'')\delta y'' - I'\, \mathrm{d}^2y'\, \mathrm{d}\delta y' \\ & + \mathrm{d}(I'\, \mathrm{d}^2y')\delta y' + \mathrm{S}\, \mathrm{d}^2(I\, \mathrm{d}^2y)\delta y, \end{aligned}$$

$$\begin{aligned} \mathrm{S}\, I\, \mathrm{d}^2z\, \delta\mathrm{d}^2z = {} & I''\, \mathrm{d}^2z''\, \mathrm{d}\delta z'' - \mathrm{d}(I''\, \mathrm{d}^2z'')\delta z'' - I'\, \mathrm{d}^2z'\, \mathrm{d}\delta z' \\ & + \mathrm{d}(I'\, \mathrm{d}^2z')\delta z' + \mathrm{S}\, \mathrm{d}^2(I\, \mathrm{d}^2z)\delta z, \end{aligned}$$

Now add these different terms to those which form the first part of the general equation of equilibrium of Article 29 and the equation of equilibrium for an inextensible and elastic string will be obtained.

48. If the coefficients of the variations δx, δy, δz which are under the integral sign S are first equated to zero, the three following indefinite equations will be obtained.

$$X\, \mathrm{d}m - \mathrm{d}\left(\frac{\lambda\, \mathrm{d}x}{\mathrm{d}s}\right) + \mathrm{d}^2(I\, \mathrm{d}^2x) = 0$$

$$Y\, \mathrm{d}m - \mathrm{d}\left(\frac{\lambda\, \mathrm{d}y}{\mathrm{d}s}\right) + \mathrm{d}^2(I\, \mathrm{d}^2y) = 0$$

$$Z\, \mathrm{d}m - \mathrm{d}\left(\frac{\lambda\, \mathrm{d}z}{\mathrm{d}s}\right) + \mathrm{d}^2(I\, \mathrm{d}^2z) = 0$$

from which λ should be eliminated which will reduce them to two and which will be sufficient to determine the equation of the string.

The first integration gives

$$\frac{\lambda\, \mathrm{d}x}{\mathrm{d}s} - \mathrm{d}(I\, \mathrm{d}^2x) = A + \int X\, \mathrm{d}m$$

$$\frac{\lambda\, \mathrm{d}y}{\mathrm{d}s} - \mathrm{d}(I\, \mathrm{d}^2y) = B + \int Y\, \mathrm{d}m$$

$$\frac{\lambda\, \mathrm{d}z}{\mathrm{d}s} - \mathrm{d}(I\, \mathrm{d}^2z) = C + \int Z\, \mathrm{d}m$$

where A, B, C are arbitrary constants and the elimination of λ will give

$$\mathrm{d}x\,\mathrm{d}(I\,\mathrm{d}^2y) - \mathrm{d}y\,\mathrm{d}(I\,\mathrm{d}^2x) = (A + \int X\,\mathrm{d}m)\mathrm{d}y - (B + \int Y\,\mathrm{d}m)\mathrm{d}x,$$

$$\mathrm{d}x\,\mathrm{d}(I\,\mathrm{d}^2z) - \mathrm{d}z\,\mathrm{d}(I\,\mathrm{d}^2x) = (A + \int X\,\mathrm{d}m)\mathrm{d}z - (C + \int Z\,\mathrm{d}m)\mathrm{d}x,$$

$$\mathrm{d}y\,\mathrm{d}(I\,\mathrm{d}^2z) - \mathrm{d}z\,\mathrm{d}(I\,\mathrm{d}^2y) = (B + \int Y\,\mathrm{d}m)\mathrm{d}z - (C + \int Z\,\mathrm{d}m)\mathrm{d}y$$

where the last equation is dependent on the two preceding equations.

These equations are again integrable and one will have

$$I(\mathrm{d}x\,\mathrm{d}^2y - \mathrm{d}y\,\mathrm{d}^2x) = F + \int(A + \int X\,\mathrm{d}m)\mathrm{d}y - \int(B + \int Y\,\mathrm{d}m)\mathrm{d}x$$

$$I(\mathrm{d}x\,\mathrm{d}^2z - \mathrm{d}z\,\mathrm{d}^2x) = G + \int(A + \int X\,\mathrm{d}m)\mathrm{d}z - \int(C + \int Z\,\mathrm{d}m)\mathrm{d}x$$

$$I(\mathrm{d}y\,\mathrm{d}^2z - \mathrm{d}z\,\mathrm{d}^2y) = H + \int(B + \int Y\,\mathrm{d}m)\mathrm{d}z - \int(C + \int Z\,\mathrm{d}m)\mathrm{d}y$$

where F, G, H are new constants.

Now we saw earlier that (Article 47)

$$I = \frac{E}{\mathrm{d}s\sqrt{\mathrm{d}^2x^2 + \mathrm{d}^2y^2 + \mathrm{d}^2z^2 - \mathrm{d}^2s^2}}$$

The square of the denominator of this quantity is

$$\begin{aligned}
&\mathrm{d}s^2(\mathrm{d}^2x^2 + \mathrm{d}^2y^2 + \mathrm{d}^2z^2) - \mathrm{d}s^2\,\mathrm{d}^2s^2 = (\mathrm{d}x^2 + \mathrm{d}y^2 + \mathrm{d}z^2)(\mathrm{d}^2x^2 + \mathrm{d}^2y^2 + \mathrm{d}^2z^2)\\
&-(\mathrm{d}x\,\mathrm{d}^2x + \mathrm{d}y\,\mathrm{d}^2y + \mathrm{d}z\,\mathrm{d}^2z)^2 =\\
&(\mathrm{d}x\,\mathrm{d}^2y - \mathrm{d}y\,\mathrm{d}^2x)^2 + (\mathrm{d}x\,\mathrm{d}^2z - \mathrm{d}z\,\mathrm{d}^2x)^2 + (\mathrm{d}y\,\mathrm{d}^2z + \mathrm{d}z\,\mathrm{d}^2y)^2
\end{aligned}$$

Therefore, if the square of the three preceding equations were summed, the following equation without the differentials would be obtained

$$\begin{aligned}
E^2 = (F &+ \int(A + \int X\,\mathrm{d}m)\mathrm{d}y - \int(B + \int Y\,\mathrm{d}m)\mathrm{d}x)^2\\
&+(G + \int(A + \int X\,\mathrm{d}m)\mathrm{d}z - \int(C + \int Z\,\mathrm{d}m)\mathrm{d}x)^2\\
&+(H + \int(B + \int Y\,\mathrm{d}m)\mathrm{d}z - \int(C + \int Z\,\mathrm{d}m)\mathrm{d}y)^2
\end{aligned}$$

and if two of these equations were divided, the following equation where the string's elasticity is eliminated would result

$$\frac{\mathrm{d}x\,\mathrm{d}^2z - \mathrm{d}z\,\mathrm{d}^2x}{\mathrm{d}x\,\mathrm{d}^2y - \mathrm{d}y\,\mathrm{d}^2x} = \frac{G + \int(A + \int X\,\mathrm{d}m)\mathrm{d}z - \int(C + \int Z\,\mathrm{d}m)\mathrm{d}x}{F + \int(A + \int X\,\mathrm{d}m)\mathrm{d}y - \int(B + \int Y\,\mathrm{d}m)\mathrm{d}x}$$

These two equations are the simplest from which to determine the elastic curve, if double curvature is to be considered.

49. It is commonly assumed that the elastic force which is opposed to the deflection of the string is inversely proportional to the radius of osculation. Therefore, by denoting this radius by ρ the elastic force will be $E = K/\rho$, where K is a constant coefficient.[92]

But it is known that $\rho = \mathrm{d}s/e$. Thus $E = Ke/\mathrm{d}s$ and the quantity I, which we assumed equal to $E/e\,\mathrm{d}s^2$ (Article 47), will become $K/\mathrm{d}s^3$ and consequently, a constant, assuming, which is permissible, that ds is constant. Thus the first three equations (Article 48) will be

$$X\,\mathrm{d}m - \mathrm{d}\frac{\lambda\,\mathrm{d}x}{\mathrm{d}s} + \frac{K\,\mathrm{d}^4x}{\mathrm{d}s^3} = 0$$

$$Y\,\mathrm{d}m - \mathrm{d}\frac{\lambda\,\mathrm{d}y}{\mathrm{d}s} + \frac{K\,\mathrm{d}^4y}{\mathrm{d}s^3} = 0$$

$$Z\,\mathrm{d}m - \mathrm{d}\frac{\lambda\,\mathrm{d}z}{\mathrm{d}s} + \frac{K\,\mathrm{d}^4z}{\mathrm{d}s^3} = 0$$

If these three equations were added together after multiplying the first by $\mathrm{d}x/\mathrm{d}s$, the second by $\mathrm{d}y/\mathrm{d}s$, and the third by $\mathrm{d}z/\mathrm{d}s$, and since

$$\frac{\mathrm{d}x}{\mathrm{d}s}\mathrm{d}\frac{\mathrm{d}x}{\mathrm{d}s} + \frac{\mathrm{d}y}{\mathrm{d}s}\mathrm{d}\frac{\mathrm{d}y}{\mathrm{d}s} + \frac{\mathrm{d}z}{\mathrm{d}s}\mathrm{d}\frac{\mathrm{d}z}{\mathrm{d}s} = \frac{1}{2}\mathrm{d}\left(\frac{\mathrm{d}x^2 + \mathrm{d}y^2 + \mathrm{d}z^2}{\mathrm{d}s^2}\right) = 0$$

the following equation will result

$$(X\,\mathrm{d}x + Y\,\mathrm{d}y + Z\,\mathrm{d}z)\frac{\mathrm{d}m}{\mathrm{d}s} + K\frac{\mathrm{d}x\,\mathrm{d}^4x + \mathrm{d}y\,\mathrm{d}^4y + \mathrm{d}z\,\mathrm{d}^4z}{\mathrm{d}s^4} = \mathrm{d}\lambda$$

Let Γ be the cross-sectional area of the string so that $\mathrm{d}m = \Gamma\,\mathrm{d}s$. The integration of the preceding equation, assuming $\mathrm{d}s$ to be constant, produces the following equation

$$\lambda = \int \Gamma(X\,\mathrm{d}x + Y\,\mathrm{d}y + Z\,\mathrm{d}z) + K\left(\frac{\mathrm{d}x\,\mathrm{d}^3x + \mathrm{d}y\,\mathrm{d}^3y + \mathrm{d}z\,\mathrm{d}^3z}{\mathrm{d}s^4} - \frac{\mathrm{d}^2x^2 + \mathrm{d}^2y^2 + \mathrm{d}^2z^2}{2\,\mathrm{d}s^4}\right)$$

This value of λ expresses the tension in the elastic member, that is, the resistance with which it reacts the force which tends to extend it, as in Article 31.

50. The simplest and most common case is the one in which the forces X, Y, Z, which are assumed acting at all points of the elastic strip, are equal to zero and the curvature of the strip results only from the forces applied at its two ends. In this case, the integrable equations of Article 48, give by replacing I by its value $K/\mathrm{d}s^3$

$$K\frac{\mathrm{d}x\,\mathrm{d}^2y - \mathrm{d}y\,\mathrm{d}^2x}{\mathrm{d}s^3} = F + Ay - Bx$$

$$K\frac{\mathrm{d}x\,\mathrm{d}^2z-\mathrm{d}z\,\mathrm{d}^2x}{\mathrm{d}s^3}=G+Az-Cx$$
$$K\frac{\mathrm{d}y\,\mathrm{d}^2z-\mathrm{d}z\,\mathrm{d}^2y}{\mathrm{d}s^3}=H+Bz-Cy$$

but another integration of these equations might be impossible in general.

When the surface of the strip is always in the same plane and taking for this plane the xy-plane and making $\mathrm{d}y=\mathrm{d}s\sin\varphi$, $\mathrm{d}x=\mathrm{d}s\cos\varphi$, the first equation which is now the only necessary equation, becomes

$$\frac{\mathrm{d}\varphi}{\mathrm{d}s}=F+A\int\sin\varphi\,\mathrm{d}s-B\int\cos\varphi\,\mathrm{d}s$$

which, after differentiation, gives

$$\frac{\mathrm{d}^2\varphi}{\mathrm{d}^2s}=A\sin\varphi-B\cos\varphi$$

Multiplying by $\mathrm{d}\varphi$ and integrating once

$$\frac{\mathrm{d}\varphi^2}{2\,\mathrm{d}s^2}=A\cos\varphi+B\sin\varphi+D$$

from which is obtained

$$\mathrm{d}s=\frac{\mathrm{d}\varphi}{\sqrt{2D+2A\cos\varphi+2B\sin\varphi}}$$

and from there

$$\mathrm{d}x=\frac{\cos\varphi\,\mathrm{d}\varphi}{\sqrt{2D+2A\cos\varphi+2B\sin\varphi}}$$

and because from the first equation $F+Ay-Bx=\mathrm{d}\varphi/\mathrm{d}s$, there results

$$y=\frac{Bx-F}{A}-\frac{1}{A}\sqrt{2D+2A\cos\varphi+2B\sin\varphi}$$

Therefore, the problem is reduced to integration of the expressions for $\mathrm{d}s$ and $\mathrm{d}x$. But the integrations depend upon the rectification of conical sections. Until now, it does not appear that research had gone farther in the general solution of the problem of the elastic curve.

51. Let us now consider the terms of the general equation which are outside the integral sign S. These terms are

$$\left(\frac{\lambda''\,\mathrm{d}x''}{\mathrm{d}s''}-\mathrm{d}(I''\,\mathrm{d}^2x'')\right)\delta x''+I''\,\mathrm{d}^2x''\,\mathrm{d}\delta x''$$

$$+(\frac{\lambda'' \, \mathrm{d}y''}{\mathrm{d}s''} - \mathrm{d}(I'' \, \mathrm{d}^2 y''))\delta y'' + I'' \, \mathrm{d}^2 y'' \, \mathrm{d}\delta y''$$
$$+(\frac{\lambda'' \, \mathrm{d}z''}{\mathrm{d}s''} - \mathrm{d}(I'' \, \mathrm{d}^2 z''))\delta z'' + I'' \, \mathrm{d}^2 z'' \, \mathrm{d}\delta z''$$
$$-(\frac{\lambda' \, \mathrm{d}x'}{\mathrm{d}s'} - \mathrm{d}(I' \, \mathrm{d}^2 x'))\delta x' - I' \, \mathrm{d}^2 x' \, \mathrm{d}\delta x'$$
$$-(\frac{\lambda' \, \mathrm{d}y'}{\mathrm{d}s'} - \mathrm{d}(I' \, \mathrm{d}^2 y'))\delta y' - I' \, \mathrm{d}^2 y' \, \mathrm{d}\delta y'$$
$$-(\frac{\lambda' \, \mathrm{d}z'}{\mathrm{d}s'} - \mathrm{d}(I' \, \mathrm{d}^2 z'))\delta z' - I' \, \mathrm{d}^2 z' \, \mathrm{d}\delta z'$$

and they should be eliminated independently of the values of $\delta x''$, $\delta y''$, etc.

1. Therefore, if the string is entirely free, the coefficients of the twelve quantities $\delta x''$, $\delta y''$, $\delta z''$, $\mathrm{d}\delta x''$, $\mathrm{d}\delta y''$, $\mathrm{d}\delta z''$, $\delta x'$, $\delta y'$, $\delta z'$, $\mathrm{d}\delta x'$, $\mathrm{d}\delta y'$, $\mathrm{d}\delta z'$ must be individually equal to zero.

   From the first integrable equations of Article 48, it is obvious that by beginning the integrations from the first end of the string, the coefficients of $\delta x'$, $\delta y'$, $\delta z'$ are equal to A, B, C, and those of $\delta x''$, $\delta y''$, $\delta z''$ become $A + \mathrm{S}\, X \,\mathrm{d}m$, $B + \mathrm{S}\, Y \,\mathrm{d}m$, $C + \mathrm{S}\, Z \,\mathrm{d}m$. Thus in this case it results that $A = 0$, $B = 0$ and $C = 0$ and also $\mathrm{S}\, X \,\mathrm{d}m = 0$, $\mathrm{S}\, Y \,\mathrm{d}m = 0$, $\mathrm{S}\, Z \,\mathrm{d}m = 0$.

   Then there also results $I'' \, \mathrm{d}^2 x'' = 0$, $I'' \, \mathrm{d}^2 y'' = 0$, $I'' \, \mathrm{d}^2 z'' = 0$, and $I' \, \mathrm{d}^2 x' = 0$, $I' \, \mathrm{d}^2 y' = 0$, $I' \, \mathrm{d}^2 z' = 0$, to eliminate the terms related to $\mathrm{d}\delta x''$, $\mathrm{d}\delta y''$, etc.; and it is clear that the second integrable equations of the same article will give $F = 0, G = 0$, $H = 0$; and $\mathrm{S}(\int X \,\mathrm{d}m.\mathrm{d}y - \int Y \,\mathrm{d}m.\mathrm{d}x) = 0$, $\mathrm{S}(\int X \,\mathrm{d}m.\mathrm{d}z \int Z \,\mathrm{d}m.\mathrm{d}x) = 0$, $\mathrm{S}(\int Y \,\mathrm{d}m.\mathrm{d}z - \int Z \,\mathrm{d}m.\mathrm{d}y) = 0$.

2. If the first end of the string were fixed, then $\delta x' = 0, \delta y' = 0, \delta z' = 0$. Consequently A, B, C will not be equal to zero, but the condition that the coefficients of $\delta x''$, $\delta y''$, $\delta z''$ are equal to zero will give $A = -\mathrm{S}\, X \,\mathrm{d}m$, $B = -\mathrm{S}\, Y \,\mathrm{d}m$, $C = -\mathrm{S}\, Z \,\mathrm{d}m$. And if the position of the tangent to this extremity was also given, it would also result that $\mathrm{d}\delta x' = 0$, $\mathrm{d}\delta y' = 0$, $\mathrm{d}\delta z' = 0$. Consequently, F, G, H would not be equal to zero, but equating to zero the coefficients of $\mathrm{d}\delta x''$, $\mathrm{d}\delta y''$, $\mathrm{d}\delta z''$ would give $F = \mathrm{S}((B + \int Y \,\mathrm{d}m)\mathrm{d}x - (A + \int X \,\mathrm{d}m)\mathrm{d}y)$, $G = \mathrm{S}((C + \int Z \,\mathrm{d}m)\mathrm{d}x - (A + \int X \,\mathrm{d}m)\mathrm{d}z)$, $H = \mathrm{S}((C + \int Z \,\mathrm{d}m)\mathrm{d}y - (B + \int Y \,\mathrm{d}m)\mathrm{d}z)$. The problem will be approached in the same manner with regard to the other end of the string.

3. Finally, if besides the forces which act on every point of the string, there were, in particular, various forces X', Y', Z', X'', Y'', and Z'' applied to both ends, it will only be necessary to add to the terms above the following relation

   $$X' \,\delta x' + Y \,\delta y' + Z \,\delta z' + X'' \,\delta x'' + Y'' \,\delta y'' + Z \,\delta z''$$

   and if there were other conditions relative to these extremities, the problem would be approached in the same fashion and from the same principles.

52. If the string were required to be doubly elastic not only with respect to extensibility but also with respect to flexibility the term $\mathrm{S}\, F\, \mathrm{d}\delta s$, that is, simply F instead of λ, where F represents the force of elasticity which resists the extension of the string (Article 42) will replace the term $\mathrm{S}\, \lambda\, \mathrm{d}\delta s$ in the general equation of equilibrium. The quantity $\mathrm{d}s$ should also be viewed as a variable in this case in the expression for δe. Consequently, one should add to the value of δe of Article 47 these two terms

$$-\frac{e\, \delta \mathrm{d}s}{\mathrm{d}s} - \frac{\mathrm{d}^2 s\, \delta \mathrm{d}^2 s}{e\, \mathrm{d}s^2}$$

Then the terms

$$-\mathrm{S}\frac{Ee}{\mathrm{d}s}\delta \mathrm{d}s - \mathrm{S}\frac{E\, \mathrm{d}^2 s}{e\, \mathrm{d}s^2}\delta \mathrm{d}^2 s$$

would have to be added to the value of $\mathrm{S}\, E\, \delta e$ of the same article. The last term is first reduced to

$$-\frac{E''\, \mathrm{d}^2 s''}{e''\, \mathrm{d}s''^2}\mathrm{d}\delta s'' + \frac{E'\, \mathrm{d}^2 s'}{e'\, \mathrm{d}s'^2}\mathrm{d}\delta s' + \mathrm{S}\, \mathrm{d}\frac{E\, \mathrm{d}^2 s}{e\, \mathrm{d}s^2}\delta \mathrm{d}s$$

Therefore, the terms

$$-\frac{E''\, \mathrm{d}^2 s''}{e''\, \mathrm{d}s''^2}\mathrm{d}\delta s'' + \frac{E'\, \mathrm{d}^2 s'}{e'\, \mathrm{d}s'^2}\mathrm{d}\delta s' + \mathrm{S}(\mathrm{d}\frac{E\, \mathrm{d}^2 s}{e\, \mathrm{d}s^2} - \frac{Ee}{\mathrm{d}s})\delta \mathrm{d}s$$

should be added to the value of $\mathrm{S}\, E\, \delta e$.

Since the last term of this expression is analogous to the term $\mathrm{S}\, F\, \delta \mathrm{d}s$, it will be possible to simplify it in a similar fashion. With respect to the two other terms, $\mathrm{d}\delta s$ should only have to be replaced by its value

$$\frac{\mathrm{d}x\, \mathrm{d}\delta x + \mathrm{d}y\, \mathrm{d}\delta y + \mathrm{d}z\, \mathrm{d}\delta z}{\mathrm{d}s}$$

obtained by marking all the letters by one or two primes.

From this point, it is easy to conclude that for the solution of the present case the same formulas as in the case where the elastic string is assumed inextensible will be obtained by only replacing λ by

$$F + \mathrm{d}\frac{E\, \mathrm{d}^2 s}{e\, \mathrm{d}s^2} - \frac{Ee}{\mathrm{d}s}$$

and by adding to the terms outside the integral sign S, the two following terms

$$\frac{E'\, \mathrm{d}^2 s'}{e'\, \mathrm{d}s'^2}\mathrm{d}\delta s' - \frac{E''\, \mathrm{d}^2 x''}{e''\, \mathrm{d}s''^2}\mathrm{d}\delta s''$$

Since in the equation of the curve the quantity λ must be eliminated, it results that the equation for the elastic strip will be the same whether it is assumed extensible or not. But the tension in the string which is expressed by λ or by F when the string is not elastic (Article 43) will be increased, in view of the elasticity E, by the quantity

$$\mathrm{d}\frac{E\rho\,\mathrm{d}^2s}{\mathrm{d}s^3} - \frac{E}{\rho}$$

with $e = \mathrm{d}s/\rho$ (Article 49).

Subsection IV
The Equilibrium of an Inextensible and Inflexible String of Given Configuration

53. Finally, let us consider the case of an inextensible and inflexible string. The sum of the moments of the forces for this case will produce the same integrable formula as in the case of Article 28, that is, $\mathrm{S}(X\,\delta x + Y\,\delta y + Z\,\delta z)\mathrm{d}m$. Then the condition of the inextensiblity of the string will give as in the same article $\delta \mathrm{d}s = 0$ and the condition of inflexibility gives $\delta e = 0$ because the tangential angle must be invariant. But these two conditions are still not sufficient in the case where the curve has double curvature, as will be seen.

In order to treat the problem in a very simple and direct fashion, I note that everything consists of requiring the different points of the curve of the string to always keep the same distances between themselves. Considering several points with the coordinates x, y, z, $x + \mathrm{d}x$, $y + \mathrm{d}y$, $z + \mathrm{d}z$, $x + 2\,\mathrm{d}x + \mathrm{d}^2x$, $y + 2\,\mathrm{d}y + \mathrm{d}^2y$, $z + 2\,\mathrm{d}z + \mathrm{d}^2z$, etc., it is clear that the squares of the distances between the first of these points and the following ones will be expressed by the quantities

$$\begin{aligned}
&\mathrm{d}x^2 + \mathrm{d}y^2 + \mathrm{d}z^2, \qquad (2\,\mathrm{d}x + \mathrm{d}^2x)^2 + (2\,dy + \mathrm{d}^2y)^2 + (2\,\mathrm{d}z + \mathrm{d}^2z)^2,\\
&(3\,\mathrm{d}x + 3\,\mathrm{d}^2x + \mathrm{d}^3x)^2 + (3\,\mathrm{d}y + 3\,\mathrm{d}^2y + \mathrm{d}^3y)^2 + (3\,\mathrm{d}z + 3\,\mathrm{d}^2z + \mathrm{d}^3z)^2\\
&\vdots
\end{aligned}$$

Let us assume in order to shorten the expressions

$$\begin{aligned}
&\mathrm{d}x^2 + \mathrm{d}y^2 + \mathrm{d}z^2 = \alpha\\
&\mathrm{d}^2x^2 + \mathrm{d}^2y^2 + \mathrm{d}^2z^2 = \beta\\
&\mathrm{d}^3x^2 + \mathrm{d}^3y^2 + \mathrm{d}^3z^2 = \gamma\\
&\vdots
\end{aligned}$$

The preceding quantities after further development will become

$$\alpha$$
$$4\alpha + 2\,\mathrm{d}\alpha + \beta$$
$$9\alpha + 9\,\mathrm{d}\alpha + 9\beta + 3(\mathrm{d}^2\alpha - 2\beta) + 3\,\mathrm{d}\beta + \gamma$$
$$\vdots$$

Therefore, the variations of these quantities must be equal to zero at every point of the curve which will give the following indefinite equations

$$\delta\alpha = 0$$
$$4\,\delta\alpha + 2\,\delta\mathrm{d}\alpha + \delta\beta = 0$$
$$9\,\delta\alpha + 9\,\delta\mathrm{d}\alpha + 3\,\delta\beta + 3\,\delta\mathrm{d}^2\alpha + 3\,\delta\mathrm{d}\beta + \delta\gamma = 0$$
$$\vdots$$

but since $\delta\alpha$ equals zero, there results $\mathrm{d}\delta\alpha = \delta\mathrm{d}\alpha = 0$, and thus $\delta\beta = 0$. Thus it will also result that $\mathrm{d}^2\delta\alpha = \delta\mathrm{d}^2\alpha = 0$, $\mathrm{d}\delta\beta = \delta\mathrm{d}\beta = 0$. Hence $\delta\gamma = 0$, and so on. Therefore, the equations of condition for the inextensibility and inflexibility of the string will be $\delta\alpha = 0$, $\delta\beta = 0$, $\delta\gamma = 0$, etc., that is, by differentiating and replacing $\delta\mathrm{d}$ by $\mathrm{d}\delta$

$$\mathrm{d}x\,\mathrm{d}\delta x + \mathrm{d}y\,\mathrm{d}\delta y + \mathrm{d}z\,\mathrm{d}\delta z = 0$$
$$\mathrm{d}^2x\,\mathrm{d}^2\delta x + \mathrm{d}^2y\,\mathrm{d}^2\delta y + \mathrm{d}^2z\,\mathrm{d}^2\delta z = 0$$
$$\mathrm{d}^3x\,\mathrm{d}^3\delta x + \mathrm{d}^3y\,\mathrm{d}^3\delta y + \mathrm{d}^3z\,\mathrm{d}^3\delta z = 0$$
$$\vdots$$

It is clear that three of these equations are sufficient to determine the three variations δx, δy, δz. Thus, it can be concluded that if the first three conditions are satisfied, all the other conditions that can be found will also be satisfied. The same conclusion could be found using the calculus, which will be shown in Article 60.

54. Thus using our method, the following general equation of equilibrium will be obtained

$$0 = \mathrm{S}(X\,\delta x + Y\,\delta y + Z\,\delta z)\mathrm{d}m + \mathrm{S}\lambda(\mathrm{d}x\,\mathrm{d}\delta x + \mathrm{d}y\,\mathrm{d}\delta y + \mathrm{d}z\,\mathrm{d}\delta z)$$
$$+S\mu(\mathrm{d}^2x\,\mathrm{d}^2\delta x + \mathrm{d}^2y\,\mathrm{d}^2\delta y + \mathrm{d}^2z\,\mathrm{d}^2\delta z) + \mathrm{S}\nu(\mathrm{d}^3x\,\mathrm{d}^3\delta x + \mathrm{d}^3y\,\mathrm{d}^3\delta y + \mathrm{d}^3z\,\mathrm{d}^3\delta z)$$

and which after the transformations described will be reduced to the following form

$$0 = \mathrm{S}(X\,\mathrm{d}m - \mathrm{d}(\lambda\,\mathrm{d}x) + \mathrm{d}^2(\mu\,\mathrm{d}^2x) - \mathrm{d}^3(\nu\,\mathrm{d}^3x))\delta x$$
$$+\mathrm{S}(Y\,\mathrm{d}m - \mathrm{d}(\lambda\,\mathrm{d}y) + \mathrm{d}^2(\mu\,\mathrm{d}^2y) - \mathrm{d}^3(\nu\,\mathrm{d}^3y))\delta y$$
$$+\mathrm{S}(Z\,\mathrm{d}m - \mathrm{d}(\lambda\,\mathrm{d}z) + \mathrm{d}^2(\mu\,\mathrm{d}^2z) - \mathrm{d}^3(\nu\,\mathrm{d}^3z))\delta z$$
$$+(\lambda''\,\mathrm{d}x'' - \mathrm{d}(\mu''\mathrm{d}^2x'') + \mathrm{d}^2(\nu''\mathrm{d}^3x''))\delta x''$$
$$+(\mu''\,\mathrm{d}^2x'' - \mathrm{d}(\nu''\,\mathrm{d}^3x'')\mathrm{d}\delta x'' + \nu''\,\mathrm{d}^3x'')\mathrm{d}^2\delta x''$$

$$
\begin{aligned}
&+(\lambda''\,\mathrm{d}y''-\mathrm{d}(\mu''\,\mathrm{d}^2y'')+\mathrm{d}^2(\nu''\,\mathrm{d}^3y''))\delta y''\\
&+(\mu''\,\mathrm{d}^2y''-\mathrm{d}(\nu''\,\mathrm{d}^3y''))\mathrm{d}\delta y''+\nu''\,\mathrm{d}^3y'')\mathrm{d}^2\delta y''\\
&+(\lambda''\,\mathrm{d}z''-\mathrm{d}(\mu''\,\mathrm{d}^2z'')+\mathrm{d}^2(\nu''\,\mathrm{d}^3z''))\delta z''\\
&+(\mu''\,\mathrm{d}^2z''-\mathrm{d}(\nu''\,\mathrm{d}^3z'')\mathrm{d}\delta z+\nu''\,\mathrm{d}^3z'')\mathrm{d}^2\delta z''\\
&-(\lambda'\,\mathrm{d}x'-\mathrm{d}(\mu'\,\mathrm{d}^2x')+\mathrm{d}^2(\nu'\mathrm{d}^3x')\delta x'\\
&-(\mu'\,\mathrm{d}^2x'-\mathrm{d}(\nu'\,\mathrm{d}^3x'))\mathrm{d}\delta x'-\nu'\mathrm{d}^3x')\mathrm{d}^2\delta x'\\
&-(\lambda'\,\mathrm{d}y'-\mathrm{d}(\mu'\,\mathrm{d}^2y')+\mathrm{d}^2(\nu'\,\mathrm{d}^3y'))\delta y'\\
&-(\mu'\,\mathrm{d}^2y'-\mathrm{d}(\nu'-\mathrm{d}^3y')\mathrm{d}\delta y'-\nu'\,\mathrm{d}^3y')\mathrm{d}^2\delta y'\\
&-(\lambda'\,\mathrm{d}z'-\mathrm{d}(\mu'\,\mathrm{d}^2z')+\mathrm{d}^2(\nu'\,\mathrm{d}^3z'))\delta z'\\
&-(\mu'\,\mathrm{d}^2z'-\mathrm{d}(\nu'\,\mathrm{d}^3z')\mathrm{d}\delta z'-\nu'\,\mathrm{d}^2z')\mathrm{d}^2\delta z'
\end{aligned}
$$

55. At the outset equate to zero the coefficients of δx, δy, δz under the integral sign S then the following three indefinite equations will result

$$
\begin{aligned}
&X\,\mathrm{d}m-\mathrm{d}(\lambda\,\mathrm{d}x)+\mathrm{d}^2(\mu\,\mathrm{d}^2x)-\mathrm{d}^3(\nu\,\mathrm{d}^3x)=0\\
&Y\,\mathrm{d}m-\mathrm{d}(\lambda\,\mathrm{d}y)+\mathrm{d}^2(\mu\,\mathrm{d}^2y)-\mathrm{d}^3(\nu\,\mathrm{d}^3y)=0\\
&Z\,\mathrm{d}m-\mathrm{d}(\lambda\,\mathrm{d}z)+\mathrm{d}^2(\mu\,\mathrm{d}^2z)-\mathrm{d}^3(\nu\,\mathrm{d}^3z)=0
\end{aligned}
$$

which, containing three undetermined variables λ, μ, ν, will only be used to determine these three quantities, so that there will be no indefinite equation between the different forces X, Y, Z which are assumed applied at every point of the rod and furthermore, the conditions of equilibrium will only depend upon the terms which are outside the integral sign S. But, since these terms contain the unknowns λ, μ, ν, the analysis should begin by determining these unknowns.

In order to achieve this result, the preceding equations must be integrated which is straightforward and from which the following three equations will be obtained

$$
\begin{aligned}
&\int X\,\mathrm{d}m-\lambda\,\mathrm{d}x+\mathrm{d}(\mu\,\mathrm{d}^2x)-\mathrm{d}^2(\nu\,\mathrm{d}^3x)=A\\
&\int Y\,\mathrm{d}m-\lambda\,\mathrm{d}y+\mathrm{d}(\mu\,\mathrm{d}^2y)-\mathrm{d}^2(\nu\,\mathrm{d}^3y)=B\\
&\int Z\,\mathrm{d}m-\lambda\,\mathrm{d}z+\mathrm{d}(\mu\,\mathrm{d}^2z)-\mathrm{d}^2(\nu\,\mathrm{d}^3z)=C
\end{aligned}
$$

where A, B, C are three arbitrary constants.

These equations give, after elimination of λ, the following three equations

$$
\begin{aligned}
&\mathrm{d}y\int X\,\mathrm{d}m-\mathrm{d}x\int Y\,\mathrm{d}m+\mathrm{d}y\,\mathrm{d}(\mu\,\mathrm{d}^2x)-\mathrm{d}x\,\mathrm{d}(\mu\,\mathrm{d}^2y)\\
&\quad-\mathrm{d}y\,\mathrm{d}^2(\nu\,\mathrm{d}^3x)+\mathrm{d}x\,\mathrm{d}^2(\nu\,\mathrm{d}^3y)=A\,\mathrm{d}y-B\,\mathrm{d}x\\
&\mathrm{d}z\int X\,\mathrm{d}m-\mathrm{d}x\int Z\,\mathrm{d}m+\mathrm{d}z\,\mathrm{d}(\mu\,\mathrm{d}^2x)-\mathrm{d}x\,\mathrm{d}(\mu\,\mathrm{d}^2z)
\end{aligned}
$$

$$-\mathrm{d}z\,\mathrm{d}^2(\nu\,\mathrm{d}^3x)+\mathrm{d}x\,\mathrm{d}^2(\nu\,\mathrm{d}^3z)=A\,\mathrm{d}z-C\,\mathrm{d}x$$

$$\mathrm{d}z\int Y\,\mathrm{d}m-\mathrm{d}y\int Z\,\mathrm{d}m+\mathrm{d}z\,\mathrm{d}(\mu\,\mathrm{d}^2y)-\mathrm{d}y\,\mathrm{d}(\mu\,\mathrm{d}^2z)$$
$$-\mathrm{d}z\,\mathrm{d}^2(\nu\,\mathrm{d}^3y)+\mathrm{d}y\,\mathrm{d}^2(\nu\,\mathrm{d}^3z)=B\,\mathrm{d}z-C\,\mathrm{d}y$$

which are also integrable. Their integrals are

$$y\int X\,\mathrm{d}m-x\int Y\,\mathrm{d}m-\int(Xy-Yx)\,\mathrm{d}m$$
$$+\mu(\mathrm{d}y\,\mathrm{d}^2x-\mathrm{d}x\,\mathrm{d}^2y)-\mathrm{d}y\,\mathrm{d}(\nu\,\mathrm{d}^3x)+\mathrm{d}x\,\mathrm{d}(\nu\,\mathrm{d}^3y)$$
$$+\nu(\mathrm{d}^2y\,\mathrm{d}^3x-\mathrm{d}^2x\,\mathrm{d}^3y)=Ay-Bx+F$$

$$z\int X\,\mathrm{d}m-x\int Z\,\mathrm{d}m-\int(Xz-Zx)\,\mathrm{d}m$$
$$+\mu(\mathrm{d}z\,\mathrm{d}^2x-\mathrm{d}x\,\mathrm{d}^2z)-\mathrm{d}z\,\mathrm{d}(\nu\,\mathrm{d}^3x)+\mathrm{d}x\,\mathrm{d}(\nu\,\mathrm{d}^3z)$$
$$+\nu(\mathrm{d}^2z\,\mathrm{d}^3x-\mathrm{d}^2x\,\mathrm{d}^3z)=Az-Cx+G$$

$$z\int Y\,\mathrm{d}m-y\int Z\,\mathrm{d}m-\int(Yz-Zy)\,\mathrm{d}m$$
$$+\mu(\mathrm{d}z\,\mathrm{d}^2y-\mathrm{d}y\,\mathrm{d}^2z)-\mathrm{d}z\,\mathrm{d}(\nu\,\mathrm{d}^3y)+\mathrm{d}y\,\mathrm{d}(\nu\,\mathrm{d}^3z)$$
$$+\nu(\mathrm{d}^2z\,\mathrm{d}^3y-\mathrm{d}^2y\,\mathrm{d}^3z)=Bz-Cy+H$$

where F, G, H are new arbitrary constants.

The last three equations will be used to determine the three quantities μ, ν and $\mathrm{d}\nu$. The first three integrable equations will give the values of λ, $\mathrm{d}\mu$, and $\mathrm{d}^2\nu$. Thus the values for all the unknowns which are contained in the terms outside the integral sign S will be found. It will suffice for this purpose to mark all the letters in the sixth equation found above by one or two primes with the exception of the arbitrary constants and to assume equal to zero in the first case the quantities affected by the integral sign $\int$ which are assumed to begin at the first end of the string and to change in the second case $\int$ by S in the same quantities in order to refer them to the other end of the string.

56. Having made these assumptions, let us see now how the conditions can result from the elimination of the terms outside the integral sign S in the general equation of equilibrium (Article 54).

At the outset, if the rod were assumed to be entirely free, the variations $\delta x'$, $\delta y'$, $\delta z'$, $\mathrm{d}\delta x'$, $\mathrm{d}\delta y'$, $\mathrm{d}\delta z'$, $\mathrm{d}^2\delta x'$, $\mathrm{d}^2\delta y'$, $\mathrm{d}^2\delta z'$, $\delta x''$, $\delta y''$, $\delta z''$, $\mathrm{d}\delta x''$, etc. will all be indeterminate. Consequently, each of their coefficients should always be equated to zero. It is obvious that for this purpose, the quantities λ', μ', ν', $\mathrm{d}\mu'$, $\mathrm{d}\nu'$, $\mathrm{d}^2\nu'$, as well as λ'', μ'', ν'', $\mathrm{d}\mu''$, $\mathrm{d}\nu''$, $\mathrm{d}^2\nu''$ are equal to zero.

Therefore, the first three integrable equations of the preceding article when referred to the first and last extremities of the string will give these six conditions

$$A=0,\quad B=0,\quad C=0,\quad \mathrm{S}X\,\mathrm{d}m=A,\quad \mathrm{S}Y\,\mathrm{d}m=B,\quad \mathrm{S}Z\,\mathrm{d}m=C$$

The last three integrals will give similarly the next six conditions

$$0 = Ay' - Bx' + F$$
$$0 = Az' - Cx' + G$$
$$0 = Bz' - Cy' + H$$
$$y''\text{S}\,X\,\text{d}m - x''\text{S}\,Y\,\text{d}m - \text{S}(Xy - Yx)\text{d}m = Ay'' - Bx'' + F$$
$$z''\text{S}\,X\,\text{d}m - x''\text{S}\,Z\,\text{d}m - \text{S}(Xz - Zx)\text{d}m = Az'' - Cx'' + G$$
$$z''\text{S}\,Y\,\text{d}m - y''\text{S}\,Z\,\text{d}m - \text{S}(Yz - Zy)\text{d}m = Bz'' - Cy'' + H$$

Therefore, $A = 0$, $B = 0$, $C = 0$, $F = 0$, $G = 0$, $H = 0$, and consequently

$$\text{S}\,X\,\text{d}m = 0, \qquad \text{S}\,Y\,\text{d}m = 0, \qquad \text{S}\,Z\,\text{d}m = 0,$$
$$\text{S}(Yy - Yx)\text{d}m = 0, \qquad \text{S}(Xz - Zx)\text{d}m = 0, \qquad \text{S}(Yz - Zy)\text{d}m = 0$$

These six conditions are the sole requirements which are necessary for the equilibrium of an inflexible rod when there is no fixed point. This is in accordance with what we said earlier (Article 25) and this is also what could have been deduced immediately from the theory given in SECTION III as we observed in the cited article.

57. Let us now assume that there is a fixed point in the rod and that this point is one of the ends of the rod. In this case, there results $\delta x' = 0$, $\delta y' = 0$, $\delta z' = 0$ so that the terms affected by these variations will disappear. Thus it will suffice to equate to zero the coefficients of $\text{d}\delta x'$, $\text{d}\delta y'$, $\text{d}\delta z'$, $\text{d}^2\delta x'$, $\text{d}^2\delta y'$, $\text{d}^2\delta z'$, as well as the coefficients of $\delta x''$, $\delta y''$, $\delta z''$, $\text{d}\delta x''$, $\text{d}\delta y''$, etc.

But it is easy to see that for this purpose it will suffice to have $\mu' = 0$, $\nu' = 0$, $\text{d}\nu' = 0$, and then $\lambda'' = 0$, $\mu'' = 0$, $\nu'' = 0$, $\text{d}\mu'' = 0$, $\text{d}\nu'' = 0$, $\text{d}^2\nu'' = 0$, as in the preceding case. And the same conditions as in the preceding article will be found with the exception that A, B, C will not be equal to zero.

Thus there will result $A = \text{S}\,X\,\text{d}m$, $B = \text{S}\,Y\,\text{d}m$, $C = \text{S}\,Z\,\text{d}m$, and then $F = Bx' - Ay'$, $G = Cx' - Az'$, $H = Cy' - Bz'$ and the other three equations will be reduced to the following

$$-\text{S}(Xy - Yx)\text{d}m = Bx' - Ay'$$
$$-\text{S}(Xz - Zx)\text{d}m = Cx' - Az'$$
$$-\text{S}(Yz - Zy)\text{d}m = Cy' - Bz'$$

that is, to the following equations

$$\text{S}(Xy - Yx)\text{d}m + x'\text{S}\,Y\,\text{d}m - y'\text{S}\,X\,\text{d}m = 0$$
$$\text{S}(Xz - Zx)\text{d}m + x'\text{S}\,Z\,\text{d}m - z'\text{S}\,X\,\text{d}m = 0$$
$$\text{S}(Yz - Zy)\text{d}m + y'\text{S}\,Z\,\text{d}m - z'\text{S}\,Y\,\text{d}m = 0$$

or, which is the same, to these equations

$$\mathrm{S}(X(y-y')-Y(x-x'))\mathrm{d}m=0$$
$$\mathrm{S}(X(z-z')-Z(x-x'))\mathrm{d}m=0$$
$$\mathrm{S}(Y(z-z')-Z(y-y'))\mathrm{d}m=0$$

These are the only conditions necessary for equilibrium and it is clear that they are equivalent to those found in Article 24.

58. If the rod were firmly attached by its first end so that not only the first point of the curve is fixed but also the tangent at this point, then it will not only result that $\delta x'=0$, $\delta y'=0$, $\delta z'=0$, but also that $\delta \mathrm{d}x'=\mathrm{d}\delta x'=0$, $\delta \mathrm{d}y'=\mathrm{d}\delta y'=0$, $\delta \mathrm{d}z'=\mathrm{d}\delta z'=0$. Consequently, all the terms affected by these quantities will disappear identically and only the terms affected by $\mathrm{d}^2\delta x'$, $\mathrm{d}^2\delta y'$, $\mathrm{d}^2\delta z'$ and by $\delta x''$, $\delta y''$, $\delta z''$, $\mathrm{d}\delta x''$, $\mathrm{d}\delta y''$, etc. will remain to be eliminated.

Thus in this case only these conditions will result

$$\nu'=0,\qquad \lambda''=0,\qquad \mu''=0,\qquad \nu''=0,$$
$$\mathrm{d}\mu''=0,\qquad \mathrm{d}\nu''=0,\qquad \mathrm{d}^2\nu''=0$$

and the constants A, B, C will retain the values

$$A=\mathrm{S}\,X\,\mathrm{d}m,\qquad B=\mathrm{S}\,Y\,\mathrm{d}m,\qquad C=\mathrm{S}\,Z\,\mathrm{d}m$$

The last three integrals of Article 55 when applied to the endpoint of the rod will give

$$F=\mathrm{S}(Yx-Xy)\mathrm{d}m,\qquad G=\mathrm{S}(Zx-Xz)\mathrm{d}m,\qquad H=\mathrm{S}(Zy-Yz)\mathrm{d}m$$

And if these same equations were applied to the other end, there would result

$$\mu'(\mathrm{d}y'\,\mathrm{d}^2\mathrm{d}x'-\mathrm{d}x'\,\mathrm{d}^2\mathrm{d}y')-\mathrm{d}\nu'(\mathrm{d}y'\,\mathrm{d}^3x'-\mathrm{d}x'\mathrm{d}^3y')=Ay'-Bx'+F$$
$$\mu'(\mathrm{d}z'\,\mathrm{d}^2\mathrm{d}x'-\mathrm{d}x'\,\mathrm{d}^2\mathrm{d}z')-\mathrm{d}\nu'(\mathrm{d}z'\,\mathrm{d}^3x'-\mathrm{d}x'\mathrm{d}^3z')=Az'-Cx'+G$$
$$\mu'(\mathrm{d}z'\,\mathrm{d}^2\mathrm{d}y'-\mathrm{d}y'\,\mathrm{d}^2\mathrm{d}z')-\mathrm{d}\nu'(\mathrm{d}z'\,\mathrm{d}^3y'-\mathrm{d}y'\mathrm{d}^3z')=Bz'-Cy'+H$$

from which after eliminating μ' and $\mathrm{d}\nu'$, the following equation will be obtained

$$A(y'\,\mathrm{d}z'-z'\,\mathrm{d}y')+B(z'\,\mathrm{d}x'-x'\,\mathrm{d}z')+C(x'\,\mathrm{d}y'-y'\,\mathrm{d}x')$$
$$+F\,\mathrm{d}z'-G\,\mathrm{d}y'-H\,\mathrm{d}x'=0$$

This equation is necessary to prevent the rod from rotating about the tangent to the first endpoint which is assumed fixed and it is easily seen that its first term will be equal to zero when the rod is rectilinear.

59. The great length of this solution using our method could be viewed as a shortcoming. Using ordinary methods, this last problem is much more difficult than the one for equilibrium of a rigid rod acted on by arbitrary forces, because the configuration that the string

must assume in order to be in equilibrium must be determined by the composition of forces while this configuration is given in the case of the rod and the equilibrium requires only that the moments of these forces be equal to zero. But if a uniform procedure is desired for all these problems with the capability to go from one problem to the other easily, as more conditions are added, it is evident that the case of an inflexible string is less simple than the case of a flexible string because the inflexibility expressed analytically requires the invariability of the respective distances between the points of the string. In this case if the curve were given, it must no longer be a result of calculation, as it is in the case of the flexible string. This is a circumstance that the analysis must reflect and which indeed it does with the three indeterminate quantities λ, μ, ν, which remain in the three indefinite equations between x, y, z of Article 55 and which have a form that is adaptable to any given curve. Therefore, these equations should not be viewed as superfluous. In addition, they are used to determine the three unknowns λ, μ, ν, on which the conditions of equilibrium depend and which express at the same time the forces which oppose the variations of the three functions α, β, γ due to the effect of the forces acting on the string.

It is true that the three indeterminates λ, μ, ν must be replaced by the three equations of condition which are founded on what the differential functions α, β, γ are assumed to express. But because of the nature of the differential calculus, the absolute values of the differentials remain indeterminate and it is only the ratio which can be given. These three conditions can only be equivalent to two, which contain the ratios of the three quantities α, β, γ and these two ratios are sufficient to determine the curve.

Indeed, based on what was demonstrated above (Article 46), it is seen that the angle of tangency made by two successive sides of the curve is expressed by $\sqrt{4\alpha\beta - \mathrm{d}\alpha^2}/(2\alpha)$ keeping the values of α, β, γ of Article 53 so that the osculating radius will be expressed by $2\alpha\sqrt{\alpha}/\sqrt{4\alpha\beta - \mathrm{d}\alpha^2}$.

If this radius is assumed given, the resulting curve has single curvature and for the curves with double curvature, it will not be difficult to prove that the second curvature coming from the angle of tangency made by the planes which successively pass by two successive elements of the curve will depend on the ratio of the three quantities α, β, γ. Therefore these three conditions with respect to the curve are reduced to the condition that the curve must be given, as was assumed.

The analysis of this problem could be expanded to the case of a surface or of a solid for which all the points would be acted on by arbitrary forces. But we are going to show how the analysis can be simplified beginning with the same equations of condition and by determining in advance from these equations the expressions for the variations of the coordinates.

Chapter IV
THE EQUILIBRIUM OF A SOLID BODY OF FINITE DIMENSION AND ARBITRARY CONFIGURATION WHERE ALL POINTS ARE LOADED BY ARBITRARY FORCES

60. Since the condition of rigidity for a body requires that all the points of this body retain the same relative position and the same distances apart, the same equations of condition as those which were found in Article 55 will be obtained between the variations δx, δy, δz, because it is obvious that by imagining an arbitrary curve within the body, it will suffice that all its points retain the same relative distances to each other whatever motion the body receives. Thus by this means, the values of these variations could immediately be determined.

To this effect, I note that by passing to the differences of the second order, it is always permissible to take one of the differences of first order as constant. Thus $\mathrm{d}x$ can be assumed constant and consequently, $\mathrm{d}^2x = 0$, $\mathrm{d}^3x = 0$, etc. so that the second and third equation of the cited article will become

$$\mathrm{d}^2y\,\mathrm{d}^2\delta y + \mathrm{d}^2z\,\mathrm{d}^2\delta z = 0, \qquad \mathrm{d}^3y\,\mathrm{d}^3\delta y + \mathrm{d}^3z\,\mathrm{d}^3\delta z = 0$$

The first of these equations gives immediately

$$\mathrm{d}^2\delta y = -\frac{\mathrm{d}^2z}{\mathrm{d}^2y}\mathrm{d}^2\delta z$$

and after differentiation

$$\mathrm{d}^3\delta y = -\frac{\mathrm{d}^2z}{\mathrm{d}^2y}\mathrm{d}^3\delta z - \left(\frac{\mathrm{d}^3z}{\mathrm{d}^2y} - \frac{\mathrm{d}^2z\,\mathrm{d}^3y}{\mathrm{d}^2y^2}\right)\mathrm{d}^2\delta z$$

After this expression is substituted in the second equation, it will be found to be divisible by $\mathrm{d}^3z - (\mathrm{d}^3y\,\mathrm{d}^2z/\mathrm{d}^2y)$ and after the division the following equation will result

$$\mathrm{d}^3\delta z - \frac{\mathrm{d}^3y}{\mathrm{d}^2y}\mathrm{d}^2\delta z = 0$$

From this last equation after integration it will result that $\mathrm{d}^2\delta z = \delta L\,\mathrm{d}^2y$, where δL is a constant. Having calculated $\mathrm{d}^2\delta z$ it will be found that $\mathrm{d}^2\delta y = -\delta L\,\mathrm{d}^2z$. Then integrating again and adding the constants $-\delta M\,\mathrm{d}x$ and $\delta N\,\mathrm{d}x$, there will result $\mathrm{d}\delta z = \delta L\,\mathrm{d}y - \delta M\,\mathrm{d}x$, $\mathrm{d}\delta y = -\delta L\mathrm{d}z + \delta N\,\mathrm{d}x$, and these values after substitution in the first equation of condition, that is, $\mathrm{d}x\,\mathrm{d}\delta x + \mathrm{d}y\,\mathrm{d}\delta y + \mathrm{d}z\,\mathrm{d}\delta z = 0$, will produce the following equation: $\mathrm{d}\delta x = -\delta N\,\mathrm{d}y + \delta M\,\mathrm{d}z$.

Finally after a third integration and by addition of the new constants $\delta\ell$, δm, δn, the following equations will be obtained

$$\delta x = \delta\ell - y\delta N + z\delta M, \qquad \delta y = \delta m + x\delta N - z\delta L, \qquad \delta z = \delta n - x\delta M + y\delta L$$

It is easy to be convinced that these expressions not only satisfy the first three equations of condition of Article 53, but also all the others that could be found and which are all contained in this general equation

$$\mathrm{d}^n x\,\mathrm{d}^n\delta x + \mathrm{d}^n y\,\mathrm{d}^n\delta y + \mathrm{d}^n z\,\mathrm{d}^n\delta z = 0$$

These are the values of δx, δy, δz for an arbitrary system of points connected in such a manner that they always keep the same distances between themselves. Thus these values will not only be used in the case of an arbitrary mobile and invariable curve in the body, but also for the case of a solid body with an arbitrary shape.

Euler was the first to find these simple and elegant formulas which express the variations of the coordinates of all the points of a solid mobile body in space. He arrived at these formulas by considerations drawn from the differential calculus but different from those that we have used and it seems to me less rigorous. Refer to the memoir in the volume of the Académie de Berlin for 1750 entitled Dècouverte d'un nouveau principe de mécanique.[93]

61. Since the preceding values of δx, δy, δz already satisfy the equations of condition of the problem, it is clear that it will be sufficient to substitute them in the formula $\mathrm{S}(X\,\delta x + Y\,\delta y + Z\,\mathrm{d}z)\mathrm{d}m$, and to equate this latter equation to zero independent of the quantities $\delta\ell$, δm, δn, δL, δM, δN, which are the only indeterminates remaining.

But since these quantities are the same at every point of the body, they should be put outside of the integral sign S in making the substitution. Consequently, this general equation of equilibrium for a solid body of arbitrary shape will be obtained

$$\begin{aligned}&\delta\ell\,\mathrm{S}\,X\,\mathrm{d}m + \delta m\,\mathrm{S}\,Y\,\mathrm{d}m + \delta n\,\mathrm{S}\,Z\,\mathrm{d}m\\&+\delta N\,\mathrm{S}(Yx - Xy)\mathrm{d}m + \delta M\,\mathrm{S}(Xz - Zx)\mathrm{d}m\\&+\delta L\,\mathrm{S}(Zy - Yz)\mathrm{d}m = 0\end{aligned}$$

from which the particular equations of equilibrium for the particular conditions of this problem will result.

62. At the outset, if the body were assumed entirely free, the six variations $\delta\ell$, δm, δn, δL, δM, δN will all be indeterminate and the quantities by which they are multiplied should also separately be equated to zero, which will give these six equations which are already known

$$\begin{aligned}&\mathrm{S}\,X\,\mathrm{d}m = 0, \qquad \mathrm{S}\,Y\,\mathrm{d}m = 0, \qquad \mathrm{S}\,Z\,\mathrm{d}m = 0,\\&\mathrm{S}(Yx - Xy)\,\mathrm{d}m = 0, \qquad \mathrm{S}(Xz - Zx)\,\mathrm{d}m = 0, \qquad \mathrm{S}(Zy - Yz)\,\mathrm{d}m = 0\end{aligned}$$

In the second place, if there were in the body a fixed point about which it has only the freedom of rotation in any direction and if the values of the coordinates x, y, z for this fixed point were called a, b, c it will result that $\delta a = 0$, $\delta b = 0$, $\delta c = 0$. Therefore

$$\delta\ell - b\,\delta N + c\,\delta M = 0, \qquad \delta m + a\,\delta N - c\,\delta L = 0, \qquad \delta n - a\,\delta M + b\,\delta L = 0$$

from which one obtains

$$\delta\ell = b\,\delta N - c\,\delta M, \qquad \delta m = c\,\delta L - a\,\delta N, \qquad \delta n = a\,\delta M - b\,\delta L$$

These expressions will be substituted in the general equation of the preceding article and putting the quantities a, b, c which are constants with respect to the different points of the body under the integral sign S, the following transformation will result

$$\begin{aligned}&\delta N\,\mathrm{S}(Y(x-a) - X(y-b))\mathrm{d}m\\ &+\delta M\,\mathrm{S}(X(z-c) - Z(x-a))\mathrm{d}m\\ &+\delta L\,\mathrm{S}(Z(y-b) - Y(z-c))\mathrm{d}m = 0\end{aligned}$$

which will give only three additional equations, that is

$$\begin{aligned}&\mathrm{S}(Y(x-a) - X(y-b))\mathrm{d}m = 0\\ &\mathrm{S}(X(z-c) - Z(x-a))\mathrm{d}m = 0\\ &\mathrm{S}(Z(y-b) - Y(z-c))\mathrm{d}m = 0\end{aligned}$$

In the third place, if there were two fixed points in the body and if f, g, h were the values of x, y, z for the second of these points, there will also result

$$\delta\ell = g\,\delta N - h\,\delta M, \qquad \delta m = h\,\delta L - f\,\delta N, \qquad \delta n = f\,\delta M - g\,\delta L$$

Thus after subtracting these expressions for $\delta\ell$, δm, δn from the preceding expressions, one will have

$$\begin{aligned}&(g-b)\delta N - (h-c)\delta M = 0, \qquad (f-a)\delta N - (h-c)\delta L = 0,\\ &(f-a)\delta M - (g-b)\delta L = 0\end{aligned}$$

The first two of these equations give

$$\delta L = \frac{f-a}{h-c}\delta N, \qquad \delta M = \frac{g-b}{h-c}\delta N$$

and because these expressions also satisfy the third equation, the variation of δN will remain indeterminate.

After making these substitutions in the transformed equations above, the following equation will result

$$\begin{aligned}&\delta N[(h-c)\mathrm{S}(Y(x-a)-X(y-b))\mathrm{d}m\\&+(g-b)\mathrm{S}(X(z-c)-Z(x-a))\mathrm{d}m\\&+(f-a)\mathrm{S}(Z(y-b)-Y(x-c))\mathrm{d}m]=0\end{aligned}$$

Thus the conditions of equilibrium will be contained in this single equation

$$\begin{aligned}&[(h-c)\mathrm{S}(Y(x-a)-X(y-b))\mathrm{d}m\\&+(g-b)\mathrm{S}(X(z-c)-Z(x-a))\mathrm{d}m\\&+(f-a)\mathrm{S}(Z(y-b)-Y(z-c))\mathrm{d}m]=0\end{aligned}$$

63. These diverse equations are related to those which we have given in SECTION III for the equilibrium of a system of isolated points and of invariant configuration. We could have immediately applied the conditions of this equilibrium to a solid body of arbitrary configuration for which all the points are acted on by given forces. But we believed that it was useful for the purpose of showing the fecundity of our methods to treat this last problem specifically and without using anything from problems already solved.

Also, if the two points of the body we assumed fixed were mobile on given lines or surfaces or even joined together in an arbitrary fashion, then one or several differential equations between the variations of the coordinates $a, b, c,\ f, g, h$ of these points will result. Substituting in place of these variations their values in $\delta\ell$, δm, δn, δL, δM, δN, from the general formulas of Article 58, there will be as many equations between these last variations by means of which some of these variations could be determined from the others. One would then substitute these values in the general equation and equate to zero each of the coefficients of the remaining variations which will give all the necessary equations for equilibrium.

The process of calculation is, as is observed, always the same. This fact should be viewed as one of the main advantages of this method.

64. The expressions found above (Article 60) for the variations δx, δy, δz show that these variations are nothing more than the motions of translation and rotation which we considered specifically in SECTION III.

Indeed, it is obvious that the terms $\delta\ell$, δm, δn which are the same for all the points of the body represent the small distances traversed by the body in the directions of the x, y, z coordinate axes by virtue of an arbitrary motion of translation. From the formulas of Article 8 of the same section it can be seen that the terms $z\,\delta M - y\,\delta N$, $x\,\delta N - z\,\delta L$, $y\,\delta L - x\,\delta M$ represent the small spaces traversed by each point of the body in these same directions by virtue of the three motions of rotation $\delta L, \delta M, \delta N$ about the three coordinate axes of x, y, z. These quantities δL, δM, δN, are associated with the quantities $\mathrm{d}\psi$, $\mathrm{d}\omega$, $\mathrm{d}\varphi$ of the cited article. Thus these expressions could have been immediately deduced from the consideration of these motions which would have been simpler but not as direct. The preceding analysis leads naturally to these expressions and proves then by a more direct and general approach than that of Article 10 of SECTION III that when the various points of a system retain their respective positions, the system can have at any instant only translational motion in space and rotation about three orthogonal axes.

SECTION VI
THE PRINCIPLES OF HYDROSTATICS

Although we have ignored the internal structure of fluids, we cannot doubt that the particles which compose them are material and that for this reason the general laws of equilibrium are as applicable to them as they are to solid bodies. Indeed, the principal property of fluids and the only one which distinguishes them from solid bodies is that all their parts have no resistance against the smallest force and can move among themselves with all possible facility, whatever their mutual action and connection. This property is easily modelled by the calculus and it follows that the laws of equilibrium for fluids do not require a separate theory but that they are only a particular case of the general theory of statics. It is from this point of view that we will consider them. However, we believe that we must begin by presenting the different principles which have been used in the past in this part of statics, which is commonly called hydrostatics, in order to complete the analysis of the principles of statics which we gave in SECTION I.

1. Again it is from Archimedes that we have the first principles for the equilibrium of fluids. His treatise *De insidentibus humido*[1] did not come down to us in the original Greek. There was only a rather imperfect Latin translation published by Tartaglia which Commandino undertook to restore and clarify with notes. This treatise was published under the direction of this learned commentator in 1565 with the title *De iis quae vehuntur in acqua*.[2]

This work, which can be viewed as one of the most precious relics of Antiquity, is divided into two books. In the first book, Archimedes formulates two principles which he considers drawn from experience and on which he bases his entire theory. The first is that the nature of fluids is such that the portions compressed are displaced by those which are more compressed and that each portion is always compressed by the weight of the column standing vertically above it. The second is that everything which is pushed upward by a fluid is always pushed along the perpendicular which passes through its center of gravity.

From the first principle, Archimedes easily demonstrates that the surface of a fluid, for which all the parts are assumed attracted toward the center of the Earth, must be spheroidal for the fluid to be in equilibrium. Then he demonstrates that a body as heavy as an equal volume of fluid must sink entirely in it because by considering two equal pyramids of the fluid assumed in equilibrium about the center of the Earth, the pyramid in which the body would be partly submerged would exert a greater pressure on the center of the Earth than the other pyramid or in general on an arbitrary spherical surface which could be imagined about this center.[3] In the same fashion, he proves that bodies lighter than an equal volume of fluid will sink in it until the submerged part displaces a volume of fluid as heavy as the entire body. From these demonstrations, he deduces the following two theorems of hydrostatics: bodies which are lighter than an equal volume of fluid are pushed upward by a force equal to the amount by which the weight of the displaced fluid exceeds the weight of the immersed body and heavier bodies lose a part of their weight equal to that of the displaced fluid.

Thereafter, Archimedes uses his second principle to establish the laws of equilibrium for floating bodies. He demonstrates that all submerged portions of a hemisphere lighter than an equal volume of fluid must necessarily assume an orientation for which the base is horizontal. His demonstration consists of showing that if the base were inclined, the total weight of the body considered concentrated at its center of gravity and the buoyant force of the fluid considered concentrated at the center of gravity of the submerged portion would always tend to rotate the body until its base became horizontal.

These are the goals of the first book. In the second book, Archimedes formulates, from the same principles, the laws of equilibrium for various solids formed by the revolution of a conic section and submerged in fluids of greater density than these bodies. He examines those cases where the conoids can remain inclined, those where they must remain horizontal and those where they must rollover or right themselves. This book is one of the most beautiful monuments to the genius of Archimedes and it contains a theory for the stability of floating bodies to which modern investigators have added little.

2. Although the pressure of a fluid on the bottom or on the walls of a vessel in which it is contained is not difficult to determine from the demonstrations of Archimedes, it is not until Stevin that this research is carried out and the paradox of hydrostatics demonstrated, that is, that a fluid can exert a pressure much greater than its own weight.[4] Stevin's theory of hydrostatics is in the third volume of the *Hypomnemata Mathematica*,[5] which was translated from Flemish by Snellius and published at Leyden in 1608. After having proved that a solid body of arbitrary shape with the same [specific] gravity as water can remain submerged at an arbitrary depth because it occupies the same volume and weighs as much as if it were water, Stevin imagines a rectangular vessel filled with water and he shows easily that its bottom must bear the entire weight of the water contained in the vessel. He then assumes that a solid body of arbitrary shape and of the same [specific] gravity as water is placed in this vessel. It is clear that the pressure will remain the same so that if a shape is assumed for the immersed solid such that it remains a column of fluid of arbitrary configuration, the pressure of the column on the base will still be the same and consequently, it will be equal to the weight of a vertical column of water which would have this same base. But Stevin observes that by assuming this solid firmly fixed in place, it cannot result in any change in the action of the water on the base of the vessel. Therefore, the pressure on this base will always be equal to the weight of the same column of water, whatever the shape of the vessel.

Stevin proceeds from there to determine the pressure of the water on the vertical or inclined walls of the vessel. He divides their surface by horizontal lines into several small parts and he shows that the pressure on each part is greater than if it were horizontal and at the elevation of its upper edge, but that at the same time the pressure is less than if it were positioned horizontally at the elevation of its lower edge. Thus by diminishing the width of the parts and by infinitely increasing their number, he proves by a limiting process that the pressure on an inclined plane, is equal to the weight of a column of fluid for which this plane will be the base and for which the height will be half the height of the vessel.

He then determines the pressure on an arbitrary part of an inclined plane and discovers that it is equal to the weight of a column of water which would be formed by applying perpendicularly to each point of this part straight lines equal to the depth of the point under water. Now that this theorem has been demonstrated for plane surfaces positioned arbitrarily, it is easy to extend it to curved surfaces and to conclude that the pressure exerted by a heavy fluid against an arbitrary surface has for its measure the weight of a fluid column which would have for a base this same surface, if necessary, converted to a plane surface and for which the heights with respect to the different points of the base would be the same as the distances from the corresponding points of the surface to the fluid level or which is the same, this pressure will be measured by the weight of a column which would have for a base the weighted surface and for a height the vertical distance from the center of gravity of this weighted surface to the top surface of the fluid.

3. The preceding theories of equilibrium and of the pressure of fluids are, as is observed, entirely independent of the general principles of statics because they are founded only on principles of experience peculiar to fluids. Hence, the approach to demonstrating the laws of hydrostatics, by deducing from empirical knowledge of some of these laws, one from all the others, has been adopted by most modern authors and this approach has made hydrostatics a science entirely different and independent of statics.

However, it is natural to attempt to combine these two sciences and to make them depend on only one principle. But among the different principles which can be used as a basis for statics and for which we have given a concise exposition in SECTION I, it is obvious that only the Principle of Virtual Velocities can be applied naturally to the equilibrium of fluids. Thus Galileo, author of this principle, has used it to demonstrate the main theorems of both statics and hydrostatics.

In his *Discorso intorno alle cose che stanno su l'acqua o che in quella si muovono*,[6] he deduces immediately from this principle the equilibrium of water in a siphon by showing that if the level of the fluid is assumed to be at the same height in the two branches, it cannot descend in one and ascend in the other and at the same time not have equal moments for the portion of the fluid which descends and for the portion which ascends. In a similar fashion, Galileo demonstrates the equilibrium of fluids with solid bodies immersed in them. It is true that these demonstrations are not very rigorous and although an attempt is made to remedy this shortcoming in the notes added to the Florentine edition of 1728, it can be asserted that they still leave a great deal to be desired. Descartes and Pascal have also used the Principle of Virtual Velocities in hydrostatics. In particular, the latter made broad use of it in his *Traité de l'equilibre des liqueurs*[7] and has used it to demonstrate the principal property of fluids which is that an arbitrary pressure applied to a point of the surface is also transferred to all other points.

4. However, these applications of the Principle of Virtual Velocities were still too hypothetical, more precisely stated, too broad to serve as a foundation for establishing a rigorous theory for the equilibrium of fluids. Thus, from that time, this principle was abandoned by most of the authors who have investigated hydrostatics and even more so by those who have undertaken to extend the frontiers of this science by investigating the laws of equilibrium of heterogeneous fluids in which all the parts are acted upon by arbitrary forces. This research is very important because of its relation to the famous question concerning the shape of the Earth.

Huygens has assumed as a principle of equilibrium in this research the perpendicularity of gravity to the fluid surface. Newton[8] began his research with the principle of the equality of the weights of central columns. Then Bouguer[9] observed that often these two principles did not give the same result and concluded that to have equilibrium in a fluid mass, the two principles had to be acting simultaneously and accordingly, would have to give the same shape to the surface of the fluid. Moreover, Clairaut[10] has demonstrated that there might exist cases where this is true but yet, there would be no equilibrium. Maclaurin has generalized Newton's principle by establishing that in a fluid mass in equilibrium, each particle must be equally compressed by all the rectilinear columns of fluid which act on this particle and terminate at the surface. Clairaut has made this principle more general yet by showing that the equilibrium of a fluid mass requires that the forces acting on all the parts of a fluid, contained in an arbitrary conduit, terminating at the surface or self-contained, equilibrate each other. Finally, he was the first to deduce from this principle, the true fundamental laws of equilibrium of a fluid mass for which all the parts are acted upon by arbitrary forces. Furthermore, he found the equations of partial differences from which these laws can be derived. This discovery has changed the science of hydrostatics and has made of it a new science.

5. Clairaut's principle is nothing less than a natural consequence of the principle of equality of pressure in all directions at a point and the same equations which hold for equilibrium in conduits can immediately be deduced from this principle. Since, by considering pressure as a force acting on each particle which can be expressed as a function of the coordinates of the particle in a fluid mass, the difference between the pressures acting on two opposed and parallel faces gives the force which tends to compress the mass of the fluid perpendicular to these faces and which must be equilibrated by the accelerating forces acting on this particle so that by projecting all these forces on the directions of the three rectangular coordinate axes and by assuming the fluid mass divided into small rectangular parallelograms having for sides elements of these coordinates, one has directly three equations of partial differences between the pressure and the given accelerating forces which are used to determine the value of the pressure and the relation which must exist between these forces. This

simple approach to finding the general laws of hydrostatics is due to Euler (Mémoires de Berlin, 1755)[11] and it is now adopted in almost every treatise on this science.

6. Thus until now the principle of the equality of pressure in every direction has been the foundation for the theory of the equilibrium of fluids and it should be said that this principle contains in effect the most simple and general property that experience has discovered in fluids in equilibrium. However, is knowledge of this property indispensable for investigating the laws of the equilibrium of fluids? Can these laws be derived directly from the nature of fluids considered as aggregates of molecules loosely held together, independent of one another and perfectly mobile in all directions? This is what I will attempt to do in the following sections using only the general principle of equilibrium which I applied heretofore solely to solid bodies. This part of my work will not only furnish one of the most beautiful applications of this principle, but it will also be used to simplify several aspects of the theory of hydrostatics.

In general, fluids are divided into two categories: incompressible fluids for which portions change shape but do so without changing volume and compressible, elastic fluids for which portions change shape and volume simultaneously and which always expand by exerting a known force which is ordinarily assumed proportional to a function of their density, for example, water, mercury, etc. belong to the first category and air, steam, etc. belong to the second.

We will begin by treating the equilibrium of incompressible fluids and later, the equilibrium of compressible, elastic fluids.

SECTION VII
THE EQUILIBRIUM OF INCOMPRESSIBLE FLUIDS

1. Let the letter m denote a mass of fluid for which all the particles are acted upon by gravity or arbitrary forces P, Q, R, etc. acting in the direction of the lines p, q, r, etc. The sum of the moments of all these forces, according to the formulations of Article 12 of SECTION IV, will be represented by the integrable formula

$$\mathrm{S}(P\,\delta p + Q\,\delta q + R\,\delta r + \cdots)\mathrm{d}m$$

which will be equal to zero for a fluid in equilibrium.

Subsection I
The Equilibrium Of A Fluid In A Very Narrow Conduit

2. Let us assume at the outset that the fluid is contained in a conduit or pipe of very small cross section and of given shape, and let us further imagine that this fluid is divided into

infinitesimal elements for which the length is ds and the cross section is ω. It is obvious that $\mathrm{d}m = \omega\,\mathrm{d}s$, since the cross section ω is assumed infinitesimal and ds is the elemental length of that portion of the fluid passing through the cross section.[12] By assuming that the fluid makes a small displacement and changes its position in the conduit by a small amount and by letting δs be the small distance that the element or particle traverses in the conduit, it is obvious that $\omega\,\delta s$ will be the quantity of fluid which will pass at the same instant through each cross section ω. Thus because of the incompressibility of the fluid, this quantity must be the same throughout so that $\omega\,\delta s = \alpha$, where the quantity α will be a constant along each of the infinitesimal elements of the conduit. Thus $\omega = \alpha/\delta s$, and consequently, $\mathrm{d}m = \alpha\,\mathrm{d}s/\delta s$ so that the formula which expresses the sum of the moments of the forces[13] will become after taking the constant quantity α outside the integral sign S

$$\frac{\alpha \mathrm{S}(P\,\delta p + Q\,\delta q + R\,\delta r + \cdots)\mathrm{d}s}{\delta s}$$

Now it is obvious that since δp, δq, δr, etc. are the variations of the lines p, q, r, etc. resulting from the variation δs, these variations must be in the same ratio to δs as the differentials $\mathrm{d}p$, $\mathrm{d}q$, $\mathrm{d}r$, etc. are to $\mathrm{d}s$ because of the given configuration of the conduit.[14] Thus there results

$$\frac{\delta p}{\delta s} = \frac{\mathrm{d}p}{\mathrm{d}s}, \qquad \frac{\delta q}{\delta s} = \frac{\mathrm{d}q}{\mathrm{d}s}, \qquad \frac{\delta r}{\delta s} = \frac{\mathrm{d}r}{\mathrm{d}s}, \qquad \ldots$$

which will reduce the preceding formula to the following form

$$\alpha\,\mathrm{S}(P\,\mathrm{d}p + Q\,\mathrm{d}q + R\,\mathrm{d}r + \cdots)$$

where the differentials $\mathrm{d}p$, $\mathrm{d}q$, $\mathrm{d}r$, etc. are taken with respect to the curvature of the conduit and the integral sign S indicates an integral taken along the entire length of the conduit.

After setting this quantity equal to zero, the following equation is obtained

$$S(P\,\mathrm{d}p + Q\,\mathrm{d}q + R\,\mathrm{d}r + \cdots) = 0$$

which expresses the general law of equilibrium of a fluid contained in a conduit of arbitrary shape.

3. If in addition to the forces P, Q, R, etc. which act on every point of the fluid, there is also at one of the ends of the conduit an external force Π' which is produced by means of a piston acting on the surface of the fluid perpendicular to the walls of the conduit, then denoting by $\delta s'$ the small distance traversed by the segment of fluid that is assumed compressed by the force Π', while the other segments traverse the various distances δs, the moment of the force Π', which will be expressed by $\Pi'\,\delta s'$, should be added to the sum of the moments of the forces P, Q, R, etc. However, if ω' denotes the cross sectional area of the conduit where the force Π' is acting, the quantity of fluid which passes through

the section ω' will be $\omega'\,\delta s'$ while the quantity of fluid passing through the other arbitrary cross sections ω will be $\omega\,\delta s$.

However, the incompressibility of the fluid requires that these quantities be everywhere the same. Thus since $\omega\,\delta s = \alpha$, it follows that $\omega'\,\delta s' = \alpha$ and consequently, $\delta s' = \alpha/\omega'$. Therefore, the total sum of the moments of the forces which act on the fluid will be represented by the formula

$$\alpha(\frac{\Pi'}{\omega'} + \mathrm{S}(P\,\mathrm{d}p + Q\,\mathrm{d}q + R\,\mathrm{d}r + \cdots))$$

so that the equation of equilibrium will be

$$\frac{\Pi'}{\omega'} + \mathrm{S}(P\,\mathrm{d}p + Q\,\mathrm{d}q + R\,\mathrm{d}r + \cdots) = 0$$

4. It is obvious that in the state of equilibrium the force Π' must be equilibrated by the fluid pressure exerted on the piston over the cross sectional area ω'. From which it follows that this force will be equal to $-\Pi'$, and consequently, equal to

$$\omega'\,\mathrm{S}(P\,\mathrm{d}p + Q\,\mathrm{d}q + R\,\mathrm{d}r + \cdots)$$

Therefore, in general, the force of the fluid exerted on each point of the piston will be expressed by the integral formula

$$S(P\,\mathrm{d}p + Q\,\mathrm{d}q + R\,\mathrm{d}r + \cdots)$$

where this integral is to be taken over the entire length of the conduit. This force will be the same if an immobile wall which seals the conduit at one end replaces the mobile piston.

5. If at the other end of the conduit there is another force Π'', produced in a similar fashion by means of a piston, the equation for the equilibrium of the fluid would be found in a similar fashion

$$\frac{\Pi'}{\omega'} + \frac{\Pi''}{\omega''} + \mathrm{S}(P\,\mathrm{d}p + Q\,\mathrm{d}q + R\,\mathrm{d}r + \cdots) = 0$$

where ω'' is the cross sectional area of the conduit at this location.

6. Thus if the fluid is only compressed by the two external forces Π' and Π'' applied to the surfaces ω' and ω'', the equilibrium equation would be

$$\frac{\Pi'}{\omega'} + \frac{\Pi''}{\omega''} = 0.$$

From which it is obvious that the two forces Π' and Π'' must be in opposite directions and at the same time mutually proportional to the areas ω' and ω'' on which these forces act.

This is a proposition which is commonly viewed as a principle drawn from experience or at least as following from the principle of the equality of pressure in every direction, which most authors writing on hydrostatics consider to be the fundamental property of fluids.

7. Knowledge of the laws of equilibrium for a fluid contained in a narrow conduit of arbitrary shape can lead to the laws of equilibrium for an arbitrary mass of fluid whether contained in a vessel or not. Indeed, if a mass of fluid is in equilibrium and if an arbitrary conduit is assumed to pass through it, the fluid contained in this conduit will also be in equilibrium, that is, independent of the other portions of the fluid. Thus the equation for equilibrium in this conduit, omitting the external forces (Article 2), will be

$$\mathrm{S}(P\,\mathrm{d}p + Q\,\mathrm{d}q + R\,\mathrm{d}r + \cdots) = 0$$

Since the configuration of the conduit can be arbitrary, the preceding equation should be independent of the configuration. From that fact, it could be immediately concluded, as Clairaut has done in his "Théorie de la Figure de la Terre", that the quantity $P\,\mathrm{d}p + Q\,\mathrm{d}q + R\,\mathrm{d}r + \cdots$ must be an exact differential. However, this conclusion could be found using solely analytical methods, and at the same time, the relations which must hold between the quantities P, Q, R, etc. could be obtained. To this effect, it is only necessary to vary the integral $\mathrm{S}(P\,\mathrm{d}p + Q\,\mathrm{d}q + R\,\mathrm{d}r + \cdots)$ by the method of variations and to assume that its variation is equal to zero.

8. Let us denote by Ψ the value of the general integral $\mathrm{S}(P\,\mathrm{d}p + Q\,\mathrm{d}q + R\,\mathrm{d}r + \cdots)$ taken over the entire length of the conduit. In addition, it follows that $\delta\Psi = 0$. Now by differentiation, the following equation results

$$\begin{aligned}\delta\Psi &= \delta[\mathrm{S}(P\,\mathrm{d}p + Q\,\mathrm{d}q + R\,\mathrm{d}r + \cdots)] = \mathrm{S}[\delta(P\,\mathrm{d}p + Q\,\mathrm{d}q + R\,\mathrm{d}r + \cdots)] \\ &= \mathrm{S}(P\,\delta\mathrm{d}p + Q\,\delta\mathrm{d}q + R\,\delta\mathrm{d}r + \cdots + \delta P\,\mathrm{d}p + \delta Q\,\mathrm{d}q + \delta R\,\mathrm{d}r + \cdots)\end{aligned}$$

Replacing $\delta\mathrm{d}$ by $\mathrm{d}\delta$ and then eliminating the double operator $\mathrm{d}\delta$ through integration by parts, the expression for $\delta\Psi$ will become

$$\begin{aligned}&\delta\Psi = P\,\delta p + Q\,\delta q + R\,\delta r + \cdots \\ &+\mathrm{S}(\delta P\,\mathrm{d}p - \mathrm{d}P\,\delta p + \delta Q\,\mathrm{d}q - \mathrm{d}Q\,\delta q + \delta R\,\mathrm{d}r - \mathrm{d}R\,\delta r + \cdots)\end{aligned}$$

where the terms which are outside of the integral sign S correspond to the limits of the integral represented by the integral sign and consequently, they correspond to the ends of the conduit where by assuming these ends fixed, the variations δp, δq, δr, etc., which correspond to them will be equal to zero. Consequently, the terms outside of the integral sign will vanish.

Now, since the quantities P, Q, R, etc. which represent the forces are or can always be assumed to be functions of p, q, r, etc., it is clear that the part of $\delta\Psi$ which is under the integral sign S cannot be reduced further. Therefore, in order to have in general $\delta\Psi = 0$, this part must be independently equal to zero and consequently, the following identity will be obtained for each point of the mass

$$\delta P\,\mathrm{d}p - \mathrm{d}P\,\delta p + \delta Q\,\mathrm{d}q - \mathrm{d}Q\,\delta q + \delta R\,\mathrm{d}r - \mathrm{d}R\,\delta r + \cdots = 0$$

By regarding the expressions of the forces P, Q, R, etc. as arbitrary functions of p, q, r, etc., the following equations written in the usual notation will be obtained

$$\mathrm{d}P = \frac{\mathrm{d}P}{\mathrm{d}p}\mathrm{d}p + \frac{\mathrm{d}P}{\mathrm{d}q}\mathrm{d}q + \frac{\mathrm{d}P}{\mathrm{d}r}\mathrm{d}r + \cdots$$

and similarly

$$\delta P = \frac{\mathrm{d}P}{\mathrm{d}p}\delta p + \frac{\mathrm{d}P}{\mathrm{d}q}\delta q + \frac{\mathrm{d}P}{\mathrm{d}r}\delta r + \cdots$$

and identically for the other differences. Substituting these expressions in the preceding equation and ordering the terms, it will assume the following form

$$\begin{aligned}
0 &= (\frac{\mathrm{d}P}{\mathrm{d}q} - \frac{\mathrm{d}Q}{\mathrm{d}p})(\delta q\,\mathrm{d}p - \mathrm{d}q\,\delta p)\\
&+ (\frac{\mathrm{d}P}{\mathrm{d}r} - \frac{\mathrm{d}R}{\mathrm{d}p})(\delta r\,\mathrm{d}p - \mathrm{d}r\,\delta p)\\
&+ (\frac{\mathrm{d}Q}{\mathrm{d}r} - \frac{\mathrm{d}R}{\mathrm{d}q})(\delta r\,\mathrm{d}q - \mathrm{d}r\,\delta q)\\
&+ \cdots
\end{aligned}$$

and should exist independently of the differences $\mathrm{d}p$, $\mathrm{d}q$, $\mathrm{d}r$, etc. and the variations δp, δq, δr, etc.

Therefore, if no relation is given between the variations p, q, r, etc., the following expressions should be put equal to zero[15]

$$\begin{aligned}
&\frac{\mathrm{d}P}{\mathrm{d}q} - \frac{\mathrm{d}Q}{\mathrm{d}p} = 0\\
&\frac{\mathrm{d}P}{\mathrm{d}r} - \frac{\mathrm{d}R}{\mathrm{d}p} = 0\\
&\frac{\mathrm{d}Q}{\mathrm{d}r} - \frac{\mathrm{d}R}{\mathrm{d}q} = 0\\
&\vdots
\end{aligned}$$

These are the equations of condition derived from the integrability of the formula $P\,\mathrm{d}p + Q\,\mathrm{d}q + R\,\mathrm{d}r + \cdots$.[16]

9. When the lines p, q, r, etc. relate to a point in space, as in the present case, they depend only on the coordinates of this point and the forces P, Q, R, etc. can always be reduced to

three in the direction of these coordinates (SECTION V, Article 7). Thus by taking p, q, r for these coordinates, either rectangular or not, and P, Q, R, etc. for the forces which act on each particle of the fluid in the direction of these same coordinates, the quantities P, Q, R viewed as functions of p, q, r must satisfy the following three equations

$$\frac{\mathrm{d}P}{\mathrm{d}q} - \frac{\mathrm{d}Q}{\mathrm{d}p} = 0, \qquad \frac{\mathrm{d}P}{\mathrm{d}r} - \frac{\mathrm{d}R}{\mathrm{d}p} = 0, \qquad \frac{\mathrm{d}Q}{\mathrm{d}r} - \frac{\mathrm{d}R}{\mathrm{d}q} = 0$$

These are the necessary conditions for the equilibrium of a fluid mass under the action of the forces P, Q, R which act on all its points.

Incidentally, up to this point the density of the fluid has not been considered or rather it has been viewed as a constant and equal to unity. However, if it is to be assumed variable, then by denoting by Γ the density of an arbitrary particle $\mathrm{d}m$, one will have (Article 2) $\mathrm{d}m = \Gamma\omega\,\mathrm{d}s$ and the quantities P, Q, R, etc. will all be multiplied by Γ. Therefore, the laws of equilibrium for fluids with variable density will be the same as the laws for the equilibrium of fluids with uniform density after multiplying the acting forces by the density of the particle on which they act, that is, by simply writing ΓP, ΓQ, ΓR, etc. instead of P, Q, R, etc.

Subsection II
Where the General Laws of Equilibrium of Incompressible Fluids are Deduced from the Nature of the Particles which Compose It

10. We are now going to determine directly from our general formula, the laws of equilibrium for incompressible fluids by viewing these types of fluids as composed of a mass of perfectly mobile particles and capable of changing shape but not volume. Let us assume for greater simplicity that all the forces which act on the particles of the fluid are reduced to three, represented by X, Y, Z and acting in the direction of the rectangular coordinates x, y, z, that is, acting towards the origin of these coordinates. We have given the means to carry out this task in Chapter I of SECTION V using the general formulas.

Denoting the mass of an arbitrary particle by $\mathrm{d}m$, the sum of the moments of the forces X, Y, Z, will be given by the integral formula

$$\mathrm{S}(X\,\delta x + Y\,\delta y + Z\,\delta z)\mathrm{d}m$$

but the volume of the particle $\mathrm{d}m$ can be represented by $\mathrm{d}x\,\mathrm{d}y\,\mathrm{d}z$. So by expressing the density by Γ it is clear that $\mathrm{d}m = \Gamma\,\mathrm{d}x\,\mathrm{d}y\,\mathrm{d}z$ and the integration will be with respect to the three variables x, y, z.

Moreover, the equation of condition expressing the incompressibility of the fluid must be considered. This equation is assumed to be expressed by $L = 0$, and will give, after differentiating according to δ, multiplying by an undetermined coefficient λ and integrating, the formula $\mathrm{S}\lambda\,\delta L$ which will be added to the preceding equation.

If there are no external forces which act on the surface of the fluid, nor conditions peculiar to the surface, the general equation of equilibrium (SECTION IV, Article 13) will simply be

$$\mathrm{S}(X\,\delta x + Y\,\delta y + Z\,\delta z)\mathrm{d}m + \mathrm{S}\,\lambda\,\delta L = 0$$

in which the integral must be taken with respect to the entire mass of the fluid.

11. The condition of incompressibility requires that the volume of each particle be invariable. So, having expressed this volume by $\mathrm{d}x\,\mathrm{d}y\,\mathrm{d}z$, it must be that $\mathrm{d}x\,\mathrm{d}y\,\mathrm{d}z =$ constant for the equation of condition and consequently, L is equal to ($\mathrm{d}x\,\mathrm{d}y\,\mathrm{d}z$ − constant) and $\delta L = \delta(\mathrm{d}x\,\mathrm{d}y\,\mathrm{d}z)$.

In order to obtain the variation $\delta(\mathrm{d}x\,\mathrm{d}y\,\mathrm{d}z)$ it seems that it will only be necessary to simply differentiate $\mathrm{d}x\,\mathrm{d}y\,\mathrm{d}z$ according to δ but at this point there is a particular consideration to be made and without which the calculation will not be rigorous. The quantity $\mathrm{d}x\,\mathrm{d}y\,\mathrm{d}z$ represents the volume of a particle as long as the shape of this particle is assumed to be a rectangular parallelepiped whose sides are parallel to the x, y, z axes of the coordinate system. This assumption is very well-founded because the fluid can be imagined divided into infinitesimal elements of arbitrary shape. However, $\delta(\mathrm{d}x\,\mathrm{d}y\,\mathrm{d}z)$ must express the variation of volume when the particles change position by an infinitesimal amount. That is, when its coordinates x, y, z, become $x + \delta x$, $y + \delta y$, $z + \delta z$ and it is clear that if in this change the element retained the shape of a rectangular parallelepiped, one would have

$$\delta(\mathrm{d}x\,\mathrm{d}y\,\mathrm{d}z) = \mathrm{d}y\,\mathrm{d}z\,\delta\mathrm{d}x + \mathrm{d}x\,\mathrm{d}z\,\delta\mathrm{d}y + \mathrm{d}x\,\mathrm{d}y\,\delta\mathrm{d}z$$

Using the principles of the Calculus of Variations, the variations $\delta\mathrm{d}x$, $\delta\mathrm{d}y$, $\delta\mathrm{d}z$ can be replaced by $\mathrm{d}\delta x$, $\mathrm{d}\delta y$, $\mathrm{d}\delta z$. However, it is necessary to recognize that since the variations $\delta x, \delta y, \delta z$ can be viewed as indeterminate and infinitesimal functions of x, y, z, only x must be considered variable in the differentiation of δx so that $\mathrm{d}\delta x$ represents the elongation of the side $\mathrm{d}x$ of the rectangular parallelepiped $\mathrm{d}x\,\mathrm{d}y\,\mathrm{d}z$ while the coordinates y and z do not vary. Therefore, according to the notation for partial differentials, $(\mathrm{d}\delta x/\mathrm{d}x)/\mathrm{d}x$ should be written in place of $\mathrm{d}\delta x$. By a similar reasoning $(\mathrm{d}\delta y/\mathrm{d}y)/\mathrm{d}y$ and $(\mathrm{d}\delta z/\mathrm{d}z)/\mathrm{d}z$ will be written instead of $\mathrm{d}\delta y$ and $\mathrm{d}\delta z$. In this fashion, with the assumption that the elemental volume $\mathrm{d}x\,\mathrm{d}y\,\mathrm{d}z$ remains rectangular after the variation, one will have

$$\delta(\mathrm{d}x\,\mathrm{d}y\,\mathrm{d}z) = \mathrm{d}x\,\mathrm{d}y\,\mathrm{d}z\left(\frac{\mathrm{d}\delta x}{\mathrm{d}x} + \frac{\mathrm{d}\delta y}{\mathrm{d}y} + \frac{\mathrm{d}\delta z}{\mathrm{d}z}\right)$$

Nothing will change, if it is assumed that the elemental volume $\mathrm{d}x\,\mathrm{d}y\,\mathrm{d}z$ became after the variation, a parallelepiped whose angles are very close to right angles. Since it is known from geometry that if a, b, c are the three sides of a parallelepiped which makes a solid angle and α, β, γ are the three angles which these sides form with each other, then the volume of the parallelepiped is expressed by the formula

$$abc\sqrt{(1 - \cos^2\alpha - \cos^2\beta - \cos^2\gamma + 2\cos\alpha\cos\beta\cos\gamma)}$$

Now the lengths of the sides become, because of the variation,

$$\mathrm{d}x(1+\frac{\mathrm{d}\delta x}{\mathrm{d}x}),\qquad \mathrm{d}y(1+\frac{\mathrm{d}\delta y}{\mathrm{d}y}),\qquad \mathrm{d}z(1+\frac{\mathrm{d}\delta z}{\mathrm{d}z})$$

and the cosines of α, β, γ become infinitesimal. Thus by substituting these values in place of a, b, c and neglecting the infinitesimal quantities of orders higher than the first, the variation of $\mathrm{d}x\,\mathrm{d}y\,\mathrm{d}z$ will give the same expression which was just found.

Although this last assumption is legitimate, we do not want to adopt it without a demonstration so that no doubt is left about the accuracy of our formulas. Thus we will investigate in a rigorous manner the variation of $\mathrm{d}x\,\mathrm{d}y\,\mathrm{d}z$ with respect to both change of position and length of each of the sides of our rectangular parallelepiped and by only assuming that these sides remain rectilinear, which is true for infinitesimal variations.

12. In order to simplify this investigation, we will begin by considering only one of the faces of the parallelepiped $\mathrm{d}x\,\mathrm{d}y\,\mathrm{d}z$, for example, the face $\mathrm{d}x\,\mathrm{d}y$ for which the four corners have the following coordinates

$$(1)\ x, y, z,\qquad (2)\ x+\mathrm{d}x, y, z,$$
$$(3)\ x, y+\mathrm{d}y, z,\qquad (4)\ x+\mathrm{d}x, y+\mathrm{d}y, z$$

Let us assume that the coordinates x, y, z of the first corner become $x+\delta x, y+\delta y, z+\delta z$, and let us further consider the variations δx, δy, δz as infinitesimal functions of x, y, z. By incrementing in turn, x and y by their differentials $\mathrm{d}x$ and $\mathrm{d}y$, the coordinates of the other three corners can be found simultaneously. Therefore, the following expressions will be obtained, using the same numbers to represent the coordinates of the four corners after variation

$$(1)\quad x+\delta x,\qquad y+\delta y,\qquad z+\delta z$$
$$(2)\quad x+\mathrm{d}x+\delta x+\frac{\mathrm{d}\delta x}{\mathrm{d}x}\mathrm{d}x,\qquad y+\delta y+\frac{\mathrm{d}\delta y}{\mathrm{d}x}\mathrm{d}x,\qquad z+\delta z+\frac{\mathrm{d}\delta z}{\mathrm{d}x}\mathrm{d}x$$
$$(3)\quad x+\delta x+\frac{\mathrm{d}\delta x}{\mathrm{d}y}\mathrm{d}y,\qquad y+\mathrm{d}y+\delta y+\frac{\mathrm{d}\delta y}{\mathrm{d}y}\mathrm{d}y,\qquad z+\delta z+\frac{\mathrm{d}\delta z}{\mathrm{d}y}\mathrm{d}y$$
$$(4)\quad \begin{cases} x+\mathrm{d}x+\delta x+\dfrac{\mathrm{d}\delta x}{\mathrm{d}x}\mathrm{d}x+\dfrac{\mathrm{d}\delta x}{\mathrm{d}y}\mathrm{d}y \\ y+\mathrm{d}y+\delta y+\dfrac{\mathrm{d}\delta y}{\mathrm{d}x}\mathrm{d}x+\dfrac{\mathrm{d}\delta y}{\mathrm{d}y}\mathrm{d}y \\ z+\mathrm{d}z+\delta z+\dfrac{\mathrm{d}\delta z}{\mathrm{d}x}\mathrm{d}x+\dfrac{\mathrm{d}\delta z}{\mathrm{d}y}\mathrm{d}y \end{cases}$$

Since these four expressions are the coordinates of the four corners of the new quadrilateral in which the rectangle $\mathrm{d}x\,\mathrm{d}y$ has been altered, it is clear that the lengths of the sides of this quadrilateral will be obtained by taking the square root of the sum of the squares of the differences of the coordinates of the two corners at the end of each side. Thus by denoting the length of the straight line which joins two corners by the two numbers assigned to the coordinates of those two corners, the following equations will be obtained

$$(1,2) = \mathrm{d}x\sqrt{(1+\frac{\mathrm{d}\delta x}{\mathrm{d}x})^2 + (\frac{\mathrm{d}\delta y}{\mathrm{d}x})^2 + (\frac{\mathrm{d}\delta z}{\mathrm{d}x})^2}$$

$$(1,3) = \mathrm{d}y\sqrt{(\frac{\mathrm{d}\delta x}{\mathrm{d}y})^2 + (1+\frac{\mathrm{d}\delta y}{\mathrm{d}y})^2 + (\frac{\mathrm{d}\delta z}{\mathrm{d}y})^2}$$

$$(3,4) = \mathrm{d}x\sqrt{(1+\frac{\mathrm{d}\delta x}{\mathrm{d}x})^2 + (\frac{\mathrm{d}\delta y}{\mathrm{d}x})^2 + (\frac{\mathrm{d}\delta z}{\mathrm{d}x})^2}$$

$$(2,4) = \mathrm{d}y\sqrt{(\frac{\mathrm{d}\delta x}{\mathrm{d}y})^2 + (1+\frac{\mathrm{d}\delta y}{\mathrm{d}y})^2 + (\frac{\mathrm{d}\delta z}{\mathrm{d}y})^2}$$

from which it is clear that the opposite sides (1,2), (3,4) are equal to each other as are the opposite sides (1,3), (2,4). Therefore, the quadrilateral is a parallelogram for which the lengths of the two contiguous sides (1,2), (1,3), after the quantities of second order under the square root sign are neglected, will be

$$(1,2) = \mathrm{d}x(1+\frac{\mathrm{d}\delta x}{\mathrm{d}x}), \qquad (1,3) = \mathrm{d}y(1+\frac{\mathrm{d}\delta y}{\mathrm{d}y})$$

13. The angle made by these two sides (1,2) and (1,3) will be found by considering the diagonal length (2,3) which is obtained by taking the square root of the sum of the squares of the differences of the respective coordinates of the corners (2) and (3), which is

$$(2,3) = \sqrt{\left(\mathrm{d}x + \frac{\mathrm{d}\delta x}{\mathrm{d}x}\mathrm{d}x - \frac{\mathrm{d}\delta x}{\mathrm{d}y}\mathrm{d}y\right)^2 + \left(\mathrm{d}y + \frac{\mathrm{d}\delta y}{\mathrm{d}y}\mathrm{d}y - \frac{\mathrm{d}\delta y}{\mathrm{d}x}\mathrm{d}x\right)^2 + \left(\frac{\mathrm{d}\delta z}{\mathrm{d}x}\mathrm{d}x - \frac{\mathrm{d}\delta z}{\mathrm{d}y}\mathrm{d}y\right)^2}$$

Denoting by α the angle in question, the triangle formed by the three sides (1,2), (1,3), (2,3) results in the following equation

$$\cos\alpha = \frac{(1,2)^2 + (1,3)^2 - (2,3)^2}{2\,(1,2)\,(1,3)}$$

Substituting into this expression the equations which have been found for (1,2), (1,3), (2,3) eliminating those terms which cancel out, and neglecting the infinitesimal quantities of the second and higher orders, one will have

$$\cos\alpha = \frac{\mathrm{d}\delta x}{\mathrm{d}y} + \frac{\mathrm{d}\delta y}{\mathrm{d}x}$$

where it is clear that the angle α differs from a right angle by infinitesimal quantities, because its cosine is infinitesimal.

14. If the same analysis is applied to the other two faces, $\mathrm{d}x\,\mathrm{d}z$ and $\mathrm{d}y\,\mathrm{d}z$ of the rectangle $\mathrm{d}x\,\mathrm{d}y\,\mathrm{d}z$, these faces will also become parallelograms and the three opposite faces will also be parallelograms, as can be demonstrated easily by geometry. Consequently, the new solid will be a parallelepiped for which the sides which form a solid angle will be

$$\mathrm{d}x(1+\frac{\mathrm{d}\delta x}{\mathrm{d}x}),\qquad \mathrm{d}y(1+\frac{\mathrm{d}\delta y}{\mathrm{d}y}),\qquad \mathrm{d}z(1+\frac{\mathrm{d}\delta z}{\mathrm{d}z})$$

and denoting by α, β, γ the angles formed by these sides, one will have

$$\cos\alpha = \frac{\mathrm{d}\delta x}{\mathrm{d}y}+\frac{\mathrm{d}\delta y}{\mathrm{d}x}$$

$$\cos\beta = \frac{\mathrm{d}\delta x}{\mathrm{d}z}+\frac{\mathrm{d}\delta z}{\mathrm{d}x}$$

$$\cos\gamma = \frac{\mathrm{d}\delta y}{\mathrm{d}z}+\frac{\mathrm{d}\delta z}{\mathrm{d}y}$$

From this result, it can be concluded that the variation of the rectangular parallelepiped $\mathrm{d}x\,\mathrm{d}y\,\mathrm{d}z$ is rigorously expressed by the formula given above in Article 11.

15. It can also be observed from this that if the variations δx, δy, δz were only functions of x, y, z, respectively, then it happens that $\cos\alpha = 0$, $\cos\beta = 0$, $\cos\gamma = 0$ such that the rectangular parallelepiped $\mathrm{d}x\,\mathrm{d}y\,\mathrm{d}z$ would remain rectangular after these variations. But since the change of configuration of this parallelepiped is only infinitesimalegin{center} and does not influence at all the magnitude of its volume, it follows that without altering the generality of the result, the variations δx, δy, δz can be assumed to be functions of x, y, z as was done in Article 31 of SECTION IV.

16. Thus since we have the exact value of $\delta(\mathrm{d}x\,\mathrm{d}y\,\mathrm{d}z)$, it will be taken for δL. Consequently, there results

$$\delta L = \mathrm{d}x\,\mathrm{d}y\,\mathrm{d}z(\frac{\mathrm{d}\delta x}{\mathrm{d}x}+\frac{\mathrm{d}\delta y}{\mathrm{d}y}+\frac{\mathrm{d}\delta z}{\mathrm{d}z})$$

After substituting this expression in the general equation of Article 10 and at the same time replacing dm by the expression $\Gamma\,\mathrm{d}x\,\mathrm{d}y\,\mathrm{d}z$, the following equation results

$$\mathrm{S}\left\{\begin{array}{l}\Gamma(X\,\delta x+Y\,\delta y+Z\,\delta z)\\ +\lambda(\frac{\mathrm{d}\delta x}{\mathrm{d}x}+\frac{\mathrm{d}\delta y}{\mathrm{d}y}+\frac{\mathrm{d}\delta z}{\mathrm{d}z})\end{array}\right\}\mathrm{d}x\,\mathrm{d}y\,\mathrm{d}z = 0$$

and it will only be necessary to eliminate the double operator $\mathrm{d}\delta$ by the method presented in Subsection II of SECTION IV.

17. Let us consider at the outset the quantity $\mathrm{S}\,\lambda(\mathrm{d}\delta x/\mathrm{d}x)\mathrm{d}x\,\mathrm{d}y\,\mathrm{d}z$ where the integral sign S indicates a triple integral with respect to x, y, z. It is clear that since the difference δx depends solely on the variations of x, it will be sufficient to consider only

the integration with respect to x. This is the reason for first giving to this quantity the form $\mathrm{S}\, dy\, dz\, \mathrm{S}\, \lambda(d\delta x/dx)dx$ and then transforming the simple integral $\mathrm{S}\, \lambda(d\delta x/dx)dx$ to $\lambda''\delta x'' - \lambda'\delta x' - \mathrm{S}\,(d\lambda/dx)\delta x\, dx$.

The quantities denoted by a prime correspond to the lower limit of the integration and those denoted by a double prime correspond to the upper limit of the integration following the adopted notation. Therefore, the quantity in question will be transformed to

$$\mathrm{S}\, dy\, dz(\lambda''\, \delta x'' - \lambda'\, \delta x') - \mathrm{S}\, dy\, dz\, \mathrm{S}\frac{d\lambda}{dx}\delta x\, dx$$

or what is the same thing $\mathrm{S}(\lambda''\, \delta x'' - \lambda'\, \delta x')dy\, dz - \mathrm{S}(d\lambda/dx)\delta x\, dx\, dy\, dz$.

In the same fashion and by a similar reasoning, the quantities $\mathrm{S}\, \lambda(d\delta y/dy)dx\, dy\, dz$ and $\mathrm{S}\, \lambda(d\delta z/dz)dx\, dy\, dz$ will be transformed to $\mathrm{S}(\lambda''\, \delta y'' - \lambda'\, \delta y')dx\, dz - \mathrm{S}(d\lambda/dy)\delta y\, dx\, dy\, dz$ and $\mathrm{S}(\lambda''\, \delta z'' - \lambda'\, \delta z')dx\, dy - \mathrm{S}(d\lambda/dz)\delta z\, dx\, dy\, dz$.

Thus after these substitutions are made, the following general equation for the equilibrium of the fluid mass will be

$$\begin{aligned}&\mathrm{S}[(\Gamma X - \frac{d\lambda}{dx})\delta x + (\Gamma Y - \frac{d\lambda}{dy})\delta y + (\Gamma Z - \frac{d\lambda}{dz})\delta z]dx\, dy\, dz\\ &+\mathrm{S}(\lambda''\, \delta x'' - \lambda'\, \delta x')dy\, dz + \mathrm{S}(\lambda''\, \delta y'' - \lambda'\, \delta y')dx\, dz\\ &+\mathrm{S}(\lambda''\, \delta z'' - \lambda'\, \delta z')dx\, dy = 0\end{aligned}$$

in which it is only necessary to equate separately to zero the coefficients of the indeterminate variations δx, δy, δz (SECTION IV, Article 16).

18. The following three equations will be obtained at the outset

$$\Gamma X - \frac{d\lambda}{dx} = 0, \qquad \Gamma Y - \frac{d\lambda}{dy} = 0, \qquad \Gamma Z - \frac{d\lambda}{dz} = 0$$

which must hold for every point of the fluid mass. Then, if the boundaries of the fluid are free, the variations $\delta x'$, $\delta y'$, $\delta z'$, $\delta x''$, $\delta y''$, $\delta z''$ which correspond to the points of the surface of the fluid will also be indeterminate and consequently, their coefficients will also equate separately to zero, which will provide $\lambda' = 0$, $\lambda'' = 0$, that is, in general $\lambda = 0$ for all points of the surface of the fluid. Furthermore, this equation will be used to determine the shape of this surface.

It will be the same for the surface of a fluid contained in an open vessel. But with respect to the part of the boundary which is supported on the sides, the variations $\delta x'$, $\delta y'$, $\delta z'$, $\delta x''$, $\delta y''$, $\delta z''$ must have ratios between them given by the shape of these sides because the fluid can only slide along the sides. And we will demonstrate below that whatever their shape, the terms which contain the variations in question will always be equal to zero individually so that there will be no condition imposed on this part of the fluid's boundary.

19. The three equations just found for the conditions of equilibrium of the fluid give

$$\frac{\mathrm{d}\lambda}{\mathrm{d}x} = \Gamma X, \qquad \frac{\mathrm{d}\lambda}{\mathrm{d}y} = \Gamma Y, \qquad \frac{\mathrm{d}\lambda}{\mathrm{d}z} = \Gamma Z$$

and since $\mathrm{d}\lambda = (\mathrm{d}\lambda/\mathrm{d}x)\mathrm{d}x + (\mathrm{d}\lambda/\mathrm{d}y)\mathrm{d}y + (\mathrm{d}\lambda/\mathrm{d}z)\mathrm{d}z$ one will have $\mathrm{d}\lambda = \Gamma(X\,\mathrm{d}x + Y\,\mathrm{d}y + Z\,\mathrm{d}z)$ and consequently, the quantity $\Gamma(X\,\mathrm{d}x + Y\,\mathrm{d}y + Z\,\mathrm{d}z)$ must be an exact differential in x, y, z. And this condition alone expresses the laws for the equilibrium of fluids.

If the quantity λ is eliminated from these equations, the following equations will be obtained

$$\frac{\mathrm{d}(\Gamma X)}{\mathrm{d}y} = \frac{\mathrm{d}(\Gamma Y)}{\mathrm{d}x}, \qquad \frac{\mathrm{d}(\Gamma X)}{\mathrm{d}z} = \frac{\mathrm{d}(\Gamma Z)}{\mathrm{d}x}, \qquad \frac{\mathrm{d}(\Gamma Y)}{\mathrm{d}z} = \frac{\mathrm{d}(\Gamma Z)}{\mathrm{d}y}$$

which are compatible with those of Article 9.

These conditions are necessary so that the fluid mass can be in equilibrium under the action of the forces X, Y, Z. When these conditions hold due to the nature of these forces, it is certain that equilibrium is possible and it remains to find the shape that the fluid mass must have in order to be in equilibrium, that is, the equation of the external boundary of the fluid.

We have seen in the preceding article that at each point of the surface, λ must be equal to zero. Thus since $\mathrm{d}\lambda = \Gamma(X\,\mathrm{d}x + Y\,\mathrm{d}y + Z\,\mathrm{d}z)$, one will have after integration

$$\lambda = \int \Gamma(X\,\mathrm{d}x + Y\,\mathrm{d}y + Z\,\mathrm{d}z) + \text{const.}$$

Consequently, the equation for the external boundary of the fluid will be

$$\int \Gamma(X\,\mathrm{d}x + Y\,\mathrm{d}y + Z\,\mathrm{d}z) = K$$

where K is an arbitrary constant. This equation will always consist of finite terms because the quantity $\Gamma(X\,\mathrm{d}x + Y\,\mathrm{d}y + Z\,\mathrm{d}z)$ is assumed to be an exact differential.

20. The quantity $X\,\mathrm{d}x + Y\,\mathrm{d}y + Z\,\mathrm{d}z$ is always an exact differential when the forces X, Y, Z are the result of one or several attractions proportional to arbitrary functions of the distances between the centers of the particles since one has in general from Article 1 of SECTION V

$$X\,\mathrm{d}x + Y\,\mathrm{d}y + Z\,\mathrm{d}z = P\,\mathrm{d}p + Q\,\mathrm{d}q + R\,\mathrm{d}r + \cdots$$

Denoting this quantity by $\mathrm{d}\Pi$, one will then have $\mathrm{d}\lambda = \Gamma\,\mathrm{d}\Pi$. Therefore, for $\mathrm{d}\lambda$ to be an exact differential, Γ must be a function of Π. Consequently, $\lambda = \int \Gamma\,\mathrm{d}\Pi$ will also be a function of Π.

Thus in this case, which is due to the essence of the problem, the shape of the surface will be represented by the equation $f(\Pi) = k$, that is, Π is equal to a constant, as if the density of the fluid were uniform. Moreover, since Π is constant at the surface and Γ is a function of Π, it follows that the density Γ must be the same at every point of the external boundary of a fluid mass in equilibrium.

Within the fluid the density can vary in an arbitrary fashion, as long as it is a function of Π. It will thus be constant wherever the value of Π is constant, so that $\Pi = h$ will in general be the equation of the layer of the same density, where h is a constant. Thus after differentiation, there results $\mathrm{d}\Pi = 0$, or $X\,\mathrm{d}x + Y\,\mathrm{d}y + Z\,\mathrm{d}z = 0$ for the general equation of these layers and it is obvious that this equation is for surfaces perpendicular to the resultant of the forces X, Y, Z which Clairaut calls "surfaces de niveau".[17] From which it follows that the density must be uniform in each layer of equal potential formed by two surfaces which are infinitesimally close to one another.

This law must hold on Earth and on the planets, assuming that these bodies were originally fluid and that they kept, during their solidification, the shape that they had assumed by virtue of the attraction of their parts combined with centrifugal force.

21. With respect to the quantity λ which we just found, it is worthwhile to remark that the term $\mathrm{S}\,\lambda\,\delta L$ of the general equation of Article 10 represents the sum of the moments of all the forces λ which tend to reduce the value of the function L (SECTION IV, Article 7) so that because $\delta L = \delta(\mathrm{d}x\,\mathrm{d}y\,\mathrm{d}z)$ (Article 11), it can be said that the force λ tends to compress each particle $\mathrm{d}x\,\mathrm{d}y\,\mathrm{d}z$ of the fluid. Consequently, this force is nothing more than the pressure applied to this particle from every direction and which it reacts by its incompressibility.

Thus, in general, the expression for the pressure at each point of the fluid mass is

$$\mathrm{S}\,\Gamma(P\,\mathrm{d}x + Q\,\mathrm{d}y + R\,\mathrm{d}z)$$

and since the quantity under the integral sign must always be integrable so that the fluid remains in equilibrium, it follows that the pressure can always be expressed by a finite function of the coordinates relative to the particle to which this pressure is applied. This is a fundamental proposition of the theory of fluids presented by Euler (SECTION VI, Article 5).

In order to present an example of the equation $\Pi =$ constant which was found earlier to represent the surface of a fluid mass in equilibrium (Article 20), we will consider the equilibrium of the sea, assuming that it covers the Earth completely. The Earth is viewed as a solid with an elliptical shape which is very close to a sphere, and each of the particles of the sea are attracted simultaneously by all the particles of the Earth and the remaining sea. Furthermore, the sea is acted on simultaneously by the centrifugal force created by the uniform rotation of the Earth about its axis.

This is an opportunity to use the formulas which we gave in Article 10 of SECTION V. We have designated by Σ the value of the function Π when the forces are the result of attractions of all the particles of a body with a given shape and we have given the expression for Σ for the case where the attraction is inversely proportional to the square of the distances and where the attracting body is an elliptical spheroid very close to a spheroidal shape. Retaining the symbols used in that article and keeping only the terms up to the second order for the eccentricities e and i, it has been found that

$$\Sigma = -m(\frac{1}{r} - \frac{e^2+i^2}{2.5r^3} + 3\frac{e^2y^2+i^2z^2}{2.5r^5})$$

where x, y, z are the rectangular coordinates of the attracted point, $r = \sqrt{(x^2+y^2+z^2)}$ is the distance from this point to the center of the spheroid and m is the mass of the spheroid which is equal to $(4\pi/3)ABC$, where A, B, C are the semi-axes of the spheroid.

If Γ is the density of the spheroid, which is assumed homogeneous, the expression for Σ must be multiplied by Γ and if the spheroid is assumed to envelope a spheroidal core for which the density is different, it will only be necessary to add the value of Σ relative to this new spheroid, multiplied by the difference in the densities. Thus by denoting by a prime the quantities corresponding to the internal spheroid and assuming its density to be $\Gamma + \Gamma'$, the total value of Σ will be

$$\Sigma = -\frac{\Gamma m + \Gamma' m'}{r} + \frac{\Gamma m(e^2+i^2) + \Gamma' m'(e'^2+i'^2)}{2.5r^3}$$
$$-3\frac{\Gamma me^2 + \Gamma' m' e'^2}{2.5r^5}y^2 - 3\frac{\Gamma mi^2 + \Gamma' m' i'^2}{2.5r^5}z^2$$

23. Let us assume that the point attracted by the spheroid is at the same time acted on by three forces represented by fx, gy, hz in the direction of the x, y, z coordinate axes away from the origin. Then their moments will be $-fx\,dx$, $-gy\,dy$ and $-hz\,dz$. The terms $-fx^2/2 - gy^2/2 - hz^2/2$ will have to be added to the quantity Σ to obtain the value of Π, which results from all the forces which act at the same point. Thus the equation of equilibrium will be

$$\Sigma - \frac{fx^2+gy^2+hz^2}{2} = \text{const.}$$

24. Now in order to apply these formulas to the problem in question, the external and internal spheroids will be assumed to be the sea and Earth, for which the densities are Γ and $\Gamma + \Gamma'$, respectively. The attracted point will be located on the surface of the sea by having its coordinates x, y, z equal to the coordinates a, b, c of the surface of the external spheroid. Thus the following equation must hold so that this surface is in equilibrium.

$$\frac{\Gamma m + \Gamma' m'}{r} - \frac{\Gamma m(e^2+i^2) + \Gamma' m'(e'^2+i'^2)}{2.5r^3} + \frac{fx^2}{2}$$
$$+(3\frac{\Gamma me^2 + \Gamma' m' e'^2}{2.5r^5} + \frac{g}{2})y^2 + (3\frac{\Gamma mi^2 + \Gamma' m' i'^2}{2.5r^5} + \frac{h}{2})z^2 = \text{const.}$$

This equation, in which $r = \sqrt{(x^2 + y^2 + z^2)}$, gives the shape of the surface. But we have assumed in the formulas of Article 10 of SECTION V that this surface is represented by the equation

$$\frac{x^2}{A^2} + \frac{y^2}{B^2} + \frac{z^2}{C^2} = 1$$

in which x, y, z have replaced a, b, c. Thus these two equations must be identical.

Using this equation, let us express the value of r as a function of y and z. In order to do this, we will substitute for x^2 in the relation $r^2 = x^2 + y^2 + z^2$, the expression

$$A^2 - \frac{A^2 y^2}{B^2} - \frac{A^2 z^2}{C^2}$$

then after replacing B^2 and C^2 by the expressions $(A^2 + e^2)$ and $(A^2 + i^2)$ (article cited), the following equation will be obtained

$$r^2 = A^2 + \frac{e^2 y^2}{A^2 + e^2} + \frac{i^2 z^2}{A^2 + i^2}$$

from which there results, after discarding powers of e and i greater than e^2 and i^2 in which we are not interested here

$$\frac{1}{r} = \frac{1}{A} - \frac{e^2 y^2 + i^2 z^2}{2A^5}$$

Thus after substituting this expression for $1/r$ as well as that for x^2 in the first equation and discarding the terms which contain e^4, i^4, e^2, i^2, etc., one will have

$$\begin{aligned}
&\frac{\Gamma m + \Gamma' m'}{A} - \frac{\Gamma m(e^2 + i^2) + \Gamma' m'(e'^2 + i'^2)}{2.5A^3} + \frac{fA^2}{2} \\
&+(3\frac{\Gamma m e^2 + \Gamma' m' e'^2}{2.5A^5} + \frac{g}{2} - \frac{fA^2}{2B^2} - \frac{(\Gamma m + \Gamma' m')e^2}{2A^5})y^2 \\
&+(3\frac{\Gamma m i^2 + \Gamma' m' i'^2}{2.5A^5} + \frac{h}{2} - \frac{fA^2}{2C^2} - \frac{(\Gamma m + \Gamma' m')i^2}{2A^5})z^2 \\
&= \text{const.}
\end{aligned}$$

Since this equation is an identity, the coefficients of the variable quantities y^2 and z^2 must be equal to zero which produces the following two equations

$$\begin{aligned}
&\frac{3\Gamma' m' e'^2}{2.5A^5} - \frac{(2\Gamma m + 5\Gamma' m')e^2}{2.5A^5} + \frac{g}{2} - \frac{fA^2}{2B^2} = 0 \\
&\frac{3\Gamma' m' i'^2}{2.5A^5} - \frac{(2\Gamma m + 5\Gamma' m')i^2}{2.5A^5} + \frac{h}{2} - \frac{fA^2}{2C^2} = 0
\end{aligned}$$

from which it is possible to determine the two eccentricities e and i of the elliptic surface of the sea.

25. It is known that centrifugal force is proportional to its distance from the axis of rotation and to the square of the angular velocity of rotation. Therefore, if the axis of rotation is taken along the semi-axis A, which is also the x-coordinate axis and if f is the centrifugal force at a distance A from the axis, fu/A is the centrifugal force applied to an arbitrary point of the spheroid where $u = \sqrt{(y^2+z^2)}$. The force fu/A directed along the line u in the positive direction will give the moment $-(fu\,du)/A$ for which the integral is $-(fu^2)/2A$, that is, $-f(y^2+z^2)/2A$, must be added to the quantity Σ to account for the centrifugal force. Thus the conditions for equilibrium of the sea can be obtained by considering the mutual attraction of all the particles of the sea and Earth and the centrifugal force resulting from the rotation of the Earth, by substituting $f = 0$, $g = f^*/A$ and $h = f^*/A$ in the two preceding equations.

Since the two constants g and h are equal, it can be seen from these equations that if the two eccentricities e' and i' of the Earth are equal, then the two eccentricities e and i for the shape of the sea are also equal to each other, so that if the Earth is a spheroid of revolution, the sea will also be spheroidal. But if the Earth is not a spheroid of revolution neither is the sea and the two equations in question will give values for the two eccentricities e and i which will be different from the eccentricities e' and i' of the Earth.

26. Also, this solution is exact only to the second order of the quantities e^2, i^2, $(e')^2$, $(i')^2$, and if terms of higher order are to be considered in the quantities Σ and r, it will generally no longer be possible to verify the equation

$$\Sigma - \frac{f^*(y^2+z^2)}{2A} = \text{const.}$$

for the surface of equilibrium. Hence, it must be concluded that this surface does not exactly have the shape of an elliptical spheroid.

I say, in general, because in the case where the spheroid is homogeneous and without a core of different density, the attractions at an arbitrary point of the surface along the three coordinates x, y, z are found to be exactly represented by the formulas

$$mLx, \qquad mMy, \qquad mNz$$

where L, M, N are functions of A, B, C given by definite integrals from which this rigorous expression for Σ can be obtained

$$\Sigma = \frac{m}{2}(Lx^2 + My^2 + Nz^2)$$

Thus since the equation of equilibrium

$$\Sigma - \frac{f^*(y^2+z^2)}{2A} = \text{const.}$$

is of the same form as the equation of a spheroid

$$\frac{x^2}{A^2} + \frac{y^2}{B^2} + \frac{z^2}{C^2} = 1$$

they can be made identical by these two conditions

$$\frac{mM - f^*}{mL} = \frac{A^2}{B^2}, \qquad \frac{mN - f^*}{mL} = \frac{A^2}{C^2}$$

because of the arbitrary constant which gives $B = C$, since the quantities M and N are similar functions of B, C and C, B. Thus they reduce to one function which is used to determine the ratio of A to B.

This is the only case at present for which a rigorous solution has been found, which by the way is due to Maclaurin,[18] so that the problem of the shape of the Earth conceived physically is exactly resolved only by assuming the spheroid to be fluid and homogeneous. In this case, the two equations found above (Article 24) give, after putting $\Gamma = 1, \Gamma' = 0$, $g = h = f^*/A$ and $e = i$, the following relation

$$\frac{2me^2}{5A^4} - f^* = 0$$

If the centrifugal force is compared to gravity which is taken as unity and which is approximately on the order of e^2, m/A^2, it is only necessary to put $m/A^2 = 1$ in order to obtain

$$\frac{2e^2}{5A^2} = f^* = 2\frac{B^2 - A^2}{5A^2}$$

from which results

$$\frac{B}{A} = \sqrt{1 + \frac{5f^*}{2}}$$

However, since $f^* = 1/288$ then $B/A = 231/230$ approximately, which has been known for a long time.

Subsection III
The Equilibrium of a Free Fluid Mass and the Solid which it Covers

27. When all the points of the fluid and solid are acted on by arbitrary forces, the particular laws for equilibrium of a fluid and a solid which is partially or completely submerged,

depend on the terms of the general equation (Article 17) which are related to the upper and lower limits of integration and which contain only double integrals. These terms give this equation at the limits of integration

$$\begin{aligned}&\mathrm{S}\,\lambda''(\delta x''\,\mathrm{d}y\,\mathrm{d}z+\delta y''\,\mathrm{d}x\,\mathrm{d}z+\delta z''\,\mathrm{d}x\,\mathrm{d}y)\\&-\mathrm{S}\,\lambda'(\delta x'\,\mathrm{d}y\,\mathrm{d}z+\delta y'\,\mathrm{d}x\,\mathrm{d}z+\delta z'\,\mathrm{d}x\,\mathrm{d}y)=0\end{aligned}$$

which must hold at all points where the fluid is contiguous to the solid.

28. Let us at the outset consider the case of a fluid mass for which the outer surface is free and which surrounds a fixed and solid core of arbitrary shape. By taking the origin of the coordinate system at a point within the core, the quantities marked by a prime will apply to the surface of the core and the quantities marked by two primes to the external surface of the fluid. Thus, for all the points of this surface, the equation $\lambda''=0$ will be obtained which gives, as already seen above (Article 19)

$$\mathrm{S}\,\Gamma(X\,\mathrm{d}x+Y\,\mathrm{d}y+Z\,\mathrm{d}z)=K$$

for the shape of this surface. Then it only remains to verify the following equation

$$\mathrm{S}\,\lambda'(\delta x'\,\mathrm{d}y\,\mathrm{d}z+\delta y'\,\mathrm{d}x\,\mathrm{d}z+\delta z'\,\mathrm{d}x\,\mathrm{d}y)=0$$

for which all the terms are related to the surface of the core.

29. Since the integration of these terms is relative to the coordinates for which the differentials are contained in the expression for the surface elements $\mathrm{d}x\,\mathrm{d}y$, $\mathrm{d}x\,\mathrm{d}z$, $\mathrm{d}y\,\mathrm{d}z$, one must begin by reducing these elements to the same form which can be done by referring them to the element of the surface to which they correspond.

Let us designate by $\mathrm{d}s^2$ the element of the surface which is related to the element $\mathrm{d}x\,\mathrm{d}y$ of the xy-plane and let us call by γ' the angle that the tangential plane makes with the xy-plane. Because of the known property of planes, the following equation will be obtained

$$\mathrm{d}x\,\mathrm{d}y=\mathrm{d}s^2\cos\gamma'$$

and the integral $\mathrm{S}\,\lambda'\,\delta z'\,\mathrm{d}x\,\mathrm{d}y$ will become $\mathrm{S}\,\lambda'\cos\gamma'\,\delta z\,\mathrm{d}s^2$ for which the domain of definition is now all the points of the fluid surface.

Similarly, if $\mathrm{d}\sigma^2$ is the element of the surface which corresponds to the element $\mathrm{d}x\,\mathrm{d}z$ of the xz-plane and if the angle that the tangential plane makes with this xz-plane is denoted by β', one will have $\mathrm{d}x\,\mathrm{d}z=\mathrm{d}\sigma^2\cos\beta'$ and the integral $\mathrm{S}\,\lambda'\,\delta y\,\mathrm{d}x\,\mathrm{d}z$ will become $\mathrm{S}\,\lambda'\cos\beta'\,\delta y'\,\mathrm{d}\sigma^2$ for which the domain of definition will also be the entire surface of the fluid.

30. I now note that although the two elements $\mathrm{d}s^2$ and $\mathrm{d}\sigma^2$ of the surface cannot be equal to each other, nevertheless, because the two integrals containing these elements are with

respect to the same surface, nothing prevents us from using the same element in the two integrals. The reason is that by the nature of the differential calculus, the absolute value of the elements is arbitrary and has no influence on the value of the integral. Thus the integral $\mathrm{S}\,\lambda' \cos\beta'\delta y'\,\mathrm{d}\sigma^2$, can be replaced by $\mathrm{S}\,\lambda' \cos\beta'\,\delta y'\,\mathrm{d}s^2$. By the same reasoning, the integral $\mathrm{S}\,\lambda'\,\delta x'\,\mathrm{d}y\,\mathrm{d}z$ can be put in the form $\mathrm{S}\,\lambda' \cos\alpha'\,\delta x'\,\mathrm{d}s^2$, by denoting by α' the angle that the tangential plane makes with the xy-plane.

On the other hand, it is obvious that the elements $\mathrm{d}x$, $\mathrm{d}y$, $\mathrm{d}z$ can always be taken so that they satisfy the following conditions

$$\mathrm{d}x\,\mathrm{d}y = \cos\gamma'\,\mathrm{d}s^2, \qquad \mathrm{d}x\,\mathrm{d}z = \cos\beta'\,\mathrm{d}s^2, \qquad \mathrm{d}y\,\mathrm{d}z = \cos\alpha'\,\mathrm{d}s^2$$

which give

$$\mathrm{d}x = \mathrm{d}s\sqrt{\left(\frac{\cos\beta'\cos\gamma'}{\cos\alpha'}\right)}$$

$$\mathrm{d}y = \mathrm{d}s\sqrt{\left(\frac{\cos\alpha'\cos\gamma'}{\cos\beta'}\right)}$$

$$\mathrm{d}z = \mathrm{d}s\sqrt{\left(\frac{\cos\alpha'\cos\beta'}{\cos\gamma'}\right)}$$

By these transformations, the equation of limits will finally become

$$\mathrm{S}\,\lambda'(\cos\alpha'\,\delta x' + \cos\beta'\,\delta y' + \cos\gamma'\,\delta z')\mathrm{d}s^2 = 0$$

where the integration has been carried out over the entire surface of the fluid contiguous to the core.

31. Let us assume that the shape of this surface is represented by the differential equation

$$A\,\mathrm{d}x' + B\,\mathrm{d}y' + C\,\mathrm{d}z' = 0$$

By denoting by α', β', γ' the angles that the tangential plane makes with the xy-, xz- and yz-planes, the following expression can be obtained using the theory of surfaces[19]

$$\cos\alpha' = \frac{A}{\sqrt{(A^2+B^2+C^2)}}$$

$$\cos\beta' = \frac{B}{\sqrt{(A^2+B^2+C^2)}}$$

$$\cos\gamma' = \frac{C}{\sqrt{(A^2+B^2+C^2)}}$$

Thus the equation of the preceding article, relative to the surface, will become

$$\mathrm{S}(\lambda'\frac{A\,\delta x' + B\,\delta y' + C\,\delta z'}{\sqrt{(A^2+B^2+C^2)}})\mathrm{d}s^2 = 0$$

Since this surface has given shape and position, the variations $\delta x'$, $\delta y'$, $\delta z'$ of the coordinates of the particles which are contiguous to the surface must have between themselves a relation dependent on the equation of the same surface. Also, having assumed this equation to be $A\,\mathrm{d}x' + B\,\mathrm{d}y' + C\,\mathrm{d}z' = 0$, the following result will also be obtained

$$A\,\delta x' + B\,\delta y' + C\,\delta z' = 0$$

which satisfies the equation of limits of the preceding article without requiring any new equation.

32. Let p' be a line perpendicular to the surface drawn from the point to which the variations $\delta x'$, $\delta y'$, $\delta z'$ correspond and to a fixed point. Since α' is the angle that the tangential plane makes with the yz-plane, this will also be the angle that the perpendicular p' to this plane makes with the x-axis which is perpendicular to the yz-plane. Similarly, β' will be the angle of this perpendicular with the y-axis and γ' will be the angle of the same perpendicular with the z-axis. Therefore, whatever the variations $\delta x'$, $\delta y'$, $\delta z'$ one will have in general from Article 7 of SECTION II, after d is replaced by δ

$$\delta p' = \cos\alpha'\,\delta x' + \cos\beta'\,\delta y' + \cos\gamma'\,\delta z'$$

and the equation of Article 30, relative to the surface of the fluid can be put in the following form

$$\mathrm{S}\,\lambda'\,\delta p'\,\mathrm{d}s^2 = 0$$

where it is seen that each element $\lambda'\,\mathrm{d}s^2\,\delta p'$ of this integral represents the moment of a force $\lambda'\,\mathrm{d}s^2$ applied to the element $\mathrm{d}s^2$ of the surface and directed along the perpendicular p' to this surface. So the integral $\mathrm{S}\,\lambda'\,\delta p'\,\mathrm{d}s^2$ will represent the sum of the moments of all the forces λ' applied to each point of the surface and acting perpendicular to this surface.

This force, equal to λ', is obviously equal to the pressure exerted by the fluid on the surface of the core and which is equilibrated by the resistance of the core. But, in general, all the terms in the equation of limits which correspond to the surface of the fluid whether this surface is free or not, can be reduced to the form $\mathrm{S}\,\lambda\,\delta p\,\mathrm{d}s^2$ and it is obvious that the pressure λ must be equal to zero at all points where the surface is free, which we have already found by other means (Article 18).

33. If the core immersed in the fluid were mobile, then the variations δx, δy, δz should be increased by the variations dependent on the change in position of the core. In order to distinguish these diverse variations, we will designate by δx, δy, δz the variations resulting directly from the displacement of the particles of the fluid relative to the core which is viewed as fixed and we will denote by $\delta\xi$, $\delta\eta$, $\delta\zeta$ the variations which depend on the displacement of the core. The latter are expressed by the following formulas which we

found in Article 60 of SECTION V.

$$\delta\xi = \delta l + z\,\delta M - y\,\delta N$$
$$\delta\eta = \delta m - z\,\delta L + x\,\delta N$$
$$\delta\zeta = \delta n + y\,\delta L - x\,\delta M$$

Hence, in the general equation of Article 17, the expressions $\delta x + \delta\xi$, $\delta y + \delta\eta$ and $\delta z + \delta\zeta$ should be put in place of δx, δy, δz. Then the terms with the variations δx, δy, δz as well as those which will have the new variations $\delta\ell$, δm, δn, δL, δM and δN should be equated to zero after they have been taken outside the integral sign S because these variations are the same for all particles of the fluid.

It is observed at the outset that the introduction of the variations $\delta\xi$, $\delta\eta$, $\delta\zeta$ does not bring any change to the equations which must hold for every point of the fluid and which result from the terms with triple integration, because by equating to zero the coefficients of δx, δy, δz in these terms, the variations $\delta\xi$, $\delta\eta$, $\delta\zeta$ disappear simultaneously. From which it follows that the general laws of equilibrium expressed by the formulas of Article 19 are independent of the state as well as the configuration of the core.

34. Thus there is nothing more to consider than the equation of limits which we have reduced to the following form in Article 30

$$\mathrm{S}\,\lambda'(\cos\alpha\,\delta x' + \cos\beta\,\delta y' + \cos\gamma\,\delta z')\mathrm{d}s^2 = 0$$

After substituting in the equation above for the variations $\delta x'$, $\delta y'$, $\delta z'$ the values $\delta x' + \delta\xi'$, $\delta y' + \delta\eta'$, $\delta z' + \delta\zeta'$ where the prime denotes the surface of the fluid contiguous to the core, this equation becomes

$$\mathrm{S}\,\lambda'(\cos\alpha\,\delta x' + \cos\beta\,\delta y' + \cos\gamma\,\delta z')\mathrm{d}s^2$$
$$+\mathrm{S}\,\lambda'(\cos\alpha\,\delta\xi' + \cos\beta\,\delta\eta' + \cos\gamma\,\delta\zeta')\mathrm{d}s^2 = 0$$

The portion of this equation which contains the variations $\delta x'$, $\delta y'$, $\delta z'$ is identically equal to zero as we have demonstrated in Article 31. The remaining portion to the left of the equal sign will also be equal to zero. The values for $\delta\xi'$, $\delta\eta'$, $\delta\zeta'$, will then be substituted and the quantities multiplied by $\delta\ell$, δm, δn and δL, δM, δN will be equated individually to zero. Thus the following six equations will result

$$\mathrm{S}\,\lambda'\cos\alpha\,\mathrm{d}s^2 = 0, \qquad \mathrm{S}\,\lambda'\cos\beta\,\mathrm{d}s^2 = 0, \qquad \mathrm{S}\,\lambda'\cos\gamma\,\mathrm{d}s^2 = 0$$
$$\mathrm{S}\,\lambda'(y'\cos\gamma - z'\cos\beta)\mathrm{d}s^2 = 0$$
$$\mathrm{S}\,\lambda'(z'\cos\alpha - x'\cos\gamma)\mathrm{d}s^2 = 0$$
$$\mathrm{S}\,\lambda'(x'\cos\beta - y'\cos\alpha)\mathrm{d}s^2 = 0$$

which are necessary for the total equilibrium of the fluid and solid.

These equations are related to those of Article 62 of SECTION V when $\mathrm{d}s^2$ is substituted for $\mathrm{d}m$ and $\lambda'\cos\alpha$, $\lambda'\cos\beta$, $\lambda'\cos\gamma$ for X, Y, Z. Indeed, λ', which is the force resulting from the pressure which acts perpendicular to the surface of the solid core, and $\lambda' \cos\alpha$, $\lambda' \cos\beta$, $\lambda' \cos\gamma$ will be the forces which result from this pressure in the direction of the x-, y- and z-coordinates and the surface must be in equilibrium since every point is acted on by these same forces.

35. But when a fluid is supported by a solid with an arbitrary configuration and both of them are submitted to arbitrary forces, it is simpler to obtain the solution of the problem directly from the fundamental equation of Article 16 by immediately substituting for δx, δy, δz, their complete values $\delta x + \delta\xi$, $\delta y + \delta\eta$, $\delta z + \delta\zeta$ (Article 33).

Since the variations δx, δy, δz are independent from the other variations $\delta\ell$, δm, etc., they will produce an equation similar to the one of Article 17 and will give the same result for the equilibrium of the fluid as in the case where the solid is assumed fixed.

With respect to the other variations $\delta\xi$, $\delta\eta$, $\delta\zeta$, it is easy to see at the outset that they will not contribute to the values of the partial differences $(\mathrm{d}\delta x)/\mathrm{d}x$, $(\mathrm{d}\delta y)/\mathrm{d}y$, $(\mathrm{d}\delta z)/\mathrm{d}z$, because the variations $\delta\ell$, δm, δn, δL, δM, δN are assumed to be independent of x, y, z.

Thus it will suffice to substitute $\delta\xi$, $\delta\eta$, $\delta\zeta$ for δx, δy, δz in the formula

$$\mathrm{S}(X\,\delta x + Y\,\delta y + Z\,\delta z)\Gamma\,\mathrm{d}x\,\mathrm{d}y\,\mathrm{d}z$$

and to equate separately to zero the quantities multiplied by each of the six variations $\delta\ell$, δm, δn, δL, δM, δN after they have been taken outside the integral sign S. It is obvious that one will have in this fashion the same equations that have been found in Chapter IV of SECTION V for the equilibrium of a solid body of which each particle $\mathrm{d}m$, which is here represented by $\Gamma\,\mathrm{d}x\,\mathrm{d}y\,\mathrm{d}z$ is acted on by arbitrary forces X, Y, Z so that the equations of equilibrium for a fluid surrounding a solid core are the same as if the fluid were solid.

36. It results from the two approaches to viewing the variations that the pressure of the fluid on the surface of the core is equivalent to the action of all the forces which act on each particle of the fluid assuming that the fluid is considered a solid and that the core is increased by the entire mass of the fluid which is assumed to be solid. Since this theorem of statics is important, we believe that we must demonstrate in a more direct manner how it can be deduced from our formulas.

Everything reduces to the demonstration that the equation

$$S(X\,\delta\xi + Y\,\delta\eta + Z\,\delta\zeta)\Gamma\,\mathrm{d}x\,\mathrm{d}y\,\mathrm{d}z = 0$$

gives the same results as the equation of limits

$$\mathrm{S}\,\lambda'(\delta\xi'\,\mathrm{d}y\,\mathrm{d}z + \delta\eta'\,\mathrm{d}x\,\mathrm{d}z + \delta\zeta'\,\mathrm{d}x\,\mathrm{d}y) = 0$$

From the conditions of equilibrium of a fluid, the following relations have been obtained (Article 19)

$$\Gamma X = \frac{\mathrm{d}\lambda}{\mathrm{d}x}, \qquad \Gamma Y = \frac{\mathrm{d}\lambda}{\mathrm{d}y}, \qquad \Gamma Z = \frac{\mathrm{d}\lambda}{\mathrm{d}z}$$

And since the values of $\delta\xi$, $\delta\eta$, $\delta\zeta$ (Article 33) are respectively independent of x, y, z, one will also have

$$\Gamma X\,\delta\xi = \frac{\mathrm{d}(\lambda\,\delta\xi)}{\mathrm{d}x}, \qquad \Gamma Y\,\delta\eta = \frac{\mathrm{d}(\lambda\,\delta\eta)}{\mathrm{d}y}, \qquad \Gamma Z\,\delta\zeta = \frac{\mathrm{d}(\lambda\,\delta\zeta)}{\mathrm{d}z}$$

Thus the first equation will become

$$\mathrm{S}(\frac{\mathrm{d}(\lambda\,\delta\xi)}{\mathrm{d}x} + \frac{\mathrm{d}(\lambda\,\delta\eta)}{\mathrm{d}y} + \frac{\mathrm{d}(\lambda\,\delta\zeta)}{\mathrm{d}z})\mathrm{d}x\,\mathrm{d}y\,\mathrm{d}z = 0$$

The first term under the integral sign is integrable with respect to x, the second with respect to y and the third with respect to z. Therefore, if these partial integrations were performed as was done in Article 17, then the following equation of limits results

$$\begin{aligned}&\mathrm{S}\,\lambda''(\delta\xi''\,\mathrm{d}y\,\mathrm{d}z + \delta\eta''\,\mathrm{d}x\,\mathrm{d}z + \delta\zeta''\,\mathrm{d}x\,\mathrm{d}y)\\ &-\mathrm{S}\,\lambda'(\delta\xi'\,\mathrm{d}y\,\mathrm{d}z + \delta\eta'\,\mathrm{d}x\,\mathrm{d}z + \delta\zeta'\,\mathrm{d}x\,\mathrm{d}y) = 0\end{aligned}$$

But λ'' equals zero (Article 23) because the external surface of the fluid is assumed free. Thus only the following equation will remain

$$\mathrm{S}\,\lambda'(\delta\xi'\,\mathrm{d}y\,\mathrm{d}z + \delta\eta'\,\mathrm{d}x\,\mathrm{d}z + \delta\zeta'\,\mathrm{d}x\,\mathrm{d}y) = 0$$

Hence, the two equations are identical.

37. When considering the variations which are dependent on the displacements of the core, the core and the fluid which covers it can be viewed as if it were one solid mass. Also, when all the points of the core are under the action of arbitrary forces, these forces should be considered in addition to those which act on the particles of the fluid and the solutions given in Chapter IV of SECTION V should be applied for the equilibrium of the mass composed of fluid and solid as if it were a continuous solid.

Subsection IV
The Equilibrium of Incompressible Fluids Contained in Vessels

38. The general equation of Article 27 for limits must be applicable at every point on the walls of the vessel in which the fluid is contained. Let us put this equation in the form

$$\begin{aligned}&\mathrm{S}(\lambda''\,\delta x'' - \lambda'\,\delta x')\mathrm{d}y\,\mathrm{d}z\\ &+\mathrm{S}(\lambda''\,\delta y'' - \lambda'\,\delta y')\mathrm{d}x\,\mathrm{d}z\\ &+\mathrm{S}(\lambda''\,\delta z'' - \lambda'\,\delta z')\mathrm{d}x\,\mathrm{d}y = 0\end{aligned}$$

and let us first consider the terms $S(\lambda''\,\delta z'' - \lambda'\,\delta z')dx\,dy$ where $\delta z''$ and $\delta z'$ are the variations of the z-coordinate so that it corresponds to two points of the fluid surface which have the same x- and y-coordinates.

It is obvious that the variations $\delta z''$ have a tendency to pull the particles away from the surface of the fluid mass and that the variations $\delta z'$, assuming that both are positive, press into the mass of fluid the particles of the opposite surface so that by giving to the latter a negative sign, the variations $\delta z''$ and $-\delta z'$ will also tend to pull the particles of the surface away from the fluid mass. Then the double integral

$$S(\lambda''\,\delta z'' - \lambda'\,\delta z')dx\,dy$$

will represent the sum of all the quantities $\lambda\,\delta z\,dx\,dy$ which correspond to all points of the surface of the fluid and for which the variations δz will be assumed to have the same effect whether they are inside or outside of the fluid. Therefore, with this condition, we can give to this integral the simpler form $S\,\lambda\,dx\,dy$. In the same fashion, and with the same conditions, the other two double integrals $S(\lambda''\,\delta y'' - \lambda'\,\delta y')dx\,dz$ and $S(\lambda''\,\delta x'' - \lambda'\,\delta x')dy\,dz$ can be reduced to the form $S\,\lambda\,\delta y\,dx\,dz$, $S\,\lambda\,\delta x\,dy\,dz$.

Therefore, the equation being considered can be reduced to the following form

$$S\,\lambda\,\delta z\,dx\,dy + S\,\lambda\,\delta y\,dx\,dz + S\,\lambda\,\delta z\,dy\,dx$$

which can also be reduced further using the approach of Article 33 to the following form

$$S\,\lambda(\cos\alpha\,\delta x + \cos\beta\,\delta y + \cos\gamma\,\delta z)ds^2 = 0$$

where α, β, γ are the angles that the plane tangent to the surface at the point corresponding to the coordinates x, y, z makes with the yz-, xz-, xy-planes. The integration of this equation must extend over the entire surface of the fluid and the variations δx, δy, δz are all assumed directed from within the fluid mass to the outside.

39. Since the variations δx, δy, δz are indeterminate at the points where the surface is free, the equation can only be satisfied if $\lambda = 0$, which will give the equation of this surface as has been observed in Article 18. If the quantities which correspond to all the other points of the surface where the fluid is contiguous to the sides of the vessel are indicated by a prime, the same equation which has been found for the surface of the core covered by a fluid (Article 30) will be used for those sides. Thus all the conclusions drawn from this equation, from the article which has just been cited down to the end of the preceding paragraph, can be applied to the sides of the vessel in which the fluid is contained, whatever its configuration and whether it remains fixed or must be in equilibrium under the pressure of the fluid and under the action of external forces which act on it in an arbitrary fashion.

SECTION VIII
The Equilibrium of Compressible and Elastic Fluids

1. As in Article 10 of the preceding section, let X, Y, Z represent the forces which act on each point of the fluid mass referred to the directions of the x-, y-, z-coordinates and directed towards the origin. The sum of their moments will be expressed at the outset as $\mathrm{S}(X\,\delta x + Y\,\delta y + Z\,\delta z)\mathrm{d}m$.

In elastic fluids there is also an internal force which is called the force of elasticity and which has a tendency to dilate or to increase their volume. Thus let ϵ be the elasticity of an arbitrary particle $\mathrm{d}m$. Since this force has a tendency to increase the volume $\mathrm{d}x\,\mathrm{d}y\,\mathrm{d}z$ of the particle, it will have or will be assumed to have for its moment the quantity $-\epsilon\,\delta(\mathrm{d}x\,\mathrm{d}y\,\mathrm{d}z)$ from Article 9 of SECTION II. I assign here a negative sign to the moment of this force because it increases the volume element dxdydz while the forces X, Y, Z decrease the values of the variables x, y, z. Thus the sum of the moments resulting from the elasticity of all the fluid mass will be expressed by $-\mathrm{S}\,\epsilon\,\delta(\mathrm{d}x\,\mathrm{d}y\,\mathrm{d}z)$.

Therefore, the total sum of the moments of the forces which act on the fluid will be

$$\mathrm{S}(X\,\delta x + Y\,\delta y + Z\,\delta z)\mathrm{d}m - \mathrm{S}\,\epsilon\delta(\mathrm{d}x\,\mathrm{d}y\,\mathrm{d}z)$$

and because there is no particular condition to fulfill here, the general equation of equilibrium will be obtained by simply equating this sum to zero.

2. Thus the equation for the equilibrium of elastic fluids will have the same form as the one which has been found in the preceding section (Article 10) for the equilibrium of incompressible fluids because in this equation $\delta L = \delta(\mathrm{d}x\,\mathrm{d}y\,\mathrm{d}z)$ (Article 11) which makes the term $\mathrm{S}\,\lambda\,\delta L$, deriving from the condition of incompressibility, entirely similar to the term $\mathrm{S}\,\epsilon\,\delta(\mathrm{d}x\,\mathrm{d}y\,\mathrm{d}z)$ resulting from the moments of the elastic forces.

It results from this development that the formulas found for the equilibrium of incompressible fluids can be immediately and without any restrictions applied to the equilibrium of elastic fluids by simply replacing the coefficient λ by $-\epsilon$, that is, by assuming that the quantity λ, which represents the pressure in an incompressible fluid, taken here with a negative sign, expresses the force of elasticity of each element in an elastic fluid.

3. The elasticity ϵ depends on the density and the temperature of each particle of the fluid and must be viewed as a known function of these two quantities. However, the density of each particle is unknown because it depends on the ratio of the mass $\mathrm{d}m$ of the particle to its volume $\mathrm{d}x\,\mathrm{d}y\,\mathrm{d}z$ and it is not possible to determine this ratio which depends on the number of elementary particles contained in the differential element $\mathrm{d}x\,\mathrm{d}y\,\mathrm{d}z$ of the fluid mass, using the differential calculus.

Therefore, the value of the elasticity is only known **a posteriori** by means of the forces which hold the fluid in equilibrium. Thus, the value of ϵ should be determined in the same fashion that λ was found in Article 19 of the preceding section.

4. By replacing λ with $-\epsilon$, the following equations will be obtained

$$\frac{\mathrm{d}\epsilon}{\mathrm{d}x} + \Gamma X = 0, \qquad \frac{\mathrm{d}\epsilon}{\mathrm{d}y} + \Gamma Y = 0, \qquad \frac{\mathrm{d}\epsilon}{\mathrm{d}z} + \Gamma Z = 0$$

which leads to

$$\mathrm{d}\epsilon + \Gamma(X\,\mathrm{d}x + Y\,\mathrm{d}y + Z\,\mathrm{d}z) = 0$$

and consequently,

$$\epsilon = \text{const.} - \mathrm{S}\,\Gamma(X\,\mathrm{d}x + Y\,\mathrm{d}y + Z\,\mathrm{d}z)$$

Thus the quantity $\Gamma(X\,\mathrm{d}x+Y\,\mathrm{d}y+Z\,\mathrm{d}z)$ must be a complete differential for the equilibrium of elastic fluids as it is for incompressible fluids. From this it can be concluded, as in Article 20 of the preceding section, that when the quantity $X\,\mathrm{d}x+Y\,\mathrm{d}y+Z\,\mathrm{d}z$ is itself a complete differential, the density Γ must be uniform at each equipotential level.

5. By designating by θ the heat which exists at every point of the fluid mass, it is ordinarily assumed for air that ϵ is proportional to $\Gamma\theta$, by neglecting other causes such as vapors, electricity, etc., which may have an effect on its elasticity.

Substituting for Γ in the equation $\mathrm{d}\epsilon + \Gamma(X\,\mathrm{d}x + Y\,\mathrm{d}y + Z\,\mathrm{d}z)$ its value $\epsilon/m\theta$, it will become

$$m\frac{\mathrm{d}\epsilon}{\epsilon} + \frac{X\,\mathrm{d}x + Y\,\mathrm{d}y + Z\,\mathrm{d}z}{\theta} = 0$$

Since heat is produced by local causes, the quantity θ will be a given function of x, y, z and in order for the preceding equation to exist, the quantity

$$\frac{X\,\mathrm{d}x + Y\,\mathrm{d}y + Z\,\mathrm{d}z}{\theta}$$

must be an exact differential.

6. Therefore, in the case where $X\,\mathrm{d}x+Y\,\mathrm{d}y+Z\,\mathrm{d}z = \mathrm{d}\Pi$ (Article 20, preceding section), θ must be a function of Π. Consequently, one will have $\mathrm{d}\theta = 0$ when $\mathrm{d}\Pi = 0$. From which it follows that the amount of heat must be constant at each equipotential level to which gravity is perpendicular. Otherwise, it will be impossible for the atmosphere to be in equilibrium. Thus in order that the air be at rest, the temperature must be equal over the entire surface of the Earth and that it varies only by ascending from one level to another in the atmosphere.

7. The equation of limits for the surface of the fluid becomes $\mathrm{S}\,\epsilon\,\delta p\,\mathrm{d}s^2 = 0$ by using the formulation of Article 32 of the preceding section. In this form the equation is readily

understood, because at the surface only the force of elasticity ϵ which acts along the line p perpendicular to the same surface must be considered. If the fluid is contained in a vessel, the variations δp are equal to zero and the equation holds, but if a part of the surface were free, then the elasticity ϵ must be equal to zero, otherwise, the fluid is no longer contained and will dissipate.

8. The elasticity ϵ in the atmosphere is proportional to the level of mercury in a barometer which we will designate by h. Let Z denote the force of gravity. Let us also take the ordinate z perpendicular to the surface of the Earth and directed from the surface upward. The equation of Article 5 will become

$$m\frac{\mathrm{d}h}{h} + \frac{Z\,\mathrm{d}z}{\theta} = 0$$

which gives after integration, taking H for the height of the barometer when $z = 0$

$$m\,\ell n\frac{H}{h} = \int \frac{Z\,\mathrm{d}z}{\theta}$$

The integral is assumed to have for a lower limit of integration the point where $z = 0$.

It is obvious here that the logarithm of the ratio of the elevations of the barometer rigorously gives only a quantity proportional to the value of the integral $\int(Z/\theta)\,\mathrm{d}z$ contained between the heights of the two stations and that to deduce the difference between the elevations of the stations, the calorific law giving θ as a function of z must be assumed known.

9. It is known that gravity decreases in an inverse proportion to the square of the distance from the center of the Earth. Thus taking r for the radius of the Earth and assuming that z is the altitude above the surface of the Earth, one has

$$Z = \frac{g}{(1+\frac{z}{r})^2}$$

where g is the acceleration of gravity at the Earth's surface. And from this

$$Z\,\mathrm{d}z = g\frac{\mathrm{d}z}{(1+\frac{z}{r})^2} = g\,\mathrm{d}x$$

and by putting

$$x = \frac{z}{1+\frac{z}{r}}$$

so that one will have

$$m\,\ell n\frac{H}{h} = g\int\frac{\mathrm{d}x}{\theta}$$

and the problem is reduced to obtaining θ as a function of x.

10. By assuming θ constant and in order to shorten the expression, by making $m\theta/g = K$, the following equation will be found

$$x = K\ell n\frac{H}{h} = K(\ell nH - \ell nh)$$

and the value of z will be obtained from the formula

$$z = \frac{x}{1 - \frac{x}{r}}$$

If for the elevations z, which are not very large the term x/r which is always a very small quantity, is neglected, one has simply that $z = x$, which gives the ordinary rule for the measure of altitude by the barometer.

The coefficient K must be determined by observation. Deluc[20] has found that this coefficient is equal to 10000 for the uniform temperature of 16-3/4° on the Reaumur scale while taking the logarithm from a table and the altitudes in toises.[21] For the other temperatures, he increased or decreased it by a 215th part for each degree above or below 16-3/4° Since the temperatures vary from one station to another, he used the arithmetic mean of the temperatures of the two stations. Since then, this rule has been improved with more exact information and by new corrections applied to the coefficient K.

11. Also, taking the uniform temperature to be the arithmetic mean between the extreme temperatures of the column of air, it is assumed that the amount of heat diminishes in an arithmetic progression. In order to see what results from this hypothesis, one will have $\theta = \Theta(1 - nz)$ or rather $\theta = \Theta(1 - nx)$ to simplify the calculation, where Θ is the temperature when $x = 0$. Substituting this value in the formula $\mathrm{d}x/\theta$, integrating and then replacing n by the value obtained from the preceding equation, one will have[22]

$$\int \frac{\mathrm{d}x}{\theta} = x\frac{\ell n\Theta - \ell n\theta}{\Theta - \theta} = \frac{x}{k}(1 - \frac{T+t}{2k} + \frac{T+Tt+t^2}{3k^2} - \cdots)$$

where $\Theta = k + T, \theta = k + t$, k is a fixed temperature, and T, t are the readings in degrees of the thermometer above this temperature.

By setting $mk/g = K$, the formula of Article 9 will give the following result, using a second order approximation for T and t

$$x = K(1 + \frac{T+t}{2k} - \frac{(T-t)^2}{12k^2}\ell n\frac{H}{h}$$

The first two terms follow Deluc's rule and the third term will almost always be negligible.

PART II

DYNAMICS

SECTION I
THE VARIOUS PRINCIPLES OF DYNAMICS

Dynamics is the science of accelerating or retarding forces and the diverse motions which they produce. This science is entirely due to modern investigators and it is Galileo who developed its basic principles. Before him, only forces applied to bodies in a state of equilibrium were considered and although the cause of the acceleration of heavy bodies and the curvilinear motion of projectiles could not be attributed to anything but the constant action of gravity, no one had been able to determine the laws of these common everyday motions with such a simple cause. Galileo was the first to make this important step and by his discovery began a new and vast development for the advancement of mechanics. This discovery is formulated and developed in the work entitled *Discorsi e dimostrazioni matematiche intorno a due nuove scienze*[1] which was published for the first time at Leyden in 1638. It did not bring him as much fame during his lifetime as the discoveries made in the skies, but this discovery constitutes the most solid and real part of the glory of this man.

The discoveries of the satellites of Jupiter, the phases of Venus, sunspots, etc. required only a telescope and diligence, but it takes an extraordinary genius to resolve and understand the laws of nature in the ever present complexity of the phenomena whose explanation had nevertheless escaped even the research of the philosophers.

Huygens, who seems to have been destined to perfect and complete most of Galileo's discoveries added to the theory of the acceleration of heavy bodies, the motion of pendulums and centripetal forces and thus prepared the way to the great discovery of universal gravitation. Mechanics later became a new science in the hands of Newton and his *Principia mathematica*, which was published for the first time in 1687, opened the epoch of this revolution.

Finally, the discovery of the infinitesimal calculus enabled geometers to reduce the laws of motion for solid bodies to analytical equations[2]and the research on forces and the motion which they produce has become the principal object of their work.

I propose here to offer a new means of facilitating this research, but before doing this, it is advantageous to discuss the principles which serve as a foundation for dynamics and then to present the order and development of the ideas which have contributed the most to the understanding and perfection of this science.

1. The theory of various types of motion and the forces of acceleration which produce them are founded on the following general laws: all motion imparted to a body is intrinsically uniform and rectilinear and different motions imparted all at once or successively to a given body are composed in such a way that the body is located at every instant at the same point in space where it should be from the composition of the individual motions if each of them existed in reality and separately in the body. It is from these two laws that consistency is developed between the known principles of the inertia force and the composition of motion. Galileo knew the first of these two principles and deduced the laws of motion of projectiles from it by combining the oblique motion, the effect of the impulse imparted to the body, with its perpendicular descent due to the action of gravity.

With respect to the laws for the acceleration of falling bodies, they can be deduced directly from the consideration of the constant and uniform action of gravity. That is, for all bodies receiving during the same instant of time equal increments of velocity in the same direction, the total velocity acquired at the end of any time must be proportional to that time. It is evident that this ratio of constant velocity to time should be proportional to the intensity of the force that gravity exerts to put the bodies in motion. This must occur in such a way that for the motion on inclined planes, this ratio must not be proportional to the absolute force of gravity as in the case of vertical motion, but to a related force which depends upon the angle of inclination of the plane. This force is determined from the laws of statics which provide an easy means of comparison between motions of bodies descending on planes of diverse inclinations.

However, it does not seem that Galileo discovered the laws of falling bodies in this manner. On the contrary, he began by considering the notion of a uniformly accelerated motion in which the velocities increase in proportion to the time. Hence, he deduced geometrically the principal properties of this type of motion and more specifically, the law by which distance increased with the square of the time. Then he convinced himself through experiments that this law actually holds in the motion of falling bodies or for those on any inclined plane. He was obliged to admit first the uncertain principle that the velocities acquired in descending equal vertical distances are always equal and it is only shortly before his death and after the publication of his *Dialogues*[3] that he discovered the demonstration of this principle when considering the relative action of gravity on an inclined plane. This demonstration has been added to later editions of this work.

2. The constant ratio existing between velocity and time or between distance and the square of the time in uniformly accelerated motions can therefore be taken as a measure of the force of acceleration which continually acts on a moving body because in effect this force can only be estimated from the effect that it produces on the body which consists of the developed velocities or the distances traversed in any given time.

Therefore, for the estimation of these forces it is sufficient to consider the motion produced in any finite or infinitely small interval of time provided the force is viewed as constant during that time. Consequently, whatever the motion of the body and its law of acceleration, by the nature of the differential calculus the action of all accelerating forces can be regarded as constant in an infinitesimal time. The value of the applied force on a body at any instant of time can always be determined by comparing the velocity developed at that instant with the duration of that instant of time or the distance that the body traverses with the square of the duration of that instant. It is not even necessary that the body actually traverses this distance. It is sufficient that it can be imagined to be traversed by a composite motion since the effect is the same in one case as in the other according to the principles of motion discussed above.

It is in this manner that Huygens discovered that the centrifugal force of a body moving in a circle with constant velocity is proportional to the square of the velocity divided by the radius of the circle.[4] He compared these forces with the gravitational force at the surface of the Earth, as can be seen from the demonstrations that he left of his theorems on centrifugal force published in 1673 at the end of the work called the *Horologium oscillatorium*.[5]

By combining this theory of centrifugal force with a theory developed later, whose author is also Huygens, and which reduces each infinitesimal segment of any curve to an arc of a circle, it was easy for him to extend it to any curve. But it was reserved for Newton to take the next step and complete the science of variable motion and the accelerating forces which generate them. This science now consists of a few basically simple differential formulas, but Newton had constantly made use of the simplified geometric method when considering the former and latter concepts and even if he sometimes used the analytical calculus, it was uniquely the series method that he used which must be distinguished from the differential method although it is easy to consider them related and to refer to them as a single principle.

The geometers who treated the subject of accelerating forces after Newton had almost always limited their efforts to generalizing these theorems and expressing them in differential notation. Thus the different centrifugal force formulas appeared which are found in various books on mechanics but they can hardly be used because they only apply to curves which are assumed to be the result of a unique force directed towards a center. Now there are general formulas for the determination of any motion produced by an arbitrary force.

3. If the motion of a body is considered and the forces producing it resolved in the direction of three orthogonal straight lines, the motions and the forces relative to each of these directions can be considered separately.[6] Because of the orthogonal nature of these directions, it is recognized that each of these partial motions can be regarded as independent of the other two and the body's motion can be changed only by the force acting in the direction of this motion. Therefore, it can be concluded that these three motions must independently obey the laws for rectilinear accelerating or retarding motions. On the other hand, in rectilinear motion the effect of accelerating force consists only of the alteration of the velocity of the body and this force must be measured by the ratio of the increase or decrease of velocity during any given instant, that is, by the derivative of velocity with respect to time. Since the velocity is defined in variable motion by the derivative of distance with respect to time, the acting force can be measured by the second differential of distance divided by the square of the differential of time which is assumed constant. Therefore, the

second differential of the distance traversed by the body or assumed traversed in each of the three orthogonal directions divided by the squared differential of time will define the accelerating force that the body will be subjected to in this same direction and therefore, be equal to the actual force assumed to act in that direction. This constitutes the very well-known principle of accelerating forces.

It is not necessary for the three directions to which the instantaneous motion of the body is referred to be absolutely fixed in space. It is sufficient that they are fixed for an instant. Hence, in motions along curved lines, one can use for each instant the following directions, one in the direction of the tangent and the other two perpendicular to the curve. In this case, the accelerating force along the tangent called the tangential force will always be used to change the absolute velocity of the body and it will be expressed by the increment of this velocity divided by the increment of time.

On the contrary, the normal forces will only change the direction of the body and will depend on the curvature of its trajectory. By reducing the normal forces to a resultant, this composite force must be in the plane of curvature and must be expressed by the square of the velocity divided by the osculating radius because at any instant the body can be regarded as moving in the osculating circle.

This is how the well-known tangent and normal forces which served for a long time to resolve the problems of the motion of bodies under the action of given forces were found. Euler's *Mechanics*[7]which was published in 1736 and which must be regarded as the first great study of mechanics where the analytical method was applied to the science of motion, is founded entirely on these formulas. But these formulas have been nearly abandoned because a simpler approach to expressing the effect of accelerating forces on mobile bodies has been found.

The simpler method consists of relating the motion of the body and the forces producing it to fixed directions in space. Then by using three rectangular coordinates, which have these same directions, for the determination of the position of the body in space, the variations of these coordinates will evidently represent the distance traversed by the body in the direction of these coordinates. Therefore, their second differentials divided by the squared differential of time which is assumed constant, will express the accelerating forces acting in the direction of these coordinates. Thus after setting these two expressions equal to the forces given by the nature of the problem, one will have three similar equations which will serve to define every property of the motion. This approach to establishing the equations for the motion of a body acted upon by any force, by reducing the motion to the rectilinear case, is preferable to all others because of its simplicity. It should have been developed first, but it seems that Maclaurin was the first to use it in his *Treatise of Fluxions*[8]which was published in English in 1742. It has now been universally adopted.

4. Using the principles which have been developed, the laws of motion of a free body acted upon by any force, can be determined provided that the body can be treated as a point.

These principles can be applied to the study of the motion of several bodies which exert on one another a mutual attraction according to a law such as one which is a known function of distance. Finally, it is not difficult to extend them to motion in a resisting medium as well as to those on any given curved surface because the resistance of the medium is nothing less than a force acting in the opposite direction to that of the motion. And whenever a body is forced to move on a given surface, there is necessarily a force acting perpendicular to the supporting surface and its unknown value can be determined according to the conditions resulting from the nature of that same surface.

However, if the motion of several bodies which act on one another by impact, by pressure or by immediate action is considered as in the case of an ordinary collision or by means of cords or rigid levers to which they are attached or in general by any other means, in this case, the question is of a higher order and the preceding principles are insufficient to resolve them. Since in this case the forces acting on the bodies are unknown, the forces of attraction that they exert on one another according to their relative positions must be known. Therefore, it is necessary to introduce a new principle which will serve to determine the force of bodies in motion according to their mass and velocity.

5. This principle states that to impart a given velocity to a mass in any direction whether it is in a state of rest or in motion, there must be a force whose magnitude is proportional to the product of the mass with its velocity and whose direction is the same as that of the velocity. The product of the mass of a body multiplied by its velocity is commonly called the quantity of motion in the body because, in effect, it is the sum of the motion of all of the parts of the body. Hence, the forces produced by the quantities of motion that they are capable of producing and reciprocally the quantity of motion of a body is measured by the force that the body is capable of exerting against an obstacle and which is called the percussion. Hence, if two inelastic bodies collide directly while coming from opposite directions with an equal quantity of motion, their forces must counterbalance and equilibrate each other and consequently, the bodies must stop and remain at rest. But if the collision takes place obliquely, it is necessary for the destruction of the motion that the forces follow the well-known law of the equilibrium of levers.

It seems that Descartes was the first to recognize the first principle which we have just discussed but he erred in applying it to the collision of bodies since he believed that the absolute quantity of motion must always be conserved.

Wallis is really the first to have a clear understanding of this principle and he has applied it successfully to develop the laws for the transfer of motion during the collision of rigid or elastic bodies as can be seen from his publications in the Philosophical Transactions of 1669 and from the third part of his treatise *de Motu*[9] published in 1671.

Just as the product of the mass and the velocity expresses the finite force of a body in motion, the product of the mass and the accelerating force which we know to be expressed by the increment of velocity divided by the increment of time will define the elementary or generating force. If this quantity is considered as the measure of the work that the body can produce due to the elementary velocity it has gained or tends to gain, it constitutes what is called pressure. But if it is regarded as the measure of the force or the necessary power to impart this velocity, it is what is called the motor force. In this manner, the pressures or motor forces will be annihilated or be in equilibrium if they are equal and opposite in direction or when they are applied to any machine, they will follow the laws of equilibrium for that machine.

6. When the bodies are joined together in such a way that they can respond freely to the impulses they receive and to the accelerating forces acting upon them, the bodies will necessarily exert on one another continuous pressures which alter their motions and render the determination of these motions difficult.

The first and the simplest of the type of problem with which the geometers have dealt is that of the center of oscillation. This celebrated problem was the object of discussion at the beginning of the last century and again from the middle of the preceding century through the efforts and tentative solutions that the greatest geometers have made to solve it. And since the immense progress made in dynamics since then is owed to those tentative solutions, I think I must present here a concise summary in order to show to what degree this science has been lifted towards the perfection that it now enjoys.

The letters of Descartes provide the first sketches of the research on the center of oscillation. In these letters it is learned that Mersenne had proposed to geometers to determine the magnitude which a body of arbitrary shape should have which when suspended from a point oscillates with the same period as a pendulum of given length with a single weight at its extremity. Descartes observes that this question is related to that of the center of gravity around which the gravity of all the parts of the body is in equilibrium in such a way that this center descends in the same manner as if the rest of the body is annihilated or concentrated at this same center. Thus for weights rotating about a fixed axis there must be a center which he calls the center of agitation[10] about which all the forces of agitation of the body are counterbalanced in such a way that this center is free of the action of these forces and can move as if all other parts of the body were annihilated or were concentrated at this same center. Consequently, all bodies in which this center is equally displaced from the axis of rotation will vibrate with the same period.

In accordance with the concept of the center of agitation, Descartes gave a general method to determine it for a body of arbitrary shape. This method consists of determining the center of gravity of the forces of agitation of all parts of the body by estimating the forces of each element of the body by the product obtained from the multiplication of the masses and the velocities which are here proportional to the distances from the axis of rotation and by assuming that all the parts of the body are projected on the plane passing through its center of gravity and its axis of rotation in a manner which maintains their distance to this axis of rotation.

The solution of Descartes became the subject of a dispute between him and Roberval.[11] The latter contended that it was good only when all the parts of the body are really or could be considered to be on the same plane passing through the axis of rotation and in other cases, only the motion perpendicular to the plane through the axis of rotation and the center of gravity should be considered. Furthermore, each particle must be referred to the point where this plane meets the direction of motion of this particle, a direction which is always perpendicular to the plane determined by this particle and the axis of rotation. But it is easy to prove that relative to the axis of rotation, the moments of the forces calculated in this fashion are always equal to those of the forces calculated according to Descartes' method.

On a firmer basis, Roberval contended that Descartes had only looked for the center of percussion about which the collisions or the moments of percussion are equal. And in order to find the real center of oscillation of a pendulum, it was also necessary to consider the action of gravity which moves the pendulum. But this research was far ahead of the mechanics of that period. The geometers continued to assume implicitly that the center of percussion is the same as the center of oscillation. Huygens was the first to conceive of the latter center from its true perspective. He has also thought that this problem must be regarded as entirely new and since he could not resolve it by existing methods he invented a new but indirect principle which has become popular since then under the name of the **Conservation des Forces Vives**.

7. A cord, considered as an inflexible line without any gravitation or mass, attached by one end to a fixed point and by the other to a small weight, which can be considered as a material point, forms what is called a simple pendulum. The laws of vibration for this pendulum depend uniquely on its length, that is to say, on the distance between the weight and the point of suspension. But if this cord is attached again to one or more weights at various distances from the suspension point, it becomes a compound pendulum whose motion must be somewhere between that of the simple pendulums obtained if each of these weights were suspended individually on the cord. Since, on the one hand, the gravitational force tends to cause all the weights to descend equally at the same time and on the other hand, the inflexibility of the cord constrains them to describe simultaneously unequal arcs proportional to their distance from the suspension point, a kind of compensation and redistribution of motion must be formed among these weights in such a manner that the weights that are closer to the point of suspension hasten the vibrations of those which are more distant and the latter retard those which are closer. In this manner, at one point of the cord where a weight is placed there will be a motion which is neither accelerated nor retarded by the other weights but it will be the same as if all the weight were suspended at that point on the cord. Therefore, this point will be the real center of oscillation of the compound pendulum and such a center can also be found in all solid bodies which oscillate about a horizontal axis regardless of its configuration.

Huygens saw that the center could not be determined in this rigorous fashion without the knowledge of the law according to which different weights of the compound pendulum mutually change the motion that gravity tends to impart to them at each instant, but instead of trying to determine this law from the fundamental principles of mechanics he was content to replace it by an indirect principle which consists of assuming that if several weights attached to a pendulum in any manner descend uniquely by the action of gravity and, if in any instant, they become detached and are separated from one another due to the velocity acquired during the descent, each of them can ascend to a height such that the common center of gravity will be found to ascend to the same height from which it had descended. In reality, Huygens did not establish this principle directly, but he deduced it from two hypotheses which he believed must be admitted as fundamental to mechanics. One of them is that the center of gravity of a system of weights can not ascend higher than the height from which it had fallen whatever the changes in the mutual relations between the bodies because otherwise perpetual motion would be possible. The other is that a compound pendulum can always ascend by itself to the same height as that from which it had descended freely. In addition, Huygens remarks that the same principle holds in the motion of weights joined in any manner as well as in the case of fluids.

It is difficult to surmise what had given the author the idea for such a principle. But it can be conjectured that he was guided by the theorem that Galileo had demonstrated on the fall of weights which either descend vertically or on inclined planes. Galileo showed that they always acquire a velocity capable of bringing them back to the same height from which they had fallen. This theorem when it is generalized and applied to the center of gravity of bodies becomes Huygens' principle.

Whatsoever it may be, this principle provides an equation between the vertical height from which the center of gravity of the system has descended in an arbitrary interval of time and the different vertical heights to which the bodies composing the system could have ascended with their acquired velocities which from Galileo's theorems are proportional to the squares of the velocities. But in a pendulum oscillating about a horizontal axis, the velocities of the various points are proportional to their distance from the axis. In this fashion, the equation can be reduced to two unknowns, one is the descent of the center of gravity of the pendulum in any interval of time and the other is the height to which a given point of this pendulum can ascend due to its acquired velocity. But the descent of the center of gravity determines that of every other point of the pendulum. Thus this result will give an equation between the height from which any given point of the pendulum has descended and that to which it can ascend due to the velocity gained from this descent. For the center of oscillation, these two heights must be equal since free bodies can always ascend to the same heights from which they have fallen. The equation shows that this equality can only take place at a point on the line perpendicular to the axis of rotation and passing through the center of gravity of the pendulum which is separated from this axis by the quantity resulting from multiplying all the elements of weight composing the pendulum by the square of their distance from the axis and dividing the sum of these products by the weight of the pendulum multiplied by the distance of its center of gravity from the same axis. This quantity will thus define the length of a simple pendulum which will have the same motion as that of the compound pendulum.

This theory of Huygens along with a large number of learned applications is presented in the *Horologium oscillatorium*. It would have been perfect if it were not based upon a precarious principle. A demonstration is necessary to put this principle beyond all criticism.

Some critical objections to this theory appeared in 1681 in the Journal des Savans of Paris. To these objections, Huygens had replied in a vague and hardly satisfactory manner. But this controversy claimed the attention of James Bernoulli and gave him the occasion to profoundly examine the theory of Huygens and to try to relate it to the basic principles of dynamics. At first, he only considers two equal weights attached to an inflexible rectilinear cord and he remarks that the velocity of the first weight, that is, the one nearest the point of suspension, acquired by describing an arbitrary arc must be less than the one it would have acquired by describing freely the same arc and at the same time, the velocity acquired by the other weight must be greater than the one it would have acquired by freely describing the same arc. The velocity lost by the first weight is thus transfered to the second and as this transfer is made by means of a mobile lever about a fixed point, it must follow the law of equilibrium of forces applied to this lever. Thus it must occur in such a manner, that the ratio of the loss of velocity of the first weight to the gain of velocity of the second is reciprocal to the length of the arm of the lever, that is, to the distance from the point of suspension. From this, and from the fact that the actual velocities of the two weights must be in direct proportion to these distances, the velocities are determined easily and consequently, the motion of the pendulum is obtained.

8. This was the first step taken towards the direct solution of this famous problem. The idea of referring to the lever the forces resulting from the velocities gained or lost by the weights is very astute and provides the key to the true theory. But James Bernoulli was mistaken when he considered the velocities acquired during an arbitrary finite time, instead he should have only considered the elementary velocities acquired during an instant of time and compared them with those that gravity would impress during the same instant. This is what l''Hôpital has since done in a work inserted in the Journal of Rotterdam in 1690. He assumes that the compound pendulum is composed of two arbitrary weights attached to an inflexible string and then he establishes equilibrium between the quantities of motion lost and gained by these weights in an arbitrary instant, that is to say, between the differences of the quantities of motion which the weights actually acquire in this instant and those that gravity tends to impress on them. He determines by this means the ratio of the instantaneous acceleration of each weight to those that gravity alone has the tendency to impress on it. Then the center of oscillation is found by seeking the point on the pendulum for which the two accelerations are equal. Ultimately, this theory is extended to a larger number of weights. But, for this, he views the first ones as successively reunited at their center of oscillation which is not any more direct, neither can it be admitted without demonstration.

This analysis induced James Bernoulli to return to his own solution and this led finally to the first direct and rigorous solution to the problem of the center of oscillation, a solution which greatly merits the attention of geometers since it contains the seed of the principle of dynamics which has become so fecund at the hands of d'Alembert.

This author considers collectively the motions that gravity impresses at each instant on the bodies which compose the pendulum and since these bodies, because of their constraints, cannot follow them, he conceives the motions that they must follow as composed of impressed motions and other motions added or subtracted which must cancel each other, and in virtue of which the pendulum must remain in equilibrium. The problem is thus reduced to the principles of statics and requires only the aid of analysis. James Bernoulli found by this means general formulas for the centers of oscillation of bodies of arbitrary shape and then he showed that these formulas were consistent with the principle of Huygens. Finally, he demonstrated the equivalence between the center of oscillation and the center of percussion. This solution had lain sketched out since 1691 in the Acta Eruditorum of Leipzig, but it was given in a complete form only in 1703, in the Mémoires of the Académie des Sciences of Paris.

9. In order not to omit anything of significance in the history of the problem of the center of oscillation, I should mention that the solution which John Bernoulli had given in the same memoirs and a similar solution which Taylor presented at the same time in the work entitled *Methodus incrementorum* was the occasion of a heated dispute between these two geometers. But, however astute the idea on which this new solution is founded and which consists of transforming at once the compound pendulum to a simple pendulum by substituting for these different weights other weights collected at the same point with fictitious masses and gravities so that they produce the same angular accelerations and the same moments with respect to the axis of rotation, and that the total gravity of the collected weights is equal to their natural gravity, nevertheless, it must be admitted that this idea is neither so natural nor so astute as the one of the equilibrium between the quantities of motion gained and lost.

A new method to resolve the same problem is also found in the *Phoronomia*[12] of Hermann published in 1716. It is based on this other principle, that the motor forces which act on the weights which comprise the pendulum must act such that the weights are moved simultaneously. These forces are equivalent to those which originate from the action of gravity. Consequently, the former forces which are to be taken acting in the opposite direction must be in equilibrium with the latter.

This principle is no different than the one by James Bernoulli, presented in a less simple manner and it is easy to relate one to the other by the principles of statics. Euler later generalized this method and used it to determine the oscillations of flexible bodies in a memoir published in 1740 in the seventh volume of the old Commentaries of St. Petersburg.

It would be too tedious to speak of the other problems of dynamics which have exercised the sagacity of the geometers after the one on the center of oscillation and before the art of solving them was reduced to fixed rules. These problems that the Bernoullis, Clairaut and Euler proposed to each other are found in the first volumes of the memoirs of St.

Petersburg and Berlin, in the memoirs of Paris for the years 1736 and 1742, in the *Oeuvres* of John Bernoulli and in the *Opuscules* of Euler. They consist of determining the motions of several heavy or weightless bodies which pull or push on one another by means of strings or rigid levers where they are firmly attached or along those they can freely slide, and having received an arbitrary impulse, are then left to themselves or constrained to move on given curves or surfaces.

The principle of Huygens was almost always used in the solution of these problems. But since this principle produces only a single equation, additional equations had to be found from the consideration of the unknown forces with which the bodies push or pull one another. These forces were considered to be elastic forces acting in the opposite sense. The consideration of these forces made it unnecessary to consider the connections between the bodies, and consequently, the laws of motion for free bodies could be applied. The conditions imposed on the motions of the various bodies which, from the nature of the problem, had to exist between the motions of the individual bodies, were used to determine the unknown forces that had been introduced in the calculation. But a particular skill was required to identify in each problem all the forces which should be considered. These factors rendered these problems challenging and a source for exciting emulation.

10. The *Traité de Dynamique* of d'Alembert, which was published in 1743, put an end to this type of challenge by giving a general and direct method to solve or at least to put into the form of equations all the problems of dynamics that can be imagined. This method reduces all the laws of the motions of bodies to those of their equilibrium and thus reduces dynamics to statics. We have already observed that the principle used by James Bernoulli in his research on the center of oscillation had the advantage of making this research dependent on the equilibrium conditions for the lever. But it was reserved for d'Alembert to conceive this principle in a general fashion and to give it all the simplicity and fecundity which it merits.

If a motion is impressed upon several bodies so that they are forced to move consistent with their mutual interaction, it is clear that these motions can be viewed as composed of those which the bodies would actually follow and of other motions which are negated, from which it follows that these latter motions must be such that the bodies following only these motions are in equilibrium.

This is the principle that d'Alembert gave in his treatise on dynamics and which he usefully applied in several problems notably the one of the precession of the equinoxes. This principle does not immediately give the necessary conditions for the solution of the problems of dynamics, but it is used to learn how to deduce them from the conditions of equilibrium. Thus by combining this principle with the ordinary principles for the equilibrium of the lever or of the composition of forces, the equations for each problem can always be found. But the difficulty of determining the forces which must be equilibrated as well as the laws of equilibrium between these forces often makes the application of this principle awkward and difficult, and the solutions obtained are almost always more complicated than if they were deduced from less simple and direct principles as is evident from the second part of the *Traité de Dynamique*.[13]

11. If the resolution of motions is to be avoided, which this principle requires, it would only be necessary to immediately establish equilibrium between the forces and the resulting motions, but taken in opposite directions. Because, if it is imagined that upon each body the motion that it must follow is impressed in the opposite direction, it is clear that the system would be at rest. Consequently, these motions should cancel those that the bodies would have and that they would have followed without their mutual interaction. Thus there must be equilibrium between all these motions or between the forces which can produce them.

This manner of reducing the laws of dynamics to those of statics is, to be sure, less direct than the one which results from the principle of d'Alembert, but it offers greater simplicity in its application. It is similar to the one of Hermann and Euler who applied it to the solution of many problems of mechanics, and it is sometimes found in treatises of mechanics under the title of the Principle of d'Alembert.

12. In PART I of this work, we reduced all the principles of statics to one general formula which gives all the laws of equilibrium of an arbitrary system of bodies acted upon by any number of forces. Thus it will also be possible to reduce all of dynamics to one general formula, because in order to apply to the motion of a system of bodies the formula defining its equilibrium, it will be sufficient to introduce forces which result from the variations of the motion of each body and which must be canceled. The development of this formula with respect to the conditions dependent on the nature of the system will give all the equations necessary for the determination of the motion of each body and it will only be necessary to integrate these equations which is a matter of analysis.

13. One of the advantages of the formula under discussion is that it provides immediately the general equations which contain the principles or theorems known under the names of the Conservation des Forces Vives, conservation of the motion of the center of gravity, conservation of the motion of rotation or principle of areas and the principle of the least quantity of action. These principles must be viewed as general results of the laws of dynamics rather than fundamental principles of this science. But, since they are often used as such in the solution of problems, we believe that we should discuss them here by explaining what they are and to which authors they are due in order not to omit anything in this preliminary exposition of the principles of dynamics.

14. The first of these four principles, the one of the Conservation des Forces Vives was discovered by Huygens, but in a form slightly different from the one which we will presently discuss. We referred to this principle earlier when we discussed the problem of the center of oscillation. The principle, as it has been used in the solution of this problem, requires the equality between the downward and upward motion of the center of gravity of several heavy bodies which simultaneously descend and which then ascend separately, ascending with the velocity that they had gained in descent. And from the known properties of the

center of gravity, the path traversed by this center in an arbitrary direction is expressed by the sum of the products of the mass of each body with the distance that it has traversed in the same direction, divided by the sum of the masses. On the other hand, from the theorems of Galileo, the vertical path traversed by a heavy body is proportional to the square of the velocity that it has gained in the free descent and with which it could rise to the same height. Thus the principle of Huygens in the motion of heavy bodies is reduced to the sum of the products of the masses by the square of the velocity at each instant is the same, whether the bodies move simultaneously in an arbitrary fashion or whether they traverse independently the same vertical heights. This is also what Huygens succinctly noted in a short work on the methods of treating the centers of oscillation by James Bernoulli and l'Hôpital.

Up to that time, this principle was only viewed as a simple theorem of mechanics. But when John Bernoulli adopted the distinction established by Leibnitz between the dead forces or pressures which act without actual motion and the live forces which accompany this motion as well as the measure of these latter forces by the product of the masses with the square of the velocities, he simply found in this principle a consequence of the theorem of live forces and a general law of nature according to which the sum of the live forces of several bodies is the same during the time that these bodies act upon one another with simple pressures, and is always equal to the simple live force which is the result of the action of the actual forces which put the bodies in motion. He thus gave to this principle the name of Conservation des Forces Vives and he used it successfully to solve some problems which had not yet been solved and which were difficult to resolve by means of the direct methods.

Daniel Bernoulli then gave a greater extension to this principle and he deduced from it the laws of motion of fluids in containers, a question which was treated before him only in a vague and arbitrary manner. Finally, he made it very general for in the Mémoires de Berlin for the year 1748, he showed how it could be applied to the motion of bodies acted upon by arbitrary mutual attractions or attracted toward fixed centers by forces proportional to various functions of the distances.

The great advantage of this principle is that it immediately gives a finite equation between the velocities of the bodies and the variables which determine their position in space so that by the nature of the problem, all these variables are reduced to one. This equation is sufficient to solve the problem of the center of oscillation completely. In general, the Conservation des Forces Vives always gives a first integral of the different differential equations of each problem which is of a great utility in many instances.

15. The second principle derives from Newton, who at the beginning of his *Principia mathematica*[14] demonstrates that the state of rest or motion of the center of gravity of several bodies is not changed by mutual interaction of these bodies whatever they might be, so that the center of gravity of the bodies which act on one another in an arbitrary fashion, either by strings or levers or by the laws of attraction, etc. without any external action or impediment is always at rest or moves in a straight line indefinitely.

Later, d'Alembert gave a broader application to this principle by showing that if each body is acted upon by a constant accelerating force which acts along parallel lines or which is directed toward a fixed point and acts with respect to distance, the center of gravity must traverse the same path as if the bodies were free, to which one can add that the motion of this center is in general the same as if all the forces applied to the bodies, whatever they might be, were each applied in its own direction.

It is obvious that this principle serves to determine the motion of the center of gravity independent of the motions of the bodies and thus it will always produce three finite equations between the coordinates of the bodies and time, which will be integrals of the differential equations of the problem.

16. The third principle[15] is much more recent than the first two and seems to have been discovered simultaneously by Euler, Daniel Bernoulli and d'Arcy, but with different formulations. According to the first two, this principle consists of the following: for the motion of several bodies about a fixed center, the sum of the products of the mass of each body with the associated velocity about the center and with the distance to the same center is always independent of the mutual action that bodies can exert upon one another and is constant as long as there is no force or external impediment. Daniel Bernoulli presented this principle in the first volume of the Mémoires of the Académie de Berlin which was published in 1746 and Euler presented it in the same year in the first volume of his *Opuscules*. It is also the same problem which inspired them, that is, the investigation of the motion of several mobile bodies in a tube with a given shape which can only rotate about a fixed point or center.

The principle of d'Arcy, as he presented it to the Académie des Sciences in the Mémoires of 1747, which were published in 1752, is that the sum of the products of the mass of each body with the swept area that its radius vector describes about a fixed center on the same plane of projection is always proportional to the time. It is clear that this principle is a generalization of the fine theorem of Newton on areas swept by arbitrary centripetal forces. In order to see the analogy or rather the identity with the one by Euler and Daniel Bernoulli, one has only to consider that the velocity of rotation is expressed by the element of the circular arc divided by the element of time and that the first of these elements multiplied by the distance to the center gives the element of the area described about this center. From this result, it is clear that this latter principle is nothing more than the differential expression of the one by d'Arcy.

This author then presented his principle in a different form which makes it closer to the former and which consists of the following: the sum of the products of the masses with their velocities and with the perpendiculars from the paths of the bodies to the center is a constant quantity.

From this point of view, he even made of it a type of metaphysical principle which he calls the Conservation of Action, to oppose or rather to substitute it for the one of the least quantity of action as if vague and arbitrary designations were the essence of the laws of nature and could, by some secret virtue, raise to final causes some simple results from the known laws of mechanics.[16]

Whatever it might be, the principle under discussion generally holds for all systems of bodies which act upon one another either through strings, inflexible lines, attraction laws, etc. and which are also acted upon by arbitrary forces directed toward a fixed center. Either the system is entirely free or it is constrained to move about this same center. The sum of the products of the masses with the areas swept about this center and projected upon an arbitrary plane, is always proportional to the time so that by projecting these areas on to three planes perpendicular to one another, three differential equations of the first order between time and the coordinates of the curves which the bodies describe are obtained and it is in these equations that the nature of the principle that we just discussed is found.

17. I finally arrive at the fourth principle which I call Least Action in analogy to the one to which Maupertuis has applied this designation and which the works of several illustrious authors have subsequently made so famous. This principle, viewed analytically, consists of the following: in the motion of bodies which act on one another, the sum of the products of the masses with the velocities and the spaces traversed is a minimum. The author deduced from it the laws of reflection and refraction of light as well as the laws governing the percussion of bodies in two memoirs read to the Académie des Sciences of Paris in 1744 and two years later at the Académie de Berlin, respectively.

However, these applications are too restrictive to be used to establish the truth of a general principle. They have also something vague and arbitrary about them which can only make the consequence which one could draw for the accuracy of the principle itself uncertain. Thus it would be wrong, it seems to me, to put this principle as it is presented on the same level with the ones we just discussed. However, there is another way to view it, more general and more rigorous and which alone deserves the attention of geometers. Euler has given the first interpretation at the end of his treatise on isoperimetric problems published at Lausanne in 1744,[17] by showing that for the trajectories described by bodies following central forces the integral of the product of the velocity by the element of the curve is always a maximum or a minimum.

The property which Euler found in the motion of isolated bodies and which seems to be confined to these bodies, I later expanded using the Conservation des Forces Vives, to the motion of any system of bodies acting upon one another in an arbitrary fashion. A new general principle derives from this effort, to wit, that the sum of the products of the masses with the integrals of the velocities multiplied by the element of the spaces traversed is always a maximum or a minimum.

This is the principle to which I improperly gave the name of Least Action and which I view not as a metaphysical principle but as a simple and general result of the laws of mechanics. In volume two of the Mémoires de Turin, the use which I made of this principle to solve several difficult problems of dynamics can be found. This principle, combined with the one of Forces Vives and further developed following the rules of the calculus of variations, gives directly all the necessary equations for the solution of each problem. From this result, a method equally simple and general derives to treat problems concerning the motion of bodies. However, this method is only a corollary to the one which is in PART II of this work and which at the same time has the advantage that it derives from the first principle of mechanics.

SECTION II
A GENERAL FORMULA OF DYNAMICS FOR THE MOTION OF A SYSTEM OF BODIES MOVED BY ARBITRARY FORCES

1. When the forces acting on a system of bodies are distributed according to the laws presented in PART I of this treatise, these forces equilibrate each other and the system is in equilibrium, but when equilibrium does not hold, the bodies must necessarily move due to all or some of the forces which act on them. The determination of the motion due to the given forces is the object of this second part.

We will consider here principally the forces of acceleration and retardation for which the action is constant, such as gravity, and which have a tendency to impress an infinitesimal velocity at each instant of time equally on all the constituent particles of the body. When these forces act freely and uniformly, they necessarily produce velocities which increase with time and the velocities created in a given time can be viewed as the effect of the most simple of this type of force and consequently, as the most obvious quantity to be used for its measure. In mechanics, the effect of the force acting alone must be assumed known and the art of this science consists of deducing uniquely the composed effects which must result from the combined and modified action of the same forces.

2. We will thus assume that for each accelerating force the velocity that it is able to impress on a mobile body by acting always in the same fashion during a given time is known and this given time will be taken as the basic unit of time. Furthermore, we will measure the accelerating force by this same velocity which, in turn, must be estimated by the space traversed by the mobile body during the same time if it moved uniformly. It is known from the theorems of Galileo that this space is always double the one that the body has really traversed from the constant action of this accelerating force.

Also, a known accelerating force can be taken as unity and to it, all the other forces can be referred. Then one should take for the basic unit of space, double the space that the same force continuing to act uniformly will have the body traverse in the time that one wants to take for the unit of time and the velocity acquired during this time by the continuous action of the force will be the unit of velocity. In this fashion, the forces, spaces, times and velocities will only be simple ratios, that is, ordinary quantities.

For example, if the gravity at the latitude of Paris is taken as the unit of the accelerating force and the time is measured in seconds, one should then take 30.196 feet at Paris as the unit of the space traversed because 15.098 feet is the distance which a body left to itself falls in one second at this latitude and the unit of velocity will be the velocity acquired by a heavy body falling from this height.

3. Once these preliminary notions are assumed, let us consider a system of bodies arranged in an arbitrary fashion and acted upon by arbitrary forces of acceleration. Let m be the mass of one of these bodies viewed as a point. For the greatest simplicity, let us refer to three rectangular coordinates x, y, z the absolute position of the same body at the end of an arbitrary time t. These coordinates are always assumed to be parallel to three fixed axes in space which intersect perpendicularly at a point called the origin of the coordinate axes.[18] They consequently express the rectilinear distances from the body to three planes passing through the same axes.

Thus since these planes are orthogonal, the coordinates x, y, z represent the distances by which the bodies in motion depart from the same planes. Consequently, the quantities $\mathrm{d}x/\mathrm{d}t$, $\mathrm{d}y/\mathrm{d}t$, $\mathrm{d}z/\mathrm{d}t$ will represent the velocities that this body has at an arbitrary instant causing it to depart from each of these planes and to move along the prolongation of the coordinates x, y, z. And these velocities, if the body was then left to itself, would remain constant in the following instants, according to the fundamental principles of the theory of motion.

But, because of the connections between the bodies and the action of the forces of acceleration which act on them, these velocities take, during the instant $\mathrm{d}t$, the increments $\mathrm{d}(\mathrm{d}x/\mathrm{d}t)$, $\mathrm{d}(\mathrm{d}y/\mathrm{d}t)$, $\mathrm{d}(\mathrm{d}z/\mathrm{d}t)$ which must be determined. These increments can be treated as new velocities impressed on each body and by dividing them by dt, one will have immediately the measure of the forces of acceleration used to produce them; since, however variable might be the action of a force, one can always by the nature of the differential calculus, view it as constant during an infinitesimal time. The velocity created by this force is also proportional to the force multiplied by the time. Consequently, the force will be expressed in terms of the velocity divided by the time.

By taking the element of time dt as constant, the forces of acceleration in question will be expressed in terms of $\mathrm{d}^2x/\mathrm{d}t^2$, $\mathrm{d}^2y/\mathrm{d}t^2$, $\mathrm{d}^2z/\mathrm{d}t^2$ and by multiplying these forces by the mass m of the body on which they act, one will have $m(\mathrm{d}^2x/\mathrm{d}t^2)$, $m(\mathrm{d}^2y/\mathrm{d}t^2)$, $m(\mathrm{d}^2z/\mathrm{d}t^2)$ as the forces applied immediately to move the body m during the time $\mathrm{d}t$ parallel to the coordinate axes x, y, z. Thus each body m of a system can be considered as acted upon by such forces. Consequently, all these forces should be equivalent to those which we assumed to be impressed on the system and for which the action is modified by the nature of the system. And the sum of their moments should always be equal to the sum of the moments of the latter from the theorem given in Article 15 of SECTION II of PART I.

4. We will use in what follows the ordinary symbol d to represent the differentials with respect to time and we will express the variations which represent the virtual velocities by the symbol δ, as we have already done in some problems of PART I.

Thus one will have $m(\mathrm{d}^2x/\mathrm{d}t^2)\delta x$, $m(\mathrm{d}^2y/\mathrm{d}t^2)\delta y$, $m(\mathrm{d}^2z/\mathrm{d}t^2)\delta z$ for the moments of the forces $m(\mathrm{d}^2x/\mathrm{d}t^2)$, $m(\mathrm{d}^2y/\mathrm{d}t^2)$, $m(\mathrm{d}^2z/\mathrm{d}t^2)$ which act in the direction of the coordinates

x, y, z and which have a tendency to increase them. The sum of their moments can then be represented by the formula

$$\mathrm{S}(\frac{\mathrm{d}^2x}{\mathrm{d}t^2}\delta x + \frac{\mathrm{d}^2y}{\mathrm{d}t^2}\delta y + \frac{\mathrm{d}^2z}{\mathrm{d}t^2}\delta z)m$$

where it is assumed that the integral operator S extends to all the bodies of the system.

5. Now let P, Q, R, etc. be given forces of acceleration which act upon each body m of the system toward the centers to which these forces are assumed directed and let p, q, r, etc. be the rectilinear distances of each of these bodies from the same centers. The differentials δp, δq, δr, etc. will represent the variations of the lines p, q, r, etc. caused by the variations δx, δy, δz of the coordinates x, y, z of the body m. But because the forces P, Q, R, etc. are assumed to have a tendency to shorten these lines, their virtual velocities must be represented by $-\delta p$, $-\delta q$, $-\delta r$, etc. (Article 3, SECTION II, PART I). Therefore, the moments of the forces mP, mQ, mR, etc. will be expressed by $-mP\,\delta p$, $-mQ\,\delta q$, $-mR\,\delta r$, etc. and the sum of the moments of these forces will be represented by $-\mathrm{S}(P\,\delta p + Q\,\delta q + R\,\delta r + \cdots)m$. Then after equating this sum to the one in the preceding article, one will have

$$\mathrm{S}(\frac{\mathrm{d}^2x}{\mathrm{d}t^2}\delta x + \frac{\mathrm{d}^2y}{\mathrm{d}t^2}\delta y + \frac{\mathrm{d}^2z}{\mathrm{d}t^2}\delta z)m = -\mathrm{S}(P\,\delta p + Q\,\delta q + R\,\delta r + \cdots)m$$

and after transposing the second member

$$\mathrm{S}(\frac{\mathrm{d}^2x}{\mathrm{d}t^2}\delta x + \frac{\mathrm{d}^2y}{\mathrm{d}t^2}\delta y + \frac{\mathrm{d}^2z}{\mathrm{d}t^2}\delta z)m + \mathrm{S}(P\,\delta p + Q\,\delta q + R\,\delta r + \cdots)m = 0$$

This is the general formula of dynamics for the motion of an arbitrary system of bodies.[19]

6. It is obvious that this formula only differs from the general formula of statics, given in SECTION II of PART I, by the terms resulting from the forces $m\mathrm{d}^2x/\mathrm{d}t^2$, $m\mathrm{d}^2y/\mathrm{d}t^2$, $m\mathrm{d}^2z/\mathrm{d}t^2$ which produce the acceleration of the body m in the direction of the prolongation of the three coordinates x, y, z. Indeed, we saw in the preceding section (Article 11) that if these forces are assumed to act in the opposite directions, that is, viewed as having a tendency to shorten the coordinates x, y, z, they must be in equilibrium with the actual forces P, Q, R, etc. which are assumed acting to shorten the lines p, q, r, etc. so that it is only necessary to add to the moments of these latter forces, the forces $m\mathrm{d}^2x/\mathrm{d}t^2$, $m\mathrm{d}^2y/\mathrm{d}t^2$, $m\mathrm{d}^2z/\mathrm{d}t^2$ for each body m to pass, at once, from the conditions of equilibrium to those of motion (Article 4, SECTION II, PART I).

7. Thus the rules that we gave in SECTION II of PART I, for the development of the general formula of statics, will also be applicable to the general formula of dynamics.

It only remains to observe, 1) that the differences which were indicated by the ordinary operator d to represent the variations will be indicated from now on by the operator δ.

2) that the differential operator d will always be relative to the time t as will the integral operator $\int$ with the exception of partial differences, where it does not matter which symbol is used. 3) that in order to represent the elements of a curve or a surface or in general of a system composed of an infinite number of particles, the differential operator D is used, which corresponds to the integral operator S. Thus when it is desired to extend the formula that we have given for equilibrium in Chapters III and IV of SECTION V of PART I to motion the operator d should be replaced everywhere by D to obtain the expression of the sum of the moments of all the forces.

8. When the motion takes place in a resisting medium, the resistance of the medium should be viewed as a force acting in the opposite direction to the motion of the body. Consequently, it can be assumed directed toward a point on the tangent.

Let us assume that the resistance is R. In order to obtain its moment $-R\,\delta r$, it is only necessary to consider that one has in general

$$r = \sqrt{(x-\ell)^2 + (y-m)^2 + (z-n)^2}$$

where ℓ, m, n are the coordinates of the center of the force R. Thus

$$\delta r = \frac{x-\ell}{r}\delta x + \frac{y-m}{r}\delta y + \frac{z-n}{r}\delta z$$

Let us take the center of the force R on the tangent to the curve described by the body and very close to the curve. For this operation, let $x - \ell = \mathrm{d}x$, $y - m = \mathrm{d}y$, $z - n = \mathrm{d}z$ which will give, taking $\mathrm{d}s$ for the element of the curve

$$\frac{x-\ell}{r} = \frac{\mathrm{d}x}{\mathrm{d}s}, \qquad \frac{y-m}{r} = \frac{\mathrm{d}y}{\mathrm{d}s}, \qquad \frac{z-n}{r} = \frac{\mathrm{d}z}{\mathrm{d}s}$$

and consequently, $\delta r = (\mathrm{d}x/\mathrm{d}s)\delta x + (\mathrm{d}y/\mathrm{d}s)\delta y + (\mathrm{d}z/\mathrm{d}s)\delta z$. If the medium is in motion, this motion should be composed with the motion of the body to obtain the direction of the force of resistance. Let us call $\mathrm{d}\alpha$, $\mathrm{d}\beta$, $\mathrm{d}\gamma$, the small spaces that the medium traverses parallel to the axes of the coordinates x, y, z while the body describes the space $\mathrm{d}s$. It will only be necessary to subtract these quantities from $\mathrm{d}x$, $\mathrm{d}y$, $\mathrm{d}z$ to obtain the relative motions and since $\mathrm{d}s = \sqrt{(\mathrm{d}x^2 + \mathrm{d}y^2 + \mathrm{d}z^2)}$, if one puts

$$\mathrm{d}\sigma = \sqrt{(\mathrm{d}x - \mathrm{d}\alpha)^2 + (\mathrm{d}y - \mathrm{d}\beta)^2 + (\mathrm{d}z - \mathrm{d}\gamma)^2}$$

there will result for this case

$$\delta r = \frac{\mathrm{d}x - \mathrm{d}\alpha}{\mathrm{d}\sigma}\delta x + \frac{\mathrm{d}y - \mathrm{d}\beta}{\mathrm{d}\sigma}\delta y + \frac{\mathrm{d}z - \mathrm{d}\gamma}{\mathrm{d}\sigma}\delta z$$

The resistance R is ordinarily a function of the velocity $\mathrm{d}s/\mathrm{d}t$. But in the case where the medium is in motion, it will be a function of the relative velocity $\mathrm{d}\sigma/\mathrm{d}t$.

In this fashion, it will be possible to apply our general formulas to motions which exist in resisting media without the need of any particular consideration for these types of motion.

9. It is important to note that the expression $d^2x\,\delta x + d^2y\,\delta y + d^2z\,\delta z$ by which the general formula of dynamics differs from the equivalent equation of statics (Article 5) is independent of the location of the coordinate axes x, y, z.

Now, let us assume that instead of these coordinates, other rectangular coordinates x', y', z' are substituted which have the same origin but are measured relative to different axes. From the formulas for the transformation of coordinates, given in Article 10 of SECTION III of PART I, one has

$$x = \alpha x' + \beta y' + \gamma z'$$
$$y = \alpha' x' + \beta' y' + \gamma' z'$$
$$z = \alpha'' x' + \beta'' y' + \gamma'' z'$$

Let us differentiate these expressions for x, y, z by considering all the coefficients α, β, γ, α', etc. as constants and the new coordinates x', y', z' as the only variables. Thus there will result

$$d^2x = \alpha\, d^2x' + \beta\, d^2y' + \gamma\, d^2z'$$
$$d^2y = \alpha'\, d^2x' + \beta'\, d^2y' + \gamma'\, d^2z'$$
$$d^2z = \alpha''\, d^2x' + \beta''\, d^2y' + \gamma''\, d^2z'$$

In a similar fashion, the following equations will be obtained

$$\delta x = \alpha\, \delta x' + \beta\, \delta y' + \gamma\, \delta z'$$
$$\delta y = \alpha'\, \delta x' + \beta'\, \delta y' + \gamma'\, \delta z'$$
$$\delta z = \alpha''\, \delta x' + \beta''\, \delta y' + \gamma''\, \delta z'$$

After substituting these values and considering the equations of condition given in the cited article above between the coefficients α, β, γ, α', etc., one will have

$$d^2x\,\delta x + d^2y\,\delta y + d^2z\,\delta z = d^2x'\,\delta x' + d^2y'\,\delta y' + d^2z'\,\delta z'$$

If the same substitutions are made in the expression for the rectilinear distances between the different bodies of a system represented by p, q, r, etc., it is easy to see that the quantities $\alpha, \beta, \gamma, \alpha'$, etc. will also disappear and that the transformed quantities will keep the same form. In effect, one has

$$p = \sqrt{(x - \mathbf{x})^2 + (y - \mathbf{y})^2 + (z - \mathbf{z})^2}$$

where x, y, z are coordinates of a body m and $\mathbf{x}, \mathbf{y}, \mathbf{z}$ the coordinates of a different body m referred to the same axes. By interchanging the axes, the former become x', y', z' and if the latter are designated by $\mathbf{x}'$, $\mathbf{y}'$, $\mathbf{z}'$, one will also have

$$\begin{aligned}
\mathbf{x} &= \alpha\mathbf{x}' + \beta\mathbf{y}' + \gamma\mathbf{z}' \\
\mathbf{y} &= \alpha'\mathbf{x}' + \beta'\mathbf{y}' + \gamma'\mathbf{z}' \\
\mathbf{z} &= \alpha''\mathbf{x}' + \beta''\mathbf{y}' + \gamma''\mathbf{z}'
\end{aligned}$$

After substituting and taking into consideration the same equations of condition, one will have

$$p = \sqrt{(x' - \mathbf{x}')^2 + (y' - \mathbf{y}')^2 + (z' - \mathbf{z}')^2}$$

and in the same fashion expressions for the analogous quantities q, r, etc. will be found.

10. It follows from this result that if the system is only acted upon by internal forces P, Q, R, etc. proportional to arbitrary functions of the distances p, q, r, etc. between the bodies and that the condition of the system only depends on the mutual arrangement of the bodies such that the equations of condition are only between the different variables, p, q, r, etc., the general formula of dynamics (Article 5) will be the same for the transformed coordinates x', y', z' as for the original coordinates x, y, z. Therefore, after having found by integration of the different equations deduced from this formula, the values of the coordinates x, y, z of each body m expressed in time, and if one took these values for x', y', z', one will have for the coordinates x, y, z the more general expressions

$$\begin{aligned}
x &= \alpha x' + \beta y' + \gamma z' \\
y &= \alpha' x' + \beta' y' + \gamma' z' \\
z &= \alpha'' x' + \beta'' y' + \gamma'' z'
\end{aligned}$$

in which the nine coefficients α, β, γ, etc. contain three indeterminate quantities since there are only six equations of condition among them. If the values x', y', z' contained all the necessary arbitrary constants to complete the different integrals, the three indeterminates in question will become part of these same arbitrary constants. But they will be able to supplement those which are missing and where the default renders the solution incomplete. Thus by means of the three new arbitrary coefficients which can be introduced at the end of the calculation, one will be free to be assumed equal to zero or to determinate quantities as many other arbitrary constants, which will often be used to facilitate and simplify the calculation.

11. Although the effects of impulsion and percussion can always be calculated in the same fashion as those of the forces of acceleration, when only the total impressed velocity is sought, one can dispense with considering these successive increases and immediately view the forces of impulsion as equivalent to the impressed motions.

Thus let P, Q, R, etc. be the forces of impulsion applied to an arbitrary body m of a system in the direction of the lines p, q, r, etc. Let us assume that the velocity imposed on this body is resolved into three component velocities represented by $\dot{x}$, $\dot{y}$, $\dot{z}$ in the directions of the coordinate axes x, y, z. There will result, as in Article 5, after substituting the velocities $\dot{x}$, $\dot{y}$, $\dot{z}$ for the forces of acceleration $\mathrm{d}^2x/\mathrm{d}t^2, \mathrm{d}^2y/\mathrm{d}t^2, \mathrm{d}^2z/\mathrm{d}t^2$, the general equation $\mathrm{S}(\dot{x}\,\delta x + \dot{y}\,\delta y + \dot{z}\delta z)m + \mathrm{S}(P\,\delta p + Q\,\delta q + R\,\delta r + \cdots) = 0$. This equation will give as many particular equations as there are independent variations after having reduced the variations indicated by δ to the smallest possible number in accordance with the conditions of the system.

SECTION III
GENERAL PROPERTIES OF MOTION DEDUCED FROM THE PRECEDING FORMULA

1. Let us consider a system of bodies having an arbitrary configuration and connected in any manner but without any fixity or obstacle hindering their motion. It is evident in this case that the constraints of the system can only depend upon the relative positions of the bodies. Consequently, the equations of condition can only contain functions of the coordinates which define the relative distances between the bodies. This consideration furnishes general independent equations for the motion of the system which depend on the nature of the system and which are analogous to those we found for the equilibrium of a system in the first paragraph of SECTION III of PART I.

Subsection I
Properties Relative to the Center of Gravity

2. Let x', y', z' be the coordinates of any given body in the system, while x, y, z represent in general the coordinates of any other body. Let us assume the following which is always permitted

$$x = x' + \xi, \qquad y = y' + \eta, \qquad z = z' + \zeta$$

It is obvious that the quantities x', y', z' will not enter into the expression for the relative distances between the bodies, but that these distances will only depend upon the different quantities ξ, η, ζ which properly express the coordinates of different bodies referred to the system corresponding to x', y', z'. Consequently, the equations of the system will only be between the variables ξ, η, ζ and will not include the variables x', y', z'.

Thus if in the general formula of dynamics (Article 5, preceding section) all the variations are reduced to $\delta x, \delta y, \delta z$ and the expressions $\delta x' + \delta\xi, \delta y' + \delta\eta, \delta z' + \delta\zeta$, are substituted for δx, δy, δz, the variations $\delta x'$, $\delta y'$, $\delta z'$ will be independent of all the others and of an arbitrary nature. In this fashion, all of the terms affected by these variations must be equated separately to zero. This results in three general equations which are independent of the particular nature of the system.

The internal forces through which the bodies act on one another and which we denote by $\bar{P}$, $\bar{Q}$, etc. as in Article 2 of SECTION III of PART I will not enter into these equations because the mutual distances $\bar{p}$, $\bar{q}$, etc. are independent of x', y', z' and the variations $\delta\bar{p}$, $\delta\bar{q}$, etc. with respect to these variables will be zero.

With regard to the external forces P, Q, R, etc., if these forces are reduced to three forces X, Y, Z directed along the x, y, z coordinates and tending to diminish them, according to the formulas given in Chapter I of SECTION V of PART I, there results $P\,\delta p + Q\,\delta q + R\,\delta r + \text{etc.} = X\,\delta x + Y\,\delta y + Z\,\delta z$, and the general formula becomes

$$\mathrm{S}(\frac{\mathrm{d}^2 x}{\mathrm{d}t^2} + X)m\,\delta x + \mathrm{S}(\frac{\mathrm{d}^2 y}{\mathrm{d}t^2} + Y)m\,\delta y + \mathrm{S}(\frac{\mathrm{d}^2 z}{\mathrm{d}t^2} + Z)m\,\delta z = 0$$

Since this equation depends only on the variations $\delta x'$, $\delta y'$, $\delta z'$ which are independent of all others, the following equation is obtained

$$\delta x'\,\mathrm{S}(\frac{\mathrm{d}^2 x}{\mathrm{d}t^2} + X)m + \delta y'\,\mathrm{S}(\frac{\mathrm{d}^2 y}{\mathrm{d}t^2} + Y)m + \delta z'\,\mathrm{S}(\frac{\mathrm{d}^2 z}{\mathrm{d}t^2} + Z)m = 0$$

from which the following three equations are immediately obtained

$$\mathrm{S}(\frac{\mathrm{d}^2 x}{\mathrm{d}t^2} + X)m = 0, \qquad \mathrm{S}(\frac{\mathrm{d}^2 y}{\mathrm{d}t^2} + Y)m = 0, \qquad \mathrm{S}(\frac{\mathrm{d}^2 z}{\mathrm{d}t^2} + Z)m = 0$$

These equations will always hold for the motion of any system of bodies when the system is entirely free.

3. Now assume that the body with coordinates x', y', z' is placed at the center of gravity of the entire system. From the known properties of this center (PART I, SECTION III, Subsection IV), we have the following equations

$$\mathrm{S}\,\xi m = 0, \qquad \mathrm{S}\,\eta m = 0, \qquad \mathrm{S}\,\zeta m = 0$$

which will give these equations after differentiation with respect to t

$$\mathrm{S}\frac{\mathrm{d}^2 \xi}{\mathrm{d}t^2}m = 0, \qquad \mathrm{S}\frac{\mathrm{d}^2 \eta}{\mathrm{d}t^2}m = 0, \qquad \mathrm{S}\frac{\mathrm{d}^2 \zeta}{\mathrm{d}t^2}m = 0$$

Thus we will have $\mathrm{S}(\mathrm{d}^2 x/\mathrm{d}t^2)m = \mathrm{S}(\mathrm{d}^2 x'/\mathrm{d}t^2)m = (\mathrm{d}^2 x'/\mathrm{d}t^2)\mathrm{S}\,m$ since x' has the same value for all the bodies and is thus independent of the operator S. Similarly, we will also have

$$\mathrm{S}\frac{\mathrm{d}^2 y}{\mathrm{d}t^2}m = \frac{\mathrm{d}^2 y'}{\mathrm{d}t^2}\mathrm{S}\,m, \qquad \mathrm{S}\frac{\mathrm{d}^2 z}{\mathrm{d}t^2}m = \frac{\mathrm{d}^2 z'}{\mathrm{d}t^2}\mathrm{S}\,m,$$

In this fashion, the three equations of the preceding article will assume the simpler form

$$\frac{\mathrm{d}^2 x'}{\mathrm{d}t^2}\mathrm{S}\,m + \mathrm{S}\,Xm = 0, \qquad \frac{\mathrm{d}^2 y'}{\mathrm{d}t^2}\mathrm{S}\,m + \mathrm{S}\,Ym = 0, \qquad \frac{\mathrm{d}^2 z'}{\mathrm{d}t^2}\mathrm{S}\,m + \mathrm{S}\,Zm = 0$$

These equations will serve as a basis to determine the motion of the center of gravity of all the bodies, independent of the particular motion of an individual body. Since the values $\mathrm{S}\,Xm$, $\mathrm{S}\,Ym$, $\mathrm{S}\,Zm$ do not contain the internal forces of the system, the motion of the center of gravity will not depend on the mutual interaction that the bodies exert on one another, but only on the accelerating forces which act on each body. This is what the general principle of the Conservation of the Motion of the Center of Gravity consists of.

This principle applies also in the case of the motion of colliding bodies because, whatever the nature of the bodies, it can always be assumed that the interaction between them is modelled, at the time of the impact, by means of a spring interposed between the bodies and which after the compression of the spring may tend to regain its original length depending on the elastic or inelastic nature of the bodies. In this fashion, the effect of the impact will be to produce forces of the same nature as those we have designated $\bar{P}$, $\bar{Q}$, etc. and which cancel out in the general formula (Article 2).

4. Otherwise, it can be seen that the equations of motion for the center of gravity are the same as those for the motion of a single body acted on simultaneously by all the accelerating forces which act on the different bodies of the system. Indeed, it is assumed that all of these bodies are combined at a point which corresponds to the x', y', z' coordinates, then in the general formula, one would have $x = x'$, $y = y'$, $z = z'$ and equating all of the terms affected by the three variations $\delta x'$, $\delta y'$, $\delta z'$ to zero, the same equations would result as those above.

From the preceding development, this general theorem results: the motion of the center of gravity of a free system of bodies having an arbitrary configuration is always the same as if the bodies were combined to make a single mass point and at the same time as if each of them were acted upon by the same accelerating forces as in their natural state.

5. This theorem is also valid when the bodies composing a free system receive only arbitrary impulses. Since by substituting $\delta x' + \delta\xi$, $\delta y' + \delta\eta$, $\delta z' + \delta\zeta$ for δx, δy, δz in the equation of Article 11 of the preceding section and by reducing the forces P, Q, R, etc. to the forces X, Y, Z, it will be shown, as in Article 2, that the variations $\delta x'$, $\delta y'$, $\delta z'$ must remain arbitrary. This will result in the following three equations

$$\mathrm{S}(m\dot{x} + X) = 0, \qquad \mathrm{S}(m\dot{y} + Y) = 0, \qquad \mathrm{S}(m\dot{z} + Z) = 0$$

If the coordinates x', y', z' are referred to the center of gravity of the system, one has, due to the properties of this center

$$x'\,\mathrm{S}\,m = \mathrm{S}\,xm, \qquad y'\,\mathrm{S}\,m = \mathrm{S}\,ym, \qquad z'\,\mathrm{S}\,m = \mathrm{S}\,zm$$

Thus by differentiating with respect to t and by putting

$$\mathrm{d}x = \dot{x}\,\mathrm{d}t, \quad \mathrm{d}y = \dot{y}\,\mathrm{d}t, \quad \mathrm{d}z = \dot{z}\,\mathrm{d}t, \quad \mathrm{d}x' = \dot{x}'\,\mathrm{d}t, \quad \mathrm{d}y' = \dot{y}'\,\mathrm{d}t, \quad \mathrm{d}z' = \dot{z}'\,\mathrm{d}t$$

and consequently

$$\dot{x}'\,\mathrm{S}\,m = \mathrm{S}\,\dot{x}m, \qquad \dot{y}'\,\mathrm{S}\,m = \mathrm{S}\,\dot{y}m, \qquad \dot{z}'\,\mathrm{S}\,m = \mathrm{S}\,\dot{z}m$$

from which

$$\dot{x}'\,\mathrm{S}\,m + \mathrm{S}\,X = 0, \qquad \dot{y}'\,\mathrm{S}\,m + \mathrm{S}\,Y = 0, \qquad \dot{z}'\,\mathrm{S}\,m + \mathrm{S}\,Z = 0$$

which shows that the velocities $\dot{x}$, $\dot{y}$, $\dot{z}$ imparted to the center of gravity are the same as if all the bodies are combined at this center and receive simultaneously the impulses X, Y, Z.

6. The general formula (Article 2), after the substitution of $\delta x' + \delta\xi$, $\delta y' + \delta\eta$, $\delta z' + \delta\zeta$ in place of δx, δy, δz and elimination of the terms affected by $\delta x'$, $\delta y'$, $\delta z'$ will be reduced to

$$\mathrm{S}(\frac{\mathrm{d}^2 x\,\delta\xi + \mathrm{d}^2 y\,\delta\eta + \mathrm{d}^2 z\,\delta\zeta}{\mathrm{d}t^2} + X\,\delta\xi + Y\,\delta\eta + Z\,\delta\zeta)m = 0$$

By substituting $x' + \xi$, $y' + \eta$, $z' + \zeta$ for x, y, z in the differentials d^2x, d^2y, d^2z and by bringing before the operator S the differentials d^2x', d^2y', d^2z', the terms affected by these differentials will be

$$\frac{\mathrm{d}^2x'}{\mathrm{d}t^2}\mathrm{S}\,\delta\xi\,m + \frac{\mathrm{d}^2y'}{\mathrm{d}t^2}\mathrm{S}\,\delta\eta\,m + \frac{\mathrm{d}^2z'}{\mathrm{d}t^2}\mathrm{S}\,\delta\zeta\,m$$

But by referring the coordinates x', y', z' to the center of gravity, we will have (Article 3)

$$\mathrm{S}\,\xi m = 0, \qquad \mathrm{S}\,\eta m = 0, \qquad \mathrm{S}\,\zeta m = 0$$

and in addition, by differentiating with respect to δ, the following equations will be obtained

$$\mathrm{S}\,\delta\xi\,m = 0, \qquad \mathrm{S}\,\delta\eta\,m = 0, \qquad \mathrm{S}\,\delta\zeta\,m = 0$$

which eliminates the above terms.

In this fashion, the general formula will be reduced to

$$\mathrm{S}(\frac{\mathrm{d}^2\xi\,\delta\xi + \mathrm{d}^2\eta\,\delta\eta + \mathrm{d}^2\zeta\,\delta\zeta}{\mathrm{d}t^2} + X\,\delta\xi + Y\,\delta\eta + Z\,\delta\zeta)m = 0$$

which is perfectly analogous to the first formula where the coordinates x, y, z which have a fixed origin in space are replaced by ξ, η, ζ which have their origin at the center of gravity.

From this result, it can be concluded that, in general, in a free system, the same equations and the same properties are obtained relative to the center of gravity as those relative to a fixed point outside of the system.

Subsection II
Properties Relative to Areas

7. Now let us consider the motion of a system about a fixed point, and let us assume further that it is entirely free to rotate in every direction about this point. If the motions of the bodies of the system with respect to one another are neglected, the rotation about each of the three axes x, y, z will give, as seen earlier in Article 8 of SECTION III of PART I, the following expressions for the variations δx, δy, δz

$$\delta x = z\,\delta\omega - y\,\delta\varphi, \qquad \delta y = x\,\delta\varphi - z\,\delta\psi, \qquad \delta z = y\,\delta\psi - x\,\delta\omega$$

where $\delta\varphi$, $\delta\omega$, $\delta\psi$ are the elementary rotations with respect to the three axes z, y, x and which must remain arbitrary.

These equations are general expressions for the variations of the coordinates of all the bodies of the system and it will only be necessary to substitute them in the formula of Article 5 of the preceding section, after reducing all the variations to δx, δy, δz and equating to zero separately the quantities containing the three indeterminates $\delta\varphi$, $\delta\omega$, $\delta\psi$.

It will be found at the outset, as in the cited article of PART I, that the variations $\delta\bar{p}$ become equal to zero and that the terms resulting from the internal forces $\bar{P}$ of the system, since they do not contain the variations $\delta\varphi$, $\delta\omega$, $\delta\psi$ will not contribute to the equations in question. It will also be found, as it is shown in the same article, that the variation δp is zero when the force P is directed toward the origin of the coordinates, and thus in this case, the force does not appear in the same equations.

By simply substituting as indicated for δx, δy, δz after having replaced the forces P, Q, R, etc. with X, Y, Z as above (Article 2), there will result, with respect to the variations $\delta\varphi$, $\delta\omega$, $\delta\psi$, the equation

$$\mathrm{S}\left\{\begin{array}{l} (\dfrac{x\,\mathrm{d}^2y - y\,\mathrm{d}^2x}{\mathrm{d}t^2} + Yx - Xy)\delta\varphi \\ +(\dfrac{z\,\mathrm{d}^2x - x\,\mathrm{d}^2z}{\mathrm{d}t^2} + Xz - Zx)\delta\omega \\ +(\dfrac{y\,\mathrm{d}^2z - z\,\mathrm{d}^2y}{\mathrm{d}t^2} + Zy - Yz)\delta\psi \end{array}\right\} m = 0$$

and since the variations $\delta\varphi$, $\delta\psi$, $\delta\omega$ are the same for all the bodies of the system, they will not appear under the integral operator S, so that the three following equations with respect to each of the variations will be obtained

$$\mathrm{S}(x\frac{\mathrm{d}^2y}{\mathrm{d}t^2} - y\frac{\mathrm{d}^2x}{\mathrm{d}t^2} + xY - yX)m = 0$$

$$\mathrm{S}(z\frac{\mathrm{d}^2x}{\mathrm{d}t^2} - x\frac{\mathrm{d}^2z}{\mathrm{d}t^2} + zX - xZ)m = 0$$

$$S(y\frac{d^2z}{dt^2} - z\frac{d^2y}{dt^2} + yZ - zY)m = 0$$

These equations will always hold whenever the system is free to rotate about each of the three axes, that is, every time that the system is capable of rotating freely in every direction about a fixed point which is the origin of the coordinate system.

It is worthwhile to observe that these equations always hold independently of the mutual action of the bodies, however this action might be exerted, even by the mutual collision of the bodies of the system, as in Article 3. For the same reason, they are, moreover, independent of the forces which would tend toward the fixed point where the origin of the coordinate system is located.

In order to have a clearer understanding of these equations, note that 1), the quantities $x\,d^2y - y\,d^2x$, $z\,d^2x - x\,d^2z$, $y\,d^2z - z\,d^2y$ are the differentials of $x\,dy - y\,dx$, $z\,dx - x\,dz$, $y\,dz - z\,dy$, which represent twice the elementary sectors described by the body m on the xy-plane, the xz-plane and the yz-plane, that is, on the planes perpendicular to the axes of z, y, x. Indeed, if in the expression $x\,dy - y\,dx$, one substitutes for x and y the expressions $\rho\cos\varphi$, $\rho\sin\varphi$, one has $\rho^2\,d\varphi$ which represents twice the area between the radius vector ρ and the next radius which makes with it the angle $d\varphi$. 2) the quantities X, Y, Z represent the forces which act on each body m, in the directions of the coordinates x, y, z and toward their origin, and which result from all the forces P, Q, R, etc. acting on this body, along arbitrary directions (SECTION II, Article 5). Thus the quantities $yX - xY$, $xZ - zX$, $zY - yZ$ express the moments of the forces which tend to make the body rotate about each of the three coordinate axes z, y, x where the word "moment" is taken in the ordinary sense, that is, the product of the force with the length of the perpendicular drawn to its direction.

9. If no external force acts on the system or if the only forces acting on the system are directed toward the point which is the origin of the coordinate system, the three preceding equations would reduce to the following

$$S(\frac{x\,d^2y - y\,d^2x}{dt^2})m = 0$$
$$S(\frac{z\,d^2x - x\,d^2z}{dt^2})m = 0$$
$$S(\frac{y\,d^2z - z\,d^2y}{dt^2})m = 0$$

which after integration with respect to the variable t and with the introduction of three arbitrary constants of integration A, B, C will result in the following equations

$$S(\frac{x\,dy - y\,dx}{dt})m = C$$
$$S(\frac{z\,dx - x\,dz}{dt})m = B$$
$$S(\frac{y\,dz - z\,dy}{dt})m = A$$

These latter equations obviously express the Principle of Areas, which we have already mentioned in SECTION I.

10. It is apropos to observe that these equations are in the form of the equations of Article 10 of the preceding section so that three new arbitrary constants can be introduced by a transformation of the coordinate axes.

Let x', y', z' be the new coordinates. Then there results

$$\mathrm{S}(\frac{x'\,\mathrm{d}y' - y'\,\mathrm{d}x'}{\mathrm{d}t})m = C'$$
$$\mathrm{S}(\frac{z'\,\mathrm{d}x' - x'\,\mathrm{d}z'}{\mathrm{d}t})m = B'$$
$$\mathrm{S}(\frac{y'\,\mathrm{d}z' - z'\,\mathrm{d}y'}{\mathrm{d}t})m = A'$$

where the quantities A', B', C' are also arbitrary constants, but different from A, B, C.

Now let us substitute in the expression $x\,\mathrm{d}y - y\,\mathrm{d}x$ for the variables x and y the variables x', y', z' given in the cited article of the same section. One will have

$$x\,\mathrm{d}y - y\,\mathrm{d}x = (\alpha\beta' - \beta\alpha')(x'\,\mathrm{d}y' - y'\,\mathrm{d}x')$$
$$+(\gamma\alpha' - \alpha\gamma')(z'\,\mathrm{d}x' - x'\,\mathrm{d}z') + (\beta\gamma' - \gamma\beta')(y'\,\mathrm{d}z' - z'\,\mathrm{d}y')$$

One will similarly find

$$z\,\mathrm{d}x - x\,\mathrm{d}z = (\beta\alpha'' - \alpha\beta'')(x'\,\mathrm{d}y' - y'\,\mathrm{d}x')$$
$$+(\alpha\gamma'' - \gamma\alpha'')(z'\,\mathrm{d}x' - x'\,\mathrm{d}z') + (\gamma\beta'' - \beta\gamma'')(y'\,\mathrm{d}z' - z'\,\mathrm{d}y')$$
$$y\,\mathrm{d}z - z\,\mathrm{d}y = (\alpha'\beta'' - \beta'\alpha'')(x'\,\mathrm{d}y' - y'\,\mathrm{d}x')$$
$$+(\gamma'\alpha'' - \alpha'\gamma'')(z'\,\mathrm{d}x' - x'\,\mathrm{d}z') + (\gamma'\beta'' - \beta'\gamma'')(y'\,dz' - z'\,\mathrm{d}y')$$

If all the terms of this equation under the operator S are integrated after multiplying them by m and dividing them by $\mathrm{d}t$ and then if the constants A, B, C, A', B', C' are substituted for the integrals affected by the operator S, the following equations will be obtained

$$C = (\alpha\beta' - \beta\alpha')C' + (\gamma\alpha' - \alpha\gamma')B' + (\beta\gamma' - \gamma\beta')A'$$
$$B = (\beta\alpha'' - \alpha\beta'')C' + (\alpha\gamma'' - \gamma\alpha'')B' + (\gamma\beta'' - \beta\gamma'')A'$$
$$A = (\alpha'\beta'' - \beta'\alpha'')C' + (\gamma'\alpha'' - \alpha'\gamma'')B' + (\gamma'\beta'' - \beta'\gamma'')A'$$

These formulas can be reduced to a simpler expression by observing that one has identically

$$(\alpha\beta' - \beta\alpha')^2 + (\beta\alpha'' - \alpha\beta'')^2 + (\alpha'\beta'' - \beta'\alpha'')^2$$
$$= (\alpha^2 + \alpha'^2 + \alpha''^2)(\beta^2 + \beta'^2 + \beta''^2) - (\alpha\beta + \alpha'\beta' + \alpha''\beta'')^2$$

a quantity which reduces to unity by virtue of the equations of condition of Article 10 of SECTION III of PART I. In addition, the following equations will be obtained

$$\alpha(\alpha'\beta'' - \beta'\alpha'') + \alpha'(\beta\alpha'' - \alpha\beta'') + \alpha''(\alpha\beta' - \beta\alpha') = 0$$
$$\beta(\alpha'\beta'' - \beta'\alpha'') + \beta'(\beta\alpha'' - \alpha\beta'') + \beta''(\alpha\beta' - \beta\alpha') = 0$$

If these equations are compared with the three equations of condition

$$\gamma^2 + \gamma'^2 + \gamma''^2 = 0, \qquad \alpha\gamma + \alpha'\gamma' + \alpha''\gamma'' = 0, \qquad \beta\gamma + \beta'\gamma' + \beta''\gamma'' = 0$$

it is easy to conclude from this comparison that one will have

$$\alpha'\beta'' - \beta'\alpha'' = \gamma, \qquad \beta\alpha'' - \alpha\beta'' = \gamma', \qquad \alpha\beta' - \beta\alpha' = \gamma''$$

The quantities γ, γ', γ'' could also be negative. But because at the intersection of the axes x', y', z' and the axes x, y, z, one must have $\alpha = 1, \beta = 0, \gamma = 0, \alpha' = 0, \beta' = 1, \gamma' = 0$, $\alpha'' = 0, \beta'' = 0, \gamma'' = 1$ (Article 11, SECTION III, PART I), this condition can only hold by taking γ'' positive and consequently, the quantities γ' and γ will also be positive.

One will find in the same fashion that

$$\gamma'\alpha'' - \alpha'\gamma'' = \beta, \qquad \alpha\gamma'' - \gamma\alpha'' = \beta', \qquad \gamma\alpha' - \alpha\gamma' = \beta''$$

$$\gamma'\beta'' - \beta'\gamma'' = \alpha, \qquad \gamma\beta'' - \beta\gamma'' = \alpha', \qquad \beta\gamma' - \gamma\beta' = \alpha''$$

and so that one will have

$$A = \alpha A' + \beta B' + \gamma C'$$
$$B = \alpha' A' + \beta' B' + \gamma' C'$$
$$C = \alpha'' A' + \beta'' B' + \gamma'' C'$$

from which one obtains from the equations of condition of Article 10 (SECTION III, PART I)

$$A' = A\alpha + B\alpha' + C\alpha''$$
$$B' = A\beta + B\beta' + C\beta''$$
$$C' = A\gamma + B\gamma' + C\gamma''$$

and

$$A^2 + B^2 + C^2 = A'^2 + B'^2 + C'^2$$

From this last equation one has in general the following result

$$(\mathrm{S}(\frac{x\,\mathrm{d}y - y\,\mathrm{d}x}{\mathrm{d}t})m)^2 + (\mathrm{S}(\frac{z\,\mathrm{d}x - x\,\mathrm{d}z}{\mathrm{d}t})m)^2 + (\mathrm{S}(\frac{y\,\mathrm{d}z - z\,\mathrm{d}y}{\mathrm{d}t})m)^2$$
$$= (\mathrm{S}(\frac{x'\,\mathrm{d}y' - y'\,\mathrm{d}x'}{\mathrm{d}t})m)^2 + (\mathrm{S}(\frac{z'\,\mathrm{d}x' - x'\,\mathrm{d}z'}{\mathrm{d}t})m)^2 + (\mathrm{S}(\frac{y'\,\mathrm{d}z' - z'\,\mathrm{d}y'}{\mathrm{d}t})m)^2$$

from which it can be concluded that the function

$$(\mathrm{S}(\frac{x\,\mathrm{d}y - y\,\mathrm{d}x}{\mathrm{d}t})m)^2 + (\mathrm{S}(\frac{z\,\mathrm{d}x - x\,\mathrm{d}z}{\mathrm{d}t})m)^2 + (\mathrm{S}(\frac{y\,\mathrm{d}z - z\,\mathrm{d}y}{\mathrm{d}t})m)^2$$

always has a value independent of the plane of projection and of the position of the axes of coordinates x, y, z in space, if these coordinates are orthogonal.

11. The expressions for A', B', C' in terms of A, B, C which we just found are similar to those for x', y', z' in terms of x, y, z of Article 9 of the preceding section. Consequently, if one puts $x' = A'$, $y' = B'$, $z' = C'$, one will have that $A = x$, $B = y$, $C = z$ and reciprocally that $x = A$, $y = B$, $z = C$ which will give $A' = x'$, $B' = y'$, $C' = z'$, that is, A, B, C and A', B', C' will be two systems of coordinates which correspond to the same point, the first being relative to the axes of x, y, z and the second to the axes of x', y', z'.

It is immediately obvious that from this result that one can put $A' = 0$, $B' = 0$, by making the axes of C' or z' pass through the point to which the coordinates A, B, C correspond and then the coordinate C' will have its largest value equal to $\sqrt{A^2 + B^2 + C^2}$. In this case, one will have

$$A = \gamma C', \qquad B = \gamma' C', \qquad C = \gamma'' C'$$

and it is easy to see that the coefficients γ, γ', γ'' will be nothing more than the cosines of the angles that the line C' makes with the axes of A, B, C.

Thus the solution of the equations

$$\mathrm{S}(\frac{x'\,\mathrm{d}y' - y'\,\mathrm{d}x'}{\mathrm{d}t})m = C'$$
$$\mathrm{S}(\frac{z'\,\mathrm{d}x' - x'\,\mathrm{d}z'}{\mathrm{d}t})m = 0$$
$$\mathrm{S}(\frac{y'\,\mathrm{d}z' - z'\,\mathrm{d}y'}{\mathrm{d}t})m = 0$$

will give the following equations

$$\mathrm{S}(\frac{x\,\mathrm{d}y - y\,\mathrm{d}x}{\mathrm{d}t})m = \gamma'' C'$$
$$\mathrm{S}(\frac{z\,\mathrm{d}x - x\,\mathrm{d}z}{\mathrm{d}t})m = \gamma' C'$$
$$\mathrm{S}(\frac{y\,\mathrm{d}z - z\,\mathrm{d}y}{\mathrm{d}t})m = \gamma C'$$

where the quantities γ, γ', γ'' are three constants such that $\gamma^2 + \gamma'^2 + \gamma''^2 = 1$ and of which two are arbitrary.

The plane perpendicular to the axis of C', when C' is a maximum, is the one called by Laplace the invariant plane. He was first to demonstrate its existence and orientation.

The orientation of this plane is easy to determine from the equations

$$A = \gamma C', \qquad B = \gamma' C', \qquad C = \gamma'' C'$$

because the quantities $\gamma, \gamma', \gamma''$ are the cosines of the angles that the axis of C' or z', which is perpendicular to the invariant plane, make with the x, y, z axes of the system. Denoting these angles by the letters ℓ, m, n, one will have since $C' = \sqrt{(A^2 + B^2 + C^2)}$

$$\cos \ell = \frac{A}{\sqrt{(A^2 + B^2 + C^2)}}$$
$$\cos m = \frac{B}{\sqrt{(A^2 + B^2 + C^2)}}$$
$$\cos n = \frac{C}{\sqrt{(A^2 + B^2 + C^2)}}$$

12. If the system is free, that is, there are no points of the system which are fixed and the origin of the coordinates x, y, z, is assumed fixed, then the location of the origin is arbitrary. Consequently, the properties of areas and moments which we have demonstrated will hold with respect to a fixed and arbitrary point taken anywhere in space.

But from what we have demonstrated in Article 6, these same properties will also hold with respect to the center of gravity of the entire system whether this center is fixed or not. Indeed, if in the three equations of Article 7, one substitutes for x, y, z the quantities $x' + \xi$, $y' + \eta$, $z' + \zeta$, by referring as in Article 3 the coordinates x', y', z' to the center of gravity of the system and taking the three equations of this last article, one will have these transformed equations

$$\mathrm{S}(\frac{\xi\,\mathrm{d}^2\eta - \eta\,\mathrm{d}^2\xi}{\mathrm{d}t^2} + \xi Y - \eta X)m = 0$$
$$\mathrm{S}(\frac{\zeta\,\mathrm{d}^2\xi - \xi\,\mathrm{d}^2\zeta}{\mathrm{d}t^2} + \zeta X - \xi Z)m = 0$$
$$\mathrm{S}(\frac{\eta\,\mathrm{d}^2\zeta - \zeta\,\mathrm{d}^2\eta}{\mathrm{d}t^2} + \eta Z - \zeta Y)m = 0$$

which are, as one observes, similar to those of Article 7 and for which the entire difference consists of the fact that in place of the coordinates x, y, z taken about a fixed point, there are the coordinates ξ, η, ζ, whose origin is at the center of gravity of the system.

Thus when the accelerating forces are zero, the following integrals will be obtained

$$\mathrm{S}(\frac{\xi\,\mathrm{d}\eta - \eta\,\mathrm{d}\xi}{\mathrm{d}t^2})m = C$$
$$\mathrm{S}(\frac{\zeta\,\mathrm{d}\xi - \xi\,\mathrm{d}\zeta}{\mathrm{d}t^2})m = B$$
$$\mathrm{S}(\frac{\eta\,\mathrm{d}\zeta - \zeta\,\mathrm{d}\eta}{\mathrm{d}t^2})m = A$$

Since these equations are analogous to those of Article 9, similar comments could be made.

13. When one of the bodies of the system is firmly fixed by some arbitrary obstacle, the formulation of Article 7 is applicable if the origin of the coordinates is put in this body.

But if two bodies of the system are assumed fixed, one will observe that the line which passes through these two bodies is a fixed axis about which the system can rotate freely and taking this axis for those of the coordinate z, one will have simply from the same article $\delta x = -y\,\mathrm{d}\varphi$, $\delta y = x\,\delta\varphi$ where $\delta\varphi$ represents an elementary rotation about this axis and which must remain indeterminate. Thus there is only one equation relative to the variation $\delta\varphi$, which will be

$$\mathrm{S}(x\frac{\mathrm{d}^2 y}{\mathrm{d}t^2} - y\frac{\mathrm{d}^2 x}{\mathrm{d}t^2} + xY - yX)m = 0$$

and when the moment $xY - yX$ of the external forces with respect to the axis of rotation is zero, one obtains after integration as in Article 9, $\mathrm{S}((x\,\mathrm{d}y - y\,\mathrm{d}x)/\mathrm{d}t)m = C$ an equation which expresses the Principle of Areas with respect to the xy-plane which is perpendicular to the axis of rotation and upon which the areas described by the bodies must be projected.

If three bodies of the system are assumed fixed, then the position of every other body in space would be determined by its distance from these three bodies. There will be no variations independent of the nature of this system and of the relative configuration of the bodies from which one can deduce general equations for the motion of an arbitrary system.

Subsection III
Properties Relative to Rotations Created by Impulsive Forces

14. When a system free to rotate in any direction about a fixed point is subjected to arbitrary impulses, the expressions for δx, δy, δz of Article 7 can also be used in the equations of Article 11 of the preceding section after reducing the forces of impulsion P, Q, R, etc. to their X, Y, Z components. By separately equating to zero the terms multiplied by the variations $\delta\varphi$, $\delta\omega$, $\delta\psi$, the following three equations will be obtained

$$\mathrm{S}\{m(x\dot{y} - y\dot{x}) + xY - yX\} = 0$$
$$\mathrm{S}\{m(z\dot{x} - x\dot{z}) + zX - xZ\} = 0$$
$$\mathrm{S}\{m(y\dot{z} - z\dot{y}) + yZ - zY\} = 0$$

for the first instant of motion produced by the impulses X, Y, Z.

In systems which are completely free, the fixed point can be taken anywhere in space and the preceding equations will always hold with respect to this point.

15. For these systems, the rotations can also be referred to three axes which pass through the center of gravity. Since, following the formulations of Article 5

$$\delta x = \delta x' + \delta\xi, \qquad \delta y = \delta y' + \delta\eta, \qquad \delta z = \delta z' + \delta\zeta$$

the variations $\delta x'$, $\delta y'$, $\delta z'$ will provide immediately the three equations relative to the motion of the center of gravity which are found in this same article.

There will then remain the equation

$$\mathrm{S}((m\dot{x}+X)\delta\xi+(m\dot{y}+Y)\delta\eta+(m\dot{z}+Z)\delta\zeta)=0$$

Referring the rotations $\delta\psi$, $\delta\omega$, $\delta\varphi$ to the coordinate axes ξ, η, ζ and considering only these rotations as in Article 7

$$\delta\xi=\zeta\,\delta\omega-\eta\,\delta\varphi,\qquad \delta\eta=\xi\,\delta\varphi-\zeta\,\delta\psi,\qquad \delta\zeta=\eta\,\delta\psi-\xi\,\delta\omega$$

and the three indeterminate variations $\delta\psi$, $\delta\omega$, $\delta\varphi$ will give the three equations

$$\mathrm{S}\{m(\xi\dot{y}-\eta\dot{x})+\xi Y-\eta X\}=0$$
$$\mathrm{S}\{m(\zeta\dot{x}-\xi\dot{z})+\zeta X-\xi Z\}=0$$
$$\mathrm{S}\{m(\eta\dot{z}-\zeta\dot{y})+\eta Z-\zeta Y\}=0$$

But recall that $\dot{x}=\dot{x}'+\dot{\xi}$, $\dot{y}=\dot{y}'+\dot{\eta}$, $\dot{z}=\dot{z}'+\dot{\zeta}$ are valid here. Thus substituting these expressions, taking outside the operator S the quantities $\dot{x}'$, $\dot{y}'$, $\dot{z}'$ which are only related to the center of gravity and noting that by the properties of this center one has

$$\mathrm{S}\,m\xi=0,\qquad \mathrm{S}\,m\eta=0,\qquad \mathrm{S}\,m\zeta=0$$

the three preceding equations will become

$$\mathrm{S}\{m(\xi\dot{n}-\eta\dot{\xi})+\xi Y-\eta X\}=0$$
$$\mathrm{S}\{m(\zeta\dot{\xi}-\xi\dot{\zeta})+\zeta X-\xi Z\}=0$$
$$\mathrm{S}\{m(\eta\dot{\zeta}-\zeta\dot{\eta})+\eta Z-\zeta Y\}=0$$

which are comparable to those of the preceding article and in which the coordinates ξ, η, ζ have their origin at the center of gravity and the velocities $\dot{\xi}$, $\dot{\eta}$, $\dot{\zeta}$ are relative to this center.

Thus the equations relative to a fixed point also hold with respect to the center of gravity of a system which is free.

16. The equations which we just found for the effect of the impulses at the first instant also hold in the following instants, if there are no accelerating forces and if the terms which depend on the impulses X, Y, Z are assumed constant. Since $\dot{x}$, $\dot{y}$, $\dot{z}$ are the velocities parallel to the x, y, z coordinate axes, one has $\mathrm{d}x=\dot{x}\,\mathrm{d}t$, $\mathrm{d}y=\dot{y}\,\mathrm{d}t$, $\mathrm{d}z=\dot{z}\,\mathrm{d}t$ and the equations of Article 9 become

$$\mathrm{S}m(x\dot{y}-y\dot{x})=C$$
$$\mathrm{S}m(z\dot{x}-x\dot{z})=B$$
$$\mathrm{S}m(y\dot{z}-z\dot{y})=A$$

which when compared to those of Article 14 result in the following equations

$$C = \mathrm{S}(yX - xY)$$
$$B = \mathrm{S}(xZ - zX)$$
$$A = \mathrm{S}(zY - yZ)$$

Thus the values of the constants A, B, C are obtained in terms of the primary impulses given to each body and it is clear that these values are nothing but the sum of the moments of these impulses with respect to the coordinate axes x, y, z.

The same situation exists for the equations relative to the center of gravity by comparing the equations of Article 12 with those of Article 15.

17. If only the motion of rotation about the three coordinate axes x, y, z is considered and if the velocities of rotation are designated by $\dot{\psi}$, $\dot{\omega}$, $\dot{\varphi}$, the variations δx, δy, δz will be proportional to the velocities $\dot{x}$, $\dot{y}$, $\dot{z}$ and the variations $\delta\psi$, $\delta\omega$, $\delta\varphi$ will be proportional at the same time to the velocities $\dot{\psi}$, $\dot{\omega}$, $\dot{\varphi}$. The formulas of Article 7 will then give

$$\dot{x} = z\dot{\omega} - y\dot{\varphi}, \qquad \dot{y} = x\dot{\varphi} - z\dot{\psi}, \qquad \dot{z} = y\dot{\psi} - x\dot{\omega}$$

The values of $\dot{x}$, $\dot{y}$, $\dot{z}$ are the only components which depend on the three rotations. In order to obtain the total values of the actual velocities $\dot{x}$, $\dot{y}$, $\dot{z}$ one must add the components which depend on the relative change of the position of the bodies of the system and which are independent of the rotations.

But when the system is invariable, which occurs for every solid body with arbitrary configuration, these components of the velocities are zero and the values of $\dot{x}$, $\dot{y}$, $\dot{z}$ reduce simply to those which we just gave. Thus these values can be substituted in the preceding equations and by taking outside the operator S the quantities $\dot{\psi}$, $\dot{\omega}$, $\dot{\varphi}$, one will have for a solid of arbitrary shape, after replacing m by Dm (Article 7, preceding section), the equations

$$\dot{\varphi}\,\mathrm{S}(x^2 + y^2)\mathrm{D}m - \dot{\psi}\,\mathrm{S}xz\,\mathrm{D}m - \dot{\omega}\,\mathrm{S}yz\,\mathrm{D}m = C$$
$$\dot{\omega}\,\mathrm{S}(x^2 + z^2)\mathrm{D}m - \dot{\psi}\,\mathrm{S}xy\,\mathrm{D}m - \dot{\varphi}\,\mathrm{S}yz\,\mathrm{D}m = B$$
$$\dot{\psi}\,\mathrm{S}(y^2 + z^2)\mathrm{D}m - \dot{\omega}\,\mathrm{S}xy\,\mathrm{D}m - \dot{\varphi}\,\mathrm{S}xz\,\mathrm{D}m = A$$

from which one can determine the velocities of the initial rotations $\dot{\psi}$, $\dot{\omega}$, $\dot{\varphi}$, produced by the impulses X, Y, Z applied at arbitrary points of the body and for which the moments about the coordinate axes x, y, z are A, B, C.

Since the angular velocities are proportional to the infinitesimal angles described simultaneously by the respective rotations, it follows from what has been demonstrated in Article 11 of SECTION III of PART I that the three velocities $\dot{\psi}$, $\dot{\omega}$, $\dot{\varphi}$ can be combined to obtain a single velocity $\dot{\theta}$ such that $\dot{\theta} = \sqrt{(\dot{\psi}^2 + \dot{\omega}^2 + \dot{\varphi}^2)}$ with which the body will actually rotate about an instantaneous axis, making with the coordinate axes x, y, z the angles λ, μ, ν, such that

$$\cos\lambda = \frac{\dot\psi}{\dot\theta}, \qquad \cos\mu = \frac{\dot\omega}{\dot\theta}, \qquad \cos\nu = \frac{\dot\varphi}{\dot\theta}$$

Thus the three preceding equations will give the orientation of the axis about which the body rotates at the first instant and the angular velocity about this axis. This axis is called *the instantaneous axis of rotation.*

18. In the following instants of time, the body will continue rotating due to its inertia and the three equations which were just found will still hold if the terms which contain the forces of impulsion X, Y, Z are viewed as constants as can be seen from Article 16. But the quantities $\mathrm{S}(x^2 + y^2)\mathrm{D}m$, $\mathrm{S}\, xy\, \mathrm{D}m$, etc. become variables because of the variation of the coordinates x, y, z during the rotation.

But a remarkable result can be drawn from these equations which is that, at any instant, the body has the same rotational motion which it would have received at this instant by the impulse from the same forces which initially put it in motion, if these forces are applied in such a manner that they would produce the same moments about the coordinate axes x, y, z.

Since these equations are nothing but the general equations for an arbitrary system of bodies of Article 16 applied to a solid body of arbitrary shape, it follows that if the system which has received the primary impulses become, from the mutual and successive action of the bodies, an invariable system or an arbitrary solid, the same equations will still hold so that the solid will have at any instant the same rotational motion which it would receive from the same primary impulses, if they were instantly applied so as to produce the same moments.

In addition, a fluid mass set in motion initially by arbitrary forces then left to itself to become solid due to the mutual attraction of its parts, will have at any instant the same rotational motion that the original forces would have produced if they were acting in the same fashion on the solid mass.

19. The three equations of Article 17 will give the values of the moments A, B, C of all the initial forces by knowing the instantaneous position of the body and its three angular velocities $\dot\psi$, $\dot\omega$, $\dot\varphi$, about the fixed coordinate axes x, y, z or the composed velocity $\dot\theta$ about the instantaneous axis with the angles λ, μ, ν that this axis makes with the fixed coordinate axes x, y, z. Conversely, with these moments, the values of the angular velocities can be deduced.

One can also see from these equations that the moments will be zero if the velocities are zero but if the moments are assumed equal to zero, it does not directly follow that the angular velocities must be equal to zero. Since by taking $A = 0$, $B = 0$, $C = 0$ one has three linear equations between $\dot\psi$, $\dot\omega$, $\dot\varphi$ and one should prove that these three equations cannot exist together unless it is assumed that $\dot\psi = 0, \dot\omega = 0, \dot\varphi = 0$.

By eliminating two of the unknowns one is left with one equation which gives the third unknown either equal to zero or to an arbitrary number but with the condition

$$\begin{aligned}&\mathrm{S}(x^2+y^2)\mathrm{D}m\,\mathrm{S}(x^2+z^2)\mathrm{D}m\,S(y^2+z^2)\mathrm{D}m\\&=\mathrm{S}(x^2+y^2)\mathrm{D}m(\mathrm{S}\,xy\,\mathrm{D}m)^2+\mathrm{S}(x^2+z^2)\mathrm{D}m(\mathrm{S}\,xz\,\mathrm{D}m)^2\\&\quad+\mathrm{S}(y^2+z^2)\mathrm{D}m(\mathrm{S}\,yz\,\mathrm{D}m)^2+2\,\mathrm{S}\,xy\,\mathrm{D}m\,\mathrm{S}\,xz\,\mathrm{D}m\,\mathrm{S}yz\,\mathrm{D}m\end{aligned}$$

and it is left to prove that this condition is impossible to satisfy, which seems very difficult. But it will be demonstrated later on (Article 31) that when the moments are zero, every rotation also disappears.

From these results, it can be concluded that it is impossible for a system of isolated points or an arbitrary fluid mass to form a solid body with a rotational motion unless the initial impulses produced a moment about the axis of rotation.

20. By the transformations presented in Article 10, the three equations of Article 17 can be transformed to similar equations in which the quantities x, y, z, A, B, C are replaced by analogous quantities x', y', z', A', B', C'.

Let us designate by $\dot{\psi}'$, $\dot{\omega}'$, $\dot{\varphi}'$ the angular velocities with respect to the new coordinate axes x', y', z'. One will also have

$$\begin{aligned}\mathrm{d}x'&=\dot{x}'\,\mathrm{d}t=(z'\dot{\omega}'-y'\dot{\varphi}')\mathrm{d}t\\\mathrm{d}y'&=\dot{y}'\,\mathrm{d}t=(x'\dot{\varphi}'-z'\dot{\psi}')\mathrm{d}t\\\mathrm{d}z'&=\dot{z}'\,\mathrm{d}t=(y'\dot{\psi}'-x'\omega')\mathrm{d}t\end{aligned}$$

and the first three equations of Article 10 will become after these substitutions and after replacing m by $\mathrm{D}m$

$$\begin{aligned}&\dot{\varphi}'\,\mathrm{S}(x'^2+y'^2)\mathrm{D}m-\dot{\psi}'\,\mathrm{S}\,x'z'\,\mathrm{D}m-\dot{\omega}'\,\mathrm{S}\,y'z'\,\mathrm{D}m=C'\\&\dot{\omega}'\,\mathrm{S}(z'^2+x'^2)\mathrm{D}m-\dot{\psi}'\,\mathrm{S}\,x'y'\,\mathrm{D}m-\dot{\varphi}'\,\mathrm{S}y'z'\,\mathrm{D}m=B'\\&\dot{\psi}'\,\mathrm{S}(y'^2+z'^2)\mathrm{D}m-\dot{\omega}'\,\mathrm{S}x'y'\,\mathrm{D}m-\dot{\varphi}'\,\mathrm{S}x'z'\,\mathrm{D}m=A'\end{aligned}$$

from which one will have from the same article

$$\begin{aligned}A'&=A\alpha+B\alpha'+C\alpha''\\B'&=A\beta+B\beta'+C\beta''\\C'&=A\gamma+B\gamma'+C\gamma''\end{aligned}$$

These equations have the advantage that the position of the axes of rotation are entirely arbitrary since they depend only on the quantities α, β, γ, α', etc. Since they are only of the first order, nothing prevents us from giving to these axes a different position from one instant to the other and to assume them fixed inside the body. Consequently, they move with the body in space. Then the quantities $\mathrm{S}(x'^2+y'^2)\mathrm{D}m$, $\mathrm{S}x'y'\,\mathrm{D}m$, etc. become constants but the quantities A', B', C' will be variable because of the variability of the quantities α, β, γ, α', etc. In the following development, we will give a direct approach to obtain these equations which are of great use in the problem of the rotation of the bodies.

21. We saw in Article 16 that the constants A, B, C express the sums of the moments of the initial impulses given to the bodies relative to the coordinate axes x, y, z. But it is easy to prove that the quantities α, α', α'' represent the cosines of the angles made by the coordinate axis x' with the directions of the axes x, y, z, that the quantities β, β', β'' represent the cosines of the angles made by the coordinate axis y' with the directions of the same axes x, y, z and that the quantities γ, γ', γ'' represent the cosines of the angles made by the coordinate axis z' with the directions of the same axes. Therefore, from what one has demonstrated in PART I on the composition of moments (SECTION III, Article 16), the three quantities A', B', C' will be the moments of the same impulses referred to the coordinate axes x', y', z', that is, to the axes of rotation fixed in the body and mobile in space. Thus one will be able to attribute to these axes the same conclusions reached in Article 19.

Subsection IV
Properties of a Freely Rotating Body of Arbitrary Shape about Fixed Axes of Rotation

22. We reserve for another chapter the complete solution of the general problem of the rotation of a solid body with an arbitrary shape. We will examine here the case where the instantaneous axis of rotation remains immobile in space or at least is always parallel to itself when the body has rectilinear motion, because this case is easily solved by the formulas of the preceding paragraph and it leads to the beautiful properties of the axes which are called the principal or natural axes of rotation.

Let us again consider the fundamental equations of Article 17. In order to shorten the expressions, let us make the following definitions

$$\ell = \mathrm{S}\, x^2\, \mathrm{D}m, \qquad m = \mathrm{S}\, y^2\, \mathrm{D}m, \qquad n = \mathrm{S}\, z^2\, \mathrm{D}m,$$

$$f = \mathrm{S}\, yz\, \mathrm{D}m, \qquad g = \mathrm{S}\, xz\, \mathrm{D}m, \qquad h = \mathrm{S}\, xy\, \mathrm{D}m$$

and then substitute for $\dot{\psi}$, $\dot{\omega}$, $\dot{\varphi}$ the values $\dot{\theta}\cos\lambda$, $\dot{\theta}\cos\mu$, $\dot{\theta}\cos\nu$, where $\dot{\theta}$ is the velocity of rotation about the instantaneous axis which makes the angles λ, μ, ν with the fixed coordinate axes x, y, z. These equations will thus become after division by $\dot{\theta}$

$$(m+n)\cos\lambda - h\cos\mu - g\cos\nu = \frac{A}{\dot{\theta}}$$
$$(\ell+n)\cos\mu - h\cos\lambda - f\cos\nu = \frac{B}{\dot{\theta}}$$
$$(\ell+m)\cos\nu - g\cos\lambda - f\cos\mu = \frac{C}{\dot{\theta}}$$

23. The six quantities ℓ, m, n, f, g, h are variables. By differentiating them and substituting for $\mathrm{d}x$, $\mathrm{d}y$, $\mathrm{d}z$ the quantities $\dot{x}\,\mathrm{d}t$, $\dot{y}\,\mathrm{d}t$, $\dot{z}\,\mathrm{d}t$ and then for $\dot{x}$, $\dot{y}$, $\dot{z}$ their values (cited article), one will obtain

$$\begin{aligned}
\mathrm{d}\ell &= 2(g\cos\mu - h\cos\nu)\dot{\theta}\,\mathrm{d}t\\
\mathrm{d}m &= 2(h\cos\nu - f\cos\lambda)\dot{\theta}\,\mathrm{d}t\\
\mathrm{d}n &= 2(f\cos\lambda - g\cos\mu)\dot{\theta}\,\mathrm{d}t\\
\mathrm{d}f &= ((m-n)\cos\lambda + g\cos\nu - h\cos\mu)\dot{\theta}\,\mathrm{d}t\\
\mathrm{d}g &= ((n-\ell)\cos\mu + h\cos\lambda - f\cos\nu)\dot{\theta}\,\mathrm{d}t\\
\mathrm{d}h &= ((\ell-m)\cos\nu + f\cos\mu - g\cos\lambda)\dot{\theta}\,\mathrm{d}t
\end{aligned}$$

These six equations added to the three of the preceding article contain the general solution but we will consider here only the case where the angles λ, μ, ν remain invariable. It is left to determine under what conditions these quantities can be constant.

24. In order to obtain this result, it is only necessary to differentiate the first three equations with this assumption and then substitute the values of the differentials $\mathrm{d}\ell$, $\mathrm{d}m$, etc. One will have after dividing by $\dot{\theta}\,\mathrm{d}t$ the three following equations

$$\begin{aligned}
&f(\cos^2\nu - \cos^2\mu) - g\cos\lambda\cos\mu + h\cos\lambda\cos\nu\\
&\quad +(m-n)\cos\mu\cos\nu = -\frac{A}{\dot{\theta}^3}\frac{\mathrm{d}\dot{\theta}}{\mathrm{d}t}\\
&f\cos\lambda\cos\mu + g(\cos^2\lambda - \cos^2\nu) - h\cos\mu\cos\nu\\
&\quad +(n-\ell)\cos\lambda\cos\nu = -\frac{B}{\dot{\theta}^3}\frac{\mathrm{d}\dot{\theta}}{\mathrm{d}t}\\
&-f\cos\lambda\cos\nu + g\cos\mu\cos\nu + h(\cos^2\mu - \cos^2\lambda)\\
&\quad +(\ell-m)\cos\lambda\cos\mu = -\frac{C}{\dot{\theta}^3}\frac{\mathrm{d}\dot{\theta}}{\mathrm{d}t}
\end{aligned}$$

If these three equations are added together, after multiplying the first equation by the quantity $\cos\lambda$, the second by $\cos\mu$, the third by $\cos\nu$, one will obtain the equation

$$0 = \left(-\frac{A\cos\lambda + B\cos\mu + C\cos\nu}{\dot{\theta}^3}\right)\frac{\mathrm{d}\dot{\theta}}{\mathrm{d}t}$$

which gives $\mathrm{d}\dot{\theta} = 0$ or better yet $A\cos\lambda + B\cos\mu + C\cos\nu = 0$. We will see later (Article 38) that the quantity $A\dot{\psi} + B\dot{\omega} + C\dot{\varphi}$ which is the same as $(A\cos\lambda + B\cos\mu + C\cos\nu)\dot{\theta}$ and which expresses the **force vive** of the body, can never be zero as long as the body is in motion.

Therefore, one must assume in general that $\mathrm{d}\dot{\theta} = 0$, and consequently, that the angular velocity $\dot{\theta}$ is constant. Then the three equations above reduce to two, which give the ratios of $\cos\lambda$, $\cos\mu$, $\cos\nu$. Since we have $\cos^2\lambda + \cos^2\mu + \cos^2\nu = 1$, these expressions will suffice to determine the values of the three cosines.

25. Let us assume $s = \cos\mu/\cos\lambda$, $u = \cos\nu/\cos\lambda$ then the three preceding equations will become, since $\mathrm{d}\dot{\theta} = 0$

$$\begin{aligned} &f(u^2 - s^2) - gs + hu + (m - n)su = 0 \\ &g(1 - u^2) - hsu + fs + (n - \ell)u = 0 \\ &h(s^2 - 1) + gsu - fu + (\ell - m)s = 0 \end{aligned}$$

the last equation gives

$$u = \frac{h(s^2 - 1) + (\ell - m)s}{f - gs}$$

After this value is substituted in the first or second equations or rather in the sum of the two which have been multiplied by g and f, respectively, in order to eliminate u^2, there will result

$$\begin{aligned} &(gh(m - n) + f(g^2 - h^2))s^3 \\ &+(g(\ell - m)(m - n) + fh(n - 2\ell + m) + g(g^2 + h^2 - 2f^2))s^2 \\ &+(f(\ell - m)(m - n) + gh(n - 2m + \ell) + f(f^2 + h^2 - 2g^2))s \\ &+fh(\ell - n) + g(f^2 - h^2) = 0 \end{aligned}$$

Since this equation is of the third degree, it will necessarily have a real root. Thus one will have a value of s and a corresponding value of u with which the position of the invariable axis with uniform rotation can be determined. But because this determination depends on the quantities ℓ, m, n, f, g, h which vary with time t, one must still prove that the variation of these quantities does not influence the value of the two quantities s and u.

26. In order to carry out this demonstration, let us denote by P, Q, R, the first terms of the three equations of Article 22. The first terms of the equations of Article 24 will be

$$\frac{\mathrm{d}P}{\dot{\theta}\,\mathrm{d}t}, \qquad \frac{\mathrm{d}Q}{\dot{\theta}\,\mathrm{d}t}, \qquad \frac{\mathrm{d}R}{\dot{\theta}\,\mathrm{d}t}$$

when $\mathrm{d}\ell$, $\mathrm{d}m$, etc., are replaced by their expressions. It is easy to see that one has, by the substitution of these expressions

$$\begin{aligned} \mathrm{d}P &= (R\cos\mu - Q\cos\nu)\dot{\theta}\,\mathrm{d}t \\ \mathrm{d}Q &= (P\cos\nu - R\cos\lambda)\dot{\theta}\,\mathrm{d}t \\ \mathrm{d}R &= (Q\cos\lambda - P\cos\mu)\dot{\theta}\,\mathrm{d}t \end{aligned}$$

From these equations, in which λ, μ, ν, $\dot{\theta}$ are constants, it is easy to see that if the values of $\mathrm{d}P/\mathrm{d}t$, $\mathrm{d}Q/\mathrm{d}t$, $\mathrm{d}R/\mathrm{d}t$ are zero when $t = 0$ or when t is equal to an arbitrary given value, the expressions for

$$\frac{\mathrm{d}^2P}{\mathrm{d}t^2}, \qquad \frac{\mathrm{d}^2Q}{\mathrm{d}t^2}, \qquad \frac{\mathrm{d}^2R}{\mathrm{d}t^2}, \qquad \frac{\mathrm{d}^3P}{\mathrm{d}t^3}, \qquad \frac{\mathrm{d}^3Q}{\mathrm{d}t^3}, \qquad \frac{\mathrm{d}^3R}{\mathrm{d}t^3}$$

and so on to infinity will also be zero for the same variation of t.

But by Taylor's theorem, it is known that the value of a function $\mathrm{d}P/\mathrm{d}t$ of t when t is replaced by $t + t'$ is equal to

$$\frac{\mathrm{d}P}{\mathrm{d}t} + \frac{\mathrm{d}^2P}{\mathrm{d}t^2}t' + \frac{\mathrm{d}^3P}{2\,\mathrm{d}t^3}t'^2 + \frac{\mathrm{d}^4P}{2\cdot 3\,\mathrm{d}t^4}t'^3 + \cdots$$

Thus if $\mathrm{d}P/\mathrm{d}t = 0$ when $t' = 0$, one will always have $\mathrm{d}P/\mathrm{d}t = 0$ whatever the value of t'. And the same formulation will hold for the values of $\mathrm{d}Q/\mathrm{d}t$ and $\mathrm{d}R/\mathrm{d}t$.

It follows from this development that if the equations of Article 25, which are nothing more than the transformation of the equations

$$\frac{\mathrm{d}P}{\mathrm{d}t} = 0, \qquad \frac{\mathrm{d}Q}{\mathrm{d}t} = 0, \qquad \frac{\mathrm{d}R}{\mathrm{d}t} = 0$$

hold for an arbitrary instant, they will hold whatever the value of the time t for the assumption that the quantities s and u are constants.

Consequently, the values of these quantities will be independent of the variability of the quantities ℓ, m, n, f, g, h so that it will be sufficient to determine the values of these latter quantities for an arbitrary position of the body with respect to the fixed x-, y- and z-axes to obtain those of the quantities s and u which define the position of the axis of rotation which must remain immobile in space or at least always parallel to itself if the body has a translational motion.

Since this axis by its nature is fixed within the body during an instant of time, it follows that it must always remain fixed because the body is assumed to rotate about it. It is obvious that if in the following instant it changed its position in the body, it would necessarily change position in space, which is contrary to the hypothesis.

27. Since the orientation of this axis in space has been found, nothing prevents us from assuming that it coincides with the x-axis since the orientation of this axis is arbitrary.

It can be assumed that $\lambda = 0$ and consequently, $\cos\lambda = 1$ which will give $s = 0$ and $u = 0$. From these results, one finds from the equations of Article 25 that $g = 0$ and $h = 0$. Thus this axis has the property that taking it for the x-axis, the values of the two integrals $\mathrm{S}\,xy\,\mathrm{D}m$, $\mathrm{S}\,xz\,\mathrm{D}m$ (Article 22) become equal to zero.

Let us now assume in our formulas that $g = 0$ and $h = 0$ and let us designate by f', ℓ', m', n' what will later become the quantities f, ℓ, m, n in this case. This assumption gives at the outset that $s = 0$ and $u = 0$ which is the preceding case. Then it also gives s and u equal to infinity and consequently, $\cos\lambda = 0$, $\lambda = 90°$. This value corresponds to the two other roots of the third-degree equation for s and consequently, to the position of the two other axes. But the first of these equations in s and u (Article 25) becomes, when g and h are zero, $f'(u^2 - s^2) + (m' - n')su = 0$ and after substituting the expressions for s and u

$$f'(\cos^2\nu - \cos^2\mu) + (m' - n')\cos\mu\cos\nu = 0$$

but by putting $\cos\lambda = 0$ in $\cos^2\lambda + \cos^2\mu + \cos^2\nu = 1$, one has the following equation $\cos\nu = \sqrt{(1-\cos^2\mu)} = \sin\mu$ and the preceding equation reduces to $\tan 2\mu = 2f'/(m'-n')$ which gives for the angle μ two values for which one is greater than the other by 90°.

Then having taken the x-axis as the primary axis of rotation, the two other axes of uniform rotation will be in the plane of the y- and z-axes and will make with the y-axis the angles μ and $\mu + 90°$, such that the three axes of rotation will be orthogonal as will be the coordinates. These two last axes can also be taken for those of y and z. Thus $\mu = 0$ and consequently, $f' = 0$ so that the value of the integral $\mathrm{S}\, yz\, \mathrm{D}m$ will also be equal to zero.

28. Therefore, there exists for each solid body, whatever its configuration and composition, and with respect to an arbitrary point of the body, three rectangular axes which intersect at this point, about which the body can freely and uniformly rotate and these three axes are determined by the following conditions

$$\mathrm{S}\, xy\, \mathrm{D}m = 0, \qquad \mathrm{S}\, xz\, \mathrm{D}m = 0, \qquad \mathrm{S}\, yz\, \mathrm{D}m = 0$$

where these axes are taken for those of the coordinates x, y, z.

When these axes pass through the center of gravity, they are called the Principal Axes after Euler, whom I am sure everyone has heard of. They are also called the natural axes of rotation or in general the principal axes, whether they pass through the center of gravity or not.

29. By having $f = 0$, $g = 0$, $h = 0$ which holds with respect to the three principal axes, one has also from the equations of Article 23, $\mathrm{d}\ell/\mathrm{d}t = 0$, $\mathrm{d}m/\mathrm{d}t = 0$, $\mathrm{d}n/\mathrm{d}t = 0$ which shows that the quantities ℓ, m, n are then maximum or minimum values. In order to distinguish the maxima from the minima, it is only necessary to investigate the values of $\mathrm{d}^2\ell/\mathrm{d}t^2, \mathrm{d}^2m/\mathrm{d}t^2, \mathrm{d}^2n/\mathrm{d}t^2$ and since $\dot\theta$ is a constant, the following equations will be found

$$\frac{\mathrm{d}^2\ell}{\mathrm{d}t^2} = 2((n-\ell)\cos^2\mu - (\ell - m)\cos^2\nu)\dot\theta^2$$
$$\frac{\mathrm{d}^2m}{\mathrm{d}t^2} = 2((\ell-m)\cos^2\nu - (m-n)\cos^2\lambda)\dot\theta^2$$
$$\frac{\mathrm{d}^2n}{\mathrm{d}t^2} = 2((m-n)\cos^2\lambda - (n-\ell)\cos^2\mu)\dot\theta^2$$

Thus if $\ell > m$ and $m > n$, the value of $\mathrm{d}^2\ell/\mathrm{d}t^2$ will always be negative, the second value $\mathrm{d}^2n/\mathrm{d}t^2$ always positive and the third $\mathrm{d}^2m/\mathrm{d}t^2$ can be positive or negative. Consequently, ℓ will always be a maximum, n a minimum and m will be neither a maximum nor a minimum. Also, it is obvious that $(\mathrm{d}^2\ell + \mathrm{d}^2m)/\mathrm{d}t^2$ will always have a negative value and $(\mathrm{d}^2m + \mathrm{d}^2n)/\mathrm{d}t^2$ will always have a positive value so that the quantity $(\ell + m)$ will always be a maximum and $(m+n)$ a minimum.

The quantities $\ell + m, \ell + n, m + n$ which express the sums of the products of each element of the body by the square of its distance to the x-, y-, and z-axes are called, following Euler, the moments of inertia of the body with respect to these axes. They are to the motion of rotation what simple masses are to translational motion because the moments of the forces of impulsion must be divided by these moments to obtain the angular velocities of rotation about the same axes.

Euler discovered formulas for the principal axes by considering the largest and smallest moments of inertia. Now they are ordinarily determined by the three following equations

$$\mathrm{S}\, xy\, \mathrm{D}m = 0, \qquad \mathrm{S}\, xz\, \mathrm{D}m = 0, \qquad \mathrm{S}\, yz\, \mathrm{D}m = 0$$

30. Since it is certain by the analysis of Article 27 that the equation for s (Article 25) has three real roots, it will always be easy to find them by comparing this equation without its second term with the known equation $x^3 - 3r^2x - 2r^3\cos\varphi = 0$ from which the three roots are

$$2r\cos\frac{\varphi}{3}, \qquad -2r\cos(60^\circ + \frac{\varphi}{3}), \qquad -2r\cos(60^\circ - \frac{\varphi}{3})$$

Thus there will be three values for s which we will designate s, s', s'', and for the corresponding values of u, there will be u, u', u''. Similarly, if one designates by λ, λ', λ'' the angles made by the three principal axes and the x-axis, by μ, μ', μ'' the angles made by them and the y-axis, and by ν, ν', ν'', those made by the same axes with the z-axis, by Articles 24 and 25 it will be found that

$$\cos\lambda = \frac{1}{\sqrt{1+s^2+u^2}}$$
$$\cos\mu = \frac{s}{\sqrt{1+s^2+u^2}}$$
$$\cos\nu = \frac{u}{\sqrt{1+s^2+u^2}}$$

and there will be similar expressions by marking the letters λ, μ, ν, s, u by one prime or two. Thus the determination of the three principal axes can always be made by these formulas for every solid body with arbitrary configuration, homogeneous or not, as long as the values of the quantities f, g, h, ℓ, m, n are known for a given orientation of the body relative to the fixed x-, y- and z-axes.

By substituting the values of $\cos\lambda$, $\cos\mu$, $\cos\nu$ in the three equations of Article 22, the expressions for the moments A, B, C will be obtained which will be required in order to rotate the body with a given constant velocity $\dot{\theta}$, about a fixed axis in space for which the position will be given by the same angles λ, μ, ν, and which at the same time will be one of the three principal axes of the body depending on whether the values for s and u are one of the three roots of the equation in s.

31. Since the three axes are always mutually orthogonal, they can always be taken for the x'-, y'- and z'-axes in the formulas of Article 20. One will thus have from the nature of these axes $\mathrm{S}\, x'y'\, \mathrm{D}m = 0$, $\mathrm{S}\, x'z'\, \mathrm{D}m = 0$, $\mathrm{S}\, y'z'\, \mathrm{D}m = 0$. If one puts

$$\ell' = \mathrm{S}\, x'^2\, \mathrm{D}m, \qquad m' = \mathrm{S}\, y'^2\, \mathrm{D}m, \qquad n' = \mathrm{S}\, z'^2\, \mathrm{D}m$$

the three equations of the cited article will assume this very simple form

$$(m' + n')\dot{\psi}' = A'$$
$$(\ell' + m')\dot{\omega}' = B'$$
$$(\ell' + n')\dot{\varphi}' = C'$$

from which the velocities of rotation $\dot{\psi}'$, $\dot{\omega}'$, $\dot{\varphi}'$ about the three principal axes will be found.

This is the time to demonstrate the proposition which we alluded to in Article 19. Indeed, by putting $A = 0, B = 0, C = 0$ one will also have (Article 20) $A' = 0$, $B' = 0$, $C' = 0$. Therefore, the preceding equations will give $\dot{\psi}' = 0, \dot{\omega}' = 0, \dot{\varphi}' = 0$ because the quantities ℓ, m, n can never be equal to zero for a three-dimensional body. From this result it must be concluded that a rotational motion cannot exist if the initial moments are zero.

If, among the three moments A', B', C' two are zero, for example B' and C', which holds when the impulse is in the $y'z'$-plane of the two angular velocities, $\dot{\omega}$, $\dot{\varphi}$ will also be equal to zero and the body will rotate about the principal axis of x' with the velocity $\dot{\psi}'$. And using the formulas of Article 20, one will have

$$A'^2 + B'^2 + C'^2 = A^2 + B^2 + C^2$$

because of the equations of condition between the quantities α, β, γ, α', etc. Thus putting $B' = 0$, $C' = 0$, one will have $A' = \sqrt{(A^2 + B^2 + C^2)}$ and consequently, the quantity will be a constant. Therefore, from the first equation, the velocity $\dot{\psi}'$ will also be constant.

32. With respect to the values of ℓ', m', n', it will be easy to deduce them from those of ℓ, m, n, f, g, h, because the expressions for x, y, z, as a function for x', y', z' in view of the equations of condition (Article 10, SECTION III, PART I), will give reciprocally

$$x' = \alpha x + \alpha' y + \alpha'' z$$
$$y' = \beta x + \beta' y + \beta'' z$$
$$z' = \gamma x + \gamma' y + \gamma'' z$$

By taking the x', y', z' axes for the principal axes, it is clear from Article 21 that the quantities α, α', α'' are identical to $\cos\lambda$, $\cos\mu$, $\cos\nu$ and similarly β, β', β'' will be identical to $\cos\lambda'$, $\cos\mu'$, $\cos\nu'$ and γ, γ', γ'' to $\cos\lambda''$, $\cos\mu''$, $\cos\nu''$. Thus by substituting the values of the cosine functions given above (Article 30), there results

$$x' = \frac{x + sy + uz}{\sqrt{1 + s^2 + u^2}}$$
$$y' = \frac{x + s'y + u'z}{\sqrt{1 + s'^2 + u'^2}}$$
$$z' = \frac{x + s''y + u''^2 z}{\sqrt{1 + s''^2 + u''^2}}$$

from which one obtains, after squaring, integrating and finally, multiplying by Dm, the following equations

$$\ell' = \frac{\ell + s^2m + u^2n + 2sh + 2ug + 2suf}{1 + s^2 + u^2}$$
$$m' = \frac{\ell + s'^2m + u'^2n + 2s'h + 2u'g + 2s'u'f}{1 + s'^2 + u'^2}$$
$$n' = \frac{\ell + s''^2m + u''^2n + 2s''h + 2u''g + 2s''u''f}{1 + s''^2 + u''^2}$$

The determination of the principal axes of various bodies is found in most treatises on mechanics. For those bodies whose shape is symmetrical, the axis of symmetry is always one of the principal axes. The other two axes can then be found from the formula of Article 27.

Subsection V
Properties Relative to Force Vive

33. In general, whatever the configuration and the connection between different bodies which compose a system, as long as this configuration is independent of time, that is, the equations of condition between the coordinates of the different bodies do not contain the time variable t, it is clear that in the general formula of dynamics one will always be able to assume the variations δx, δy, δz equal to the differentials dx, dy, dz which represent the real distances traversed by the bodies in the instant dt, while the variations in question must represent the arbitrary distances that the bodies could traverse during the same instant with respect to their mutual arrangement.

This assumption is only particular and therefore, it can only produce one equation. But since it is independent of the configuration of the system, it has the advantage of giving a general equation for the motion of any arbitrary system.

Then substituting in the formula of Article 5 (preceding section) for the variations δx, δy, δz the differentials dx, dy, dz and consequently, also the differentials dp, dq, dr, etc. for the variations δp, δq, δr, etc., which depend on δx, δy, δz, the following general equation for any arbitrary system will be obtained

$$\mathrm{S}(\frac{\mathrm{d}x\,\mathrm{d}^2x + \mathrm{d}y\,\mathrm{d}^2y + \mathrm{d}z\,\mathrm{d}^2z}{\mathrm{d}t^2} + P\,\mathrm{d}p + Q\,\mathrm{d}q + R\,\mathrm{d}r + \cdots)m = 0$$

34. In the case where the quantity $P\,\mathrm{d}p + Q\,\mathrm{d}q + R\,\mathrm{d}r +$ etc. is integrable, which holds when the forces P, Q, R, etc. are directed to fixed centers or to bodies of the same system and are functions of the distances p, q, r, etc., by setting

$$P\,\mathrm{d}p + Q\,\mathrm{d}q + R\,\mathrm{d}r + \cdots = \mathrm{d}\Pi$$

the preceding equation becomes

$$S(\frac{\mathrm{d}x\,\mathrm{d}^2x + \mathrm{d}y\,\mathrm{d}^2y + \mathrm{d}z\,\mathrm{d}^2z}{\mathrm{d}t^2} + \mathrm{d}\Pi)m = 0$$

for which the integral is

$$S(\frac{\mathrm{d}x^2 + \mathrm{d}y^2 + \mathrm{d}z^2}{2\,\mathrm{d}t^2} + \Pi)m = H$$

in which H designates an arbitrary constant equal to the value of the first term of the equation at a given instant.

This last equation expresses the principle known by the name of the **Conservation des Forces Vives**. Indeed, the expression $(\mathrm{d}x^2 + \mathrm{d}y^2 + \mathrm{d}z^2)$, which is the square of the distance that the body traverses during the instant $\mathrm{d}t$, gives $(\mathrm{d}x^2 + \mathrm{d}y^2 + \mathrm{d}z^2)/\mathrm{d}t^2$ as the square of its velocity and $[(\mathrm{d}x^2 + \mathrm{d}y^2 + \mathrm{d}z^2)/\mathrm{d}t^2]m$ as its **force vive**. Thus $\mathrm{S}((\mathrm{d}x^2 + \mathrm{d}y^2 + \mathrm{d}z^2)/\mathrm{d}t^2)\,m$ will be the sum of the **forces vives** of all the bodies or the **force vive** of the entire system. It is obvious from the cited equation that this **force vive** is equal to the quantity $2H - 2\mathrm{S}\,\Pi m$, which depends simply on the accelerating forces acting on the bodies and not on their mutual connections so that the **force vive** of the system has at every instant the same magnitude that the bodies would have gained if, acted upon by the same forces, they were each freely moved on the line they described. This conception is what has given the name of **Conservation des Forces Vives** to this property of motion.

35. This principle also holds when the motions of the bodies are referred to the center of gravity. If the three coordinates of the center of gravity are denoted by x', y', z' as above (Article 3) and if one sets $x = x' + \xi$, $y = y' + \eta$, $z = z' + \zeta$, the coordinates ξ, η, ζ will have their origin at the center of gravity. Thus one will have

$$\begin{aligned}\mathrm{S}(\frac{\mathrm{d}x^2 + \mathrm{d}y^2 + \mathrm{d}z^2}{2\,\mathrm{d}t^2})m &= \frac{\mathrm{d}x'^2 + \mathrm{d}y'^2 + \mathrm{d}z'^2}{2\,\mathrm{d}t^2}\mathrm{S}m \\ &+\frac{\mathrm{d}x'}{\mathrm{d}t}S\frac{\mathrm{d}\xi}{\mathrm{d}t}m + \frac{\mathrm{d}y'}{\mathrm{d}t}S\frac{\mathrm{d}\eta}{\mathrm{d}t}m + \frac{\mathrm{d}z'}{\mathrm{d}t}S\frac{\mathrm{d}\zeta}{\mathrm{d}t}m \\ &+\mathrm{S}\frac{\mathrm{d}\xi^2 + \mathrm{d}\eta^2 + \mathrm{d}\zeta^2}{2\,\mathrm{d}t^2}m\end{aligned}$$

By the nature of the center of gravity, one has (Article cited),

$$\mathrm{S}\frac{\mathrm{d}\xi}{\mathrm{d}t}m = 0, \qquad \mathrm{S}\frac{\mathrm{d}\eta}{\mathrm{d}t}m = 0, \qquad \mathrm{S}\frac{\mathrm{d}\zeta}{\mathrm{d}t}m = 0$$

Thus after the preceding equation is differentiated and subtracted from the equation of Article 33, one will have

$$\begin{aligned}&\frac{\mathrm{d}x'\,\mathrm{d}^2x' + \mathrm{d}y'\,\mathrm{d}^2y' + \mathrm{d}z'\,\mathrm{d}^2z'}{\mathrm{d}t^2}\mathrm{S}\,m + \mathrm{S}(\frac{\mathrm{d}\xi\,\mathrm{d}^2\xi + \mathrm{d}\eta\,\mathrm{d}^2\eta + \mathrm{d}\zeta\,\mathrm{d}^2\zeta}{\mathrm{d}t^2})m \\ &+\mathrm{S}(P\,\mathrm{d}p + Q\,\mathrm{d}q + R\,\mathrm{d}r + \cdots)m = 0\end{aligned}$$

Let us replace the quantity $P\,\mathrm{d}p+Q\,\mathrm{d}q+R\,\mathrm{d}r+$etc. by its equivalent $X\,\mathrm{d}x+Y\,\mathrm{d}y+Z\,\mathrm{d}z$, and substitute for $\mathrm{d}x$, $\mathrm{d}y$, $\mathrm{d}z$ the values $\mathrm{d}x' + \mathrm{d}\xi$, $\mathrm{d}y' + \mathrm{d}\eta$, $\mathrm{d}z' + \mathrm{d}\zeta$. The last equation will simplify in view of the differential equations of Article 3, to the following equation

$$\mathrm{S}(\frac{\mathrm{d}\xi\,\mathrm{d}^2\xi + \mathrm{d}\eta\,\mathrm{d}^2\eta + \mathrm{d}\zeta\,\mathrm{d}^2\zeta}{\mathrm{d}t^2})m + \mathrm{S}(X\,\mathrm{d}\xi + Y\,\mathrm{d}\eta + Z\,\mathrm{d}\zeta)m = 0$$

which is analogous to the equation of Article 33, but where the quantity $X\,\mathrm{d}\xi+Y\,\mathrm{d}\eta+Z\,\mathrm{d}\zeta$ will be integrable as long as the forces are directed towards the same bodies of the system and proportional to a function of the distances. In this case, one will have

$$\mathrm{S}(\frac{\mathrm{d}\xi^2 + \mathrm{d}\eta^2 + \mathrm{d}\zeta^2}{2\,\mathrm{d}t^2} + \Pi)m = H$$

an equation which expresses the **Conservation des Forces Vives** with respect to the center of gravity.

36. Also, it is not so much the principle of **forces vives** as those of the center of gravity and areas which hold whatever the action that the bodies of the system might exert on one another even through impact, since all the internal forces disappear from the equations which express these two principles.

The equation of the **Conservation des Forces Vives** contain all the terms resulting from the **forces vives**, external as well as internal, and is independent only of the action of the bodies resulting from their mutual action. Thus this principle holds in the motion of inelastic fluids, as long as they form a continuous mass and there is no point of impact between their parts. And if the quantity of **force vive** is the same before and after the impact of elastic bodies, this is the case where one assumes that the bodies bounce back after the impact in the same state they were in prior to the impact so that the terms $\int P\,\mathrm{d}p$ of the expression for Π, which originate from the forces P due to the restoring forces of the bodies and for which the value is greatest when the compression is at its peak, then decrease by an equal degree during the restitution and become equal to zero at the end of the impact. It is only with this hypothesis that the **Conservation des Forces Vives** holds in the impact of elastic bodies.

In all other cases, when there is a sudden change in the velocities of some of the bodies of the system, the total **force vive** is diminished by the quantity of **force vive** due to the accelerating forces which could produce these changes. This quantity can always be estimated from the sum of the masses multiplied by the squares of the velocities which these masses have lost or are assumed to have lost in the sudden changes of the actual velocities of the bodies. This result is the theorem which Carnot found for the impact of hard[20] bodies.

37. One can also assume, in the equation of Article 11 of the preceding section that the variations δx, δy, δz are proportional to the velocities $\dot{x}$, $\dot{y}$, $\dot{z}$ which the bodies receive from the impulse. Thus the following equation will be obtained

$$\mathrm{S}\{m(\dot{x}^2 + \dot{y}^2 + \dot{z}^2) + X\dot{x} + Y\dot{y} + Z\dot{z}\} = 0$$

in which the expression $\mathrm{S}\, m(\dot{x}^2 + \dot{y}^2 + \dot{z}^2)$ represents the **forces vives** of the entire system.

When this equation is combined with the three equations of Article 14, it has the property of maximis and minimis relative to the line about which the system rotates in the first instant after it has received an arbitrary impulse. This line can be called the axis of instantaneous rotation.

If α, β, γ denote the parts of the velocities $\dot{x}, \dot{y}, \dot{z}$ which depend on the respective change of position of the bodies of the system and if they are added to those which result from the rotations (Article 17), the complete expressions for $\dot{x}, \dot{y}, \dot{z}$ expressed as follows will be obtained

$$\dot{x} = z\dot{\omega} - y\dot{\varphi} + \alpha, \qquad \dot{y} = x\dot{\varphi} - z\dot{\psi} + \beta, \qquad \dot{z} = y\dot{\psi} - x\dot{\omega} + \gamma$$

Now assume that these expressions are differentiated, considering as variables only $\dot{\psi}$, $\dot{\omega}$, $\dot{\varphi}$ and if these differentials are denoted by the operator δ, one will have

$$\delta\dot{x} = z\,\delta\dot{\omega} - y\,\delta\dot{\varphi}, \qquad \delta\dot{y} = x\,\delta\dot{\varphi} - z\,\delta\dot{\psi}, \qquad \delta\dot{z} = y\,\delta\dot{\psi} - x\,\delta\dot{\omega}$$

But after the three equations of Article 14 are multiplied respectively by $\delta\dot{\varphi}$, $\delta\dot{\omega}$, $\delta\dot{\psi}$ then added together and placed under the operator S, the differentials $\delta\dot{\varphi}$, $\delta\dot{\omega}$, $\delta\dot{\psi}$ which are the same for all the bodies, give after substitution of the preceding expressions

$$\mathrm{S}\{m(\dot{x}\,\delta\dot{x} + \dot{y}\,\delta\dot{y} + \dot{z}\,\delta\dot{z}) + X\,\delta\dot{x} + Y\,\delta\dot{y} + Z\,\delta\dot{z}\} = 0$$

But the equation for the **force vive** found above after differentiation relative to δ gives

$$\mathrm{S}\{2m(\dot{x}\,\delta\dot{x} + \dot{y}\,\delta\dot{y} + \dot{z}\,\delta\dot{z}) + X\,\delta\dot{x} + Y\,\delta\dot{y} + Z\,\delta\dot{z}\} = 0$$

Therefore, from the comparison of these two equations, the following equation results

$$\mathrm{S}\, m(\dot{x}\,\delta\dot{x} + \dot{y}\,\delta\dot{y} + \dot{z}\,\delta\dot{z}) = 0$$

and consequently, $\delta\,\mathrm{S}\, m(\dot{x}^2 + \dot{y}^2 + \dot{z}^2) = 0$ which shows that the **force vive** which the system gains from the impulse is always a maximum or a minimum with respect to the rotations about the three axes. Since these three rotations are composed of an unique rotation about the instantaneous axis, it follows that the position of this axis is always such that the **force vive** of the entire system has the smallest or largest value among the possible rotations about this axis.

Euler demonstrated this property of the instantaneous axis of rotation for solid bodies with arbitrary configuration. It is clear, from the preceding analysis, that it is a general result for a system of bodies linked together in an invariant or variant fashion when these bodies receive arbitrary impulses.

38. When the system is a solid body which can rotate freely about a point and which is not acted upon by any accelerating force, it can be concluded from the combination of the equation of **forces vives** with the one of areas that it is a relation worthy of note because of its simplicity. This property between the velocities of rotation $\dot{\psi}$, $\dot{\omega}$, $\dot{\varphi}$, about three fixed axes of coordinates x, y, z has not been mentioned by anyone, as far as I know. In this case, one has simply (Article 17)

$$\begin{aligned} dx &= \dot{x}\,dt = (z\dot{\omega} - y\dot{\varphi})dt \\ dy &= \dot{y}\,dt = (x\dot{\varphi} - z\dot{\psi})dt \\ dz &= \dot{z}\,dt = (y\dot{\psi} - x\dot{\omega})dt \end{aligned}$$

If after multiplying the last three equations of Article 9 by $\dot{\varphi}$, $\dot{\omega}$, $\dot{\psi}$, they are added together and put under the operator S and then if dx/dt, dy/dt, dz/dt are substituted for their values, the following equation results

$$S(\frac{dx^2 + dy^2 + dz^2}{dt^2})m = A\dot{\psi} + B\dot{\omega} + C\dot{\varphi}$$

but the equation of Article 34 gives, when $H = 0$

$$S(\frac{dx^2 + dy^2 + dz^2}{dt^2})m = H$$

Thus one will have $A\dot{\psi} + B\dot{\omega} + C\dot{\varphi} = 2H$ where A, B, C are the moments of the initial forces of impulsion and H is an arbitrary constant which must necessarily be positive.

If in this equation the expressions of Article 11, $\gamma C'$, $\gamma' C'$, $\gamma'' C'$, or $C' \cos \ell$, $C' \cos m$, $C' \cos n$, are substituted for A, B, C and for $\dot{\psi}$, $\dot{\omega}$, $\dot{\varphi}$, those of Article 17, $\dot{\theta} \cos \lambda$, $\dot{\theta} \cos \mu$, $\dot{\theta} \cos \nu$ are substituted, one will have

$$\dot{\theta}(\cos \ell \cos \lambda + \cos m \cos \mu + \cos n \cos \nu) = \frac{2H}{C'}$$

In this formula, ℓ, m, n, are the angles which the axis perpendicular to the invariable plane makes with the fixed axes of x, y, z and λ, μ, ν are the angles which the instantaneous axis of the combined rotation, for which the velocity is $\dot{\theta}$, makes with the same axes. Thus if one calls the angle that the instantaneous axis of rotation makes with the axis perpendicular to the invariable plane, one will have, from the well-known formula

$$\cos\sigma = \cos\ell\cos\lambda + \cos m\cos\mu + \cos n\cos\nu$$

and consequently, $\dot\theta\cos\sigma = 2H/C'$ where the quantity $2H/C'$ is a constant which depends on the initial state and which gives a ratio between the real velocity of rotation at each instant and the position of the axis of rotation relative to the invariable plane independent of the configuration of the body.

Also, if the xy-plane is taken in such a manner that it passes through the center of the body and contains the straight line along which the impulsion takes place, the constants A and B will be zero (Article 16) and the general equation found above will be reduced to $C\dot\varphi = 2H$, which shows that the velocity of rotation about the z-axis, that is, parallel to the plane containing the impulse, always remains the same.

Subsection VI
Properties Relative to Least Action

39. We are now going to consider the fourth principle, that is, the Principle of Least Action.

Denoting by the letter u the velocity of each body m of the system, one has $u^2 = (\mathrm{d}x^2 + \mathrm{d}y^2 + \mathrm{d}z^2)/\mathrm{d}t^2$ and the equation of **force vive** (Article 34) becomes $\mathrm{S}(u^2/2 + \Pi)m = H$ which, after differentiation with respect to the operator δ, gives $\mathrm{S}(u\,\delta u + \delta\Pi)m = 0$. But since Π is a function of p, q, r, etc., the following equation can be obtained

$$\delta\Pi = P\,\delta p + Q\,\delta q + R\,\delta r + \cdots$$

Thus

$$\mathrm{S}(P\,\delta p + Q\,\delta q + R\,\delta r + \cdots)m = -\mathrm{S}\,mu\,\delta u$$

and this equation always holds as long as $P\,\mathrm{d}p + Q\,\mathrm{d}q + R\,\mathrm{d}r + \cdots$ is an integrable quantity and the relations between the bodies are independent of time. It would not hold if either of these conditions were not met.

If the preceding equation is substituted in the general formula of dynamics (SECTION II, Article 5), it will become

$$\mathrm{S}(\frac{\mathrm{d}^2x}{\mathrm{d}t^2}\delta x + \frac{\mathrm{d}^2y}{\mathrm{d}t^2}\delta y + \frac{\mathrm{d}^2z}{\mathrm{d}t^2}\delta z - u\,\delta u)m = 0$$

But the equation $\mathrm{d}^2x\,\delta x + \mathrm{d}^2y\,\delta y + \mathrm{d}^2z\,\delta z = \mathrm{d}(\mathrm{d}x\,\delta x + \mathrm{d}y\,\delta y + \mathrm{d}z\,\delta z) - \mathrm{d}x\,\mathrm{d}\delta x - \mathrm{d}y\,\mathrm{d}\delta y - \mathrm{d}z\,\mathrm{d}\delta z$ holds, because the operators d and δ represent differences or variations entirely independent of one another which means that the quantities $\mathrm{d}\delta x$, $\mathrm{d}\delta y$, $\mathrm{d}\delta z$ must be the same as the quantities $\delta \mathrm{d}x, \delta \mathrm{d}y, \delta \mathrm{d}z$. Also, it is obvious that $\mathrm{d}x\,\delta \mathrm{d}x + \mathrm{d}y\,\delta \mathrm{d}y + \mathrm{d}z\,\delta \mathrm{d}z = \frac{1}{2}\delta(\mathrm{d}x^2 + \mathrm{d}y^2 + \mathrm{d}z^2)$. Thus the following equation will be obtained

$$\mathrm{d}^2x\,\delta x + \mathrm{d}^2y\,\delta y + \mathrm{d}^2z\,\delta z = \mathrm{d}(\mathrm{d}x\,\delta x + \mathrm{d}y\,\delta y + \mathrm{d}z\,\delta z) - \tfrac{1}{2}\delta(\mathrm{d}x^2 + \mathrm{d}y^2 + \mathrm{d}z^2)$$

Let the quantity s represent the path or curve described by the body m during the time t. One will have $\mathrm{d}s = \sqrt{\mathrm{d}x^2 + \mathrm{d}y^2 + \mathrm{d}z^2}$ and $\mathrm{d}t = \mathrm{d}s/u$. Therefore

$$\mathrm{d}^2x\,\delta x + d^2y\,\delta y + d^2z\,\delta z = \mathrm{d}(\mathrm{d}x\,\delta x + \mathrm{d}y\,\delta y + \mathrm{d}z\,\delta z) - \mathrm{d}s\,\delta \mathrm{d}s$$

from which

$$\frac{\mathrm{d}^2x}{\mathrm{d}t^2}\delta x + \frac{\mathrm{d}^2y}{\mathrm{d}t^2}\delta y + \frac{\mathrm{d}^2z}{\mathrm{d}t^2}\delta z = \frac{\mathrm{d}(\mathrm{d}x\,\delta x + \mathrm{d}y\,\delta y + \mathrm{d}z\,\delta z)}{\mathrm{d}t^2} - \frac{u^2\,\delta \mathrm{d}s}{\mathrm{d}s}$$

Thus the general formula under consideration will become

$$\mathrm{S}\Big(\frac{\mathrm{d}(\mathrm{d}x\,\delta x + \mathrm{d}y\,\delta y + \mathrm{d}z\,\delta z)}{\mathrm{d}t^2} - \frac{u^2\,\delta \mathrm{d}s}{\mathrm{d}s} - u\,\delta u\Big)m = 0$$

or by multiplying all the terms by the constant element $\mathrm{d}t = \mathrm{d}s/u$ and observing that $u\,\delta \mathrm{d}s + \mathrm{d}s\,\delta u = \delta(u\,\mathrm{d}s)$, the following equation results

$$\mathrm{S}\Big(\frac{\mathrm{d}(\mathrm{d}x\,\delta x + \mathrm{d}y\,\delta y + \mathrm{d}z\,\delta z)}{\mathrm{d}t} - \delta(u\,\mathrm{d}s)\Big)m = 0$$

Since the integral sign S has no relation to the differential operators d and δ, the order of the operations can be reversed. Then the preceding equation will assume the following form

$$\frac{\mathrm{d}\Big(S(\mathrm{d}x\,\delta x + \mathrm{d}y\,\delta y + \mathrm{d}z\,\delta z)\Big)m}{\mathrm{d}t} - \delta(\mathrm{S}\,mu\,\mathrm{d}s) = 0$$

Let us integrate with respect to the differential operator d and denote this integration by the ordinary integral sign $\int$ in order to obtain

$$\frac{\mathrm{S}(\mathrm{d}x\,\delta x + \mathrm{d}y\,\delta y + \mathrm{d}z\,\delta z)m}{\mathrm{d}t} - \int \delta(\mathrm{S}\,mu\,\mathrm{d}s) = \text{const.}$$

Since the integral sign $\int$ in the expression $\int \delta(\mathrm{S}\,mu\,\mathrm{d}s)$ applies only to the variables u and s and since it has no relation to the operators S and δ, it is clear that this expression is the same as the following expression $\delta(\mathrm{S}\,m \int u\,\mathrm{d}s)$. If it is assumed that at the points where the integrals $\int u\, ds$ begin one has $\delta x = 0$, $\delta y = 0$, $\delta z = 0$, then the arbitrary constant must be zero because the initial value of the equation at these points is zero. Thus there results for this case

$$\delta(\mathrm{S}\, m \int u\, \mathrm{d}s) = \frac{\mathrm{S}(\mathrm{d}x\, \delta x + \mathrm{d}y\, \delta y + \mathrm{d}z\, \delta z)m}{\mathrm{d}t}$$

Moreover, if it is assumed that the variations δx, δy, δz are also zero for the points where the integrals $\int u\, \mathrm{d}s$ end, one will simply have $\delta(\mathrm{S}\, m \int u\, \mathrm{d}s) = 0$, that is, the variations of the quantity $\mathrm{S}\, m \int u\, \mathrm{d}s$ will be equal to zero. Consequently, this quantity will be a maximum or a minimum.

From this formulation, the following general theorem results: In the motion of an arbitrary system of bodies acted upon by mutual forces of attraction or directed toward fixed centers and proportional to arbitrary functions of distance, the paths described by the different bodies and their velocities are such that the sum of the products of each mass and the integral of the velocity multiplied by the differential element of the path is a maximum or a minimum, where the first and the last points of each path are viewed as given so that the variations of the coordinates corresponding to these points are zero. This is the theorem which was discussed at the end of SECTION I under the title of the Principle of Least Action.

40. This theorem does not only express a very remarkable property of the motion of bodies, it can also be used to determine this motion. Indeed, because the formula Sm$\int$uds must be a maximum or a minimum, one has only to analyze by the method of variations the conditions which give it this property and by using the general equation of the **Conservation des Forces Vives**, all the equations necessary to determine the motion of each body will always be found. Since for a maximum or a minimum, the variation must be zero, and one has $\delta(\mathrm{S}\, m \int u\, \mathrm{d}s) = 0$. And from this result, by applying in reversed order the operations given above, the same general formula with which one began will be found.

In order to make this method more comprehensible, we will briefly discuss it here. The condition of maximum or minimum gives, in general, $\delta(\mathrm{S}\, m \int u\, \mathrm{d}s) = 0$ and putting the differential operator δ after the integral signs S and $\int$ (which is obviously permissible because of the nature of these various operations), one will have the equation $\mathrm{S}\, m \int \delta(u\, \mathrm{d}s) = 0$, or better yet, by performing the differentiation by δ

$$\mathrm{S}\, m \int (\mathrm{d}s\, \delta u + u\, \delta \mathrm{d}s) = 0$$

At the outset, let us consider the first part of the equation $\mathrm{S}\, m \int \mathrm{d}s\, \delta u$. By replacing $\mathrm{d}s$ by its equivalent $u\, \mathrm{d}t$, it becomes $\mathrm{S}\, m \int u\, \delta u\, \mathrm{d}t$ or after changing the order of the integral signs S and $\int$ which are absolutely independent of one another, $\int \mathrm{d}t\, \mathrm{S}\, m u\, \delta u$. But the general equation of the principle of **force vive** gives (Article 34) $\mathrm{S}\,u^2m = 2H - 2\,\mathrm{S}(\Pi m)$, where $\mathrm{d}\Pi = P\,\mathrm{d}p + Q\,\mathrm{d}q + R\,\mathrm{d}r + \cdots$. Thus after differentiating by δ, one will have

$$\mathrm{S}\,u\,\delta u\,m = -\mathrm{S}\,\delta\Pi m = -\mathrm{S}(P\,\delta p + Q\,\delta q + R\,\delta r + \cdots)m$$

because Π is assumed an algebraic function of p, q, r, etc., the variation $\delta\Pi$ is the same as $\mathrm{d}\Pi$, if d is replaced by δ. Hence, the quantity $\mathrm{S}\,m\int \mathrm{d}s\,\delta u$ will be reduced to this form

$$-\int \mathrm{d}t\,\mathrm{S}(P\,\delta p + Q\,\delta q + R\,\delta r + \cdots)m$$

Secondly, let us consider the second part of the equation, $\mathrm{S}\,m\int u\,\delta \mathrm{d}s$ and substitute for $\mathrm{d}s$ its expression in rectangular coordinates or in other arbitrary variables. Using the rectangular coordinates x, y, z one has $\mathrm{d}s = \sqrt{\mathrm{d}x^2 + \mathrm{d}y^2 + \mathrm{d}z^2}$. Then after differentiating with respect to δ, one obtains

$$\delta \mathrm{d}s = \frac{\mathrm{d}x\,\delta \mathrm{d}x + \mathrm{d}y\,\delta \mathrm{d}y + \mathrm{d}z\,\delta \mathrm{d}z}{\mathrm{d}s}$$

or better yet, by transposing the symbols d, δ and writing $\mathrm{d}\delta$ instead of $\delta\mathrm{d}$ which is always permitted because of the independence of these operations

$$\delta \mathrm{d}s = \frac{\mathrm{d}x\,d\delta x + \mathrm{d}y\,\mathrm{d}\delta y + \mathrm{d}z\,\mathrm{d}\delta z}{\mathrm{d}s}$$

After the substitution of this expression and replacing $\mathrm{d}s/u$ by $\mathrm{d}t$, one will then have

$$\int u\,\delta \mathrm{d}s = \int \frac{\mathrm{d}x\,\mathrm{d}\delta x + \mathrm{d}y\,\mathrm{d}\delta y + \mathrm{d}z\,\mathrm{d}\delta z}{\mathrm{d}t}$$

Since it is here under the integral sign $\int$ that the differentials of the variations δx, δy, δz are found, one must eliminate them by the well-known method of integration by parts, according to the principles of the method of variations. Thus one will transform the quantity $\int \mathrm{d}x\,\mathrm{d}\delta x/\mathrm{d}t$ to the following equivalent quantity $(\mathrm{d}x/\mathrm{d}t)\delta x - \int \delta x\,\mathrm{d}(\mathrm{d}x/\mathrm{d}t)$ and assuming that the two terms of the curve are given so that the coordinates of the two limits of the integral do not vary, one will have simply

$$\int \frac{\mathrm{d}x\,\mathrm{d}\delta x}{\mathrm{d}t} = -\int \delta x\,\mathrm{d}\left(\frac{\mathrm{d}x}{\mathrm{d}t}\right)$$

Similarly, it will be found that

$$\int \frac{\mathrm{d}y\,\mathrm{d}\delta y}{\mathrm{d}t} = -\int \delta y\,\mathrm{d}\left(\frac{\mathrm{d}y}{\mathrm{d}t}\right)$$

and similarly

$$\int \frac{\mathrm{d}z\,\mathrm{d}\delta z}{\mathrm{d}t} = - \int \delta z\,\mathrm{d}\left(\frac{\mathrm{d}z}{\mathrm{d}t}\right)$$

so that the following transformed equation results

$$\int u\,\delta \mathrm{d}s = - \int \left(\delta x\,\mathrm{d}\left(\frac{\mathrm{d}x}{\mathrm{d}t}\right) + \delta y\,\mathrm{d}\left(\frac{\mathrm{d}y}{\mathrm{d}t}\right) + \delta z\,\mathrm{d}\left(\frac{\mathrm{d}z}{\mathrm{d}t}\right)\right)$$

After transposing the integral signs S and $\int$, and assuming $\mathrm{d}t$ to be constant, the quantity $\mathrm{S}\,m \int u\,\delta \mathrm{d}s$ will become

$$- \int \mathrm{d}t\,\mathrm{S}\left(\delta x\,\mathrm{d}\left(\frac{\mathrm{d}x}{\mathrm{d}t^2}\right) + \delta y\,\mathrm{d}\left(\frac{\mathrm{d}y}{\mathrm{d}t^2}\right) + \delta z\,\mathrm{d}\left(\frac{\mathrm{d}z}{\mathrm{d}t^2}\right)\right) m$$

The equation for the maximum or minimum is then

$$\int \mathrm{d}t\,\mathrm{S}\left\{\begin{array}{l} P\,\delta p + Q\,\delta q + R\,\delta r + \cdots \\ +\delta x\,\mathrm{d}\left(\frac{\mathrm{d}x}{\mathrm{d}t^2}\right) + \delta y\,\mathrm{d}\left(\frac{\mathrm{d}y}{\mathrm{d}t^2}\right) + \delta z\,\mathrm{d}\left(\frac{\mathrm{d}z}{\mathrm{d}t^2}\right) \end{array}\right\} m = 0$$

which holds in general for all possible variations when the quantity under the integral sign is zero at each instant. One has then the indefinite equation

$$\mathrm{S}\left(P\,\delta p + Q\,\delta q + R\,\delta r + \cdots + \delta x\,\mathrm{d}\left(\frac{\mathrm{d}x}{\mathrm{d}t^2}\right) + \delta y\,\mathrm{d}\left(\frac{\mathrm{d}y}{\mathrm{d}t^2}\right) + \delta z\,\mathrm{d}\left(\frac{\mathrm{d}z}{\mathrm{d}t^2}\right)\right) m = 0$$

which is identical to the general formula of dynamics (SECTION II, Article 5) and consequently, it provides all the necessary equations for the solution of the problem.

41. In place of the coordinates x, y, z, it is possible to use other arbitrary unknowns as long as it is possible to express the element of the curve $\mathrm{d}s$ as a function of these quantities. For example, if the radius vector or the rectilinear distance from the origin of coordinates is denoted by ρ, then employing two angles, where the angle ψ is the inclination of the radius vector to the xy-plane and the second angle φ is the angle of projection of the component of the radius vector in the xy-plane on the x-axis, one has the following relations: $z = \rho \sin\psi$, $y = \rho\cos\psi\sin\varphi$, $x = \rho\cos\psi\cos\varphi$. From these relations it is found that $\mathrm{d}s^2 = \mathrm{d}x^2 + \mathrm{d}y^2 + \mathrm{d}z^2 = \mathrm{d}\rho^2 + \rho^2(\mathrm{d}\psi^2 + \cos^2\psi\,\mathrm{d}\varphi^2)$, which is an expression which can also be found directly from geometrical considerations. Now differentiating by δ and replacing $\delta\mathrm{d}$ by $\mathrm{d}\delta$, one has the relation $\mathrm{d}s\,\delta\mathrm{d}s = \mathrm{d}\rho\,\mathrm{d}\delta\rho + \rho(\mathrm{d}\psi^2 + \cos^2\psi\,\mathrm{d}\varphi^2)\delta\rho +$

$\rho^2(\mathrm{d}\psi\,\mathrm{d}\delta\psi - \sin\psi\cos\psi\,\mathrm{d}\varphi^2\,\delta\psi + \cos^2\psi\,\mathrm{d}\varphi\,\mathrm{d}\delta\varphi)$ and after dividing by $\mathrm{d}t = \mathrm{d}s/u$, and integrating, the following equation results

$$\int u\,\delta\mathrm{d}s = \int \frac{\mathrm{d}\rho\,\mathrm{d}\delta\rho + \rho(\mathrm{d}\psi^2 + \cos^2\psi\,\mathrm{d}\varphi^2)\delta\rho}{\mathrm{d}t}$$
$$+ \int \frac{\rho^2(\mathrm{d}\psi\,\mathrm{d}\delta\psi - \sin\psi\cos\psi\,\mathrm{d}\varphi^2\,\delta\psi + \cos^2\psi\,\mathrm{d}\varphi\,\mathrm{d}\delta\varphi)}{\mathrm{d}t}$$

The double operator $\mathrm{d}\delta$ can be eliminated from under the integral sign $\int$ by integration by parts. The terms outside of the integral sign $\int$, which could contain variations can be ignored, because these variations are with respect to the end-points of the integral and are zero by the assumption that the first and last points of the trajectories described by the bodies are given and invariable. One will thus have the following transformation

$$\int u\,\delta\mathrm{d}s = -\int \mathrm{d}u\,\delta s = -\int [(\mathrm{d}\frac{\mathrm{d}\rho}{\mathrm{d}t} - \rho\frac{\mathrm{d}\psi^2 + \cos^2\psi\,\mathrm{d}\varphi^2}{\mathrm{d}t})\delta\rho$$
$$+(\frac{\rho^2\sin\psi\cos\psi\,\mathrm{d}\varphi^2}{\mathrm{d}t} + \mathrm{d}\frac{\rho^2\,\mathrm{d}\psi}{\mathrm{d}t})\delta\psi + \mathrm{d}\frac{\cos^2\psi\,\mathrm{d}\varphi}{\mathrm{d}t}\delta\varphi]$$

Consequently, the equation of maxima and minima will be

$$\int \mathrm{d}t\,\mathrm{S}\left\{\begin{array}{l} P\,\delta p + Q\,\delta q + R\,\delta r + \cdots \\ +(\mathrm{d}\frac{\mathrm{d}\rho}{\mathrm{d}t^2} - \rho\frac{\mathrm{d}\psi^2 + \cos^2\psi\,\mathrm{d}\varphi^2}{\mathrm{d}t^2})\delta\rho + \\ (\frac{\rho^2\sin\psi\cos\psi\,\mathrm{d}\varphi^2}{\mathrm{d}t^2} + \mathrm{d}\frac{\rho^2\,\mathrm{d}\psi}{\mathrm{d}t^2})\delta\psi + \mathrm{d}\frac{\cos^2\psi\,\mathrm{d}\varphi}{\mathrm{d}t^2}\delta\varphi \end{array}\right\} m = 0$$

After equating to zero the quantity which is under the integral sign $\int$, one will have an indefinite equation similar to the one of the preceding article but which, instead of the variations δx, δy, δz, will contain the variations $\delta\rho$, $\delta\varphi$, $\delta\psi$ and from it, the necessary equations for the solution of the problem are derived by first reducing the variations to the smallest possible number, then developing individual equations for the terms affected by each of the remaining variations.

Using other unknowns will lead to different formulas, but one can always be certain to have in each case the most simple formulas that the nature of the selection of unknowns can provide. For additional information, the reader should refer to the second volume of the Mémoires of the Académie de Turin[21] where this method has been used to solve various problems of mechanics.

42. Also, since $\mathrm{d}s = u\,\mathrm{d}t$, the formula $\mathrm{S}\,m\int u\,\mathrm{d}s$, which is a maximum or a minimum, can also be put in the form $\mathrm{S}\,m\int u^2\,\mathrm{d}t$ or $\int \mathrm{d}t\,\mathrm{S}\,mu^2$ in which $\mathrm{S}\,mu^2$ expresses the **force vive** of the entire system at an arbitrary instant.[22] Thus the principle in question is properly reduced to the following: the sum of the instantaneous **force vive** of all the bodies from the time they depart from given points to the time they arrive at other given points is a

maximum or a minimum. It could be called with more reason, the principle of the largest or smallest **force vive**. And this manner of conceiving it would have the advantage of generality with respect to motion as well as to equilibrium, since we have seen in Article 22 of SECTION III of PART I, that the **force vive** of a system is always a maximum or a minimum in the state of equilibrium.

SECTION IV
DIFFERENTIAL EQUATIONS FOR THE SOLUTION OF ALL PROBLEMS OF DYNAMICS

1. The formula to which the entire theory of dynamics has been reduced in SECTION II requires only further development to give all the necessary equations for the solution of any problem of this science. But this development which is only a matter of pure calculation can still be simplified in several ways by means which we will present in this section.

Since the problem consists of reducing the various variables which are contained in the formula to the smallest possible number by means of the equations of condition provided by the nature of each problem, one of the main objectives is to substitute for these variables functions of other variables. This objective is always easy to fulfill by ordinary methods. But there is one particular approach which has the advantage of leading directly to the simplest transformation for the proposed formula.

2. The formula is composed of two different parts which must be considered separately. The first part contains the terms

$$\mathrm{S}(\frac{\mathrm{d}^2x}{\mathrm{d}t^2}\delta x + \frac{\mathrm{d}^2y}{\mathrm{d}t^2}\delta y + \frac{\mathrm{d}^2z}{\mathrm{d}t^2}\delta z)m$$

which results solely from the inertia forces of the bodies. The second part is composed of the terms

$$\mathrm{S}(P\,\delta p + Q\,\delta q + R\,\delta r + \cdots)m$$

and is due to the accelerating forces P, Q, R, etc. which are assumed to act effectively on each body in the directions of the lines p, q, r, etc. and which have a tendency to shorten these lines. The sum of these two quantities, when equated to zero, constitutes the general formula of dynamics (SECTION II, Article 5).

3. At the outset, consider the expression $\mathrm{d}^2x\,\delta x + \mathrm{d}^2y\,\delta y + \mathrm{d}^2z\,\delta z$ It is clear that if the following expression is added to the preceding expression

$$\mathrm{d}x\,\mathrm{d}\delta x + \mathrm{d}y\,\mathrm{d}\delta y + \mathrm{d}z\,\mathrm{d}\delta z$$

the sum is integrable. After integration there results

$$\mathrm{d}x\,\delta x + \mathrm{d}y\,\delta y + \mathrm{d}z\,\delta z$$

from which it follows that

$$\mathrm{d}^2x\,\delta x + \mathrm{d}^2y\,\delta y + \mathrm{d}^2z\,\delta z = \mathrm{d}(\mathrm{d}x\,\delta x + \mathrm{d}y\,\delta y + \mathrm{d}z\,\delta z)$$
$$-\mathrm{d}x\,\mathrm{d}\delta x - \mathrm{d}y\,\mathrm{d}\delta y - \mathrm{d}z\,\mathrm{d}\delta z$$

Since it has been shown from known principles that the double operator $\mathrm{d}\delta$ is equivalent to its inversion $\delta\mathrm{d}$, the expression

$$\mathrm{d}x\,\mathrm{d}\delta x + \mathrm{d}y\,\mathrm{d}\delta y + \mathrm{d}z\,\mathrm{d}\delta z$$

can be written in the form $\mathrm{d}x\,\delta\mathrm{d}x + \mathrm{d}y\,\delta\mathrm{d}y + \mathrm{d}z\,\delta\mathrm{d}z$ which is equivalent to $\frac{1}{2}\delta(\mathrm{d}x^2 + \mathrm{d}y^2 + \mathrm{d}z^2)$. Thus the following simplification is obtained

$$\mathrm{d}^2x\,\delta x + \mathrm{d}^2y\,\delta y + \mathrm{d}^2z\,\delta z = \mathrm{d}(\mathrm{d}x\,\delta x + \mathrm{d}y\,\delta y + \mathrm{d}z\,\delta z) - \frac{1}{2}\delta(\mathrm{d}x^2 + \mathrm{d}y^2 + \mathrm{d}z^2)$$

from which it is clear that in order to evaluate the expression

$$\mathrm{d}^2x\,\delta x + \mathrm{d}^2y\,\delta y + \mathrm{d}^2z\,\delta z$$

it suffices to differentiate the two following expressions which contain only first differences $\mathrm{d}x\,\delta x + \mathrm{d}y\,\delta y + \mathrm{d}z\,\delta z$ and $\mathrm{d}x^2 + \mathrm{d}y^2 + \mathrm{d}z^2$. The first expression should be differentiated with respect to d and the second with respect to δ.

4. Let us thus assume that it is a question of substituting in place of the variables x, y, z known functions of other variables ξ, ψ, φ, etc. After differentiating these functions, expressions of the following form will be obtained

$$\mathrm{d}x = A\,\mathrm{d}\xi + B\,\mathrm{d}\psi + C\,\mathrm{d}\varphi + \cdots$$
$$\mathrm{d}y = A'\,\mathrm{d}\xi + B'\,\mathrm{d}\psi + C'\,\mathrm{d}\varphi + \cdots$$
$$\mathrm{d}z = A''\,\mathrm{d}\xi + B''\,\mathrm{d}\psi + C''\,\mathrm{d}\varphi + \cdots$$

in which A, A', A'', B, B', B'', etc. will be known functions of the same variables ξ, ψ, φ, etc. and the values of δx, δy, δz will also be expressed in the same manner, after replacing the operator d by δ.

If these substitutions are made in the quantity $\mathrm{d}x\,\delta x + \mathrm{d}y\,\delta y + \mathrm{d}z\,\delta z$, it will assume this form

$$F\,\mathrm{d}\xi\,\delta\xi + G(\mathrm{d}\xi\,\delta\psi + \mathrm{d}\psi\,\delta\xi) + H\,\mathrm{d}\psi\,\delta\psi + I(\mathrm{d}\xi\,\delta\varphi + \mathrm{d}\varphi\,\delta\xi) + \cdots$$

where F, G, H, I, etc. will be finite functions of ξ, ψ, φ, etc.

Then by replacing δ by d, the expression for $\mathrm{d}x^2 + \mathrm{d}y^2 + \mathrm{d}z^2$ will also be obtained, which will be

$$F\,\mathrm{d}\xi^2 + 2G\,\mathrm{d}\xi\,\mathrm{d}\psi + H\,\mathrm{d}\psi^2 + 2I\,\mathrm{d}\xi\,\mathrm{d}\varphi + \cdots$$

Now if the first of these two quantities is differentiated by d, the following differential will be obtained

$$\begin{aligned}
&\mathrm{d}(F\,\mathrm{d}\xi)\delta\xi + F\,\mathrm{d}\xi\,\mathrm{d}\delta\xi + \mathrm{d}(G\,\mathrm{d}\xi)\delta\psi \\
&+\mathrm{d}(G\,\mathrm{d}\psi)\delta\xi + G\,\mathrm{d}\xi\,\mathrm{d}\delta\psi + G\,\mathrm{d}\psi\,\mathrm{d}\delta\xi \\
&+\mathrm{d}(H\,\mathrm{d}\psi)\delta\psi + H\,\mathrm{d}\psi\,\mathrm{d}\delta\psi + \cdots
\end{aligned}$$

If the second equation is differentiated by δ, there will result

$$\begin{aligned}
&\delta F\,\mathrm{d}\xi^2 + 2F\,\mathrm{d}\xi\,\delta\mathrm{d}\xi + 2\,\delta G\mathrm{d}\xi\,\mathrm{d}\psi + 2G\,\mathrm{d}\psi\,\delta\,\mathrm{d}\xi \\
&+2G\,\mathrm{d}\xi\,\delta\,\mathrm{d}\psi + \delta H\,\mathrm{d}\psi^2 + 2H\,\mathrm{d}\psi\,\delta\,\mathrm{d}\psi + \cdots
\end{aligned}$$

If one-half of this last differential is subtracted from the first and recalling that $\mathrm{d}\delta$ and $\delta\mathrm{d}$ are equivalent, there results

$$\begin{aligned}
&\mathrm{d}(F\,\mathrm{d}\xi)\delta\xi - \frac{1}{2}\delta F\,\mathrm{d}\xi^2 + \mathrm{d}(G\,\mathrm{d}\xi)\delta\psi + \mathrm{d}(G\,\mathrm{d}\psi)\delta\xi \\
&-\frac{1}{2}\delta G\,\mathrm{d}\xi\,\mathrm{d}\psi + \mathrm{d}(H\,\mathrm{d}\psi)\delta\psi - \frac{1}{2}\delta H\,\mathrm{d}\psi^2 + \cdots
\end{aligned}$$

for the transformed expression of the quantity $\mathrm{d}^2x\,\delta x + \mathrm{d}^2y\,\delta y + \mathrm{d}^2z\,\delta z$.

But it is obvious that this value can be deduced immediately from the last differential, by dividing all the terms by 2, changing the signs of those which do not contain the double operator $\delta\mathrm{d}$ and deleting in the others the d after the δ to apply it to the quantities which multiply the double differential affected by $\delta\mathrm{d}$. Thus the term $\delta F\,\mathrm{d}\xi^2$ gives $-\frac{1}{2}\delta F\,\mathrm{d}\xi^2$, the term $2F\,\mathrm{d}\xi\,\delta\mathrm{d}\xi$ will give $\mathrm{d}(F\,\mathrm{d}\xi)\delta\xi$, the term $2\delta G\,\mathrm{d}\xi\,\mathrm{d}\psi$ will give $-\delta G\,\mathrm{d}\xi\,\mathrm{d}\psi$, the term $2G\,\mathrm{d}\psi\,\delta\mathrm{d}\xi$ will give $\mathrm{d}(G\,\mathrm{d}\psi)\delta\xi$ and similarly for the others.

5. From this formulation, it follows that the quantity $\frac{1}{2}(\mathrm{d}x^2 + \mathrm{d}y^2 + \mathrm{d}z^2)$ is transformed to

$$\begin{aligned}
&\mathrm{d}^2x\,\delta x + \mathrm{d}^2y\,\delta y + \mathrm{d}^2z\,\delta z = (-\frac{\delta\phi}{\delta\xi} + \mathrm{d}\frac{\delta\phi}{\delta\,\mathrm{d}\xi})\delta\xi \\
&+(-\frac{\delta\phi}{\delta\psi} + \mathrm{d}\frac{\delta\phi}{\delta\mathrm{d}\psi})\delta\psi + (-\frac{\delta\phi}{\delta\varphi} + \mathrm{d}\frac{\delta\phi}{\delta\mathrm{d}\varphi})\delta\varphi + \cdots
\end{aligned}$$

by substituting ξ, ψ, φ, etc. for the values of x, y, z and by denoting by ϕ the function of ξ, ψ, φ, etc. and $\mathrm{d}\xi$, $\mathrm{d}\psi$, $\mathrm{d}\varphi$, etc. Now $\delta\phi/\delta\xi$ denotes, according to common usage, the coefficient of $\delta\xi$ in the difference $\delta\phi$, and $\delta\phi/\delta\mathrm{d}\xi$ denotes the coefficient of $\delta\mathrm{d}\xi$ in the same difference and similarly for the remaining quantities.

6. What was just found by a particular approach could also have been found more simply and generally by the principles of the Method of Variations.[23]

Let ϕ be an arbitrary function of x, y, z, etc., $\mathrm{d}x$, $\mathrm{d}y$, $\mathrm{d}z$, d^2x, d^2y, d^2z, etc., which becomes a function of ξ, ψ, φ, etc., $\mathrm{d}\xi$, $\mathrm{d}\psi$, $\mathrm{d}\varphi$, etc., $\mathrm{d}^2\xi$, $\mathrm{d}^2\psi$, $\mathrm{d}^2\varphi$, etc., after the substitution of the values of x, y, z, etc. expressed as a function of ξ, ψ, φ, etc. Differentiating with respect to δ one will have this equivalent equation

$$\delta\phi = \frac{\delta\phi}{\delta x}\delta x + \frac{\delta\phi}{\delta \mathrm{d}x}\delta \mathrm{d}x + \frac{\delta\phi}{\delta \mathrm{d}^2 x}\delta \mathrm{d}^2 x + \cdots$$
$$+\frac{\mathrm{d}\phi}{\delta y}\delta y + \frac{\delta\phi}{\delta \mathrm{d}y}\delta \mathrm{d}y + \frac{\delta\phi}{\delta \mathrm{d}^2 y}\delta \mathrm{d}^2 y + \cdots$$
$$+\frac{\delta\phi}{\delta z}\delta z + \frac{\delta\phi}{\delta \mathrm{d}z}\delta \mathrm{d}z + \frac{\delta\phi}{\delta \mathrm{d}^2 z}\delta \mathrm{d}^2 z + \cdots$$
$$= \frac{\delta\phi}{\delta \xi}\delta \xi + \frac{\delta\phi}{\delta \psi}\delta \psi + \frac{\delta\phi}{\delta \varphi}\delta \varphi + \cdots$$
$$+\frac{\delta\phi}{\delta \mathrm{d}\xi}\delta \mathrm{d}\xi + \frac{\delta\phi}{\delta \mathrm{d}\psi}\delta \mathrm{d}\psi + \frac{\delta\phi}{\delta \mathrm{d}\varphi}\delta \mathrm{d}\varphi + \cdots$$
$$+\frac{\delta\phi}{\delta \mathrm{d}^2\xi}\delta \mathrm{d}^2\xi + \frac{\delta\phi}{\delta \mathrm{d}^2\psi}\delta \mathrm{d}^2\psi + \frac{\delta\phi}{\delta \mathrm{d}^2\varphi}\delta \mathrm{d}\varphi + \cdots$$

If the double operators $\delta\mathrm{d}$, $\delta\mathrm{d}^2$, etc. are replaced by their equivalents $\mathrm{d}\delta$, $\mathrm{d}^2\delta$, etc. and subsequently, integrated with respect to d and then by the integration by parts all the double operators $\mathrm{d}\delta$, $\mathrm{d}^2\delta$, etc. are eliminated under the integral sign $\int$ which is related to the differential operator d, an equation of the following form will result

$$\int (A\,\delta x + B\,\delta y + C\,\delta z + \cdots) + Z = \int (A'\,\delta\xi + B'\,\delta\psi + C'\,\delta\varphi + \cdots) + Z'$$

in which

$$A = \frac{\delta\phi}{\delta x} - \mathrm{d}\frac{\delta\phi}{\delta \mathrm{d}x} + \mathrm{d}^2\frac{\delta\phi}{\delta \mathrm{d}^2 x} - \cdots$$
$$B = \frac{\delta\phi}{\delta y} - \mathrm{d}\frac{\delta\phi}{\delta \mathrm{d}y} + \mathrm{d}^2\frac{\delta\phi}{\delta \mathrm{d}^2 y} - \cdots$$
$$C = \frac{\delta\phi}{\delta z} - \mathrm{d}\frac{\delta\phi}{\delta \mathrm{d}z} + \mathrm{d}^2\frac{\delta\phi}{\delta \mathrm{d}^2 z} - \cdots$$
$$\vdots$$
$$A' = \frac{\delta\phi}{\delta \xi} - \mathrm{d}\frac{\delta\phi}{\delta \mathrm{d}\xi} + \mathrm{d}^2\frac{\delta\phi}{\delta \mathrm{d}^2 \xi} - \cdots$$
$$B' = \frac{\delta\phi}{\delta \psi} - \mathrm{d}\frac{\delta\phi}{\delta \mathrm{d}\psi} + \mathrm{d}^2\frac{\delta\phi}{\delta \mathrm{d}^2 \psi} - \cdots$$
$$C' = \frac{\delta\phi}{\delta \varphi} - \mathrm{d}\frac{\delta\phi}{\delta \mathrm{d}\varphi} + \mathrm{d}^2\frac{\delta\phi}{\delta \mathrm{d}^2 \varphi} - \cdots$$
$$\vdots$$
$$Z = (\frac{\delta\phi}{\delta \mathrm{d}x} - \mathrm{d}\frac{\delta\phi}{\delta \mathrm{d}^2 x} + \cdots)\delta x + \frac{\delta\phi}{\delta \mathrm{d}^2 x}\mathrm{d}\delta x + \cdots$$
$$+(\frac{\delta\phi}{\delta \mathrm{d}y} - \mathrm{d}\frac{\delta\phi}{\delta \mathrm{d}^2 y} + \cdots)\delta y + \frac{\delta\phi}{\delta \mathrm{d}^2 y}\mathrm{d}\delta y + \cdots$$
$$+(\frac{\delta\phi}{\delta \mathrm{d}z} - \mathrm{d}\frac{\delta\phi}{\delta \mathrm{d}^2 z} + \cdots)\delta z + \frac{\delta\phi}{\delta \mathrm{d}^2 z}\mathrm{d}\delta z + \cdots$$

$$\vdots$$

$$Z' = (\frac{\delta\phi}{\delta \mathrm{d}\xi} - \mathrm{d}\frac{\delta\phi}{\delta \mathrm{d}^2\xi} + \cdots)\delta\xi + \frac{\delta\phi}{\delta \mathrm{d}^2\xi}\mathrm{d}\delta\xi + \cdots$$
$$+(\frac{\delta\phi}{\delta \mathrm{d}\psi} - \mathrm{d}\frac{\delta\phi}{\delta \mathrm{d}^2\psi} + \cdots)\delta\psi + \frac{\delta\phi}{\delta \mathrm{d}^2\psi}\delta\mathrm{d}\psi + \cdots$$
$$+(\frac{\delta\phi}{\delta \mathrm{d}\varphi} - \mathrm{d}\frac{\delta\phi}{\delta \mathrm{d}^2\varphi} + \cdots)\delta\varphi + \frac{\delta\phi}{\delta \mathrm{d}^2\varphi}\mathrm{d}\delta\varphi + \cdots$$

$$\vdots$$

Now differentiating again and transposing will produce the following equation

$$A\,\delta x + B\,\delta y + C\,\delta z + \cdots - A'\,\delta\xi - B'\,\delta\psi - C'\,\delta\varphi - \cdots = \mathrm{d}Z' - \mathrm{d}Z$$

which must be an identify and exist whatever the variations or differences indicated by the operator δ.

Thus since the second member of this equation is an exact differential with respect to the differential operator d, the first member must also be exact with respect to the same operator and independent of the operator δ. But this is not possible because the terms of this first member contain simply the variations δx, δy, δz, etc., $\delta\xi$, $\delta\psi$, etc. and not the differentials of these variations.

From which it follows that in order for the equation to be valid, it is necessary that the two members are each equal to zero, which will lead to the following two equations

$$A\,\delta x + B\,\delta y + C\,\delta z + \cdots = A'\,\delta\xi + B'\,\delta\psi + C'\,\delta\varphi + \cdots$$
$$\mathrm{d}Z = \mathrm{d}Z'$$

These equations can be useful in different situations.

For example, let $\Phi = \frac{1}{2}(\mathrm{d}x^2 + \mathrm{d}y^2 + \mathrm{d}z^2)$, one will have $\delta\Phi/\delta x = 0$, $\delta\Phi/\delta\mathrm{d}x = \mathrm{d}x$, $\delta\Phi/\delta\mathrm{d}^2x = 0$, etc. and similarly for the other quantities. Thus

$$A = -\mathrm{d}^2x, \qquad B = -\mathrm{d}^2y, \qquad C = -\mathrm{d}^2z$$

then, since Φ contains only differences of the first order, there will simply result

$$A' = \frac{\delta\Phi}{\delta\xi} - \mathrm{d}\frac{\delta\Phi}{\delta\mathrm{d}\xi}$$
$$B' = \frac{\delta\Phi}{\delta\psi} - \mathrm{d}\frac{\delta\Phi}{\delta\mathrm{d}\psi}$$
$$C' = \frac{\delta\Phi}{\delta\varphi} - \mathrm{d}\frac{\delta\Phi}{\delta\mathrm{d}\varphi} \quad \cdots$$

Thus the following equation will result

$$-\mathrm{d}^2x\,\delta x - \mathrm{d}^2y\,\delta y - \mathrm{d}^2z\,\delta z = (\frac{\delta\Phi}{\delta\xi} - \mathrm{d}\frac{\delta\Phi}{\delta \mathrm{d}\xi})\delta\xi$$

$$+(\frac{\delta\Phi}{\delta\psi} - \mathrm{d}\frac{\delta\Phi}{\delta \mathrm{d}\psi})\delta\psi + (\frac{\delta\Phi}{\delta\varphi} - \mathrm{d}\frac{\delta\Phi}{\delta \mathrm{d}\varphi})\delta\varphi + \cdots$$

which agrees with the results of Article 5.

7. A result of the preceding development is that in order to have the value of the quantity

$$\mathrm{S}(\frac{\mathrm{d}^2x}{\mathrm{d}t^2}\delta x + \frac{\mathrm{d}^2y}{\mathrm{d}t^2}\delta y + \frac{\mathrm{d}^2z}{\mathrm{d}t^2}\delta z)m$$

as a function of ξ, ψ, φ, etc. it suffices to determine the value of the quantity

$$\mathrm{S}(\frac{\mathrm{d}x^2 + \mathrm{d}y^2 + \mathrm{d}z^2}{2\,\mathrm{d}t^2})m$$

as a function of ξ, ψ, φ, etc., and their differentials, because by denoting this function by T, one will immediately have the transformed equation

$$(\mathrm{d}\frac{\delta T}{\delta \mathrm{d}\xi} - \frac{\delta T}{\delta\xi})\delta\xi + (\mathrm{d}\frac{\delta T}{\delta \mathrm{d}\psi} - \frac{\delta T}{\delta\psi})\delta\psi + (\mathrm{d}\frac{\delta T}{\delta \mathrm{d}\varphi} - \frac{\delta T}{\delta\varphi})\delta\varphi + \cdots$$

This transformation will also hold if among the new variables there appears even the time t, as long as it is viewed as a constant, that is, as long as $\delta t = 0$.[24]

Moreover, it is easy to see that a similar transformation will also hold in the case where the variations $\delta\xi$, $\delta\psi$, $\delta\varphi$, etc. will not be exact differentials as long as they represent undetermined quantities and the variation δT is of the form

$$\delta T = \frac{\delta T}{\delta\xi}\delta\xi + \frac{\delta T}{\mathrm{d}\delta\xi}\mathrm{d}\delta\xi + \frac{\delta T}{\delta\psi}\delta\psi + \frac{\delta T}{\mathrm{d}\delta\psi}\mathrm{d}\delta\psi + \cdots$$

whatever the form of the coefficients $\delta T/\delta\xi$, $\delta T/\mathrm{d}\delta\xi$, $\delta T/\delta\psi$, etc.

8. Also, it is advantageous to note that if the expression for T contains a term $\mathrm{d}A$ which is the complete differential of a function A in which one of the variables such as ξ, enters only in a finite form, this term will not contribute anything to the preceding transformation relative to this variable. Since by defining

$$T = \mathrm{d}A = \frac{\mathrm{d}A}{\mathrm{d}\xi}\mathrm{d}\xi + \frac{\mathrm{d}A}{\mathrm{d}\psi}\mathrm{d}\psi + \cdots$$

there results

$$\frac{\delta T}{\delta \mathrm{d}\xi} = \frac{\mathrm{d}A}{\mathrm{d}\xi}, \qquad \frac{\delta T}{\delta \xi} = \frac{\delta \dfrac{\mathrm{d}A}{\mathrm{d}\xi}}{\mathrm{d}\xi}\mathrm{d}\xi + \frac{\delta \dfrac{\mathrm{d}A}{\mathrm{d}\psi}}{\delta \xi}\mathrm{d}\psi + \cdots$$

$$= \frac{\mathrm{d}^2 A}{\mathrm{d}\xi^2}\mathrm{d}\xi + \frac{\mathrm{d}^2 A}{\mathrm{d}\xi\,\mathrm{d}\psi}\mathrm{d}\psi + \cdots = \mathrm{d}\frac{\mathrm{d}A}{\mathrm{d}\xi}$$

Thus the expression $\mathrm{d}(\delta T/\delta \mathrm{d}\xi) - \delta T/\delta\xi$, which is the coefficient of $\delta\xi$, will become equal to $\mathrm{d}(\mathrm{d}A/\mathrm{d}\xi) - \mathrm{d}(\mathrm{d}A/\mathrm{d}\xi) = 0$.

It follows that if the expression for T contained a term of the form $B\,\mathrm{d}A$, where A is a function of ξ, ψ, etc. without $\mathrm{d}\xi$, and B is an arbitrary function without ξ, this term would simply give, relative to the variation of ξ, the term $\mathrm{d}B(\delta A/\delta\xi)$.

By expressing the term $B\,\mathrm{d}A$ in the form $\mathrm{d}(BA) - A\,\mathrm{d}B$, it is obvious at the outset that the term $\mathrm{d}(BA)$ will not contribute relative to the variation of ξ, since AB contains ξ without $\mathrm{d}\xi$. Then, since $\mathrm{d}B$ contains neither ξ nor $\mathrm{d}\xi$, and A contains ξ without $\mathrm{d}\xi$, it is clear that by defining $T = A\,\mathrm{d}B$, one has $\delta T/\delta\mathrm{d}\xi = 0$ and $\delta T/\delta\xi = -(\delta A/\delta\xi)\mathrm{d}B$ in such a way that the coefficient of $\delta\xi$ will be reduced to $(\delta A/\delta\xi)\mathrm{d}B$.

9. With respect to the quantity $P\,\delta p + Q\,\delta q + R\,\delta r +$ etc., it is always easy to simplify it with respect to ξ, ψ, φ, etc., because it is only a question of simplifying separately the expressions for the lengths p, q, r, etc. and for the forces P, Q, R, etc. But this operation becomes even easier when the forces are such that the sum of the moments, that is, the quantity $P\,\mathrm{d}p + Q\,\mathrm{d}q + R\,\mathrm{d}r +$ etc. is integrable, which is, as we have already observed, properly a property of its nature.

Assuming as in Article 34 of SECTION III, that

$$\mathrm{d}\Pi = P\,\mathrm{d}p + Q\,\mathrm{d}q + R\,\mathrm{d}r + \cdots$$

the quantity Π will be expressed as a finite function of p, q, r, etc. Consequently, one will also have

$$\delta\Pi = P\,\delta p + Q\,\delta q + R\,\delta r + \cdots$$

After multiplying this equation by m and composing the sum of all the bodies of the system, one will have

$$\mathrm{S}(P\,\delta p + Q\,\delta q + R\,\delta r + \cdots)m = \mathrm{S}\,\delta\Pi m = \delta(\mathrm{S}\,\Pi m)$$

since the integral operator S is independent of the operator δ.

It will only be necessary to analyze the quantity after it is expressed as a function of ξ, ψ, φ, etc. which only requires the substitution of the quantities of x, y, z for ξ, ψ, φ, etc. in the expressions for p, q, r, etc. (Article 1, SECTION II, Part I) and for this particular value of $\mathrm{S}\,\Pi m$ which is denoted by the letter V, one will immediately have

$$\delta V = \frac{\mathrm{d}V}{\mathrm{d}\xi}\delta\xi + \frac{\mathrm{d}V}{\mathrm{d}\psi}\delta\psi + \frac{\mathrm{d}V}{\mathrm{d}\varphi}\delta\varphi + \cdots$$

10. In this fashion, the general formula of dynamics (Article 2) will be transformed to the following equation

$$\Xi\,\delta\xi + \Psi\,\delta\psi + \Phi\,\delta\varphi + \cdots = 0$$

in which one will have

$$\begin{aligned}
\Xi &= \mathrm{d}\frac{\delta T}{\delta \mathrm{d}\xi} - \frac{\delta T}{\delta\xi} + \frac{\delta V}{\delta\xi}\\
\Psi &= \mathrm{d}\frac{\delta T}{\delta \mathrm{d}\psi} - \frac{\delta T}{\delta\psi} + \frac{\delta V}{\delta\psi}\\
\Phi &= \mathrm{d}\frac{\delta T}{\delta \mathrm{d}\varphi} - \frac{\delta T}{\delta\varphi} + \frac{\delta V}{\delta\varphi}\\
&\vdots
\end{aligned}$$

assuming

$$T = \mathrm{S}(\frac{\mathrm{d}x^2 + \mathrm{d}y^2 + \mathrm{d}z^2}{2\,\mathrm{d}t^2})m, \qquad V = \mathrm{S}\,\Pi m$$

and

$$\mathrm{d}\Pi = P\,\mathrm{d}p + Q\,\mathrm{d}q + R\,\mathrm{d}r + \cdots$$

If the bodies m and m' of the system, viewed as mass points and for which the mutual distance is p, were attracted with an accelerating force represented by P, which is a function of p, it is easy to see that the moment of this force would be expressed by $mm'P\,\mathrm{d}p$. Hence, it is only necessary to add to the value of V the quantity $mm'\int P\,\mathrm{d}p$ and so on if there are in the system other forces of mutual attraction.

In general, if the system contained arbitrary forces F, G, etc. which have a tendency to reduce the value of the quantities f, g, etc., one would have $F\,\delta f$, $G\,\delta g$, etc., for the moments of these forces (Article 9, SECTION II, Part I). By considering F as a function of f and G as a function of g, etc., one should add to the value of V as many terms of the form $\int F\,\mathrm{d}f$, $\int G\,\mathrm{d}g$, etc., as there are corresponding forces.

However, if in the choice of the new variables ξ, ψ, φ, etc., the equations of condition given by the nature of the assumed system are such that these variables were now entirely independent of one another and consequently, the variations $\delta\xi$, $\delta\psi$, $\delta\varphi$, etc. remain absolutely indeterminate, one will immediately have the particular equations $\Xi = 0$, $\Psi = 0$, $\Phi = 0$, etc. The particular equation will be used to determine the motion of the system, because the number of these equations are the same as the number of variables ξ, ψ, φ, etc. on which the configuration of the system depends at every instant.

11. But even if the problem can always be brought to this state, since it is only a question of using the equations of condition to eliminate as many variables as possible and then to take ξ, ψ, φ, etc. as the remaining variables, nevertheless, there may be some cases where this method is too laborious and where it may be judicious to retain a greater number of these variables in order not to overly complicate the calculation. Then, the equations of condition which are not yet satisfied will have to be used to eliminate in the general formula some of the variations $\delta\xi$, $\delta\psi$, etc. But, instead of the actual elimination, the method of multipliers which was presented in SECTION IV of PART I could be used.

Let $L = 0$, $M = 0$, $N = 0$, etc. be the equations in question, reduced to functions of ξ, ψ, φ, etc. such that L, M, N, etc., are given functions of these variables. One will add to the first member of the general formula (preceding article) the quantity $\lambda\,\delta L + \mu\,\delta M + \nu\,\delta N +$ etc. in which λ, μ, ν, etc. are indeterminate coefficients. Then the variations $\delta\xi$, $\delta\psi$, $\delta\varphi$, etc. can be viewed as independent and arbitrary.

Thus the general equation will be obtained

$$\Xi\,\delta\xi + \Psi\,\delta\psi + \Phi\,\delta\varphi + \cdots + \lambda\,\delta L + \mu\,\delta M + \nu\,\delta N + \cdots = 0$$

which has to be examined independently of the variations $\delta\xi$, $\delta\psi$, $\delta\varphi$, etc. and which will give the following particular equations for the motion of the system

$$\begin{aligned}
&\Xi + \lambda\frac{\delta L}{\delta\xi} + \mu\frac{\delta M}{\delta\xi} + \nu\frac{\delta N}{\delta\xi} + \cdots = 0\\
&\Psi + \lambda\frac{\delta L}{\delta\psi} + \mu\frac{\delta M}{\delta\psi} + \nu\frac{\delta N}{\delta\psi} + \cdots = 0\\
&\Phi + \lambda\frac{\delta L}{\delta\varphi} + \mu\frac{\delta M}{\delta\varphi} + \nu\frac{\delta N}{\delta\varphi} + \cdots = 0\\
&\quad\vdots
\end{aligned}$$

from which the unknowns λ, μ, ν, etc. shall be eliminated, which will diminish the number of equations by as much, but by adding the equations of condition, which must necessarily be done, there will be as many equations as there are variables.

12. Since these equations can assume different forms, more or less simple and more importantly, more or less appropriate for the integration, the form in which they are first presented matters a great deal. It is perhaps one of the main advantages of our approach that it always produces equations for each problem in the simplest form with respect to the variables which are used and it also enables one to determine in advance those variables

whose application can facilitate the integration the most. Here are for this purpose some general principles from which one recognizes the application to the solution of various problems.

It is clear from the formulas which we developed that the differential terms of the equations for the motion of an arbitrary system of bodies derive uniquely from the quantity T which expresses the sum of all the quantities $[(\mathrm{d}x^2+\mathrm{d}y^2+\mathrm{d}z^2)/(2\,\mathrm{d}t^2)]m$ relative to the various bodies. Each finite variable, such as ξ, which enters in the expression for T produces the term $-\delta T/\delta\xi$ and each differential variable such as $\delta\xi$, produces the term $\mathrm{d}(\delta T/\delta \mathrm{d}\xi)$. From which, it is clear at the outset that the terms in question can not contain any other variable function with the exception of those which will be in the expression for T. Consequently, if, by using sines and cosines of angles, which occur naturally in the solution of several problems, it happens that the sines and cosines disappear from the function T, then it will only contain the differentials of these angles and the terms in question will also only contain these same differentials. Thus there will always be a gain in simplicity in the equations of the problem by using these types of solutions.

For example, if in place of the two coordinates x and y, one uses the radius vector r, drawn from the origin of the same coordinates and making with the x-axis the angle φ, one will have $x = r\cos\varphi$, $y = r\sin\varphi$, and after differentiation $\mathrm{d}x = \cos\varphi\,\mathrm{d}r - r\sin\varphi\,\mathrm{d}\varphi$, $\mathrm{d}y = \sin\varphi\,\mathrm{d}r + r\cos\varphi\,\mathrm{d}\varphi$. Thus $\mathrm{d}x^2+\mathrm{d}y^2 = \mathrm{d}r^2 + r^2\,\mathrm{d}\varphi^2$ is a very simple expression which contains neither sine functions nor cosine functions of φ, but only its differential $\mathrm{d}\varphi$. In this fashion, the quantity $\mathrm{d}x^2+\mathrm{d}y^2+\mathrm{d}z^2$ will become $r^2\,\mathrm{d}\varphi^2+\mathrm{d}r^2+\mathrm{d}z^2$.

One could also substitute in place of r and z a new radius vector ρ with the angle ψ which this radius makes with r, which is the projection and which will give $r = \rho\cos\psi$, $z = \rho\sin\psi$. Consequently, $\mathrm{d}r^2+\mathrm{d}z^2 = \mathrm{d}\rho^2+\rho^2\,\mathrm{d}\psi^2$ such that the quantity $\mathrm{d}x^2+\mathrm{d}y^2+\mathrm{d}z^2$ would be transformed to the following equation $\rho^2(\cos^2\psi\,\mathrm{d}\varphi^2+\mathrm{d}\psi^2)+\mathrm{d}\rho^2$. It is clear that ρ will be the radius drawn from the origin of the coordinate system to the point in space where the body m is located and ψ will be the inclination of this radius on the xy-plane. Furthermore, the variable φ will be the angle of the projection of this radius on the same plane with the x-axis. Thus, the following relations will result as in Article 4 of SECTION III

$$x = \rho\cos\psi\cos\varphi, \qquad y = \rho\cos\psi\sin\varphi, \qquad z = \rho\sin\psi$$

Finally, other substitutions could be used at will. When the system is composed of several bodies, they could be immediately located relative to one another by relative coordinates. The circumstances of each problem will always indicate the approach which is most appropriate. Then, after having found after a substitution one or some of the equations of the problem, the others could be deduced from other substitutions, which will give new means to diversify these equations and to find the simplest and easiest way to integrate them.

13. The other terms of the equations in question depend upon the accelerating forces that are assumed acting on the bodies and on the equations of condition which must hold between the variables relative to the position of the bodies in space.

When the forces P, Q, R, etc. act toward fixed centers or toward bodies of the same system, and are proportional to arbitrary functions of the distance between them, as this exists in nature, the quantity V which expresses the sum of the quantities $m \int (P\,\mathrm{d}p + Q\,\mathrm{d}q + R\,\mathrm{d}r + \text{etc.})$ for all the bodies m of the system, will be an algebraic function of the distances and will give for each variable ξ of which it is composed, a finite term of the form $\delta V/\delta\xi$.

Similarly, the equations of condition $L = 0$, $M = 0$, etc. will give for the same variable ξ the terms $\lambda(\delta L/\delta\xi)$, $\mu(\delta M/\delta\xi)$, etc. and in the same fashion for the others. Thus it will only be necessary to add to the value of V the quantities λL, μM, etc. viewing λ, μ, etc. as constants in the differentiation by δ.

Thus if some of the variables which are in the function T are not in V or in L, M, etc., the equations with respect to these variables will only contain differential terms and the integration will be much easier, especially if these variables only appear in T in a differential form. This will hold when the bodies are attracted toward centers and the distances to these centers and the angles described about them can be used for coordinates.

14. An integration which is always possible when the forces are functions of distance and when the functions T, V, L, M, etc. do not contain the finite variable t, is the one which will result in the principle of the **Conservation des Forces Vives**. Although we have already shown how this principle results from our general formula of dynamics (SECTION III, Article 43), it will be worthwhile to show that the particular equations deduced from this formula always furnish an integrable equation which is the one of the **Conservation des Forces Vives**.

These equations, considered in all their generality, are each of the form (Article 11)

$$\mathrm{d}\frac{\delta T}{\delta \mathrm{d}\xi} - \frac{\delta T}{\delta\xi} + \frac{\delta V}{\delta\xi} + \lambda\frac{\delta L}{\delta\xi} + \mu\frac{\delta M}{\delta\xi} + \cdots = 0$$

If they were added together after each equation is multiplied by the differentials of $\mathrm{d}\xi$, $\mathrm{d}\psi$, etc., respectively and if one insures that the quantities V, L, M, etc., are independent of t by the assumption of the algebraic functions of the variables ξ, ψ, etc., it is clear that the following equation will be obtained

$$\begin{aligned}&(\mathrm{d}\frac{\delta T}{\delta \mathrm{d}\xi} - \frac{\delta T}{\delta\xi})\mathrm{d}\xi + (\mathrm{d}\frac{\delta T}{\delta \mathrm{d}\psi} - \frac{\delta T}{\delta\psi})\mathrm{d}\psi + \cdots\\&+\mathrm{d}V + \lambda\,\mathrm{d}L + \mu\,\mathrm{d}M + \cdots = 0\end{aligned}$$

but since $L = 0$, $M = 0$, etc. are the equations of condition, it will generally result that $\mathrm{d}L = 0$, $\mathrm{d}M = 0$, etc. Consequently, the preceding equation will be reduced to

$$(\mathrm{d}\frac{\delta T}{\delta \mathrm{d}\xi} - \frac{\delta T}{\delta \xi})\mathrm{d}\xi + \cdots + \mathrm{d}V = 0$$

But one has

$$\mathrm{d}\xi\, \mathrm{d}\frac{\delta T}{\delta \mathrm{d}\xi} = \mathrm{d}(\frac{\delta T}{\delta \mathrm{d}\xi}\mathrm{d}\xi) - \frac{\delta T}{\delta \mathrm{d}\xi}\mathrm{d}^2\xi$$

and since T is an algebraic function of the variables ξ, ψ, etc. and their differentials $\mathrm{d}\xi$, $\mathrm{d}\psi$, etc. independent of t, one will have

$$\mathrm{d}T = \frac{\delta T}{\delta \xi}\mathrm{d}\xi + \frac{\delta T}{\delta \mathrm{d}\xi}\mathrm{d}^2\xi + \frac{\delta T}{\delta \psi}\mathrm{d}\psi + \frac{\delta T}{\delta \mathrm{d}\psi}\mathrm{d}^2\psi + \cdots$$

Thus the equation will become

$$\mathrm{d}(\frac{\delta T}{\delta \mathrm{d}\xi}\mathrm{d}\xi + \frac{\delta T}{\delta \mathrm{d}\psi}\mathrm{d}\psi + \cdots) - \mathrm{d}T + \mathrm{d}V = 0$$

which is obviously integrable and for which the integral is

$$\frac{\delta T}{\delta \mathrm{d}\xi}\mathrm{d}\xi + \frac{\delta T}{\delta \mathrm{d}\psi}\mathrm{d}\psi + \cdots - T + V = \text{const.}$$

Now since $T = \mathrm{S}((\mathrm{d}x^2 + \mathrm{d}y^2 + \mathrm{d}z^2)/2\,\mathrm{d}t^2)m$, it is obvious that whatever variables are substituted for x, y, z, the resulting function will have to be homogeneous and of two dimensions relative to the differences of these variables. Thus from the well-known theorem, one will have

$$\frac{\delta T}{\delta \mathrm{d}\xi}\mathrm{d}\xi + \frac{\delta T}{\delta \mathrm{d}\psi}\mathrm{d}\psi + \cdots = 2T$$

Therefore, the derived integral will simply be $T + V$ = constant which expresses the principle of the **Conservation des Forces Vives** (SECTION III, Article 34).

If the quantity V were not an algebraic function, one will not have $\mathrm{d}V = (\delta V/\delta\xi)\mathrm{d}\xi +$ etc. and if the quantities T, L, M, etc. also contained the variable t, then the differentials $\mathrm{d}T$, $\mathrm{d}L$, $\mathrm{d}M$, etc. will also contain the terms $(\delta T/\delta t)\mathrm{d}t$, $(\delta L/\delta t)\mathrm{d}t$, $(\delta M/\delta t)\mathrm{d}t$, etc. Thus the simplifications which made the equation integrable would not hold and consequently, neither would the principle of the **Conservation des Forces Vives**.

15. Although the theorem on homogeneous functions which we just discussed is demonstrated in various works and consequently, it could be assumed to be well-known, the

demonstration to be presented here is so simple that I am compelled to present it. If the function F is a homogeneous function of various variables x, y, etc. and of degree n, it is clear that by replacing x, y, etc. with ax, ay, etc., it will necessarily become $a^n F$ whatever the quantity a. Therefore, by defining $a = 1 + \alpha$, and treating α as an infinitesimal quantity, the infinitesimal increase in F due to the infinitesimal increases αx, αy, etc. of x, y, etc. will be $n\alpha F$. But by varying x, y, etc. in the quantities αx, αy, one will have in general for the variation of F, $(\delta F/\delta x)\alpha x + (\delta F/\delta y)\alpha y +$ etc. Thus equating these two expressions for the increase in F and dividing by α, one will have[25]

$$nF = \frac{\delta F}{\delta x}x + \frac{\delta F}{\delta y}y + \cdots$$

16. The integral of the principle of the **Conservation des Forces Vives** is of great use in the solution of the problems of mechanics, especially if the function T contains only the differential of one variable which is not found in the function V, because this integral can then be used to determine this variable and to eliminate it from the differential equations.

Integrals which are related to the Conservation of the Motion of the Center of Gravity and to the Principle of Areas which were formulated in a general manner in SECTION III, will be present in the solution of each problem, if one is careful in the choice of variables, to separate the absolute motion of the system from the relative motion of the bodies as we have shown in the cited section.

The other integrals will depend on the nature of the differential equations for each problem and a general rule to find them cannot be given. However, there is a rather general case which is always amenable to a complete solution in finite terms. This is the case in which the system moves about its position of equilibrium with very small oscillations. We will devote an entire section to this problem because of its importance.

17. When the system for which the motion is to be found is composed of an infinite number of particles or elements for which the aggregation forms a finite mass with a variable shape, an analysis similar to the one we proposed in Subsection II of SECTION IV of PART I must be used. But in place of the operator d, which we have used (from Article 11) to designate the differences of the variables relative to the various elements of the system, one should substitute the operator D which corresponds to the integral operator S, relative to the entire system in order to be able to retain the other operator d for the differences relative to time, to which it has been applied in SECTION II.

Thus by denoting by the letter m the entire mass and by $\mathrm{D}m$ one of its elements, one should replace m by $\mathrm{D}m$ in the expressions for T and V of Article 10.

If to each element of the body the forces F, G, etc. are applied, which have a tendency to diminish the quantities f, g, etc. for which these forces are functions, the expressions $\mathrm{S}\int F\,\mathrm{d}f$, $\mathrm{S}\int G\,\mathrm{d}g$, etc. should be added to the value of V.

If equations of condition $L = 0$, $M = 0$, etc. exist, which must hold at every point of the mass m, one should put S$\lambda\,\delta L$, S$\mu\,\delta M$, etc. in place of $\lambda\,\delta L$, $\mu\,\delta M$, etc. in the formulas of Article 11.

Since the quantities f, g, etc. as well as the quantities L, M, etc. are capable of containing the differences of the variables relative to the operator D, one should then eliminate the double signs δD, δD^2, etc. by the well-known operation of integration by parts such that only the simple variations indicated by δ remain under the integral operator S and the terms outside the operator S will refer uniquely to the limits of the integrals.

Finally, the forces and the equations of condition relative to the determined points of the mass m should be considered. Moreover, they should be taken into consideration in the general formula. However, they will only give the terms independent of the integral operator S.

By equating their coefficients to zero, the variations which remain under the integral operator S will give an equal number of indefinite equations for the motion of each element of the system and the variations outside the system will give determinate equations for various points of the system.

SECTION V
A GENERAL METHOD OF APPROXIMATION FOR THE PROBLEMS OF DYNAMICS BASED ON THE VARIATION OF ARBITRARY CONSTANTS

Since the general equations presented in the preceding section are of second order, integrations which often exceed the known methods of analysis are still required. Consequently, it is necessary to use approximate methods and our formulas furnish the most tractable approach to reach this goal.

1. Every approximate solution to the proposed problem retains the form of the exact solution in which the elements or quantities regarded as very small are neglected. This solution expresses the first approximation and then it is improved by successively incorporating the neglected quantities.

For problems of mechanics which can only be resolved by approximate methods, the first solution is ordinarily found by considering only the primary forces acting on the bodies. In order to extend this solution to the secondary forces which are called perturbing forces, the simplest approach is to keep the form of the first solution by considering as variables the arbitrary constants contained in it. If the quantities which were neglected and which we want to take into account, are very small, the new variables will be nearly constant and the ordinary methods of approximation could be applied. Thus the difficulty is reduced to finding the equations between these variables.

The general method of varying the arbitrary constants resulting from the integration of differential equations is well-known, so that these integrals are also correct for the same equations augmented by given terms.[26] But the form given in the preceding section (Article 10) to the general equations of dynamics has the advantage of furnishing a relation between the variations of the arbitrary constants that the integration must produce, which simplifies remarkably the formulas of these variations in the problems where they express the effect of the perturbing forces. We will first demonstrate this relation. Then we will give the simplest equations to determine the variations of the arbitrary constants in the problems in question.

Subsection I
Where A General Relation Between the Variations of Arbitrary Constants is Deduced from the Equations of the Preceding Section

2. Let there be an arbitrary system of bodies m acted upon by accelerating forces P, Q, R, etc. which act toward arbitrary, fixed or moveable centers and which are proportional to arbitrary functions of their distances p, q, r, etc. to these centers.

Let us assume that from the consideration of the equations of condition of the system, the coordinates x, y, z of each body have been expressed as functions of other variables ξ, ψ, φ, etc. which are entirely independent of one another and which are sufficient to determine the configuration of the system at every instant.

The equations of Article 10 of the preceding section will describe the motion of the entire system and it is easy to see that these equations will be of second order with respect to the variables ξ, ψ, φ etc. such that the total values of these variables which are found by integration and which are expressed as a function of time t, will contain twice as many arbitrary constants as there are variables. Since these constants must remain arbitrary, they can be varied at will. Consequently, it is possible to differentiate the equations in question with respect to these constants, which are assumed contained in the expressions for the variables ξ, ψ, φ, etc.

3. For simplicity, let us define the following relations $\mathrm{d}\xi = \xi'\,\mathrm{d}t$, $\mathrm{d}\psi = \psi'\,\mathrm{d}t$, $\mathrm{d}\varphi = \varphi'\,\mathrm{d}t$, etc. Then the quantity T will become a function of ξ, ψ, φ, etc. and of ξ', ψ', φ', etc. If the forces act toward fixed centers or toward bodies of the same system, the quantity V will be a simple function of ξ, ψ, φ, etc. In this case, by defining $Z = T - V$, one will have

$$\frac{\delta T}{\delta \mathrm{d}\xi} = \frac{\delta Z}{\delta \xi'\,\mathrm{d}t}, \qquad \frac{\delta T}{\delta \mathrm{d}\psi} = \frac{\delta Z}{\delta \psi'\,\mathrm{d}t}, \qquad \frac{\delta T}{\delta \mathrm{d}\varphi} = \frac{\delta Z}{\delta \varphi'\,\mathrm{d}t}, \qquad \ldots$$

where the operator δ can be replaced by d, since it is used only to represent partial differences.

Thus the differential equations of motion of the system (Article 10, preceding section) after multiplication by $\mathrm{d}t$ will be reduced to this simpler form

$$\mathrm{d}\frac{\mathrm{d}Z}{\mathrm{d}\xi'} - \frac{\mathrm{d}Z}{\mathrm{d}\xi}\mathrm{d}t = 0, \qquad \mathrm{d}\frac{\mathrm{d}Z}{\mathrm{d}\psi'} - \frac{\mathrm{d}Z}{\mathrm{d}\psi}\mathrm{d}t = 0, \qquad \mathrm{d}\frac{\mathrm{d}Z}{\mathrm{d}\varphi'} - \frac{\mathrm{d}Z}{\mathrm{d}\varphi}\mathrm{d}t = 0, \qquad \ldots$$

4. Let us differentiate these equations by the operator δ, which we will regard as uniquely relative to the variations of the arbitrary constants which are assumed contained in the expressions for the variables ξ, ψ, φ, etc. of which Z is a function. Because the operator d which affects the terms $\mathrm{d}(\mathrm{d}Z/\mathrm{d}\xi')$, $\mathrm{d}(\mathrm{d}Z/\mathrm{d}\psi')$, etc. is only relative to the variable t which represents time, it is possible, following the principles of the Calculus of Variations, to replace the double operator $\delta\mathrm{d}$ by $\mathrm{d}\delta$ so that the equations

$$\mathrm{d}\delta\frac{\mathrm{d}Z}{\mathrm{d}\xi'} - \delta\frac{\mathrm{d}Z}{\mathrm{d}\xi}\,\mathrm{d}t = 0, \quad \mathrm{d}\delta\frac{\mathrm{d}Z}{\mathrm{d}\psi'} - \delta\frac{\mathrm{d}Z}{\mathrm{d}\psi}\,\mathrm{d}t = 0, \quad \mathrm{d}\delta\frac{\mathrm{d}Z}{\mathrm{d}\varphi'} - \delta\frac{\mathrm{d}Z}{\mathrm{d}\varphi}\mathrm{d}t = 0, \quad \cdots$$

will be obtained.

Similarly, if in order to represent different variations of the same arbitrary constants, the operator Δ is used, the following equations result

$$\mathrm{d}\Delta\frac{\mathrm{d}Z}{\mathrm{d}\xi'} - \Delta\frac{\mathrm{d}Z}{\mathrm{d}\xi}\,\mathrm{d}t = 0, \quad \mathrm{d}\Delta\frac{\mathrm{d}Z}{\mathrm{d}\psi'} - \Delta\frac{\mathrm{d}Z}{\mathrm{d}\psi}\,\mathrm{d}t = 0, \quad \mathrm{d}\Delta\frac{\mathrm{d}Z}{\mathrm{d}\varphi'} - \Delta\frac{\mathrm{d}Z}{\mathrm{d}\varphi}\mathrm{d}t = 0, \quad \cdots$$

5. After multiplying the first equations by $\Delta\xi$, $\Delta\psi$, $\Delta\varphi$, etc. respectively and subtracting from their sum the last equations multiplied by $\delta\xi$, $\delta\psi$, $\delta\varphi$, etc. respectively, the following equation will be obtained

$$\begin{aligned}
&\Delta\xi\,\mathrm{d}\delta\frac{\mathrm{d}Z}{\mathrm{d}\xi'} + \Delta\psi\,\mathrm{d}\delta\frac{\mathrm{d}Z}{\mathrm{d}\psi'} + \Delta\varphi\,\mathrm{d}\delta\frac{\mathrm{d}Z}{\mathrm{d}\varphi'} + \cdots \\
&-\delta\xi\,\mathrm{d}\Delta\frac{\mathrm{d}Z}{\mathrm{d}\xi'} - \delta\psi\,\mathrm{d}\Delta\frac{\mathrm{d}Z}{\mathrm{d}\psi'} - \delta\varphi\,\mathrm{d}\Delta\frac{\mathrm{d}Z}{\mathrm{d}\varphi'} + \cdots \\
&-(\Delta\xi\,\delta\frac{\mathrm{d}Z}{\mathrm{d}\xi} + \Delta\psi\,\delta\frac{\mathrm{d}Z}{\mathrm{d}\psi} + \Delta\varphi\,\delta\frac{\mathrm{d}Z}{\mathrm{d}\varphi} + \cdots)\mathrm{d}t \\
&+(\delta\xi\,\Delta\frac{\mathrm{d}Z}{\mathrm{d}\xi} + \delta\psi\,\Delta\frac{\mathrm{d}Z}{\mathrm{d}\psi} + \delta\varphi\,\Delta\frac{\mathrm{d}Z}{\mathrm{d}\varphi} + \cdots)\mathrm{d}t = 0
\end{aligned}$$

Now

$$\Delta\xi\,\mathrm{d}\delta\frac{\mathrm{d}Z}{\mathrm{d}\xi'} = \mathrm{d}(\Delta\xi\,\delta\frac{\mathrm{d}Z}{\mathrm{d}\xi'}) - \mathrm{d}\Delta\xi\,\delta\frac{\mathrm{d}Z}{\mathrm{d}\xi'}$$

but

$$\mathrm{d}\Delta\xi = \Delta\mathrm{d}\xi = \Delta\xi'\,\mathrm{d}t$$

since $\mathrm{d}\xi = \xi'\,\mathrm{d}t$ by hypothesis. Thus

$$\Delta\xi\,\mathrm{d}\delta\frac{\mathrm{d}Z}{\mathrm{d}\xi'} = \mathrm{d}(\Delta\xi\,\delta\frac{\mathrm{d}Z}{\mathrm{d}\xi'}) - \Delta\xi'\,\delta\frac{\mathrm{d}Z}{\mathrm{d}\xi'}\mathrm{d}t$$

Similarly, the following equation will be obtained

$$\delta\xi\,\mathrm{d}\Delta\frac{\mathrm{d}Z}{\mathrm{d}\xi'} = \mathrm{d}(\delta\xi\,\Delta\frac{\mathrm{d}Z}{\mathrm{d}\xi'}) - \delta\xi'\,\Delta\frac{\mathrm{d}Z}{\mathrm{d}\xi'}\mathrm{d}t$$

and similarly for the remaining formulas.

By means of these transformations, the preceding equation will be of the following form

$$\begin{aligned}
&\mathrm{d}\left\{\begin{array}{l}\Delta\xi\,\delta\dfrac{\mathrm{d}Z}{\mathrm{d}\xi'} + \Delta\psi\,\delta\dfrac{\mathrm{d}Z}{\mathrm{d}\psi'} + \Delta\varphi\,\delta\dfrac{\mathrm{d}Z}{\mathrm{d}\varphi'} + \cdots \\ -\delta\xi\,\Delta\dfrac{\mathrm{d}Z}{\mathrm{d}\xi'} - \delta\psi\,\Delta\dfrac{\mathrm{d}Z}{\mathrm{d}\psi'} - \delta\varphi\,\Delta\dfrac{\mathrm{d}Z}{\mathrm{d}\varphi'} - \cdots\end{array}\right\} \\
&- \left\{\begin{array}{l}\Delta\xi\,\delta\dfrac{\mathrm{d}Z}{\mathrm{d}\xi} + \Delta\psi\,\delta\dfrac{\mathrm{d}Z}{\mathrm{d}\psi} + \Delta\varphi\,\delta\dfrac{\mathrm{d}Z}{\mathrm{d}\varphi} + \cdots \\ +\Delta\xi'\,\delta\dfrac{\mathrm{d}Z}{\mathrm{d}\xi'} + \Delta\psi'\,\delta\dfrac{\mathrm{d}Z}{\mathrm{d}\psi'} + \Delta\varphi'\,\delta\dfrac{\mathrm{d}Z}{\mathrm{d}\varphi'} + \cdots\end{array}\right\} \mathrm{d}t \\
&+ \left\{\begin{array}{l}\delta\xi\,\Delta\dfrac{\mathrm{d}Z}{\mathrm{d}\xi} + \delta\psi\,\Delta\dfrac{\mathrm{d}Z}{\mathrm{d}\psi} + \delta\varphi\,\Delta\dfrac{\mathrm{d}Z}{\mathrm{d}\varphi} + \cdots \\ +\delta\xi'\,\Delta\dfrac{\mathrm{d}Z}{\mathrm{d}\xi'} + \delta\psi'\,\Delta\dfrac{\mathrm{d}Z}{\mathrm{d}\psi'} + \delta\varphi'\,\Delta\dfrac{\mathrm{d}Z}{\mathrm{d}\varphi'} + \cdots\end{array}\right\} \mathrm{d}t = 0
\end{aligned}$$

6. But if the expressions $\delta(\mathrm{d}Z/\mathrm{d}\xi)$, $\delta(\mathrm{d}Z/\mathrm{d}\xi')$, etc. as well as the similar expressions $\Delta(\mathrm{d}Z/\mathrm{d}\xi)$, $\Delta(\mathrm{d}Z/\mathrm{d}\xi')$, etc. are developed and if Z is viewed as a function of ξ, ψ, φ, etc. and of ξ', ψ', φ', etc., it is easy to see that the terms multiplied by dt in the preceding equation cancel one another. In effect, there will result

$$\begin{aligned}
\delta\frac{\mathrm{d}Z}{\mathrm{d}\xi} &= \frac{\mathrm{d}^2Z}{\mathrm{d}\xi^2}\delta\xi + \frac{\mathrm{d}^2Z}{\mathrm{d}\xi\,\mathrm{d}\psi}\delta\psi + \cdots + \frac{\mathrm{d}^2Z}{\mathrm{d}\xi\,\mathrm{d}\xi'}\delta\xi' + \frac{\mathrm{d}^2Z}{\mathrm{d}\xi\,\mathrm{d}\psi'}\delta\psi' + \cdots \\
\delta\frac{\mathrm{d}Z}{\mathrm{d}\psi} &= \frac{\mathrm{d}^2Z}{\mathrm{d}\xi\,\mathrm{d}\psi}\delta\xi + \frac{\mathrm{d}^2Z}{\mathrm{d}\psi^2}\delta\psi + \cdots + \frac{\mathrm{d}^2Z}{\mathrm{d}\psi\,\mathrm{d}\xi'}\delta\xi' + \frac{\mathrm{d}^2Z}{\mathrm{d}\psi\,\mathrm{d}\psi'}\delta\psi' + \cdots \\
&\vdots \\
\delta\frac{\mathrm{d}Z}{\mathrm{d}\xi'} &= \frac{\mathrm{d}^2Z}{\mathrm{d}\xi\,\mathrm{d}\xi'}\delta\xi + \frac{\mathrm{d}^2Z}{\mathrm{d}\psi\,\mathrm{d}\xi'}\delta\psi + \cdots + \frac{\mathrm{d}^2Z}{\mathrm{d}\xi'^2}\delta\xi' + \frac{\mathrm{d}^2Z}{\mathrm{d}\xi\,\mathrm{d}\psi'}\delta\psi' + \cdots \\
\delta\frac{\mathrm{d}Z}{\mathrm{d}\psi'} &= \frac{\mathrm{d}^2Z}{\mathrm{d}\xi\,d\psi'}\delta\xi + \frac{\mathrm{d}^2Z}{\mathrm{d}\psi\,\mathrm{d}\psi'}\delta\psi + \cdots + \frac{\mathrm{d}^2Z}{\mathrm{d}\xi'\,d\psi'}\delta\xi' + \frac{\mathrm{d}^2Z}{\mathrm{d}\psi'^2}\delta\psi' + \cdots \\
&\vdots
\end{aligned}$$

which, after ordering the terms with respect to the partial differences of Z, lead to the following formulation

$$\begin{aligned}
&\Delta\xi\,\delta\frac{\mathrm{d}Z}{\mathrm{d}\xi} + \Delta\psi\,\delta\frac{\mathrm{d}Z}{\mathrm{d}\psi} + \cdots + \Delta\xi'\,\delta\frac{\mathrm{d}Z}{\mathrm{d}\xi'} + \Delta\psi'\,\delta\frac{\mathrm{d}Z}{\mathrm{d}\psi'} + \cdots \\
&= \frac{\mathrm{d}^2Z}{\mathrm{d}\xi^2}\Delta\xi\,\delta\xi + \frac{\mathrm{d}^2Z}{\mathrm{d}\xi\,\mathrm{d}\psi}(\Delta\xi\,\delta\psi + \Delta\psi\,\delta\xi) + \frac{\mathrm{d}^2Z}{\mathrm{d}\psi^2}\Delta\psi\,\delta\psi + \cdots
\end{aligned}$$

$$
\begin{aligned}
&+\frac{\mathrm{d}^2 Z}{\mathrm{d}\xi\,\mathrm{d}\xi'}(\Delta\xi\,\delta\xi' + \Delta\xi'\,\delta\xi) + \frac{\mathrm{d}^2 Z}{\mathrm{d}\xi\,\mathrm{d}\psi'}(\Delta\xi\,\delta\psi' + \Delta\psi'\,\delta\xi) + \cdots \\
&+\frac{\mathrm{d}^2 Z}{\mathrm{d}\psi\,\mathrm{d}\xi'}(\Delta\psi\,\delta\xi' + \Delta\xi'\,\delta\psi) + \frac{\mathrm{d}^2 Z}{\mathrm{d}\psi\,\mathrm{d}\psi'}(\Delta\psi\delta\psi' + \Delta\psi'\,\delta\psi) + \cdots \\
&+\frac{\mathrm{d}^2 Z}{\mathrm{d}\xi'^2}\Delta\xi'\,\delta\xi' + \frac{\mathrm{d}^2 Z}{\mathrm{d}\xi'\,\mathrm{d}\psi'}(\Delta\xi'\,\delta\psi' + \Delta\psi'\,\delta\xi') + \frac{\mathrm{d}^2 Z}{\mathrm{d}\psi'^2}\Delta\psi'\,\delta\psi' + \cdots \\
&\ \vdots
\end{aligned}
$$

By interchanging the operators δ and Δ, the formulation of a similar expression results

$$
\delta\xi\,\Delta\frac{\mathrm{d}Z}{\mathrm{d}\xi} + \delta\psi\,\Delta\frac{\mathrm{d}Z}{\mathrm{d}\psi} + \cdots + \delta\xi'\,\Delta\frac{\mathrm{d}Z}{\mathrm{d}\xi'} + \delta\psi'\,\Delta\frac{\mathrm{d}Z}{\mathrm{d}\psi'} + \cdots
$$

It should be noted that this interchange does not produce a change in the preceding formulation, from which it follows that the two expressions are identical. Since their signs are different in the above equation, they must cancel one another.

7. Thus one will have simply the equation

$$
\mathrm{d}\left\{
\begin{aligned}
&\Delta\xi\,\delta\frac{\mathrm{d}Z}{\mathrm{d}\xi'} + \Delta\psi\,\delta\frac{\mathrm{d}Z}{\mathrm{d}\psi'} + \Delta\varphi\,\delta\frac{\mathrm{d}Z}{\mathrm{d}\varphi'} + \cdots \\
&-\delta\xi\,\Delta\frac{\mathrm{d}Z}{\mathrm{d}\xi'} - \delta\psi\,\Delta\frac{\mathrm{d}Z}{\mathrm{d}\psi'} - \delta\varphi\,\Delta\frac{\mathrm{d}Z}{\mathrm{d}\varphi'} - \cdots
\end{aligned}
\right\} = 0
$$

in which Z can be replaced by T, since $Z = T - V$ and since V must not contain the variables ξ', ψ', φ', etc. (Article 3). It is clear from this equation that the quantity

$$
\begin{aligned}
&\Delta\xi\,\delta\frac{\mathrm{d}T}{\mathrm{d}\xi'} + \Delta\psi\,\delta\frac{\mathrm{d}T}{\mathrm{d}\psi'} + \Delta\varphi\,\delta\frac{\mathrm{d}T}{\mathrm{d}\varphi'} + \cdots \\
&-\delta\xi\,\Delta\frac{\mathrm{d}T}{\mathrm{d}\xi'} - \delta\psi\,\Delta\frac{\mathrm{d}T}{\mathrm{d}\psi'} - \delta\varphi\,\Delta\frac{\mathrm{d}T}{\mathrm{d}\varphi'} - \cdots
\end{aligned}
$$

is necessarily always constant with respect to the time t, to which the differentials indicated by the operator d are related. Consequently, if the expressions for the variables ξ, ψ, φ, etc. are expressed as a function of t and the arbitrary constants deduced from the equations of any problem of mechanics are substituted, the variable t will cancel out, whatever the variations which may be imposed on these constants in the quantities affected by the operators δ and Δ. This is a new and very remarkable property of the function T, which represents the **force vive** of the entire system and which furnishes a general criterion to judge the accuracy of a solution found by any arbitrary method. But the principal application of this formula is to the variation of the arbitrary constants in the problems of mechanics as we will presently demonstrate.

Subsection II
Where the Simplest Differential Equations Are Developed to Determine the Variations of the Arbitrary Constants due to Perturbing Forces

8. Let us now assume that after having resolved the problem presented in the differential equations of Article 3 by the complete integration of these equations, it remains to solve the same problem but with the addition of new forces applied to the same system acting toward fixed or mobile centers in an arbitrary manner and proportional to functions of the distances to these centers. These new forces, which can be treated as forces perturbing the motion of the system and are of a nature similar to the forces P, Q, R, etc. on which the function V depends, will add to this function an analogous function which we will designate by $-\Omega$. Therefore, it only remains to replace V by $V - \Omega$ in the equations of Article 10 (preceding section) and consequently, $Z - \Omega$ in place of Z in the terms of Article 3 which contains the partial differences of Z with respect to ξ, ψ, φ, etc. in order to obtain the equations of the new problem, which will be

$$\mathrm{d}\frac{\mathrm{d}Z}{\mathrm{d}\xi'} - \frac{\mathrm{d}Z}{\mathrm{d}\xi}\mathrm{d}t = \frac{\mathrm{d}\Omega}{\mathrm{d}\xi}\mathrm{d}t$$

$$\mathrm{d}\frac{\mathrm{d}Z}{\mathrm{d}\psi'} - \frac{\mathrm{d}Z}{\mathrm{d}\psi}\mathrm{d}t = \frac{\mathrm{d}\Omega}{\mathrm{d}\psi}\mathrm{d}t$$

$$\mathrm{d}\frac{\mathrm{d}Z}{\mathrm{d}\varphi'} - \frac{\mathrm{d}Z}{\mathrm{d}\varphi}\mathrm{d}t = \frac{\mathrm{d}\Omega}{\mathrm{d}\varphi}\mathrm{d}t$$

$$\vdots$$

9. If it is assumed that the expressions for the variables ξ, ψ, φ, etc. are known as functions of t and the arbitrary constants, one can in the case where the second members of these equations are zero, by keeping these same expressions and by making their arbitrary constants variable, also make them satisfy all these equations. The object of the following analysis is to present the simplest formulas to determine these constants which are now considered variables.

At the outset, it should be noted that since the number of these constants is twice the number of variables ξ, ψ, φ, etc. which was noted earlier (Article 2) that, as a consequence, twice the number of equations must be satisfied. They can also be constrained by a number of conditions equal to the number of variables.

The simplest conditions and at the same time, the most appropriate to the problem, are that the values of $\mathrm{d}\xi/\mathrm{d}t$, $\mathrm{d}\psi/\mathrm{d}t$, $\mathrm{d}\varphi/\mathrm{d}t$, etc. also retain the same form as if the constants did not vary. In this fashion, not only the spaces traversed by the bodies but also their velocities are determined by similar formulas whether the arbitrary constants remain invariant when there are no perturbing forces or whether they become variables from the effect of these forces. These conditions will also have the advantage that the differential equations between the new variables will be reduced to first order so that there will be twice as many equations but all of the first order.

10. By using, as in Article 4, the operator δ to designate the differentials due solely to the variation of the arbitrary constants while the operator d will be related only to the differentials of the time t, the conditions of which we just spoke will be expressed by the equations $\delta\xi = 0$, $\delta\psi = 0$, $\delta\varphi = 0$, ... in which it must be noted that all the arbitrary constants must simultaneously become variable so that the operator δ will indicate in the following development the simultaneous variation of all arbitrary constants instead of as in the formulas of Article 4 and what follows, where the same operator denoted in general the differentials relative to the variation of all the constants or only to some of them at will. This remark applies as well for the other operator Δ.

Therefore, by varying all of them, the differentials of ξ, ψ, φ, etc. will simply be $\mathrm{d}\xi$, $\mathrm{d}\psi$, $\mathrm{d}\varphi$, etc. or $\xi'\,\mathrm{d}t$, $\psi'\,\mathrm{d}t$, $\varphi'\,\mathrm{d}t$, etc. as if only time varied.

Thus in the equations of Article 8, the function Z will be the same, whether the arbitrary constants are assumed variable or not. But by considering these constants variable, the differences $\mathrm{d}(\mathrm{d}Z/\mathrm{d}\xi')$, $\mathrm{d}(\mathrm{d}Z/\mathrm{d}\psi')$, $\mathrm{d}(\mathrm{d}Z/\mathrm{d}\varphi')$, etc. should be augmented by the terms $\delta(\mathrm{d}Z/\mathrm{d}\xi')$, $\delta(\mathrm{d}Z/\mathrm{d}\psi')$, $\delta(\mathrm{d}Z/\mathrm{d}\varphi')$, etc. resulting from the variation of the constants.

On the other hand, in view of the assumption that the functions of t and the constants which represent the values of ξ, ψ, φ, etc. satisfy identically the same equations without their second members, in the case where these constants do not vary, whatever their value, it is clear that the terms

$$\mathrm{d}\frac{\mathrm{d}Z}{\mathrm{d}\xi'} - \frac{\mathrm{d}Z}{\mathrm{d}\xi}\mathrm{d}t, \qquad \mathrm{d}\frac{\mathrm{d}Z}{\mathrm{d}\psi'} - \frac{\mathrm{d}Z}{\mathrm{d}\psi}\mathrm{d}t, \qquad \mathrm{d}\frac{\mathrm{d}Z}{\mathrm{d}\varphi'} - \frac{\mathrm{d}Z}{\mathrm{d}\varphi}\mathrm{d}t, \qquad \cdots$$

cancel one another and consequently, they can be eliminated.

Thus for the variation of the arbitrary constants, the following equations will be obtained

$$\delta\frac{\mathrm{d}Z}{\mathrm{d}\xi'} = \frac{\mathrm{d}\Omega}{\mathrm{d}\xi}\mathrm{d}t, \qquad \delta\frac{\mathrm{d}Z}{\mathrm{d}\psi'} = \frac{\mathrm{d}\Omega}{\mathrm{d}\psi}\mathrm{d}t, \qquad \delta\frac{\mathrm{d}Z}{\mathrm{d}\varphi'} = \frac{\mathrm{d}\Omega}{\mathrm{d}\varphi}\mathrm{d}t, \qquad \cdots$$

which must be combined with the above equations $\delta\xi = 0$, $\delta\psi = 0$, $\delta\varphi = 0$, etc.

Since these equations are twice the number of those of the variables ξ, ψ, φ, etc. and consequently, of the same number as the arbitrary constants (Article 2), they will serve to determine all the constants which have become variable.

11. The equations which we have found after multiplication by $\Delta\xi$, $\Delta\psi$, $\Delta\varphi$, etc., respectively and after adding them together, give

$$\Delta\xi\,\delta\frac{\mathrm{d}Z}{\mathrm{d}\xi'} + \Delta\psi\,\delta\frac{\mathrm{d}Z}{\mathrm{d}\psi'} + \Delta\varphi\,\delta\frac{\mathrm{d}Z}{\mathrm{d}\varphi'} + \cdots$$
$$= \left(\frac{\mathrm{d}\Omega}{\mathrm{d}\xi}\Delta\xi + \frac{\mathrm{d}\Omega}{\mathrm{d}\psi}\Delta\psi + \frac{\mathrm{d}\Omega}{\mathrm{d}\varphi}\Delta\varphi + \cdots\right)\mathrm{d}t$$

Here $\Delta\xi$, $\Delta\psi$, $\Delta\varphi$, etc. express, as in Article 4, the differentials of the functions ξ, ψ, φ, etc. taken by varying only the arbitrary constants in an arbitrary fashion whether all or only some of them are varied simultaneously.

By considering Ω as a function of ξ, ψ, φ, etc., one will have, after differentiating with respect to Δ

$$\Delta\Omega = \frac{d\Omega}{d\xi}\Delta\xi + \frac{d\Omega}{d\psi}\Delta\psi + \frac{d\Omega}{d\varphi}\Delta\varphi + \cdots$$

Then there will result

$$\Delta\Omega\, dt = \Delta\xi\,\delta\frac{dZ}{d\xi'} + \Delta\psi\,\delta\frac{dZ}{d\psi'} + \Delta\varphi\,\delta\frac{dZ}{d\varphi'} + \cdots$$

Let us subtract from the second member of this equation the quantity

$$\delta\xi\,\Delta\frac{dZ}{d\xi'} + \delta\psi\,\Delta\frac{dZ}{d\psi'} + \delta\varphi\,\Delta\frac{dZ}{d\varphi'} + \cdots$$

which is zero because of the equations of condition $\delta\xi = 0$, $\delta\psi = 0$, $\delta\varphi = 0$, etc. Then this general formula results

$$\begin{aligned}\Delta\Omega\, dt &= \Delta\xi\,\delta\frac{dZ}{d\xi'} + \Delta\psi\,\delta\frac{dZ}{d\psi'} + \Delta\varphi\,\delta\frac{dZ}{d\varphi'} + \cdots\\ &\quad - \delta\xi\,\Delta\frac{dZ}{d\xi'} - \delta\psi\,\Delta\frac{dZ}{d\psi'} - \delta\varphi\,\Delta\frac{dZ}{d\varphi'} - \cdots\\ &= \Delta\xi\,\delta\frac{dT}{d\xi'} + \Delta\psi\,\delta\frac{dT}{d\psi'} + \Delta\varphi\,\delta\frac{dT}{d\varphi'} + \cdots\\ &\quad - \delta\xi\,\Delta\frac{dT}{d\xi'} - \delta\psi\,\Delta\frac{dT}{d\psi'} - \delta\varphi\,\Delta\frac{dT}{d\varphi'} - \cdots\end{aligned}$$

after replacing Z with T as in Article 7.

One observes that the second member of the preceding equation is the same function which we saw earlier to be independent of the time t (Article 7). From which it follows that after having substituted the values of ξ, ψ, φ, etc. as functions of t and the arbitrary constants, the variable t could be put equal to zero or to some arbitrary value.

12. Therefore, if it is assumed, which is always permissible, that these functions as well as those which represent the values of $dT/d\xi'$, $dT/d\psi'$, $dT/d\varphi'$, etc. are expanded in a series of increasing powers of t in the following fashion

$$\begin{aligned}\xi &= \alpha + \alpha' t + \alpha'' t^2 + \alpha'' t^3 + \cdots\\ \psi &= \beta + \beta' t + \beta'' t^2 + \beta''' t^3 + \cdots\\ \varphi &= \gamma + \gamma' t + \gamma'' t^2 + \gamma''' t^3 + \cdots\\ &\vdots\\ \frac{dT}{d\xi'} &= \lambda + \lambda' t + \lambda'' t^2 + \lambda''' t^3 + \cdots\\ \frac{dT}{d\psi'} &= \mu + \mu' t + \mu' t^2 + \mu''' t^3 + \cdots\end{aligned}$$

$$\frac{\mathrm{d}T}{\mathrm{d}\varphi'} = \nu + \nu' t + \nu'' t^2 + \nu''' t^3 + \cdots$$
$$\vdots$$

and if these expressions are substituted in the second member of the equation of the preceding article, one could put $t = 0$, which will reduce them to the first terms, α, β, γ, etc., λ, μ, ν, etc.

Thus this equation will be reduced to the form

$$\Delta\Omega\,\mathrm{d}t = \Delta\alpha\,\delta\lambda + \Delta\beta\,\delta\mu + \Delta\gamma\,\delta\nu + \cdots - \Delta\lambda\,\delta\alpha - \Delta\mu\,\delta\beta - \Delta\nu\,\delta\gamma - \cdots$$

13. The quantities α, β, γ, etc., λ, μ, ν, etc. can only be functions of arbitrary constants that the double integration introduces in the finite expressions for the variables ξ, ψ, φ, etc. and they can also be taken for these same constants.

Indeed, the arbitrary constants which give to the solution of a problem of mechanics all the generality that it can have are the initial values of the variables and those of their first differentials, that is, the values of ξ, ψ, φ, etc. and of $\mathrm{d}\xi/\mathrm{d}t$, $\mathrm{d}\psi/\mathrm{d}t$, $\mathrm{d}\varphi/\mathrm{d}t$, etc. when $t = 0$. These values are in the expressions of ξ, ψ, φ, etc., in which we have adopted α, β, γ, etc., α', β', γ', etc. But T is a given function of ξ, ψ, φ, etc. and $\xi' = \mathrm{d}\xi/\mathrm{d}t$, $\psi' = \mathrm{d}\psi/\mathrm{d}t$, $\varphi' = \mathrm{d}\varphi/\mathrm{d}t$, etc. and it is clear that by making $t = 0$ in the functions $\mathrm{d}T/\mathrm{d}\xi'$, $\mathrm{d}T/\mathrm{d}\psi'$, $\mathrm{d}T/\mathrm{d}\varphi'$, etc., which reduce them to λ, μ, ν, etc., the constants λ, μ, ν, etc. will be the same functions of the constants α, β, γ, etc., α', β', γ', etc., then the quantities $\mathrm{d}T/\mathrm{d}\xi'$, $\mathrm{d}T/\mathrm{d}\psi'$, $\mathrm{d}T/\mathrm{d}\varphi'$, etc. are functions of the variables ξ, ψ, φ, etc., ξ', ψ', φ', etc. Consequently, instead of immediately taking α', β', γ', etc. for the arbitrary constants, one can take λ, μ, ν, etc. which depend on them. Thus one will have α, β, γ, etc., λ, μ, ν, etc. for the arbitrary constants in the expressions for ξ, ψ, φ, etc. The number of these constants will be precisely twice the number of variables ξ, ψ, φ, etc.

In this fashion, the differential $\Delta\Omega$ in which the operator Δ affects only the arbitrary constants contained in Ω, that is, the values ξ, ψ, φ, etc. which contain these constants, will become

$$\Delta\Omega = \frac{\mathrm{d}\Omega}{\mathrm{d}\alpha}\Delta\alpha + \frac{\mathrm{d}\Omega}{\mathrm{d}\beta}\Delta\beta + \frac{\mathrm{d}\Omega}{\mathrm{d}\gamma}\Delta\gamma + \cdots + \frac{\mathrm{d}\Omega}{\mathrm{d}\lambda}\Delta\lambda + \frac{\mathrm{d}\Omega}{\mathrm{d}\mu}\Delta\mu + \frac{\mathrm{d}\Omega}{\mathrm{d}\nu}\Delta\nu + \cdots$$

Substituting this relation for the first member of the equation of the preceding equation and ordering the terms with respect to the differences denoted by Δ, there results

$$\begin{aligned}&(\frac{\mathrm{d}\Omega}{\mathrm{d}\alpha}\mathrm{d}t - \delta\lambda)\Delta\alpha + (\frac{\mathrm{d}\Omega}{\mathrm{d}\beta}\mathrm{d}t - \delta\mu)\Delta\beta + (\frac{\mathrm{d}\Omega}{\mathrm{d}\gamma}\mathrm{d}t - \delta\nu)\Delta\gamma + \cdots\\ &+(\frac{\mathrm{d}\Omega}{\mathrm{d}\lambda}\mathrm{d}t + \delta\alpha)\Delta\lambda + (\frac{\mathrm{d}\Omega}{\mathrm{d}\mu}\mathrm{d}t + \delta\beta)\Delta\mu + (\frac{\mathrm{d}\Omega}{\mathrm{d}\nu}\mathrm{d}t + \delta\gamma)\Delta\nu + \cdots = 0\end{aligned}$$

Since it is possible to assign to the differences $\Delta\alpha$, $\Delta\beta$, etc., denoted by the operator Δ, an arbitrary value, it happens that the equation is verified independent of these differences, which produces a number of particular equations such as

$$\frac{d\Omega}{d\alpha}dt = \delta\lambda, \qquad \frac{d\Omega}{d\beta}dt = \delta\mu, \qquad \frac{d\Omega}{d\gamma}dt = \delta\nu, \qquad \cdots$$

$$\frac{d\Omega}{d\lambda}dt = -\delta\alpha, \qquad \frac{d\Omega}{d\mu}dt = -\delta\beta, \qquad \frac{d\Omega}{d\nu}dt = -\delta\gamma, \qquad \cdots$$

14. The differences indicated by the operator δ are properly the differentials of the arbitrary constants which have become variables (Article 10). Now since these differentials can also be referred to the time t, it is permissible and even convenient to change the δ to d, and for the determination of the new variables α, β, γ, etc., λ, μ, ν, etc., the following equations are available

$$\frac{d\alpha}{dt} = -\frac{d\Omega}{d\lambda}, \qquad \frac{d\beta}{dt} = -\frac{d\Omega}{d\mu}, \qquad \frac{d\gamma}{dt} = -\frac{d\Omega}{d\nu}, \qquad \cdots$$

$$\frac{d\lambda}{dt} = \frac{d\Omega}{d\alpha}, \qquad \frac{d\mu}{dt} = \frac{d\Omega}{d\beta}, \qquad \frac{d\nu}{dt} = \frac{d\Omega}{d\gamma}, \qquad \cdots$$

which are, as can be seen, in a very simple form. Thus they represent the simplest solution to the problem of the variation of arbitrary constants.

15. Since the function Ω contains the quantities α, β, γ, etc., λ, μ, ν, etc., they should also be viewed as variables in the partial differences of this function. But when the value of Ω, which depends on perturbing forces, is assumed very small, it is clear that the variations of these quantities will also be very small, and to the first approximation, they can be viewed as constants in the partial differences of Ω, and only their variability considered in the following approximations.

Let us denote by a, b, c, etc., ℓ, m, n, etc., the constant parts of α, β, γ, etc., λ, μ, ν, etc., and by α', β', γ', etc., λ', μ', ν', etc., their variable parts, which being of the same order as the quantity Ω will necessarily be very small, and let O be the value of Ω in which α, β, γ, etc., λ, μ, ν, etc. are replaced by a, b, c, etc., ℓ, m, n, etc.

Thus the following relations result

$$\alpha = a + \alpha', \qquad \beta = b + \beta', \qquad \gamma = c + \gamma', \qquad \cdots$$

$$\lambda = \ell + \lambda', \qquad \mu = m + \mu', \qquad \nu = n + \nu', \qquad \cdots$$

and after further development, there will result

$$\Omega = O + \frac{dO}{da}\alpha' + \frac{dO}{db}\beta' + \frac{dO}{dc}\gamma' + \cdots$$

$$+\frac{dO}{d\ell}\lambda' + \frac{dO}{dm}\mu' + \frac{dO}{dn}\nu' + \cdots$$

$$\vdots$$

The differential equations of the preceding article provide

$$d\alpha' = -\frac{d\Omega}{d\ell}dt, \qquad d\beta' = -\frac{d\Omega}{dm}dt, \qquad d\gamma' = -\frac{d\Omega}{dn}dt, \qquad \cdots$$

$$d\lambda' = \frac{d\Omega}{da}dt, \qquad d\mu' = \frac{d\Omega}{db}dt, \qquad d\nu' = \frac{d\Omega}{dc}dt, \qquad \cdots$$

because it is obvious that the partial differences relative to α, β, γ, etc., λ, μ, ν, etc., can be referred to the analogous quantities a, b, c, etc., ℓ, m, n, etc.

For the first approximation, one will have $\Omega = O$, where O is a simple function of t. Thus after integration, the following equations result

$$\alpha' = -\int \frac{dO}{d\ell}dt, \qquad \beta' = -\int \frac{dO}{dm}dt, \qquad \gamma' = -\int \frac{dO}{dn}dt, \qquad \cdots$$

$$\lambda' = \int \frac{dO}{da}dt, \qquad \mu' = \int \frac{dO}{db}dt, \qquad \nu' = \int \frac{dO}{dc}dt, \qquad \cdots$$

By substituting these expressions in the equation for Ω, one will have for the second approximation

$$\Omega = O + \frac{dO}{d\ell}\int \frac{dO}{da}dt - \frac{dO}{da}\int \frac{dO}{d\ell}dt$$
$$+ \frac{dO}{dm}\int \frac{dO}{db}dt - \frac{dO}{db}\int \frac{dO}{dm}dt$$

and similarly for the others.

16. There is an important remark to be made here. If the function O contains the time only within the arguments of sines and cosines, it is obvious that the expression for Ω will only contain, in the first approximation, the same sines and cosines. But one could question it if, in the following approximation, it did not contain terms where the time t would be outside the sines and cosines and which, by continually increasing, would increase the value of Ω infinitely. Consequently, it would make the approximation inaccurate.

In order to remove this doubt, we will note that similar terms could only come from the constant part of Ω, that is, free of all sine or cosine functions containing the time t.

Thus let A be this part which will be a function of the arbitrary constants α, β, γ, etc., λ, μ, ν, etc. Hence, the function O will contain a similar function of a, b, c, etc., ℓ, m, n, etc., which we will denote also by the letter A.

By substituting A for O in the expression for Ω of the preceding article, the portion of Ω due to the constant A in the second approximation will be obtained and this portion will be

$$A + \frac{dA}{d\ell}\frac{dA}{da}t - \frac{dA}{da}\frac{dA}{d\ell}t + \frac{dA}{dm}\frac{dA}{db}t - \frac{dA}{db}\frac{dA}{dm}t$$

$$\vdots$$

where it is clear that the terms containing t cancel one another.

Thus one can be certain that the second approximation does not give any term in Ω which increases with the time t. But it remains to be seen if there could be any in the following approximations.

In addition, the same constant term A could also give terms in Ω multiplied by t, which can be combined with variable terms of the same function Ω. But then t, which would be free of sine and cosine functions, would be simultaneously multiplied by sine and cosine functions of angles proportional to time. The same thing would happen if the coefficient of t in the sine and cosine functions was a function of the arbitrary constants α, β, γ, etc., because then the partial derivatives of Ω with respect to these constants will put t outside the sine and cosine functions. But, in general, when the successive approximations produce terms of this form in which the sine and cosine functions will be multiplied by the angle which is in these sine and cosine functions, these types of terms are always the result of the development of other sine and cosine functions, and one can avoid them by directly integrating the differential equations between the arbitrary constants which have become variables.

17. Although the arbitrary constants that we have used are those which are the most natural to use and which give the most simple results, it often happens that the different integrations introduce in their place different constants which can only be a function of the former.

We will designate, in general, the arbitrary constants which are assumed to enter in the expressions of the variables ξ, ψ, φ, etc., by a, b, c, f, g, etc. of which the number must be twice the number of variables. And to obtain the relation between these new constants and the former, it suffices to assume $t = 0$ in the expressions for the functions ξ, ψ, φ, etc., $\mathrm{d}T/\mathrm{d}\xi'$, $\mathrm{d}T/\mathrm{d}\psi'$, $\mathrm{d}T/\mathrm{d}\varphi'$, etc. and to equate the results to the quantities α, β, γ, etc., λ, μ, ν, etc. In this fashion, one will have as many equations between these different constants as needed to determine the values of a, b, c, f, g, etc. as functions of α, β, γ, etc., λ, μ, ν, etc.

Thus let us assume these functions to be known and their differentiation will give us immediately

$$\mathrm{d}a = \frac{\mathrm{d}a}{\mathrm{d}\alpha}\mathrm{d}\alpha + \frac{\mathrm{d}a}{\mathrm{d}\beta}\mathrm{d}\beta + \frac{\mathrm{d}a}{\mathrm{d}\gamma}\mathrm{d}\gamma + \cdots$$

$$+\frac{\mathrm{d}a}{\mathrm{d}\lambda}\mathrm{d}\lambda+\frac{\mathrm{d}a}{\mathrm{d}\mu}\mathrm{d}\mu+\frac{\mathrm{d}a}{\mathrm{d}\nu}\mathrm{d}\nu+\cdots$$

Thus substituting the expressions found above (Article 14) for $\mathrm{d}\alpha$, $\mathrm{d}\beta$, etc. and after dividing by $\mathrm{d}t$, one will obtain

$$\begin{aligned}\frac{\mathrm{d}a}{\mathrm{d}t}&=\frac{\mathrm{d}a}{\mathrm{d}\lambda}\frac{\mathrm{d}\Omega}{\mathrm{d}\alpha}+\frac{\mathrm{d}a}{\mathrm{d}\mu}\frac{\mathrm{d}\Omega}{\mathrm{d}\beta}+\frac{\mathrm{d}a}{\mathrm{d}\nu}\frac{\mathrm{d}\Omega}{\mathrm{d}\gamma}+\cdots\\&\quad-\frac{\mathrm{d}a}{\mathrm{d}\alpha}\frac{\mathrm{d}\Omega}{\mathrm{d}\lambda}-\frac{\mathrm{d}a}{\mathrm{d}\beta}\frac{\mathrm{d}\Omega}{\mathrm{d}\mu}-\frac{\mathrm{d}a}{\mathrm{d}\gamma}\frac{\mathrm{d}\Omega}{\mathrm{d}\nu}+\cdots\end{aligned}$$

It will be the same for the values of $\mathrm{d}b/\mathrm{d}t$, $\mathrm{d}c/\mathrm{d}t$, etc. for which it is only necessary to replace a in the preceding equation by b, and then by c, etc.

18. But these formulas still contain the partial differences of Ω relative to the constants α, β, γ, etc. and they have to be changed to partial differences with respect to a, b, c, etc. which is easy to do by known operations.

In effect, since Ω is assumed to be a function of a, b, c, etc. and these quantities are themselves functions of α, β, γ, etc., λ, μ, ν, etc., one has immediately from the algorithm of partial differences

$$\begin{aligned}\frac{\mathrm{d}\Omega}{\mathrm{d}\alpha}&=\frac{\mathrm{d}\Omega}{\mathrm{d}a}\frac{\mathrm{d}a}{\mathrm{d}\alpha}+\frac{\mathrm{d}\Omega}{\mathrm{d}b}\frac{\mathrm{d}b}{\mathrm{d}\alpha}+\frac{\mathrm{d}\Omega}{\mathrm{d}c}\frac{\mathrm{d}c}{\mathrm{d}\alpha}+\cdots\\\frac{\mathrm{d}\Omega}{\mathrm{d}\beta}&=\frac{\mathrm{d}\Omega}{\mathrm{d}a}\frac{\mathrm{d}a}{\mathrm{d}\beta}+\frac{\mathrm{d}\Omega}{\mathrm{d}b}\frac{\mathrm{d}b}{\mathrm{d}\beta}+\frac{\mathrm{d}\Omega}{\mathrm{d}c}\frac{\mathrm{d}c}{\mathrm{d}\beta}+\cdots\\&\;\;\vdots\end{aligned}$$

Then it only remains to substitute these expressions in those for $\mathrm{d}a/\mathrm{d}t$, $\mathrm{d}b/\mathrm{d}t$, etc. of the preceding article.

By making these substitutions and ordering the terms with respect to the partial differences of Ω, it is clear at the outset that the coefficient of $\mathrm{d}\Omega/\mathrm{d}a$ is zero in the expression for $\mathrm{d}a/\mathrm{d}t$, that the one of $\mathrm{d}\Omega/\mathrm{d}b$ is zero in the expression for $\mathrm{d}b/\mathrm{d}t$, etc.

Then, if to represent the expression for $\mathrm{d}a/\mathrm{d}t$, the following formula is used

$$\frac{\mathrm{d}a}{\mathrm{d}t}=(a,b)\frac{\mathrm{d}\Omega}{\mathrm{d}b}+(a,c)\frac{\mathrm{d}\Omega}{\mathrm{d}c}+(a,f)\frac{\mathrm{d}\Omega}{\mathrm{d}f}+\cdots$$

one will have

$$\begin{aligned}(a,b)&=\frac{\mathrm{d}a}{\mathrm{d}\lambda}\frac{\mathrm{d}b}{\mathrm{d}\alpha}+\frac{\mathrm{d}c}{\mathrm{d}\mu}\frac{\mathrm{d}b}{\mathrm{d}\beta}+\frac{\mathrm{d}a}{\mathrm{d}\nu}\frac{\mathrm{d}b}{\mathrm{d}\gamma}+\cdots\\&\quad-\frac{\mathrm{d}a}{\mathrm{d}\alpha}\frac{\mathrm{d}b}{\mathrm{d}\lambda}-\frac{\mathrm{d}a}{\mathrm{d}\beta}\frac{\mathrm{d}b}{\mathrm{d}\mu}-\frac{\mathrm{d}a}{\mathrm{d}\gamma}\frac{\mathrm{d}b}{\mathrm{d}\nu}-\cdots\\(a,c)&=\frac{\mathrm{d}a}{\mathrm{d}\lambda}\frac{\mathrm{d}c}{\mathrm{d}\alpha}+\frac{\mathrm{d}a}{\mathrm{d}\mu}\frac{\mathrm{d}c}{\mathrm{d}\beta}+\frac{\mathrm{d}a}{\mathrm{d}\nu}\frac{\mathrm{d}c}{\mathrm{d}\gamma}+\cdots\end{aligned}$$

$$- \frac{da}{d\alpha}\frac{dc}{d\lambda} - \frac{da}{d\beta}\frac{dc}{d\mu} - \frac{da}{d\gamma}\frac{dc}{d\nu} - \cdots$$

$$\vdots$$

To obtain the value of db/dt, it is only necessary to interchange in these formulas a by b, and b by a, and noting that one has $(b, a) = -(a, b)$; there will result

$$\frac{db}{dt} = -(a, b)\frac{d\Omega}{da} + (b, c)\frac{d\Omega}{dc} + (b, f)\frac{d\Omega}{df} + \cdots$$

$$\vdots$$

$$(b, c) = \frac{db}{d\lambda}\frac{dc}{d\alpha} + \frac{db}{d\mu}\frac{dc}{d\beta} + \frac{db}{d\nu}\frac{dc}{d\gamma} + \cdots$$
$$- \frac{db}{d\alpha}\frac{dc}{d\lambda} - \frac{db}{d\beta}\frac{dc}{d\mu} - \frac{db}{d\gamma}\frac{dc}{d\nu} - \cdots$$

$$\vdots$$

In general, if k represents any of the arbitrary constants a, b, c, f, etc. and if it is noted that the value of the symbols represented by two brackets become zero when the two letters within the brackets are identical and that they simply change sign when one changes the order of the letters, the general formulas will be obtained

$$\frac{dk}{dt} = (k, a)\frac{d\Omega}{da} + (k, b)\frac{d\Omega}{db} + (k, c)\frac{d\Omega}{dc} + \cdots$$

$$(k, a) = \frac{dk}{d\lambda}\frac{da}{d\alpha} + \frac{dk}{d\mu}\frac{da}{d\beta} + \frac{dk}{d\nu}\frac{da}{d\gamma} + \cdots$$
$$- \frac{dk}{d\alpha}\frac{da}{d\lambda} - \frac{dk}{d\beta}\frac{da}{d\mu} - \frac{dk}{d\gamma}\frac{da}{d\nu} - \cdots$$

$$\vdots$$

19. The principal use of these formulas is in the theory of planetary motion where they are used to calculate the effect of perturbations by reducing the problem to the variation of arbitrary constants which are elements of the primary motion. They are useful principally to determine the variations that the astronomers call secular[27] because they have very long periods and are independent of those which exist in the fundamental variables.

Since the equations of Article 18 do not contain functions of time other than the partial differences of the function Ω, if one investigates by the resolution of series or by some other method the portion A of the function Ω which is independent of time t and contains only the arbitrary constants a, b, c, etc., it will suffice to substitute in these equations A instead of Ω. One will have directly the equations between the quantities a, b, c, etc. which are now variables and the time t which will be used to determine their secular variations because they are free of sine and cosine functions.

Subsection III
Where an Important Property of the Quantity which Expresses the Force Vive in a System Acted Upon by Perturbing Forces is Demonstrated.

20. The arbitrary constants, for which we just gave the variations, depend on the nature of each problem and can only be determined in particular cases. However, there is one which holds for all problems where V is solely a function of ξ, ψ, φ, etc. This is the case where the integrations must be added to t, because the differential equations contain only the element $\mathrm{d}t$. It is clear that in the finite expressions of the variables as a function of t one can always put t plus an arbitrary constant in place of t.

Let us denote this constant by K and refer to it the differences affected by the operator Δ in the general formula of Article 11. Then one will have

$$\Delta\Omega = \frac{\mathrm{d}\Omega}{\mathrm{d}K}\Delta K, \qquad \Delta\xi = \frac{\mathrm{d}\xi}{\mathrm{d}K}\Delta K, \qquad \Delta\psi = \frac{\mathrm{d}\psi}{\mathrm{d}K}\Delta K, \qquad \ldots$$

But since ξ, ψ, φ, etc., are functions of $t + K$, it is obvious that we will have $\mathrm{d}\xi/\mathrm{d}K = \mathrm{d}\xi/\mathrm{d}t = \xi'$, and similarly, $\mathrm{d}\psi/\mathrm{d}K = \mathrm{d}\psi/\mathrm{d}t = \psi'$, $\mathrm{d}\varphi/\mathrm{d}K = \mathrm{d}\varphi/\mathrm{d}t = \varphi'$, etc. Thus

$$\Delta\xi = \xi'\,\Delta K, \qquad \Delta\psi = \psi'\,\Delta K, \qquad \Delta\varphi = \varphi'\,\Delta K, \qquad \ldots$$

For the same reason, one will have

$$\Delta\frac{\mathrm{d}Z}{\mathrm{d}\xi'} = -\frac{\mathrm{d}\dfrac{\mathrm{d}Z}{\mathrm{d}\xi'}}{\mathrm{d}t}\Delta K, \qquad \Delta\frac{\mathrm{d}Z}{\mathrm{d}\psi'} = \frac{\mathrm{d}\dfrac{\mathrm{d}Z}{\mathrm{d}\psi'}}{\mathrm{d}t}\Delta K, \qquad \ldots$$

But the differential equations of Article 3 give

$$\frac{\mathrm{d}\dfrac{\mathrm{d}Z}{\mathrm{d}\xi'}}{\mathrm{d}t} = \frac{\mathrm{d}Z}{\mathrm{d}\xi}, \qquad \frac{\mathrm{d}\dfrac{\mathrm{d}Z}{\mathrm{d}\psi'}}{\mathrm{d}t} = \frac{\mathrm{d}Z}{\mathrm{d}\psi}, \qquad \ldots$$

Thus one will have

$$\Delta\frac{\mathrm{d}Z}{\mathrm{d}\xi'} = \frac{\mathrm{d}Z}{\mathrm{d}\xi}\Delta K, \qquad \Delta\frac{\mathrm{d}Z}{\mathrm{d}\psi'} = \frac{\mathrm{d}Z}{\mathrm{d}\psi}\Delta K, \qquad \ldots$$

In this fashion, the general formula of Article 11 will become after these substitutions and division by ΔK

$$\begin{aligned}\frac{\mathrm{d}\Omega}{\mathrm{d}K}\mathrm{d}t &= \xi'\,\delta\frac{\mathrm{d}Z}{\mathrm{d}\xi'} + \psi'\,\delta\frac{\mathrm{d}Z}{\mathrm{d}\psi'} + \varphi'\,\delta\frac{\mathrm{d}Z}{\mathrm{d}\varphi'} + \cdots \\ &\quad - \frac{\mathrm{d}Z}{\mathrm{d}\xi}\delta\xi - \frac{\mathrm{d}Z}{\mathrm{d}\psi}\delta\psi - \frac{\mathrm{d}Z}{\mathrm{d}\varphi}\delta\varphi - \cdots\end{aligned}$$

Now one has

$$\xi'\,\delta\frac{dZ}{d\xi'} + \psi'\,\delta\frac{dZ}{d\psi'} + \varphi'\,\delta\frac{dZ}{d\varphi'} + \cdots$$
$$= \delta(\xi'\frac{dZ}{d\xi'} + \psi'\frac{dZ}{d\psi'} + \varphi'\frac{dZ}{d\varphi'} + \cdots)$$
$$-\frac{dZ}{d\xi'}\delta\xi' - \frac{dZ}{d\psi'}\delta\psi' - \frac{dZ}{d\varphi'}\delta\varphi' - \cdots$$

and since Z is assumed to be a function of ξ, ψ, φ, etc. and of ξ', ψ', φ', etc., one will have

$$\delta Z = \frac{dZ}{d\xi}\delta\xi + \frac{dZ}{d\psi}\delta\psi + \frac{dZ}{d\varphi}\delta\varphi + \cdots$$
$$+ \frac{dZ}{d\xi'}\delta'\xi + \frac{dZ}{d\psi'}\delta\psi' + \frac{dZ}{d\varphi'}\delta\varphi' + \cdots$$

Thus the preceding equation will become

$$\frac{d\Omega}{dK}dt = \delta(\xi'\frac{dZ}{d\xi'} + \psi'\frac{dZ}{d\psi'} + \varphi'\frac{dZ}{d\varphi'} + \cdots - Z)$$

where the second member must be a function of the arbitrary constants and independent of t.

21. In effect, if Z is replaced by $T - V$ and ξ', ψ', φ', etc., by $d\xi/dt$, $d\psi/dt$, $d\varphi/dt$, etc. (Article 3), it is easy to see that the quantity

$$\xi'\frac{dZ}{d\xi'} + \psi'\frac{dZ}{d\psi'} + \varphi'\frac{dZ}{d\varphi'} + \cdots - Z$$

will be the same as the quantity

$$\frac{\delta T}{\delta d\xi}d\xi + \frac{\delta T}{\delta d\psi}d\psi + \frac{\delta T}{\delta d\varphi}d\varphi + \cdots - T + V$$

which we saw earlier is equal to a constant and which is reduced to $T + V$ (preceding section, Article 14), from which derives the equation $T + V = H$, which expresses the **Conservation des Forces Vives** for the system.

Thus by taking H for one of the arbitrary constants, one will have for its variation resulting from the perturbing forces contained in the function Ω, this very simple formula $dH = (d\Omega/dK)dt$.

22. This formula could also be obtained by a shorter path. Indeed, if one again considered the equations of Article 8 and added them together after having multiplied them respectively by $d\xi$, $d\psi$, $d\varphi$, etc. and if one integrated by using the same simplifications which we used in Article 14 of the preceding section, the following equation would be obtained directly

$$T + V = H + \int(\frac{d\Omega}{d\xi}d\xi + \frac{d\Omega}{d\psi}d\psi + \frac{d\Omega}{d\varphi}d\varphi + \cdots)$$

in which the quantity which is under the integral sign is in general not integrable, since the function Ω, because of the mobility that one can assume for the centers of the perturbing forces, is assumed to contain besides the variables ξ, ψ, φ, etc. other variables independent of the former.

In the case where there are no perturbing forces, one has simply $T + V = H$. But it is obvious that one can keep this form for the integral which has just been found by making the quantity H variable and by doing so $\mathrm{d}H = (\mathrm{d}\Omega/\mathrm{d}\xi)\mathrm{d}\xi + (\mathrm{d}\Omega/\mathrm{d}\psi)\mathrm{d}\psi + (\mathrm{d}\Omega/\mathrm{d}\varphi)\mathrm{d}\varphi + \cdots$ but it is clear that the quantity

$$\frac{\mathrm{d}\Omega}{\mathrm{d}\xi}\mathrm{d}\xi + \frac{\mathrm{d}\Omega}{\mathrm{d}\psi}\mathrm{d}\psi + \frac{\mathrm{d}\Omega}{\mathrm{d}\varphi}\mathrm{d}\varphi + \cdots$$

is nothing else but the differential of Ω by varying only the quantities ξ, ψ, φ, etc. which depend on the fundamental differential equations and which are assumed to be known functions of $t + K'$, denoting K' as in Article 20 the constant which can always be added to the variable t. Thus since the variables ξ, ψ, φ, etc. vary only with the time t, it is easy to see that the quantity in question will be the same as $(\mathrm{d}\Omega/\mathrm{d}K')\mathrm{d}t$. Consequently, one will have, as above, the equation $\mathrm{d}H/\mathrm{d}t = \mathrm{d}\Omega/\mathrm{d}K'$.

23. This equation can also be put in the form $\mathrm{d}H/\mathrm{d}t = (\mathrm{d}\Omega/\mathrm{d}t)$ as long as in the partial difference of Ω one varies t whenever it is contained in the expressions of the variables ξ, ψ, φ, etc. It results from this formula that if the function does contain the time t only in the sine and cosine functions, as occurs in the theory of planetary motion, the expression for $(\mathrm{d}\Omega/\mathrm{d}t)$ will only contain periodic terms since all the constant terms of Ω will disappear after differentiation with respect to t. Thus, in the first approximation, where one views the arbitrary constants in the function Ω as absolutely invariable, the integral for $(\mathrm{d}\Omega/\mathrm{d}t)\mathrm{d}t$, that is, the value of H will not contain terms such as Nt which increase with the time t. We earlier saw (Article 16), that the second approximation cannot contribute to Ω any term which is not periodic.

Therefore, the same conclusion with respect to the variation of H will also hold in the second approximation.

24. The quantity T expresses the **force vive** of the system and it is equal to $H - V$. When the system is not disturbed by some other perturbing force, the quantity H is constant and the force vive depends only on the accelerating forces contained in the expression for V as seen in Article 34 of SECTION III. This quantity becomes variable when there are perturbing forces. Consequently, the **force vive** will be altered by the action of these forces. But, by what we just demonstrated, one sees that its alterations can only be periodic, if the expression for the perturbing forces is periodic, at least in the first two approximations. This result is of great importance to the calculus of perturbations.

SECTION VI
THE VERY SMALL OSCILLATIONS OF AN ARBITRARY SYSTEM OF BODIES

The differential equations describing the motion of an arbitrary system of bodies are always integrable in the case where the bodies depart by a very small amount from their positions of equilibrium. Thus one can then determine the laws of oscillation of the entire system. The general analysis of this case, which is very vast, and the solution of some of the significant related problems are the object of this section.

Subsection I
A General Solution of the Problem of Very Small Oscillations of a System of Bodies about their Points of Equilibrium

1. Let a, b, c, represent the values of the rectangular coordinates x, y, z of each body m of the assumed system about the position of equilibrium. Since it is assumed that the system, in its motion, departs very little from its configuration of equilibrium, one will have in general

$$x = a + \alpha, \qquad y = b + \beta, \qquad z = c + \gamma$$

where the variables α, β, γ are always very small quantities. Consequently, it will suffice to consider only the first order of these quantities in the differential equations of motion. The same conditions will hold for the other analogous quantities, which will be distinguished by primes of order one, two, etc. relative to the different bodies m', m'', etc. of the same system.

Let us consider at the outset the equations of condition which must derive from the nature of the system and which can be represented by $L = 0$, $M = 0$, etc. where L, M, etc. are given algebraic functions of the coordinates x, y, z, x', y', etc. Since the configuration of equilibrium is one among the many that the system could have, it results that the same equations $L = 0$, $M = 0$, etc. must hold. If it is assumed that x, y, z, x', etc. become a, b, c, a', etc. it is easy to conclude that these equations could not be a function of the time t.

Let A, B, etc. be the variables which become L, M, etc. when x, y, z, x', etc. become a, b, c, a', etc. It is clear that by substituting for x, y, z, x', etc. the values $a + \alpha$, $b + \beta$, $c + \gamma$, $a' + \alpha'$, etc. one will obtain because of the minuteness of α, β, α', etc.

$$L = A + \frac{dA}{da}\alpha + \frac{dA}{db}\beta + \frac{dA}{dc}\gamma + \frac{dA}{da'}\alpha' + \cdots$$
$$M = B + \frac{dB}{da}\alpha + \frac{dB}{db}\beta + \frac{dB}{dc}\gamma + \frac{dB}{da'}\alpha' + \cdots$$

and so on.

Therefore, 1. one has $A = 0$, $B = 0$, etc. relative to the state of equilibrium; 2. one will have the equations

$$\begin{aligned} &\frac{dA}{da}\alpha + \frac{dA}{db}\beta + \frac{dA}{dc}\gamma + \frac{dA}{da'}\alpha' + \cdots = 0 \\ &\frac{dB}{da}\alpha + \frac{dB}{db}\beta + \frac{dB}{dc}\gamma + \frac{dB}{da'}\alpha' + \cdots = 0 \\ &\vdots \end{aligned}$$

which will give the relations which must hold between the variables α, β, γ, α', etc.

By neglecting at the outset the very small quantities of the second and higher orders, linear equations will result from which the values of some of these variables can be determined in terms of the others. Then, using these approximate values, it will be possible to find more accurate expressions by taking into account the second and higher order terms, as one desires. Thus the values of some of the variables α, β, γ, α', etc. will be expressed by a function in terms of the other variables. These remaining variables will be absolutely mutually independent.

In most of the cases, by considering the conditions of the problem, the number of coordinates can be reduced immediately by substitution of rational and whole functions of other mutually independent variables which are very small quantities whose value is zero in the state of equilibrium.

Thus, in general, it will be assumed that one has

$$\begin{aligned} x &= a + a1\xi + a2\psi + a3\varphi + \cdots + a'1\xi^2 + \cdots \\ y &= b + b1\xi + b2\psi + b3\varphi + \cdots + b'1\xi^2 + \cdots \\ z &= c + c1\xi + c2\psi + c3\varphi + \cdots + c'1\xi^2 + \cdots \end{aligned}$$

and so on for the other coordinates x', y', etc. The quantities $a, b, c, a1, b1$, etc. are constants and the quantities ξ, ψ, φ, etc. are variables, very small and zero at equilibrium.

2. It is only a question of making these substitutions in the expressions for T and V of Article 10 of SECTION IV and it will suffice to take into account the second-order terms in order to obtain linear differential equations. At the outset, it is clear that the expression for T will be of the form

$$\begin{aligned} T = &\frac{(1)\,d\xi^2 + (2)\,d\psi^2 + (3)\,d\varphi^2 + \cdots}{2\,dt^2} \\ &+ \frac{(1,2)\,d\xi\,d\psi + (1,3)\,d\xi\,d\varphi + (2,3)\,d\psi\,d\varphi + \cdots}{dt^2} \end{aligned}$$

and if it is assumed in order to shorten the expressions that

$$\begin{aligned} (1) &= S(a1^2 + b1^2 + c1^2)m \\ (2) &= S(a2^2 + b2^2 + c2^2)m \end{aligned}$$

$$(3) = \mathrm{S}(a3^2 + b3^2 + c3^2)m$$
$$\vdots$$
$$(1,2) = \mathrm{S}(a1a2 + b1b2 + c1c2)m$$
$$(1,3) = \mathrm{S}(a1a3 + b1b3 + c1c3)m$$
$$(2,3) = \mathrm{S}(a2a3 + b2b3 + c2c3)m$$
$$\vdots$$

where the operator S denotes integrations or sums relative to all the various bodies m of the system and at the same time, independent of the variables ξ, ψ, φ, etc. as well as of the time t.

Then, if one denotes by F the algebraic function Π, replacing x, y, z by a, b, c, it is clear that the general value of Π, can be expressed as

$$\begin{aligned} F &+ \frac{\mathrm{d}F}{\mathrm{d}a}(a1\xi + a2\psi + a3\varphi + \cdots) \\ &+ \frac{\mathrm{d}F}{\mathrm{d}b}(b1\xi + b2\psi + b3\varphi + \cdots) \\ &+ \frac{\mathrm{d}F}{\mathrm{d}c}(c1\xi + c2\psi + c3\varphi + \cdots) \\ &+ \frac{\mathrm{d}^2F}{2\,\mathrm{d}a^2}(a1\xi + a2\psi + a3\varphi + \cdots)^2 \\ &+ \frac{\mathrm{d}^2F}{\mathrm{d}a\,\mathrm{d}b}(a1\xi + a2\psi + a3\varphi + \cdots)(b1\xi + b2\psi + b3\varphi + \cdots) \\ &+ \frac{\mathrm{d}^2F}{2\mathrm{d}b^2}(b1\xi + b2\psi + b3\varphi + \cdots)^2 \\ &\vdots \end{aligned}$$

where it suffices to consider the second order terms of ξ, ψ, φ, etc.

After multiplying this function by m and integrating according to the operator S, one will have in general

$$\begin{aligned} V &= H + H1\xi + H2\psi + H3\varphi + \cdots \\ &+ \frac{[1]\xi^2 + [2]\psi^2 + [3]\varphi^2 + \cdots}{2} \\ &+ [1,2]\xi\psi + [1,3]\xi\varphi + [2,3]\psi\varphi, \cdots + \cdots \end{aligned}$$

$$H = \mathrm{S}\,Fm$$
$$H1 = \mathrm{S}(\frac{\mathrm{d}F}{\mathrm{d}a}a1 + \frac{\mathrm{d}F}{\mathrm{d}b}b1 + \frac{\mathrm{d}F}{\mathrm{d}c}c1)m$$
$$H2 = \mathrm{S}(\frac{\mathrm{d}F}{\mathrm{d}a}a2 + \frac{\mathrm{d}F}{\mathrm{d}b}b2 + \frac{\mathrm{d}F}{\mathrm{d}c}c2)m$$
$$H3 = \mathrm{S}(\frac{\mathrm{d}F}{\mathrm{d}a}a3 + \frac{\mathrm{d}F}{\mathrm{d}b}b3 + \frac{\mathrm{d}F}{\mathrm{d}c}c3)m$$

$$\vdots$$

$$[1] = \mathrm{S}(\frac{\mathrm{d}^2F}{\mathrm{d}a^2}a1^2 + \frac{\mathrm{d}^2F}{\mathrm{d}b^2}b1^2 + \frac{\mathrm{d}^2F}{\mathrm{d}c^2}c1^2$$
$$+2\frac{\mathrm{d}^2F}{\mathrm{d}a\,\mathrm{d}b}a1b1 + 2\frac{\mathrm{d}^2F}{\mathrm{d}a\,\mathrm{d}c}a1c1 + 2\frac{\mathrm{d}^2F}{\mathrm{d}b\,\mathrm{d}c}b1c1)m$$
$$[2] = \mathrm{S}(\frac{\mathrm{d}^2F}{\mathrm{d}a^2}a2^2 + \frac{\mathrm{d}^2F}{\mathrm{d}b^2}b2^2 + \frac{\mathrm{d}^2F}{\mathrm{d}c^2}c2^2$$
$$+2\frac{\mathrm{d}^2F}{\mathrm{d}a\,\mathrm{d}b}a2b2 + 2\frac{\mathrm{d}^2F}{\mathrm{d}a\,\mathrm{d}c}a2c2 + 2\frac{\mathrm{d}^2F}{\mathrm{d}b\,\mathrm{d}c}b2c2)m$$
$$[3] = \mathrm{S}(\frac{\mathrm{d}^2F}{\mathrm{d}a^2}a3^2 + \frac{\mathrm{d}^2F}{\mathrm{d}b^2}b3^2 + \frac{\mathrm{d}^2F}{\mathrm{d}c^2}c3^2$$
$$+2\frac{\mathrm{d}^2F}{\mathrm{d}a\,\mathrm{d}b}a3b3 + 2\frac{\mathrm{d}^2F}{\mathrm{d}a\,\mathrm{d}c}a3c3 + 2\frac{\mathrm{d}^2F}{\mathrm{d}b\,\mathrm{d}c}b3c3)m$$
$$[1,2] = \mathrm{S}(\frac{\mathrm{d}^2F}{\mathrm{d}a^2}a1a2 + \frac{\mathrm{d}^2F}{\mathrm{d}b^2}b1b2 + \frac{\mathrm{d}^2F}{\mathrm{d}c^2}c1c2$$
$$+\frac{\mathrm{d}^2F}{\mathrm{d}a\,\mathrm{d}b}(a1b2 + a2b1) + \frac{\mathrm{d}^2F}{\mathrm{d}a\,\mathrm{d}c}(a1c2 + a2c1) + \frac{\mathrm{d}^2F}{\mathrm{d}b\,\mathrm{d}c}(b1c2 + b2c1))m$$
$$[1,3] = \mathrm{S}(\frac{\mathrm{d}^2F}{\mathrm{d}a^2}a1a3 + \frac{\mathrm{d}^2F}{\mathrm{d}b^2}b1b3 + \frac{\mathrm{d}^2F}{\mathrm{d}c^2}c1c3$$
$$+\frac{\mathrm{d}^2F}{\mathrm{d}a\,\mathrm{d}b}(a1b3 + a3b1 + \frac{\mathrm{d}^2F}{\mathrm{d}a\,\mathrm{d}c}(a1c3 + a3c1) + \frac{\mathrm{d}^2F}{\mathrm{d}b\,\mathrm{d}c}(b1c3 + b3c1))m$$
$$[2,3] = \mathrm{S}(\frac{\mathrm{d}^2F}{\mathrm{d}a^2}a2a3 + \frac{\mathrm{d}^2F}{\mathrm{d}b^2}b2b3 + \frac{\mathrm{d}^2F}{\mathrm{d}c^2}c2c3$$
$$+\frac{\mathrm{d}^2F}{\mathrm{d}a\,\mathrm{d}b}(a2b3 + a3b2) + \frac{\mathrm{d}^2F}{\mathrm{d}a\,\mathrm{d}c}(a2c3 + a3c2) + \frac{\mathrm{d}^2F}{\mathrm{d}b\,\mathrm{d}c}(b2c3 + b3c2))m$$

$$\vdots$$

3. Thus having the quantities T and V expressed as functions of the mutually independent variables ξ, ψ, φ, etc. there will be no equation of condition to enforce, since the quantity T contains only the differentials of the variables. Thus one will have immediately for the motion of the system, the following equations

$$\mathrm{d}\frac{\delta T}{\delta \mathrm{d}\xi} + \frac{\delta V}{\delta \xi} = 0$$
$$\mathrm{d}\frac{\delta T}{\delta \mathrm{d}\psi} + \frac{\mathrm{d}V}{\delta \psi} = 0$$
$$\mathrm{d}\frac{\delta T}{\delta \mathrm{d}\varphi} + \frac{\delta V}{\delta \varphi} = 0$$
$$\vdots$$

where it is clear that the number of equations is equal to the number of variables.

These equations must hold in the state of equilibrium, because if the system reaches this state it would remain there. But, at equilibrium, one always has $x = a$, $y = b$, $z = c$, $x' = a'$, etc. by hypothesis. Therefore, $\xi = 0$, $\psi = 0$, $\varphi = 0$, etc. and $\mathrm{d}\xi/\mathrm{d}t = 0$, $\mathrm{d}\psi/\mathrm{d}t = 0$, etc. and $\mathrm{d}^2\xi/\mathrm{d}t^2 = 0$, etc. Thus the terms $\mathrm{d}(\delta T/\delta \mathrm{d}\xi)$, $\mathrm{d}(\delta T/\delta \mathrm{d}\psi)$, etc. will be zero and the terms $\delta V/\delta\xi$, $\delta V/\delta\psi$, $\delta V/\delta\varphi$, etc. reduce to $H1$, $H2$, $H3$, etc. Consequently, one will have

$$H1 = 0, \qquad H2 = 0, \qquad H3 = 0, \qquad \cdots$$

These are the necessary conditions if a, b, c, a', etc. are to assume the values of x, y, z, x', etc. in the state of equilibrium, as one intends.

Indeed, it is obvious that

$$\mathrm{d}V = \mathrm{S}(P\,\mathrm{d}p + Q\,\mathrm{d}q + R\,\mathrm{d}r + \cdots)m$$

expresses the sum of the moments of all the forces Pm, Qm, Rm, etc. applied to all the bodies m of the system. These moments must cancel each other in the state of equilibrium. Thus from the general formula given in SECTION II of PART I, one should have $\mathrm{d}V = 0$ with respect to each of the independent variables. Consequently,

$$\frac{\delta V}{\delta\xi} = 0, \qquad \frac{\delta V}{\delta\psi} = 0, \qquad \frac{\delta V}{\delta\varphi} = 0, \qquad \cdots$$

will be the conditions of equilibrium which are assumed to correspond to $\xi = 0$, $\psi = 0$, $\varphi = 0$, etc. One will have $H1 = 0$, $H2 = 0$, $H3 = 0$, etc. so that the first orders of the variables ξ, ψ, φ, etc. in the expression for V will always disappear.

Then substituting in the general equations the expressions of T and V and putting $H1, H2, H3$, etc. equal to zero, one will have for the motion of the system

$$\begin{aligned}
0 &= (1)\frac{\delta^2\xi}{\mathrm{d}t^2} + (1,2)\frac{\mathrm{d}^2\psi}{\mathrm{d}t^2} + (1,3)\frac{\mathrm{d}^2\varphi}{\mathrm{d}t^2} + \cdots \\
&\quad + [1]\xi + [1,2]\psi + [1,3]\varphi + \cdots \\
0 &= (2)\frac{\mathrm{d}^2\psi}{\mathrm{d}t^2} + (1,2)\frac{\mathrm{d}^2\xi}{\mathrm{d}t^2} + (2,3)\frac{\mathrm{d}^2\varphi}{\mathrm{d}t^2} + \cdots \\
&\quad + [2]\psi + [1,2]\xi + [2,3]\varphi + \cdots \\
0 &= (3)\frac{\mathrm{d}^2\varphi}{\mathrm{d}t^2} + (1,3)\frac{\mathrm{d}^2\xi}{\mathrm{d}t^2} + (2,3)\frac{\mathrm{d}^2\psi}{\mathrm{d}t^2} + \cdots \\
&\quad + [3]\varphi + [1,3]\xi + [2,3]\psi + \cdots \\
&\ \ \vdots
\end{aligned}$$

Since these equations have a linear form with constant coefficients they can be easily integrated and usually by established methods.

4. At the outset, it can be assumed that the variables in these types of equations, have constant ratios between themselves, that is, that one has $\psi = f\xi, \varphi = g\xi$, etc. With these substitutions, the equations will become

$$((1) + (1,2)f + (1,3)g + \cdots)\frac{\mathrm{d}^2\xi}{\mathrm{d}t^2} + ([1] + [1,2]f + [1,3]g + \cdots)\xi = 0$$

$$((2)f + (1,2) + (2,3)g + \cdots)\frac{\mathrm{d}^2\xi}{\mathrm{d}t^2} + ([2]f + [1,2] + [2,3]g + \cdots)\xi = 0$$

$$((3)g + (1,3) + (2,3)f + \cdots)\frac{\mathrm{d}^2\xi}{\mathrm{d}t^2} + ([3]g + [1,3] + [2,3]f + \cdots)\xi = 0$$

which give $(\mathrm{d}^2\xi/\mathrm{d}t^2) + k\xi = 0$, by making

$$\begin{aligned} k &= \frac{[1] + [1,2]f + [1,3]g + \cdots}{(1) + (1,2)f + (1,3)g + \cdots} \\ &= \frac{[2]f + [1,2] + [2,3]g + \cdots}{(2)f + (1,2) + (2,3)g + \cdots} \\ &= \frac{[3]g + [1,3] + [2,3]f + \cdots}{(3)g + (2,3) + (2,3)f + \cdots} \end{aligned}$$

The number of these equations is, as one observes, equal to the number of unknowns f, g, etc., k. Consequently, the unknowns are determined exactly since by keeping for the first member the term k and multiplying it by the denominator of the second, linear equations in f, g, etc. will be obtained. One will be able to eliminate them by known methods. It is not difficult to see from the general formulas of elimination that the resulting expression for k will be of equal degree to those of the equations. Consequently, the degree will be equal to the proposed differential equations so that one will have for k a similar number of different values which when substituted in the expressions for f, g, etc. will give the corresponding values of these quantities.

Now integrating the equation $(d^2\xi/\mathrm{d}t^2) + k\xi = 0$, gives the solution

$$\xi = E\sin(\sqrt{k}\,t + \epsilon)$$

where E, ϵ, are arbitrary constants. Thus since one has assumed $\psi = f\xi, \varphi = g\xi$, etc. the expressions for ψ, φ, etc. can be calculated.

This is only a particular solution, but at the same time it is also a two-fold, three-fold, etc. solution, depending on the number of values for k. Consequently, by superimposing these terms, one will obtain the general solution since on one side of the equality sign, the sum of the particular values of ξ, ψ, φ, etc. will also satisfy the differential equations because of their linear form and on the other side, this sum will contain twice as many arbitrary constants as terms. Consequently, there will be as many as the complete integrals can admit.

Let us denote by k', k'', k''', etc. the different values of k, that is, the roots of the equation in k and by f', g', etc. f'', g'', etc. f''', g''', etc. the corresponding values of f, g, etc. By taking a similar number of arbitrary coefficients E'', E''', etc. and arbitrary angles ϵ', ϵ'', ϵ''', etc. one will have the complete expressions for ξ, ψ, φ, etc.

$$\begin{aligned}
\xi &= E' \sin(\sqrt{k'}\, t + \epsilon') + E'' \sin(\sqrt{k''}\, t + \epsilon'') + E''' \sin(\sqrt{k'''}\, t + \epsilon''') + \cdots \\
\psi &= f'E' \sin(\sqrt{k'}\, t + \epsilon') + f''E'' \sin(\sqrt{k''}\, t + \epsilon'') + f'''E''' \sin(\sqrt{k'''}\, t + \epsilon''') + \cdots \\
\varphi &= g'E' \sin(\sqrt{k'}\, t + \epsilon') + g''E'' \sin(\sqrt{k''}\, t + \epsilon'') + g'''E''' \sin(\sqrt{k'''}\, t + \epsilon''') + \cdots \\
&\vdots
\end{aligned}$$

in which the arbitrary constants E', E'', E''', etc. ϵ', ϵ'', ϵ''', etc. will depend on the values of ξ, ψ, φ, etc. and $\mathrm{d}\xi/dt$, $\mathrm{d}\psi/\mathrm{d}t$, $\mathrm{d}\varphi/dt$, etc. when t is equal to zero and the system is in its initial state.

Indeed, if in the developed expressions for ξ, ψ, φ, etc. one puts $t = 0$ and assumes as given the values of ξ, ψ, φ, etc. one will have linear equations between the unknowns $E' \sin \epsilon'$, $E'' \sin \epsilon''$, etc. from which it will be possible to determine each of the unknowns. Similarly, if one puts $t = 0$ in the differentials of the same expressions and if one regards as given the values of $\mathrm{d}\xi/\mathrm{d}t$, $\mathrm{d}\psi/\mathrm{d}t$, $\mathrm{d}\varphi/\mathrm{d}t$, etc. one will have a second system of linear equations between $E' \cos \epsilon'$, $E'' \cos \epsilon''$, etc. which will serve for their determination. From this result, one will easily obtain the values of E', E'', etc. as well as those of $\tan \epsilon'$, $\tan \epsilon''$, etc. and finally, those of the angles ϵ', ϵ'', etc.

But there is a simpler method to determine directly these unknowns and without the bother of elimination.

5. By adding together the differential equations of Article 3 after having multiplied the second equation by f, the third by g and similarly for the others, and making the following definitions in order to shorten the expressions

$$\begin{aligned}
p &= (1) + (1,2)f + (1,3)g + \cdots \\
P &= [1] + [1,2]f + [1,3]g + \cdots \\
q &= (2)f + (1,2) + (2,3)g + \cdots \\
Q &= [2]f + [1,2] + [2,3]g + \cdots \\
r &= (3)g + (1,3) + (2,3)f + \cdots \\
R &= [3]g + [1,3] + [2,3]f + \cdots \\
&\vdots
\end{aligned}$$

one has the equation

$$\begin{aligned}
0 = {} & p\frac{\mathrm{d}^2\xi}{\mathrm{d}t^2} + q\frac{\mathrm{d}^2\psi}{\mathrm{d}t^2} + r\frac{\mathrm{d}^2\varphi}{\mathrm{d}t^2} + \cdots \\
& + P\xi + Q\psi + R\varphi + \cdots
\end{aligned}$$

But the equation of Article 4 gives

$$P = kp, \qquad Q = kq, \qquad R = kr, \qquad \cdots$$

Therefore, the preceding equation will assume the form

$$0 = \frac{\mathrm{d}^2(r\xi + q\psi + r\varphi + \cdots)}{\mathrm{d}t^2} + (p\xi + q\psi + r\varphi + \cdots)k$$

which has as its integral

$$p\xi + q\psi + r\varphi + \cdots = L\sin(\sqrt{k}t + \lambda)$$

where L and λ are two arbitrary constants.

This equation must also hold for all the various values of k which we have denoted by k', k'', etc. and which result from the same equations of condition. Thus in the same fashion designating by p', p'', etc. q', q'', etc. the corresponding values of p, q, etc. and using different arbitrary constants denoted by L', L'', etc., λ', λ'', etc., one will obtain the following equations

$$\begin{aligned}
p'\xi + q'\psi + r'\varphi + \cdots &= L'\sin(\sqrt{k'}\,t + \lambda')\\
p''\xi + q''\psi + r''\varphi + \cdots &= L''\sin(\sqrt{k''}\,t + \lambda'')\\
p'''\xi + q'''\psi + r'''\varphi + \cdots &= L'''\sin(\sqrt{k'''}\,t + \lambda''')
\end{aligned}$$

These equations would equally serve to determine the values of ξ, ψ, φ, etc. and it is obvious that these values should be the same as those which we found above (Article 6) since they both result from the same differential equations. Thus by substituting the values found in the cited article in the preceding equations, they should become completely identical.

From this result, it is easy to conclude that for the first equation, one will have

$$\lambda' = \epsilon', \qquad L' = (p' + f'q' + g'r' + \cdots)E'$$

then

$$p' + f''q' + g''r' + \cdots = 0, \qquad p' + f'''q' + g'''r' + \cdots = 0, \qquad \cdots$$

and one will have similarly for the second equation

$$\lambda'' = \epsilon'', \qquad L'' = (p'' + f''q'' + g''r'' + \cdots)E'$$

then

$$p'' + f'q'' + g'r'' + \cdots = 0, \qquad p'' + f'''q'' + g'''r'' + \cdots = 0, \qquad \cdots$$

and similarly for the others.

Thus substituting in the above equations for λ', L', λ'', δ'', λ''', δ''', etc. the values which have just been found, one will obtain the following

$$E' \sin(\sqrt{k'}\, t + \epsilon') = \frac{p'\xi + q'\psi + r'\varphi + \cdots}{p' + q' f' + r' g' + \cdots}$$
$$E'' \sin(\sqrt{k''}\, t + \epsilon'') = \frac{p''\xi + q''\psi + r''\varphi + \cdots}{p'' + q'' f'' + r'' g'' + \cdots}$$
$$E''' \sin(\sqrt{k'''}\, t + \epsilon''') = \frac{p'''\xi + q'''\psi + r'''\varphi + \cdots}{p''' + q''' f''' + r''' g''' + \cdots}$$
$$\vdots$$

which are the counterparts of those of Article 4.

Now the determination of the arbitrary constants E', E'', etc., ϵ', ϵ'', ϵ''', etc. presents no additional difficulty because 1° by assuming $t = 0$, the first members of the preceding equations become $E' \sin \epsilon'$, $E'' \sin \epsilon''$, etc., and the second members are all known, assuming the values of ξ, ψ, φ, etc. given for the first instants; 2° by differentiating the same equations and then assuming $t = 0$, the first members will be $\sqrt{k'}\, E' \cos \epsilon'$, $\sqrt{k''}\, E'' \cos \epsilon''$, etc. and all the second ones will also be known, regarding as given the quantities $\mathrm{d}\xi/\mathrm{d}t$, $\mathrm{d}\psi/\mathrm{d}t$, $\mathrm{d}\varphi/\mathrm{d}t$, etc. when $t = 0$. Therefore, ...

6. Thus the solution of the problem is uniquely reduced to the determination of the quantities k, f, g, h, etc. and we have seen in Article 4 that this determination depends on the solution of the following equations

$$pk - P = 0, \qquad qk - Q = 0, \qquad rk - R = 0$$

keeping the expressions p, q, r, etc., P, Q, R, etc. of Article 5.

Now if one denotes by A what is to become the quantity T by changing in it $\mathrm{d}\xi/\mathrm{d}t$, $\mathrm{d}\psi/\mathrm{d}t$, $\mathrm{d}\varphi/\mathrm{d}t$, etc. to e, f, g, etc. and by P what is to become the part of the quantity V where the variables ξ, ψ, φ, etc. form together two dimensions, similarly changing these variables to e, f, g, etc., it is easy to see and one could even convince oneself **a priori** that the following equations will be obtained

$$p = \frac{\mathrm{d}A}{\mathrm{d}e}, \qquad q = \frac{\mathrm{d}A}{\mathrm{d}f}, \qquad r = \frac{\mathrm{d}A}{\mathrm{d}g}, \qquad \cdots$$
$$P = \frac{\mathrm{d}B}{\mathrm{d}e}, \qquad Q = \frac{\mathrm{d}B}{\mathrm{d}f}, \qquad R = \frac{\mathrm{d}B}{\mathrm{d}g}, \qquad \cdots$$

after making e equal to one.

Therefore, in general, if one puts $Ak - B = K$, the equations for the determination of the unknowns k, f, g, etc. will be

$$\frac{\mathrm{d}K}{\mathrm{d}e} = 0, \qquad \frac{\mathrm{d}K}{\mathrm{d}f} = 0, \qquad \frac{\mathrm{d}K}{\mathrm{d}g} = 0, \qquad \cdots$$

assuming e to be equal to one. Thus since the quantity K is immediately formed from the quantities T and V, it will be possible to find directly the equations in question without having need to deduce them from the differential equations of the motion of the system.

I now note that since K is a homogeneous function of two dimensions of e, f, g, etc., one will have from the property of these types of functions which is demonstrated in Article 15 of SECTION IV

$$2K = e\frac{dK}{de} + f\frac{dK}{df} + g\frac{dK}{dg} + \cdots$$

Thus one will also obtain K is equal to zero. Consequently, the unknowns f, g, h, etc. must be such that not only the quantity K is zero but also that each of its differentials relative to these unknowns are also zero. From which it follows that the quantity K viewed as a function of these unknowns, dependent on the equation $K = 0$, shall be a maximum or a minimum.

If at the outset one puts e equal to one and then replaces with $K = 0$ the equation $dK/de = 0$, one will obtain, from the determination of the unknowns f, g, h, etc., the equations

$$K = 0, \qquad \frac{dK}{df} = 0, \qquad \frac{dK}{dg} = 0, \qquad \cdots$$

Therefore, if the value of f is calculated first from the equation $dK/df = 0$, and then substituted in the equation $K = 0$, one changes this equation to $K' = 0$. Taking the derivative with respect to g and forming the equation $dK'/dg = 0$, and similarly, substituting the value of g obtained from this last equation again in $K' = 0$, now denoting the resulting equation by $K'' = 0$, formulate again the equation $dK''/dh = 0$ and so on. By this means one will be able to obtain a final equation which will not contain the unknowns f, g, h, etc., but only the quantity k and which will be the sought after equation k whose roots have been called k', k'', k''', etc.

This equation can even be reduced to a general formula by considering that since the quantities f, g, h, etc. compose in the value of K only two orders, the quantity $(2K\,d^2K - dK^2)/df^2$ will necessarily be independent of f, its differential relative to f is $2K\,d^3K/df^2$ and consequently, zero. Such that it is possible to put $K' = (2K\,d^2K - dK^2)/df^2$ and since in this quantity K' the remaining unknowns g, h, etc., only include the second order, one could similarly put $K'' = (2K'\,d^2K' - dK'^2)/dg^2$ and so on. The last of the quantities K, K', K'', etc., all equated to zero, will be the desired equation in k. It is true that this equation can go to a degree higher than needed because of the extraneous factors introduced in the equations K'', K''', etc., but if by developing these equations one takes care to eliminate successively these factors and then to only take for the values of K'', etc. their first factors, thus simplified, the final equation will be reduced by itself to the form and to a degree for which it is applicable.

With respect to the values of f, g, etc., one determines them in succession from the equations $\mathrm{d}K/\mathrm{d}f = 0$, $\mathrm{d}K'/\mathrm{d}g = 0$, etc. starting with the last one and working backward to the first by successive substitution of the determined values.

7. Since the preceding solution is founded on the assumption that the variables ξ, ψ, φ, etc. are very small, one must, for it to be legitimate that this assumption hold effectively, require that the roots k', k'', etc. are all real, positive and unequal, such that the time t, which increases to infinity, is always contained in the sine and cosine functions. If some of these roots were to become negative or imaginary, they would introduce in the corresponding sine or cosine functions, real exponentials, and if they were to become simply equals, they would introduce algebraic powers of the arc. This can be shown to be true by known methods by replacing in the first case, the sine or cosine functions with the equivalent imaginary exponential expressions and by assuming in the second case that the equal roots differ by indeterminate infinitesimal quantities. But since the development of these cases would serve no useful purpose for the present investigation, we will not elaborate on it.

If the condition which requires the coefficients of t to be real and unequal numbers holds, it is obvious that the largest values of ξ, η, etc. will be less than the sums of the quantities E', E'', E''', etc. and the quantities $f'E', f''E'', f'''E'''$, etc. assuming all these quantities are positive. Consequently, if these different sums are very small, it will be certain that the values of the variables will always be small also.

But since the coefficients E', E'', E''', etc. are arbitrary and depend uniquely on the initial displacement of the system, it is possible that the variables ξ, ψ, etc. will remain very small, yet among the quantities $\sqrt{k'}$, $\sqrt{k''}$, etc., some would be imaginary or equal because it is sufficient for this result that the corresponding quantities E', E'', etc. are zero, which will make the terms which increase with time t disappear. Then, the solution without being exact or general will yet in the particular case be so when the preceding condition prevails.

8. There are methods to determine if a given equation, whatever its degree, has all real roots or not, and to judge, in the case where the roots are real, of their sign and inequality. But the application of these methods is always a little difficult. Here are some simple and general rules for a large number of cases which will be used to determine the form of the roots in question.

Beginning with the equation $K = 0$ or $Ak - B = 0$ (Article 6), one has $k = B/A$. Now it is easy to convince oneself that the quantity A always has necessarily a positive value, as long as f, g, etc., are real quantities. Since the function T, from which it results by changing $\mathrm{d}\xi/\mathrm{d}t$, $\mathrm{d}\psi/\mathrm{d}t$, $\mathrm{d}\varphi/\mathrm{d}t$, etc. to $1, f, g$, etc. (article cited), is composed of the sum of several squares multiplied by coefficients which are necessarily positive. Thus if the quantity B is also always positive, which holds when the portion of the function V, where the variables ξ, ψ, φ, etc. together form two dimensions, is reducible to the same form as the function T, because the quantity B also results from this part of V, after changing ξ, ψ, φ, etc. by $1, f, g$, etc., it is certain that the values of k, that is, the roots of the equation for k, will always be positive whenever they are real.

On the contrary, if the quantity B is always negative which will be the same when it is composed of several squared quantities multiplied by negative coefficients, the real values of K will always be negative. In the latter case, the solution can not be the applicable one since the roots of the equation for k can only be imaginary or negative real numbers, the expressions of the variables ξ, ψ, etc. will necessarily contain the time t outside of the sine and cosine functions.

In the first case, where B is positive, it is obvious that if the roots are real, they are necessarily positive and although it may perhaps be difficult to demonstrate directly that they must all be real, one can convince oneself by other means that they must be so.

Since the principle of the Conservation des Forces Vives, which we have demonstrated in Subsection V of SECTION III, gives the equation $T + V$ = constant (Article 14, preceding section), which always holds because T and V are functions independent of t (Article 2). Now if one denotes by V' the part of V which contains the terms of two dimensions so that $V = H + V'$, because $H1 = 0, H2 = 0, H3 = 0$, etc. (Article 3), one will obtain

$$T + H + V' = \text{const.} = (T) + H + (V')$$

where (T) and (V') denote the initial values of T and V'. Therefore, $T + V'$ will be equal to $(T) + (V')$.

Since T is by its form always a positive quantity, if V' is also positive, one will necessarily have V' greater than zero and $(T) + (V')$ greater than zero so that the value of V' and also those of the variables ξ, ψ, φ, etc. will be contained in the given boundaries and only dependent on the initial state. These variables will not be able to contain the time t outside the sine and cosine functions, because then they could increase infinitely. When the value of B is constantly positive, the value of V' is also always positive. Consequently, the roots of the equation for k will necessarily all be real, positive and unequal (Article 7) and the solution will always be acceptable.

In this case, the state of equilibrium from which the system has been displaced will be stable since the system will always return to it or will always have the tendency to return after some very small oscillations. At least, it will only be able to depart from it by a very small amount.

9. It is in this fashion that we demonstrated at the end of SECTION III on statics (Articles 23 and so on) that when the function Π is a minimum in the state of equilibrium, this state is stable because it is easy to see that the function denoted by Π in Article 21 of the cited section is the same as the one we represent here by V, since both are the integral of the total moments of the forces acting on the various bodies of the system, the sum of which must be zero at equilibrium. And since one has $V = H + V'$, and V' contains only the variables ξ, ψ, φ etc. to only the second dimension, it results that V will be a minimum or a maximum, depending on whether the value of V' will be positive or negative, if these variables are given arbitrary values. Thus the equilibrium will necessarily be stable for the case where V (preceding article) is a minimum.

On the contrary, for the case in which V is a maximum, since the quantity V' is always negative, the quantity B will also be negative because by putting $\psi = f\xi$, $\varphi = g\xi$, etc. the value of V' becomes $\xi^2 B$ (Article 6) and from what we have demonstrated in the preceding article, the expressions of the variables will necessarily contain terms where t will be outside the sine and cosine functions. Thus the equilibrium will not be stable because if the system is displaced by however small an amount, it will always depart more from it. The second part of this theorem given in the cited section on statics could not have been demonstrated earlier because of the lack of the necessary principles. We delayed the demonstration until we reached dynamics and the one we just gave leaves nothing to be desired.

10. Also, between the two states of absolute stability and instability, in which the system departs from equilibrium by an arbitrarily small amount, the system will tend to either return to equilibrium by itself or to depart more and more from the equilibrium state. There can be states of conditional and relative stability. In these states, the return to equilibrium will depend on the initial displacement of the system. Because, if some of the values of $\sqrt{k}$ are imaginary, the corresponding terms of the values of the variables will contain segments of a circle and the equilibrium will not be stable in general. But, if the coefficients of these terms become zero, which depends on the initial state of the system, the segments of the circle will disappear and the equilibrium can then be viewed as stable, at least with respect to this particular state.

11. When all the values of $\sqrt{k}$ are real and unequal numbers and consequently, the equilibrium is stable, the expressions of all the variables will be composed of as many terms of the form

$$E \sin(\sqrt{k}t + \epsilon)$$

as there are variables.

This term represents the very small and isochronous oscillations of a simple pendulum for which the length is g/k, where g is the force of gravity. Therefore, the oscillations of the different bodies of the system can be viewed as composed of simple oscillations analogous to those of pendulums for which the lengths would be g/k', g/k'', g/k''', etc.

Since the coefficients E', E'', etc. are arbitrary and depend only on the initial state of the system, it can always be assumed for this state that all the coefficients, except one arbitrary coefficient, are zero. Then, all the bodies of the system will make simple oscillations, analogous to those of a common pendulum and one observes that the same system is capable of as many simple oscillations as there are mobile bodies. Therefore, in general, the arbitrary oscillations of a system will only be composed of all the simple oscillations which could take place due to the nature of the system.

Daniel Bernoulli observed the composition of simple and isochronous oscillations in the motion of a vibrating string loaded with several small weights and he viewed it as representative of a general law of all the small and reciprocal motions which can exist in an arbitrary system of bodies. A single case, such as the one of the vibrating string, does not suffice to establish such a law, but the analysis which we just gave does establish this law in a certain and general fashion and shows that however irregular the small oscillations which are observed in nature may seem, they can always be reduced to simple oscillations for which the number will be equal to those of the oscillating bodies in the same system.

This is a consequence of the nature of linear equations to which the motions of the bodies which compose an arbitrary system are reduced when these motions are very small.

12. If the values of the quantities $\sqrt{k'}$, $\sqrt{k''}$, $\sqrt{k'''}$, etc. are incommensurable, it is clear that the period of these oscillations will also be incommensurable and consequently, the system will never be able to regain its initial position.

If these quantities are to one another as number to number, and their largest common measure is μ, it is easily seen that the system will always return to the same position at the end of the time $\theta = 2\pi/\mu$ where π is an angle of 180 degrees. Thus θ will be the period of the composed oscillation of the entire system.

13. The solution which we just gave requires that the coordinates can be expressed by a series expansion of functions of very small variables which are zero in the state of equilibrium as we have assumed in Article 3.

This is always possible as we have observed when the equations of condition, reduced to a series, contain the first orders of the variables which are assumed very small, because these terms give at the outset equations whose solutions are rational and then the method of series can always be used to obtain increasingly exact rational solutions.

Yet it can happen that the terms of the first dimension are missing in one or several of the equations of condition, which will happen for example, if in the equation $L = 0$, the values of the coordinates for equilibrium are such that they make not only L equal to zero, but also each of its first differentials, because then one will have $\mathrm{d}A/\mathrm{d}a = 0$, $\mathrm{d}A/\mathrm{d}b = 0$, etc. and the equation $L = 0$ will not contain the second and higher powers of α, β, γ, α', etc. (Article 1). In this case, if the coordinates are reduced to functions of the independent variables, these functions could no longer be rational and the differential equations will be neither linear nor even rational. Therefore, the assumption of very small displacements of the system will not be used to simplify the solution of the problem or rather will not make it susceptible to the general methods which we have presented.

To resolve these types of questions in the most simple manner, the equations of condition will be neglected at the outset where the first dimensions of the variables will not be found. Thus one will be able to obtain expressions for T and V in the same form as in Article 2. Then, it will be necessary to add to this value of V the first members of the equations of condition which have not yet been considered, each multiplied by an indeterminate coefficient which is assumed constant in the differentiation by δ. It will suffice in the terms resulting from the equations of condition to take into account the lowest dimensions of these very small variables. From these terms the differential equations will be found in the ordinary fashion and it will be necessary to eliminate from these equations the indeterminate coefficients.

If these equations of condition were of the second degree and if the indeterminate coefficients could be assumed constant, the value of V would still be of the same form as in the general solution. Consequently, it could also be applied to this case. The coefficients would then be determined in such a manner that the equations of condition are satisfied. Thus it will always be possible to begin by adopting this assumption. The next step is to determine if the values which are obtained for the variables can satisfy the equations of condition, in which case, the assumption would be legitimate and the solution exact. Otherwise, an attempt should be made to integrate the differential equations by particular methods.

Subsection II
The Oscillations of a Linear System of Bodies

14. When the bodies which compose the proposed system are arranged with respect to each other in a uniform and regular manner, the calculations can be simplified and general and symmetrical formulas can be found by using the notation and algorithms of finite differences. We will give an example by investigating the case where an arbitrary number of bodies lying on a straight or curved line oscillate because of arbitrary applied forces combined with their mutual interaction.

Let x, y, z be the rectangular coordinates of one of the bodies of the system, which we will denote by $\mathrm{D}m$, where the upper case letter D denotes the finite difference (SECTION IV, Article 17). One will obtain at the outset

$$T = \mathrm{S}\frac{\mathrm{d}x^2 + \mathrm{d}y^2 + \mathrm{d}z^2}{2\,\mathrm{d}t^2}\mathrm{D}m$$

where the operator S represents the sums relative to the entire system.

The function V must contain the sum $\mathrm{S}\,\Pi\,\mathrm{D}m$ resulting from the accelerating forces P, Q, R, etc., which are assumed in such a fashion that one has

$$\Pi = \int (P\,\mathrm{d}p + Q\,\mathrm{d}q + R\,\mathrm{d}r + \cdots)$$

This function must also contain the sum $\mathrm{S}\int \Phi\,\mathrm{dD}s$, assuming that Φ is the force with which two neighboring bodies separated by a distance $\mathrm{D}s$ attract one another and that this force is a function of the same distance $\mathrm{D}s$ such that $\mathrm{S}\int \Phi\,\mathrm{dD}s$ is an integrable quantity for which differentiation by δ is $\Phi\,\delta\mathrm{D}s$. This force Φ, which we assume to be a function of $\mathrm{D}s$ can vary from one body to another and consequently, it will also be a function of the number or of the quantity which represents the location of each body in the series of all the bodies, and to which is referred the summing operator S. If the bodies, instead of attracting each other were repulsing one another, the quantity Φ should be taken negative.

Thus there results $V = \mathrm{S}\Pi\,\mathrm{D}m + \mathrm{S}\int \Phi\,\mathrm{dD}s$ and consequently

$$\delta V = \mathrm{S}\,\delta\Pi\,\mathrm{D}m + \mathrm{S}\Phi\,\delta\mathrm{D}s$$

It is worthwhile to note that this expression for δV would be the same if the bodies were joined together such that their respective distances were invariant. One would have in this case the equation of condition $\delta\mathrm{D}s = 0$, which would give in the expression for δV the term $\mathrm{S}\,\lambda\,\delta\mathrm{d}S$ (article cited).

15. By expressing the element $\mathrm{D}s$ by the finite differences of x, y, z, it is clear that one will have

$$Ds = \sqrt{\mathrm{D}x^2 + \mathrm{D}y^2 + \mathrm{D}z^2}$$

then after differentiating by δ there results

$$\delta\mathrm{D}s = \frac{\mathrm{D}x\,\delta\mathrm{D}x + \mathrm{D}y\,\delta\mathrm{D}y + \mathrm{D}z\,\delta\mathrm{D}z}{\mathrm{D}s}$$

After substituting this expression and by defining $\Phi/\mathrm{D}s = \Psi$, a function of $\mathrm{D}s$ in order to shorten the expression, the following equation results

$$\delta V = \mathrm{S}\,\delta\Pi\,\mathrm{D}m + \mathrm{S}\,\Psi(\mathrm{D}x\,\delta\mathrm{D}x + \mathrm{D}y\,\delta\mathrm{D}y + \mathrm{D}z\,\delta\mathrm{D}z)$$

Since the operators D and δ are independent of one another, one can substitute $\delta\mathrm{D}$ for $\mathrm{D}\delta$ to obtain

$$\delta V = \mathrm{S}\,\delta\Pi\,\mathrm{D}m + \mathrm{S}\,\Psi(\mathrm{D}x\,\mathrm{D}\delta x + \mathrm{D}y\,\mathrm{D}\delta y + \mathrm{D}z\,\mathrm{D}\delta z)$$

Also, the operator D before the operator δ can be eliminated by integrating the finite difference expressions by parts.

16. In effect, one has in general

$$\mathrm{D}(xy) = x\,\mathrm{D}y + y\,\mathrm{D}x + \mathrm{D}x\,\mathrm{D}y = (x + \mathrm{D}x)\mathrm{D}y + y\,\mathrm{D}x = x_{\prime}\,\mathrm{D}y + y\,\mathrm{D}x$$

denoting by $x_{\prime}$, the term which follows x in the series of consecutive terms x, $x + \mathrm{D}x$, etc. Therefore, by changing from differences to sums, one will obtain

$$\mathrm{S}\,y\,\mathrm{D}x = xy - \mathrm{S}\,x_{\prime}\,\mathrm{D}y$$

In the same fashion, it is possible to obtain $\mathrm{S}\,\mathbf{y}\,\mathrm{D}^2 x = y\,\mathrm{D}x - x_{\prime}\,\mathrm{D}y + \mathrm{S}\,x_{\prime\prime}\,\mathrm{D}^2 y$ and similarly x, $x_{\prime}$, $x_{\prime\prime}$, etc. are the terms which follow each other in the same series.

In order to complete these summations, the terms outside the operator S should be referred to the endpoint of the finite integral $\mathrm{S}\,y\,\mathrm{D}x$ and the same terms referred to the first point should be subtracted from it. Thus using the subscripts zero and i to represent the first and last point, one will have these complete summations

$$\mathrm{S}\,y\,\mathrm{D}x = x_i y_i - x_0 y_0 - \mathrm{S}\,x_{\prime}\,\mathrm{D}y$$
$$\mathrm{S}\,y\,\mathrm{D}^2 x = y_i\,\mathrm{D}x_i - x_{i+1}\,\mathrm{D}y_i$$
$$-y_0\,\mathrm{D}x_0 + x_{\prime}\,\mathrm{D}y_0 + \mathrm{S}\,\mathbf{x}_{\prime\prime}\,\mathrm{D}y$$
$$\vdots$$

When the operator S indicates the total sum of a number of given terms, it is clear that one can, in place of the terms $x_{\prime}\,\mathrm{D}y$, $x_{\prime\prime}\,\mathrm{D}y$ under the operator S, take the preceding terms that are denoted by $x\,\mathrm{D}_{\prime}y$, $x\,\mathrm{D}_{\prime\prime}y$, etc. by marks of one prime, of two, etc. located at the left, the terms ${}_{\prime\prime}y$, ${}_{\prime}y$ which are before y in the infinite series ${}_{\prime\prime}y$, ${}_{\prime}y$, y, $y_{\prime}$, $y_{\prime\prime}$, etc.

17. Granting the preceding development, let us substitute δx in place of x in the preceding formulas and $\Psi\,\mathrm{D}x$ in place of y to obtain the following transformations

$$\mathrm{S}\,\Psi\,\mathrm{D}x\,\mathrm{D}\delta x = (\Psi\,\mathrm{D}x\,\delta x)_i - (\Psi\,\mathrm{D}x\,\delta x)_0 - \mathrm{S}\,\delta x\,\mathrm{D}_{\prime}(\Psi\,\mathrm{D}x)$$

and similarly

$$\mathrm{S}\,\Psi\,\mathrm{D}y\,\mathrm{D}\delta y = (\Psi\,\mathrm{D}y\,\delta y)_i - (\Psi\,\mathrm{D}y\,\delta y)_0 - \mathrm{S}\,\delta y\,\mathrm{D}_{\prime}(\Psi\,\mathrm{D}y)$$
$$\mathrm{S}\,\Psi\,\mathrm{D}z\,\mathrm{D}\delta z = (\Psi\,Dz\delta z)_i - (\Psi\,\mathrm{D}z\,\delta z)_0 - \mathrm{S}\,\delta z\,\mathrm{D}_{\prime}(\Psi\,\mathrm{D}z)$$

and one will make these substitutions in the expression for δV.

If the first and last body are assumed fixed, the variations δx, δy, δz and δx_i, δy_i, δz_i, which are related, will be zero. At the outset, we will adopt this assumption which simplifies the formulas and consequently, we will have

$$\delta V = \mathrm{S}\,\delta\Pi\,\mathrm{D}m - \mathrm{S}\,\delta x\,\mathrm{D}_{\prime}(\Psi\,\mathrm{D}x) - \mathrm{S}\,\delta y\,\mathrm{D}_{\prime}(\Psi\,\mathrm{D}y) - \mathrm{S}\,\delta z\,\mathrm{D}_{\prime}(\Psi\,\mathrm{D}z)$$

In general, since the variations must always vanish, if the first or last body or both were not fixed, one should assume the value of Ψ is equal to zero at the beginning or at the end. One would thus have, since $\Psi = \Phi/\mathrm{D}s$, the condition to fulfill which is that $\Phi_0 = 0$ or $\Phi_i = 0$, if the first or last body is assumed mobile. If both were mobile, one would have the two conditions $\Phi_0 = 0$ and $\Phi_i = 0$.

18. After the variation δV is reduced to this simple formula, the general equations of Article 10 of SECTION IV as functions of the variables x, y, z for each of the bodies

of the system, will give for these variables the three following equations, in which Φ is reintroduced in place of Ψ Ds

$$\frac{\mathrm{d}^2x}{\mathrm{d}t^2}\mathrm{D}m + \frac{\delta\Pi}{\delta x}\mathrm{D}m - D_{\prime}(\frac{\Phi\,\mathrm{D}x}{\mathrm{D}s}) = 0$$
$$\frac{\mathrm{d}^2y}{\mathrm{d}t^2}\mathrm{D}m + \frac{\delta\Pi}{\delta y}\mathrm{D}m - D_{\prime}(\frac{\Phi\,\mathrm{D}y}{\mathrm{D}s}) = 0$$
$$\frac{\mathrm{d}^2z}{\mathrm{d}t^2}\mathrm{D}m + \frac{\delta\Pi}{\delta z}\mathrm{D}m - \mathrm{D}_{\prime}(\frac{\Phi\,\mathrm{D}z}{\mathrm{D}s}) = 0$$

These equations are accurate, whatever the motion of the bodies. When the motions are very small, the equations are simplified and become linear, as we saw earlier (Subsection I).

19. Let us assume that in the state of equilibrium the system coordinates x, y, z become a, b, c and that they appear in the displacement $a + \xi$, $b + \eta$, $c + \zeta$, where the quantities ξ, η, ζ are very small. The function Π will become

$$\Pi + \frac{\mathrm{d}\Pi}{\mathrm{d}a}\xi + \frac{\mathrm{d}\Pi}{\mathrm{d}b}\eta + \frac{\mathrm{d}\Pi}{\mathrm{d}c}\zeta$$

Thus by treating Π henceforth as a simple function of a, b, c, the three partial differences $\delta\Pi/\delta x$, $\delta\Pi/\delta y$, $\delta\Pi/\delta z$ can be expressed as

$$\frac{\mathrm{d}\Pi}{\mathrm{d}a} + (\frac{\mathrm{d}^2\Pi}{\mathrm{d}a^2}\xi + \frac{\mathrm{d}^2\Pi}{\mathrm{d}a\,\mathrm{d}b}\eta + \frac{\mathrm{d}^2\Pi}{\mathrm{d}a\,\mathrm{d}c}\zeta)$$
$$\frac{\mathrm{d}\Pi}{\mathrm{d}b} + (\frac{\mathrm{d}^2\Pi}{\mathrm{d}a\,\mathrm{d}b}\xi + \frac{\mathrm{d}^2\Pi}{\mathrm{d}b^2}\eta + \frac{\mathrm{d}^2\Pi}{\mathrm{d}b\,\mathrm{d}c}\zeta)$$
$$\frac{\mathrm{d}\Pi}{\mathrm{d}c} + (\frac{\mathrm{d}^2\Pi}{\mathrm{d}a\,\mathrm{d}c}\xi + \frac{\mathrm{d}^2\Pi}{\mathrm{d}b\,\mathrm{d}c}\eta + \frac{\mathrm{d}^2\Pi}{\mathrm{d}c^2}\zeta)$$

With the same substitutions of $a + \xi$, $b + \eta$, $c + \zeta$, instead of x, y, z, the differences Dx, Dy, Dz will become Da + Dξ, Db + Dη, Dc + Dζ. With respect to the quantity Φ, which is assumed a function of Ds, if the following definition is made in order to shorten the expression $\mathrm{D}f = \sqrt{\mathrm{D}a^2 + \mathrm{D}b^2 + \mathrm{D}c^2}$, one will immediately obtain

$$\mathrm{D}s = \mathrm{D}f + \frac{\mathrm{D}a}{\mathrm{D}f}\mathrm{D}\xi + \frac{\mathrm{D}b}{\mathrm{D}f}\mathrm{D}\eta + \frac{\mathrm{D}c}{\mathrm{D}f}\mathrm{D}\zeta$$

then if F denotes what will become the function Φ after changing Ds to Df and in addition, if the following definition is made

$$\frac{\mathrm{d}F}{\mathrm{d}(\mathrm{D}f)} = \frac{F'}{\mathrm{D}f}$$

the development will lead to

$$\Phi = F + F'(\frac{\mathrm{D}a}{\mathrm{D}f}\frac{\mathrm{D}\xi}{\mathrm{D}f} + \frac{\mathrm{D}b}{\mathrm{D}f}\frac{\mathrm{D}\eta}{\mathrm{D}f} + \frac{\mathrm{D}c}{\mathrm{D}f}\frac{\mathrm{D}\zeta}{\mathrm{D}f})$$

and as a consequence

$$\frac{\Phi}{Ds} = \frac{F}{Df} + \frac{F' - F}{Df}(\frac{Da}{Df}\frac{D\xi}{Df} + \frac{Db}{Df}\frac{d\eta}{Df} + \frac{Dc}{Df}\frac{D\zeta}{Df})$$

20. Now make these substitutions in the three equations found above and since in the state of equilibrium the variables ξ, η, ζ are assumed zero, these equations will verify this assumption. Thus the constant terms will mutually cancel which will give at the outset the three equations of condition

$$\frac{d\Pi}{da}Dm - D_{\prime}(\frac{F\,Da}{Df}) = 0$$
$$\frac{d\Pi}{db}Dm - D_{\prime}(\frac{F\,Db}{Df}) = 0$$
$$\frac{d\Pi}{dc}Dm - D_{\prime}(\frac{F\,Dc}{Df}) = 0$$

These equations will give the values that the coordinates a, b, c, must have in the state of equilibrium. It is obvious that they represent in a general fashion the equations which were found in SECTION V of PART I for the equilibrium of several bodies joined by an extensible or inextensible string.

21. After the following definitions are made

$$G = F - F'$$
$$a' = \frac{Da}{Df}, \qquad b' = \frac{Db}{Df}, \qquad c' = \frac{Dc}{Df}$$

the following three equations between the variables ξ, η, ζ and t will be obtained

$$\frac{D^2\xi}{dt^2}Dm + (\frac{d^2\Pi}{da^2}\xi + \frac{d^2\Pi}{da\,db}\eta + \frac{d^2\Pi}{da\,dc}\zeta)Dm$$
$$-D_{\prime}\left[\frac{F\,D\xi}{Df} - Ga'(\frac{a'\,D\xi}{Df} + \frac{b'\,D\eta}{Df} + \frac{c'\,D\zeta}{Df})\right] = 0$$
$$\frac{d^2\eta}{dt^2}Dm + (\frac{d^2\Pi}{da\,db}\xi + \frac{d^2\Pi}{db^2}\eta + \frac{d^2\Pi}{db\,dc}\zeta)Dm$$
$$-D_{\prime}\left[\frac{F\,D\eta}{Df} - Gb'(\frac{a'\,D\xi}{Df} + \frac{b'\,D\eta}{Df} + \frac{c'\,D\zeta}{Df})\right] = 0$$
$$\frac{d^2\zeta}{dt^2}Dm + (\frac{d^2\Pi}{da\,dc}\xi + \frac{d^2\Pi}{db\,dc}\eta + \frac{d^2\Pi}{dc^2}\zeta)Dm$$
$$-D_{\prime}\left[\frac{F\,D\zeta}{Df} - Gc'(\frac{a'\,D\xi}{Df} + \frac{b'\,D\eta}{Df} + \frac{c'\,D\zeta}{Df})\right] = 0$$

These are the equations which will be used to determine the oscillations of the system which are assumed to be very small. They are of the type commonly called "of finite differences" and "infinitely small". Since they have constant coefficients, they can be treated according to the general method discussed in the preceding paragraph.

22. The equations of Article 20, which contain the conditions of equilibrium, produce by passing from differences to sums

$$\frac{F\,\mathrm{D}a}{\mathrm{D}f} = \mathrm{S}\frac{\mathrm{d}\Pi}{\mathrm{d}a}\mathrm{D}m + A$$
$$\frac{F\,\mathrm{D}b}{\mathrm{D}f} = \mathrm{S}\frac{\mathrm{d}\Pi}{\mathrm{d}b}\mathrm{D}m + B$$
$$\frac{F\,\mathrm{D}c}{\mathrm{D}f} = \mathrm{S}\frac{\mathrm{d}\Pi}{\mathrm{d}c}\mathrm{D}m + C$$

where A, B, C are three arbitrary constants. From the preceding equations, one has immediately

$$F = \sqrt{(\mathrm{S}\frac{\mathrm{d}\Pi}{\mathrm{d}a}\mathrm{D}m + A)^2 + (\mathrm{S}\frac{\mathrm{d}\Pi}{\mathrm{d}b}\mathrm{D}m + B)^2 + (\mathrm{S}\frac{\mathrm{d}\Pi}{\mathrm{d}c}\mathrm{D}m + C)^2}$$

When the quantity F is a given function of $\mathrm{D}f$, which holds when it is assumed that the bodies are mutually attracted or repulsed by a force Φ which is a function of the distances $\mathrm{D}s$, the preceding expression for F will give the value of $\mathrm{D}f$ which must hold in the state of equilibrium.

But when the distances $\mathrm{D}s$ are assumed given and invariant, the quantity Φ, which consists of the multiplier λ (Article 14), is unknown and must be determined from the preceding formula. In this case, one has $\mathrm{D}s = \mathrm{D}f$ and consequently (Article 19)

$$\frac{\mathrm{D}a}{\mathrm{D}f}\mathrm{D}\xi + \frac{\mathrm{D}b}{\mathrm{D}f}\mathrm{D}\eta + \frac{\mathrm{D}c}{\mathrm{D}f}\mathrm{D}\zeta = 0$$

which simplifies the equations of the preceding article.

23. The essence of the method of Article 4 consists of assuming that each variable is expressed by the same function of t, each multiplied by a different quantity.

If this function is designated by θ and in addition, if the following definitions are introduced

$$\xi = \theta X, \qquad \eta = \theta Y, \qquad \zeta = \theta Z$$

after having substituted these values in the equations of Article 21, it will be easy to see that in order to verify these equations, it is necessary to determine the variable θ from an equation of the form $\mathrm{d}^2\theta/\mathrm{d}t^2 + k\theta = 0$ because by replacing $\mathrm{d}^2\theta/\mathrm{d}t^2$ by its value $-k\theta$ and dividing all the terms by θ, the following three finite difference equations will be obtained

$$kX\,\mathrm{D}m = (\frac{\mathrm{d}^2\Pi}{\mathrm{d}a^2}X + \frac{\mathrm{d}^2\Pi}{\mathrm{d}a\,\mathrm{d}b}Y + \frac{\mathrm{d}^2\Pi}{\mathrm{d}a\,\mathrm{d}c}Z)\mathrm{D}m$$
$$-\mathrm{D}_{\prime}\left[\frac{F\,\mathrm{D}X}{\mathrm{D}f} - Ga'(\frac{a'\,\mathrm{D}X}{\mathrm{D}f} + \frac{b'\,\mathrm{D}Y}{\mathrm{D}f} + \frac{c'\,\mathrm{D}Z}{\mathrm{D}f})\right]$$

$$kY\,\mathrm{D}m = (\frac{\mathrm{d}^2\Pi}{\mathrm{d}a\,\mathrm{d}b}X + \frac{\mathrm{d}^2\Pi}{\mathrm{d}b^2}Y + \frac{\mathrm{d}^2\Pi}{\mathrm{d}b\,\mathrm{d}c}Z)\mathrm{D}m$$
$$-\mathrm{D}_{\prime}\left[\frac{F\,\mathrm{D}Y}{\mathrm{D}f} - Gb'(\frac{a'\,\mathrm{D}X}{\mathrm{D}f} + \frac{b'\,\mathrm{D}Y}{\mathrm{D}f} + \frac{c'\,\mathrm{D}Z}{\mathrm{D}f})\right] = 0$$
$$kZ\,\mathrm{D}m = (\frac{\mathrm{d}^2\Pi}{\mathrm{d}a\,\mathrm{d}c}X + \frac{\mathrm{d}^2\Pi}{\mathrm{d}b\,\mathrm{d}c}Y + \frac{\mathrm{d}^2\Pi}{\mathrm{d}c^2}Z)\mathrm{D}m$$
$$-\mathrm{D}_{\prime}\left[\frac{F\,\mathrm{D}Z}{\mathrm{D}f} - Gc'(\frac{a'\,\mathrm{D}X}{\mathrm{D}f} + \frac{b'\,\mathrm{D}Y}{\mathrm{D}f} + \frac{c'\,\mathrm{D}Z}{\mathrm{D}f})\right] = 0$$

24. The equation for θ is easily integrated. It gives

$$\theta = E\sin(\sqrt{k}\,t + \epsilon)$$

where E and ϵ are two arbitrary constants.

With respect to the equations which are functions of X, Y, Z, they are in general only integrable in finite terms by known methods when they have constant coefficients. But if the finite differences indicated by D are developed, they assume the following form (Article 16)

$$AX_{\prime} + BY_{\prime} + CZ_{\prime} + A'X + B'Y + C'Z + A''_{\prime}X + B''_{\prime}Y + C''_{\prime}Z = 0$$

where the coefficients A, B, C, A', B', etc. are constant or variable but independent of t and the quantity k enters only through the values of A', B', C', and then only to the first order.

Now if one designates by X_0, X_1, X_2, X_3, etc. the consecutive values of X, beginning with the first which corresponds to the first body of the system, and similarly by Y_0, Y_1, Y_2, Y_3, etc. Z_0, Z_1, Z_2, Z_3, etc. the consecutive values of Y and Z, and then substitutes successively these values in the three equations reduced to the preceding form, it is easy to see that the first three will give the values of X_2, Y_2, Z_2 as linear functions of X_0, Y_0, Z_0, X_1, Y_1, Z_1. The substitution of the three following values will give X_3, Y_3, Z_3, as linear functions of $X_2, Y_2, Z_2, X_1, Y_1, Z_1$, which by the substitution of the values X_2, Y_2, Z_2, will also become linear functions of $X_0, Y_0, Z_0, X_1, Y_1, Z_1$, and so on.

Therefore, in general, the values of $X_{n+1}, Y_{n+1}, Z_{n+1}$ will be of the form

$$AX_0 + BY_0 + CZ_0 + A'X_1 + B'Y_1 + C'Z_1$$

and using the calculus, it is easy to be certain that the quantities A, B, C, will be rational and complete functions of k of order $n-2$, and that the quantities A', B', C', are similar functions of the order $n-1$.

We have assumed (Article 17) that the first and last bodies of the system were fixed. The first body corresponds to the index 0 and if n indicates the number of mobile bodies, then the last body, which must be fixed, will have the index $n + 1$. Thus there should result

$$X_0 = 0, \qquad Y_0 = 0, \qquad Z_0 = 0, \qquad X_{n+1} = 0, \qquad Y_{n+1} = 0, \qquad Z_{n+1} = 0$$

which will give between X_1, Y_1, Z_1, three linear equations of the form $A'X_1 + B'Y_1 + C'Z_1 = 0$ in which the coefficients A', B', C' are rational and complete functions of k to the nth degree. By eliminating the quantities X_1, Y_1, Z_1, one obtains an equation in k of degree $3n$, a number of unknowns X, Y, Z and consequently, an equation which has $3n$ roots.

The same equations will give the ratios between the three quantities X_1, Y_1, Z_1, so that the value of any one of these quantities can be selected. Because these ratios will be expressed by rational functions of k, the value of the three quantities X_1, Y_1, Z_1, can be expressed by rational and complete functions of k. This implies that the unknowns X, Y, Z will also be expressed in general by known rational and complete functions of k.

25. We will denote by k', k'', k''', etc. $k^{(3n)}$ the different roots of the equation of k for which the solution must be assumed known and we will similarly denote by X', X'', X''', etc., Y', Y'', Y''', etc., Z', Z'', Z''', etc., the corresponding values of the quantities X, Y, Z, which result from the substitution of these different roots in place of k.

Therefore, since it was found (Articles 23, 24) that

$$\xi = XE\sin(\sqrt{k}\,t + \epsilon), \qquad \eta = YE\sin(\sqrt{k}\,t + \epsilon), \qquad \zeta = ZE\sin(\sqrt{k}\,t + \epsilon)$$

by successively substituting the different values of k and by taking various arbitrary constants E and ϵ, one will have as many particular values of ξ, η, ζ, for which the sum will give the complete solution for these variables by the properties of the linear equations.

The particular values of ξ, η, ζ, are analogous to those which represent the small oscillations of a pendulum for which the length would be g/k (Article 11), as long as k is a real and positive quantity. The motion of each body will be compounded of as many similar oscillations as there are different values for k such that if all these values are incommensurable to each other, it will be impossible for the system to regain its initial position unless the values of ξ, η, ζ are reduced to the particular values which are only one of the roots k. In this case, by putting $t = 0$ in the preceding formulas, one will have $XE\sin\epsilon$, $YE\sin\epsilon$, $ZE\sin\epsilon$, for the values of ξ, η, ζ, and $XE\cos\epsilon$, $YE\cos\epsilon$, $ZE\cos\epsilon$ for those of $\mathrm{d}\xi/\mathrm{d}t$, $\mathrm{d}\eta/\mathrm{d}t$, $\mathrm{d}\zeta/\mathrm{d}t$. Thus, in order for this case to hold, the initial displacements ξ, η, ζ, as well as the initial velocities $\mathrm{d}\xi/\mathrm{d}t$, $\mathrm{d}\eta/\mathrm{d}t$, $\mathrm{d}\zeta/\mathrm{d}t$ will have to be proportional to X, Y, Z and there will be as many ways to satisfy these conditions as there are different values of k.

26. If the different arbitrary constants are designated by upper primes, the following equations will be obtained

$$\xi = X'E'\sin(\sqrt{k'}\,t + \epsilon') + X''E''\sin(\sqrt{k''}\,t + \epsilon'') + X'''E'''\sin(\sqrt{k'''}\,t + \epsilon''') + \cdots$$

$$\eta = Y'E'\sin(\sqrt{k'}\,t+\epsilon') + Y''E''\sin(\sqrt{k''}\,t+\epsilon'') + Y'''E'''\sin(\sqrt{k'''}\,t+\epsilon''') + \cdots$$
$$\zeta = Z'E'\sin(\sqrt{k'}\,t+\epsilon') + Z''E''\sin(\sqrt{k''}\,t+\epsilon'') + Z'''E'''\sin(\sqrt{k'''}\,t+\epsilon''') + \cdots$$

as complete values of the variables ξ, η, ζ which represent the oscillations of each body of the given system whatever their initial state.

These values can be represented in a simpler manner by using the symbol Σ to express the sum of all the values corresponding to the different values of k. Thus one will have

$$\xi = \Sigma(XE\sin(\sqrt{k}\,t+\epsilon)), \quad \eta = \Sigma(YE\sin(\sqrt{k}\,t+\epsilon)), \quad \zeta = \Sigma(ZE\sin(\sqrt{k}\,t+\epsilon))$$

and the particular expressions of the variables $\xi_1, \eta_1, \zeta_1, \xi_2, \eta_2, \zeta_2$, etc. will be obtained for each body of the system by changing in the preceding formulas X, Y, Z by X_1, Y_1, Z_1, X_2, Y_2, Z_2, etc. and taking for E and ϵ different arbitrary constants E_1, E_2, etc. ϵ_1, ϵ_2, etc. which depend on the initial state of the system.

27. In order to determine these constants in the simplest fashion, consider again the equations in ξ, η, ζ, of Article 21 and add them together, after having multiplied the first equation by X, the second by Y and the third by Z. Then consider the sum of all the equations composed in this fashion relative to all the bodies of the system and denote this sum by the operator S. If care is taken that this operator is independent of the operator d for the relative differences of t, the following equation will be obtained

$$\begin{aligned}
&\frac{\mathrm{d}^2\mathrm{S}(X\xi + Y\eta + Z\zeta)\mathrm{D}m}{\mathrm{d}t^2}\\
&+\mathrm{S}\left(\frac{\mathrm{d}^2\Pi}{\mathrm{d}a^2}X + \frac{\mathrm{d}^2\Pi}{\mathrm{d}a\,\mathrm{d}b}Y + \frac{\mathrm{d}^2\Pi}{\mathrm{d}a\,\mathrm{d}c}Z\right)\xi\,\mathrm{D}m\\
&+\mathrm{S}\left(\frac{\mathrm{d}^2\Pi}{\mathrm{d}a\,\mathrm{d}b}X + \frac{\mathrm{d}^2\Pi}{\mathrm{d}b^2}Y + \frac{\mathrm{d}^2\Pi}{\mathrm{d}b\,\mathrm{d}c}Z\right)\eta\,\mathrm{D}m\\
&+\mathrm{S}\left(\frac{\mathrm{d}^2\Pi}{\mathrm{d}a\,\mathrm{d}c}X + \frac{\mathrm{d}^2\Pi}{\mathrm{d}b\,\mathrm{d}c}Y + \frac{\mathrm{d}^2\Pi}{\mathrm{d}c^2}Z\right)\zeta\,\mathrm{D}m\\
&-\mathrm{S}\,X\,\mathrm{D}_{\prime}\left[\frac{F\,\mathrm{D}\xi}{\mathrm{D}f} - Ga'\left(\frac{a'\,\mathrm{D}\xi}{\mathrm{D}f} + \frac{b'\,\mathrm{D}\eta}{\mathrm{D}f} + \frac{c'\,\mathrm{D}\zeta}{\mathrm{D}f}\right)\right]\\
&-\mathrm{S}\,Y\,\mathrm{D}_{\prime}\left[\frac{F\,\mathrm{D}\eta}{\mathrm{D}f} - Gb'\left(\frac{a'\,\mathrm{D}\xi}{\mathrm{D}f} + \frac{b'\,\mathrm{D}\eta}{\mathrm{D}f} + \frac{c'\,\mathrm{D}\zeta}{\mathrm{D}f}\right)\right]\\
&-\mathrm{S}\,Z\,\mathrm{D}_{\prime}\left[\frac{F\,\mathrm{D}\zeta}{\mathrm{D}f} - Gc'\left(\frac{a'\,\mathrm{D}\xi}{\mathrm{D}f} + \frac{b'\,\mathrm{D}\eta}{\mathrm{D}f} + \frac{c'\,\mathrm{D}\zeta}{\mathrm{D}f}\right)\right] = 0
\end{aligned}$$

In this equation, the terms which contain the differences indicated by D under the operator S are capable of further simplification using procedures similar to those of integration by parts and for which we have given the approach in Article 16. For this, let us consider in general an arbitrary term of the form $\mathrm{S}\,X\,\mathrm{D}_1(V\,\mathrm{D}\xi)$, following the procedures of the cited article and by taking care that the quantities X and ξ are zero at the beginning and end of the integrations indicated by D (Article 24), the following equation results

$$\mathrm{S}\,X\,\mathrm{D}_1(V\,\mathrm{D}\xi) = -\mathrm{S}\,V\,\mathrm{D}\xi\,\mathrm{D}X = \mathrm{S}\,\xi_1(V\,\mathrm{D}X)$$

But $\mathrm{S}\,\xi_1\,\mathrm{D}(V\,\mathrm{D}X)$ is the same as $\mathrm{S}\,\xi\,\mathrm{D}_1(V\,\mathrm{D}X)$, taking instead of the term $\xi_1\,\mathrm{D}(V\,\mathrm{D}X)$ the one just before it.

Therefore, in general, one will have $\mathrm{S}\,X\,\mathrm{D}_1(V\,\mathrm{D}\xi) = \mathrm{S}\,\xi\,\mathrm{D}_1(V\,\mathrm{D}X)$ and it will be the same for similar terms. In this fashion, the preceding equation will assume the form

$$\frac{\mathrm{d}^2\mathrm{S}(X\xi + Y\eta + Z\zeta)\mathrm{D}m}{\mathrm{d}t^2} + \mathrm{S}((X)\Pi + (Y)\eta + (Z)\zeta) = 0$$

in which the quantities indicated by (X), (Y), (Z) will contain the same terms which comprise the second members of the equations of Article 23 such that these equations will give

$$(X) = kX\,\mathrm{D}m, \qquad (Y) = kY\,\mathrm{D}m, \qquad (Z) = kZ\,\mathrm{D}m$$

from which it follows that the above equation will become

$$\frac{\mathrm{d}^2\mathrm{S}(X\xi + Y\eta + Z\zeta)\mathrm{D}m}{\mathrm{d}t^2} + k\,\mathrm{S}(X\xi + Y\eta + Z\zeta)\mathrm{D}m = 0$$

which after integration gives immediately

$$\mathrm{S}(X\xi + Y\eta + Z\zeta)\mathrm{D}m = L\sin(\sqrt{k}\,t + \lambda)$$

where L and λ are two arbitrary constants.

28. It is easy to see, from the nature of the calculation that if k is replaced by one of the roots of the equation for k which we have designated by k', k'', k''', etc. (Article 25) we shall have an identical result similar to the one obtained for the expressions of ξ, η, ζ, given in Article 26 so that by substituting these same expressions in the preceding equation, it should become absolutely identical for all the values of k.

Thus the identical equation will be obtained

$$\mathrm{S}\left\{\begin{array}{l} X\Sigma(XE\sin(\sqrt{k}\,t + \epsilon) \\ +Y\Sigma(YE\sin(\sqrt{k}\,t + \epsilon) \\ +Z\Sigma(ZE\sin(\sqrt{k}\,t + \epsilon) \end{array}\right\}\mathrm{D}m = L\sin(\sqrt{k}\,t + \gamma)$$

for each of the values k', k'', k''', etc. of k. Since this identity must hold independent of the value of t, it will not be difficult to be convinced that all the terms which will contain the same arc $\sqrt{k}\,t$ shall be identical in the first and second members of the equation. From which it follows immediately that λ will necessarily be equal to ϵ for all values of λ and ϵ.

Then if attention is paid to the value of the integral signs S and Σ, for which the former, S, represents the sum of the quantities under the sign which belongs to all the bodies of the system and which we have designated by subscript letter indices (Article 24) and for which the latter, Σ, represents the sum of similar quantities which correspond to all the roots k', k'', k''', etc., $k^{(3n)}$ and which we designate by upper primes (Article 25), one will find by the comparison of terms with the same sines, the equation

$$E\,\mathrm{S}(X^2+Y^2+Z^2)\mathrm{D}m = L$$

Thus in general one will have

$$E\sin(\sqrt{k}\,t+\epsilon) = \frac{L\sin(\sqrt{k}\,t+\epsilon)}{\mathrm{S}(X^2+Y^2+Z^2)\mathrm{D}m}$$

and consequently, by Article 27

$$E\sin(\sqrt{k}\,t+\epsilon) = \frac{\mathrm{S}(X\xi+Y\eta+Z\zeta)\mathrm{D}m}{\mathrm{S}(X^2+Y^2+Z^2)\mathrm{D}m}$$

and equation which will hold for all values of k.

29. When $t=0$, $\xi=\alpha$, $\eta=\beta$, $\zeta=\gamma$, $\mathrm{d}\xi/\mathrm{d}t=\dot\alpha$, $\mathrm{d}\eta/\mathrm{d}t=\dot\beta$, $\mathrm{d}\zeta/\mathrm{d}t=\dot\gamma$, these six quantities are given by the initial state of the system. Thus if they are introduced in the preceding equation and in its differential with respect to t and by setting $t=0$, the following values for the arbitrary constants will be obtained

$$E\sin\epsilon = \frac{\mathrm{S}(X\alpha+Y\beta+Z\gamma)\mathrm{D}m}{\mathrm{S}(X^2+Y^2+Z^2)\mathrm{D}m}$$

$$E\cos\epsilon = \frac{\mathrm{S}(X\dot\alpha+Y\dot\beta+Z\dot\gamma)\mathrm{D}m}{\sqrt{k}\,\mathrm{S}(X^2+Y^2+Z^2)\mathrm{D}m}$$

Finally, if these values are substituted in the expressions for ξ, η, ζ of Article 26, one will have

$$\xi = \Sigma\Big(\frac{X\,\mathrm{S}(X\alpha+Y\beta+Z\gamma)\mathrm{D}m}{\mathrm{D}\,\mathrm{S}(X^2+Y^2+Z^2)\mathrm{D}m}\cos\sqrt{k}\,t\Big) + \Sigma\Big(\frac{X\,\mathrm{S}(X\dot\alpha+Y\dot\beta+Z\dot\gamma)\mathrm{D}m}{\sqrt{k}\,\mathrm{S}(X^2+Y^2+Z^2)\mathrm{D}m}\sin\sqrt{k}\,t\Big)$$

$$\eta = \Sigma\Big(\frac{Y\,\mathrm{S}(X\alpha+Y\beta+Z\gamma)\mathrm{D}m}{\mathrm{S}(X^2+Y^2+Z^2)\mathrm{D}m}\cos\sqrt{k}\,t\Big) + \Sigma\Big(\frac{Y\,\mathrm{S}(X\dot\alpha+Y\dot\beta+Z\dot\gamma)\mathrm{D}m}{\sqrt{k}\,\mathrm{S}(X^2+Y^2+Z^2)\mathrm{D}m}\sin\sqrt{k}\,t\Big)$$

$$\zeta = \Sigma\Big(\frac{Z\,\mathrm{S}(X\alpha+Y\beta+Z\gamma)\mathrm{D}m}{\mathrm{S}(X^2+Y^2+Z^2)\mathrm{D}m}\cos\sqrt{k}\,t\Big) + \Sigma\Big(\frac{Z\,\mathrm{S}(X\dot\alpha+Y\dot\beta+Z\dot\gamma)\mathrm{D}m}{\sqrt{k}\,\mathrm{S}(X^2+Y^2+Z^2)\mathrm{D}m}\sin\sqrt{k}\,t\Big)$$

These formulas, remarkable for their generality as well as for their simplicity, contain the solution to several problems for which the analysis would be very difficult by other methods. We will apply them to two problems already solved in different works, but in a more or less unfinished fashion.

Subsection III
An Application of the Preceding Formulas to the Vibrations of a Tensioned String Loaded by Several Bodies and to the Oscillations of an Inextensible String Loaded by an Arbitrary Number of Weights and Suspended by Either One or Both Ends

30. The expressions which were found for the variables ξ, η, ζ are greatly simplified when, in the differential equations of Article 21, the variables in question are separated. Then the variables X, Y, Z are also separated in the equations of finite differences of Article 23. Each of these equations will provide a particular equation in k of degree m by Article 24. If k, k_1, k_2, represent the values of k which correspond to the quantities X, Y, Z given by these three equations and if the designations of the preceding article are kept, the expressions for ξ, η, ζ will be reduced in the present case to the following equations

$$\xi = \Sigma(\frac{X\,\mathrm{S}\,X\alpha\,Dm}{\mathrm{S}\,X^2\,Dm}\cos\sqrt{k}\,t) + \Sigma(\frac{X\,\mathrm{S}\,X\dot{\alpha}\,Dm}{\mathrm{S}\,X^2\,Dm\sqrt{k}}\sin\sqrt{k}\,t)$$

$$\eta = \Sigma(\frac{Y\,\mathrm{S}\,Y\beta\,Dm}{\mathrm{S}\,Y^2\,Dm}\cos\sqrt{k_1}\,t) + \Sigma(\frac{Y\,\mathrm{S}\,Y\dot{\beta}\,Dm}{\mathrm{S}\,Y^2\,Dm\sqrt{k_1}}\sin\sqrt{k_1}\,t)$$

$$\zeta = \Sigma(\frac{Z\,\mathrm{S}\,Z\gamma\,Dm}{\mathrm{S}\,Z^2\,Dm}\cos\sqrt{k_2}\,t) + \Sigma(\frac{Z\,\mathrm{S}\,Z\dot{\gamma}\,Dm}{\mathrm{S}\,Z^2\,Dm\sqrt{k_2}}\sin\sqrt{k_2}\,t)$$

31. This case is applicable primarily when the bodies in a state of equilibrium are assumed to be in a straight line. Because if this line is taken for the x-axis, the ordinates b and c will become zero as well as their differences Db, Dc, and the equations of condition of Article 20 require that one has $\mathrm{d}\Pi/\mathrm{d}b = 0$, $\mathrm{d}\Pi/\mathrm{d}c = 0$, that is, that the perpendicular forces to the axis are zero.

Thus one will also have $\mathrm{d}^2\Pi/\mathrm{d}a\,\mathrm{d}b = 0$, $\mathrm{d}^2\Pi/\mathrm{d}a\,\mathrm{d}c = 0$, etc. and the equations of Article 21 will become because of the fact that $a' = 1$, $b' = 0$, $c' = 0$, and $G = F - F'$

$$\frac{\mathrm{d}^2\xi}{\mathrm{d}t^2}Dm + \frac{\mathrm{d}^2\Pi}{\mathrm{d}a^2}\xi - \mathrm{D}_1(\frac{F'\,\mathrm{D}\xi}{\mathrm{D}f}) = 0$$

$$\frac{\mathrm{d}^2\eta}{\mathrm{d}t^2}Dm - D_1(\frac{F\,\mathrm{D}\eta}{\mathrm{D}f}) = 0$$

$$\frac{\mathrm{d}^2\zeta}{\mathrm{d}t^2}Dm - D_1(\frac{F\,\mathrm{D}\zeta}{\mathrm{D}f}) = 0$$

Consequently, the equations of Article 23 will be reduced to the following equations

$$(k - \frac{\mathrm{d}\Pi}{\mathrm{d}a^2})X\,Dm + \mathrm{D}_1(\frac{F'\,\mathrm{D}X}{\mathrm{D}f}) = 0$$

$$kY\,\mathrm{D}m + \mathrm{D}_1\left(\frac{F\,\mathrm{D}Y}{\mathrm{D}f}\right) = 0$$

$$kZ\,\mathrm{D}m + \mathrm{D}_1\left(\frac{F\,\mathrm{D}Z}{\mathrm{D}f}\right) = 0$$

in which it is observed that the variables are separated in such a manner that each can be determined individually.

The undetermined constant k can then be different in these three equations and each of them will give an equation of the nth degree for the determination of this constant. Thus the formulas of the preceding article will result.

32. Because, in this case, $\mathrm{D}b = 0$, $\mathrm{D}c = 0$, one will have $\mathrm{D}f = \mathrm{D}a$ (Article 19) and the equations of equilibrium (Article 22) will give $F = \mathrm{S}(\mathrm{d}\Pi/\mathrm{d}a)\mathrm{D}m + A$.

In order to obtain the expression for the quantity F' (Article 19), the expression for F should be known as a function of $\mathrm{D}f$ or $\mathrm{D}a$ and the expression for F' should be deduced as a function of F by differentiation.

For example, if it is assumed that $\Phi = K(\mathrm{D}s)^m$, one will obtain $F = K(\mathrm{D}f)^m$, and from this result $F' = mK(\mathrm{D}f)^m = mF$.

In the case where all extraneous forces are disregarded, one would have $\mathrm{d}\Pi/\mathrm{d}a = 0$, which gives $F = A$ and consequently, F is a constant for all bodies. But the value of F' can vary from one body to the next unless the interval $\mathrm{D}a$ between the successive bodies is also the same for all bodies. In the last case, the quantities F and F' will be two constants which can be determined **a posteriori** without knowing the form of the function Φ.

This case is the one of a taut string or rope whose two ends are fixed and which is loaded with an arbitrary number of equally-spaced bodies. Then the quantity F expresses the tension of the string or the weight which can produce it. But, the quantity F' cannot be derived from F without knowing the elastic law of the string.

This problem, which goes by the name of "the problem of the vibrating string", merits a special examination because it is amenable to a general solution as much as it is intimately related to the famous problem of the vibration of sonorous strings.[28]

33. We will assume that all the bodies $\mathrm{D}m$ which load the string are mutually equal and without weight and that the intervals $\mathrm{D}f$ or $\mathrm{D}a$ which separate them in the state of equilibrium are also equal.

Since n represents the number of mobile bodies, if one designates by M the entire mass or the sum of all the masses $\mathrm{D}m$ including the last one which is assumed fixed and by ℓ the length of the string in the state of equilibrium, it is clear that one will have $\mathrm{D}m = M/(n+1)$ and $\mathrm{D}f = \mathrm{D}a = \ell/(n+1)$. The three equations for X, Y, Z of Article 31 will become

$$\frac{\ell M k}{(n+1)^2 F'}X + \mathrm{D}_1^2 X = 0$$

$$\frac{\ell Mk}{(n+1)^2 F}Y + D_1^2 Y = 0$$
$$\frac{\ell Mk}{(n+1)^2 F}Z + D_1^2 Z = 0$$

which, since they are similar to each other, will suffice to solve the first equation and then it would only be necessary to replace F' by F to obtain the solution of the other two equations.

34. Let r be the index of the rank that an arbitrary term X holds in the series for X. In general, we will designate this term by X_r and the term preceding ${}_{\prime}X$ will be X_{r-1}. Thus the first equation will be

$$\frac{\ell Mk}{(n+1)^2 F'}X_r + D^2 X_{r-1} = 0$$

Let us assume in order to solve this equation that

$$X_r = H\sin(r\varphi + e)$$

where H and e are two arbitrary constants. From the known formulas for the multiplication of angles, the following equation will be obtained

$$D^2 X_{r-1} = X_{r+1} - 2X_r + X_{r-1} = -4H\sin(r\varphi + e)\sin^2\frac{\varphi}{2}$$

and after substitution of the values in the preceding equation and division by X_r

$$\frac{\ell Mk}{(n+1)^2 F'} - 4\sin^2\frac{\varphi}{2} = 0$$

which gives

$$\sqrt{k} = 2(n+1)\sqrt{\frac{F'}{\ell M}}\sin\frac{\varphi}{2}$$

But there are (Article 24) two conditions to fulfill: $X_0 = 0$ and $X_{n+1} = 0$. The first condition gives $e = 0$, the second gives $\sin(n+1)\varphi = 0$ from which one obtains $(n+1)\varphi = \rho\pi$, where π is an angle of 180° and ρ an arbitrary integer. Therefore, one will have $\varphi = \rho\pi/n+1$. Consequently, by putting $H = 1$, which is permissible, one will obtain in general

$$X_r = \sin r\frac{\rho\pi}{n+1}$$

and the same expression will be obtained for Y_r and Z_r which will be substituted in place of X, Y, Z in the expressions for ξ, η, ζ of Article 30.

If the same value of φ is substituted in the expression for $\sqrt{k}$ which was found above, the following equation results

$$\sqrt{k} = 2(n+1)\sqrt{\frac{F'}{\ell M}}\sin\frac{\rho\pi}{2(n+1)}$$

where all the integers from 0 to n inclusively can be used to replace ρ, because $\rho = n+1$ gives X, Y, Z equal to zero and above $n+1$, the sines of $\rho\pi/2(n+1)$ are the same. Thus there will be as many different values of k as there are mobile bodies. These values will be the roots of the equation for k.

By replacing F' with F, the values of the roots k_1 and k_2 of the two other equations for k will be obtained.

Thus these substitutions will be made in the general formulas of Article 30 and it should be noted that the integral operator S must only correspond to the indices of order r, from $r = 1$ to $r = n$, and that the integral sign Σ must correspond to the indices ρ of the different roots from $\rho = 1$ to $\rho = n$.

With regard to the value of $\text{S}\,X^2\,\text{D}m = \text{D}m\,\text{S}\,X^2$, since $\varphi = \rho\pi/(n+1)$, one will have the following sum

$$\begin{aligned}\text{S}\,X^2 &= \sin^2\varphi + \sin^2 2\varphi + \sin^2 3\varphi + \cdots + \sin^2 n\varphi\\ &= \frac{1}{2}n - \frac{1}{2}(\cos 2\varphi + \cos 4\varphi + \cos 6\varphi + \cdots + \cos 2n\varphi)\\ &= \frac{1}{2}n - \frac{1}{2}(\frac{\cos 2n\varphi - \cos 2(n+1)\varphi}{2(1-\cos 2\varphi)} - \frac{1}{2}) = \frac{n+1}{2}\end{aligned}$$

Similarly, the following equation results $\text{S}\,Y^2 = \text{S}\,Z^2 = (n+1)/2$.

35. Since the values of k are incommensurable to each other, the string will never be able to regain its initial position, unless the expressions for ξ, η, ζ, are reduced to only one term (Article 25). In this case, by replacing X, Y, Z and k in the formulas of the cited article by the values that have just been found and making the following definitions in order to shorten the expressions

$$h' = \sqrt{\frac{F'}{\ell M}}, \qquad h = \sqrt{\frac{F}{\ell M}}$$

the following expressions will be obtained in which the angle φ has been retained rather than using the equivalent expression $\rho\pi/(n+1)$

$$\xi = E\sin r\varphi\sin(h't\sin\frac{\varphi}{2} + \epsilon)$$

$$\eta = E \sin r\varphi \sin(ht \sin \frac{\varphi}{2} + \epsilon)$$
$$\zeta = E \sin r\varphi \sin(ht \sin \frac{\varphi}{2} + \epsilon)$$

but the initial values α, β, γ, $\dot{\alpha}, \dot{\beta}, \dot{\gamma}$, which correspond to $t = 0$ must be proportional to $\sin r\varphi$. This is the known solution in which it is assumed that the bodies perform only simple and isochronous oscillations.

36. In order to obtain general expressions applicable to an initial arbitrary state, one must use the formulas of Article 30 by substituting the values found above (Article 34). For additional clarity, we will apply to the variables ξ, η, ζ a subscripted index r to indicate the order of the body to which they refer and with respect to the quantities α, β, γ, $\dot{\alpha}$, $\dot{\beta}$, $\dot{\gamma}$, and X, Y, Z which are under the summation sign S, we will use the letter s instead of r since this letter refers only to the symbol S which indicates that one must take the sum of all the terms related to the values of S from 0 to n.

Thus the following general formula will result

$$\xi_r = \Sigma \frac{2 \sin r\varphi}{n+1} \left\{ \begin{array}{l} \mathrm{S}\, \alpha_s \sin s\varphi \cos(2(n+1)h't \sin \frac{\varphi}{2}) \\ +\mathrm{S}\, \dot{\alpha}_s \sin s\varphi \dfrac{\sin(2(n+1)h't \sin \frac{\varphi}{2})}{2(n+1)h' \sin \frac{\varphi}{2}} \end{array} \right\}$$

and to obtain the expressions for η_r and ζ_r, it will only be necessary to change h' to h, α, $\dot{\alpha}$ to β, $\dot{\beta}$ and $\dot{\gamma}$ to γ.

The variables ξ_r, represent the longitudinal displacements of the bodies from the straight line or axis which passes through the two fixed ends of the string. The variables η_r, ζ_r represent their transverse or lateral displacements, i.e. in a direction perpendicular to the axis, which are the only displacements which have been considered heretofore in the solution of the problem of vibrating strings.

Considering the summation symbol Σ, one will recall that it expresses the sum of all the quantities under this sign which correspond to $\rho = 1, 2, 3$, etc., in from which one sees that the displacements of each body, longitudinal as well as transverse, will be composed in general of as many particular displacements analogous to those of the different pendulums for which the lengths would be

$$\frac{g}{4(n+1)^2 h'^2 \sin^2 \frac{\varphi}{2}}$$

or

$$\frac{g}{4(n+1)^2 h^2 \sin^2 \frac{\varphi}{2}}$$

as there are mobile bodies where g is the force of gravity.

In order to insure that the values of h and h' are real numbers, the quantities F and F' must be positive (Article 35). Thus, according to the hypothesis of Article 32, the index m must be positive. If the bodies were repulsing one another, F would be a negative quantity and then the index m would also be negative. Moreover, one must have $\beta = 0$, $\dot{\beta} = 0$, $\gamma = 0$, $\dot{\gamma} = 0$ to have the transverse displacements η and ζ equal to zero.

37. There is an important observation to be made about the general expression for ξ, which we have just found. Although we assumed that the number of mobile bodies n is given and that the string, for which the length is also given, is fixed at its two ends, the calculation is not stymied by these assumptions. The expression in question gives a value of ξ for every body located on the same straight line of which the order would be expressed by an arbitrary integer number r positive or negative.

Indeed, since this number r appears only in the $\sin r\varphi$ term, it is obvious that one can give it whatever value one chooses and it can be seen at the same time that because $\varphi = \rho\pi/n+1$, this sine function will not change value if one replaces r with $2\lambda(n+1)+r$ and will simply become negative if one changes r by $2\lambda(n+1)-r$ where λ is an arbitrary integer number, positive or negative. From which it follows that by imagining according to the intent of the calculation that the string is extended indefinitely from one end to the other and that it is loaded, along its entire length by equal bodies placed at equal intervals from one another, the motions of these bodies will be such that one will always have

$$\xi_{2\lambda}(n+1) \pm r = \pm\xi_r$$

It is easy to see that the formula $2\lambda(n+1) \pm r$ can represent all the integer positive or negative numbers, assuming r lies between 0 and $n+1$. Because if one divides an arbitrary integer repetitively by $2(n+1)$ until the remainder which is positive or negative is less than $n+1$, which is always possible and if one takes λ for the quotients and $\pm r$ for the remainder, this number will be represented by $2\lambda(n+1)\pm r$. Thus the values of ξ, relative to an arbitrary body located on the same straight line, at whatever distance desired from the origin of the axis ℓ, will always be reduced to the value of ξ for one of the bodies located on this axis.

Since the relation we just found between the different values of ξ is general whatever the number r, if $\lambda(n+1)+r$ is substituted for r and if the lower signs are taken, it becomes

$$\xi_{\lambda(n+1)} - r = -\xi_{\lambda(n+1)} + r$$

From which it is easy to conclude that if one imagines the indefinite length of the string divided into equal parts along the axis ℓ of the given string, the values of ξ, in each of these parts, will be the same at equal distance from the points of partition but of different signs in the contiguous parts. If, therefore, one represents the values of ξ of all the bodies located on the axis ℓ by the ordinates of the angles of a polygon described on this axis, one will only have to move this polygon alternatively and symmetrically below and above the axis prolonged from the sides to the infinite such that the sides which end at the points of partition are the same but located in opposite directions and in the same direction. Thus one will have at each instant, the values of ξ for all the bodies, which are assumed distributed on the same straight line, extended infinitely by the ordinates of the angles of this polygon composed of an infinite number of branches. These values will be zero at each point of the partition such that the bodies located at these points will be immobile by themselves. The analysis must satisfy the condition that the two ends of the given string are fixed.

What we just demonstrated with respect to the variables ξ, also holds for the differentials $\mathrm{d}\xi/\mathrm{d}t$ because by differentiating the expression of ξ with respect to t, one has an expression for $\mathrm{d}\xi_1/\mathrm{d}t$ to which one can apply the same reasoning.

Thus the values of α and $\dot{\alpha}$, which represent those of ξ and $\mathrm{d}\xi/\mathrm{d}t$ at the first instant and which are arbitrary for all the bodies located on the axis ℓ will be represented by a similar construction in the stretching of the string of indefinite length.

Since the expressions for the other two variables η and ζ differ from those of ξ only by the initial values β, $\dot{\beta}$ and γ, $\dot{\gamma}$ which take the place of α, $\dot{\alpha}$, the same results will also hold with respect to these variables.

38. Thus it can be concluded in general that if a tensioned string of arbitrary length is loaded by equal bodies placed at equal intervals from one another and that after having divided this string in numerous equal parts where each division is between two bodies, all the bodies with the exception of those which are the points of partition are distributed simultaneously such that the disturbance is the same, but in opposite directions for those which are at equal distances from each point of partition. The bodies located at the points of partition remain immobile and each part of the string will have the same motion as if it were isolated and its two ends were absolutely fixed.

It can be concluded from this that for a tensioned string of length ℓ, fixed at its two ends, loaded by n bodies and divided into v equal parts where v is a divisor of $n+1$, if the initial state is such that the bodies placed at the points of partition are not disturbed and those which are at an equal distance from each side of a point of partition were disturbed equally but in opposite directions, the string will oscillate as if the points of partition were fixed and the string had only a length ℓ/v.

39. The separation of the variables in the equations for ξ, η, ζ, can still take place without having to assume that the bodies are distributed on a straight line in the state of equilibrium, if it is assumed that their mutual distances do not vary during the motion. We remarked in Article 14 that this case depends on the same general formulas by viewing the quantity Θ, and consequently, also the quantity F as indeterminate. Furthermore, we saw in Article 22 that we will have the following equation of condition

$$\frac{\mathrm{D}a}{\mathrm{D}f}\mathrm{D}\xi + \frac{\mathrm{D}b}{\mathrm{D}f}\mathrm{D}\eta + \frac{\mathrm{D}c}{\mathrm{D}f}\mathrm{D}\zeta = 0$$

which eliminates in the general equations of Article 21 all the terms multiplied by G.

By considering only the gravity of the bodies and taking the axes of abscissas x and a directed vertically upward, one will have $\mathrm{d}\Pi/\mathrm{d}a$ equal to the accelerating force of gravity that we will designate by g, and moreover, $\mathrm{d}\Pi/\mathrm{d}b = 0$, $\mathrm{d}\Pi/\mathrm{d}c = 0$ and the equations of the cited article will become

$$\frac{\mathrm{d}^2\xi}{\mathrm{d}t^2}\mathrm{D}m - \mathrm{D}_1\left(\frac{F\,\mathrm{D}\xi}{\mathrm{D}f}\right) = 0$$
$$\frac{\mathrm{d}^2\eta}{\mathrm{d}t^2}\mathrm{D}m - \mathrm{D}_1\left(\frac{F\,\mathrm{D}\eta}{\mathrm{D}f}\right) = 0$$
$$\frac{\mathrm{d}^2\zeta}{\mathrm{d}t^2}\mathrm{D}m - \mathrm{D}_1\left(\frac{F\,\mathrm{D}\zeta}{\mathrm{D}f}\right) = 0$$

where the variables are separated.

The value of F will be (Article 22)

$$F = \sqrt{(g\mathrm{S}\,\mathrm{D}m + A)^2 + B^2 + C^2}$$

The equation in X, Y, Z will thus become (Article 23)

$$kX\,\mathrm{D}m + \mathrm{D}_1\left(\frac{F\,\mathrm{D}X}{\mathrm{D}f}\right) = 0$$
$$kY\,\mathrm{D}m + \mathrm{D}_1\left(\frac{F\,\mathrm{D}Y}{\mathrm{D}f}\right) = 0$$
$$kZ\,\mathrm{D}m + \mathrm{D}_1\left(\frac{F\,\mathrm{D}Z}{\mathrm{D}f}\right) = 0$$

which are, as one observes, exactly the same so that it can be assumed that $X = Y = Z$, since the arbitrary constants, by which these quantities can differ are determined from the same conditions, will become identical. Therefore, the values of ξ, η, ζ given in the general formulas of Article 30 can only differ by the initial values α, β, γ, $\dot{\alpha}, \dot{\beta}, \dot{\gamma}$, which can be arbitrary.

Thus the problem reduces to finding the general expression for X. But this is a solution which could not be found by known methods. This is the case of an inextensible string loaded with several weights and fixed firmly at its extremities.

40. When the string is only fixed at one of its ends, which we will take for the upper extremity, the lowest body will be free and according to Article 17, the value of Φ or F will be zero at the lower end. By assuming this end as the origin of the abscissas, which we assume directed upward and where the origin of the sum $\mathrm{S}\,\mathrm{D}m$ begins, the value of F will be zero if one has $A = 0$, $B = 0$, $C = 0$. One will thus have $F = g\,\mathrm{S}\,\mathrm{D}m$.

Since one has in this case $\mathrm{d}\Pi/\mathrm{d}a = g$, $\mathrm{d}\Pi/\mathrm{d}b = 0$, $\mathrm{d}\Pi/\mathrm{d}c = 0$, the equations of Article 22 will give $\mathrm{D}a = \mathrm{D}f$, $\mathrm{D}b = 0$, $\mathrm{D}c = 0$, that is, the ordinates b and c will be constant such that one will have, for the state of equilibrium, a straight line parallel to the vertical axis of the abscissas a. Therefore, one can put $b = 0$, $c = 0$, by taking for the vertical axis, the vertical which passes through the point of suspension of the string.

This case, which considers the very small oscillations of a string suspended from a fixed point and loaded with an arbitrary number of weights, is also amenable to a general solution when the weights are all mutually equal and placed at equal distances from one another.

41. In the latter case, by denoting the number of bodies by n, M the sum of their masses Dm and ℓ the length of the string, one has D$m = M/n$, Df = D$a = \ell/n$ and moreover, if one denotes by r the number of bodies, beginning from the lowest to the one which is related to the variables ξ, η, ζ, one will obtain S D$m = (r-1)$D$m = ((r-1)M)/n$ and from this expression, the following expression results $F = (g(r-1)M)/n$.

The equation in X of Article 39 after multiplication by ℓ/gm will become, by replacing X by X_r and observing that $,X$ becomes X_{r-1}, and $X,$ becomes X_{r+1},

$$\frac{\ell k}{gn}X_r + \mathrm{D}((r-1)\mathrm{D}X_{r-1}) = 0$$

by performing the differentiations indicated by the operator D, according to the formula of Article 16.

$$\frac{\ell k}{gn}X_r + (X_{r+1} - X_r) + (r-1)(X_{r-1} - 2X_1 + X_{r+1}) = 0$$

This equation because of the variable coefficient r cannot be treated in the same fashion as those which give the ordinary recurring series. But, one can successively deduce from it the values of X_2, X_3, etc.

For this, one has only to put this equation in the following form, where $h = \ell k/gn$

$$X_{r+1} = \frac{2r-h-1}{r}X_r - \frac{r-1}{rX_{r-1}}$$

From this expression by successively putting $r = 1, 2, 3$, etc., one will have

$$X_2 = (1-h)X_1$$
$$X_3 = \frac{3-h}{2}X_2 - \frac{1}{2}X_1 = (1 - 2h + \frac{h^2}{2})X_1$$
$$X_4 = \frac{5-h}{3}X_3 - \frac{2}{3}X_2 = (1 - 3h + \frac{3h^2}{2} - \frac{h^3}{2.3})X_1$$
$$X_5 = (1 - 4h + \frac{6h^2}{2} - \frac{4h^3}{2.3} + \frac{h^4}{2.3.4})X_1$$

and so on, such that one will have in general

$$X_{r+1} = (1 - rh + \frac{r(r-1)}{4}h^2 - \frac{r(r-1)(r-2)}{4.9}h^3 + \cdots)X_1$$

Since the upper end of the string must be fixed, one can assume that it corresponds to the body for which the order is $n+1$. Thus one will have $X_{r+1} = 0$ which gives the following equation, after replacing h by its value $\ell k/gn$

$$1 - \frac{\ell k}{g} + \frac{(n-1)\ell^2 k^2}{4ng^2} - \frac{(n-1)(n-2)\ell^2 k^3}{4.9gn^2 g^3} + \cdots = 0$$

which will be, with respect to k, of the nth degree and consequently, will give the n values of k which we will designate in general by $k^{(\rho)}$.

42. Thus one will only have to substitute in the formulas of Article 30 the preceding expression for X_r in place of X, Y, Z, and those of $k^{(\rho)}$ in place of k and then to perform the summations indicated by the symbols S and Σ. But one must observe that in the present case where one assumes $\mathrm{D}b = 0$ and $\mathrm{D}c = 0$ (Article 40), the equation of condition of Article 39 gives $\mathrm{D}\xi = 0$ and consequently, ξ is equal to a constant for all the bodies, but it could be a function of t. Therefore, one will have for the beginning of the motion α and $\dot{\alpha}$ both equal to constants. But since the first body is assumed fixed, the initial values α and $\dot{\alpha}$ are zero for this body. Thus they will also be zero for all the others. Consequently, the general expression of the variable ξ will become zero. This holds by neglecting, as we have done, the squares and the powers of higher order of the variables ξ, η, ζ which are assumed very small. In effect, the equation $\mathrm{D}s = \mathrm{D}f$ of Article 19 gives since $\mathrm{D}s^2 = \mathrm{D}x^2 + \mathrm{D}y^2 + \mathrm{D}z^2$ and $\mathrm{D}b = 0, \mathrm{D}c = 0$, $\mathrm{D}a^2 = (\mathrm{D}a + \mathrm{D}\xi)^2 + \mathrm{D}\eta^2 + \mathrm{D}\zeta^2$, from which one obtains

$$\mathrm{D}\xi = -\frac{\mathrm{D}\eta^2 + \mathrm{D}\zeta^2}{2\,\mathrm{D}a}$$

such that the variables ξ will be of the second order with respect to η and ζ.

Let us now designate this function of r by Φr

$$1 - (r-1)\frac{\ell k^{(\rho)}}{gn} + \frac{(r-1)(r-2)}{4}\left(\frac{\ell k^{(\rho)}}{gn}\right)^2$$
$$-\frac{(r-1)(r-2)(r-3)}{4.9}\left(\frac{\ell k^{(\rho)}}{gn}\right)^3 + \cdots$$

and let us put in the general expressions for the variable η of Article 30, similarly to what was done in Article 36, η_r instead of η and Φr instead of Y in the terms which are outside the sign S. But in those which are under this sign, we will replace r by s and put $\beta_s, \dot{\beta}_s$, in place of β and $\dot{\beta}$. Therefore, one will obtain, for an arbitrary body for which the rank is r and increasing

$$n_r = \Sigma\left(\frac{\Phi r\, \mathrm{S}(\beta_s \Phi_s)}{\mathrm{S}(\Phi s)^2}\cos\sqrt{k^{(\rho)}}\,t\right)$$
$$+\Sigma\left(\frac{\Phi r\, \mathrm{S}(\dot{\beta}_s \Phi_s)}{\mathrm{S}(\Phi s)^2\sqrt{k^{(\rho)}}}\sin\sqrt{k^{(\rho)}}\,t\right)$$

where the operator S expresses the sum of all the terms which correspond to $s = 1, 2, 3$, etc. n, and the sign Σ represents the sum of the terms which correspond to $\rho = 1, 2, 3$, etc., n, assuming that $k^{(1)}, k^{(2)}, k^{(3)}$, etc., $k^{(n)}$ are the roots of the equation in $k^{(\rho)}$ represented by $\Phi(n+1) = 0$.

One will have an entirely similar expression for the variable ζ_r, by simply replacing β_s, $\dot{\beta}_s$, by γ_s, $\dot{\gamma}_s$.

The problem of the infinitesimal oscillations of a string loaded with an arbitrary number of equal weights is therefore completely resolved. It only remains to determine the roots of the equation in $k^{(\rho)}$, which does not seem possible in general.

43. Also, although one can not determine these roots, nevertheless, one can be certain that they must all be real, positive and unequal. Otherwise, the values of ξ, η, ζ would contain terms which would increase with time, which is impossible since it is obvious from the nature of the problem that the oscillations of the string must always be very small if the initial values of ξ, η, ζ are very small.

If one assumes that the quantity g, which is the force of gravity, to be negative, that is, acting in the opposite direction, the contrary result will hold. Since this will be the case when the point of suspension of the vertical string is located at its lower end, the string will flip over as soon as it is displaced from the vertical position. In effect, by making g negative in the equation for k, all its terms will become positive so that it can not have imaginary or real and negative roots.

These results can also be found **a priori** by the principles established in Article 8. This determination can be used to show the accuracy of these principles. Indeed, if one considers the condition of inextensibility of the string, which gives (preceding article) by taking the sums counted from the lowest body

$$\xi = \xi_1 - \mathrm{S}\frac{\mathrm{D}\eta^2 + \mathrm{D}\zeta^2}{2\,\mathrm{D}a}$$

the expression for V will simply be $\mathrm{S}\,\Pi\,\mathrm{D}m$ and one will have $\Pi = gx = ga + g\xi$.

Since the most elevated body which corresponds to $n+1$ is assumed fixed, the value of ξ must be zero in this position. Thus one will have

$$\xi_1 = (\mathrm{S}\frac{\mathrm{D}\eta^2 + \mathrm{D}\zeta^2}{2\,\mathrm{D}a})$$

assuming that the sum contained between two brackets is the total sum. Therefore, one will have

$$\xi = \mathrm{S}'\frac{\mathrm{D}\eta^2 + \mathrm{D}\zeta^2}{2\,\mathrm{D}a}$$

where the operator S′ denotes the sums taken in the opposite direction starting with the most elevated body and which are the differences of the total sum and of the partial sums denoted by S, which must start from the lowest body where the origin of abscissas is located.

One will thus obtain

$$V = g\,\mathrm{S}a\,\mathrm{D}m + g\,\mathrm{S}\,\mathrm{D}m\,\mathrm{S}'\frac{\mathrm{D}\eta^2 + \mathrm{D}\zeta^2}{2\,\mathrm{D}a}$$

from which it can be seen that the part of V which contains the second orders of the variables η and ζ which are now independent, is necessarily always positive and consequently, the roots of the equation for k will always be real, positive and unequal. The opposite result is obtained if a negative value is given to g.

Subsection IV
The Sonorous Vibrations of Strings, Viewed as Tensioned Cords, Loaded by an Infinite Number of Small Weights Placed Infinitesimally Close to One Another and About the Discontinuity of Arbitrary Functions

44. The general solution that we have given to the problem of sonorous strings holds whatever the number n of mobile bodies and in addition, whatever their initial state. Consequently, it must also apply to the case where the number n would become infinitely large and the intervals between the bodies would become infinitesimal but such that the length of the string would remain the same. Then the motion of each body will be represented by an infinite series, for which the sum will be equivalent to a finite function different from each of its terms. This is the case of a sonorous string with uniform thickness. It is customary to solve it directly by the differential calculus. However, it is interesting to show how it can be deduced from the general solution, even more so, since by this means one will be certain of having a solution applicable to any shape that the string can assume at the beginning of its motion.

45. At the outset, we will remark that by assuming n infinite, the value of $\sqrt{k}$ (Article 34) becomes $\sqrt{(F'/\ell M)}\rho\pi$, since the final limit of $2(n+1)\sin(\rho\pi/2(n+1)$ is $\rho\pi$ so that the roots of the equation in k which were all incommensurable to each other as long as the number of mobile bodies is finite, all become commensurable when the number of bodies is infinite, having for common measure $\pi\sqrt{F'/(\ell M)}$ in the longitudinal displacements ξ and $\pi\sqrt{F/(\ell M)}$ in the transverse displacements η and ζ. From which it follows that the string will always regain its initial configuration with respect to the axis after a time equal to $2\sqrt{\ell M/F}$, whatever its initial state might be.

It is true that since the number ρ can become infinite there will be instances where it would not be possible to assume $2(n+1)\sin(\rho\pi/2(n+1)) = \rho\pi$, but because this is possible only after an infinite number of terms in the infinite series indicated by Σ, it follows from the known properties of these series that these particular cases are not an exception to the general result.

One can also convince oneself of this claim directly. In the case where n is infinite, the finite differences marked by D become infinitesimal. Therefore, the equation for X of Article 33 becomes, after replacing D with d and $n+1$ by the value $\ell/\mathrm{d}a$

$$\frac{Mk}{\ell F'}X + \frac{\mathrm{d}^2X}{\mathrm{d}a^2} = 0$$

which, after integration, gives

$$X = H\sin(a\sqrt{\frac{Mk}{\ell F'}} + \epsilon)$$

Since X must be zero when $a = 0$ or when $a = \ell$, the two extremities of the string are fixed. The first condition gives $\epsilon = 0$ and the second $\ell\sqrt{Mk/(\ell F')} = \rho\pi$, from which one obtains $\sqrt{k} = \rho\pi\sqrt{F'/(\ell M)}$, as above.

Therefore, since the string always returns to its initial state in this case, there is no need to assume that it makes only simple oscillations similar to those of a pendulum, as in Article 35, because whatever its initial state, it is certain that its vibrations will always be isochronous to each other and synchronous to those of a simple pendulum of length g/k. However, the law of these vibrations will be different from those of the vibrations of pendulums and will depend on the initial state of the string.

In order to understand this law, one must find the general expressions for ξ, η, ζ in the case where n is infinite. This is what we will examine.

46. Let us substitute $\rho\pi/n+1$ in place of φ and $\rho\pi/2(n+1)$ in place of $\sin\varphi/2$, in the general formula of Article 36 assuming n to be infinite and instead of the indices r and s which denote the rank of the bodies to which the variables ξ and α belong let us use, which is simpler, the parts of the axis or the abscissas which correspond to these bodies, by denoting by x the abscissa relative to ξ and by a the abscissa relative to α and $\dot{\alpha}$. Since the total length of the string is assumed equal to ℓ, one will have

$$\frac{r}{n+1} = \frac{x}{\ell}, \qquad \frac{s}{n+1} = \frac{a}{\ell}, \qquad n+1 = \frac{\ell}{\mathrm{D}a}$$

and the formula in question will give this general expression for the longitudinal displacements ξ

$$\xi = 2\Sigma\sin\frac{\rho\pi x}{\ell}(A^{(\rho)}\cos(\rho\pi h't) + \dot{A}^{(\rho)}\frac{\sin(\rho\pi h't}{\rho\pi h'})$$

where

$$A^{(\rho)} = \mathrm{S}(\sin\frac{\rho\pi a}{\ell}\frac{\alpha\,\mathrm{D}a}{\ell)}$$

$$\dot{A}^{(\rho)} = \mathrm{S}(\sin\frac{\rho\pi a}{\ell}\frac{\dot{\alpha}\,\mathrm{D}a}{\ell)}$$

The symbol Σ denotes here an infinite series whose terms correspond to $\rho = 1, 2, 3$, etc. to infinity and the integral sign S denotes other infinite series with terms corresponding to all the values of a, Da, 2Da, 3Da, etc., to infinity, since Da is infinitesimal. One will have similar expressions for the transverse displacements η and ζ, by replacing h' with h and α, $\dot{\alpha}$ by β, $\dot{\beta}$, and γ, $\dot{\gamma}$.

47. Daniel Bernoulli, by generalizing the solution of the problem of sonorous strings given by Taylor, arrived at a formula similar to the one above, but in which the coefficients $\dot{A}^{(\rho)}$ were zero and the coefficients $A^{(\rho)}$ simply denoted arbitrary constants dependent on the initial configuration of the string (Mémoires de Berlin, 1753). He believed that he was able to explain the harmonic sounds which a sonorous string makes with the fundamental sound from the different terms of his formula. Our formula in which the coefficients are expressed by the initial values α, $\dot{\alpha}$, puts us in a position to understand this explanation which has been adopted by several authors following Bernoulli's lead.

Indeed, it is easy to see that the fundamental sound of the string will be given by the first or the first two terms of the series which corresponds to $\rho = 1$, and that the successive harmonic sounds, that is, the octave, the twelfth, the double octave, the seventeenth, etc., will be given by the following terms which correspond to $\rho = 2, 3, 4, 5$, etc. Therefore, it must be assumed that the coefficients $A^{(1)}$, $\dot{A}^{(1)}$ are much larger than all the others taken together and that the following coefficients

$$A^{(2)}, \quad A^{(3)}, \quad A^{(4)}, \quad \ldots, \quad \dot{A}^{(2)}, \quad \dot{A}^{(3)}, \quad \dot{A}^{(4)}, \quad \ldots$$

form a series which is quickly convergent in order to have the fundamental sound dominate all the others and to have the first of the harmonics heard simultaneously. But from the way in which these coefficients depend on the initial values α and $\dot{\alpha}$, it is obvious that this assumption is inadmissible if the initial state of the string is arbitrary. It is also clear that in most cases, these coefficients will form a divergent series which will not prevent the string from making isochronous or vibrations of equal duration, the sole and necessary condition to produce a tone.

48. Although the formulas of Article 46 rigorously give the motion of the string after an arbitrary time t, the infinite series contained in these formulas nevertheless prevent the representation of this motion in a neat and sensible manner. But by viewing the general formula of Article 36 from a different perspective, a simple and uniform construction can be obtained to determine the configuration of the string at every instant whatever its initial state might be.

Let us again consider this formula and put it in the following form, which is permissible because of the independence of the summation signs S and Σ

$$\xi_r = \mathrm{S}\,\alpha_s \Sigma \left[\frac{2 \sin r\varphi}{n+1} \sin s\varphi \cos(2(n+1)h't \sin \frac{\varphi}{2}) \right]$$

$$+ \mathrm{S}\dot{\alpha}_s\Sigma\left[\frac{2\sin r\varphi}{n+1}\sin s\varphi\frac{\sin(2(n+1)h't\sin\frac{\varphi}{2})}{(n+1)h'\sin\frac{\varphi}{2}}\right]$$

At the outset, we will derive from this formula a result which will be very useful to us. Since it has been assumed that α is the initial value of ξ (Article 29), this formula must reduce to α_r, when we put $t = 0$ in the preceding expression for ξ_r. Consequently, one must have this identity

$$\alpha_r = \mathrm{S}\,\alpha_s\Sigma\frac{2\sin r\varphi}{n+1}\sin s\varphi$$

It is evident that the second member of this equation cannot be reduced to α_r, unless one has in general

$$\Sigma\frac{2\sin r\varphi}{n+1}\sin s\varphi = 0$$

as long as s is not equal to r. When s equals r, one has

$$\Sigma\frac{2\sin r\varphi}{n+1}\sin r\varphi = 1$$

where φ is equal to $\rho\pi/(n+1)$, and the summation sign Σ is with respect to the consecutive values 1, 2, 3, etc., n of ρ, which gives a series formed from the products of the sines of the angles which are multiples of $\rho\pi/(n+1)$ and $s\pi/(n+1)$ for which the sum must always be zero in the first case and equal to $n + 1/2$ in the second. These results can also be demonstrated directly by the known formulas for the sum of these types of series.

In these formulas, r and s are assumed to be arbitrary integers between 0 and $n + 1$. But since $\varphi = \rho\pi/(n+1)$, where ρ is also an integer, if one replaces r by $2\lambda(n+1)\pm r$ where λ is an arbitrary positive or negative integer, one will have $\sin(2\lambda(n+1)\pm r)\varphi = \pm\sin r\varphi$. Consequently, in general the following equation will be obtained

$$\Sigma(\frac{2\sin(2\lambda(n+1)\pm r)\varphi}{n+1}\sin s\varphi) = \pm 1 \quad \text{or} \quad 0$$

depending on whether s will be equal to r or not.

The formula $2\lambda(n+1)\pm r$ can represent all the positive or negative integers as we have seen in Article 37. Thus having an arbitrary integer N, one can put $N = 2\lambda(n+1)\pm r$, which will give $r = \pm(N - 2\lambda(n+1))$ and in general, whatever the value of N, the following equation will be obtained

$$\Sigma\frac{\sin N\varphi\sin s\varphi}{n+1} = \pm\frac{1}{2} \quad \text{or} \quad 0$$

depending on whether s is equal to $\pm(N - 2\lambda(n+1))$ or not, where s is an integer between 0 and $n+1$.

49. Now that this problem has been formulated, there is another factor which must be taken into account. Since the expression for ξ_r is composed of two parts, in which the first contains the initial values of α, for the variable ξ, and in which the second contains the initial values of $\dot{\alpha}$ for the differentials $\mathrm{d}\xi/\mathrm{d}t$, we will consider these two parts separately. The first part will be designated by ξ_r' and the second by ξ_r'', such that one will have $\xi_r = \xi_r' + \xi_r''$.

By assuming n infinite, the angle $\varphi = \rho\pi/n+1$ becomes infinitesimal and $\sin\varphi/2$ is reduced to $\varphi/2$ (Article 46). After making these substitutions in the expression for ξ_r', one will obtain (Article 48) the following expression

$$\xi_r' = \mathrm{S}\,\alpha_s \Sigma \frac{2}{n+1} \sin r\varphi \sin s\varphi \cos(n+1)h't\varphi$$

and after further developing the product $\sin r\varphi \cos(n+1)h't\varphi$, the following equation results

$$\begin{aligned}\xi_r' &= \mathrm{S}\,\alpha_s \Sigma(\frac{\sin(r+(n+1)h't)\varphi}{n+1} \sin s\varphi) \\ &+ \mathrm{S}\,\alpha_s \Sigma(\frac{\sin(r-(n+1)h't)\varphi}{n+1} \sin s\varphi)\end{aligned}$$

Since n is assumed to be an infinitely large number, it will always be possible to view the number $(n+1)h't$ as an integer whatever might be the number expressed by $h't$.

Thus by replacing in the last formula of the preceding article $N = r + (n+1)h't$, one will obtain

$$\mathrm{S}\,\alpha_s \Sigma(\frac{\sin(r+(n+1)h't)\varphi}{n+1} \sin s\varphi) = \pm\frac{1}{2}\alpha_s$$

where $s = \pm(r + (n+1)h't - 2\lambda(n+1)$ and putting $N = r - (n+1)h't$, one will similarly have

$$\mathrm{S}\,\alpha_s \Sigma \frac{\sin(r-(n+1)h't)\varphi}{n+1} \sin s\varphi = \pm\frac{1}{2}\alpha_s$$

where

$$s' = \pm(r - (n+1)h't - 2\lambda'(n+1)$$

and where λ and λ' are arbitrary integers or zeros.

Thus adding these two values, one will have simply

$$\xi'_r = \frac{1}{2}(\pm\alpha_s \pm \alpha_{s'})$$

where the plus or minus signs of α_s and $\alpha_{s'}$ correspond to those of the values of s and s'.

50. But instead of the indices r and s which denote the ranking of the bodies to which correspond the variables ξ and α, it is easier to use the segments of the string contained between the first fixed end and these same bodies.

As in Article 46, let us designate by x the segment of the line where the abscissa corresponds to ν, and by a the segment which corresponds to α where the length of the string is ℓ. Then one will obtain $r/n+1 = x/\ell$, $s/n+1 = a/\ell$ and similarly, $s'/n+1 = a'/\ell$ which will give

$$r = \frac{(n+1)x}{\ell}, \qquad s = \frac{(n+1)a}{\ell}, \qquad s' = \frac{(n+1)a'}{\ell}$$

and instead of $\xi'_r, \alpha_s, \alpha_{s'}$, one can simply write $\xi'_x, \alpha_a, \alpha_{a'}$.

After substituting the values of r, s, s' in the formulas for s and s' of the preceding article and multiplying and dividing by ℓ and $n+1$, respectively, one will obtain

$$a = \pm(x + \ell h't - 2\lambda\ell)$$
$$a' = \pm(x - \ell h't - 2\lambda'\ell)$$
$$\xi'_x = \frac{1}{2}(\pm\alpha_a \pm \alpha_{a'})$$

where the plus or minus signs of α_a and $\alpha_{a'}$ correspond to those of a and a'. The signs will be determined as well as the values of a and a' from the condition that the values are positive and less than ℓ.

51. Let us represent by A and A' the values of $\pm\alpha_a$ and $\pm\alpha_{a'}$ such that one has in general $\xi'_x = (A + A')/2$. Thus

1. If $x + \ell h't$ is between 0 and ℓ, one will take $a = x + \ell h't$ and $A = +\alpha_a$.
2. If $x + \ell h't$ is between ℓ and 2ℓ, one will take $a = -(x + \ell h't - 2\ell)$ and $A = -\alpha_a$.
3. If $x + \ell h't$ is between 2ℓ and 3ℓ, one will take $a = x + \ell h't - 2\ell$ and $A = \alpha_a$.

...

Similarly,

1. If $x - \ell h't$ is between ℓ and 0, one will take $a' = x - \ell h't$ and $A' = \alpha_a$.
2. If $x - \ell h't$ is between 0 and $-\ell$, one will take $a' = -(x - \ell h't)$ and $A' = -\alpha_{a'}$.
3. If $x - \ell h't$ is between $-\ell$ and -2ℓ, one will take $a' = x - \ell h't + 2\ell$ and $A' = \alpha_{a'}$.

...

It is clear that these different cases are reduced to determining the abscissas a or a' by adding or subtracting from the abscissa x the line $\ell h't$ such that when it passes either end of the axis ℓ it is bent backward and reflected by impediments located at its two ends. The corresponding ordinate α_a or $\alpha_{a'}$ should be taken as positive if the number of reflections is even, or negative if this number is odd.

52. But it is still simpler to continue the curve of α on the same axis ℓ, prolonged from both ends such that one has directly the ordinates α_a and $\alpha_{a'}$ which correspond to the abscissas $x + \ell h't$ and $x - \ell h't$.

For this result, after having described on the axis ℓ the polygon having an infinite number of sides or the curve for which the coordinates are α_x, for an arbitrary abscissa x and which will be given by the initial values of the displacements ξ_x of all the points of the string, it will only be necessary to move this same curve alternately below and above the same axis prolonged indefinitely on both sides such that a continuous curve consisting of equal segments located symmetrically about the axis and joined by the same ends results for which the ordinates taken at equal distances from each side of each of the two ends of the axis ℓ are always equal to each other and of opposite sign.

By taking for this curve the ordinates which correspond to the abscissas $x + \ell h't$ and $x - \ell h't$, the values of A and A' will be obtained and the variable ξ'_x will be given, after an arbitrary length of time t, by the formula

$$\xi'_x = \frac{1}{2}(\alpha_{x+\ell h't} + \alpha_{x-\ell h't})$$

One could have immediately deduced the continuation of the curve which represents the values of α from the general demonstration given in Article 37 by assuming that the string, instead of being terminated by two fixed points, continues on both sides to infinity. The polygon which we hypothesized in this article will become here a continuous curve which when applied to the first instant of the motion will be the curve of the values of α prolonged to infinity.

53. Let us now consider the second part of ξ_r, which we denote by ξ''_r and which is represented by the following formula (Article 46)

$$\xi''_r = \mathrm{S}\,\dot{\alpha}_s \Sigma \left[\frac{2 \sin r\varphi}{n+1} \sin s\varphi \frac{\sin(2(n+1)h't \sin \frac{\varphi}{2})}{2(n+1)h' \sin \frac{\varphi}{2}} \right]$$

One must begin by removing from the denominator $\sin(\varphi/2)$, to make it similar to the formulation of ξ'_r and amenable to the same simplifications. For this reason, I take the difference as $\mathrm{D}\xi''_r$ and since the index r is only contained in $\sin r\varphi$, it will suffice to denote this sine function by the operator D.

Now, from known theorems, one has

$$\mathrm{D} \sin r\varphi = \sin(r+1)\varphi - \sin r\varphi = 2 \sin \frac{\varphi}{2} \cos(r + \frac{1}{2})\varphi$$

Therefore, after substituting this value in the expression for $D\xi''_r$ the following equation will result

$$D\xi''_r = \frac{1}{(n+1)h'} S\dot{\alpha}_s \Sigma \left[\frac{2\cos(r+\frac{1}{2})\varphi}{(n+1)} \sin s\varphi \sin(2(n+1)h't \sin\frac{\varphi}{2})\right]$$

For the case of n infinite, if one sets $\sin(\varphi/2) = \varphi/2$, and develops the product $\cos(r + \frac{1}{2})\rho \sin(n+1)h't\varphi$, one will have

$$\begin{aligned} D\xi''_r &= \frac{1}{(n+1)h'} S\dot{\alpha}_s \Sigma \frac{\sin(r+(n+1)h't+\frac{1}{2})\varphi}{n+1} \sin s\varphi \\ &- \frac{1}{(n+1)h'} S\dot{\alpha}_s \Sigma \frac{\sin(r-(n+1)h't+\frac{1}{2})\varphi}{n+1} \sin s\varphi \end{aligned}$$

This expression for $D\xi''_r$ is composed of two parts similar to those of ξ'_r (Article 49). Therefore, the same argumentation can be applied and thereby, it will be reduced to a similar formulation.

Thus having drawn on the axis ℓ the polygon with an infinite number of sides or the curve for which the ordinates for each abscissa x are $\dot{\alpha}_\alpha$, and which will be given by the initial velocities $\dot{\alpha}$, one will transport it alternately below and above the same axis prolonged to infinity on both sides such that one has a continuous curve similar to the one of the preceding article. Then by putting ℓ/Da or ℓ/Dx in place of $n+1$ and neglecting as a small quantity in comparison to x the term $1/(2(n+1))$ one will find

$$D\xi''_x = \frac{Dx}{2\ell h'}(\dot{\alpha}_{x+\ell h't} - \dot{\alpha}_{x-\ell h't})$$

and passing from differences to sums, the following equation will be found

$$\xi''_x = \frac{1}{2\ell h'} S(\dot{\alpha}_{x+\ell h't} - \dot{\alpha}_{x-\ell h't})Dx$$

54. These sums or integrals represent, as one has seen, areas of the curve for which the ordinates are $\dot{\alpha}$ and these areas must begin only at the points where $x = 0$ and where the abscissas are $\ell h't$ and $-\ell h't$. But it is easier to have them start at the common origin of abscissas, which is the first inner extremity of the axis ℓ. In order to do this, one must subtract from the area which begins at this point and which corresponds to the abscissa $x + \ell h't$, the area which corresponds to the abscissa $\ell h't$ so that the remaining area begins only at the point where $x = 0$. With respect to the area which corresponds to the abscissa $x - \ell h't$, one should add to it, the area corresponding to $-\ell h't$ to bring the origin of this area to the origin of abscissas.

Let us denote in general by $(\int \dot{\alpha}\, dx)\alpha$ each area which begins at this origin and which corresponds to an arbitrary abscissa x. According to what was just said, one will have in the expression for ξ''_x

$$\mathrm{S}\,\dot\alpha_{x+\ell h't}\mathrm{D}x = (\int \dot\alpha\,\mathrm{d}x)_{x+\ell h't} - (\int \dot\alpha\,\mathrm{d}x)_{\ell h't}$$

$$\mathrm{S}\,\dot\alpha_{x-\ell h't}\mathrm{D}x = (\int \dot\alpha\,\mathrm{d}x)_{x-\ell h't} - (\int \dot\alpha\,\mathrm{d}x)_{-\ell h't}$$

Thus substituting these values and observing that in general

$$(\int \dot\alpha\,\mathrm{d}x)_{\ell h't} + (\int \dot\alpha\,\mathrm{d}x)_{-\ell h't} = 0$$

because from the nature of the curve for $\dot\alpha$, the ordinates which correspond to equal abscissas, but of opposite signs, are also equal and of opposite sign such that one has always $\dot\alpha_{\ell h't} + \dot\alpha_{-\ell h't} = 0$.

Thus one will simply have (preceding article)

$$\xi''_x = \frac{1}{2\ell h'}((\int \dot\alpha\,\mathrm{d}x)_{x+\ell h't} - (\int \dot\alpha\,\mathrm{d}x)_{x-\ell h't})$$

55. Finally, after combining the values of ξ'_x and ξ''_x, and after an arbitrary length of time t, the following general expression for ξ_x, will be obtained

$$\xi_x = \frac{1}{2}(\alpha_{x+\ell h't} + \alpha_{x-\ell h't}) + \frac{1}{2\ell h'}((\int \dot\alpha\,\mathrm{d}x)_{x+\ell h't} - (\int \dot\alpha\,\mathrm{d}x)_{x-\ell h't})$$

One will have similar expressions for the variables η_x, ζ_x, by only replacing h' with h and α, $\dot\alpha$ with β, $\dot\beta$ and γ, $\dot\gamma$, and by assuming that one has drawn in the same manner the curves corresponding to the initial values β, $\dot\beta$ and $\gamma, \dot\gamma$.

Thus having the longitudinal displacements for ξ_x and the lateral displacements η_x, ζ_x at each point of the string which correspond to the abscissa x taken along the axis, one will know the state of the string after an arbitrary length of time t spent after the beginning of the motion and since the initial values α, β, γ as well as $\dot\alpha, \dot\beta, \dot\gamma$, are absolutely arbitrary, one sees that nothing can limit this solution as long as the curves formed from these values have a continuous curvature and do not form any finite angles, which could produce jumps in the expression for the velocities and accelerating forces.

It has been assumed (Article 35) that $h = \sqrt{F/(\ell M)}$, $h' = \sqrt{F'/(\ell M)}$, where ℓ is the length of the string and M is the total mass of all the weights with which it is loaded (Article 33). Thus M will be the mass or the weight of the entire string which is assumed of uniform thickness so that if one calls P its specific weight per unit length of string which depends on the density and cross-section of the string, one will have $M = \ell P$. Consequently, the following equations result

$$h = \frac{1}{\ell}\sqrt{\frac{F}{P}}, \qquad h' = \frac{1}{\ell}\sqrt{\frac{F'}{P}},$$

With respect to the quantities F and F', we have seen that they are two constants of which F expresses the tension in the string and consequently, is proportional to the weight which puts it in tension. But F' depends on the (constitutive) law of the material, relative to the deformation of the string (Article 32).

56. If one examines the nature of the curves which represent the values of α and $\dot{\alpha}$, it is easy to see that the ordinates, which are a distance 2ℓ apart, will always be equal and of the same sign and that the areas which end at these ordinates will also be mutually equal because each area which corresponds to an interval 2ℓ, taken at an arbitrary location of the axis prolonged to infinity, is always zero, since it is composed of two equal parts but of opposite sign.

It follows from this that the value of ξ_x will remain the same if one increases the time t by the quantity $2/h'$ or by an arbitrary multiple of this quantity. Thus the longitudinal displacements of the string will become the same after an interval of time equal to $2/h'$ or $2\ell\sqrt{P/F}$. This is the duration of the longitudinal vibrations.

It will be the same for the values of η_x and ζ_x by replacing h' with h, that is, F' by F. Thus the duration of the transverse vibrations will be $2\ell\sqrt{P/F}$.

All the authors up to the present time who have treated the vibrations of sonorous strings have only considered transverse vibrations and they have found for their duration the same formula which we just gave.

With respect to longitudinal vibrations, Chladni[29] is the sole author, as far as I know, who has mentioned it in his interesting treatise on acoustics in Subsection 45. He gives a method to produce them on a violin string and he observes that the tone they give is not the same as those of transverse oscillations, from which it follows that F' is different from F. Consequently, in the very realistic hypothesis that the elastic force with which each element of the string resists being elongated or shortened is proportional to the exponent m of this element, one has $\Phi = K(\mathrm{D}s)^m$ (Article 14). Then the exponent m must be different from unity (Article 32). And if, as Chladni seems to imply, the longitudinal is always higher than the transverse tone, one should have $F' > F$ and consequently, $m > \ell$.

57. We have seen (Article 36) that a tensioned string, of length ℓ and loaded with n bodies, can move as if it has only a length ℓ/ν, where ν is a divisor of $n+1$. When n is an infinitely large number, ν can be an arbitrary integer number. Thus a sonorous string of length ℓ can oscillate as a string for which the length will be ℓ/ν, that is, an aliquot of ℓ and the duration of its oscillations will be reduced to $(2\ell/\nu)\sqrt{P/F'}$ for longitudinal oscillations and to $(2\ell/\nu)\sqrt{P/F}$ for transverse oscillations.

Indeed, if the initial and arbitrary values α and $\dot{\alpha}$ are such that the curves or the locations of these values on the ℓ-axis, cross this axis in two or ν equal parts and that the segments which correspond to these parts are the same but located alternately above and below the axis, such that at equal distances from both sides of each of these points of intersection, the ordinates are equal and of opposite sign. If these curves are then prolonged to infinity, according to the formulation of Article 49, they will have the same shape as if they were from a string of which the length will only be ℓ/ν and the general expression ξ_x (Article 52) shows that the values of ξ which correspond to the points of intersection are always zero such that the string in its longitudinal oscillations will be partitioned by itself in as many equal parts which will each oscillate as if their ends were fixed.

It will be the same for transverse oscillations represented by the variables η and ζ.

58. Since the tone which a sonorous string produces depends only on the duration of its isochronous oscillations which, for the same tensioned string are proportional to its length, it follows that a string partitioned into aliquot parts[30] will produce tones which will be to the fundamental tone in which the oscillation is total as the fractions which express these parts are to unity. Thus if the string is partitioned into two, three, four, etc. equal parts, the tones will be expressed by the fractions 1/2, 1/3, 1/4, etc. and consequently, will be with respect to the octave, the twelfth, the double octave, the seventeenth, etc. of the fundamental tone.

The various tones that the same string can produce are called "harmonics" and they can be produced at will by lightly pressing the string during its vibration at one of the points of division which are called the "nodes of vibration", after Sauveur[31] who was the first to explain by means of nodes the harmonic sounds of the marine trumpet and of other instruments in the Mémoires of the Académie des Sciences for 1701. Wallis had already observed them in strings vibrating at the octave, the twelfth, the double octave, etc. below another string which was made to vibrate and which vibrates by dividing naturally into two, three, four, etc. equal parts, of which each would give the same tone as the string which is made to vibrate. (Refer to Chapter 107 of his algebra.[32])

59. Theory and experience are in good agreement on the production of harmonic sounds, but it is not easy to explain what is called by Rameau,[33] who has made of it the basis of his system, the resonance of a sonorous body and which consists of the superposition of harmonics on the fundamental tone of any string that is made to vibrate in an arbitrary fashion.

If these harmonics are indeed produced by the same string, at the same time as the fundamental tone, it must be assumed that the string produces simultaneously whole and partial vibrations and that the effective vibrations are composed of these different vibrations as if the entire motion is capable of being composed or regarded as composed of several component motions.[34]

We have already seen above (Article 47) that one can not explain in a plausible manner the coexistence of harmonic tones with Daniel Bernoulli's formula. In addition, the series which could give these different sounds disappears from the formula when one assumes the number of bodies infinite. Furthermore, for each point of the string a simple and uniform isochronous law results, which immediately and simply depends on the initial state as we just demonstrated.

Moreover, if the multiple resonance of strings by composed vibrations is to be explained by any means, one should consider the initial configuration, for example, as formed of different curves superimposed on one another such that one serves as the axis for the next one and for which the first only forms a wave along the entire length of the string. The second forms two equal waves located symmetrically which divide the axis in two equal parts while the third forms three equal waves which divide the axis in three equal parts and so on.

Thus the vibrations of the string can be viewed as composed of entire vibrations along the entire length of the string and of vibrations which correspond only to half of the string, to a third, to a fourth, etc., of the string. Since this superposition of curves and vibrations is only hypothetical, the consequences which could be deduced relative to the coexistence of harmonic sounds would be entirely precarious.

60. Let us return to the general formula found in Article 55. Since the quantities $\alpha_{x+\ell h't}$ and $\alpha_{x-\ell h't}$ are the coordinates of a given curve, which corresponds to the abscissas $x + \ell h't$ and $x - \ell h't$, they can be represented by functions of these abscissas with the same form. Thus by designating an indeterminate function by the letter F, one will have

$$\alpha_{x+\ell h't} = F(x + \ell h't), \qquad \alpha_{x-\ell h't} = F(x - \ell h't)$$

Similarly, by taking another function designated by the letter f, one can put

$$(\int \alpha\, \mathrm{d}x)_{x+\ell h't} = f(x + \ell h't), \qquad (\int \alpha\, \mathrm{d}x)_{x-\ell h't} = f(x - \ell h't)$$

Thus the expression for ξ_x (Article 55) can be put in the following form

$$\xi_x = \frac{F(x + \ell h't) + F(x - \ell h't)}{2} + \frac{f(x + \ell h't) - f(x - \ell h't)}{2\ell h'}$$

in which the functions marked by the letters F and f are arbitrary, since they depend on the initial state of the string.

This expression can even be reduced to a simpler form by observing that $F(x + \ell h't)/2 + f(x + \ell h't)/(2\ell h')$ properly represents a function of $x + \ell h't$ which one can indicate by the function Φ and thus that $F(x - \ell h't)/2 - f(x + \ell h't)/(2\ell h')$ only represents one function of $x - \ell h't$, but different from the preceding one and which can be indicated by another function Ψ.

In this manner, the general expression of ξ will become simply

$$\xi = \Phi(x + \ell h't) + \Psi(x - \ell h't)$$

61. One can arrive directly at this expression using the differential equation which determines the variable ξ (Article 31). This equation, by putting $\mathrm{d}\Pi/\mathrm{d}a = 0$ and F' constant,

as in Article 32 and changing the operator D of the finite differences to the operator d of the infinitesimal differences, becomes

$$\frac{\mathrm{d}^2\xi}{\mathrm{d}t^2}\mathrm{d}m - F'\,\mathrm{d}(\frac{\mathrm{d}\xi}{\mathrm{d}f}) = 0$$

Now if one puts $\mathrm{d}f = \mathrm{d}x$, $\mathrm{d}m = (M\,\mathrm{d}x)/\ell$ and $h' = \sqrt{F'/(\ell M)}$, this equation becomes

$$\frac{\mathrm{d}^2\xi}{\mathrm{d}t^2} - \ell^2 h'^2 \frac{\mathrm{d}^2\xi}{\mathrm{d}x^2} = 0$$

which is to the partial differences of the second order between the three variables ξ, x and t, and which has for its complete integral

$$\xi = \Phi(x + \ell h't) + \Psi(x - \ell h't)$$

where the symbols Φ and Ψ denote two arbitrary functions.

These functions must be determined by the initial state of the string and by the conditions that its two ends are fixed. If they are decomposed into two other functions denoted by the letters F and f, and such that $\Phi = F/2 + f/2\ell h'$ and $\Psi = F/2 - f/2\ell h'$, in such a way that one has

$$\xi = \frac{F(x + \ell h't) + F(x - \ell h't)}{2} + \frac{f(x + \ell h't) - f(x - \ell h't)}{2\ell h'}$$

as we have already deduced from our formulation. The first condition will give by putting $t = 0$, $\xi = Fx = \alpha$ and $\mathrm{d}\xi/\mathrm{d}t = f'x = \dot{\alpha}$ from which one obtains $fx = \int \dot{\alpha}\,\mathrm{d}x$. Thus the values of the functions Fx and fx along the entire length ℓ of the string result immediately by means of the initial values α and $\dot{\alpha}$.

The conditions of fixity at the ends of the string give $\xi = 0$ when $x = 0$ and when $x = \ell$, whatever the value of t. By separately adding to these two conditions the two functions F and f, which is permissible, one has for the first equation

$$F(-\ell h't) = -F(\ell h't), \qquad F(\ell + \ell h't) = -F(\ell - \ell h't)$$

and for the second

$$f(-\ell h't) = f(\ell h't), \qquad f(\ell + \ell h't) = f(\ell - \ell h't)$$

which after differentiation results in the following equations

$$-f'(-\ell h't) = f(\ell h't), \qquad f'(\ell + \ell h't) = -f'(\ell - \ell h't)$$

where one recognizes that the conditions of the function f' are the same as those of the function F.

These conditions determine the values of the function Fx, $f'x$ for the abscissas x negative or greater than ℓ from the values of these functions for the abscissas between 0 and ℓ. It is easy to see that from this development the formulations given in Articles 52 and 53 result.

If instead of the longitudinal displacements ξ, the transverse displacements η and ζ are considered, one has the same differential equation. Consequently, the same integral and the same formulations result by replacing only h' with h and $\dot{\alpha}$, α with $\dot{\beta}$, β or γ, $\dot{\gamma}$.

These formulations are similar to those given by Euler to determine the shape of the string at an arbitrary instant from its initial shape by neglecting the velocities impressed at the beginning of the motion. But it must be noted that because they are founded here only on functions which represent the integrals of the equations of partial differences, they cannot have a greater domain of applicability than the nature of the functions contain whether algebraic or transcendental. But since the differential equation is the same for all the points of the string and for all the instants of its motion, the relation it represents must constantly and uniformly exist between the variables whatever the domain of definition giving it. Consequently, although the arbitrary functions are by themselves of undetermined form, nevertheless, when this form is given by the initial state of the string for some domain of definition, it is natural to conclude that it must remain the same for the entire domain of definition of the function and that it is not permissible to change it in order to adapt it to the conditions which depend on the assumed fixity of the ends of the string.

Also, d'Alembert, to whom is owed the discovery of this integral of arbitrary functions, has always maintained that the formulation which results from it is only legitimate when the initial curve is such that it has by its nature branches alternately equal and similar, all contained in the same equation so that the same function can represent this curve to infinity with all its branches. Euler, on the contrary, while adopting the analytical solution of d'Alembert, believed that it was enough to transport the initial curve alternately above and below the axis to infinity to form a continuous curve without considering whether its different branches could be dependent through the same equation and constrained to the law of continuity of analytical functions. The reader should refer to the Mémoires de Berlin for the years 1747, 1748 and Volumes I and IV of the *Opuscules* of d'Alembert.

62. The formulas which describe the motion of a tensioned string, loaded by an infinite number of identical bodies are straight-forward, since the motion of each body is determined by an individual equation. It is obvious that if these formulas can be applied to the motion of a string with uniform thickness by assuming the number of bodies to be infinite and their respective distance from one another to be infinitesimal, then the law which will result for the vibrations of the string will be entirely independent of its initial state. And if this law happens to be the same as the one which is deduced from the consideration of arbitrary functions, it will prove that these functions can be of arbitrary form, continuous or discontinuous, as long as they represent the initial state of the string. This is the approach

that I used to demonstrate Euler's formulation which was then supported by inadequate proofs in the first volume of the Mémoires de Turin. The analysis which I used is, minus some simplifications which I added afterward, the same as the one I just gave and I believe that it will not be out of place in this treatise because it leads directly to the rigorous solution of one of the most interesting problems of mechanics.

The generality of the arbitrary functions and their independence of the law of continuity which has been demonstrated for the integral of the equation relative to the vibrations of sonorous strings, provides the justification to use these functions, in the same fashion, in the integrals of the other equation of partial differences. I even showed, in the second volume of the cited memoirs, how one could integrate several of these equations without consideration of arbitrary functions and still arrive at the same solutions which could be found by means of these functions taken over their entire domain of definition.

The principle of discontinuous functions is generally accepted for the integrals of all the equations of partial differences and the constructions which Monge has given for a great number of these equations, added to his theory for the generation of surfaces by arbitrary functions, does not leave any doubt about the use of discontinuous functions in problems which depend on equations of this type.

63. There is one observation worthy of note, which is that the same expression $\xi = \Phi(x+kt)+\Psi(x-kt)$ which satisfies the equation of partial differences $(\mathrm{d}^2\xi/\mathrm{d}t^2) - k^2(\mathrm{d}^2\xi/\mathrm{d}x^2) = 0$ also satisfies the same equation of finite differences which can be represented by $(\mathrm{D}_t^2\xi/\mathrm{D}t^2) - k^2(\mathrm{D}_t^2\xi/\mathrm{D}x^2) = 0$ if it is assumed that $\mathrm{D}x = k\,\mathrm{D}t$ and $\mathrm{D}t$ is constant. Indeed, one has, by varying x only

$$D_t^2\Phi(x+kt) = \Phi(x+\mathrm{D}x+kt) - 2\Phi(x+kt) + \Phi(x-\mathrm{D}x+kt)$$

and by varying t only

$$D_t^2\Phi(x+kt) = \Phi(x+kt+k\,\mathrm{D}t) - 2\Phi(x+kt) + \Phi(x+kt-k\,\mathrm{D}t)$$

expressions which become equal by putting $\mathrm{D}x = k\,\mathrm{D}t$. The same result will be found for the function $\Psi(x-kt)$.

If the quantities are infinitesimal, the condition $\mathrm{d}x = k\,\mathrm{d}t$ disappears and the integral always holds. The reason is that then the expression $(\mathrm{d}^2\xi/\mathrm{d}t^2)$, which seems to represent the second difference of ξ, divided by the square of the difference of t is nothing more than a symbol which expresses a simple function of t derived from the basic function ξ but different from this function, which is entirely independent of the value of $\mathrm{d}t$. It is also the same for the expressions $\mathrm{d}^2\xi/\mathrm{d}a^2$, with respect to x. It is in this change of functions that the passage from the finite to the infinitesimal occurs which is the essence of the differential calculus.

64. I would like to make an observation at this point which is often very useful. Its purpose is a new method of interpolation which results from the formulas of Article 48.

We have seen that the formula

$$\frac{2}{n+1}\Sigma(\sin(\frac{r\rho\pi}{n+1})\,\mathrm{S}\sin(\frac{s\rho\pi}{n+1})\alpha_s)$$

becomes equal to α_r when $r = 1, 2, 3, \ldots, n$. Therefore, if there are a series of quantities $\alpha_1, \alpha_2, \alpha_3, \ldots, \alpha_n$, for which the number is n, one will be able to represent by the preceding formula an arbitrary intermediate term for which the order will be marked by an arbitrary number r, integer or not, since by successively putting $r = 1, 2, 3, \ldots, n$, the formula gives $\alpha_1, \alpha_2, \alpha_3, \ldots, \alpha_n$.

The summation sign S represents the sum of all the terms which correspond to $s = 1, 2, 3, \ldots, n$ and the integral sign Σ the sum of all the terms which correspond to $\rho = 1, 2, 3, \ldots, n$, where the quantity π is the sum of two right angles.

Let us assume that only one term α_1 is given, then put $n = 1, s = 1, \rho = 1$, and we will arrive at the general expression for α_r

$$\alpha_r = \alpha_1 \sin\frac{r\pi}{2}$$

Let $n = 2$ and assume that the two terms α_1, α_2, are given. The following general equation will be obtained by putting $s = 1, 2, \rho = 1, 2$

$$\alpha_r = \frac{2}{3}(A'\sin\frac{r\pi}{3} + A''\sin\frac{2r\pi}{3})$$

by defining

$$A' = \alpha_1 \sin\frac{\pi}{3} + \alpha_2 \sin\frac{2\pi}{3})$$
$$A'' = \alpha_1 \sin\frac{2\pi}{3} + \alpha_2 \sin\frac{4\pi}{3}$$

Let $n = 3$ and assume that the given terms are $\alpha_1, \alpha_2, \alpha_3$. The following general equation will be obtained by putting $s = 1, 2, 3$ and $\rho = 1, 2, 3$.

$$\alpha_r = \frac{2}{4}(A'\sin\frac{r\pi}{4} + A''\sin\frac{2r\pi}{4} + A'''\sin\frac{3r\pi}{4})$$

where the coefficients A', A'', A''' are determined by the following formulas

$$A' = \alpha_1 \sin\frac{\pi}{4} + \alpha_2 \sin\frac{2\pi}{4} + \alpha_3 \sin\frac{3\pi}{4}$$
$$A'' = \alpha_1 \sin\frac{2\pi}{4} + \alpha_2 \sin\frac{4\pi}{4} + \alpha_3 \sin\frac{6\pi}{4}$$
$$A''' = \alpha_1 \sin\frac{3\pi}{4} + \alpha_2 \sin\frac{6\pi}{4} + \alpha_3 \sin\frac{9\pi}{4}$$

and so on for any given value of n.

In the usual method of interpolation, a parabolic curve of the form

$$y = a + bx + cx^2 + dx^3 + \cdots$$

is assumed and passed through the ordinates which represent the given numbers. In the preceding method, instead of a parabolic curve, a function of the following form is assumed

$$y = A' \sin(\frac{\pi x}{a}) + A'' \sin(\frac{2\pi x}{a}) + A''' \sin(\frac{3\pi x}{a}) + \cdots$$

There are many cases where this assumption is preferable since it conforms to the nature of the problem to a greater degree.

A
LVIGI LAGRANGE
LA PATRIA

MÉCANIQUE
ANALYTIQUE,

Par J. L. Lagrange, de l'Institut des Sciences, Lettres et Arts, du Bureau des Longitudes; Grand-Officier de la Légion-d'Honneur, etc., etc., etc.

NOUVELLE ÉDITION,

REVUE ET AUGMENTÉE PAR L'AUTEUR.

TOME SECOND.

PARIS,

Mme Ve COURCIER, IMPRIMEUR-LIBRAIRE POUR LES MATHÉMATIQUES.

1815.

TABLE OF CONTENTS

PART II. DYNAMICS

Part II

DYNAMICS

SECTION VII
THE MOTION OF A SYSTEM OF FREE BODIES TREATED AS MASS POINTS AND ACTED UPON BY FORCES OF ATTRACTION

All systems of bodies acting on one another and for which the motion can be determined by the laws of mechanics can only belong to one of three specific classes because their interaction can only occur in three different ways, that is, by forces of attraction when the bodies are free or by the members which link them or yet by immediate impact. Our planetary system belongs to the first class and consequently, all related problems are of primary importance among the problems of dynamics. These problems are the object of this section.

For systems of this class, where the bodies are assumed to move freely, it is very easy to find the equations of motion because it is only a matter of reducing all the forces to three mutually perpendicular directions and then to equate, using the principle of accelerating forces, the force along each of these directions to the component of the velocity in the same direction divided by the time element. Nevertheless, the application of the formulas given in SECTION IV is always preferable because they give directly, without a resolution of forces at the outset, the simplest differential equations, whatever the system of coordinates used to determine the location of the bodies and even when the bodies are constrained to move along given surfaces or lines instead of being free to move.

We begin by discussing the formulas which we will use.

1. Let m, m', m'', etc., be the masses of different bodies treated as mass points and let x, y, z be the rectangular coordinates of the body m; x', y', z' those of the body m'; and so on. All of the coordinates are taken with respect to the same fixed axes in space. Thus

$$T = m\frac{\mathrm{d}x^2 + \mathrm{d}y^2 + \mathrm{d}z^2}{2\,\mathrm{d}t^2} + m'\frac{\mathrm{d}x'^2 + \mathrm{d}y'^2 + \mathrm{d}z'^2}{2\,\mathrm{d}t^2} + \cdots$$

and if in place of the rectangular coordinates x, y, z other coordinates are used such as ξ, η, ζ, it is only necessary to substitute for the quantities x, y, z in the formula $\mathrm{d}x^2 + \mathrm{d}y^2 + \mathrm{d}z^2$ those of ξ, η, ζ. If, in addition, the rectangular coordinates x', y', z' are to be transformed to ξ', η', ζ', it is only necessary to substitute for the values of x', y', z' in the formula $(\mathrm{d}x')^2 + (\mathrm{d}y')^2 + (\mathrm{d}z')^2$ those of ξ', η', ζ' and so on. In this manner, the quantity T will become a function of the variables $\xi, \eta, \zeta, \xi', \eta', \zeta'$, etc. and their first differences.

Now let R, Q, P, etc. be the forces with which each point of the mass m tends towards fixed or free centers at distances denoted by r, q, p, etc. which are also functions of ξ, η, ζ since they are expressed as functions of x, y, z. Let

$$\delta\Pi = R\,\delta r + Q\,\delta q + P\,\delta p + \cdots$$

where $\delta\Pi$ can be either an exact or non-exact differential. Let us also denote by the same letters marked with one, two, or more primes the same quantities relative to the bodies m', m'', etc. so that

$$\delta V = m\,\delta\Pi + m'\,\delta\Pi' + m''\,\delta\Pi'' + \cdots$$

If there were, besides the forces directed towards given centers, forces of mutual attraction between the masses of the bodies m and m', the quantity $mm'R\,\delta r$ should be added to δV where r is the distance between these bodies treated as points and R is the force of attraction which could be a function of distance. Similar quantities should be added for the other bodies which mutually attract each other.

Because the bodies are assumed to be free, the coordinates which determine their position in space are independent and each of them, such as ξ, will give an equation of the form

$$\mathrm{d}\frac{\delta T}{\delta \mathrm{d}\xi} - \frac{\delta T}{\delta\xi} + \frac{\delta V}{\delta\xi} = 0$$

2. When the quantities $\delta\Pi$, $\delta\Pi'$, etc. are exact differentials, which is always the case when the forces are proportional to arbitrary functions of the distances to the centers of attraction and which is the case commonly found in nature, it will be simpler to begin with the integrals Π, Π', etc., which are

$$\begin{aligned}
\Pi &= \int R\,\mathrm{d}r + \int Q\,\mathrm{d}q + \int P\,\mathrm{d}p + \cdots \\
\Pi' &= \int R'\,\mathrm{d}r' + \int Q'\,\mathrm{d}q' + \int P'\,\mathrm{d}p' + \cdots \\
&\vdots
\end{aligned}$$

Then the quantity V will become

$$V = m\Pi + m'\Pi' + m''\Pi'' + \cdots$$

for which, after transformation to a function of the variables $\xi, \eta, \zeta, \xi', \eta', \zeta'$, etc., it will be easy to find the partial derivatives $\delta V/\delta\xi$, $\delta V/\delta\eta$, etc. by differentiation. In this case, if the functions T and V do not contain the time coordinate t explicitly, the following relation will always hold

$$T + V = H$$

where H is an arbitrary constant. This relation expresses the principle of the **Conservation des Forces Vives.**

Chapter I
THE MOTION OF A BODY TREATED AS A POINT AND ATTRACTED BY FORCES PROPORTIONAL TO A FUNCTION OF THE DISTANCE TO A FIXED CENTER AND IN PARTICULAR, ON THE MOTION OF PLANETS AND COMETS ABOUT THE SUN

3. When only the motion of a free body is considered, its mass m can be assumed to be equal to unity, and one will simply have

$$T = \frac{\mathrm{d}x^2 + \mathrm{d}y^2 + \mathrm{d}z^2}{2\,\mathrm{d}t^2}, \qquad V = \Pi$$

where $\delta V = R\,\delta r + Q\,\delta q + P\,\delta p$. In this case, whatever the three independent coordinates which determine the position of the bodies in space, they provide three differential equations of the following form

$$\mathrm{d}\frac{\delta T}{\delta \mathrm{d}\xi} - \frac{\delta T}{\delta \xi} + \frac{\delta V}{\delta \xi} = 0$$

to which the following first order equation $T + V = H$ can be added. This last equation will take the place of one of the three differential equations.

If the motion takes place in a resisting medium where R denotes the resistance, the following terms (SECTION II, Article 8) would have to be added to the value of δV

$$R(\frac{\mathrm{d}x}{\mathrm{d}s}\delta x + \frac{\mathrm{d}y}{\mathrm{d}s}\delta y + \frac{\mathrm{d}z}{\mathrm{d}s}\delta z)$$

but the equation $T + V = H$ would no longer hold.

4. Let us assume that the body m is attracted towards a fixed center by a force R, which is a function of the distance r between the body and its center, then there will result

$$V = \int R\,\mathrm{d}r$$

Let us take the distance r for one of the coordinates of the body and for the two other coordinates let us take the angle between the radius vector r and the xy-plane and the angle between the projection of r on this plane and the x-axis. After placing the origin of the rectangular coordinates x, y, z at the origin of the radius r, such that $r = \sqrt{x^2 + y^2 + z^2}$, the following relations will be easily found[1]

$$x = r \cos\psi \cos\varphi, \qquad y = r\cos\psi \sin\varphi, \qquad z = r \sin\psi$$

and then

$$T = \frac{r^2(\cos^2\psi\,\mathrm{d}\varphi^2 + \mathrm{d}\psi^2) + \mathrm{d}r^2}{2\,\mathrm{d}t^2}, \qquad V = \int R\,\mathrm{d}r$$

Therefore, the following three differential equations with respect to r, ψ, φ will result

$$\begin{aligned}
&\mathrm{d}\frac{\delta T}{\delta \mathrm{d}r} - \frac{\delta T}{\delta r} + \frac{\delta V}{\delta r} = 0\\
&\mathrm{d}\frac{\delta T}{\delta \mathrm{d}\psi} - \frac{\delta T}{\delta \psi} + \frac{\delta V}{\delta \psi} = 0\\
&\mathrm{d}\frac{\delta T}{\delta \mathrm{d}\varphi} - \frac{\delta T}{\delta \varphi} + \frac{\delta V}{\delta \varphi} = 0
\end{aligned}$$

which will become

$$\begin{aligned}
&\frac{\mathrm{d}^2 r}{\mathrm{d}t^2} - \frac{r(\cos^2\psi\,\mathrm{d}\varphi^2 + \mathrm{d}\psi^2)}{\mathrm{d}t^2} + R = 0\\
&\mathrm{d}\frac{r^2\,\mathrm{d}\psi}{\mathrm{d}t^2} + \frac{r^2\sin\psi\cos\psi\,\mathrm{d}\varphi^2}{\mathrm{d}t^2} = 0\\
&\mathrm{d}\frac{r^2\cos^2\psi\,\mathrm{d}\varphi}{\mathrm{d}t^2} = 0
\end{aligned}$$

and the equation $T + V = H$ will readily give the following integral equation

$$\frac{r^2(\cos^2\psi\,\mathrm{d}\varphi^2 + \mathrm{d}\psi^2) + \mathrm{d}r^2}{\mathrm{d}t^2} + 2\int R\,\mathrm{d}r = 2H$$

where H is an arbitrary constant.

5. The third equation of the differential equations can be integrated separately. The integral is $r^2\cos^2\psi\,\mathrm{d}\varphi/\mathrm{d}t = C$, where C is an arbitrary constant. The second equation becomes integrable after substituting the expression $C/(r^2\cos^2\psi)$ for $\mathrm{d}\varphi/\mathrm{d}t$ obtained from the integration of the third equation and after multiplying by $2r^2\,\mathrm{d}\psi$. The integral is then

$$\frac{r^4\,\mathrm{d}\psi^2}{\mathrm{d}t^2} + \frac{C^2}{\cos^2\psi} = E^2$$

where E is a new arbitrary constant.

At the outset, note that if ψ and $\mathrm{d}\psi/\mathrm{d}t$ are simultaneously assumed to be equal to zero in this integral at a given time, they will always be zero, because if $\psi = 0$ and $\mathrm{d}\psi/\mathrm{d}t = 0$ for an instant, the last equation becomes $C^2 = E^2$, and then, after substituting C^2 for E^2, it will become

$$\frac{r^4\,\mathrm{d}\psi^2}{\mathrm{d}t^2} + C^2\tan^2\psi = 0$$

which can only hold if the two following equations are satisfied

$$\psi = 0, \qquad \frac{d\psi}{dt} = 0$$

This assumption requires that the body move for an instant in the xy-plane, which is always possible, because the orientation of this plane is arbitrary. Then the body will continue to move in the same plane and describe a planar curve, that is, a line with a single curvature. This can also be demonstrated directly by integration of the differential equation.

After substituting the expression obtained from the first integral for dt, the equation becomes

$$\frac{C^2\, d\psi^2}{\cos^4\psi\, d\varphi^2} + \frac{C^2}{\cos^2\psi} = E^2$$

Let $\psi = 0$ when $d\psi/d\varphi = \tan i$, then

$$E^2 = C^2 + C^2\tan^2 i = \frac{C^2}{\cos^2 i}$$

and the last equation will become

$$\frac{d\psi^2}{\cos^4\psi\, d\varphi^2} = \frac{1}{\cos^2 i} - \frac{1}{\cos^2\psi} = \tan^2 i - \tan^2\psi$$

from which there results

$$d\varphi = \frac{d\psi}{\cos^2\psi\sqrt{\tan^2 i - \tan^2\psi}}$$

an equation whose integral is

$$\varphi - h = \sin^{-1}\left(\frac{\tan\psi}{\tan i}\right)$$

or

$$\tan\psi = \tan i \sin(\varphi - h)$$

where h is the value of φ when $\psi = 0$.

This equation shows that $\varphi - h$ and ψ are the two sides of a triangle[2] in which the angle i is opposite the side ψ. Therefore, since the arc $\varphi - h$ lies in the xy-plane and the arc ψ is always perpendicular to this plane, it follows that the arc which joins these two, and which forms the hypothenuse of the triangle, will make a constant angle i with the base $\varphi - h$. Consequently, this arc will pass through the extremities of all the arcs ψ and all the radii r will be in the plane of the same arc, which will be the plane of the orbit of the body, for which the angle with the xy-plane will be the constant angle i, and for which the angle between the intersection with this same plane and the x-axis will be the angle h.

If, in order to fix these ideas, the xy-plane is taken for the ecliptic, φ will be the longitude with respect to the ecliptic, ψ the latitude, h the longitude of the node of the orbit and i its inclination.

6. Let us now consider the integral

$$\frac{r^2(\cos^2\psi\,\mathrm{d}\varphi^2 + \mathrm{d}\psi^2) + \mathrm{d}r^2}{\mathrm{d}t^2} + 2\int R\,\mathrm{d}r = 2H$$

After replacing $\mathrm{d}\psi$ with the expression for $\mathrm{d}\varphi$ found above, it becomes

$$\frac{r^2\cos^4\psi\,\mathrm{d}\varphi^2}{\cos^2 i\,\mathrm{d}t^2} + \frac{\mathrm{d}r^2}{\mathrm{d}t^2} + 2\int R\,\mathrm{d}r = 2H$$

which can then be combined with the other integral

$$\frac{r^2\cos^2\psi\,\mathrm{d}\varphi}{\mathrm{d}t} = C$$

If the expression for $\mathrm{d}\varphi$ obtained from this last equation, is substituted and with the following definition $C/\cos i = D$, the integral equation will now become

$$\frac{\mathrm{d}r^2}{\mathrm{d}t^2} + \frac{D^2}{r^2} + 2\int R\,\mathrm{d}r = 2H$$

which can be solved for the differential of t

$$\mathrm{d}t = \frac{\mathrm{d}r}{\sqrt{2H - 2\int R\,\mathrm{d}r - D^2/r^2}}$$

The expressions for t as a function of r and reciprocally, for r as a function of t, are obtained from the integration of this equation.

7. The variable φ is then obtained from the equation

$$\mathrm{d}\varphi = \frac{D\cos i\,\mathrm{d}t}{r^2\cos^2\psi}$$

or since the plane of the angle φ is arbitrary, if this plane is considered to be coincident with the plane of the orbit by setting $i = 0$ it will result that $\psi = 0$ (Article 5). Consequently, $\mathrm{d}\varphi = D\,\mathrm{d}t/r^2$ and in this case, the angle $\mathrm{d}\varphi$ will be the angle made by the radius r in the plane of the orbit. Thus, if this angle is denoted in general by $\mathrm{d}\Phi$ the following relation will be obtained

$$\mathrm{d}\Phi = \frac{D\,\mathrm{d}t}{r^2}$$

and if the expression above for $\mathrm{d}t$ is substituted, the integral equation becomes

$$\mathrm{d}\Phi = \frac{D\,\mathrm{d}r}{r^2\sqrt{2H - 2\int R\,\mathrm{d}r - D^2/r^2}}$$

The integral of this equation will give the quantity Φ as a function of r and reciprocally the quantity r as a function of Φ.

Then the expression for φ as a function of Φ will be obtained from the equation

$$\mathrm{d}\Phi = \frac{\cos^2\psi\,\mathrm{d}\varphi}{\cos i}$$

which after substituting for $\cos\psi$ the value obtained from the equation

$$\tan\psi = \tan i \sin(\varphi - h)$$

which was developed earlier in this section, it will become

$$\mathrm{d}\Phi = \frac{\mathrm{d}\varphi}{\cos i[1 + \tan^2 i \sin^2(\varphi - h)]} = \frac{\cos i\,\mathrm{d}(\tan(\varphi - h))}{\cos^2 i + \tan^2(\varphi - h)}$$

from which the following equation is obtained after integration

$$\Phi + k = \tan^{-1}\left(\frac{\tan(\varphi - h)}{\cos i}\right)$$

where k is an arbitrary constant. From this result

$$\tan(\varphi - h) = \cos i \tan(\Phi + k)$$

which is the equation which demonstrates that $(\Phi + k)$ is the hypothenuse of the same triangle for which the base is $(\varphi - h)$ and the adjacent angle is i (Article 5) and in addition, the side opposite to i is ψ.

It is clear from the preceding results that $(\Phi + k)$ is the angle described by the radius r in the plane of the orbit and whose origin is the line of intersection of this plane with the xy-plane, that $(\varphi - h)$ is the angle described by the projection of this radius on the same plane, and that i is the inclination of the plane of the orbit on the fixed xy-plane.

8. Thus the problem is solved because it depends only on the integration of two independent equations between t, Φ and r. The six arbitrary constants required to complete the integration of the three differential equations between r, φ and ψ will be i, h, D and H, and the two constants that the integration will introduce in the expressions for t and Φ.

In the preceding solution, we have taken for coordinates the radius vector and the two angles of longitude and latitude, in order to follow the conventions of the astronomers. Thus this solution has the advantage of providing directly most of the theorems which cannot ordinarily be found except by the methods of spherical trigonometry. However, from the analytical point of view, it is simpler if the original rectangular coordinates are retained. It is useful to demonstrate this claim for the new formulas which will be used in the following developments.

9. By taking x, y, z as the three independent variables, the general formulas of Article 3 will immediately give the three differential equations

$$\frac{\mathrm{d}^2 x}{\mathrm{d}t^2} + R\frac{x}{r} = 0, \qquad \frac{\mathrm{d}^2 y}{\mathrm{d}t^2} + R\frac{y}{r} = 0, \qquad \frac{\mathrm{d}^2 z}{\mathrm{d}t^2} + R\frac{z}{r} = 0$$

and the integral equation

$$\frac{\mathrm{d}x^2 + \mathrm{d}y^2 + \mathrm{d}z^2}{2\,\mathrm{d}t^2} + \int R\,\mathrm{d}r = H$$

By eliminating R from the three differential equations, three integrable equations result immediately, for which the integrals are

$$\frac{x\,\mathrm{d}y - y\,\mathrm{d}x}{\mathrm{d}t} = C, \qquad \frac{z\,\mathrm{d}x - x\,\mathrm{d}z}{\mathrm{d}t} = B, \qquad \frac{y\,\mathrm{d}z - z\,\mathrm{d}y}{\mathrm{d}t} = A$$

where C, B, A are arbitrary constants. The first constant C is identical to the constant in the equation $(r^2 \cos^2 \psi d\,\varphi)/\mathrm{d}t = C$ of Article 5, because this latter equation is only a transformation of the equation $(x\,\mathrm{d}y - y\,\mathrm{d}x)/\mathrm{d}t = C$ after the expressions for x, y, z of Article 4 have been substituted.

The three integrals correspond to those given for a system of bodies in Article 9 of SECTION III from which they could have been obtained.

10. By adding together the squares of the last three equations and using a well-known method for simplification, the following equation results

$$\begin{aligned}&(x\,\mathrm{d}y - y\,\mathrm{d}x)^2 + (z\,\mathrm{d}x - x\,\mathrm{d}z)^2 + (y\,\mathrm{d}z - z\,\mathrm{d}y)^2\\ &= (x^2 + y^2 + z^2)(\mathrm{d}x^2 + \mathrm{d}y^2 + \mathrm{d}z^2) - (x\,\mathrm{d}x + y\,\mathrm{d}y + z\,\mathrm{d}z)^2\end{aligned}$$

Consequently, the following equation results

$$\frac{r^2(\mathrm{d}x^2+\mathrm{d}y^2+\mathrm{d}z^2)-r^2\,\mathrm{d}r^2}{\mathrm{d}t^2}=A^2+B^2+C^2$$

which becomes, after substituting for $(\mathrm{d}x^2+\mathrm{d}y^2+\mathrm{d}z^2)$ its equivalent expression which is obtained from the first integral, and denoting for the sake of brevity

$$A^2+B^2+C^2=D^2$$

the following equation

$$2r^2\Big(H-\int R\,\mathrm{d}r\Big)-\frac{r^2\,\mathrm{d}r^2}{\mathrm{d}t^2}=D^2$$

The following equation is immediately obtained from the preceding equation

$$\mathrm{d}t=\frac{\mathrm{d}r}{\sqrt{2\big(H-\int R\,\mathrm{d}r\big)-D^2/r^2}}$$

as in Article 6.

If the same equations are added together after multiplication of the first equation by z, the second by y and the third by x, the result will be

$$Cz+By+Ax=0$$

which is the expression for a plane passing through the origin of the coordinate system and it shows that the orbit described by the body is a planar curve about the center of force.

11. Let ξ and η be the rectangular coordinates of this curve with the ξ-axis taken along the line of intersection of the plane of the curve with the xy-plane. Also let, as in Article 5, i represent the angle made by these two planes and let h be the angle of that same line of intersection with the x-axis. The two quantities i and h will be constant and from known formulas for the transformation of coordinates the following equations will result

$$x=\xi\cos h-\eta\cos i\sin h$$
$$y=\xi\sin h+\eta\cos i\cos h$$
$$z=\eta\sin i$$

These expressions, after substitution in the same equations, will give the equations

$$\frac{\xi\,\mathrm{d}\eta-\eta\,\mathrm{d}\xi}{\mathrm{d}t}\cos i=C$$
$$\frac{\eta\,\mathrm{d}\xi-\xi\,\mathrm{d}\eta}{\mathrm{d}t}\sin i\cos h=B$$
$$\frac{\xi\,\mathrm{d}\eta-\eta\,\mathrm{d}\xi}{\mathrm{d}t}\sin i\sin h=A$$

By adding together the squares of these equations and extracting its root, one will have (Article 6)

$$\frac{\xi\,\mathrm{d}\eta - \eta\,\mathrm{d}\xi}{\mathrm{d}t} = \sqrt{A^2 + B^2 + C^2} = \frac{C}{\cos i} = D$$

so that the expressions for the constants A, B, C will be

$$C = D\cos i, \qquad B = -D\sin i\cos h, \qquad A = D\sin i\sin h$$

It is obvious that, by designating $(\Phi + k)$, as in Article 7, as the angle that the radius r makes with the line of intersection of the plane of the orbit and the fixed xy-plane, the resulting equations will be

$$\xi = r\cos(\Phi + k), \qquad \eta = r\sin(\Phi + k)$$

and the last of the preceding equations will become

$$r^2\,\mathrm{d}\Phi = D\,\mathrm{d}t$$

which gives the known theorem of sectors $\int r^2\,\mathrm{d}\Phi$ proportional to the time t.

After substituting the value of $\mathrm{d}t$, one will have

$$\mathrm{d}\Phi = \frac{D\,\mathrm{d}r}{r^2\sqrt{2H - 2\int R\,\mathrm{d}r - D^2/r^2}}$$

as in the cited article.

Thus, the problem is again reduced to the integration of two independent equations which are a function of t, Φ and r and which were found earlier (Articles 6 and 7). But this integration depends upon the expression for the central force R, which is a function of the radius r.

12. It is clear from these equations that the radius will be a maximum or a minimum, either relative to time t or to the angle Φ, when it is determined from the following equation

$$2H - 2\int R\,\mathrm{d}r - \frac{D^2}{r^2} = 0$$

Let us assume that in integrating these equations, the second members of the integrals are taken such that they start at the point where r is a minimum and that the origin of the angle Φ is also at this point. The angle k will then be the angle made by the radius passing through the same point of the line of intersection of the orbit with the fixed plane (Article 7). The constant k when added to the constant that the integration will add to t and to the constants A, B, C and H or to D, i, h and H will complete the number of six arbitrary constants that the integration of the three differential equations in x, y, z and t must provide.

13. If now

$$X = r\cos\Phi, \qquad Y = r\sin\Phi$$

it is clear that X and Y will be the rectangular coordinates of the curve in its plane with the same origin as the radius r and with the abscissa X directed towards the point where r is a minimum. And, if these quantities are substituted in the expressions for ξ and η of Article 11, one will obtain

$$\xi = X\cos k - Y\sin k, \qquad \eta = Y\cos k + X\sin k$$

Let us substitute these expressions in those for x, y, z of the same article. Also, in order to simplify the expressions, let us make the following definitions

$$\begin{aligned}
\alpha &= \cos k\cos h - \sin k\sin h\cos i\\
\beta &= -\sin k\cos h - \cos k\sin h\cos i\\
\alpha_1 &= \cos k\sin h + \sin k\cos h\cos i\\
\beta_1 &= -\sin k\sin h + \cos k\cos h\cos i\\
\alpha_2 &= \sin k\sin i\\
\beta_2 &= \cos k\sin i
\end{aligned}$$

from which the following equations will be obtained

$$\begin{aligned}
x &= \alpha X + \beta Y = r(\alpha\cos\Phi + \beta\sin\Phi)\\
y &= \alpha_1 X + \beta_1 Y = r(\alpha_1\cos\Phi + \beta_1\sin\Phi)\\
z &= \alpha_2 X + \beta_2 Y = r(\alpha_2\cos\Phi + \beta_2\sin\Phi)
\end{aligned}$$

These expressions have the advantage that the quantities dependent on the motion in the orbit are separated from the quantities which uniquely depend upon the orientation of the orbit relative to the fixed xy-plane.

These expressions for x, y, z conform to the general theory presented in Article 10 of SECTION II and they could have been immediately deduced.

Indeed, by considering the motion in the orbit first, the coordinates X and Y are obtained while the third coordinate Z is equal to zero. These coordinates which only contain three arbitrary constants can be viewed as particular values of the general coordinates x, y, z.

Then they can be obtained by means of the coefficients α, β, α_1, etc., which contain the three other constants.

14. If instead of considering the motion of the body in the orbit, this motion is referred to an arbitrary plane, with the three coordinates X, Y, Z, which also contains only three arbitrary constants, one would then have from the same theory, the general expressions

$$x = \alpha X + \beta Y + \gamma Z$$
$$y = \alpha_1 X + \beta_1 Y + \gamma_1 Z$$
$$z = \alpha_2 X + \beta_2 Y + \gamma_2 Z$$

and since it was found in Article 10 of SECTION III that

$$\gamma = \alpha_1\beta_2 - \beta_1\alpha_2, \qquad \gamma_1 = \beta\alpha_2 - \alpha\beta_2, \qquad \gamma_2 = \alpha\beta_1 - \beta\alpha_1$$

one would have

$$\gamma = \sin h \sin i, \qquad \gamma_1 = -\cos h \sin i, \qquad \gamma_2 = \cos i$$

These expressions for α, β, γ, α_1, etc., which contain the three arbitrary values k, h, i, satisfy in a general manner the six equations of condition given in Article 10 of SECTION III of PART I, which are

$$\alpha^2 + \alpha_1^2 + \alpha_2^2 = 1, \qquad \beta^2 + \beta_1^2 + \beta_2^2 = 1$$

$$\gamma^2 + \gamma_1^2 + \gamma_2^2 = 1, \qquad \alpha\beta + \alpha_1\beta_1 + \alpha_2\beta_2 = 0$$

$$\alpha\gamma + \alpha_1\gamma_1 + \alpha_2\gamma_2 = 0, \qquad \beta\gamma + \beta_1\gamma_1 + \beta_2\gamma_2 = 0$$

With the general formulas for the motion of a body attracted towards a fixed point, it only remains to apply them to the motions of planets and comets. This is the object of the following sections.

Subsection I
The Motion Of Planets And Comets About The Sun Which Is Assumed Fixed

15. In our world system, the force of attraction is inversely proportional to the square of the distance. Let $R = g/r^2$, where g is the force with which a planet is attracted toward the Sun from a unit distance, which will give $\int R\,\mathrm{d}r = -g/r$.

After substituting this expression in the equation between Φ and r (Article 11), the quantity under the integral sign becomes

$$2H + \frac{2g}{r} - \frac{D^2}{r^2}$$

which can be put in the form

$$2H + \frac{g^2}{D^2} - \left(\frac{D}{r} - \frac{g}{D}\right)^2$$

Then the second member of the equation will express the differential of the angle for which the cosine is

$$\frac{\frac{D}{r} - \frac{g}{D}}{\sqrt{2H + \frac{g^2}{D^2}}}$$

such that by integrating, adding the arbitrary constant K to Φ, and using the cosines of the arcs, one will have

$$\frac{D}{r} - \frac{g}{D} = \sqrt{2H + \frac{g^2}{D^2}} \cos(\Phi + K)$$

It is clear that the minimum value of r is obtained when the angle $(\Phi + K)$ is zero. Therefore, as we have assumed in Article 12, the angle Φ is to be measured from the point which corresponds to a minimum for r and where the value of K is found to be zero.

Then if the following definitions are made in order to simplify the resulting equation

$$b = \frac{D^2}{g}, \qquad e = \sqrt{1 + \frac{2HD^2}{g^2}}$$

the following equation will be obtained

$$r = \frac{b}{1 + e \cos \Phi}$$

This is the polar equation of a conic section for which b is the parameter, e is the eccentricity, that is, the ratio of the distance between the foci to the length of the major axis, r is the radius vector from one of the foci, and Φ is the angle that the radius vector makes with the segment of the major axis from this focus to the nearest vertex.

Since the largest and smallest values of r are $b/(1 - e)$ and $b/(1 + e)$, their half sum will be $b/(1 - e^2)$; that is, the average distance, which we will designate by a, such that we will have $b = a(1 - e^2)$ and if in place of b and e, the expressions for D and H are substituted, one will have

$$\frac{1}{a} = \frac{1 - e^2}{b} = -\frac{2H}{2g}$$

from which it is clear that the constant H must be negative so that the orbit is an ellipse. If it were zero, the axis $2a$ would be infinite and the orbit would be a parabola. But, if it were positive, the axis $2a$ would be negative,[3] and the orbit would be a hyperbola. In the first case, the value of the eccentricity e will be less than unity. It will be equal to unity in the second case and greater than unity in the third case.

There is also another hypothesis defining the law of attraction which gives an elliptical orbit. This is the hypothesis that the attraction is directly proportional to distance. But because it is not applicable to the planets, it will not be considered further. For additional discussion of this case, the reader should refer to the *Principia* of Newton and the works where his theories of analysis have been discussed.

16. Let us now return to the equation which gives t as a function of r (Article 10) and let us substitute for $\int R\,\mathrm{d}r$ the expression $(-g/r)$, in place of D^2 substitute $gb = ga(1-e^2)$, and for $2H$ substitute the expression $-g/a$. The equation will now become

$$\mathrm{d}t = \frac{r\,\mathrm{d}r}{\sqrt{ga}\sqrt{e^2 - (1 - r/a)^2}}$$

The equation $1 - r/a = e\cos\theta$ can be solved for r, which gives

$$r = a(1 - e\cos\theta)$$

Thus the above equation becomes

$$\mathrm{d}t = \sqrt{\frac{a^3}{g}}\,(1 - e\cos\theta)\mathrm{d}\theta$$

and after integrating and introducing an arbitrary constant of integration c

$$t - c = \sqrt{\frac{a^3}{g}}(\theta - e\sin\theta)$$

This equation will give θ as a function of t and since r has already been obtained as a function of θ one will have after the substitution, r as a function of t.

If the same substitution is made in the equation between Φ and r of Article 11, one will have

$$\mathrm{d}\Phi = \frac{\mathrm{d}\theta\sqrt{1-e^2}}{1 - e\cos\theta}$$

for which the integral is

$$\Phi = \sin^{-1}\left(\frac{\sin\theta\sqrt{1-e^2}}{1 - e\cos\theta}\right) + \text{const.}$$

But the expression for Φ as a function of θ can be obtained without another integration by a simple comparison of the expressions for r, which gives the equation

$$\frac{b}{1+e\cos\Phi}=a(1-e\cos\theta)$$

from which it is found, since $b=a(1-e^2)$, that

$$\cos\Phi=\frac{\cos\theta-e}{1-e\cos\theta},\qquad \sin\Phi=\frac{\sin\theta}{1-e\cos\theta}\sqrt{1-e^2}$$

and from which

$$\tan\frac{\Phi}{2}=\sqrt{\left(\frac{1+e}{1-e}\right)}\tan\frac{\theta}{2}$$

It is clear from these formulas that when the angle θ is increased by 360 degrees, the radius r remains the same but the angle Φ is also increased by 360 degrees. Therefore, the planet returns to the same point after having completed an entire revolution. But since the angle θ has increased by 360 degrees, the time t will increase by $360^\circ\sqrt{a^3/g}$. This is the time required for the planet to return to the same point in its orbit. Hence, this amount of time is called the planet's **period**. The **period** depends only on the major axis $2a$ and it would be the same if the planet described a circle for which the radius is the average distance a. In this case, one would have

$$e=0,\qquad t-c=\theta\sqrt{\frac{a^3}{g}},\qquad \theta=\Phi$$

Thus the time would be proportional to the described angles. If one assumed that the quantity g is equal to unity and that the average distance a from the Earth is unity, the times would be represented by the same angles that the Earth would describe if it moved in a circle for which the average distance would be the radius with a velocity equal to unity. The motion, in this circle, is called by astronomers the **average motion** of the Earth or Sun and it is to this motion that they commonly relate the motions of the other planets.

17. For a hyperbolic orbit, the major axis a is negative and the angle θ is imaginary. In order to apply the above formulas to this case, let

$$a=-A,\qquad \theta=\frac{\Theta}{\sqrt{-1}}$$

One can obtain from known formulas, where i is the number for which the value of the hyperbolic logarithm[4] is equal to unity, the following equations

$$\sin\theta=\frac{\mathrm{i}^{\Theta}-\mathrm{i}^{-\Theta}}{2\sqrt{-1}},\qquad \cos\theta=\frac{\mathrm{i}^{\Theta}+\mathrm{i}^{-\Theta}}{2}$$

and the equations of the preceding article will become

$$t - c = \sqrt{\frac{A^3}{g}}\left(\Theta - e\frac{\mathrm{i}^{\Theta} - \mathrm{i}^{-\Theta}}{2}\right)$$

and in addition

$$\tan\frac{\Phi}{2} = \sqrt{\frac{e+1}{e-1}}\,\frac{\mathrm{i}^{\frac{1}{2}\Theta} - \mathrm{i}^{-\frac{1}{2}\Theta}}{\mathrm{i}^{\frac{1}{2}\Theta} + \mathrm{i}^{-\frac{1}{2}\Theta}}$$

since e is greater than unity.

18. The equation $r(1 + e\cos\Phi) = b$, found in Article 15, gives after substituting the quantity X (Article 13) for $r\cos\Phi$

$$X = \frac{b - r}{e} = \frac{a(1 - e^2) - r}{e}$$

After substituting for r its equivalent in terms of θ, that is, $a(1 - e\cos\theta)$, the following equation results

$$X = a(\cos\theta - e)$$

and since $Y = \sqrt{r^2 - X^2}$, one will have

$$Y = a\sqrt{1 - e^2}\sin\theta$$

which are very simple expressions which can be substituted in the general expressions for x, y, z of the same article.

Thus, it will only be a question of substituting the expression for θ as a function of t, obtained from the equation given in Article 16 in order to obtain the three coordinates as functions of time.

19. The angle θ that we have introduced in place of t is what is called in astronomy the **eccentric anomaly**, and which corresponds to the **average anomaly** $(t - c)\sqrt{g/a^3}$ and to the **true anomaly** Φ. But astronomers are accustomed to compute these angles from the vertex of the ellipse, the farthest point on the orbit from the focus where the Sun is assumed to be located and which is called the **aphelion** or **apsid superior**, instead of in the preceding formulas where they are assumed computed from the vertex of the nearest point on the orbit from this same focus, which is called the **perihelion** or **apsid inferior**. In order to relate them to the **aphelion**, it is only necessary to add 180 degrees or, which is the same, to change the sign of the quantity e. But by taking the origin of the anomalies at the **perihelion**, there is the advantage that the formulas are equally applicable to planets

for which the eccentricities are rather small and to comets for which the eccentricities are almost equal to unity because their major axis is very large while this parameter keeps a finite value.

20. It remains for us to determine θ as a function of t, that is, the **eccentric anomaly** from the **average anomaly**. This problem is known as the **Problem of Kepler**, because he was the first to propose it and the first to search for a solution. Since the equation between t and θ is transcendental, it is impossible to obtain, in general, the expression for θ as a function of t in finite form. But by assuming the eccentricity e to be very small, a relation can be obtained using a series which is more or less convergent. In order to obtain this expression in the simplest manner, we will use the general formula which we demonstrated earlier[5] for the solution by series of an arbitrary equation.

Let there be an equation of the form

$$u = \theta - f(\theta)$$

where $f(\theta)$ is an arbitrary function of θ, then

$$\theta = u + f(u) + \frac{\mathrm{d}(f(u))^2}{2\,\mathrm{d}u} + \frac{\mathrm{d}^2(f(u))^3}{2.3\,\mathrm{d}u^2} + \cdots$$

In general, if the expression for an arbitrary function of θ, denoted by $F(\theta)$, is sought, note that

$$F'(\theta) = \frac{\mathrm{d}F(\theta)}{\mathrm{d}\theta}$$

so that one will have

$$F(\theta) = F(u) + f(u)F'(u) + \frac{\mathrm{d}[(f(u))^2 F'(u)]}{2\,\mathrm{d}u} + \frac{\mathrm{d}^2[(f(u))^3 F'(u)]}{2.3\,\mathrm{d}u^2} + \cdots$$

21. In order to apply this formula to the equation of Article 16, let us put

$$f(\theta) = e\sin\theta, \qquad u = (t-c)\sqrt{\frac{g}{a^3}}$$

then the following equation is obtained immediately

$$\theta = u + e\sin u + e^2\frac{\mathrm{d}\sin^2 u}{2\,\mathrm{d}u} + e^3\frac{\mathrm{d}^2\sin^3 u}{2.3\,\mathrm{d}u^2} + \cdots$$

where only the indicated differentiations remain to be carried out. But in order to obtain the simplest expressions, it is necessary to first develop the powers of the sine function in terms of sine and cosine functions of angles which are a multiple of u.

Thus we have

$$\sin\theta = \sin u + e\sin u\cos u + e^2\frac{d(\sin^2 u\cos u)}{2\,du} + e^3\frac{d^2(\sin^3 u\cos u)}{2.3\,du^2} + \cdots$$

$$\cos\theta = \cos u - e\sin^2 u - e^2\frac{d\sin^3 u}{2\,du} - e^3\frac{d^2\sin^4 u}{2.3\,du^2} - \cdots$$

$$\tan\theta = \tan u + e\frac{\sin u}{\cos^2 u} + e^2\frac{d\left(\dfrac{\sin^2 u}{\cos^2 u}\right)}{2\,du} + e^3\frac{d^2\left(\dfrac{\sin^3 u}{\cos^2 u}\right)}{2.3\,du^2} + \cdots$$

By the formulas of Articles 16 and 17, we will also have

$$r = a(1 - e\cos u + e^2\sin^2 u + e^3\frac{d\sin^3 u}{2\,du} + e^4\frac{d^2\sin^4 u}{2.3\,du^2} - \cdots$$

$$r^n = a^n\left\{\begin{array}{l}(1-e\cos u)^n + ne^2\sin^2 u(1-e\cos u)^{n-1} \\ +\dfrac{ne^3\,d\sin^3 u(1-e\cos u)^{n-1}}{2\,du} + \cdots\end{array}\right\}$$

$$X = a[\cos u - e(1+\sin^2 u) - e^2\frac{d\sin^3 u}{2\,du} - e^3\frac{d^2\sin^4 u}{2.3\,du^3} - \cdots]$$

$$Y = a\sqrt{1-e^2}\left\{\begin{array}{l}\sin u + e\sin u\cos u + e^2\dfrac{d(\sin^2 u\cos u)}{2\,du} \\ +e^3\dfrac{d^2(\sin^3 u\cos u)}{2.3\,du^2} + \cdots\end{array}\right\}$$

$$\tan\frac{\Phi}{2} = \sqrt{\left(\frac{1+e}{1-e}\right)}\left\{\begin{array}{l}\tan\dfrac{u}{2} + 2e\sin^2\dfrac{u}{2}\dfrac{\sin^2 u}{1+cosu} + \dfrac{e^2}{2}\dfrac{d\left(\dfrac{\sin^2 u}{1+\cos u}\right)}{2\,du} \\ +e^3\dfrac{d^2\left(\dfrac{\sin^3 u}{1+\cos u}\right)}{2.3\,du^2} + \cdots\end{array}\right\}$$

22. The expression for the angle Φ could be obtained from the series which gives the tangent of the angle. But, it would be difficult in this fashion to have a series for which the law is known. In order to obtain the series, the expression for the angle Φ will first be obtained from the equation

$$\tan\frac{\Phi}{2} = \sqrt{\frac{1+e}{1-e}}\tan\frac{\theta}{2}$$

which can be done elegantly by using imaginary exponentials. Thus the following transformation will be obtained after defining i as the number for which the hyperbolic logarithm is equal to unity[6]

$$\frac{i^{\frac{\Phi}{2}\sqrt{-1}} - i^{-\frac{\Phi}{2}\sqrt{-1}}}{i^{\frac{\Phi}{2}\sqrt{-1}} + i^{-\frac{\Phi}{2}\sqrt{-1}}} = \sqrt{\frac{1+e}{1-e}}\frac{i^{\frac{\theta}{2}\sqrt{-1}} - i^{-\frac{\theta}{2}\sqrt{-1}}}{i^{\frac{\theta}{2}\sqrt{-1}} + i^{-\frac{\theta}{2}\sqrt{-1}}}$$

The preceding equation can be reduced to

$$\frac{\mathrm{i}^{\Phi\sqrt{-1}}-1}{\mathrm{i}^{\Phi\sqrt{-1}}+1}=\sqrt{\frac{1+e}{1-e}}\frac{\mathrm{i}^{\theta\sqrt{-1}}-1}{\mathrm{i}^{\theta\sqrt{-1}}+1}$$

from which we will obtain, after setting $\sqrt{(1+e)/(1-e)}=\epsilon$

$$\mathrm{i}^{\Phi\sqrt{-1}}=\frac{(1+\epsilon)\mathrm{i}^{\theta\sqrt{-1}}+1-\epsilon}{(1-\epsilon)\mathrm{i}^{\theta\sqrt{-1}}+1+\epsilon}$$

or rather, by assuming

$$E=\frac{\epsilon-1}{\epsilon+1}=\frac{e}{1+\sqrt{1-e^2}}$$

one will have

$$\mathrm{i}^{\Phi\sqrt{-1}}=\mathrm{i}^{\theta\sqrt{-1}}\frac{1-E\mathrm{i}^{-\theta\sqrt{-1}}}{1-E\mathrm{i}^{\theta\sqrt{-1}}}$$

Let us now take the logarithms of the two members. After dividing by the quantity $\sqrt{-1}$, the following equation will be obtained

$$\Phi=\theta+\frac{1}{\sqrt{-1}}\ln(1-E\mathrm{i}^{-\theta\sqrt{-1}})-\frac{1}{\sqrt{-1}}\ln(1-E\mathrm{i}^{\theta\sqrt{-1}})$$

After reducing the logarithms of the second member to a series and then substituting for the imaginary exponentials the corresponding real sine functions, the following series[7] will be obtained

$$\Phi=\theta+2E\sin\theta+\frac{2E^2}{2}\sin 2\theta+\frac{2E^3}{3}\sin 3\theta+\cdots$$

Then it only remains to substitute for θ its expression in terms of u. Thus if in order to shorten the expression the following definition is made

$$U=\cos u+E\cos 2u+E^2\cos 3u+\cdots$$

one will have

$$\Phi=u+2E\sin u+\frac{2E^2}{2}\sin 2u+\frac{2E^3}{3}\sin 3u+\cdots$$
$$+2eEU\sin u+2e^2E\frac{\mathrm{d}(U\sin^2 u)}{2\,\mathrm{d}u}+2e^3E\frac{\mathrm{d}^2(U\sin^3 u)}{2.3\,\mathrm{d}u^2}+\cdots$$

It is possible to reduce the expression for U to a finite form which gives

$$U = \frac{\cos u - E}{1 - 2E\cos u + E^2} = \frac{(1 - \sqrt{1+E^2})\cos u - e}{2(1 - e\cos u)}$$

The advantage of these formulas is that they give the sum of the series which is a new result.

23. By taking the xy-plane for the fixed ecliptic and by assuming the x-axis directed towards the first point of *Aries*,[8] the angle φ is what is called the longitude of the planet, the angle h is the longitude of the node, the angle Ψ is the latitude. It is clear that the angle $(\Phi + k)$, for which $(\varphi - h)$ is the projection on the ecliptic, will be the longitude in the orbit beginning at the node or what is called the *argument of latitude* and the equation (Article 7)

$$\tan(\varphi - h) = \cos i \tan(\Phi + k)$$

which gives the angle φ as a function of Φ, can be solved when the inclination is small by using series and the method of imaginary exponentials presented earlier. Then it is only necessary to replace in the expression for Φ as a function of θ, the quantity $(\varphi - h)$ for $\Phi/2$, $\Phi + k$ for $\theta/2$, and $\cos i$ for ϵ, which will give

$$E = \frac{\cos i - 1}{\cos i + 1} = -\left(\frac{\sin\frac{i}{2}}{\cos\frac{i}{2}}\right)^2 = -\tan^2\frac{i}{2}$$

and one will have

$$\varphi - h = (\Phi + k) - (\tan\frac{i}{2})^2 \sin 2(\Phi + k) + \frac{1}{2}(\tan\frac{i}{2})^4 \sin 4(\Phi + k) - \frac{1}{3}(\tan\frac{i}{2})^6 \sin 6(\Phi + k) + \cdots$$

The equation which gives Ψ as a function of φ (Article 5)

$$\tan\psi = \tan i \sin(\varphi - h)$$

could then be solved in the same manner, but a less elegant series would result. At the outset, one would have the equation in imaginary exponentials, where i is the number for which the hyperbolic logarithm[9] is equal to unity

$$\frac{\mathrm{i}^{\psi\sqrt{-1}} - \mathrm{i}^{-\psi\sqrt{-1}}}{\mathrm{i}^{\psi\sqrt{-1}} + \mathrm{i}^{-\psi\sqrt{-1}}} = \tan i \frac{\mathrm{i}^{(\varphi-h)\sqrt{-1}} - \mathrm{i}^{(\varphi-h)\sqrt{-1}}}{2}$$

from which the following equation can be obtained

$$\mathrm{i}^{2\psi\sqrt{-1}} = \frac{1 + \frac{\tan \mathrm{i}}{2}[\mathrm{i}^{(\varphi-h)\sqrt{-1}} - \mathrm{i}^{-(\varphi-h)\sqrt{-1}}]}{1 - \frac{\tan \mathrm{i}}{2}[\mathrm{i}^{(\varphi-h)\sqrt{-1}} - \mathrm{i}^{-(\varphi-h)\sqrt{-1}}]}$$

and by taking the logarithm of this expression

$$\begin{aligned}\psi = & \frac{\tan \mathrm{i}}{2\sqrt{-1}}[\mathrm{i}^{(\varphi-h)\sqrt{-1}} - \mathrm{i}^{-(\varphi-h)\sqrt{-1}}] \\ & + \frac{\tan^3 i}{3.8\sqrt{-1}}[\mathrm{i}^{(\varphi-h)\sqrt{-1}} - \mathrm{i}^{-(\varphi-h)\sqrt{-1}}]^3 \\ & + \frac{\tan^5 i}{5.32\sqrt{-1}}[\mathrm{i}^{(\varphi-h)\sqrt{-1}} - \mathrm{i}^{-(\varphi-h)\sqrt{-1}}]^5 + \cdots\end{aligned}$$

Finally, by expanding the powers of the imaginary exponentials and substituting for them the corresponding sine functions, there will result

$$\begin{aligned}\psi = & \tan i \sin(\varphi - h) + \frac{\tan^3 i}{3.4}[\sin 3(\varphi \quad h) - 3\sin(\varphi - h)] \\ & + \frac{\tan^5 i}{5.16}[\sin 5(\varphi - h) - 5\sin 3(\varphi - h) + 10\sin(\varphi - h)] + \cdots\end{aligned}$$

The series which we just found are only convergent because the eccentricity e or the inclination i is very small and therefore are only applicable to elliptic orbits not much different from the circle and with small inclination, such as the planets and their satellites. The sole exception would be the planet Pallas,[10] one of the four small planets for which the inclination to the ecliptic is about 34 degrees and for which $\tan^2 i/2$ is still a rather small fraction, such that the series for φ as a function of Φ will converge very rapidly, but the series for Ψ as a function of φ will converge more slowly.

24. Besides the case where the eccentricity e is very small, Kepler's problem is still solveable analytically for the case where the eccentricity is not much different from unity, which is the case for orbits which are almost parabolic, such as those of comets. In this case, the semi-major axis a will become very large and the equation of Article 15

$$\frac{1}{a} = \frac{1 - e^2}{b}$$

where b is the semi-parameter, gives

$$e = \sqrt{1 - \frac{b}{a}} = 1 - \frac{b}{2a} - \frac{b^2}{8a^2} + \cdots$$

The equation between t and θ (Article 16) which is of the form

$$(t - c)\sqrt{\frac{g}{a^3}} = \theta - e \sin \theta$$

shows that when a is very large, θ becomes very small so that $\sin\theta$ can be expanded as a function of θ

$$\sin\theta = \theta - \frac{\theta^3}{2.3} + \frac{\theta^5}{2.3.4.5} - \cdots$$

After making these substitutions in the preceding equation, the following equation will be obtained

$$(t-c)\sqrt{\frac{g}{a^3}} = \frac{\theta^3}{2.3} - \frac{\theta^5}{2.3.4.5} + \cdots + \frac{b}{2a}(\theta - \frac{\theta^3}{2.3} + \cdots) + \frac{b^2}{8a^2}(\theta - \cdots) + \cdots$$

where it is seen that the quantity θ is of the order of $1/\sqrt{a}$. Thus, if $\theta = \Theta/\sqrt{a}$ and if the approximation is only carried out to the order $1/a$, one will have

$$(t-c)\sqrt{g} = \frac{b}{2}\Theta + \frac{1}{2.3}\Theta^3 + \frac{1}{a}(\frac{b^2}{8}\Theta - \frac{b}{4.3}\Theta^3 - \frac{1}{2.3.4.5}\Theta^5) + \cdots$$

Using the same simplification techniques, one will find

$$r = \frac{1}{2}(b+\Theta^2) + \frac{1}{4a}(\frac{b^2}{2} - b\Theta^2 - \frac{1}{2.3}\Theta^4) + \cdots$$

$$\tan\frac{\Phi}{2} = \frac{1}{\sqrt{b}}\Theta - \frac{\sqrt{b}}{4a}(\Theta + \frac{1}{3b}\Theta^2)$$

$$X = \frac{1}{2}(b-\Theta^2) + \frac{b^2}{8a}(1 + \frac{1}{3}\Theta^4)$$

$$Y = \sqrt{b}\Theta - \frac{\sqrt{b}}{2.3a}\Theta^3$$

Let T be the value of Θ when a increases without bound, which is the case for the parabola. Then T can be determined as a function of t from the following equation of the third degree

$$T^3 + 3bT = 6(t-c)\sqrt{g}$$

which gives

$$T = \sqrt[3]{3(t-c)\sqrt{g} + \sqrt{9(t-c)^2 g + b^3}} + \sqrt[3]{3(t-c)\sqrt{g} - \sqrt{9(t-c)^2 g + b^3}}$$

and if one puts

$$T' = \frac{-\frac{b^2T}{4} + \frac{bT^3}{3} + \frac{T^5}{3.4.5}}{b + \frac{T^3}{3}}$$

the following equation will be obtained

$$\Theta = T + \frac{T'}{a} + \cdots$$

and from this result we have

$$r = \frac{1}{2}(b + T^2) + \frac{1}{a}\left(\frac{b^2}{8} - \frac{bT^2}{4} - \frac{T^4}{12} + TT'\right)$$

$$\tan\frac{\Phi}{2} = \frac{T}{\sqrt{b}} - \frac{1}{a}\left(\frac{T\sqrt{b}}{4} + \frac{T^3}{12\sqrt{b}} - \frac{T'}{\sqrt{b}}\right)$$

$$X = \frac{1}{2}(b - T^2) + \frac{1}{a}\left(\frac{b^2}{8} + \frac{b^2T^4}{24} - TT'\right)$$

$$Y = T\sqrt{b} + \frac{1}{a}\left(\frac{T^3\sqrt{b}}{6} - T'\sqrt{b}\right)$$

But the irrational nature of these expressions when they are written as a function of T will always prevent these formulas from being of much use in the analytical calculus of parabolic or nearly parabolic orbits.

25. It is useful to note that relative to parabolic motion the time required to travel over an arbitrary segment of the parabola can be determined from a rather simple formula which contains only the sum of the radius vectors corresponding to the two ends and the chord subtended by this segment.

By having the quantity a infinite and $\Theta = \tau\sqrt{b}$, the preceding formulas give

$$6(t - c)\sqrt{g} = b\sqrt{b}(3\tau + \tau^3), \qquad \tau = \tan\frac{\Phi}{2},$$
$$2r = b(1 + \tau^2), \qquad 2X = b(1 - \tau^2), \qquad Y = b\tau$$

Let us denote the same quantities at a different point on the parabola by a prime. Then the difference $(t' - t)$, or the time spent to travel over a segment of the parabola lying between two given points, will be expressed by the following formula

$$6(t' - t)\sqrt{g} = b\sqrt{b}(3 + \tau^2 + \tau\tau' + \tau'^2)(\tau' - \tau)$$

But one has

$$X = b - r, \qquad Y = \sqrt{2br - b^2}$$

and if the chord which joins the two ends of the radii r and r' is denoted by ν, the following equation will result

$$\nu^2 = (X' - X)^2 + (Y' - Y)^2 = (r' - r)^2 + (\sqrt{2br' - b^2} - \sqrt{2br - b^2})^2$$

In order to shorten this expression, let us put

$$U^2 = \nu^2 - (r' - r)^2$$

so that

$$U = \sqrt{2br' - b^2} - \sqrt{2br - b^2}$$

an equation which is needed in order to calculate the value of b.

After eliminating the square roots and ordering the terms according to the powers of b, the following equation results

$$b^2[(r' - r)^2 + U^2] - bU^2(r' + r) + \frac{U^4}{4} = 0$$

from which the following equation can be obtained

$$b = \frac{U^2(r' + r + \sqrt{4r'r - U^2})}{2[(r' - r)^2 + U^2]}$$

or rather, by multiplying numerator and denominator by $r' + r - \sqrt{(4r'r - U^2)}$

$$b = \frac{U^2}{2(r' + r - \sqrt{4r'r - U^2})}$$

Now, one has

$$\tau = \frac{\sqrt{2br - b^2}}{b}$$

Therefore

$$\tau' - \tau = \frac{U}{b}, \qquad \tau^2 + \tau'^2 + \tau\tau' = \frac{3(r + r')}{b} - 3 - \frac{U^2}{2b^2}$$

and thus

$$6(t' - t)\sqrt{g} = \frac{U}{2b\sqrt{b}}[6b(r' + r) - U^2]$$

After substituting the value of b, this quantity becomes

$$[2(r + r') + \sqrt{4rr' - U^2}]\sqrt{2(r' + r) - 2\sqrt{4rr' - U^2}}$$

Finally, by replacing U^2 by the expression given earlier and putting $r + r' = s$, one will have

$$t' - t = \frac{(2s + \sqrt{s^2 - \nu^2})\sqrt{2s - 2\sqrt{s^2 - \nu^2}}}{6\sqrt{g}}$$

an expression which can be put in the following simpler form

$$t' - t = \frac{(s+\nu)^{\frac{3}{2}} - (s-\nu)^{\frac{3}{2}}}{6\sqrt{g}}$$

which can be verified by taking the squares.

26. This elegant formula was given first by Euler, in the seventh volume of the *Miscellanea Berolinensis*. It could have been deduced from the tenth lemma of the Third Book of the *Principia* by putting in analytical terms the geometrical construction used by Newton to determine the uniform velocity with which the chord of a segment of a parabola would be traversed in the same time that the arc would be traversed by a comet and by observing that in the parabola the half sum of the radius vectors which end at the extremities of an arbitrary arc is always equal to the radius vector which ends at the top of the diameter drawn from the middle of the chord parallel to the axis added to the portion of this diameter between the arc and the chord. From this result and from the ninth lemma, the magnitude of this last radius is obtained, expressed in terms of the chord and the sum of the radius vectors corresponding to its two extremities.

It will be shown later how the same formula can be extended to elliptical or hyperbolic motion.

27. Finally, the equation between θ and t is always solvable by approximate methods when the time t, is assumed to be very small. Then the variable θ, and consequently, all the variables which it depends upon, is to be expressed in series and ordered in powers of t. The series will converge more quickly for smaller values of t. But, in this case, it is simpler to extract the solution directly from the differential equations in x, y, z and t of Article 9 by setting $R = g/r^2$.

By viewing the variables x, y, z as functions of t and by assuming that they become $x + x'$, $y + y'$, and $z + z'$, when t becomes $t + t'$, there results from known theorems that

$$\begin{aligned}
x' &= \frac{dx}{dt}t' + \frac{d^2x}{dt^2}\frac{t'^2}{2} + \frac{d^3x}{dt^3}\frac{t'^3}{2.3} + \cdots \\
y' &= \frac{dy}{dt}t' + \frac{d^2y}{dt^2}\frac{t'^2}{2} + \frac{d^3y}{dt^3}\frac{t'^3}{2.3} + \cdots \\
z' &= \frac{dz}{dt}t' + \frac{d^2z}{dt^2}\frac{t'^2}{2} + \frac{d^3z}{dt^3}\frac{t'^3}{2.3} + \cdots
\end{aligned}$$

and it is only necessary to substitute the values of the differentials of x, y, z obtained from the three following equations

$$\frac{d^2x}{dt^2}+\frac{gx}{r^3}=0,\qquad \frac{d^2y}{dt^2}+\frac{gy}{r^3}=0,\qquad \frac{d^2z}{dt^2}+\frac{gz}{r^3}=0$$

to which can be added, in order to simplify the calculation, the equation for r of Article 10

$$2Hr^2+2gr-\frac{r^2\,dr^2}{dt^2}=D^2$$

and which, after differentiation and division by $2r\,dr$, becomes

$$2H+\frac{g}{r}-\frac{d(r\,dr)}{dt^2}=0$$

After differentiating this equation again and putting $s=r\,dr/dt$, in order to shorten the expression, the following equation will be obtained

$$\frac{d^2s}{dt^2}+\frac{gs}{r^3}=0$$

which has the same form as the preceding equations.

Thus after successive differentiations and substitutions, the following equations will be obtained

$$\begin{aligned}
\frac{d^2x}{dt^2}&=-\frac{gx}{r^3},\qquad \frac{d^3x}{dt^3}=\frac{3gs}{r^5}x-\frac{g}{r^3}\frac{dx}{dt}\\
\frac{d^4x}{dt^4}&=\left(\frac{3g}{r^5}\frac{ds}{dt}-\frac{3.5gs^3}{r^7}+\frac{g^2}{r^6}\right)x+\frac{2.3gs}{r^5}\frac{dx}{dt}\\
\frac{d^5x}{dt^5}&=\left(-\frac{3.3.5g}{r^7}\frac{s\,ds}{dt}+\frac{3.5.7gs^3}{r^9}-\frac{3.5g^2s}{r^8}\right)x\\
&\quad+\left(\frac{3.3g}{r^5}\frac{ds}{dt}-\frac{3.3.5gs^2}{r^7}+\frac{g^2}{r^6}\right)\frac{dx}{dt}
\end{aligned}$$

and so on.

Similar expressions will be obtained for the differentials of y and z after first replacing x by y and then by z.

28. With these substitutions and because in these formulas the quantities x, y, z and their differentials are with respect to the initial time t', if t' is replaced by t and if $\mathbf{x}, \mathbf{y}, \mathbf{z}, \mathbf{r}$ and s designate the values of x, y, z, r and s at time $t=0$, and if in order to shorten the expressions, the following quantities are defined

$$T=1-\frac{g}{r^3}\frac{t^2}{2}+\frac{3gs^2}{r^5}\frac{t^3}{2.3}+\left(\frac{3g}{r^5}\frac{ds}{dt}-\frac{3.5gs^2}{r^7}+\frac{g^2}{r^6}\right)\frac{t^4}{2.3.4}$$

$$+\left(-\frac{3.3.5g}{r^7}\frac{s\,\mathrm{d}s}{\mathrm{d}t}+\frac{3.5.7gs^3}{r^9}-\frac{3.5g^2s}{r^8}\right)\frac{t^5}{2.3.4.5}+\cdots$$

$$V=t-\frac{g}{r^3}\frac{t^3}{2.3}+\frac{2.3gs}{r^5}\frac{t^4}{2.3.4}+\left(\frac{3.3g}{r^5}\frac{\mathrm{d}s}{\mathrm{d}t}-\frac{3.3.5gs^2}{r^7}+\frac{g^2}{r^6}\right)\frac{t^5}{2.3.4.5}+\cdots$$

one will have these expressions

$$x=\mathbf{x}T+\frac{\mathrm{d}x}{\mathrm{d}t}V,\qquad y=\mathbf{y}T+\frac{\mathrm{d}y}{\mathrm{d}t}V,\qquad z=\mathbf{z}T+\frac{\mathrm{d}z}{\mathrm{d}t}V$$

With respect to the constants s and $\mathrm{d}s/\mathrm{d}t$ which are contained in these expressions, it is useful to note that they immediately reduce to the constants D and H, on which depend the parameters a, b, c of the elliptical orbit, as we saw in Article 8. Since, by referring the two equations in r of the preceding article to the origin of the time t, one has

$$\frac{(r\,\mathrm{d}r)^2}{\mathrm{d}t^2}-2gr=2Hr^2-D^2,\qquad \mathrm{d}\frac{r\,\mathrm{d}r}{\mathrm{d}t^2}-\frac{g}{r}=2H$$

that is

$$s^2=2gr+2Hr^2-D^2,\qquad \frac{\mathrm{d}s}{\mathrm{d}t}=2H+\frac{g}{r}$$

and after substituting for H and D^2 their values $-g/2a$ and gb (Article 15), the following equation will result

$$\frac{\mathrm{d}s}{\mathrm{d}t}=g\left(\frac{1}{a}-\frac{1}{r}\right),\qquad s^2=g\left(2r-\frac{r^2}{a}-b\right)$$

from which one has

$$\frac{1}{a}=\frac{1}{r}+\frac{\mathrm{d}s}{g\,\mathrm{d}t},\qquad b=2r-\frac{r^2}{a}-\frac{s^2}{g}$$

It is clear from this development that the quantities T and V depend only upon the shape of the orbit and not upon the orientation of its plane.

29. Because the quantity $r\,\mathrm{d}r/\mathrm{d}t$ or s is determined by a differential equation similar to those which determine x, one will also have for this quantity a similar expression by changing only x and $\mathrm{d}x/\mathrm{d}t$ to s and $\mathrm{d}s/\mathrm{d}t$. Thus

$$s=\frac{r\,\mathrm{d}r}{\mathrm{d}t}=sT+\frac{\mathrm{d}s}{\mathrm{d}t}V$$

From this result, after integrating and adding the constant r_0^2

$$r^2=r_0^2+2s\int T\,\mathrm{d}t+\frac{2\,\mathrm{d}s}{\mathrm{d}t}\int V\,\mathrm{d}t$$

where the integrals must be taken such that they are zero at time $t = 0$. After substituting the values of T and V, and ordering the terms with respect to the powers of t, there will result

$$r^2 = r_0^2 + 2st + \frac{\mathrm{d}s}{\mathrm{d}t}t^2 - \frac{gs}{r^3}\frac{t^3}{3} + \left(\frac{3gs^2}{r^5} - \frac{g}{r^3}\frac{\mathrm{d}s}{\mathrm{d}t}\right)\frac{t^4}{3.4}$$
$$+ \left(\frac{9g}{r^5}\frac{s\,\mathrm{d}s}{\mathrm{d}t} - \frac{15gs^3}{r^7} + \frac{g^2s}{r^6}\right)\frac{t^5}{3.4.5} + \cdots$$

This expression for r^2 must be identical to the one which would give the values of x, y, z because $r^2 = x^2 + y^2 + z^2$. Thus there results

$$r^2 = (\mathbf{x}^2 + \mathbf{y}^2 + \mathbf{z}^2)T^2 + 2\frac{\mathbf{x}\,\mathrm{d}\mathbf{x} + \mathbf{y}\,\mathrm{d}\mathbf{y} + \mathbf{z}\,\mathrm{d}\mathbf{z}}{\mathrm{d}t}TV + \frac{\mathrm{d}\mathbf{x}^2 + \mathrm{d}\mathbf{y}^2 + \mathrm{d}\mathbf{z}^2}{\mathrm{d}t^2}V^2$$

But

$$\mathbf{x}^2 + \mathbf{y}^2 + \mathbf{z}^2 = r^2, \qquad \frac{\mathbf{x}\,\mathrm{d}\mathbf{x} + \mathbf{y}\,\mathrm{d}\mathbf{y} + \mathbf{z}\,\mathrm{d}\mathbf{z}}{\mathrm{d}t} = \frac{r\,\mathrm{d}r}{\mathrm{d}t} = s$$

and $(\mathrm{d}\mathbf{x}^2 + \mathrm{d}\mathbf{y}^2 + \mathrm{d}\mathbf{z}^2)/\mathrm{d}t^2 = 2H + 2g/r$ (from Art. 9) $= \mathrm{d}(r\,\mathrm{d}r)/\mathrm{d}t + g/r$ (from Art. 27) $= \mathrm{d}s/\mathrm{d}t + g/r$ so that there results

$$r^2 = r_0^2T^2 + 2sTV + \left(\frac{\mathrm{d}s}{\mathrm{d}t} + \frac{g}{r}\right)V^2$$

an expression which is the same as the one developed earlier.

Subsection II
The Determination Of The Elements Of Elliptical Or Parabolic Motion

30. In the theory of planetary motion, the six constant quantities which are used to determine the shape of the orbit, its location with respect to a fixed plane called the plane of the ecliptic and the time at which the planet is at the aphelion or perihelion are called the *elements of the motion*.

Let a be, as in the preceding paragraph, the semi-major axis or the average distance and let b be the semi-parameter. These two elements determine the configuration of the orbit and if e is called the eccentricity or rather the ratio of the distance between the two focii to the major axis, one has

$$b = a(1 - e^2)$$

and consequently

$$e = \sqrt{1 - \frac{b}{a}}$$

In addition, let c correspond to the time t in which the planet reaches the point of perihelion. This element, in conjunction with the two preceding elements, will be used to determine elliptical motion, independent of the orientation of the orbit in space.

In order to determine the orientation, let k be the longitude of perihelion measured from the line of nodes, that is, the angle that the segment of the major axis which corresponds to perihelion makes with the line of intersection of the plane of the orbit and a fixed plane. This element determines the orientation of the ellipse on the plane of the orbit.

Finally, let i be the inclination of this plane on the fixed plane to which it is referred and which is ordinarily taken in astronomy as the ecliptic. We take it in our formulas for the plane with coordinates x and y, and let h be the longitude of the node, that is, the angle that the intersection of the two planes makes with a fixed line, which astronomers assume is directed toward the first point of *Aries*[11] , and which we take for the x-axis.

These six quantities a, b, c, h, i and k are the elements which need to be determined from whatever conditions are available for a given elliptical motion.

31. The simplest example of this problem is when the position of the body, its velocity and its direction at any given instant are known. In this case, the given parameters are the values of $x, y, z, \mathrm{d}x/\mathrm{d}t, \mathrm{d}y/\mathrm{d}t$ and $\mathrm{d}z/\mathrm{d}t$ for a given instant; values that we will denote by the boldface letters **x, y, z, dx/dt, dy/dt, dz/dt** and which will have to be used to express the six elements a, b, c, k, h and i.

At the outset, Article 9 gives, by substituting $-g/r$ in place of $\int R\,\mathrm{d}r$ and by replacing $x, y, z, r, \mathrm{d}x/\mathrm{d}t, \mathrm{d}y/\mathrm{d}t, \mathrm{d}z/\mathrm{d}t$ by **x, y, z, r, dx/dt, dy/dt, dz/dt**

$$
\begin{aligned}
A &= \mathbf{z}\frac{\mathbf{dy}}{\mathbf{dt}} - \mathbf{y}\frac{\mathbf{dz}}{\mathbf{dt}}\\
B &= \mathbf{z}\frac{\mathbf{dx}}{\mathbf{dt}} - \mathbf{x}\frac{\mathbf{dz}}{\mathbf{dt}}\\
C &= \mathbf{x}\frac{\mathbf{dy}}{\mathbf{dt}} - \mathbf{y}\frac{\mathbf{dx}}{\mathbf{dt}}\\
2H &= (\frac{\mathbf{dx}}{\mathbf{dt}})^2 + (\frac{\mathbf{dy}}{\mathbf{dt}})^2 + (\frac{\mathbf{dz}}{\mathbf{dt}})^2 - \frac{2\mathrm{g}}{\mathrm{r}}
\end{aligned}
$$

and Articles 11 and 15 give

$$
A = D\sin i\sin h, \qquad B = D\sin i\cos h, \qquad C = D\cos i,
$$

$$
D = \sqrt{gb}, \qquad H = -\frac{g}{2a}
$$

Thus from these formulas the values of the semi-major axis a and the semi-parameter b, from which the eccentricity $e = \sqrt{1 - b/a}$ is obtained and the angles h and i, will be immediately obtained. It only remains to find the quantities c and k.

32. It is useful to observe that the values of a and b can be reduced to a simpler form. Indeed, it is clear that $(\mathbf{x}'^2 + \mathbf{y}'^2 + \mathbf{z}'^2)$ is the square of the initial velocity which if it is denoted by u will lead to

$$\frac{1}{a} = \frac{2}{r} - \frac{u^2}{g}$$

from which it is obvious that the major axis of the conic section and consequently, also the periodic time (Article 16), depends only on the initial distance from the body to the focus of attraction and on the velocity of projection.

With respect to the parameter $2b$ in Article 11, the quantity D has been reduced to the form $r^2 \mathrm{d}\Phi/dt$, where $\mathrm{d}\Phi$ is the angle described by the radius r at the instant dt so that $\mathbf{r}\,\mathrm{d}\Phi$ is the smallest arc described by the same radius and consequently, $\mathbf{r}\,\mathrm{d}\Phi/dt$ is the velocity of rotation perpendicular to this radius to the body about the focus.

If the velocity of rotation is denoted by v, the following equation is obtained

$$\frac{r^2\,\mathrm{d}\Phi}{\mathrm{d}t} = \mathbf{r}v = \sqrt{gb}$$

and consequently

$$b = \frac{r^2 v^2}{g}$$

Thus the parameter $2b$ depends only on the radius r and on the part of the velocity u with which the body has the tendency to rotate about the focus toward which it is attracted.

33. In order to find the value of the element c, which determines the time of passage at perihelion, one will note that this constant was included in the calculations by integration which gave the value of r as a function of t (Article 16).

Therefore, if ζ is the value of θ at the time $t = 0$, one will have by the formulas of the cited article, after setting $t = 0$, which changes r to $\mathbf{r}$ and θ to ζ

$$-c = \sqrt{\frac{a^3}{g}}\,(\zeta - e\sin\zeta), \qquad r = a(1 - e\cos\zeta)$$

Thus after elimination of ζ, the value of c as a function of $\mathbf{r}$ will be found because a and e are already known.

Finally, in order to determine the last element k, which was also included by the integration between $\mathbf{r}$ and Φ (Article 15), one will note at the outset that after changing x, y to $\mathbf{x}$, $\mathbf{y}$ (Article 4) and referring the angle φ to the origin of time t

$$\frac{\mathbf{y}}{\mathbf{x}} = \tan\varphi$$

Then from Article 7 there results

$$\tan\Phi = \frac{\tan(\varphi - h)}{\cos i}$$

Now since h and i are already known, one will obtain from the intermediary angle φ, the angle Φ as a function of **x** and **y**. From this result, the variable k is obtained from the equation of Article 15 with respect to the instant where $t = 0$

$$\cos(\varphi - k) = \frac{1}{c}(\frac{b}{r} - 1)$$

34. If two points on the body's orbit and the time spent in orbit between the two points were known, one would also have six given values for the coordinates which correspond to these two points on the orbit and the six elements will be determined by the values of these coordinates. But the transcendental expression for the time will prevent a general and algebraic solution of the problem from being obtained. One will only be able to obtain a solution by approximation by using the formulas of Article 28, provided the interval of time between the two points on the orbit is small enough.

Let $\mathbf{x}, \mathbf{y}, \mathbf{z}$ be the three coordinates of the first point on the orbit, and let $\mathbf{x}', \mathbf{y}', \mathbf{z}'$ be those of the second point. By taking t for the time spent by the body between these two points, one will generally have (Article 28)

$$\mathbf{x}' = \mathbf{x}T + \frac{\mathrm{d}\mathbf{x}}{dt}V, \qquad \mathbf{y} = -\mathbf{y}T + \frac{\mathrm{d}\mathbf{y}}{dt}V, \qquad \mathbf{z} = -\mathbf{z}T + \frac{\mathrm{d}\mathbf{z}}{dt}V$$

Let us assume that the precision of the calculation is to be carried only to the third power of t, then

$$T = 1 - \frac{g}{r^3}\frac{t^2}{2} + \frac{3gs}{r^5}\frac{t^3}{6}, \qquad V = t - \frac{g}{r^3}\frac{t^3}{6}$$

Because the expression for T contains the constant

$$s = \frac{r\,\mathrm{d}r}{\mathrm{d}t} = \frac{\mathbf{x}\,\mathrm{d}\mathbf{x} + \mathbf{y}\,\mathrm{d}\mathbf{y} + \mathbf{z}\,\mathrm{d}\mathbf{z}}{\mathrm{d}t}$$

it will be determined by adding together the three preceding equations, after having multiplied the first by **x**, the second by **y** and the third by **z**. Thus the following equation will result

$$\mathbf{x}\mathbf{x}' + \mathbf{y}\mathbf{y}' + \mathbf{z}\mathbf{z}' = \mathbf{r}^2T + \mathrm{s}V$$

from which the value of **s** is taken which will then be substituted in the expression for T.

Since the values of T and V are known, the same equations will give the values of the differentials $\mathbf{dx}/dt$, $\mathbf{dy}/dt$ and $\mathbf{dz}/dt$. Thus the problem will be reduced to the preceding case.

35. Finally, if the three radius vectors $\mathbf{r}, \mathbf{r}', \mathbf{r}''$ were known and the times t and t' spent between the points located by $\mathbf{r}$ and $\mathbf{r}'$ and by $\mathbf{r}$ and $\mathbf{r}''$ were known, the orbit could still be determined by the formulas of Article 29 by assuming the times t and t' to be sufficient.

Because by setting $t = 0$ in the expression for r^2 and carrying in the series the values of r'^2 and r''^2 only to t^2 and t'^2, one will have

$$r^2 = \mathbf{r}^2, \qquad r'^2 = \mathbf{r}^2 + (2t - \frac{gt^3}{3r^3})s + t^2 \frac{ds}{dt}$$

$$r''^2 = \mathbf{r}^2 + (2t' - \frac{gt'^3}{3r^3})s + t'^2 \frac{ds}{dt}$$

equations from which the values $\mathbf{r}$, $\mathbf{s}$ and ds/dt will be obtained. The last two equations give immediately, using the formulas of Article 19

$$\frac{1}{a} = \frac{1}{r} - \frac{ds}{g\,dt}, \qquad b = 2\mathbf{r} - \frac{r^2}{a} - \frac{s^2}{g}$$

then one will have the angle Π between the radius r and the radius of perihelion, using the formula (Article 15)

$$\cos \Pi = \frac{b - r}{re}$$

where

$$e = \sqrt{1 - \frac{b}{a}}$$

If the orbit were a parabola, one would have a increasing without bound, and consequently, $ds/dt = -g/r$. Then, it would suffice to know the two lengths r and r'. The first value will give the value of $\mathbf{r}$, and the second the value of s from the equation

$$r'^2 = \mathbf{r}^2 - \frac{gt^2}{r} + (2t - \frac{gt^3}{3r^3})s$$

36. The elements of the planets are well-known. It is from the observations of longitude and latitude that they were determined and the minuteness of their eccentricities and of their inclinations on the plane of the ecliptic have contributed greatly to facilitate the determination.

By taking the ecliptic for the xy-plane, the angles φ and ψ (Article 4) represent, the longitude and latitude of the body, respectively. We have given in Articles 22 and 23 for the expressions of φ and ψ as a function of t, series which converge well when the eccentricity e and the inclination i are very small. By taking six observations, three of longitude and three of the corresponding latitude, six equations will generally be obtained from which the six elements can be determined, at least for the Sun and the Moon which rotate about the Earth.

For the other planets which rotate about the Sun, the calculation is a little more complicated because the observations give immediately longitude and latitude observed from the Earth, which are called geocentric. But by assuming that the motion of the Sun is known one can always deduce an equation from each of these observations so that six observations will suffice as a minimum to determine these six elements.

This problem is especially important for comets for which the elements, when they appear, are entirely unknown. Therefore, excluding Newton who was the first to attempt to solve this problem, there have been few geometers and astronomers who have attempted to solve it. Because they could not establish as for the planets the approximation based on the minuteness of the eccentricity and the inclination, they all assumed that the interval of time between the observations is very small and they have given more or less exact methods to deduce the elements of the comets from three observed values of longitude and three of latitude. Since the method I proposed in the *Mémoires de Berlin* in 1783 seems to me to give the most direct and general solution to the problem of comets, I believe I can give it here but somewhat simplified and with a new commentary. It will present an important application of the principal formulas that we have developed in the preceding paragraph.

Subsection III
The Determination Of The Orbits Of Comets

37. At an arbitrary instant, let R be the distance from the comet to the Earth, and let ℓ, m, n be the cosines of the angles made by the line or visual radius R with the three orthogonal axes assumed fixed in space. One will have $R\ell, Rm, Rn$ for the three rectangular coordinates of the comet, parallel to the axes and having for origin the center of the Earth. The quantity R will be unknown but the three quantities ℓ, m, n will be known from observation of the comet and should satisfy the condition

$$\ell^2 + m^2 + n^2 = 1$$

because it must be true from the hypothesis that

$$R^2 = (R\ell)^2 + (Rm)^2 + (Rn)^2$$

Similarly, let ρ, λ, μ, ν, be the corresponding quantities relative to the Sun so that $\rho\lambda, \rho\mu, \rho\nu$ are the rectangular coordinates of the position of the Sun with respect to the Earth and parallel to the same axes. These quantities must be assumed known from the calculation of the position of the Sun at the same instant of observation of the comet and one will then have the condition

$$\lambda^2 + \mu^2 + \nu^2 = 1$$

Finally, let x, y, z be the rectangular coordinates of the position of the comet with respect to the Sun, parallel to the same axes and let r be the radius vector of its orbit about the Sun. It is clear that the following three equations will be obtained

$$x = R\ell - \rho\lambda, \qquad y = Rm - \rho\mu, \qquad z = Rn - \rho\nu$$

and since $r^2 = x^2 + y^2 + z^2$, one will have

$$r^2 = R^2 + \rho^2 - 2R\rho(\ell\lambda + m\mu + n\nu)$$

But, it is known that the expression $(\ell\lambda + m\mu + n\nu)$ is the expression for the cosine of the angle formed by the two radii R and ρ, emanating from the common center of the Earth and directed, one towards the comet and the other towards the Sun, so that if this angle is denoted by σ, one will have

$$r^2 = R^2 - 2R\rho\cos\sigma + \rho^2$$

Therefore, if three observations of a comet are made at known intervals three similar systems of equations will be obtained where each contain a new unknown ρ. From the properties of the parabola[12] three additional equations will be obtained.

38. The simplest approach to reaching this goal is to use the formula given in Article 25 from which the time taken by the comet to describe an arbitrary arc defined by the chord f of the arc and by the sum of the radius vectors which extend to its two ends and are independent of all the elements of the orbit because the three intervals of time between the three observations, taken two at a time, will give the three equations sought.

If the letters which designate analogous quantities in the second observation are denoted by a prime, we will have

$$r'^2 = R'^2 - 2R'\rho'\cos\sigma' + \rho'^2, \qquad s = r + r'$$

For the chord u of the arc traversed by the comet in the time interval between the two observations, it is clear that the following equation will be obtained

$$u^2 = (x' - x)^2 + (y' - y)^2 + (z' - z)^2 = r'^2 + r^2 - 2(xx' + yy' + zz')$$

The following equation results if the expressions for x, y, z and x', y', z' are substituted

$$xx' + yy' + zz' = RR'(\ell\ell' + mm' + nn')$$
$$+\rho\rho'(\lambda\lambda' + \mu\mu' + \nu\nu') - R\rho'(\ell\lambda' + m\mu' + n\nu') - R'\rho(\ell'\lambda + m'\mu + n'\nu)$$

Now from known theorems the expression $(\ell\ell' + mm' + nn')$ must represent the cosine of the angle between the two radii R and R' emanating from the center of the Earth and directed towards the two positions of the comet for the two observations. Similarly, $(\lambda\lambda' + \mu\mu' + \nu\nu')$ will be the cosine of the angle made at the center of the Earth by the two radii ρ and ρ' directed towards two positions of the Sun and similarly for similar expressions.

Therefore, if for additional clarity, it is imagined that the two apparent positions of the comet are marked on the surface of the sphere by the letters C and C$'$, and similarly, the apparent positions on the Sun by the letters S and S$'$ and if the four points C, C$'$, S, S$'$ are connected by large circles it is obvious that the arcs CS, C$'$S$'$, will represent the angles which we have denoted by σ and σ' and that the arcs CC$'$ and SS$'$ will represent the angles for which the cosines are

$$(\ell\ell' + mm' + nn'), \qquad (\lambda\lambda' + \mu\mu' + \nu\nu')$$

and finally, the arcs CS$'$ and C$'$S will represent the angles for which the cosines are

$$(\ell\lambda' + m\mu' + n\nu'), \qquad (\ell'\lambda + m'\mu + n'\nu)$$

Thus, by considering the spheroidal quadrilateral CC$'$SS$'$, which is assumed to be given by the two observations of the comet and by the two calculated positions of the Sun, one will have

$$r^2 = R^2 - 2R\rho\cos(\mathrm{CS}) + \rho^2$$
$$r'^2 = R'^2 - 2R'\rho'\cos(\mathrm{C'S'}) + \rho'^2$$
$$u^2 = r^2 + r'^2 - 2RR'\cos(\mathrm{CC'}) - 2\rho\rho'\cos(\mathrm{SS'})$$
$$+ 2R\rho'\cos(\mathrm{CS'}) + 2R'\rho\cos(\mathrm{C'S})$$

Therefore, since the difference $(t' - t)$ of the times t and t' which correspond to the two observations, that is, the interval of time, is assumed given, one will have from the formula of the cited article, the following equation

$$t' - t = \frac{(r + r' + u)^{\frac{3}{2}} - (r + r' - u)^{\frac{3}{2}}}{6\sqrt{g}}$$

in which the only unknowns are the two distances R and R'.

If there is a third observation, for which analogous quantities are denoted by the same letters marked by two primes, one will have by comparison of the first and third observations, a second similar equation in which the letters marked by a prime in the preceding equation will possess two primes and which will contain only the two unknowns R and R''.

In the same fashion, a third similar equation will be obtained by comparison of the second and third observations, by marking by a single prime in the first equation all the letters which have no primes and by two primes all those which have one prime. Therefore, the third equation will contain only the same unknowns R' and R'' so that the three equations will contain only the three unknowns R, R', R''. Thus the three unknowns R, R', R'' can be determined. But, although these equations are in a rather simple form, their solution presents almost insurmountable difficulties since the unknowns are mixed together and are present in different radicals.

39. Also, if in some fashion the values of the radii R and R' could be found, the semi-parameter b will immediately be obtained which, for the parabola, is equal to twice the perihelion distance from the formula of Article 25

$$b = \frac{u^2 - (r' - r)^2}{2[r + r' - \sqrt{(r + r')^2 - u^2}]}$$

and since we have in general (Article 10) the equation $Cz + Ax + By = 0$ in which $A/C = \sin h \tan i$, $B/C = -\cos h \tan i$ (Article 11), one will have these two equations

$$z - (x \sin h + y \cos h) \tan i = 0$$
$$z' - (x' \sin h + y' \cos h) \tan i = 0$$

from which one will easily obtain the values of $\tan i$ and $\tan h$, which will give the orientation of the plane of the parabola with respect to the plane which will have been chosen for the x- and y-axes.

It should be noted that by means of these equations, which depend on the fact that the orbit of the comet is assumed to be in a plane which passes through the Sun, the three unknowns can be reduced to two.

Indeed, if the following definitions are made in order to shorten the expressions

$$L = \tan i \sin h, \qquad M = \tan i \cos h$$

after substituting the values of x, y and z, the following equation will be obtained

$$Rn - \rho\nu = L(R\ell - \rho\lambda) + M(Rm - \rho\mu)$$

and from which the following equation can be derived

$$R = \rho\frac{\nu - \lambda L - \mu M}{n - \ell L - mM}$$

and in a similar fashion, the expressions for R' and R'', where we have indicated by a prime or by two primes the letters, with the exception of L and M, which are the same for all observations.

In this manner, the three unknowns R, R', R'' will be reduced to two, L and M, so that only two equations will be needed for their determination, which simplifies somewhat the solution of the problem.

40. In order to further simplify the expression, it does not appear that there are other means than to assume the time intervals between the observations small enough so that several terms can be neglected because they are of no consequence, which will first give an approximate solution which can be made more exact later with additional terms. This is also the approach used in all the solutions which have been given for this problem.

By applying this hypothesis to the preceding solution, the chord u will become very small and by keeping only the first two members under the radicals which are in the expression for the interval of time $t' - t$, between the first two observations, one will have

$$t' - t = \frac{u\sqrt{r + r'}}{2\sqrt{g}}$$

which gives

$$u^2(r + r') = 4g(t' - t)^2$$

and there will not be any other radicals in this equation with the exception of those which are in the expressions for r and r'. But the equations between the three unknowns R, R', R'', or between the two expressions L and M, will still be too complicated to be used successfully.

It can be concluded from these results that these unknowns are not the best suited for use in the preceding problem. When only an approximate solution is required, it is simpler to use the formulas that were given in Article 28 for the case where the time t is assumed to be very small.

41. In order to apply these formulas to the determination of the orbits of comets, it is only necessary to substitute for x, y, z the expressions given in Article 37. Thus one will have generally

$$R\ell - \rho\lambda = \mathbf{x}T + \frac{\mathrm{d}\mathbf{x}}{\mathrm{d}t}V$$

$$Rm - \rho\mu = \mathbf{y}T + \frac{\mathrm{d}\mathbf{y}}{\mathrm{d}t}V$$

$$Rn - \rho\nu = \mathbf{z}T + \frac{\mathrm{d}\mathbf{z}}{\mathrm{d}t}V$$

where the quantities $\mathbf{x}, \mathbf{y}, \mathbf{z}$, $\mathrm{d}\mathbf{x}/\mathrm{d}t$, $\mathrm{d}\mathbf{y}/\mathrm{d}t$, $\mathrm{d}\mathbf{z}/\mathrm{d}t$ correspond to the origin of the time t and are viewed as constants, and where T and V are rational functions of t and of the constants r, $\mathrm{d}r/\mathrm{d}t$, $\mathrm{d}^2r/\mathrm{d}t^2$.

Since the origin for the time t is arbitrary, it can be fixed at the instant of the first observation. Now by setting $t = 0$, one has

$$T = 1, \qquad V = 0$$

Thus the following system of equations is obtained for the first observation

$$\begin{aligned} R\ell - \rho\lambda &= \mathbf{x} \\ Rm - \rho\mu &= \mathbf{y} \\ Rn - \rho\nu &= \mathbf{z} \end{aligned}$$

For the second observation at a time t after the first observation, one will have after marking by a prime the letters R, ℓ, m, n, ρ, λ, μ, ν a second system of equations

$$\begin{aligned} R'\ell' - \rho'\lambda' &= \mathbf{x}T + \frac{\mathrm{d}\mathbf{x}}{\mathrm{d}t}V \\ R'm' - \rho'\mu' &= \mathbf{y}T + \frac{\mathrm{d}\mathbf{y}}{\mathrm{d}t}V \\ R'n' - \rho'\nu' &= \mathbf{z}T + \frac{\mathrm{d}\mathbf{z}}{\mathrm{d}t}V \end{aligned}$$

Similar equations will be obtained for the third observation at a time t' after the first observation after marking by two primes the letters marked by one prime in the preceding equations and by a prime the letters T and V to indicate that the variable t must be changed to t'. Therefore, the following system of equations will be obtained

$$\begin{aligned} R''\ell'' - \rho''\lambda'' &= \mathbf{x}T' + \frac{\mathrm{d}\mathbf{x}}{\mathrm{d}t}V' \\ R''m'' - \rho''\mu'' &= \mathbf{y}T' + \frac{\mathrm{d}\mathbf{y}}{\mathrm{d}t}V' \\ R''n'' - \rho''\nu'' &= \mathbf{z}T' + \frac{\mathrm{d}\mathbf{z}}{\mathrm{d}t}V' \end{aligned}$$

The two constants $\mathbf{x}$ and $\mathbf{dx}/dt$ can be eliminated from each of the three systems. And after making the following definitions in order to shorten the expressions

$$TV' - VT' = V''$$

one will have

$$(R\ell - \rho\lambda)V'' - (R'\ell' - \rho'\lambda')V' + (R''\ell'' - \rho''\lambda'')V = 0$$

Similarly, eliminating the two constants $\mathbf{y}$ and $\mathbf{dy}/dt$ from the second equations of the three systems of equations will give

$$(Rm - \rho\mu)V'' - (R'm' - \rho'\mu')V' + (R''\ell'' - \rho''\lambda'')V = 0$$

and the elimination of the constants **z** and d**z**/dt from the last equations of these systems will similarly give

$$(Rn-\rho\nu)V''-(R'n'-\rho'\nu')V'+(R''n''-\rho''\nu'')V=0$$

From these three equations, the expressions for R, R' and R'' can be obtained

$$R=\frac{\rho\Gamma V''-\rho'\Gamma'V'-\rho''\Gamma''V}{GV''}$$
$$R'=-\frac{\rho\Gamma_1V''-\rho'\Gamma_1'V'+\rho''\Gamma_1''V}{GV'}$$
$$R''=\frac{\rho\Gamma_2V''-\rho'\Gamma_2'V'+\rho''\Gamma_2''V}{GV}$$

and in order to shorten the expressions it is assumed that

$$G=\ell m'n''+mn'\ell''+n\ell'm''-\ell n'm''-m\ell'n''-nm'\ell''$$

and by denoting by $\Gamma, \Gamma', \Gamma''$ what becomes of G when ℓ, m, n are changed to λ, μ, ν, then to λ', μ', ν', and finally to λ'', μ'', ν'', respectively; similarly, by denoting by Γ_1, Γ_1', Γ_1'' and by $\Gamma_2, \Gamma_2', \Gamma_2''$ what G becomes by changing in the same manner the quantities ℓ', m', n', and ℓ'', m'', n''.

Now the three observations also give (Articles 37 and 38) the equations

$$R^2-2R\rho\cos(\mathrm{CS})+\rho^2=r^2$$
$$R'^2-2R'\rho'\cos(\mathrm{C'S'})+\rho'^2=r'^2$$
$$R''^2-2R''\rho''\cos(\mathrm{C''S''})+\rho''^2=r''^2$$

Thus if the preceding values of R, R', R'' were substituted, three equations will be obtained which will only contain known quantities, with the quantities V, V', V'' and r, r', r'', which are given as functions of time, and the three constants r, s and ds/dt on which the elements of the orbit depend (Article 28 and 29), so that the three constants can be determined.

42. By carrying the approximation only to the fourth power of t, one has

$$V=t-g\frac{t^3}{6r^3}+gs\frac{t^4}{4r^5}+\cdots$$

and similarly

$$V'=t'-g\frac{t'^3}{6r^3}+gs\frac{t'^4}{4r^5}+\cdots$$

and since $V''=TV'-VT'$, one will find

$$V''=t'-t-g\frac{(t'-t)^3}{6r^3}+g\frac{(t'-t)^3(t+t')^5}{4r^5}+\cdots$$

By making these substitutions in the expressions for R, R', R'' and assuming that $t' = mt$, where the coefficient m is given by the ratio between the two intervals of time between the three observations, it is clear that the fourth dimension of t will disappear from the division and thus it will suffice to consider the third observation in the expressions of r' and r''.

But one has, in general, for the approximation to the fourth power of t

$$r^2 = \mathbf{r}^2 + 2st + \frac{\mathrm{d}s}{\mathrm{d}t}t^2 - \frac{gs}{3r^3}t^3 + \cdots$$

Now we have assumed that the first observation corresponds to $t = 0$ and that the two following observations correspond to times t and $t' = mt$. Thus one will have

$$r^2 = \mathbf{r}^2, \qquad r'^2 = \mathbf{r}^2 + 2st + \frac{\mathrm{d}s}{\mathrm{d}t}t^2 - \frac{gs}{3r^3}t^3$$

$$r''^2 = \mathbf{r}^2 + 2mst + m^2\frac{\mathrm{d}s}{\mathrm{d}t}t^2 + m^2\frac{gs}{3r^3}t^3$$

If these substitutions are made in the last three equations of the preceding article and if the terms which contain powers of t greater than 3 are omitted, one will have three equations between the unknowns $\mathbf{r}$, s and $\mathrm{d}s/\mathrm{d}t$, in which the last two will only appear in a linear form so that it will be very easy to eliminate them and to reduce the problem to one equation in $\mathbf{r}$. This represents the main advantage of the method proposed here.

If one wanted to extend the approximation and consider a larger number of terms in the expressions for V, V', V'', r'^2, r''^2, equations would result which would be non-linear in the unknowns s and $\mathrm{d}s/\mathrm{d}t$ which would render their elimination more difficult and the final equation still more complicated.

43. In order to give a method of calculation, we will consider just the third power of t and t' in the expressions for V, V', V'', which will cause the terms which include the unknown s to disappear. For greater simplicity, let $g = 1$, by taking the average distance from the Earth to the Sun as unity and by representing the time by the median motions of the Sun (Article 23). Now after assuming that $t' = mt$, we will have

$$V = t - \frac{t^3}{6r^3}, \qquad V' = mt - \frac{m^3t^3}{6r^3}$$

$$V'' = (m-1)t - \frac{(m-1)^3t^3}{6r^3}$$

The expressions for R, R', R'' will assume the form

$$R = \frac{6Pr^3 - Qt^2}{[6(m-1)r^3 - (m-1)^3t^2]G}$$

$$R' = \frac{6P_1r^3 - Q_1t^2}{[6(m-1)r^3 - (m-1)^3t^2]G}$$

$$R'' = \frac{6P_2r^3 - Q_2t^2}{[6(m-1)r^3 - (m-1)^3t^2]G}$$

where it has been assumed in order to shorten the expressions that

$$P = (m-1)\rho\Gamma - m\rho'\Gamma' + \rho''\Gamma''$$
$$Q = (m-1)^3\rho\Gamma - m^3\rho\Gamma' + \rho''\Gamma''$$

and denoting by P_1, Q_1 and P_2, Q_2 what becomes of P and Q after replacing $\Gamma, \Gamma', \Gamma''$ by $\Gamma_1, \Gamma_1', \Gamma_1''$ and by $\Gamma_2, \Gamma_2', \Gamma_2''$.

These expressions for R, R', R'' which are the distances from the comet to the Earth in the three observations, contain, as is seen, only the single unknown radius vector of the comet from the first observation. Thus if the expression for R is substituted in the equation (Article 25)

$$R^2 + 2R\rho\cos(\text{CS}) + \rho^2 = \mathbf{r}^2$$

the result will be an equation in $\mathbf{r}$, which will be to the eighth power and the problem will be reduced to the solution of this equation.

After having found the expression for $\mathbf{r}$, those of R' and R'' will be obtained from the preceding formulas. Then, from the formulas of Article 42 the expressions for the three radius vectors r, r', r'', as well as those of the coordinates $\mathbf{x}, \mathbf{y}, \mathbf{z}$ and their differentials $\mathrm{d}\mathbf{x}/dt$, $\mathrm{d}\mathbf{y}/dt$ and $\mathrm{d}\mathbf{z}/dt$ will be obtained. The orbit can be determined by the formulas of Subsection II or, if it is preferred, from the known trigonometric formulas and from the three distances R, R', R'' from the comet to the Earth.

44. The expressions of the distances R, R', R'' can be simplified by the following consideration: since the Earth and comet move about the Sun with the same attracting force from this star, if the letters ξ, η, ζ denote the rectangular coordinates of the Earth from the Sun when $t = 0$ and if the letters Θ, Υ denote what becomes of the functions T and V when the elements of the orbit of the comet are changed to those of the Earth, then as in Article 28, the three following equations will be obtained

$$-\rho\lambda = \xi\Theta + \frac{\mathrm{d}\xi}{\mathrm{d}t}\Upsilon$$
$$-\rho\mu = \eta\Theta + \frac{\mathrm{d}\eta}{\mathrm{d}t}\Upsilon$$
$$-\rho\nu = \zeta\Theta + \frac{\mathrm{d}\zeta}{\mathrm{d}t}\Upsilon$$

because, after having denoted (Article 24) by $\rho\lambda, \rho\mu, \rho\nu$ the rectangular coordinates of the position of the Sun with respect to the Earth, one will have $-\rho\lambda, -\rho\mu, -\rho\nu$ for those of the Earth with respect to the Sun.

Since the sole difference between these equations and those in Article 41 is that $\mathbf{x}, \mathbf{y}, \mathbf{z}, T, V$ have been changed to $\xi, \eta, \zeta, \Theta, \Upsilon$ and that R is zero, it is clear that analogous results will be obtained by making the same changes in those that we found in the preceding article. Therefore, since the expressions of R, R', R'', given at the end of this article do not contain other quantities dependent on the elements of the orbit other than the radius vector r, if one changes r to ρ, the radius vector of the orbit of the Earth, there will result

$$R = 0, \qquad R' = 0, \qquad R'' = 0$$

from which the following equations are obtained

$$6P = \frac{Qt^2}{\rho^3}, \qquad 6P_1 = \frac{Q_1t^2}{\rho^3}, \qquad 6P_2 = \frac{Q_2t^2}{\rho^3}$$

These expressions are now to be substituted in the same expressions for R, R', R'' and neglecting in the denominator the very small term of second order $(m-1)^3t^2$ in comparison to the finite term $6(m-1)r^3$, the following simplified expressions will be obtained

$$R = \frac{Qt^2}{6(m-1)G}(\frac{1}{\rho^3} - \frac{1}{r^3})$$
$$R' = \frac{Q_1t^2}{6(m-1)G}(\frac{1}{\rho^3} - \frac{1}{r^3})$$
$$R'' = \frac{Q_2t^2}{6(m-1)G}(\frac{1}{\rho^3} - \frac{1}{r^3})$$

Therefore, if the expression for R is substituted in the equation

$$R^2 - 2R\rho\cos(\text{CS}) + \rho^2 - \mathbf{r}^2 = 0$$

and if, in order to shorten the expression, the following definition is made

$$\frac{Qt^2}{6(m-1)G} = K$$

which is a quantity known entirely from observation and by multiplying it by $\rho^6\mathbf{r}^6$, one obtains the equation

$$K^2(\mathbf{r}^3 - \rho^3)^2 - 2K\rho^4\mathbf{r}^3(\mathbf{r}^3 - \rho^3)\cos(\text{CS}) + \rho^6\mathbf{r}^6(\rho^2 - \mathbf{r}^2) = 0$$

where the unknown r will increase to the eighth order. However, since this equation is divisible by $\mathbf{r} - \rho$, it will only be of the seventh degree after the division.

The reduction of the degree of the equation in $\mathbf{r}$ is a consequence of having described the motion of the Earth in a similar fashion to the motion of the comet, with approximate formulas where the powers of t larger than the third were neglected. It would not have occurred if the value of R of the preceding article was used where the positions of the Sun are assumed exact and are determined from tables.

45. The preceding equation can be reduced to a rather simple construction. Draw from a given point two straight lines making an angle equal to the arc CS which is the apparent distance from the comet to the Sun in the first observation. The first observation is equal to K/ρ^3 and the second is equal to ρ. The problem will be to find on the first line a point such that the ratio between the part included between this point and the end of the same line with the total length of this line is equal to the ratio of the cube of the second line with the cube of the line which joins this one and the point in question. Then this last line will be equal to $\mathbf{r}$ and the part of the first line intercepted between the given point and the point to be found will be equal to R. From this construction, one will have the proportion[13]

$$\frac{K}{\rho^3} - R : \frac{K}{\rho^3} :: \rho^3 : \mathbf{r}^3$$

which gives

$$R = K(\frac{1}{\rho^3} - \frac{1}{\mathbf{r}^3})$$

and then

$$\mathbf{r} = \sqrt{\rho^2 - 2\rho R \cos(\mathrm{CS}) + R^2}$$

from which the above equation in r results.

I believe that Lambert is the first to have reduced the problem of comets, viewed in an approximate but accurate fashion,[14] to a single equation with only one unknown. He succeeded by an ingenious consideration which is founded on the degree that the apparent location of the comet in the second observation deviates from the large circle obtained by its apparent locations in the first and third observations. The determination of this deviation led him straight to a construction which is analogous to the one which we just gave and which is reduced to an equation in $\mathbf{r}$ of the seventh degree. Additional information can be found in the *Mémoires de l'Académie de Berlin* for the year 1771.

Thus since the values of $\mathbf{r}$ and R are known, one will have

$$R' = \frac{Q_1}{Q} R, \qquad R'' = \frac{Q_2}{Q} R$$

and the two equations (Articles 40 and 41)

$$R'^2 - 2R'\rho' \cos(\mathrm{C'S'}) + \rho'^2 = r'^2 = \mathbf{r}^2 + (2t - \frac{t^3}{3r^3})s + t^2 \frac{\mathrm{d}s}{\mathrm{d}t}$$

$$R''^2 - 2R''\rho'' \cos(\mathrm{C''S''}) + \rho''^2 = r''^2 = \mathbf{r}^2 + (2mt - \frac{m^3 t^3}{3r^3})s + m^2 t^2 \frac{\mathrm{d}s}{\mathrm{d}t}$$

will give the values of the constants s and $\mathrm{d}s/\mathrm{d}t$, from which the values of the elements a and b of the orbit by the formulas of Article 28 will be obtained where $2a$ is the major axis and $2b$ is the parameter.

46. If a parabolic orbit is assumed, the quantity a will be infinite, which will give $\mathrm{d}s/\mathrm{d}t = g/r$. In this case, the last two equations will only contain the unknown s, which after it is eliminated will result in a new equation in R which will have a root in common with the one already found. This fact will simplify the search for this root.

If the first hypothesis is that the comet has a parabolic orbit, it will be preferable to have the solution depend only on this last equation because it has the advantage that it is free of the quantity G, which is very small and of the third order when the intervals of time t and t' or mt are very small compared to the first order. It will be seen later that the errors in observation from which this quantity results can have a large influence.

If the following definitions are made in order to shorten the expressions

$$\frac{m(Q_1)^2 - (Q_2)^2}{Q^2} = M, \qquad \frac{mQ_1\rho' \cos(\mathrm{C'S'}) - Q_2\rho'' \cos(\mathrm{C''S''})}{Q} = N$$

and neglecting the terms containing t^3 in the coefficients of ρ, the elimination of this quantity will give the following equation in R

$$MR^2 - 2NR + m\rho'^2 - \rho''^2 - (m-1)\mathbf{r}^2 + m(m-1)(\frac{gt^2}{r}) = 0$$

which after it is combined with the equation

$$R^2 - 2R\rho\cos(\mathrm{CS}) + \rho^2 - \mathbf{r}^2 = 0$$

will give after elimination of R an equation in $\mathbf{r}$ of the sixth degree. If in the combined equation, the square of the term $m(m-1)gt^2/r$, which is of the fourth order, is neglected, then the final equation will be only of the fifth degree. One could even, for the first approximation, neglect this term, which is only of the second order. Then the final equation will only be of the fourth degree and could be directly solved by known methods.

The value of $\mathbf{r}$ will give those of R, R', R'' and from these, the values of s will be found using the formulas of the preceding article. Since a is assumed infinite, one will have

$$b = 2\mathbf{r} - s^2$$

where the semi-parameter b becomes double the perihelic distance of the comet.

47. After having reduced the problem of comets to final equations with one unknown, it remains to examine the quantities which are assumed to be known. These quantities are:

1. The three radii ρ, ρ', ρ'', which represent the distances from the Sun to the Earth in the three observations and which must be calculated using tables.
2. The quantities $G, \Gamma, \Gamma', \Gamma''$, $\Gamma_1, \Gamma_1', \Gamma_1''$, $\Gamma_2, \Gamma_2', \Gamma_2''$, on which the values of P, Q, P_1, Q_1, P_2, Q_2 depend (Articles 41 and 43). These must be determined from the three observations of the comet and from the calculation of the positions of the Sun. But they can be reduced to simpler expressions which make the determinations much easier.

Let us begin with the quantity G from which the others are only derived. One has (Article 41)

$$G = \ell m'n'' + mn'\ell'' + n\ell'm'' - \ell n'm'' - m\ell'n'' - nm'\ell''$$

The square of this expression can be put in the following form

$$\begin{aligned} G^2 = {} & (\ell^2 + m^2 + n^2)(\ell'^2 + m'^2 + n'^2)(\ell''^2 + m''^2 + n''^2) \\ & + 2(\ell\ell' + mm' + nn')(\ell\ell'' + mm'' + nn'')(\ell'\ell'' + m'm'' + n'n'') \\ & - (\ell^2 + m^2 + n^2)(\ell'\ell'' + m'm'' + n'n'')^2 \\ & - (\ell'^2 + m'^2 + n'^2)(\ell\ell'' + mm'' + nn'')^2 \\ & - (\ell''^2 + m''^2 + n''^2)(\ell\ell' + mm' + nn')^2 \end{aligned}$$

This expression can be shown to be correct by developing the preceding equation. Because of the nature of the quantities ℓ, m, n, ℓ', m', n', ℓ'', m'', n'' (Article 37), one has

$$\ell^2 + m^2 + n^2 = 1, \qquad \ell'^2 + m'^2 + n'^2 = 1, \qquad \ell''^2 + m''^2 + n''^2 = 1$$

Then, in order to shorten the expressions, make the following definitions

$$\begin{aligned} L &= \ell\ell' + mm' + nn' \\ L' &= \ell\ell'' + mm'' + nn'' \\ L'' &= \ell'\ell'' + m'm'' + n'n'' \end{aligned}$$

after which, one will have

$$G^2 = 1 + 2LL'L'' - L^2 - L'^2 - L''^2$$

It was noted earlier (Article 38) that the quantity $(\ell\ell' + mm' + nn')$ is equal to the cosine of the angle between the two radii R and R' directed toward the comet in the first two observations. We have designated this angle by the side CC$'$ of the spherical triangle CC$'$C$''$, assumed drawn on the sphere by joining the three apparent locations of the comet with arcs of large circles in the three observations. This triangle is completely defined by the observations of the comet wherever they have been made. We can view as known its three sides CC′, CC″, C′C″, as well as the angles C, C′, C″, which are opposite the sides C′C″, CC″, CC′, respectively.

Thus one will have

$$L = \cos(\mathrm{CC}')$$

and similarly

$$L' = \cos(\mathrm{CC}''), \qquad L'' = \cos(\mathrm{C}'\mathrm{C}'')$$

and the expression for the quantity G^2 will become

$$\begin{aligned} G^2 &= 1 + 2\cos(\mathrm{CC}')\cos(\mathrm{CC}'')\cos(C'C'') \\ &\quad - \cos^2(\mathrm{CC}') - \cos^2(\mathrm{CC}'') - \cos^2(\mathrm{C}'\mathrm{C}'') \end{aligned}$$

This expression for G^2 can be simplified even further because it is easy to convince oneself from the development of the terms that it is the same as

$$[\cos(\mathrm{CC}' + \mathrm{CC}'') - \cos(C'C'')][\cos(\mathrm{C}'\mathrm{C}'') - \cos(\mathrm{CC}' - \mathrm{CC}'')]$$

which after transformation by well-known methods becomes

$$\begin{aligned} G^2 &= -4\sin(\frac{\mathrm{CC}' + \mathrm{CC}'' + C'C''}{2})\sin(\frac{\mathrm{CC}' + \mathrm{CC}'' - C'C''}{2}) \\ &\quad \times \sin(\frac{\mathrm{CC}' - \mathrm{CC}'' + C'C''}{2})\sin(\frac{\mathrm{C}'\mathrm{C}'' + \mathrm{CC}'' - \mathrm{CC}'}{2}) \end{aligned}$$

which is a very useful formula for trigonometric calculation.

If the angles of the same triangle are used, a simpler expression for the quantity G can still be obtained because from known formulas

$$\cos(\mathrm{C}'\mathrm{C}'') = \cos(\mathrm{CC}')\cos(\mathrm{CC}'') + \sin(\mathrm{CC}')\sin(\mathrm{CC}'')\cos\mathrm{C}$$

If this equation is substituted in the first expression for G^2, one will have after simplifying the expression

$$G^2 = \sin^2(\mathrm{CC}')\sin^2(\mathrm{CC}'')\sin^2\mathrm{C}$$

and consequently, after extracting the square root

$$G = \sin(\mathrm{CC}')\sin(\mathrm{CC}'')\sin\mathrm{C}$$

In the same fashion, it will be found that

$$G = \sin(\mathrm{C'C''})\sin(\mathrm{C'C})\sin\mathrm{C'} = \sin(\mathrm{C''C})\sin(\mathrm{C'C''})\sin\mathrm{C''}$$

It is easy to prove that the quantity G is nothing more than the volume, taken six times, of the triangular pyramid for which the apex is at the center of the sphere whose radius is assumed equal to unity and which rests on the spherical triangle CC′C″, that is, which has for base the rectilinear triangle made by the chords of the three arcs CC′, CC″, C′C″. If one of the faces of this pyramid is considered, for example, the one which has for base the chord of the arc CC′, the area of this isosceles triangle will be $\frac{1}{2}\sin(\mathrm{CC'})$. Then, if the adjacent face which has for a base the chord of the arc CC″ is considered, it is clear that the mutual inclination of these two faces will be equal to the angle C of the spherical triangle. Consequently, the perpendicular drawn from the angle C″ on the first face will be equal to $\sin(\mathrm{CC''})\sin\mathrm{C}$. This perpendicular becomes the height of the pyramid, assuming it is resting[15] on the first face which is equal to $\frac{1}{2}\sin(\mathrm{CC''})$. Therefore, the volume of the pyramid will be equal to

$$\frac{1}{6}\sin(\mathrm{CC'})\sin(\mathrm{CC''})\sin\mathrm{C}$$

and consequently, it is equal to $G/6$.

48. In general, we will denote by the symbol (CC′C″) a function of the sides of the angles of any spherical triangle CC′C″ by which we have expressed the quantity G.

Having marked on a globe the three apparent positions C, C′, C″ of the comet given by the three observations and constructing the spherical triangle CC′C″, one will have immediately

$$G = (\mathrm{CC'C''})$$

Now, if the three locations of the Sun S, S′, S″, from the three observations are placed on the same globe and if these locations are joined with those of the comet by arcs of great circles, three different spheroidal triangles SC′C″, S′C′C″, etc., will be formed and it is easy to see, from what we said in Article 40, relative to the quantities $\Gamma, \Gamma', \Gamma'', \Gamma_1, \Gamma'_1, \Gamma''_1, \Gamma_2, \Gamma'_2, \Gamma''_2$ that the first three will be given by similar functions of the triangles SC′C″, S′C′C″, S″C′C″ and that the three others will be given by similar functions of the triangles CSC″, CS′C″, CS″C″, and finally that the last three will be given by similar functions of the triangles CC′S, CC′S′, CC′S″. Thus one will have, using the same notation

$$\begin{aligned} &\Gamma = (\mathrm{SC'C''}), && \Gamma' = (\mathrm{S'C'C''}), && \Gamma'' = (\mathrm{S''C'C''}) \\ &\Gamma_1 = (\mathrm{CSC''}), && \Gamma'_1 = (\mathrm{CS'C''}), && \Gamma''_1 = (\mathrm{CS''C''}) \\ &\Gamma_2 = (\mathrm{CC'S}), && \Gamma'_2 = (\mathrm{CC'S'}), && \Gamma''_2 = (\mathrm{CC'S''}) \end{aligned}$$

These quantities only depend, as is obvious, on the mutual position of the apparent locations of the comet and Sun, and because they are the only ones to be in the equations which determine the absolute elements of the orbit, our analysis has the advantage of separating the determination of these elements from the other elements which can be called **relative** because they correspond to the orientation of the orbit in space.

49. It should still be noted that the expressions that we just gave hold whatever the position of the apparent locations of the comet and Sun. But from our viewpoint, the locations of the comet are not far from one another and thus the arcs CC′, C′C″ will be very small and the angle C′ between these two arcs will not be much different from a right angle. It would be equal to a right angle if the Earth and comet described, between the first and third observations, straight lines because then the three apparent locations of the comet would be in the same great circle. The sines of CC′, C′C″ and of C′ will therefore be very small and the quantity

$$G = \sin(\mathrm{CC}') \sin(\mathrm{C}'\mathrm{C}'') \sin \mathrm{C}'$$

will also be very small of the third order. But the quantities

$$\Gamma = \sin(\mathrm{SC}') \sin(\mathrm{SC}'') \sin \mathrm{S}$$
$$\Gamma' = \sin(\mathrm{S}'\mathrm{C}') \sin(\mathrm{S}'\mathrm{C}'') \sin \mathrm{S}'$$
$$\cdots\cdots\cdots$$

will only be of the first order. Also, since only quantities dependent upon the apparent locations of the comet are contained in the expression for G instead of the quantities Γ, Γ', etc., they only depend upon the location of the Sun, which since they are given by Tables can be viewed as exact. It follows that the value of the quantity G will always be more susceptible to error than those of the quantities Γ, Γ', etc., and it will have to be avoided as much as possible, as we have demonstrated in Article 46.

50. Finally, we will note that because the observations of a comet ordinarily give its vertical ascension and declination, this information can be immediately used by assuming that the three axes to which are referred the radii R, R', R'' directed toward the comet, and the radii ρ, ρ', ρ'' directed toward the Sun, are directed, the first toward the equinox of spring, the second at a right angle on the plane of the equator and according to the order of signs, the third toward the boreal pole of the equator. Then by denoting by the letter a the vertical ascension of the comet, d its declination in the first observation and similarly α the vertical ascension of the Sun, δ its declination at the same instant, it is easy to see that one will have

$$\ell = \sin a \cos d, \qquad m = \cos a \cos d, \qquad n = \sin d$$
$$\lambda = \sin \alpha \cos \delta, \qquad \mu = \cos \alpha \cos \delta, \qquad \nu = \sin \delta$$

From these equations, one will have (cited article)

$$\cos(\mathrm{CS}) = \ell\lambda + m\mu + n\nu = \cos(a - \alpha) \cos d \cos \delta + \sin d \sin \delta$$

and similarly

$$\cos(C'S') = \cos(a' - \alpha')\cos d' \cos\delta' + \sin d' \sin\delta'$$
$$\cos(C''S'') = \cos(a'' - \alpha'')\cos d'' \cos\delta'' + \sin d'' \sin\delta''$$

by marking, as was done previously, with one and two primes the analogous quantities which refer to the second and third observations.

In the same fashion, one will have

$$\cos(CC') = \cos(a - a')\cos d \cos d' + \sin d \sin d'$$
$$\cos(SS') = \cos(\alpha - \alpha')\cos\delta \cos\delta' + \sin\delta \sin\delta'$$
$$\cos(CS') = \cos(a - \alpha')\cos d \cos\delta' + \sin d \sin\delta'$$

and similarly for the other sines.

If, then the same values of $\ell, m, n, \ell', m', n', \ell'', m'', n''$, were substituted in the expression for G, one will have

$$\begin{aligned} G &= \cos d \cos d' \sin d'' \sin(a' - a) \\ &\quad - \cos d \cos d'' \sin d' \sin(a'' - a) + \cos d' \cos d'' \sin d \sin(a'' - a') \\ &= \cos d \cos d' \cos d'' \begin{bmatrix} \sin(a' - a)\tan d'' - \sin(a'' - a)\tan d' \\ + \sin(a'' - a')\tan d \end{bmatrix} \end{aligned}$$

and from it the values of $\Gamma, \Gamma', \Gamma''$ can be deduced replacing a and d with α and δ, and α' and δ' with α'' and δ''; those of $\Gamma_1, \Gamma_1', \Gamma_1''$ by making the same substitutions for a' and d' and those of $\Gamma_2, \Gamma_2', \Gamma_2''$, by making the same changes on a'' and d''. Thus one will have

$$\Gamma = \cos\delta \cos d' \cos d'' \begin{bmatrix} \sin(a' - \alpha)\tan d'' - \sin(a'' - \alpha)\tan d' \\ + \sin(a'' - a')\tan\delta \end{bmatrix}$$

$$\Gamma_1 = \cos d \cos\delta \cos d'' \begin{bmatrix} \sin(\alpha - a)\tan d'' - \sin(a'' - a)\tan\delta \\ + \sin(a'' - \alpha)\tan d \end{bmatrix}$$

$$\Gamma_2 = \cos d \cos d' \cos\delta \begin{bmatrix} \sin(a' - a)\tan\delta - \sin(\alpha - a)\tan d' \\ + \sin(\alpha - a')\tan\delta \end{bmatrix}$$

and to obtain the expressions for $\Gamma', \Gamma_1', \Gamma_2'$ and for $\Gamma'', \Gamma_1'', \Gamma_2''$, it will only be necessary to mark in the expressions for $\Gamma, \Gamma_1, \Gamma_2$, the letters α and δ with one and two primes.

It is unnecessary to observe here that if instead of the rectilinear ascensions and declinations, the data were the longitudes and latitudes, it will only be necessary to substitute the latter for the former in the same formulas. The orbit will then be referenced to the ecliptic instead of the equator.

51. After having calculated these values, one will compute those of the quantities Q, Q_1, Q_2, by the formula of Article 42 and if the method of Article 44 is used as the

shorter one, the final equation in **r**, for which the resolution will not be difficult by reducing it, for a first approximation, to the fourth degree, will be immediately obtained.

If the intervals between observations were equal, one would have $t' = 2t$, and consequently, $m = 2$ which would give

$$Q = \rho\Gamma - 8\rho'\Gamma' + \rho''\Gamma'' = -6\rho'\Gamma' + \Delta^2\rho\Gamma$$

after denoting by the characteristic Δ^2 the second differences of the quantities $\rho\Gamma$, $\rho'\Gamma'$, $\rho''\Gamma''$ in which there are only quantities relative to the Sun which vary. But since it is assumed that the observations are close to one another, the differences of these quantities will be very small and consequently, the second difference $\Delta^2\rho\Gamma$ will be very small of the second order and can be neglected in comparison with the finite quantity $-6\rho'\Gamma'$, which will reduce the value of Q to this single quantity. The same simplifications can be made on the analogous quantities Q_1, Q_2, so that one will simply have

$$Q = -6\rho'\Gamma', \qquad Q_1 = -6\rho'\Gamma'_1, \qquad Q_2 = -6\rho'\Gamma'_2$$

which will even further shorten the calculation of the first approximation.

With respect to the measurement of time, since the time is to be represented by the average motion of the Sun, if one wants to express it in average days, it will suffice to multiply the number of days and the parts of days by the angle of the average motion of the Sun in one day, reduced to parts of the radius. This angle is $59'8.3''$ and gives in parts of the radius the number 0.0172021 by which one will have to multiply the intervals of time t, reduced to average days.

Chapter II
THE VARIATION OF THE ELEMENTS OF ELLIPTICAL ORBITS PRODUCED BY AN IMPULSE OR BY ACCELERATING FORCES

52. One of the first and best results of Newton's theory on the system of the world is that all the orbits of celestial bodies are of the same nature and only differ by the force of projection that these bodies can be assumed to have received at their creation. It follows that if a planet or comet receives an external arbitrary impulse, its orbit would be disturbed but only the elements, which are the arbitrary constants of the equations, would change. Thus it is in this fashion that the circular or elliptical orbit of a planet could become parabolic or even hyperbolic, which will transform the planet into a comet.

It is the same for all the problems of mechanics. Since the arbitrary constants introduced by integration depend only on the initial state of the system, which can be defined to be any instant, if these bodies are assumed to receive arbitrary impulses during their motion, the velocities produced by these impulses are added to the velocities already acquired by the bodies. Hence, they can be viewed as initial velocities which will only change the values of the constants.

If instead of finite impulses which act only for an instant, infinitesimal impulses were assumed, but which act continuously, these same constants will become variables and will be used to determine the effect of these forces, which must be viewed as perturbing forces. Then, one will have the problem for which we gave a general solution in SECTION V and which we will apply here to the orbits of planets.

Subsection I
The Change Produced In The Elements Of The Orbit Of A Planet When The Planet Is Assumed To Be Subjected To An Arbitrary Impulse

53. We have already seen in Subsection II of the preceding chapter how all the elements of the elliptical motion of a planet can be expressed, by functions of the coordinates x, y, z and of their differentials $\mathrm{d}x/\mathrm{d}t$, $\mathrm{d}y/\mathrm{d}t$, $\mathrm{d}z/\mathrm{d}t$, which express the velocities in the directions of these coordinates. Thus if it is assumed that a planet during its motion receives at an arbitrary point in its orbit an impulse which affects the velocities in the direction of the x, y, z coordinates and with a tendency to increase them, it will only be necessary to replace $\mathrm{d}x/\mathrm{d}t$, $\mathrm{d}y/\mathrm{d}t$, $\mathrm{d}z/\mathrm{d}t$ by $(\mathrm{d}x/\mathrm{d}t) + \dot{x}$, $(\mathrm{d}y/\mathrm{d}t) + \dot{y}$, $(\mathrm{d}z/\mathrm{d}t) + \dot{z}$ and one will have the elements of the new orbit followed by the planet after the impulse.

If, instead of the rectangular coordinates x, y, z, one takes, as in Article 5, the radius vector r with the angles ψ and φ for which ψ is the inclination of r on the fixed xy-plane and φ is the angle made by the projection of r on this plane with the fixed x-axis, the equations of the orbit become simpler.

Indeed, by substituting $r\cos\psi\cos\varphi$, $r\cos\psi\sin\varphi$ and $r\sin\psi$ in place of x, y, z, the elements a, b, h and i become

$$\frac{1}{a} = \frac{2}{r} - \frac{r^2(\cos^2\psi\,\mathrm{d}\varphi^2 + \mathrm{d}\psi^2) + \mathrm{d}r^2}{g\,\mathrm{d}t^2}$$

$$b = \frac{r^4(\cos^2\psi\,\mathrm{d}\varphi^2 + \mathrm{d}\psi^2)}{g\,\mathrm{d}t^2}$$

$$\tan h = \frac{\sin\varphi\,\mathrm{d}\psi - \sin\psi\cos\psi\cos\varphi\,\mathrm{d}\varphi}{\cos\varphi\,\mathrm{d}\psi - \sin\psi\cos\psi\sin\varphi\,\mathrm{d}\varphi}$$

$$\tan i = \frac{\sqrt{\mathrm{d}\psi^2 + \sin^2\psi\cos^2\psi\,\mathrm{d}\varphi^2}}{\cos^2\psi\,\mathrm{d}\varphi}$$

In these formulas, the differential expressions $\mathrm{d}r/\mathrm{d}t$, $r\,\mathrm{d}\varphi/\mathrm{d}t$ and $r\,\mathrm{d}\psi/\mathrm{d}t$ represent the velocities in the direction of the radius r, in a direction perpendicular to this radius and parallel to the plane of projection, and in a direction perpendicular to this same plane.

54. For greater simplicity, let us take the plane of projection in the plane of the orbit and let us assume that the velocity developed by the impulse is resolved into three components,

one along the radius r, the other perpendicular to this radius in the plane of the orbit and the third perpendicular to this plane. If the first is designated by r, the second by $r\varphi$ and the third by $r\psi$ one will have the elements of the new orbit following the impulse, by substituting in the preceding expressions for $\mathrm{d}r$, $\mathrm{d}\varphi$, and $\mathrm{d}\psi$ the quantities $\mathrm{d}r + \dot{r}\,\mathrm{d}t$, $\mathrm{d}\varphi + \dot{\varphi}\,\mathrm{d}t$, $\mathrm{d}\psi + \dot{\psi}\,\mathrm{d}t$ and by setting $\psi = 0$, $\mathrm{d}\psi = 0$. Then the position of the new orbit will be with respect to the plane of the original orbit.

Let A, B, H, and I represent what becomes of the elements a, b, h, and i for the new orbit. Then one will have

$$\frac{1}{A} = \frac{2}{r} - \frac{r^2[(\mathrm{d}\varphi + \dot{\varphi}\,\mathrm{d}t)^2 + \dot{\psi}^2\,\mathrm{d}t^2] + (\mathrm{d}r + \dot{r}\,\mathrm{d}t)^2}{g\,\mathrm{d}t^2}$$

$$B = \frac{r^4[(\mathrm{d}\varphi + \dot{\varphi}\,\mathrm{d}t)^2 + \dot{\psi}\,\mathrm{d}t^2]}{g\,\mathrm{d}t^2}$$

$$\tan I = \frac{\dot{\psi}\,\mathrm{d}t}{\mathrm{d}\varphi + \dot{\varphi}\,\mathrm{d}t}$$

$$\tan H = \frac{\sin\varphi}{\cos\varphi} = \tan\varphi$$

The last equation shows that $H = \varphi$. Indeed, it is clear that the node between the new orbit and the original orbit must be at the point where the impulse is applied.

Also, if ψ and $\mathrm{d}\psi/\mathrm{d}t$ are put equal to zero in the expressions for the original elements a and b, one has

$$\frac{1}{a} = \frac{2}{r} - \frac{r^2\,\mathrm{d}\varphi^2 + \mathrm{d}r^2}{g\,\mathrm{d}t^2}, \qquad b = \frac{r^4\,\mathrm{d}\varphi^2}{g\,\mathrm{d}t^2}$$

and from this expression, we obtain the following equations

$$\frac{\mathrm{d}\varphi}{\sqrt{g}\,\mathrm{d}t} = \sqrt{\frac{b}{r^2}}, \qquad \frac{\mathrm{d}r}{\sqrt{g}\,\mathrm{d}t} = \sqrt{\frac{2}{r} - \frac{1}{a} - \frac{b}{r^2}}$$

By substituting these expressions, one will have the elements of the new orbit expressed in terms of those of the original orbit and of the velocities $\dot{r}, r\dot{\varphi}, r\dot{\psi}$ produced by the impulse.

55. Let us now assume that the impulse necessary to change the original elements a and b to A and B and to have the new orbit inclined to the original one by an angle I is needed. It will suffice to obtain the expressions $\dot{\varphi}, \dot{\psi}, \dot{r}$ as functions of A, B, I, and a, b, r. The formulas we just found give

$$\dot{\psi} = \frac{\sqrt{gB}\sin I}{r^2}$$

$$\dot{\varphi} = \frac{\sqrt{gB}\cos I - \sqrt{gb}}{r^2}$$

$$\dot{r} = \sqrt{g}\sqrt{\frac{2}{r} - \frac{1}{A} - \frac{B}{r^2}} - \sqrt{g}\sqrt{\frac{2}{r} - \frac{1}{a} - \frac{b}{r^2}}$$

Let u be the velocity resulting from the impulse and let α, β, γ be the angles that the direction of the impulse makes with three axes of which one is the prolonged radius r, the other the perpendicular in the plane of the original orbit and in the direction of the motion of the planet and the third perpendicular to the same plane. From the principle of resolution, one will have $u\cos\alpha, u\cos\beta, u\cos\gamma$ for the three component velocities along the axes, which are also those that we have designated by $\dot{r}$, $r\dot{\varphi}$ and $r\dot{\psi}$. Therefore, one will have

$$u\cos\alpha = \dot{r}, \qquad u\cos\beta = r\dot{\varphi}, \qquad u\cos\gamma = r\dot{\psi}$$

from which, and with $\cos^2\alpha + \cos^2\beta + \cos^2\gamma = 1$, the following equation is obtained

$$u = \sqrt{\dot{r}^2 + r^2\dot{\varphi}^2 + r^2\dot{\psi}^2}$$

If, in order to shorten the expressions, the following definitions are made

$$F = \sqrt{\frac{2}{r} - \frac{1}{A} - \frac{B}{r^2}}, \qquad f = \sqrt{\frac{2}{r} - \frac{1}{a} - \frac{b}{r^2}}$$

one will have

$$u = \sqrt{g\left(\frac{4}{r} - \frac{1}{A} - \frac{1}{a} - 2\frac{\sqrt{Bb}}{r^2}\cos I - 2Ff\right)}$$

$$\cos\alpha = \frac{F - f}{u}\sqrt{g}$$

$$\cos\beta = \frac{\sqrt{B}\cos I - \sqrt{b}}{ur}\sqrt{g}$$

$$\cos\gamma = \frac{\sqrt{B}\sin I}{ur}\sqrt{g}$$

But if the direction of the impulse is to be referred to two other axes located in the plane of the original orbit, where one would be perpendicular and the other tangent to this orbit, then denoting by ϵ the angle made by the perpendicular to the orbit and the radius vector r and for which the tangent is expressed by $\mathrm{d}r/r\,\mathrm{d}\varphi$, the velocities impressed along these two axes will be

$$\dot{r}\cos\epsilon - r\dot{\varphi}\sin\epsilon, \qquad \dot{r}\sin\epsilon + r\dot{\varphi}\cos\epsilon$$

The velocity along the third axis perpendicular to the plane of the orbit will remain the same. If the angles that the directions of the impulse make with these new axes are designated by α' and β', we will have that

$$u\cos\alpha' = \dot{r}\cos\epsilon - r\dot{\varphi}\sin\epsilon$$

$$u\cos\beta' = \dot{r}\sin\epsilon + r\dot{\varphi}\cos\epsilon$$

now one has

$$\tan\epsilon = \frac{\mathrm{d}r}{r\,\mathrm{d}\varphi} = \frac{fr}{\sqrt{b}}$$

from which the following equation is obtained, after substituting the expression for f

$$\sin\epsilon = \frac{f}{\sqrt{\dfrac{2}{r} - \dfrac{1}{a}}}, \qquad \cos\epsilon = \frac{\sqrt{b}}{r\sqrt{\dfrac{2}{r} - \dfrac{1}{a}}}$$

and from this result, one will have

$$\cos\alpha' = \frac{F\sqrt{b} - f\sqrt{B}\cos I}{ur\sqrt{\dfrac{2}{r} - \dfrac{1}{a}}}\sqrt{g}$$

$$\cos\beta' = \left\{\frac{r^2Ff + \sqrt{Bb}\cos I}{ur^2\sqrt{\dfrac{2}{r} - \dfrac{1}{a}}} - \frac{\sqrt{\dfrac{2}{r} - \dfrac{1}{a}}}{u}\right\}\sqrt{g}$$

where it is to be noted that

$$\sqrt{g}\sqrt{\frac{2}{r} - \frac{1}{a}}$$

is the velocity in the original orbit.

With respect to the alternate signs of the radicals which are in these formulas, it should be noted that:

1. Since f is the value of $\mathrm{d}r/g\,\mathrm{d}t$ which expresses the velocity along the radius r in the original orbit, the quantity F will express the velocity along this radius in the new orbit. Therefore, the quantities will be taken positive or negative according to whether the velocities they represent have the tendency to increase or decrease the radius r, that is, to push or pull the body towards or away from the focus.
2. Since $\sqrt{b}/r$ is equal to $r^2\,\mathrm{d}\varphi/g\,\mathrm{d}t$, which is the circular velocity about the focus in the original orbit, and in the same fashion $\sqrt{B}/r$ represents the circular velocity in the new orbit, then $(\sqrt{B}/r)\cos I$ will be the circular velocity with respect to the plane of the original orbit. Thus by taking $\sqrt{b}$ positive, the other radical $\sqrt{B}$ can be taken positive or negative, depending on whether the new orbit with respect to the plane of the original orbit will have the same or opposite sense in this orbit, that is, whether the motion in the new orbit will be in the same direction or whether it will be in the opposite direction, relative to the motion in the original orbit.

56. If these formulas are to be used for planets and comets, simply put $g = 1$, taking the average distance from the Earth to the Sun as unity and the average velocity of the Earth in its orbit as unity. This velocity is about 7 leagues, where a league is 25 to the degree per second.[16] The velocity of a number 24 cannonball, at the muzzle of the cannon, is about 1400 feet or 233 toise per second, which is about the velocity of a point on the equator in the diurnal motion of the Earth, which is 238 toise per second. Therefore, if in order to obtain a better understanding of our estimate, we take the velocity of a number 24 cannonball for unity, which is about one tenth of a league [per second], the velocity of the Earth in its orbit will be expressed by the number 70. Consequently, the value of the impulse created velocity u should be multiplied by 70.

Let us determine what the largest value of u could be.

By calling e the eccentricity of the original orbit and by φ the true anomaly which corresponds to the radius r, one has (Article 15)

$$r = \frac{b}{1 + e\cos\varphi} = \frac{a(1 - e^2)}{1 + e\cos\varphi}$$

therefore

$$\frac{1}{a} = \frac{1 - e^2}{r(1 + e\cos\varphi)}$$

Thus the smallest value of $1/a$ will be $(1 - e)/r$ and similarly the smallest value of $1/A$ will be $(1 - E)/r$, where E denotes the eccentricity of the new orbit. Thus the largest value of the terms $4/r - 1/A - 1/a$ will be $(2 + E + e)/r$, and this expression will also hold for hyperbolic orbits where the value of E and e will be greater than unity.

With the same formulas, one has $b/r = 1 + e\cos\varphi$ for which the largest value is $1 + e$, the largest value of B/r will similarly be $1 + E$. Therefore, the largest value of $\sqrt{Bb/r^2}$ will be

$$\frac{\sqrt{(1 + E)(1 + e)}}{r}$$

but it is easy to prove that

$$\sqrt{(1 + E)(1 + e)} < 1 + \frac{E + e}{2}$$

because the difference of their squares is $\frac{1}{4}(E - e)^2$. Thus the following equation will always be obtained

$$\frac{2\sqrt{Bb}}{r^2} < \frac{2 + E + e}{r}$$

It still remains to search for the largest values of f and F. Because the smallest values of $1/a$ and b/r^2 are $(1-e)/r$, the largest value of f will be $\sqrt{2e/r}$, and similarly, the largest value of F will be $\sqrt{2E/r}$.

Therefore, since in the expressions for u, the quantities $\sqrt{b}$, $\sqrt{B}$, f and F can be positive or negative, by taking as positive the terms which contain these radicals, and also by giving to $\cos I$ its largest value, one will have

$$u < \sqrt{\frac{4 + 2(E+e) + 4\sqrt{Ee}}{r}}$$

This limit will become $\sqrt{6/r}$ when the original orbit is circular or nearly circular like those of the planets and the new orbit is parabolic like those of the comets.

57. The principal circumstances surrounding the motion of the planets about the Sun give credence to the idea that they have a common origin. This is not true of comets. They have no common properties outside of the fact that their motion is a parabola or in general, a conic section, and furthermore, they seem to have been randomly thrown into space.

Can it not be assumed that the cause which created our planets has at the same time created a larger number of others located beyond Saturn and describing similar orbits, such as Uranus, but for which several could have become comets after bursting from an internal explosion. When a planet is broken into two or more parts, each of the parts will receive an impulse from the force of the explosion which will make it describe an orbit different from the planet's orbit. For this orbit to be parabolic, it will suffice[17] that the velocity impressed by the explosion does not exceed $70\sqrt{6/r}$ times the velocity of a cannonball. For Saturn, one has $r = 9$ and for Uranus, $r = 19$. Assuming that $r = 24$, it will require a velocity less than 35 times the velocity of a cannonball which can be produced by a handful of powder.

The hypothesis of a planet shattered by an internal explosion has already been proposed by Olbers to explain the near equality of the elements of four new planets. What might confirm it are the variations which are observed in the light reflected from these planets and which, by indicating a rotational motion, indicate simultaneously that their configurations are not solids of revolution as are the other planets. Consequently, they can not be fluid but they must have been solid before they became planets.

If the original orbit is assumed circular and the orbit after the explosion is assumed to be elliptical but not much different from a circle and only slightly inclined to the plane of the original orbit, and if only the first dimensions of the eccentricity E and the sine of the inclination I are considered, one has

$$u = \frac{\sqrt{E^2(\sin^2\Phi + \frac{1}{4}\cos^2\Phi) + \sin^2 I}}{\sqrt{r}}$$

$$\cos\alpha = \frac{E\sin\Phi}{u\sqrt{r}}, \qquad \cos\beta = \frac{E\cos\Phi}{2u\sqrt{r}}, \qquad \cos\gamma = \frac{\sin I}{u\sqrt{r}}$$

where the angle Φ is formed by the radius r and the radius of perihelion.

Since the eccentricities and inclinations of the planets do not follow any law among themselves and have only their small size in common, it could be assumed that the orbits of the planets were originally circular and that they became elliptical and inclined due to the effect of small internal explosions. Indeed, if a small part m of mass from the total mass M of a planet is detached and ejected with a velocity V capable of making it a comet, the planet would have only received a small velocity $mV/(M - m)$ in an opposite direction which would have changed its circular orbit to an elliptical and inclined orbit, such as the orbits of our planets and the same impulse could also have produced some changes in its rotation, as we will see later.

Subsection II
Variation Of The Elements Of Planetary Motion Produced By Perturbing Forces

58. Let us now assume that the impulses which change the arbitrary constants are infinitesimal and continuous. These constants will become variable and in this manner, it will be possible to reduce the effect of perturbing forces on planets to variations of the elements of their orbits.

Let X, Y, Z be the perturbing forces resolved in the directions of the rectangular coordinates x, y, z and having a tendency to increase the coordinates. These forces will create during the instant dt the small velocities $X\,\mathrm{d}t, Y\,\mathrm{d}t, Z\,\mathrm{d}t$ which should be added to the velocities $\mathrm{d}x/\mathrm{d}t, \mathrm{d}y/\mathrm{d}t, \mathrm{d}z/\mathrm{d}t$ in the expression for each of the elements a, b, c, etc. as in Article 52. But because the added velocities are infinitesimal they will only produce in the elements infinitesimal variations which can be determined by the differential calculus.

In order to shorten the expressions, let us make the following definitions

$$\frac{\mathrm{d}x}{\mathrm{d}t} = x', \qquad \frac{\mathrm{d}y}{\mathrm{d}t} = y', \qquad \frac{\mathrm{d}z}{\mathrm{d}t} = z'$$

Each of the elements will be expressed by a given function of $x, y, z,\ x', y', z'$. Let a be one of these elements. One will have its variation $\mathrm{d}a$ by augmenting x', y', z' of the infinitesimal quantities $X\,\mathrm{d}t, Y\,\mathrm{d}t, Z\,\mathrm{d}t$. Thus one will have

$$\mathrm{d}a = \left(\frac{\mathrm{d}a}{\mathrm{d}x'}X + \frac{\mathrm{d}a}{\mathrm{d}y'}Y + \frac{\mathrm{d}a}{\mathrm{d}z'}Z\right)\mathrm{d}t$$

and similar equations will be obtained for the other elements of the orbit b, c, h, i, k.

In order to use these equations, one should substitute for the variables x, y, z, x', y', z' their values as functions of t, a, b, c, etc., given by the formulas found in Chapter I. Thus there will be as many equations of the first order between the time t and the elements a, b, c, etc., which become variables, as there are elements and it will suffice to integrate them.

If it is desired to introduce directly the perturbing forces in the equations of the original orbit (Article 4), it would suffice to add the quantities X, Y, Z, respectively to the terms $R(\mathrm{d}r/\mathrm{d}x)$, $R(\mathrm{d}r/\mathrm{d}y)$, $R(\mathrm{d}r/\mathrm{d}z)$ of these equations. Thus the preceding equations between the new variables a, b, c can be viewed as transformations of the equations in x, y, z. But these transformations would be of little use for the general solution of the problem. Their great use is when a rigorous solution is impossible and the perturbing forces are very small. They provide a means of approximation which we presented in a general manner in SECTION V.

59. This approximation, which is based on the variation of the elements, is mainly applicable to the elliptical orbits of planets, if they are perturbed by the actions of other planets. Geometers have used it often in the theory of planets and comets. It can be said that it is the observations themselves which made it known before it was discovered by the calculus. This approximation has the advantage of retaining the elliptical shape of the orbits and at the same time, of keeping the ellipse invariant during an infinitesimal time in such a way that not only the location of the planet, but also its velocity and direction,[18] are not affected by the instantaneous variation of the elements.

Indeed, by viewing the coordinates x, y, z as functions of time and the elements a, b, c, etc. as variables, one has after differentiation

$$\mathrm{d}x = \frac{\mathrm{d}x}{\mathrm{d}t}\mathrm{d}t + \frac{\mathrm{d}x}{\mathrm{d}a}\mathrm{d}a + \frac{\mathrm{d}x}{\mathrm{d}b}\mathrm{d}b + \frac{\mathrm{d}x}{\mathrm{d}c}\mathrm{d}c + \cdots$$

It is easy to prove that the part which contains the variations $\mathrm{d}a$, $\mathrm{d}b$, etc. becomes zero after substitution of the value of $\mathrm{d}a$ given above and of the similar values of $\mathrm{d}b$, $\mathrm{d}c$, etc. because by making these substitutions in the terms $(\mathrm{d}x/\mathrm{d}a)\mathrm{d}a + (\mathrm{d}x/\mathrm{d}b)\mathrm{d}b + \cdots$ and ordering with respect to the quantities X, Y, Z, one will have

$$\begin{aligned}
&\left(\frac{\mathrm{d}x}{\mathrm{d}a}\frac{\mathrm{d}a}{\mathrm{d}x'} + \frac{\mathrm{d}x}{\mathrm{d}b}\frac{\mathrm{d}b}{\mathrm{d}x'} + \frac{\mathrm{d}x}{\mathrm{d}c}\frac{\mathrm{d}c}{\mathrm{d}x'} + \cdots\right) X\,\mathrm{d}t \\
+ &\left(\frac{\mathrm{d}x}{\mathrm{d}a}\frac{\mathrm{d}a}{\mathrm{d}y'} + \frac{\mathrm{d}x}{\mathrm{d}b}\frac{\mathrm{d}b}{\mathrm{d}y'} + \frac{\mathrm{d}x}{\mathrm{d}c}\frac{\mathrm{d}c}{\mathrm{d}y'} + \cdots\right) Y\,\mathrm{d}t \\
+ &\left(\frac{\mathrm{d}x}{\mathrm{d}a}\frac{\mathrm{d}a}{\mathrm{d}z'} + \frac{\mathrm{d}x}{\mathrm{d}b}\frac{\mathrm{d}b}{\mathrm{d}z'} + \frac{\mathrm{d}x}{\mathrm{d}c}\frac{\mathrm{d}c}{\mathrm{d}z'} + \cdots\right) Z\,\mathrm{d}t
\end{aligned}$$

But by viewing x, y, z, x', y', z' at the outset as functions of a, b, c, h, i, k and subsequently, a, b, c, etc. as functions of x, y, z, x', etc. the following equations result

$$\begin{aligned}
\mathrm{d}x &= \frac{\mathrm{d}x}{\mathrm{d}a}\mathrm{d}a + \frac{\mathrm{d}x}{\mathrm{d}b}\mathrm{d}b + \frac{\mathrm{d}x}{\mathrm{d}c}\mathrm{d}c + \frac{\mathrm{d}x}{\mathrm{d}h}\mathrm{d}h + \cdots \\
\mathrm{d}y &= \frac{\mathrm{d}y}{\mathrm{d}a}\mathrm{d}a + \frac{\mathrm{d}y}{\mathrm{d}b}\mathrm{d}b + \frac{\mathrm{d}y}{\mathrm{d}c}\mathrm{d}c + \frac{\mathrm{d}y}{\mathrm{d}h}\mathrm{d}h + \cdots \\
&\cdots\cdots\cdots \\
\mathrm{d}a &= \frac{\mathrm{d}a}{\mathrm{d}x}\mathrm{d}x + \frac{\mathrm{d}a}{\mathrm{d}y}\mathrm{d}y + \frac{\mathrm{d}a}{\mathrm{d}z}\mathrm{d}z + \frac{\mathrm{d}a}{\mathrm{d}x'}\mathrm{d}x' + \cdots
\end{aligned}$$

$$db = \frac{db}{dx}dx + \frac{db}{dy}dy + \frac{db}{dz}dz + \frac{db}{dx'}dx' + \cdots$$

$$\dots\dots\dots$$

After substituting in the expressions for dx, dy, etc., the values of da, db, etc., one will have identical equations. Consequently, the terms with dx', dy', dz' in the expressions for dx, dy, dz, must be zero, which will give with respect to dx, the identical equations

$$\frac{dx}{da}\frac{da}{dx'} + \frac{dx}{db}\frac{db}{dx'} + \frac{dx}{dc}\frac{dc}{dx'} + \cdots = 0$$
$$\frac{dx}{da}\frac{da}{dy'} + \frac{dx}{db}\frac{db}{dy'} + \frac{dx}{dc}\frac{dc}{dy'} + \cdots = 0$$
$$\frac{dx}{da}\frac{da}{dz'} + \frac{dx}{db}\frac{db}{dz'} + \frac{dx}{dc}\frac{dc}{dz'} + \cdots = 0$$

Thus, one will simply have $dx = (dx/dt)dt$, and one will find in the same manner that $dy = (dy/dt)dt$, $dz = (dz/dt)dt$ as if the constants a, b, c, h, etc., did not vary.

60. If the perturbing forces are the result of the attractions of other fixed or mobile bodies and if these attractions are proportional to functions of the distance, the sum of the integrals of each force multiplied by the element of its distance to the center of attraction can be denoted by $-\Omega$ as in Article 8 of SECTION V. Furthermore, if the quantity Ω is viewed as a function of x, y, z, the forces X, Y, Z will have the form

$$X = \frac{d\Omega}{dx}, \qquad Y = \frac{d\Omega}{dy}, \qquad Z = \frac{d\Omega}{dz}$$

The quantity Ω is taken as positive because it is assumed that the forces X, Y, Z have the tendency to increase the distances x, y, z, while in the function $-\Omega$, the perturbing forces, directed toward the centers, are assumed to have the tendency to shorten the distances from the bodies to these centers.

In this case, which is the case for natural phenomena, the variations of the elements a, b, c can be expressed in a simpler manner by using instead of the partial differences of Ω relative to x, y, z its partial differences relative to a, b, c, etc., after the substitution of the values of x, y, z in t and a, b, c, etc. It is this consideration which gave birth to the new theory of the variation of arbitrary constants.

If x, y, z are viewed as functions of a, b, c, etc., one will have

$$\frac{d\Omega}{dx} = \frac{d\Omega}{da}\frac{da}{dx} + \frac{d\Omega}{db}\frac{db}{dx} + \frac{d\Omega}{dc}\frac{dc}{dx} + \cdots$$
$$\frac{d\Omega}{dy} = \frac{d\Omega}{da}\frac{da}{dy} + \frac{d\Omega}{db}\frac{db}{dy} + \frac{d\Omega}{dc}\frac{dc}{dy} + \cdots$$
$$\frac{d\Omega}{dz} = \frac{d\Omega}{da}\frac{da}{dz} + \frac{d\Omega}{db}\frac{db}{dz} + \frac{d\Omega}{dc}\frac{dc}{dz} + \cdots$$

and after the equations are substituted in the expression for da of Article 58 for X, Y, Z, it will become

$$\mathrm{d}a = \left(\frac{\mathrm{d}a}{\mathrm{d}x'}\frac{\mathrm{d}a}{\mathrm{d}x} + \frac{\mathrm{d}a}{\mathrm{d}y'}\frac{\mathrm{d}a}{\mathrm{d}y} + \frac{\mathrm{d}a}{\mathrm{d}z'}\frac{\mathrm{d}a}{\mathrm{d}z}\right)\frac{\mathrm{d}\Omega}{\mathrm{d}a}\mathrm{d}t$$
$$+ \left(\frac{\mathrm{d}a}{\mathrm{d}x'}\frac{\mathrm{d}b}{\mathrm{d}x} + \frac{\mathrm{d}a}{\mathrm{d}y'}\frac{\mathrm{d}b}{\mathrm{d}y} + \frac{\mathrm{d}a}{\mathrm{d}z'}\frac{\mathrm{d}b}{\mathrm{d}z}\right)\frac{\mathrm{d}\Omega}{\mathrm{d}b}\mathrm{d}t$$
$$+ \left(\frac{\mathrm{d}a}{\mathrm{d}x'}\frac{\mathrm{d}c}{\mathrm{d}x} + \frac{\mathrm{d}a}{\mathrm{d}y'}\frac{\mathrm{d}c}{\mathrm{d}y} + \frac{\mathrm{d}a}{\mathrm{d}z'}\frac{\mathrm{d}c}{\mathrm{d}z}\right)\frac{\mathrm{d}\Omega}{\mathrm{d}c}\mathrm{d}t$$

The terms multiplied by dΩ/da can be eliminated from this expression by noting that Ω does not contain the variables x', y', z' and one has

$$\frac{\mathrm{d}\Omega}{\mathrm{d}x'} = \frac{\mathrm{d}\Omega}{\mathrm{d}a}\frac{\mathrm{d}a}{\mathrm{d}x'} + \frac{\mathrm{d}\Omega}{\mathrm{d}b}\frac{\mathrm{d}b}{\mathrm{d}x'} + \frac{\mathrm{d}\Omega}{\mathrm{d}c}\frac{\mathrm{d}c}{\mathrm{d}x'} + \cdots = 0$$
$$\frac{\mathrm{d}\Omega}{\mathrm{d}y'} = \frac{\mathrm{d}\Omega}{\mathrm{d}a}\frac{\mathrm{d}a}{\mathrm{d}y'} + \frac{\mathrm{d}\Omega}{\mathrm{d}b}\frac{\mathrm{d}b}{\mathrm{d}y'} + \frac{\mathrm{d}\Omega}{\mathrm{d}c}\frac{\mathrm{d}c}{\mathrm{d}y'} + \cdots = 0$$
$$\frac{\mathrm{d}\Omega}{\mathrm{d}z'} = \frac{\mathrm{d}\Omega}{\mathrm{d}a}\frac{\mathrm{d}a}{\mathrm{d}z'} + \frac{\mathrm{d}\Omega}{\mathrm{d}b}\frac{\mathrm{d}b}{\mathrm{d}z'} + \frac{\mathrm{d}\Omega}{\mathrm{d}c}\frac{\mathrm{d}c}{\mathrm{d}z'} + \cdots = 0$$

Therefore, if from the expression for da the following quantity is subtracted

$$\left(\frac{\mathrm{d}\Omega}{\mathrm{d}x'}\frac{\mathrm{d}a}{\mathrm{d}x} + \frac{\mathrm{d}\Omega}{\mathrm{d}y'}\frac{\mathrm{d}a}{\mathrm{d}y} + \frac{\mathrm{d}\Omega}{\mathrm{d}z'}\frac{\mathrm{d}a}{\mathrm{d}z}\right)\mathrm{d}t$$

which is zero, one will have

$$\mathrm{d}a = \left(\frac{\mathrm{d}a}{\mathrm{d}x'}\frac{\mathrm{d}b}{\mathrm{d}x} + \frac{\mathrm{d}a}{\mathrm{d}y'}\frac{\mathrm{d}b}{\mathrm{d}y} + \frac{\mathrm{d}a}{\mathrm{d}z'}\frac{\mathrm{d}b}{\mathrm{d}z} - \frac{\mathrm{d}a}{\mathrm{d}x}\frac{\mathrm{d}b}{\mathrm{d}x'} - \frac{\mathrm{d}a}{\mathrm{d}y}\frac{\mathrm{d}b}{\mathrm{d}y'} - \frac{\mathrm{d}a}{\mathrm{d}z}\frac{\mathrm{d}b}{\mathrm{d}z'}\right)\frac{\mathrm{d}\Omega}{\mathrm{d}b}\mathrm{d}t$$
$$+ \left(\frac{\mathrm{d}a}{\mathrm{d}x'}\frac{\mathrm{d}c}{\mathrm{d}x} + \frac{\mathrm{d}a}{\mathrm{d}y'}\frac{\mathrm{d}c}{\mathrm{d}y} + \frac{\mathrm{d}a}{\mathrm{d}z'}\frac{\mathrm{d}c}{\mathrm{d}z} - \frac{\mathrm{d}a}{\mathrm{d}x}\frac{\mathrm{d}c}{\mathrm{d}x'} - \frac{\mathrm{d}a}{\mathrm{d}y}\frac{\mathrm{d}c}{\mathrm{d}y'} - \frac{\mathrm{d}a}{\mathrm{d}z}\frac{\mathrm{d}c}{\mathrm{d}z'}\right)\frac{\mathrm{d}\Omega}{\mathrm{d}c}\mathrm{d}t$$
$$\cdots\cdots\cdots$$

This expression for da seems to be more complicated than the original formula from which we started. But, on the other hand, it has the great advantage that the coefficients of partial differences dΩ/db, dΩ/dc, etc. become independent of the time t after substitution for the values of x, y, z, x', y', z' by t and a, b, c, etc., given by the elliptical motion of the planet as can be checked by differentiation by making the time t variable in the coefficients.

61. Indeed, since a is assumed to be a function of t, x, y, z, x', y', z' and that x, y, z, x', y', z', also vary with t such that dx/dt = x', dy/dt = y', dz/dt = z' and dx'/dt = $-$dV/dx, dy'/dt = $-$dV/dy, dz'/dt = $-$dV/dz, from the differential equations of the problem (Article 4), it follows that one will have after differentiating with respect to t

$$\mathrm{d}\frac{\mathrm{d}a}{\mathrm{d}x} = \left\{\begin{array}{l} \dfrac{\mathrm{d}^2a}{\mathrm{d}x\,\mathrm{d}t} + \dfrac{\mathrm{d}^2a}{\mathrm{d}x^2}x' + \dfrac{\mathrm{d}^2a}{\mathrm{d}x\,\mathrm{d}y}y' + \dfrac{\mathrm{d}^2a}{\mathrm{d}x\,\mathrm{d}z}z' \\ -\dfrac{\mathrm{d}^2a}{\mathrm{d}x\,\mathrm{d}x'}\dfrac{\mathrm{d}V}{\mathrm{d}x} - \dfrac{\mathrm{d}^2a}{\mathrm{d}x\,\mathrm{d}y'}\dfrac{\mathrm{d}V}{\mathrm{d}y} - \dfrac{\mathrm{d}^2a}{\mathrm{d}x\,\mathrm{d}z'}\dfrac{\mathrm{d}V}{\mathrm{d}z} \end{array}\right\}\mathrm{d}t$$

Since a is one of the arbitrary constants introduced by the integration of the same equations, its differential relative to t must become identically zero for the same values of $\mathrm{d}x'/\mathrm{d}t, \mathrm{d}y'/\mathrm{d}t, \mathrm{d}z'/\mathrm{d}t$. Thus one will have

$$\frac{\mathrm{d}a}{\mathrm{d}t}+\frac{\mathrm{d}a}{\mathrm{d}x}x'+\frac{\mathrm{d}a}{\mathrm{d}y}y'+\frac{\mathrm{d}a}{\mathrm{d}z}z'-\frac{\mathrm{d}a}{\mathrm{d}x'}\frac{\mathrm{d}V}{\mathrm{d}x}-\frac{\mathrm{d}a}{\mathrm{d}y'}\frac{\mathrm{d}V}{\mathrm{d}y}-\frac{\mathrm{d}a}{\mathrm{d}z'}\frac{\mathrm{d}V}{\mathrm{d}z}=0$$

an identity which exists, consequently, by varying individually x, y, z, x', y', z'.

Let us vary x. One will also have

$$\begin{aligned}&\frac{\mathrm{d}^2a}{\mathrm{d}x\,\mathrm{d}t}+\frac{\mathrm{d}^2a}{\mathrm{d}x^2}x'+\frac{\mathrm{d}^2a}{\mathrm{d}x\,\mathrm{d}y}y'+\frac{\mathrm{d}^2a}{\mathrm{d}x\,\mathrm{d}z}z'-\frac{\mathrm{d}^2a}{\mathrm{d}x\,\mathrm{d}x'}\frac{\mathrm{d}V}{\mathrm{d}x}-\frac{\mathrm{d}^2a}{\mathrm{d}x\,\mathrm{d}y'}\frac{\mathrm{d}V}{\mathrm{d}y}\\&-\frac{\mathrm{d}^2a}{\mathrm{d}x\,\mathrm{d}z'}\frac{\mathrm{d}V}{\mathrm{d}z}-\frac{\mathrm{d}a}{\mathrm{d}x'}\frac{\mathrm{d}^2V}{\mathrm{d}x^2}-\frac{\mathrm{d}a}{\mathrm{d}y'}\frac{\mathrm{d}^2V}{\mathrm{d}x\,\mathrm{d}y}-\frac{\mathrm{d}a}{\mathrm{d}z'}\frac{\mathrm{d}^2V}{\mathrm{d}x\,\mathrm{d}z}=0\end{aligned}$$

Then the value of the differential of $\mathrm{d}a/\mathrm{d}x$ reduces to

$$\mathrm{d}\frac{\mathrm{d}a}{\mathrm{d}x}=\left(\frac{\mathrm{d}a}{\mathrm{d}x'}\frac{\mathrm{d}^2V}{\mathrm{d}x^2}+\frac{\mathrm{d}a}{\mathrm{d}y'}\frac{\mathrm{d}^2V}{\mathrm{d}x\,\mathrm{d}y}+\frac{\mathrm{d}a}{\mathrm{d}z'}\frac{\mathrm{d}^2V}{\mathrm{d}x\,\mathrm{d}z}\right)\mathrm{d}t$$

It will be found in the same manner that

$$\mathrm{d}\frac{\mathrm{d}a}{\mathrm{d}y}=\left(\frac{\mathrm{d}a}{\mathrm{d}x'}\frac{\mathrm{d}^2V}{\mathrm{d}x\,\mathrm{d}y}+\frac{\mathrm{d}a}{\mathrm{d}y'}\frac{\mathrm{d}^2V}{\mathrm{d}y^2}+\frac{\mathrm{d}a}{\mathrm{d}z'}\frac{\mathrm{d}^2V}{\mathrm{d}y\,\mathrm{d}z}\right)\mathrm{d}t$$

$$\mathrm{d}\frac{\mathrm{d}a}{\mathrm{d}z}=\left(\frac{\mathrm{d}a}{\mathrm{d}x'}\frac{\mathrm{d}^2V}{\mathrm{d}x\,\mathrm{d}z}+\frac{\mathrm{d}a}{\mathrm{d}y'}\frac{\mathrm{d}^2V}{\mathrm{d}x\,\mathrm{d}z}+\frac{\mathrm{d}a}{\mathrm{d}z'}\frac{\mathrm{d}^2V}{\mathrm{d}z^2}\right)\mathrm{d}t$$

In turn, one will have

$$\mathrm{d}\frac{\mathrm{d}a}{\mathrm{d}x'}=\left\{\begin{aligned}&\frac{\mathrm{d}^2a}{\mathrm{d}x'\,\mathrm{d}t}+\frac{\mathrm{d}^2a}{\mathrm{d}x\,\mathrm{d}x'}x'+\frac{\mathrm{d}^2a}{\mathrm{d}y\,\mathrm{d}x'}y'+\frac{\mathrm{d}^2a}{\mathrm{d}z\,\mathrm{d}x'}z'\\&-\frac{\mathrm{d}^2a}{\mathrm{d}x'^2}\frac{\mathrm{d}V}{\mathrm{d}x}-\frac{\mathrm{d}^2a}{\mathrm{d}x'\,\mathrm{d}y'}\frac{\mathrm{d}V}{\mathrm{d}y}-\frac{\mathrm{d}^2a}{\mathrm{d}x'\,\mathrm{d}z'}\frac{\mathrm{d}V}{\mathrm{d}z}\end{aligned}\right\}\mathrm{d}t$$

But by varying x' in the identical equation $\mathrm{d}a=0$ and noting that the variables x', y', z' are assumed not to be contained in the function V, one has

$$\begin{aligned}&\frac{\mathrm{d}^2a}{\mathrm{d}x'\,\mathrm{d}t}+\frac{\mathrm{d}a}{\mathrm{d}x}+\frac{\mathrm{d}^2a}{\mathrm{d}x\,\mathrm{d}x'}x'+\frac{\mathrm{d}^2a}{\mathrm{d}y\,\mathrm{d}x'}y'+\frac{\mathrm{d}^2a}{\mathrm{d}z\,\mathrm{d}x'}z'\\&-\frac{\mathrm{d}^2a}{\mathrm{d}x'^2}\frac{\mathrm{d}V}{\mathrm{d}x}-\frac{\mathrm{d}^2a}{\mathrm{d}x'\,\mathrm{d}y'}\frac{\mathrm{d}V}{\mathrm{d}y}-\frac{\mathrm{d}^2a}{\mathrm{d}x'\,\mathrm{d}z'}\frac{\mathrm{d}V}{\mathrm{d}z}=0\end{aligned}$$

Therefore, one will have more simply

$$\mathrm{d}\frac{\mathrm{d}a}{\mathrm{d}x'}=-\frac{\mathrm{d}a}{\mathrm{d}x}\mathrm{d}t$$

and in the same manner it will be found that

$$\mathrm{d}\frac{\mathrm{d}a}{\mathrm{d}y'} = -\frac{\mathrm{d}a}{\mathrm{d}y}\mathrm{d}t, \qquad \mathrm{d}\frac{\mathrm{d}a}{\mathrm{d}z'} = -\frac{\mathrm{d}a}{\mathrm{d}z}\mathrm{d}t$$

Similar expressions will be found for the differentials $\mathrm{d}(\mathrm{d}b/\mathrm{d}x), \mathrm{d}(\mathrm{d}b/\mathrm{d}y), \mathrm{d}(\mathrm{d}b/\mathrm{d}z)$, $\mathrm{d}(\mathrm{d}b/\mathrm{d}x'), \mathrm{d}(\mathrm{d}b/\mathrm{d}y'), \mathrm{d}(\mathrm{d}b/\mathrm{d}z')$ by only changing a to b and similarly for the other similar quantities.

Now if the coefficient $(\mathrm{d}\Omega/\mathrm{d}b)\mathrm{d}t$ is differentiated in the expression of Article 60 and if the values we just found were substituted for the differentials of $\mathrm{d}a/\mathrm{d}x, \mathrm{d}a/\mathrm{d}y, \mathrm{d}a/\mathrm{d}z$, $\mathrm{d}a/\mathrm{d}x', \mathrm{d}a/\mathrm{d}y'$, $\mathrm{d}a/\mathrm{d}z'$ one will first see that the terms which contain the differentials of $\mathrm{d}a/\mathrm{d}x', \mathrm{d}b/\mathrm{d}x'$, $\mathrm{d}a/\mathrm{d}y', \mathrm{d}b/\mathrm{d}y'$, $\mathrm{d}a/\mathrm{d}z', \mathrm{d}b/\mathrm{d}z'$, will cancel each other and the terms which contain the differentials of $\mathrm{d}a/\mathrm{d}x$, $\mathrm{d}b/\mathrm{d}x$, etc. which are ordered with respect to the partial differences of V will also cancel each other in each of the coefficients of the partial differences.

From this result, it can be concluded that the coefficient of $\mathrm{d}\Omega/\mathrm{d}b$ in the expression for da will be constant with respect to the time t and can not be a function of a, b, c, etc. after the substitution of the values of x, y, z, x', y', z' for a, b, c, etc. and t such that the variable t will disappear by itself. It will suffice to substitute the values of x, y, z, x', y', z', which correspond at $t = 0$, at an arbitrary value of t.

In the same manner, it will be proved that t will disappear from the other coefficients of the partial differences of Ω in the same expression for $\mathrm{d}a$. Thus the variation of a will be represented by a formula which will only contain partial differences of Ω with respect to b, c, etc., each multiplied by a function of a, b, c, etc. without t. It will be the same for the variations of the other arbitrary constants b, c, h, etc.

This important result that we have found **a posteriori** is only a particular case of the general theory of the variation of arbitrary constants that we presented in Subsection II of SECTION V and we could have deduced it immediately from this theory. But we believe that it is useful to show how it can be obtained starting from the formulas which directly give the variation of the elements resulting from the perturbing forces and in addition, how these variations have a simple and elegant form by the reduction of the perturbing forces to the partial differences of the same function, relative to these same elements viewed as variables.

62. We have assumed in Article 60 that the forces X, Y, Z could be expressed by partial differences of the same function Ω, relative to x, y, z. This hypothesis simplifies the calculations but it is not absolutely necessary for accuracy because the differential equations are always independent of the nature of the accelerating forces of the mobile body. It is only a question of knowing what should be substituted in the partial differences of Ω relative to the arbitrary constants a, b, c, etc. But these constants are contained in the function Ω

only because they are contained in the expressions for x, y, z which Ω is assumed to be a function of. Thus one will have

$$\frac{\mathrm{d}\Omega}{\mathrm{d}a} = \frac{\mathrm{d}\Omega}{\mathrm{d}x}\frac{\mathrm{d}x}{\mathrm{d}a} + \frac{\mathrm{d}\Omega}{\mathrm{d}y}\frac{\mathrm{d}y}{\mathrm{d}a} + \frac{\mathrm{d}\Omega}{\mathrm{d}z}\frac{\mathrm{d}z}{\mathrm{d}a}$$

and after substituting X, Y, Z for $\mathrm{d}\Omega/\mathrm{d}x$, $\mathrm{d}\Omega/\mathrm{d}y$, $\mathrm{d}\Omega/\mathrm{d}z$, one will have

$$\frac{\mathrm{d}\Omega}{\mathrm{d}a} = X\frac{\mathrm{d}x}{\mathrm{d}a} + Y\frac{\mathrm{d}y}{\mathrm{d}a} + Z\frac{\mathrm{d}z}{\mathrm{d}a}$$

whatever the values of X, Y, Z. It will be the same for $\mathrm{d}\Omega/\mathrm{d}b$, $\mathrm{d}\Omega/\mathrm{d}c$, etc. after replacing a by b, c, etc.

In general, if the operator δ denotes the variation of Ω relative to the arbitrary constants a, b, c, etc., one will have

$$\delta\Omega = X\,\delta x + Y\,\delta y + Z\,\delta z$$

and if it is assumed that the perturbing forces are R, Q, P, etc. directed toward centers for which the respective distances are r, q, p, etc. the following equations are obtained

$$-\mathrm{d}\Omega = R\,\mathrm{d}r + Q\,\mathrm{d}q + P\,\mathrm{d}p + \cdots$$

and it also follows that

$$-\delta\Omega = R\,\delta r + Q\,\delta q + P\,\delta p + \cdots$$

A negative sign is given to $\mathrm{d}\Omega$ because the forces R, Q, P, etc. are assumed to have a tendency to shorten the lines r, q, p, etc., while the forces X, Y, Z are assumed to have a tendency to lengthen the lines x, y, z, as we have already observed in Article 60.[19]

63. In order to apply the general formulas of Article 18 of the cited section to the elements of a planet, it is only necessary that the coordinates x, y, z which are independent be taken for the variables ξ, ψ, φ, and because there is only one mobile body for which the mass m can be assumed equal to unity, one will have simply, as in Article 3

$$T = \frac{\mathrm{d}x^2 + \mathrm{d}y^2 + \mathrm{d}z^2}{2\,\mathrm{d}t^2} = \frac{x'^2 + y'^2 + z'^2}{2}$$

Thus

$$\frac{\mathrm{d}T}{\mathrm{d}x'} = x', \qquad \frac{\mathrm{d}T}{\mathrm{d}y'} = y', \qquad \frac{\mathrm{d}T}{\mathrm{d}z'} = z',$$

Therefore, the constants α, β, γ and λ, μ, ν which represent the values of x, y, z and x', y', z' when $t = 0$ (SECTION V, Article 12), will be here $\mathbf{x}$, $\mathbf{y}$, $\mathbf{z}$, $\mathbf{x}'$, $\mathbf{y}'$, $\mathbf{z}'$ (Article 31) and the variations of the elements a, b, c, etc., will have the form

$$\mathrm{d}a = \left[(a,b)\frac{\mathrm{d}\Omega}{\mathrm{d}b} + (a,c)\frac{\mathrm{d}\Omega}{\mathrm{d}c} + \cdots\right]\mathrm{d}t$$

$$\mathrm{d}b = \left[-(a,b)\frac{\mathrm{d}\Omega}{\mathrm{d}a} + (b,c)\frac{\mathrm{d}\Omega}{\mathrm{d}c} + \cdots\right]\mathrm{d}t$$

$$\cdots$$

The coefficients represented by the symbols (a, b), (a, c), etc. can be expressed as

$$(a,b) = \frac{\mathrm{d}a}{\mathrm{d}\mathbf{x}'}\frac{\mathrm{d}b}{\mathrm{d}\mathbf{x}} + \frac{\mathrm{d}a}{\mathrm{d}\mathbf{y}'}\frac{\mathrm{d}b}{\mathrm{d}\mathbf{y}} + \frac{\mathrm{d}a}{\mathrm{d}\mathbf{z}'}\frac{\mathrm{d}b}{\mathrm{d}\mathbf{z}} - \frac{\mathrm{d}a}{\mathrm{d}\mathbf{x}}\frac{\mathrm{d}b}{\mathrm{d}\mathbf{x}'} - \frac{\mathrm{d}a}{\mathrm{d}\mathbf{y}}\frac{\mathrm{d}b}{\mathrm{d}\mathbf{y}'} - \frac{\mathrm{d}a}{\mathrm{d}\mathbf{z}}\frac{\mathrm{d}b}{\mathrm{d}\mathbf{z}'}$$

$$(a,c) = \frac{\mathrm{d}a}{\mathrm{d}\mathbf{x}'}\frac{\mathrm{d}c}{\mathrm{d}\mathbf{x}} + \frac{\mathrm{d}a}{\mathrm{d}\mathbf{y}'}\frac{\mathrm{d}c}{\mathrm{d}\mathbf{y}} + \frac{\mathrm{d}a}{\mathrm{d}\mathbf{z}'}\frac{\mathrm{d}c}{\mathrm{d}\mathbf{z}} - \frac{\mathrm{d}a}{\mathrm{d}\mathbf{x}}\frac{\mathrm{d}c}{\mathrm{d}\mathbf{x}'} - \frac{\mathrm{d}a}{\mathrm{d}\mathbf{y}}\frac{\mathrm{d}c}{\mathrm{d}\mathbf{y}'} - \frac{\mathrm{d}a}{\mathrm{d}\mathbf{z}}\frac{\mathrm{d}c}{\mathrm{d}\mathbf{z}'}$$

$$\cdots$$

It is obvious that the expressions for $\mathrm{d}a$, $\mathrm{d}b$, etc., coincide with those that we found above (Article 60) but instead of the letters x, y, z there are the letters $\mathbf{x}$, $\mathbf{y}$, $\mathbf{z}$ which represent the values of x, y, z when t is equal to zero or to an arbitrary value because the beginning of the time t is arbitrary. Since the coefficients (a, b), (a, c), etc., are independent of t, the quantities a, b, c, etc., must be the same functions of x, y, z, x', y', z' rather than of $\mathbf{x}$, $\mathbf{y}$, $\mathbf{z}$, $\mathbf{x}'$, $\mathbf{y}'$, $\mathbf{z}'$.

64. Since the quantities $\mathbf{x}$, $\mathbf{y}$, $\mathbf{z}$, $\mathbf{x}'$, $\mathbf{y}'$, $\mathbf{z}'$ are also arbitrary constants, they can be used in place of the six constants a, b, c, h, i, k. Thus after changing a to $\mathbf{x}$, b to $\mathbf{x}'$, c to $\mathbf{y}$, h to $\mathbf{y}'$ one will have

$$(\mathbf{x}, \mathbf{x}') = -1, \qquad (\mathbf{y}, \mathbf{y}') = -1, \qquad (\mathbf{z}, \mathbf{z}') = -1$$

and all the other coefficients $(\mathbf{x}, \mathbf{y})$, $(\mathbf{x}, \mathbf{z})$, $(\mathbf{x}', \mathbf{y})$, etc. will be equal to zero such that the variations of $\mathbf{x}$, $\mathbf{y}$, $\mathbf{z}$, $\mathbf{x}'$, $\mathbf{y}'$, $\mathbf{z}'$ will be represented by the very simple formulas

$$\mathrm{d}\mathbf{x} = -\frac{\mathrm{d}\Omega}{\mathrm{d}\mathbf{x}'}\mathrm{d}t, \qquad \mathrm{d}\mathbf{y} = -\frac{\mathrm{d}\Omega}{\mathrm{d}\mathbf{y}'}\mathrm{d}t, \qquad \mathrm{d}\mathbf{z} = -\frac{\mathrm{d}\Omega}{\mathrm{d}\mathbf{z}'}\mathrm{d}t$$

$$\mathrm{d}\mathbf{x}' = \frac{\mathrm{d}\Omega}{\mathrm{d}\mathbf{x}}\mathrm{d}t, \qquad \mathrm{d}\mathbf{y}' = \frac{\mathrm{d}\Omega}{\mathrm{d}\mathbf{y}}\mathrm{d}t, \qquad \mathrm{d}\mathbf{z}' = \frac{\mathrm{d}\Omega}{\mathrm{d}\mathbf{z}}\mathrm{d}t$$

which also are the result of those equations which we obtained directly in Article 14 of SECTION V. Therefore, there would always be an advantage to use these constants rather than the constants a, b, c, etc.

But whatever the constants a, b, c, etc., they can only be a function of the constants $\mathbf{x}$, $\mathbf{y}$, $\mathbf{z}$, $\mathbf{x}'$, etc. Therefore, the latter constants can be viewed as functions of the former. Thus the following equation will be obtained

$$\frac{d\Omega}{da} = \frac{d\Omega}{d\mathbf{x}}\frac{d\mathbf{x}}{da} + \frac{d\Omega}{d\mathbf{y}}\frac{d\mathbf{y}}{da} + \frac{d\Omega}{d\mathbf{z}}\frac{d\mathbf{z}}{da} + \frac{d\Omega}{d\mathbf{x}'}\frac{d\mathbf{x}'}{da} + \frac{d\Omega}{d\mathbf{y}'}\frac{d\mathbf{y}'}{da} + \frac{d\Omega}{d\mathbf{z}'}\frac{d\mathbf{z}'}{da}$$

and after substituting the values of $d\Omega/d\mathbf{x}$, $d\Omega/d\mathbf{y}$, etc. of the preceding article, one will have

$$\frac{d\Omega}{da}dt = \frac{d\mathbf{x}}{da}d\mathbf{x}' + \frac{d\mathbf{y}}{da}d\mathbf{y}' + \frac{d\mathbf{z}}{da}d\mathbf{z}' - \frac{d\mathbf{x}'}{da}d\mathbf{x} - \frac{d\mathbf{y}'}{da}d\mathbf{y} - \frac{d\mathbf{z}'}{da}d\mathbf{z}$$

Since $\mathbf{x}$, $\mathbf{y}$, $\mathbf{z}$, $\mathbf{x}'$, etc. are functions of a, b, c, etc., there will also result

$$\begin{aligned} d\mathbf{x} &= \frac{d\mathbf{x}}{da}da + \frac{d\mathbf{x}}{db}db + \frac{d\mathbf{x}}{dc}dc + \cdots \\ d\mathbf{x}' &= \frac{d\mathbf{x}'}{da}da + \frac{d\mathbf{x}'}{db}db + \frac{d\mathbf{x}'}{dc}dc + \cdots \\ &\cdots \end{aligned}$$

After substituting these expressions and ordering the terms with respect to the variations da, db, dc, etc. the following equation will be obtained

$$\frac{d\Omega}{da}dt = [a,b]db + [a,c]dc + [a,h]dh + \cdots$$

where the symbols $[a,b]$, $[a,c]$, etc. are defined by the following formulas

$$\begin{aligned} [a,b] &= \frac{d\mathbf{x}}{da}\frac{d\mathbf{x}'}{db} + \frac{d\mathbf{y}}{da}\frac{d\mathbf{y}'}{db} + \frac{d\mathbf{z}}{da}\frac{d\mathbf{z}'}{db} - \frac{d\mathbf{x}'}{da}\frac{d\mathbf{x}}{db} - \frac{d\mathbf{y}'}{da}\frac{d\mathbf{y}}{db} - \frac{d\mathbf{z}'}{da}\frac{d\mathbf{z}}{db} \\ [a,c] &= \frac{d\mathbf{x}}{da}\frac{d\mathbf{x}'}{dc} + \frac{d\mathbf{y}}{da}\frac{d\mathbf{y}'}{dc} + \frac{d\mathbf{z}}{da}\frac{d\mathbf{z}'}{dc} - \frac{d\mathbf{x}'}{da}\frac{d\mathbf{x}}{dc} - \frac{d\mathbf{y}'}{da}\frac{d\mathbf{y}}{dc} - \frac{d\mathbf{z}'}{da}\frac{d\mathbf{z}}{dc} \\ &\cdots \end{aligned}$$

Similarly, one will have, since $[b,a] = -[a,b]$

$$\begin{aligned} \frac{d\Omega}{db}dt &= -[a,b]da + [b,c]dc + [b,h]dh + \cdots \\ [b,c] &= \frac{d\mathbf{x}}{db}\frac{d\mathbf{x}'}{dc} + \frac{d\mathbf{y}}{db}\frac{d\mathbf{y}'}{dc} + \frac{d\mathbf{z}}{db}\frac{d\mathbf{z}'}{dc} - \frac{d\mathbf{x}'}{db}\frac{d\mathbf{x}}{dc} - \frac{d\mathbf{y}'}{db}\frac{d\mathbf{y}}{dc} - \frac{d\mathbf{z}'}{db}\frac{d\mathbf{z}}{dc} \end{aligned}$$

and similarly, by simply changing the quantities a, b, c, h, i, k, between themselves, taken two at a time, and by observing that in general

$$[b,a] = -[a,b]$$

such that the value of the symbols changes sign with a permutation of the two quantities they contain.

If the values of the symbols indicated by brackets are compared to those marked by parentheses (Article 63), a remarkable analogy is noted. It is that they are expressed in the same manner as the partial differences of a, b, c, etc. relative to **x, x′, y**, etc., or of **x, x′, y** relative to a, b, c, etc.

65. I found the last formulas directly in my first memoir on the variation of arbitrary constants[20] and they also result immediately from the formula of Article 12 of SECTION V, which after making the substitutions indicated above (Article 61) is reduced to

$$\Delta\Omega\, dt = \Delta\mathbf{x}\,\delta\mathbf{x}' + \Delta\mathbf{y}\,\delta\mathbf{y}' + \Delta\mathbf{z}\,\delta\mathbf{z}' - \Delta\mathbf{x}'\,\delta\mathbf{x} - \Delta\mathbf{y}'\,\delta\mathbf{y} - \Delta\mathbf{z}'\,\delta\mathbf{z}$$

In this formula, the differences marked by δ must refer to the variations of all the arbitrary constants, a, b, c, etc. But the differences marked by Δ can refer to the variation of each of these constants in particular (Article 10, cited section). Thus by applying the operator Δ to a, b, c, etc. successively, one will have

$$\frac{d\Omega}{da}dt = \frac{d\mathbf{x}}{da}\delta\mathbf{x}' + \frac{d\mathbf{y}}{da}\delta\mathbf{y}' + \frac{d\mathbf{z}}{da}\delta\mathbf{z}' - \frac{d\mathbf{x}'}{da}\delta\mathbf{x} - \frac{d\mathbf{y}'}{da}\delta\mathbf{y} - \frac{d\mathbf{z}'}{da}\delta\mathbf{z}$$

and similar equations will be obtained by replacing a with b, c, etc.

But one has

$$\delta\mathbf{x} = \frac{d\mathbf{x}}{da}da + \frac{d\mathbf{x}}{db}db + \frac{d\mathbf{x}}{dc}dc + \cdots$$
$$\delta\mathbf{x}' = \frac{d\mathbf{x}'}{da}da + \frac{d\mathbf{x}'}{db}db + \frac{d\mathbf{x}'}{dc}dc + \cdots$$

and similarly for $\delta\mathbf{y}, \delta\mathbf{y}', \delta\mathbf{z}, \delta\mathbf{z}'$. By making these substitutions, one has for $d\Omega/da$, $d\Omega/db$, etc., the same formulas found above.

An important consequence derives from these formulas. It is that the variation of the function Ω is always zero as long as it depends upon the elements a, b, c, etc. Indeed, if in the differential

$$\frac{d\Omega}{da}da + \frac{d\Omega}{db}db + \frac{d\Omega}{dc}dc + \cdots$$

the values of $d\Omega/da, d\Omega/db$, etc. are replaced by da/dt, db/dt, etc., it will be found that all the terms cancel each other, which is a very remarkable result.

66. Since in the solution to the main problem, there is no consideration that the perturbing forces must give the values of the variables x, y, z as a function of t with arbitrary constants a, b, c, etc., it can be assumed at the outset that $t = 0$ in these expressions and in those of their differentials as a function of t and then their partial differentials relative to a, b, c, etc.

can be taken. It is then easy to obtain the coefficients of the differences da, db, dc, etc. in the expressions for $(\mathrm{d}\Omega/\mathrm{d}a)\mathrm{d}t$, $(\mathrm{d}\Omega/\mathrm{d}b)\mathrm{d}t$, etc. and it is only a question of finding these differences by linear eliminations, as I have done in the cited memoir for the elements of planets.

To this end, the formulas of Article 63 seem to have the advantage of giving directly the same differences but they require that we find at the outset the expressions for the arbitrary constants a, b, c, etc., from the variables x, y, z and their differentials, which, in several instances, can only be obtained from elimination of an order higher than the linear ones. Then, after having taken their partial differences relative to x, y, z, x', y', z', one must replace these variables as a function of a, b, c, etc. because in the final analysis, the coefficients $(a, b), (a, c)$, etc., must become functions of a, b, c, etc., without t, which is the essence and strength of this analysis.

After all, after having given in Subsection I very simple expressions for the coordinates x, y, z, as functions of t and a, b, c, h, i, k, we will apply the formulas of the last article to deduce the variations of the elements a, b, c, etc. as we have done in the cited memoir, because the calculations with these formulas acquire a simplicity and elegance which would certainly not exist with the other formulas.

67. Let us consider again the expressions for x, y, z given in Article 13

$$x = \alpha X + \beta Y, \qquad y = \alpha_1 X + \beta_1 Y, \qquad z = \alpha_2 X + \beta_2 Y$$

in which (Article 17)

$$X = a(\cos\theta - e), \qquad Y = a\sqrt{1 - e^2}\sin\theta$$

where the angle θ is defined by the following equation (Article 16)

$$t - c = \sqrt{\frac{a^3}{g}}\,(\theta - e\sin\theta)$$

These formulas have the advantage that the three elements of the orbit, a, b, c, are only found in the variable quantities X and Y and consequently, they are separated from the three other elements h, i, k which depend upon the orientation of the orbit and for which the coefficients α, β, α', etc. are functions (Article 13).

Let us consider at the outset the formula

$$\frac{\mathrm{d}x}{\mathrm{d}a}\frac{\mathrm{d}x'}{\mathrm{d}b} + \frac{\mathrm{d}y}{\mathrm{d}a}\frac{\mathrm{d}y'}{\mathrm{d}b} + \frac{\mathrm{d}z}{\mathrm{d}a}\frac{\mathrm{d}z'}{\mathrm{d}b} - \frac{\mathrm{d}x'}{\mathrm{d}a}\frac{\mathrm{d}x}{\mathrm{d}b} - \frac{\mathrm{d}y'}{\mathrm{d}a}\frac{\mathrm{d}y}{\mathrm{d}b} - \frac{\mathrm{d}z'}{\mathrm{d}a}\frac{\mathrm{d}z}{\mathrm{d}b}$$

In addition, let us substitute in it the expressions for x, y, z given above. After defining

$$X' = \frac{\mathrm{d}X}{\mathrm{d}t}, \qquad Y' = \frac{\mathrm{d}Y}{\mathrm{d}t}$$

one will have for x', y', z', the same expressions where the quantities X and Y will be marked by a prime. Since the constants a and b are only in X and Y, one will have

$$\frac{dx}{da} = \alpha \frac{dX}{da} + \beta \frac{dY}{da}, \qquad \frac{dx'}{da} = \alpha \frac{dX'}{da} + \beta \frac{dY'}{da},$$

$$\frac{dx}{db} = \alpha \frac{dX}{db} + \beta \frac{dY}{db}, \qquad \frac{dx'}{db} = \alpha \frac{dX'}{db} + \beta \frac{dY'}{db}$$

and after replacing α, β, with α_1, β_1, and α_2, β_2, the expressions for dy/da, dy'/da, etc. will be obtained.

These different expressions are to be substituted in the preceding formula and with due respect to the equations of condition

$$\alpha^2 + \alpha_1^2 + \alpha_2^2 = 1, \qquad \beta^2 + \beta_1^2 + \beta_2^2 = 1, \qquad \alpha\beta + \alpha_1\beta_1 + \alpha_2\beta_2 = 0$$

which exist between the coefficients α, β, α_1, etc., (Article 14), this formula will be reduced to the form

$$\frac{dX}{da}\frac{dX'}{db} + \frac{dY}{da}\frac{dY'}{db} - \frac{dX'}{da}\frac{dX}{db} - \frac{dY'}{da}\frac{dY}{db}$$

where it is clear that the quantities α, β, α', etc., which depend on the orientation of the orbit have disappeared.

A similar result will be obtained for the partial differentials with respect to c, and it will only be necessary to replace a by b and c in the preceding formula.

Therefore, if one substitutes in the expressions for X, Y, X', Y' their expressions in terms of t and then if t is put equal to zero or to an arbitrary determined quantity and furthermore, if one designates by X, Y, X', Y' what X, Y, X', Y' became, one will have (Article 64)

$$[a, b] = \frac{dX}{da}\frac{dX'}{db} + \frac{dY}{da}\frac{dY'}{db} - \frac{dX'}{da}\frac{dX}{db} - \frac{dY'}{da}\frac{dY}{db}$$

Similarly, the values of $[a, c]$, $[b, c]$ will be obtained by replacing b with c and a with b in the partial differences.

68. But one has

$$X = a(\cos\theta - e), \qquad Y = a\sqrt{1 - e^2}\sin\theta$$

thus since $X' = dX/dt$, $Y' = dY/dt$ one will have

$$X' = -a\sin\theta\frac{d\theta}{dt}, \qquad Y' = a\sqrt{1 - e^2}\cos\theta\frac{d\theta}{dt}$$

But the equation

$$(t-c)\sqrt{\frac{g}{a^3}} = \theta - e\sin\theta$$

gives after differentiation

$$\frac{\mathrm{d}\theta}{\mathrm{d}t} = \frac{\sqrt{\frac{g}{a^3}}}{1 - e\cos\theta}$$

Therefore, one will have

$$X' = -\sqrt{\frac{g}{a}}\frac{\sin\theta}{1-e\cos\theta}, \qquad Y' = \sqrt{\frac{g(1-e^2)}{a}}\frac{\cos\theta}{1-e\cos\theta}$$

Now it is necessary to differentiate these formulas by varying the three constants a, e, c. Let us denote by the operator δ the variations relative to these constants. Thus there results at the outset

$$\delta\theta = \frac{(t-c)d\sqrt{\frac{g}{a^3}} + \sin\theta\,\mathrm{d}e - \sqrt{\frac{g}{a^3}}\,\mathrm{d}e}{1-e\cos\theta}$$

and there follows

$$\begin{aligned}
\delta X &= -a\sin\theta\,\delta\theta + \cos\theta\,\mathrm{d}a - \mathrm{d}(ae)\\
\delta Y &= a\sqrt{1-e^2}\cos\theta\,\delta\theta + \sin\theta\,\mathrm{d}(a\sqrt{1-e^2})\\
\delta X' &= -\sqrt{\frac{g}{a}}\frac{\cos\theta - e}{(1-e\cos\theta)^2}\delta\theta\\
&\quad - \sqrt{\frac{g}{a}}\frac{\sin\theta\cos\theta}{(1-e\cos\theta)^2}\mathrm{d}e - \frac{\sin\theta}{1-e\cos\theta}\mathrm{d}\sqrt{\frac{g}{a}}\\
\delta Y' &= -\sqrt{\frac{g}{a}}\sqrt{1-e^2}\frac{\sin\theta}{(1-e\cos\theta)^2}\delta\theta\\
&\quad + \sqrt{\frac{g}{a}}\sqrt{1-e^2}\frac{\cos^2\theta}{(1-e\cos\theta)^2}\mathrm{d}e\\
&\quad + \frac{\cos\theta}{1-e\cos\theta}\mathrm{d}\left(\sqrt{\frac{g}{a}}\sqrt{1-e^2}\right)
\end{aligned}$$

It is possible to put here $t = 0$ but it is simpler to put $t = c$, which also gives $\theta = 0$. Thus after replacing **X, Y** with X, Y there will result

$$\delta\theta = -\sqrt{\frac{g}{a^3}}\frac{\mathrm{d}e}{(1-e)}$$

$$\delta X = (1-e)\mathrm{d}a - a\,\mathrm{d}e$$

$$\delta Y = a\sqrt{1-e^2}\delta\theta = -\sqrt{\frac{g}{a}}\frac{\sqrt{1-e^2}}{1-e}\mathrm{d}c$$

$$\delta X' = -\sqrt{\frac{g}{a}}\frac{\delta\theta}{1-e} = \frac{g}{a^2}\frac{\mathrm{d}c}{(1-e)^2}$$

$$\delta Y' = \sqrt{\frac{g}{a}}\sqrt{1-e^2}\frac{\mathrm{d}e}{(1-e)^2} + \frac{\mathrm{d}\left(\sqrt{\frac{g}{a}}\sqrt{1-e^2}\right)}{1-e}$$

$$= \mathrm{d}\left(\sqrt{\frac{g}{a}}\frac{\sqrt{(1-e^2)}}{1-e}\right) = \frac{\sqrt{1-e^2}}{1-e}\mathrm{d}\sqrt{\frac{g}{a}} + \sqrt{\frac{g}{a}}\frac{\mathrm{d}e}{(1-e)\sqrt{1-e^2}}$$

69. The quantity e, which is the semi-eccentricity, has been retained here. But, if instead of using the quantity e the semi-parameter $b = a(1-e^2)$ is used, then after differentiation, the following equation is obtained

$$\mathrm{d}e = \frac{(1-e^2)\mathrm{d}a - \mathrm{d}b}{2ae}$$

and the expressions for δX and $\delta Y'$, which contain $\mathrm{d}e$, will become

$$\delta X = -\frac{-(1-e^2)\mathrm{d}a + \mathrm{d}b}{2e}$$

$$\delta Y' = \sqrt{\frac{g}{a}}\frac{\sqrt{1-e^2}}{2ae}\mathrm{d}a - \sqrt{\frac{g}{a}}\frac{\mathrm{d}b}{2ae(1-e)\sqrt{1-e^2}}$$

From these expressions the following partial difference equations are found

$$\frac{\mathrm{d}X}{\mathrm{d}a} = -\frac{(1-e)^2}{2e}, \qquad \frac{\mathrm{d}X}{\mathrm{d}b} = \frac{1}{2e}, \qquad \frac{\mathrm{d}X}{\mathrm{d}c} = 0$$

$$\frac{\mathrm{d}Y}{\mathrm{d}a} = 0, \qquad \frac{\mathrm{d}Y}{\mathrm{d}b} = 0, \qquad \frac{\mathrm{d}Y}{\mathrm{d}c} = -\sqrt{\frac{g}{a}}\sqrt{\frac{1-e^2}{1-e}}$$

$$\frac{\mathrm{d}X'}{\mathrm{d}a} = 0, \qquad \frac{\mathrm{d}X'}{\mathrm{d}b} = 0, \qquad \frac{\mathrm{d}X'}{\mathrm{d}c} = \frac{g}{a^2}\frac{1}{(1-e)^2}$$

$$\frac{\mathrm{d}Y'}{\mathrm{d}a} = \sqrt{\frac{g}{a}}\frac{\sqrt{1-e^2}}{2ae}, \qquad \frac{\mathrm{d}Y'}{\mathrm{d}b} = -\sqrt{\frac{g}{a}}\frac{1}{2ae(1-e)\sqrt{1-e^2}}, \qquad \frac{\mathrm{d}Y'}{\mathrm{d}c} = 0$$

After substitution of these expressions in the equations of $[a, b]$, $[a, c]$, $[b, c]$ of Article 67, there results

$$[a, b] = 0, \qquad [a, c] = -\frac{g}{2a^2e} + \frac{g(1-e^2)}{2a^2e(1-e)} = \frac{g}{2a^2}$$

$$[b, c] = \frac{g}{2a^2e}\left[\frac{1}{(1-e)^2} - \frac{\sqrt{1-e^2}}{(1-e)^2\sqrt{1-e^2}}\right] = 0$$

The same results are obtained by replacing b with e, if the eccentricity instead of the parameter is to be retained.

70. Let us consider the formula

$$\frac{\mathrm{d}x}{\mathrm{d}a}\frac{\mathrm{d}x'}{\mathrm{d}h}+\frac{\mathrm{d}y}{\mathrm{d}a}\frac{\mathrm{d}y'}{\mathrm{d}h}+\frac{\mathrm{d}z}{\mathrm{d}a}\frac{\mathrm{d}z'}{\mathrm{d}h}-\frac{\mathrm{d}x'}{\mathrm{d}a}\frac{\mathrm{d}x}{\mathrm{d}h}-\frac{\mathrm{d}y'}{\mathrm{d}a}\frac{\mathrm{d}y}{\mathrm{d}h}-\frac{\mathrm{d}z'}{\mathrm{d}a}\frac{\mathrm{d}z}{\mathrm{d}h}$$

Since the quantity h is only found in the coefficients α, β, α_1, etc., which do not contain a, one will have

$$\frac{\mathrm{d}x}{\mathrm{d}h}=\frac{\mathrm{d}\alpha}{\mathrm{d}h}X+\frac{\mathrm{d}\beta}{\mathrm{d}h}Y, \qquad \frac{\mathrm{d}x'}{\mathrm{d}h}=\frac{\mathrm{d}\alpha}{\mathrm{d}h}X'+\frac{\mathrm{d}\beta}{\mathrm{d}h}Y'$$

and after replacing α, β with α_1, β_1 and with α_2, β_2, the expressions for $\mathrm{d}y/\mathrm{d}h$, $\mathrm{d}y'/\mathrm{d}h$, $\mathrm{d}z/\mathrm{d}h$, $\mathrm{d}z'/\mathrm{d}h$ will be obtained. With respect to the partial differences relative to a, they will be the same as in the preceding article.

By making these substitutions, it is useful to note that since identical equations of condition are differentiated, one will have

$$\alpha\,\mathrm{d}\alpha+\alpha_1\,\mathrm{d}\alpha_1+\alpha_2\,\mathrm{d}\alpha_2=0, \qquad \beta\,\mathrm{d}\beta+\beta_1\,\mathrm{d}\beta_1+\beta_2\,\mathrm{d}\beta_2=0$$
$$\alpha\,\mathrm{d}\beta+\alpha_1\,\mathrm{d}\beta_1+\alpha_2\,\mathrm{d}\beta_2=-\beta\,\mathrm{d}\alpha-\beta_1\,\mathrm{d}\alpha_1-\beta_2\,\mathrm{d}\alpha_2$$

so that in order to shorten the expressions, the following definition is made

$$\beta\,\mathrm{d}\alpha+\beta_1\,\mathrm{d}\alpha_1+\beta_2\,\mathrm{d}\alpha_2=\mathrm{d}\chi$$

The differential expression[21] $\mathrm{d}\chi$ is used although the expression for $\mathrm{d}\chi$ is not a complete differential. The formula

$$\frac{\mathrm{d}x}{\mathrm{d}a}\frac{\mathrm{d}x'}{\mathrm{d}h}+\frac{\mathrm{d}y}{\mathrm{d}a}\frac{\mathrm{d}y'}{\mathrm{d}h}+\frac{\mathrm{d}z}{\mathrm{d}a}\frac{\mathrm{d}z'}{\mathrm{d}h}$$

will be reduced to the form

$$\left(X'\frac{\mathrm{d}Y}{\mathrm{d}a}-Y'\frac{\mathrm{d}X}{\mathrm{d}a}\right)\frac{\mathrm{d}\chi}{\mathrm{d}h}$$

and the formula

$$\frac{\mathrm{d}x'}{\mathrm{d}a}\frac{\mathrm{d}x}{\mathrm{d}h}+\frac{\mathrm{d}y'}{\mathrm{d}a}\frac{\mathrm{d}y}{\mathrm{d}h}+\frac{\mathrm{d}z'}{\mathrm{d}a}\frac{\mathrm{d}z}{\mathrm{d}h}$$

to this similar form

$$\left(X\frac{\mathrm{d}Y'}{\mathrm{d}a}-Y\frac{\mathrm{d}X'}{\mathrm{d}a}\right)\frac{\mathrm{d}\chi}{\mathrm{d}h}$$

Therefore, by subtracting the second of these quantities from the first and observing that

$$X'\,\mathrm{d}Y + Y\,\mathrm{d}X' = \mathrm{d}(X'Y), \qquad Y'\,\mathrm{d}X + X\,\mathrm{d}Y' = \mathrm{d}(XY')$$

one will have for the transformation of the formula in question which contains partial differences relative to a and h

$$\frac{\mathrm{d}(YX' - XY')}{\mathrm{d}a}\frac{\mathrm{d}\chi}{\mathrm{d}h}$$

and similar equations will be obtained by replacing a with b and c, and then h with i and k.

71. It remains for us to consider the formulas where there are only the partial differences relative to h, i and k. Since these quantities are only contained in the coefficients α, β, α_1, etc. these coefficients will become variable.

The differentials of these coefficients reduce to a very simple form, if the analogous coefficients γ, γ', γ'', etc. are used. Furthermore, the equations of condition between the different coefficients must be considered (Article 14).

Indeed, if it is assumed that

$$\gamma\,\mathrm{d}\alpha + \gamma_1\,\mathrm{d}\alpha_1 + \gamma_2\,\mathrm{d}\alpha_2 = \mathrm{d}\pi$$
$$\gamma\,\mathrm{d}\beta + \gamma_1\,\mathrm{d}\beta_1 + \gamma_2\,\mathrm{d}\beta_2 = \mathrm{d}\sigma$$

and the three equations

$$\alpha\,\mathrm{d}\alpha + \alpha_1\,\mathrm{d}\alpha_1 + \alpha_2\,\mathrm{d}\alpha_2 = 0$$
$$\beta\,\mathrm{d}\alpha + \beta_1\,\mathrm{d}\alpha_1 + \beta_2\,\mathrm{d}\alpha_2 = \mathrm{d}\chi$$
$$\gamma\,\mathrm{d}\alpha + \gamma_1\,\mathrm{d}\alpha_1 + \gamma_2\,\mathrm{d}\alpha_2 = \mathrm{d}\pi$$

are added together, after having multiplied the first by α, the second by β and the third by γ, one will have

$$\mathrm{d}\alpha = \beta\,\mathrm{d}\chi + \gamma\,\mathrm{d}\pi$$

If they are multiplied by $\alpha_1, \beta_1, \gamma_1$ and by $\alpha_2, \beta_2, \gamma_2$ and then added together one has in a similar fashion

$$\mathrm{d}\alpha_1 = \beta_1\,\mathrm{d}\chi + \gamma_1\,\mathrm{d}\pi$$
$$\mathrm{d}\alpha_2 = \beta_2\,\mathrm{d}\chi + \gamma_2\,\mathrm{d}\pi$$

In a similar fashion, the three equations

$$\alpha\,\mathrm{d}\beta + \alpha_1\,\mathrm{d}\beta_1 + \alpha_2\,\mathrm{d}\beta_2 = -\mathrm{d}\chi$$
$$\beta\,\mathrm{d}\beta + \beta_1\,\mathrm{d}\beta_1 + \beta_2\,\mathrm{d}\beta_2 = 0$$
$$\gamma\,\mathrm{d}\beta + \gamma_1\,\mathrm{d}\beta_1 + \gamma_2\,\mathrm{d}\beta_2 = \mathrm{d}\sigma$$

will give

$$\mathrm{d}\beta = -\alpha\,\mathrm{d}\chi + \gamma\,\mathrm{d}\sigma$$
$$\mathrm{d}\beta_1 = -\alpha_1\,\mathrm{d}\chi + \gamma_1\,\mathrm{d}\sigma$$
$$\mathrm{d}\beta_2 = -\alpha_2\,\mathrm{d}\chi + \gamma_2\,\mathrm{d}\sigma$$

Finally, the three equations

$$\alpha\,\mathrm{d}\gamma + \alpha_1\,\mathrm{d}\gamma_1 + \alpha_2\,\mathrm{d}\gamma_2 = -\,\mathrm{d}\pi$$
$$\beta\,\mathrm{d}\gamma + \beta_1\,\mathrm{d}\gamma_1 + \beta_2\,\mathrm{d}\gamma_2 = -\,\mathrm{d}\sigma$$
$$\gamma\,\mathrm{d}\gamma + \gamma_1\,\mathrm{d}\gamma_1 + \gamma_2\,\mathrm{d}\gamma_2 = 0$$

will similarly give

$$\mathrm{d}\gamma = -\alpha\,\mathrm{d}\pi - \beta\,\mathrm{d}\sigma$$
$$\mathrm{d}\gamma_1 = -\alpha_1\,\mathrm{d}\pi - \beta_1\,\mathrm{d}\sigma$$
$$\mathrm{d}\gamma_2 = -\alpha_2\,\mathrm{d}\pi - \beta_2\,\mathrm{d}\sigma$$

72. By means of these formulas, there will result

$$\frac{\mathrm{d}x}{\mathrm{d}h} = X\frac{\mathrm{d}\alpha}{\mathrm{d}h} + Y\frac{\mathrm{d}\beta}{\mathrm{d}h} = (\beta X - \alpha Y)\frac{\mathrm{d}\chi}{\mathrm{d}b} + \gamma\left(X\frac{\mathrm{d}\pi}{\mathrm{d}h} + Y\frac{\mathrm{d}\sigma}{\mathrm{d}h}\right)$$

and marking the quantities α, β, γ by one or two primes, the expressions for $\mathrm{d}y/\mathrm{d}h$ and $\mathrm{d}z/\mathrm{d}h$ will be obtained. In order to obtain those of $\mathrm{d}x'/\mathrm{d}h$, $\mathrm{d}y'/\mathrm{d}h$, $\mathrm{d}z'/\mathrm{d}h$, it is only necessary to mark with one prime the quantities **X** and **Y**. Similarly, for the partial differences relative to i and k where h will be replaced with i and k.

After making these substitutions and with due consideration of the equations of condition, the formula

$$\frac{\mathrm{d}x}{\mathrm{d}h}\frac{\mathrm{d}x'}{\mathrm{d}i} + \frac{\mathrm{d}y}{\mathrm{d}h}\frac{\mathrm{d}y'}{\mathrm{d}i} + \frac{\mathrm{d}z}{\mathrm{d}h}\frac{\mathrm{d}z'}{\mathrm{d}i} - \frac{\mathrm{d}x'}{\mathrm{d}h}\frac{\mathrm{d}x}{\mathrm{d}i} - \frac{\mathrm{d}y'}{\mathrm{d}h}\frac{\mathrm{d}y}{\mathrm{d}i} - \frac{\mathrm{d}z'}{\mathrm{d}h}\frac{\mathrm{d}z}{\mathrm{d}i}$$

will be reduced to

$$\left(X\frac{\mathrm{d}\pi}{\mathrm{d}h} + Y\frac{\mathrm{d}\sigma}{\mathrm{d}h}\right)\left(X'\frac{\mathrm{d}\pi}{\mathrm{d}i} + Y'\frac{\mathrm{d}\sigma}{\mathrm{d}i}\right)$$
$$-\left(X'\frac{\mathrm{d}\pi}{\mathrm{d}h} + Y'\frac{\mathrm{d}\sigma}{\mathrm{d}h}\right)\left(X\frac{\mathrm{d}\pi}{\mathrm{d}i} + Y\frac{\mathrm{d}\sigma}{\mathrm{d}i}\right)$$
$$= (XY' - YX')\left(\frac{\mathrm{d}\pi}{\mathrm{d}h}\frac{\mathrm{d}\sigma}{\mathrm{d}i} - \frac{\mathrm{d}\pi}{\mathrm{d}i}\frac{\mathrm{d}\sigma}{\mathrm{d}h}\right)$$

Similar formulas will be obtained by replacing h and i with k.

Since the coefficients $\alpha, \beta, \gamma, \alpha_1$, etc., are functions of three elements h, i, k (Articles 13 and 14), the three quantities $\mathrm{d}\chi$, $\mathrm{d}\pi$, $\mathrm{d}\sigma$, which are introduced in the preceding formulas must also be functions of the same elements. If in the expressions for these three quantities, the expressions of α, β, etc. given in the cited articles were substituted, it will be found after some very easy manipulations

$$\mathrm{d}\chi = \mathrm{d}k + \cos i\,\mathrm{d}h$$
$$\mathrm{d}\pi = -\cos k \sin i\,\mathrm{d}h + \sin k\,\mathrm{d}i$$
$$\mathrm{d}\sigma = \sin k \sin i\,\mathrm{d}h + \cos k\,\mathrm{d}i$$

But these quantities are not only used to simplify the computation, they also represent in a very simple way the instantaneous variations of the orientation of the orbit. Indeed, since the xy-plane to which we have referred the inclination i and the longitude h of the node is arbitrary, it can coincide at an instant with the plane of the orbit by assuming $i = 0$. Then, there results

$$\mathrm{d}\chi = \mathrm{d}k + \mathrm{d}h, \qquad \mathrm{d}\pi = \sin k\,\mathrm{d}i, \qquad \mathrm{d}\sigma = \cos k\,\mathrm{d}i$$

In this case, $(h + k)$ will be the angle that the major axis of the ellipse makes with a fixed line. Consequently, $(\mathrm{d}h + \mathrm{d}k)$ or $\mathrm{d}\chi$ will be the elementary rotation of the major axis of the orbit in its plane.

The elementary angle $\mathrm{d}i$ will be the inclination between two successive orientations of the plane of the orbit which became mobile and the angle h will be the longitude of the nodes described by these two orientations, measured on the same plane so that if $\mathrm{d}i'$ and h' designate these two elements, one will have

$$\mathrm{d}i' = \sqrt{\mathrm{d}\pi^2 + \mathrm{d}\sigma^2}$$

and

$$\tan h' = \frac{\mathrm{d}\pi}{\mathrm{d}\sigma}$$

and thus the instantaneous variations of the orientation of the orbit is determined by the three elements $\mathrm{d}\chi$, $\mathrm{d}\pi$, $\mathrm{d}\varphi$ independent of any plane of projection.

73. It is now very easy to find the values of the other coefficients represented by the symbols $[a, h]$, $[b, h]$, etc. It will only be necessary to substitute for $XY' - YX'$, that is for $(X\,\mathrm{d}Y - Y\,\mathrm{d}X)/\mathrm{d}t$, its expression which is equal to D (Article 11) and equal to $\sqrt{(gb)}$ (Article 15) and $\mathrm{d}\chi$, $\mathrm{d}\pi$, $\mathrm{d}\sigma$ with their expressions in h, i, k of the preceding article. But instead of the element k, we will retain the element χ,[22] which denotes the angle that the major axis of the orbit traverses by rotating about its mobile plane. This is properly the motion of aphelion or of perihelion in the same plane. Thus there results

$$\chi = k + \int \cos i\,\mathrm{d}h$$

therefore

$$k = \chi - \int \cos i \, dh$$

and it follows that

$$\frac{d\chi}{d\chi} = 1, \qquad \frac{d\pi}{dh} = -\cos k \sin i, \qquad \frac{d\pi}{di} = \sin k,$$
$$\frac{d\sigma}{dh} = \sin k \sin i, \qquad \frac{d\sigma}{di} = \cos k$$

and all the other partial differences will be zero, which gives

$$\frac{d\pi}{dh}\frac{d\sigma}{di} - \frac{d\pi}{di}\frac{d\sigma}{dh} = -\sin i$$

From this result, one will have

$$[a, h] = 0, \qquad [a, i] = 0, \qquad [a, \chi] = 0,$$

$$[b, h] = 0, \qquad [b, i] = 0, \qquad [b, \chi] = -\tfrac{1}{2}\sqrt{\frac{g}{b}},$$

$$[c, h] = 0, \qquad [c, i] = 0, \qquad [c, \chi] = 0,$$

$$[h, i] = \sqrt{gb} - \sin i, \qquad [h, \chi] = 0, \qquad [i, \chi] = 0$$

74. These values, added to those that we have already found (Article 69), will finally give

$$\frac{d\Omega}{da}dt = \frac{g}{2a^2}dc, \qquad \frac{d\Omega}{db}dt = -\tfrac{1}{2}\sqrt{\frac{g}{b}}d\chi,$$
$$\frac{d\Omega}{dc}dt = -\frac{g}{2a^2}da, \qquad \frac{d\Omega}{dh}dt = -\sqrt{gb}\sin i \, di,$$
$$\frac{d\Omega}{di}dt = \sqrt{gb}\sin i \, dh, \qquad \frac{d\Omega}{d\chi}dt = \tfrac{1}{2}\sqrt{\frac{g}{b}}db$$

from which result these very simple expressions of the variations of the elliptical elements

$$da = -\frac{2a^2}{g}\frac{d\Omega}{dc}dt, \qquad dc = \frac{2a^2}{g}\frac{d\Omega}{da}dt,$$
$$db = \frac{2\sqrt{b}}{\sqrt{g}}\frac{d\Omega}{d\chi}dt, \qquad d\chi = -\frac{2\sqrt{b}}{\sqrt{g}}\frac{d\Omega}{db}dt,$$
$$dh = \frac{1}{\sqrt{gb}\sin i}\frac{d\Omega}{di}dt, \qquad di = -\frac{1}{\sqrt{gb}\sin i}\frac{d\Omega}{dh}$$

75. The formulas would be more complex if instead of the semi-parameter b, the semi-eccentricity e is used. Then since $b = a(1 - e^2)$ one would have

$$XY' - YX' = \sqrt{ga(1 - e^2)}$$

which would give

$$[a, \chi] = -\frac{\sqrt{g(1 - e^2)}}{2\sqrt{a}}, \qquad [e, \chi] = \frac{e\sqrt{ga}}{\sqrt{1 - e^2}}$$

and the values of $(\mathrm{d}\Omega/\mathrm{d}a)\mathrm{d}t$, $(\mathrm{d}\Omega/\mathrm{d}e)\mathrm{d}t$, $(\mathrm{d}\Omega/\mathrm{d}\chi)\mathrm{d}t$ would become

$$\frac{\mathrm{d}\Omega}{\mathrm{d}a}\mathrm{d}t = \frac{g}{2a^2}\mathrm{d}c - \frac{\sqrt{g(1 - e^2)}}{2\sqrt{a}}\mathrm{d}\chi$$

$$\frac{\mathrm{d}\Omega}{\mathrm{d}e}\mathrm{d}t = \frac{e\sqrt{ga}}{\sqrt{1 - e^2}}\mathrm{d}\chi$$

$$\frac{\mathrm{d}\Omega}{\mathrm{d}\chi}\mathrm{d}t = \frac{\sqrt{g(1 - e^2)}}{2\sqrt{a}}\mathrm{d}a - \frac{e\sqrt{ga}}{\sqrt{1 - e^2}}\mathrm{d}e$$

from which one obtains after substituting for $\mathrm{d}a$ its value given above

$$\mathrm{d}c = \frac{2a^2}{g}\frac{\mathrm{d}\Omega}{\mathrm{d}a}\mathrm{d}t + \frac{a(1 - e^2)}{g}\frac{\mathrm{d}\Omega}{e\,\mathrm{d}e}\mathrm{d}t$$

$$\mathrm{d}e = -\frac{\sqrt{1 - e^2}}{\sqrt{ga}}\frac{\mathrm{d}\Omega}{e\,\mathrm{d}\chi}\mathrm{d}t - \frac{a(1 - e^2)}{ge}\frac{\mathrm{d}\Omega}{\mathrm{d}c}\mathrm{d}t$$

$$\mathrm{d}\chi = \frac{\sqrt{1 - e^2}}{\sqrt{ga}}\frac{\mathrm{d}\Omega}{e\,\mathrm{d}e}\mathrm{d}t$$

which are substituted in place of the expressions for $\mathrm{d}c$, $\mathrm{d}b$, $\mathrm{d}\chi$ of the preceding article while the other expressions remain the same.

From these formulas, the forces perturbing the motion of a planet can be obtained by making variable the quantities which, without these forces, would be constants. Although one may in this manner determine all the inequalities resulting from these perturbations, it is mainly for the inequalities which are called **secular**[23] that the formulas which we gave are useful. Since these inequalities are independent of the periods relative to the motion of the planets, they essentially affect their elements and produce variations which are either increasing with the time or periodic, but with their own periods and of long duration.

76. In order to determine the **secular** variations, it is only necessary to substitute for Ω the non-periodic part of this function, that is, the first term of the expansion of Ω in a series of sines and cosines of angles dependent on the mean motion of the perturbed planet and the perturbing planets. Since Ω is only a function of the elliptical coordinates of these planets, which can always be reduced to a series of sines and cosines of angles proportional to the anomalies and to the mean longitudes as long as the eccentricities and inclinations are not

large, the function Ω can always be developed in a series of the same type and the first term without a sine or cosine will be the only one which can give **secular** equations.

Let us designate by (Ω) the first term of Ω, which will be a simple function of the elements a, b, c, e, h, i of the perturbed planet and of similar elements for the perturbing planets. It is clear that the element c which is dependent on the time t will not be in it. Thus, by substituting h for Ω, one will have for the **secular** variations the formulas

$$\mathrm{d}a = 0, \qquad \mathrm{d}c = \frac{2a^2}{g}\frac{(\mathrm{d}\Omega)}{\mathrm{d}a}\mathrm{d}t + \frac{a(1-e^2)}{ge}\frac{(\mathrm{d}\Omega)}{\mathrm{d}e},$$

$$\mathrm{d}e = -\frac{\sqrt{1-e^2}}{\sqrt{ga}}\frac{(\mathrm{d}\Omega)}{e\,\mathrm{d}\chi}\mathrm{d}t, \qquad \mathrm{d}\chi = \frac{\sqrt{1-e^2}}{\sqrt{ga}}\frac{(\mathrm{d}\Omega)}{e\,\mathrm{d}e}\mathrm{d}t,$$

$$\mathrm{d}h = \frac{1}{\sqrt{gb}}\frac{(\mathrm{d}\Omega)}{\sin i\,\mathrm{d}i}, \qquad \mathrm{d}i = -\frac{1}{\sqrt{gb}}\frac{(\mathrm{d}\Omega)}{\sin i\,\mathrm{d}h}$$

where $b = a(1-e^2)$.

77. The equation $\mathrm{d}a = 0$ shows that the semi-major axis, where the average distance a is not subject to any secular variation, is only a particular case of the general theorem that we have demonstrated in Article 23 of SECTION V. Because the quantity **H** of this article is the same as the quantity **H** of Article 3 and of the preceding section, one sees with Article 15 that **H**$= -ga/2$. Thus one must apply to the average distance of planets the results that we have found on the value of the active force of an arbitrary system (SECTION V, Subsection III).

The variation de produces an alteration of the average motion since $u(t-c)\sqrt{g/a^3}$ is the average anomaly, that is, the angle of the average motion starting from perihelion (Article 19), this anomaly will be subject to a variation expressed by $-\sqrt{g/a^3}$ since $\mathrm{d}a = 0$. If the variation $\mathrm{d}\chi$ is added instead to the perihelion of the orbit, one will have $\mathrm{d}\chi - \sqrt{g/a^3}\,\mathrm{d}e$ for the **secular** variation of the average longitude which we will designate by $\mathrm{d}\lambda$. Thus there will result

$$\mathrm{d}\lambda = \mathrm{d}\chi - \sqrt{\frac{g}{a^3}}\mathrm{d}e = -2\sqrt{\frac{a}{g}}\frac{\mathrm{d}(\Omega)}{\mathrm{d}a}\mathrm{d}t + \frac{\sqrt{1-e^2}}{\sqrt{ga}}\frac{e}{1+\sqrt{1-e^2}}\frac{\mathrm{d}(\Omega)}{\mathrm{d}e}$$

because

$$1-\sqrt{1-e^2} = \frac{e^2}{1+\sqrt{1-e^2}}$$

78. When the eccentricity e is very small, the expressions for de and $\mathrm{d}\chi$ have a shortcoming: to wit, their denominators contain the very small quantity e. But it is easy to circumvent this shortcoming by substituting in place of e and χ the transformed quantities $e\sin\chi$ and $e\cos\chi$.

Indeed, if one defines

$$m = e\sin\chi, \qquad n = e\cos\chi$$

there results

$$dm = \sin\chi\,\mathrm{d}e + e\cos\chi\,\mathrm{d}\chi, \qquad dn = \cos\chi\,\mathrm{d}e - e\sin\chi\,\mathrm{d}\chi$$

Therefore, after substituting the values of de and dχ

$$dm = \frac{\sqrt{1-e^2}}{\sqrt{ga}}\left[\cos\chi\frac{\mathrm{d}(\Omega)}{\mathrm{d}e} - \sin\chi\frac{\mathrm{d}(\Omega)}{e\,\mathrm{d}\chi}\right]\mathrm{d}t$$

$$dn = -\frac{\sqrt{1-e^2}}{\sqrt{ga}}\left[\sin\chi\frac{\mathrm{d}(\Omega)}{\mathrm{d}e} + \cos\chi\frac{\mathrm{d}(\Omega)}{e\,\mathrm{d}\chi}\right]\mathrm{d}t$$

But by viewing (Ω) as a function of e and χ and also as a function of m and n, one has

$$\frac{\mathrm{d}(\Omega)}{\mathrm{d}\chi}\mathrm{d}\chi + \frac{\mathrm{d}(\Omega)}{\mathrm{d}e}\mathrm{d}e = \frac{\mathrm{d}(\Omega)}{dm}dm + \frac{\mathrm{d}(\Omega)}{dn}dn$$

an identity, which, after substitution of the values of dm and dn, produces these two equations

$$\frac{\mathrm{d}(\Omega)}{\mathrm{d}\chi} = \frac{\mathrm{d}(\Omega)}{dm}e\cos\chi - \frac{\mathrm{d}(\Omega)}{dn}e\sin\chi$$

$$\frac{\mathrm{d}(\Omega)}{\mathrm{d}e} = \frac{\mathrm{d}(\Omega)}{dm}\sin\chi + \frac{\mathrm{d}(\Omega)}{dn}\cos\chi$$

Thus after making these substitutions, the following equations result

$$dm = \frac{\sqrt{1-e^2}}{\sqrt{ga}}\frac{\mathrm{d}(\Omega)}{dn}\mathrm{d}t, \qquad dn = -\frac{\sqrt{1-e^2}}{\sqrt{ga}}\frac{\mathrm{d}(\Omega)}{dm}\mathrm{d}t$$

which can be used in place of those which give the expressions for de and dχ (Article 75).

Similar transformations can be applied to the last equations which give the expressions for dh and di.

To this effect, let

$$p = \sin i\sin h, \qquad q = \sin i\cos h$$

One will find by an analogous procedure that

$$\mathrm{d}p = \frac{\cos i}{\sqrt{gb}}\frac{\mathrm{d}(\Omega)}{\mathrm{d}q}\mathrm{d}t, \qquad \mathrm{d}q = -\frac{\cos i}{\sqrt{gb}}\frac{\mathrm{d}(\Omega)}{\mathrm{d}p}\mathrm{d}t$$

79. The perturbing forces which are considered in the theory of planets are due to the attraction of other planets and we will later give the expression for Ω which results from

this attraction. But the resistance they would be subject to from a nearly frictionless fluid in which they would be assumed to be immersed could also be viewed as a perturbing force.[24] In this case, by taking R for the resistance and following the procedures of Article 8 of SECTION II we have that

$$\delta r = \frac{\mathrm{d}x}{\mathrm{d}s}\delta x + \frac{\mathrm{d}y}{\mathrm{d}s}\delta y + \frac{\mathrm{d}z}{\mathrm{d}s}\delta z$$

assuming the resisting fluid at rest.

From this equation, the following terms (Article 62) result in the expression for $\delta\Omega$

$$-R\,\delta r = -R\left(\frac{\mathrm{d}x\delta x + \mathrm{d}y\,\delta y + \mathrm{d}z\,\delta z}{\mathrm{d}s}\right)$$

The resistance is ordinarily assumed to be proportional to the square of the velocity, which is represented by $\mathrm{d}s/\mathrm{d}t$, and to the density of the medium which is denoted by Γ. Thus the terms for the resistance in the expression of $\delta\Omega$ will be

$$-\frac{\Gamma\,\mathrm{d}s(\mathrm{d}x\,\delta x + \mathrm{d}y\,\delta y + \mathrm{d}z\,\delta z)}{\mathrm{d}t^2}$$

In order to evaluate the quantity $\mathrm{d}x\,\delta x + \mathrm{d}y\,\delta y + \mathrm{d}z\,\delta z$, it will only be necessary to use the formulas of Article 67 and 70 and to note that the operator d refers to the time t which only appears in the expressions for **X** and **Y**. Furthermore, the operator δ must refer to the arbitrary constants a, b, etc., which are contained in the expressions for **X** and **Y** and in the coefficients α, β, α_1, etc.

Thus after replacing d by δ in the expressions for $\mathrm{d}\alpha$, $\mathrm{d}\beta$, etc. there results

$$\mathrm{d}x = \alpha\,\mathrm{d}X + \beta\,\mathrm{d}Y, \qquad \mathrm{d}y = \alpha_1\,\mathrm{d}X + \beta_1\,\mathrm{d}Y, \qquad \mathrm{d}z = \alpha_2\,\mathrm{d}X + \beta_2\,\mathrm{d}Y$$
$$\begin{aligned}\delta x &= \alpha\,\delta X + \beta\,\delta Y + X(\beta\,\delta X + \gamma\,\delta\pi) + Y(-\alpha\,\delta X + \gamma\,\delta\sigma)\\ &= \alpha(\delta X - Y\,\delta\chi) + \beta(\delta Y + X\,\delta\chi) + \gamma(X\,\delta\pi + Y\,\delta\sigma)\\ \delta y &= \alpha_1(\delta X - Y\,\delta\chi) + \beta_1(\delta Y + X\,\delta\chi) + \gamma_1(X\,\delta\pi + Y\,\delta\sigma)\\ \delta z &= \alpha_2(\delta X - Y\,\delta\chi) + \beta_2(\delta Y + X\,\delta\chi) + \gamma_2(X\,\delta\pi + Y\,\delta\sigma)\end{aligned}$$

From this result, and with due consideration of the equations of condition between the coefficients α, β, γ, α_1, etc. (Article 14), the following equation results

$$\mathrm{d}x\,\delta x + \mathrm{d}y\,\delta y + \mathrm{d}z\,\delta z = \mathrm{d}X\,\delta X + \mathrm{d}Y\,\delta Y + (X\,\delta Y - Y\,\delta X)\delta\chi$$

and if we substitute for X and Y, the expressions $r\cos\Phi$, $r\sin\Phi$ (Article 13), one will have

$$\begin{aligned}&\mathrm{d}X\,\delta X + \mathrm{d}Y\,\delta Y = \mathrm{d}r\,\delta r + r^2\,\mathrm{d}\Phi\,\delta\Phi\\ &X\,\mathrm{d}Y - Y\,\mathrm{d}X = r^2\,\mathrm{d}\Phi\\ &\mathrm{d}s = \sqrt{\mathrm{d}X^2 + \mathrm{d}Y^2} = \sqrt{\mathrm{d}r^2 + r^2\,\mathrm{d}\Phi^2}\end{aligned}$$

Therefore, the terms to add to $\delta\Omega$, because of the resistance of the medium, will be given by

$$-\frac{\Gamma\sqrt{\mathrm{d}r^2 + r^2\,\mathrm{d}\Phi^2}(\mathrm{d}r\,\delta r + r^2\,\mathrm{d}\Phi\,\delta\Phi + r^2\,\mathrm{d}\Phi\,\delta\chi)}{\mathrm{d}t^2}$$

where one will only have to substitute for r and Φ their expressions as a function of t, given by the formulas of Articles 21 and 22, being careful to note that the operator d refers to the variable t and that the operator δ refers to the arbitrary constants.

Chapter III
THE MOTION OF A BODY ATTRACTED TO TWO FIXED CENTERS BY FORCES INVERSELY PROPORTIONAL TO THE SQUARES OF THE DISTANCE

80. Although this problem does not have any application to the system of the world where all the centers of attraction are in motion, it is, nevertheless, rather interesting from an analytical point of view and consequently, merits being treated in particular with some detail.

Let us assume that an isolated body is attracted simultaneously toward two fixed centers by forces proportional to arbitrary functions of distance.

As in Article 4, let one of the centers be at the origin of the coordinates and let R be the attracting force. For the other center, let us assume that its position is determined by the coordinates a, b, c, parallel to x, y, z. Also, let Q be its attracting force and q the distance from the body to this center. It is clear that one will have

$$q = \sqrt{(x-a)^2 + (y-b)^2 + (z-c)^2}$$

and after substituting for x, y, z the quantities r, ψ, φ (Article 4)

$$q = \sqrt{(r^2 - 2r[(a\cos\varphi + b\sin\varphi)\cos\psi + c\sin\psi] + h^2)}$$

where $h = \sqrt{a^2 + b^2 + c^2}$ is the distance between the two centers.

It is clear that the expression for T will be the same as in the problem of Chapter I, but the value of V will be augmented by the term $\int Q\,\mathrm{d}q$. Because Q is a function of q, and q is a function of r, φ, ψ, this term will give in the derivatives $\delta V/\delta\psi$, $\delta V/\delta\varphi$, $\delta V/\delta r$ the following terms, namely, $Q(\mathrm{d}q/\mathrm{d}\psi)$, $Q(\mathrm{d}q/\mathrm{d}\varphi)$, $Q(\mathrm{d}q/\mathrm{d}r)$ which consequently should be added to the first members of the differential equations of the cited article.

One will thus have for the motion of a body attracted toward two centers by the forces R and Q, the following three equations

$$(1)\qquad \frac{\mathrm{d}^2 r}{\mathrm{d}t^2} - \frac{r(\cos^2\psi\,\mathrm{d}\varphi^2 + \mathrm{d}\psi^2)}{\mathrm{d}t^2} + R + Q\frac{\mathrm{d}q}{\mathrm{d}r} = 0$$

$$(2) \qquad \frac{d(r^2\,\mathrm{d}\psi)}{\mathrm{d}t^2} + \frac{r^2 \sin\psi \cos\psi\,\mathrm{d}\varphi^2}{\mathrm{d}t^2} + Q\frac{\mathrm{d}q}{\mathrm{d}\psi} = 0$$

$$(3) \qquad \frac{d(r^2 \cos^2\psi\,\mathrm{d}\varphi)}{\mathrm{d}t^2} + Q\frac{\mathrm{d}q}{\mathrm{d}\varphi} = 0$$

If the body were attracted simultaneously toward other centers, one would only have to add to these equations similar terms for each of the centers.

The equation $T + V = H$ will give a fourth equation, which is an integral of the preceding equations

$$\frac{r^2(\cos^2\psi\,\mathrm{d}\varphi^2 + \mathrm{d}\psi^2) + \mathrm{d}r^2}{2\,\mathrm{d}t^2} + \int R\,\mathrm{d}r + \int Q\,\mathrm{d}q = 2H$$

and it is clear that after multiplying the three preceding equations by $\mathrm{d}\psi, \mathrm{d}\varphi, \mathrm{d}r$, respectively and adding them together, an integrable equation will be obtained for which the integral is the one we just presented.

One derives from this equation the following result

$$\frac{r^2(\cos^2\psi\,\mathrm{d}\varphi^2 + \mathrm{d}\psi^2)}{\mathrm{d}t^2} = 4H - 2\int R\,\mathrm{d}r - 2\int Q\,\mathrm{d}q - \frac{\mathrm{d}r^2}{\mathrm{d}t^2}$$

which after substituting this equation in the first equation multiplied by r, will reduce to

$$\frac{\mathrm{d}^2 r^2}{2\,\mathrm{d}t^2} + Rr + 2\int R\,\mathrm{d}r + Qr\frac{\mathrm{d}q}{\mathrm{d}r} + 2\int Q\,\mathrm{d}q = 4H$$

Now since

$$q^2 = r^2 + h^2 - 2r[(a\cos\varphi + b\sin\varphi)\cos\psi + c\sin\psi]$$

one will have by varying r

$$q\frac{\mathrm{d}q}{\mathrm{d}r} = r - (a\cos\varphi + b\sin\varphi)\cos\psi - c\sin\psi = r - \frac{r^2 + h^2 - q^2}{2r} = \frac{r^2 + q^2 - h^2}{2r}$$

Thus after substituting the expression for $\mathrm{d}q/\mathrm{d}r$, one will finally have

$$\frac{\mathrm{d}^2 r^2}{2\,\mathrm{d}t^2} + Rr + 2\int R\,\mathrm{d}r + Q\frac{r^2 + q^2 - h^2}{2q} + 2\int Q\,\mathrm{d}q = 4H$$

This equation has the advantage that it only contains the two variables r and q, and it indicates at the same time that there must be a similar equation between q and r which is obtained by simply changing r to q as well as R to Q. Since it makes no difference whether the motion of the body is referred to either one or the other of the two fixed centers, it is clear that by referring it to the center of the force Q, it will be found that with an analysis similar to the preceding one

$$\frac{\mathrm{d}^2q^2}{2\,\mathrm{d}t^2} + Qq + 2\int Q\,\mathrm{d}q + R\frac{r^2+q^2-h^2}{2r} + 2\int R\,\mathrm{d}r = 4H$$

Consequently, with these two equations the two radii r and q can be determined directly.

I now note that it can, without reducing the generality, be assumed that the two coordinates a and b of the center of forces Q are zeroes, which is identical to placing the axis of the z-coordinate on the line which joins the two centers. With this assumption, one will have $c = h$, and the quantity q will become

$$\sqrt{r^2 - 2hr\sin\psi + h^2}$$

which, since it no longer contains φ, one will have

$$\frac{\mathrm{d}q}{\mathrm{d}\varphi} = 0$$

Consequently, the third differential equation will be reduced to

$$\frac{\mathrm{d}(r^2\cos^2\psi\,\mathrm{d}\varphi)}{\mathrm{d}t^2} = 0$$

for which the integral is

$$\frac{r^2\cos^2\psi\,\mathrm{d}\varphi}{\mathrm{d}t} = B$$

where B is an arbitrary constant. From this last equation, there results

$$\frac{\mathrm{d}\varphi}{\mathrm{d}t} = \frac{B}{r^2\cos^2\psi}$$

But one has

$$\sin\psi = \frac{r^2+h^2-q^2}{2hr}$$

thus

$$\cos\psi = \frac{\sqrt{4h^2r^2-(r^2+h^2-q^2)^2}}{2hr}$$

Consequently, after substituting this expression, one will have

$$\frac{d\varphi}{dt} = \frac{4Bh^2}{4h^2r^2 - (r^2 + h^2 - q^2)^2}$$

such that if r and q are known as a function of t, one will also have φ as a function of t.

But, since $\sin\psi$ and $d\varphi/dt$ are already given by r and q, it is clear that the fourth equation can be reduced so that it only contains r and q, and then $r^2\, d\psi^2$ will necessarily be a complete integral of the two equations above in r and q because of the arbitrary constant B. Indeed, one will have

$$r^2\, d\psi^2 = \frac{[(r^2 + q^2 - h^2)dr - 2rq\, dq]^2}{4h^2r^2 - (r^2 + h^2 - q^2)^2}$$

and after adding dr^2 and reducing, it will become

$$r^2\, d\psi^2 + dr^2 = 4\frac{q^2r^2\, dr^2 + r^2q^2\, dq^2 - (r^2 + q^2 - h^2)rq\, dr\, dq}{4h^2r^2(r^2 + h^2 - q^2)^2}$$

Moreover, one has

$$\frac{r^2\cos^2\psi\, d\varphi^2}{dt^2} = \frac{4B^2}{4h^2r^2 - (r^2 + h^2 - q^2)^2}$$

Thus after making these substitutions in the fourth equation and by cancelling the denominator, one will have this integral

$$\text{(a)} \qquad \begin{cases} 2\dfrac{q^2r^2\, dr^2 + r^2q^2\, dq^2 - (r^2 + q^2 - h^2)rq\, dr\, dq}{dt^2} + 2B^2 \\ +[4h^2r^2 - (r^2 + h^2 - q^2)^2](\displaystyle\int R\, dr + \int Q\, dq - 2H) = 0 \end{cases}$$

and it is easy to see now, from the form of this equation, which results from the two equations in r and q, multiplied respectively by

$$2q^2\, d(r^2) - (r^2 + q^2 - h^2)d(q^2), \qquad 2r^2\, d(q^2) - (r^2 + q^2 - h^2)d(r^2)$$

added together and then integrated. It would have been rather difficult to discover this integral **a priori**.

81. In order to complete the solution, another integral must still be obtained from the same equations. But it can only be obtained for particular values of R and Q.

If it is assumed, which is in the nature of the problem, that

$$R = \frac{\alpha}{r^2}, \qquad Q = \frac{\beta}{q^2}$$

it will then be found that these equations, one multiplied by $\mathrm{d}(q^2)$ and the other by $\mathrm{d}(r^2)$ give an integrable sum for which the integral is

$$\text{(b)} \qquad \begin{cases} \dfrac{\mathrm{d}(r^2)\,\mathrm{d}(q^2)}{2\,\mathrm{d}t^2} - \dfrac{\alpha(3r^2 + q^2 - h^2)}{r} - \dfrac{\beta(3q^2 + r^2 - h^2)}{q} \\ = 4H(r^2 + q^2) + 2C \end{cases}$$

where C is a new arbitrary constant.

This equation multiplied by $r^2 + q^2 - h^2$ and added to the integral (a) found previously gives, with the present hypothesis, a reduced equation of the form

$$\text{(c)} \qquad \begin{cases} \dfrac{q^2(\mathrm{d}(r^2))^2 + r^2(\mathrm{d}(q^2))^2}{2\,\mathrm{d}t^2} - 2\alpha r(r^2 + 3q^2 - h^2) - 2\beta q(q^2 + 3r^2 - h^2) \\ = 2H(r^4 + q^4 + 6r^2q^2 - h^4) + 2C(r^2 + q^2 - h^2) - 2B^2 \end{cases}$$

The same equation after being multiplied by $2rq$ and then added or subtracted with this one, will give this double equation

$$\text{(d)} \qquad \begin{cases} \dfrac{(q\,\mathrm{d}(r^2) \pm r\,\mathrm{d}(q^2))^2}{4\,\mathrm{d}t^2} - \alpha[(r \pm q)^3 - h^2(r \pm q)] - \beta[(q \pm r)^3 - h^2(q \pm r)] \\ = H[(r \pm q)^4 - h^4] + C(r \pm q)^2 - B^2 \end{cases}$$

By defining $r + q = s$, $r - q = u$, one will have the two following equations

$$\text{(e)} \qquad \begin{cases} \dfrac{(s^2 - u^2)^2\,\mathrm{d}s^2}{16\,\mathrm{d}t^2} - (\alpha + \beta)s^3 + h^2(\alpha + \beta)s = H(s^4 - h^4) + Cs^2 - B^2 \\ \dfrac{(s^2 - u^2)^2\,\mathrm{d}u^2}{16\,\mathrm{d}t^2} - (\alpha - \beta)u^3 + h^2(\alpha - \beta)u = H(u^4 - h^4) + Cu^2 - B^2 \end{cases}$$

from which one first obtains this equation where the variables are separated

$$\text{(f)} \qquad \begin{cases} \dfrac{\mathrm{d}s}{\sqrt{Hs^4 + (\alpha + \beta)s^3 + Cs^2 - h^2(\alpha + \beta)s - Hh^4 - B^2}} \\ = \dfrac{\mathrm{d}u}{\sqrt{Hu^4 + (\alpha - \beta)u^3 + Cu^2 - h^2(\alpha - \beta)u - Hh^4 - B^2}} \end{cases}$$

and then

$$\text{(g)} \qquad \mathrm{d}t = \begin{cases} \dfrac{s^2\,\mathrm{d}s}{4\sqrt{Hs^4+(\alpha+\beta)s^3+Cs^2-h^2(\alpha+\beta)s-Hh^4-B^2}} \\ -\dfrac{u^2\,\mathrm{d}u}{4\sqrt{Hu^4+(\alpha-\beta)u^3+Cu^2-h^2(\alpha+\beta)u^2-Hh^4-B^2}} \end{cases}$$

If the same substitutions are used in the expression for $\mathrm{d}\varphi/\mathrm{d}t$ found above, one will have

$$\frac{\mathrm{d}\varphi}{\mathrm{d}t} = -\frac{4Bh^2}{(s^2-h^2)(u^2-h^2)} = \frac{4Bh^2}{s^2-u^2}\left(\frac{1}{s^2-h^2} - \frac{1}{u^2-h^2}\right)$$

and after substituting the expression for $\mathrm{d}t$

$$\text{(h)} \qquad \mathrm{d}\varphi = \begin{cases} \dfrac{Bh^2\,\mathrm{d}s}{(s^2-h^2)\sqrt{Hs^4+(\alpha+\beta)s^3+Cs^2-h^2(\alpha+\beta)s-Hh^4-B^2}} \\ -\dfrac{Bh^2\,\mathrm{d}u}{(u^2-h^2)\sqrt{Hu^4+(\alpha-\beta)u^3+Cu^2-h^2(\alpha-\beta)u-Hh^4-B^2}} \end{cases}$$

If each of these differentials could be integrated, one would obtain first an equation between s and u, then t and φ as functions of s and u will be obtained. Thus one will have q and from there t and φ as functions of r. And since

$$\sin\psi = \frac{r^2+h^2-q^2}{2hr}$$

one will also have ψ as function of r. But because these differentials are with respect to the rectification of conic sections, they can only be integrated by approximation and the best method seems to be the one I gave elsewhere[25] for the integration of all the differentials which contain a square root where the variable is of the fourth order under the integral sign.

82. If, besides the two forces α/r^2 and β/q^2 which attract the body toward the two fixed centers, there was a third force proportional to the distance which attracts it toward the point located in the middle of the line which joins the two centers, it is clear that this force could be resolved into two components, both directed toward the same points and also proportional to the distances. Thus, in this case, one will have

$$R = \frac{\alpha}{r^2} + 2\gamma r, \qquad Q = \frac{\beta}{q^2} + 2\gamma q$$

and it will be found that the integral (b) would also hold in this case. But one should add to its first member the terms

$$\gamma[5r^2q^2 + \tfrac{3}{2}(r^4+q^4) - h^2(r^2+q^2)]$$

and then one will have to add to the first member of the equation (c) the terms

$$\frac{\gamma}{2}[r^6 + q^6 + 15r^2q^2(r^2 + q^2) - h^2(r^4 + q^4 + 6r^2q^2)]$$

and consequently, to the first member of equation (d), the following terms

$$\frac{\gamma}{4}[(r \pm q)^6 - h^2(r \pm q)^4]$$

so that it is only necessary to add to the polynomials in s and u under the integral sign in equations (e), (f) and (g), the following terms, respectively

$$-\frac{\gamma}{4}(s^6 - h^2s^4), \qquad -\frac{\gamma}{4}(u^6 - h^2u^4)$$

which incidently does not make the solution more complicated.

83. Although it is impossible to integrate, in general, equation (f) between s and u, and consequently, to obtain a finite relation between these two variables, nevertheless, one can obtain two particular integrals represented by s = constant and u = constant.

Indeed, if one represents this equation in general by

$$\frac{\mathrm{d}s}{\sqrt{S}} = \frac{\mathrm{d}u}{\sqrt{U}}$$

it is clear that it will also hold by making ds and du equal to zero, as long as the denominators $\sqrt{S}$ or $\sqrt{U}$ are also equal to zero at the same time and of the same order.

In order to determine the necessary conditions in this case, begin by making the following definition

$$s = f + \omega$$

where f is a constant and ω is an infinitesimal quantity. If the quantity F denotes what becomes of S when s is replaced by f, the term $\mathrm{d}s/\sqrt{S}$ will become

$$\frac{\mathrm{d}\omega}{\sqrt{\left(F + \frac{\mathrm{d}F}{\mathrm{d}f}\omega + \frac{\mathrm{d}^2F}{2\,\mathrm{d}f^2}\omega^2 + \cdots\right)}}$$

Then, in order to have the same dimension for ω in the numerator and denominator, one must have

$$F = 0, \qquad \frac{\mathrm{d}F}{\mathrm{d}f} = 0$$

then, since ω is an infinitesimal quantity, the differential in question will be reduced to

$$\frac{\mathrm{d}\omega}{\omega\sqrt{\dfrac{\mathrm{d}^2F}{2\,\mathrm{d}f^2}}}$$

for which the integral is

$$\frac{1}{\sqrt{\dfrac{\mathrm{d}^2F}{2\,\mathrm{d}f^2}}}\ell n\left(\frac{\omega}{k}\right)$$

where k is an arbitrary constant. Thus if one puts $\omega = 0$ and simultaneously, if one puts $k = 0$, the value of $\ell n(\omega/k)$ will become indeterminate and the equation will always remain indeterminate whatever the value that the other member $\int \mathrm{d}u/\sqrt{U}$ acquires. But it is known and it is obvious by itself that

$$F = 0, \qquad \frac{\mathrm{d}F}{\mathrm{d}f} = 0$$

are the conditions which make f the square root of the equation $F = 0$. From this result,[26] it follows that, in general, if the polynomial S has one or more square roots, each of these roots will give a particular value of s. It will be the same for the polynomial U.

Now, it is clear that the equation $s = f$ or $r + q = f$ represents an ellipse where the two foci are the two centers of the radii r and q, and for which the major axis is equal to f. Similarly, the equation $u = g$ or $r - q = g$ represent a hyperbola for which the two foci are at the same centers and for which the major axis is g.

Thus, the particular solutions which we just discussed give ellipses or hyperbolas described about the centers of the forces α/r^2 and β/q^2 taken as foci. Since the polynomials S and V contain the three arbitrary constants A, B, C, dependent on the direction and the initial velocity of the body, it is clear that one will always be able to take these elements such that the body traverses a given ellipse or hyperbola about given foci. Thus the same conic section which can be described by a force directed toward one of the foci and acting inversely proportional to the square of the distances or directed to the center and acting proportional to the distances can also be described by three similar forces pointing toward the two foci and the center, which is very remarkable.[27]

84. If there were only one center toward which the body was attracted by the force α/r^2, the case of an elliptical orbit is obtained which was resolved in Chapter I. In this case, one will have $\beta = 0$ and $\gamma = 0$. The two polynomials S and U will become similar and will not exceed the fourth degree. The equations (f), (g) and (h) of Article 81 will then be integrable by known methods and the motion of the body will be determined by formulas

in s and u, that is, by the distances to the two centers, of which one, the one for which the attraction is zero, can be located arbitrarily. These formulas are only for pure curiosity but there is one case where they simplify and give a remarkable result, that is, the one where the center of zero attraction is located on the perimeter of the ellipse.

In order to obtain this case, the constants B and C will be determined in such a manner that since the radius q is zero, the other radius r is equal to h, which is the distance between the two centers. Consequently, the variables $s = r + q$ and $u = r - q$ both become equal to h. Equations (e) of Article 81 are perfect for this determination.

Since $s = u = h$, the first of these equations gives

$$B^2 = Ch^2$$

then, if the difference of these equations is divided by $s - u$, and if the equality $s = u = h$ is enforced, since $\beta = 0$ there will result

$$-3\alpha h^2 + \alpha h^2 = 4Hh^3 + 2Ch$$

from which one obtains

$$C = -\alpha h - 2Hh^2$$

With the substitution of these expressions, the polynomial

$$Hs^4 + \alpha s^3 + Cs^2 - \alpha h^2 s - Hh^4 - B^2$$

becomes

$$H(s^4 - 2s^2h^2 + h^4) + \alpha(s^3 - s^2h^2 - sh^2 + h^3)$$

which can be reduced to the form

$$H(s + h)^2(s - h)^2 + \alpha(s + h)(s - h)^2$$

It will be the same for the polynomial in u.

But, from Article 15, one has in this case

$$\alpha = g, \qquad H = -\frac{g}{2a}$$

where a is the semi-major axis of the ellipse. Thus, the equations (f) and (g) will become

$$\frac{\mathrm{d}s}{(s-h)\sqrt{g(s+h)-\dfrac{g}{2a}(s+h)^2}}=\frac{\mathrm{d}u}{(u-h)\sqrt{g(u+h)-\dfrac{g}{2a}(u+h)^2}}$$

$$\mathrm{d}t=\frac{s^2\,\mathrm{d}s}{4(s-h)\sqrt{g(s+h)-\dfrac{g}{2a}(s+h)^2}}-\frac{u^2\,\mathrm{d}u}{4(u-h)\sqrt{g(u+h)-\dfrac{g}{2a}(u+h)^2}}$$

and if the first equation is multiplied by h^2, and then subtracted from the last equation, and if the numerators and denominators are divided by $(s-h)$ and $(u-h)$ respectively, one will have

$$\mathrm{d}t=\frac{(s+h)\,\mathrm{d}s}{4\sqrt{g}\sqrt{s+h-\dfrac{(s+h)^2}{2a}}}-\frac{(u+h)\,\mathrm{d}u}{4\sqrt{g}\sqrt{u+h-\dfrac{(u+h)^2}{2a}}}$$

an expression which has the advantage that it contains only the major axis $2a$.

85. If the following relation is defined

$$\int\frac{z\,\mathrm{d}z}{\sqrt{z-\dfrac{z^2}{2a}}}=f(z)$$

where the integral begins where z has an arbitrary given value. If s and u are replaced by the values $(p+q)$ and $(p-q)$, one will obtain by integration

$$4t\sqrt{g}=f(h+p+q)-f(h+p-q)$$

where it is seen that $t=0$ when $q=0$, however the integration is carried out.

But, since p is the radius vector which starts from one focus, q is the radius vector which starts from the other focus taken at a point on the ellipse, and for which the distance to the focus is h, it is clear that h and p will be two radius vectors and that q will be the chord of the arc between these two radii. Consequently, the preceding expression for t will be the time the mobile body requires to describe this arc of the ellipse, which will be given by the sum of the radius vectors $(h+p)$, by the chord q, and by the major axis $2a$.

The integral designated by the function $f(z)$ depends on the arcs of a circle or of logarithms, whether a is positive or negative. But when the axis $2a$ is very large, this function is reduced to a rapidly convergent series. One has then

$$f(z)=\frac{2}{3}z^{\frac{3}{2}}+\frac{z^{\frac{5}{2}}}{5.2a}+\frac{3z^{\frac{7}{2}}}{4.7.4a^2}+\cdots$$

The first term gives the expression for time in the parabola and one has

$$4t\sqrt{g} = \tfrac{2}{3}(h+p+q)^{\frac{3}{2}} - \tfrac{2}{3}(h+p-q)^{\frac{3}{2}}$$

which coincides with the equation found in Article 25. The remainder of the series gives the difference in time required to traverse an arc of a parabola and an arc of an ellipse or hyperbola having the same chord u and the same sum s for the radius vectors.

This elegant property[28] for motion in conic sections was found by Lambert, who has given an ingenious demonstration in his treatise entitled *Insigniores orbitae cometarum proprietates*. Also, the reader should refer to the *Mémoires de l'Académie de Berlin* for the year 1778.

The problem which we just solved was first solved by Euler for the case where there are only two fixed centers which attract inversely proportional to the square of the distances and where the body moves in a plane containing the two centers (*Mémoires de Berlin* for 1760). His solution is specifically remarkable for the skill he has shown in using various substitutions to reduce the differential equations to the first order and to integration. These differential equations could not be solved by known methods because of their complexity.

By giving a different form to these equations, I obtained directly the same results and I was even able to generalize them to the case where the curve is not in the same plane and where there is also a force proportional to the distance and directed toward a fixed center located in the middle of the two other centers. The reader should refer to the Fourth Volume of the old *Mémoires de Turin*, from which the preceding analysis is taken and in which the case is found where one of the centers is moving towards infinity and the force directed towards this center becomes uniform and acts along parallel lines. It is surprising that in this case the solution is not greatly simplified. Only the radicals, which enter in the denominators of the individual equations, contain only the third degree of these variables rather than the fourth degree, which also makes their integration dependent on the rectification of conic sections.

Chapter IV
THE MOTION OF TWO OR MORE FREE BODIES WHICH ATTRACT ONE ANOTHER AND IN PARTICULAR, ON THE MOTION OF PLANETS ABOUT THE SUN AND THE SECULAR VARIATION OF THEIR ELEMENTS

86. When several bodies attract one another with forces proportional to their masses and as functions of the distance between them, their motions are expressed by the general formulas of Articles 1 and 2, by taking the bodies themselves for the centers of attraction.

Let m, m', m'', etc. be the masses of the bodies and $x, y, z,\ x', y', z',\ x'', y'', z''$, etc. their rectangular coordinates referred to fixed axes in space. The quantity T will be, as in Article 1

$$T = m\frac{\mathrm{d}x^2 + \mathrm{d}y^2 + \mathrm{d}z^2}{2\,\mathrm{d}t^2} + m'\frac{\mathrm{d}x'^2 + \mathrm{d}y'^2 + \mathrm{d}z'^2}{2\,\mathrm{d}t^2} + m''\frac{\mathrm{d}x''^2 + \mathrm{d}y''^2 + \mathrm{d}z''^2}{2\,\mathrm{d}t^2} + \cdots$$

Let ρ', ρ'', ρ''', etc. be the distances from the bodies m', m'', m''', etc. to the body m, and let R', R'', etc. be functions of these distances, proportional to the attractions.

Also, let $\rho_{,}''$, $\rho_{,}'''$, etc. be the distances from the bodies m'', m''', etc. to the body m', and let $R_{,}''$, $R_{,}'''$, etc. be functions of these distances, proportional to the attractions.

Similarly, let $\rho_{,,}'''$, $\rho_{,,}^{\mathrm{IV}}$, etc. be the distances from the bodies m''', m^{IV}, etc. to the body m'' and let $R_{,,}'''$, $R_{,,}^{\mathrm{IV}}$, etc. be the functions of the distances proportional to the attractions.

And so on. One will have

$$\begin{aligned}
\rho' &= \sqrt{(x'-x)^2 + (y'-y)^2 + (z'-z)^2}\\
\rho'' &= \sqrt{(x''-x)^2 + (y''-y)^2 + (z''-z)^2}\\
\rho_{,}'' &= \sqrt{(x''-x')^2 + (y''-y')^2 + (z''-z')^2}\\
\rho_{,}''' &= \sqrt{(x'''-x')^2 + (y'''-y')^2 + (z'''-z')^2}\\
&\ \vdots
\end{aligned}$$

and the quantity V (Article 2) will be

$$\begin{aligned}
V &= m\Big(m'\int R'\,\mathrm{d}\rho' + m''\int R''\,\mathrm{d}\rho'' + m'''\int R'''\,\mathrm{d}\rho''' + \cdots\Big)\\
&+ m'\Big(m''\int R_{,}''\,\mathrm{d}\rho_{,}'' + m'''\int R_{,}'''\,\mathrm{d}\rho_{,}''' + \cdots\Big) + m''\Big(m'''\int R_{,,}'''\,\mathrm{d}\rho_{,,}''' + \cdots\Big)\\
&\ \vdots
\end{aligned}$$

But, whatever independent coordinates are adopted, one will always have with respect to each of them, such as ξ, an equation of canonical form

$$\mathrm{d}\frac{\delta T}{\delta \mathrm{d}\xi} - \frac{\delta T}{\delta \xi} + \frac{\delta V}{\delta \xi} = 0$$

Since, in the systems that we are considering, there are no fixed points, the location of the origin of the coordinates can be taken arbitrarily. Then one will always have, as seen in SECTION III, the three finite integrals relative to areas and finally, the integral of the **forces vives** $T + V = H$.

In this fashion, one will obtain the absolute motion of bodies in space. But since the solution of this problem is only important for the motion of planets and it is only their relative motion with respect to an immobile Sun, which is of interest to astronomy, it remains to show how the general equations of absolute motion of bodies of a system can be transformed to relative motions.

Subsection I
General Equations For The Relative Motion Of Bodies Which Attract One Another

87. Let us assume that it is desired to know the relative motions of the bodies m', m'', etc. with respect to the body m. Let us designate by ξ', η', ζ' the rectangular coordinates of the body m', with respect to the body m by taking the latter for the origin of coordinates. Similarly, let ξ'', η'', ζ'' be the rectangular coordinates of the body m'' with respect to the same body m and so on. The problem will be to find a general formula which only contains these coordinates.

At the outset, it is obvious that

$$\begin{array}{lll} x' = x + \xi', & y' = y + \eta', & z' = z + \zeta' \\ x'' = x + \xi'', & y'' = y + \eta'', & z'' = z + \zeta'' \\ \ldots\ldots\ldots\ldots & & \end{array}$$

$$\begin{aligned} \rho' &= \sqrt{\xi'^2 + \eta'^2 + \zeta'^2} \\ \rho'' &= \sqrt{\xi''^2 + \eta''^2 + \zeta''^2} \\ &\ldots\ldots\ldots\ldots \end{aligned}$$

$$\begin{aligned} \rho''_{\prime} &= \sqrt{(\xi'' - \xi')^2 + (\eta'' - \eta')^2 + (\zeta'' - \zeta')^2} \\ \rho'''_{\prime} &= \sqrt{(\xi''' - \xi')^2 + (\eta''' - \eta')^2 + (\zeta''' - \zeta')^2} \\ &\ldots\ldots\ldots\ldots \end{aligned}$$

$$\begin{aligned} \rho'''_{\prime\prime} &= \sqrt{(\xi''' - \xi'')^2 + (\eta''' - \eta'')^2 + (\zeta''' - \zeta'')^2} \\ &\ldots\ldots\ldots\ldots \end{aligned}$$

and the quantity T will become

$$\begin{aligned} T &= (m + m' + m'' + \cdots)\frac{\mathrm{d}x^2 + \mathrm{d}y^2 + \mathrm{d}z^2}{2\,\mathrm{d}t^2} \\ &+ \frac{\mathrm{d}x(m'\,\mathrm{d}\xi' + m''\,\mathrm{d}\xi'' + \cdots) + \mathrm{d}y(m'\,\mathrm{d}\eta' + m''\,\mathrm{d}\eta'' + \cdots) + \mathrm{d}z(m'\,\mathrm{d}\zeta' + m''\,\mathrm{d}\zeta'' + \cdots)}{\mathrm{d}t^2} \\ &+ m'\frac{\mathrm{d}\xi'^2 + \mathrm{d}\eta'^2 + \mathrm{d}\zeta'^2}{2\,\mathrm{d}t^2} + m''\frac{\mathrm{d}\xi''^2 + \mathrm{d}\eta''^2 + \mathrm{d}\zeta''^2}{2\,\mathrm{d}t^2} + \cdots \end{aligned}$$

Since the variables x, y, z no longer appear in the quantity V after these substitutions, and the variables are not in T in a finite form, one will have with respect to these same variables, the equations

$$\mathrm{d}\frac{\delta T}{\delta \mathrm{d}x} = 0, \qquad \mathrm{d}\frac{\delta T}{\delta \mathrm{d}y} = 0, \qquad \mathrm{d}\frac{\delta T}{\delta \mathrm{d}z} = 0$$

which gives

$$\frac{\delta T}{\delta \mathrm{d}x} = \alpha, \qquad \frac{\delta T}{\delta \mathrm{d}y} = \beta, \qquad \frac{\delta T}{\delta \mathrm{d}z} = \gamma$$

where α, β, γ are arbitrary constants.

Thus the three following equations are obtained

$$(m+m'+m''+\cdots)\frac{dx}{dt}+m'\frac{d\xi'}{dt}+m''\frac{d\xi''}{dt}+\cdots=\alpha$$
$$(m+m'+m''+\cdots)\frac{dy}{dt}+m'\frac{d\eta'}{dt}+m''\frac{d\eta''}{dt}+\cdots=\beta$$
$$(m+m'+m''+\cdots)\frac{dz}{dt}+m'\frac{d\zeta'}{dt}+m''\frac{d\zeta''}{dt}+\cdots=\gamma$$

where the quantities α, β, γ are constants.

Now, if the expressions for dx/dt, dy/dt, dz/dt were substituted in the preceding expression for T and if in order to shorten the expressions, the following definitions are made

$$X = m'\xi' + m''\xi'' + m'''\xi''' + \cdots$$
$$Y = m'\eta' + m''\eta'' + m'''\eta''' + \cdots$$
$$Z = m'\zeta' + m''\zeta'' + m'''\zeta''' + \cdots$$
$$M = m + m' + m'' + m''' + \cdots$$

one will obtain

$$T=\frac{\alpha^2+\beta^2+\gamma^2}{2M}-\frac{dX^2+dY^2+dZ^2}{2M\,dt^2}$$
$$+m'\frac{d\xi'^2+d\eta'^2+d\zeta'^2}{2\,dt^2}+m''\frac{d\xi''^2+d\eta''^2+d\zeta''^2}{2\,dt^2}+\cdots$$

88. Since the variables ξ', η', ζ', ξ'', etc. are independent and the quantity T does not contain these variables in a finite form, one will immediately have with respect to each of them the equations

$$m'\left(\frac{d^2\xi'}{dt^2}-\frac{d^2X}{M\,dt^2}\right)+\frac{dV}{d\xi'}=0,\qquad m''\left(\frac{d^2\xi''}{dt^2}-\frac{d^2X}{M\,dt^2}\right)+\frac{dV}{d\xi''}=0,\qquad\cdots$$
$$m'\left(\frac{d^2\eta'}{dt^2}-\frac{d^2Y}{M\,dt^2}\right)+\frac{dV}{d\eta'}=0,\qquad m''\left(\frac{d^2\eta''}{dt^2}-\frac{d^2Y}{M\,dt^2}\right)+\frac{dV}{d\eta''}=0,\qquad\cdots$$
$$m'\left(\frac{d^2\zeta'}{dt^2}-\frac{d^2Z}{M\,dt^2}\right)+\frac{dV}{d\zeta'}=0,\qquad m''\left(\frac{d^2\zeta''}{dt^2}-\frac{d^2Z}{M\,dt^2}\right)+\frac{dV}{d\zeta''}=0,\qquad\cdots$$

If the first equations with respect to the variables ξ', ξ'', etc., were added together, there will result

$$\frac{m}{M}\frac{d^2X}{dt^2}+\frac{dV}{d\xi'}+\frac{dV}{d\xi''}+\cdots=0$$

which gives

$$\frac{d^2X}{dt^2} = -\frac{M}{m}\left(\frac{dV}{d\xi'} + \frac{dV}{d\xi''} + \cdots\right)$$

and similarly, it will be found that by adding the second and the third equations

$$\frac{d^2Y}{dt^2} = -\frac{M}{m}\left(\frac{dV}{d\eta'} + \frac{dV}{d\eta''} + \cdots\right)$$
$$\frac{d^2Z}{dt^2} = -\frac{M}{m}\left(\frac{dV}{d\zeta'} + \frac{dV}{d\zeta''} + \cdots\right)$$

expressions which can be substituted in the preceding equations.

Thus one will have for the motion of the body m' about the body m, the three equations

$$m'\frac{d^2\xi'}{dt^2} + \frac{dV}{d\xi'} + \frac{m'}{m}\left(\frac{dV}{d\xi'} + \frac{dV}{d\xi''} + \cdots\right) = 0$$
$$m'\frac{d^2\eta'}{dt^2} + \frac{dV}{d\eta'} + \frac{m'}{m}\left(\frac{dV}{d\eta'} + \frac{dV}{d\eta''} + \cdots\right) = 0$$
$$m'\frac{d^2\zeta'}{dt^2} + \frac{dV}{d\zeta'} + \frac{m'}{m}\left(\frac{dV}{d\zeta'} + \frac{dV}{d\zeta''} + \cdots\right) = 0$$

and similar equations will be obtained for the motion of the bodies m'', m''', etc., about the body m by only changing the quantities marked by two, three, etc., primes.

There remains to substitute the expression for V and to take its partial differences with respect to the different variables. But this substitution can be simplified by the following consideration.

89. Let us denote by U the sum of all the terms of the quantity V which contain the distances $\rho''_{\prime}$, $\rho'''_{\prime}$, etc. $\rho'''_{\prime\prime}$, etc. and let us note that the expressions for these distances remain the same if the coordinates ξ', ξ'', ξ''', etc. are augmented by the same arbitrary quantity which is present in all the expressions. From this result, it follows that by varying the same coordinates by an infinitesimal quantity, the variation of U will be zero. Thus, the following equation will result

$$\frac{dU}{d\xi'} + \frac{dU}{d\xi''} + \frac{dU}{d\xi'''} + \cdots = 0$$

It will be found in the same fashion, since the same property holds with respect to the coordinates η', η'', η''', etc. and to the coordinates $\zeta', \zeta'', \zeta'''$, etc., that

$$\frac{dU}{d\eta'} + \frac{dU}{d\eta''} + \frac{dU}{d\eta'''} + \cdots = 0$$
$$\frac{dU}{d\zeta'} + \frac{dU}{d\zeta''} + \frac{dU}{d\zeta'''} + \cdots = 0$$

Therefore, since

$$V = m(m' \int R' \,\mathrm{d}\rho' + m'' \int R'' \,\mathrm{d}\rho'' + \cdots) + U$$

ρ' contains only ξ', η', ζ'; ρ'' contains only ξ'', η'', ζ'' and so on, the first equation will become after these substitutions and after dividing it by m'

$$\frac{\mathrm{d}^2\xi'}{\mathrm{d}t^2} + (m+m')R'\frac{\mathrm{d}\rho'}{\mathrm{d}\xi'} + m''R''\frac{\mathrm{d}\rho''}{\mathrm{d}\xi''} + \cdots + \frac{\mathrm{d}U}{m'\,\mathrm{d}\xi'} = 0$$

But in the quantity U there are only terms which contain $\rho_{\prime}''$, $\rho_{\prime}'''$, etc., which depend in turn on the variables ξ', η', ζ' (Article 86). Thus the expression for U can be reduced to

$$U = m'(m'' \int R_{\prime}'' \,\mathrm{d}\rho_{\prime}'' + m''' \int R_{\prime}''' \,\mathrm{d}\rho_{\prime}''' + \cdots)$$

After substituting the value of $\mathrm{d}U/\mathrm{d}\xi'$ in the preceding equation, it changes to

$$\frac{\mathrm{d}^2\xi'}{\mathrm{d}t^2} + (m+m')R\frac{\mathrm{d}\rho'}{\mathrm{d}\xi'} + m''R_{\prime}''\frac{\mathrm{d}\rho_{\prime}''}{\mathrm{d}\xi'} + m'''R_{\prime}'''\frac{\mathrm{d}\rho_{\prime}'''}{\mathrm{d}\xi'} + \cdots$$
$$+ m''R''\frac{\mathrm{d}\rho''}{\mathrm{d}\xi''} + m'''R'''\frac{\mathrm{d}\rho'''}{\mathrm{d}\xi'''} + \cdots = 0$$

and in the same fashion there will be obtained

$$\frac{\mathrm{d}^2\eta'}{\mathrm{d}t^2} + (m+m')R'\frac{\mathrm{d}\rho'}{\mathrm{d}\eta'} + m''R_{\prime}''\frac{\mathrm{d}\rho_{\prime}''}{\mathrm{d}\eta'} + m'''R_{\prime}'''\frac{\mathrm{d}\rho_{\prime}'''}{\mathrm{d}\eta'} + \cdots$$
$$+ m''R''\frac{\mathrm{d}\rho''}{\mathrm{d}\eta''} + m'''R'''\frac{\mathrm{d}\rho'''}{\mathrm{d}\eta'''} + \cdots = 0$$

$$\frac{\mathrm{d}^2\zeta'}{\mathrm{d}t^2} + (m+m')R'\frac{\mathrm{d}\rho'}{\mathrm{d}\zeta'} + m''R_{\prime}''\frac{\mathrm{d}\rho_{\prime}''}{\mathrm{d}\zeta'} + m'''R_{\prime}'''\frac{\mathrm{d}\rho_{\prime}'''}{\mathrm{d}\zeta'} + \cdots$$
$$+ m''R''\frac{\mathrm{d}\rho''}{\mathrm{d}\zeta''} + m'''R'''\frac{\mathrm{d}\rho'''}{\mathrm{d}\zeta'''} + \cdots = 0$$

90. These equations can be reduced to the general form, which has the advantage that it is also applicable to arbitrary coordinates.

If the first equation is multiplied by $\delta\xi'$, the second by $\delta\eta'$, the third by $\delta\zeta'$, and then all are added together, one will have first the differential part

$$\frac{\mathrm{d}^2\xi'}{\mathrm{d}t^2}\delta\xi' + \frac{\mathrm{d}^2\eta'}{\mathrm{d}t^2}\delta\eta' + \frac{\mathrm{d}^2\zeta'}{\mathrm{d}t^2}\delta\zeta'$$

which after transforming the coordinates ξ', η', ζ', to other independent coordinates ξ, ψ, φ, will give for the terms multiplied by $\delta\xi$, the formula (SECTION IV, Article 7)

$$\left(\mathrm{d}\frac{\delta T'}{\delta \mathrm{d}\xi} - \frac{\delta T'}{\delta\xi}\right)\delta\xi$$

after defining

$$T' = \frac{\mathrm{d}\xi'^2 + \mathrm{d}\eta'^2 + \mathrm{d}\zeta'^2}{2\,\mathrm{d}t^2}$$

With respect to the terms which contain the forces R', R'', etc., it is easy to see that after replacing the operator δ by d, all of the terms are integrable with respect to the variables ξ', η', ζ'. The integral will first contain the terms

$$(m + m')\int R'\,\mathrm{d}\rho' + m''\int R''_{\prime}\,\mathrm{d}\rho''_{\prime} + m'''\int R'''_{\prime}\,\mathrm{d}\rho'''_{\prime} + \cdots$$

then it will contain the terms

$$\left(m''R''\frac{\mathrm{d}\rho''}{\mathrm{d}\xi''} + m'''R'''\frac{\mathrm{d}\rho'''}{\mathrm{d}\xi'''} + \cdots\right)\xi'$$
$$+\left(m''R''\frac{\mathrm{d}\rho''}{\mathrm{d}\eta''} + m'''R'''\frac{\mathrm{d}\rho'''}{\mathrm{d}\eta'''} + \cdots\right)\eta'$$
$$+\left(m''R''\frac{\mathrm{d}\rho''}{\mathrm{d}\zeta''} + m'''R'''\frac{\mathrm{d}\rho'''}{\mathrm{d}\zeta'''} + \cdots\right)\zeta'$$

But one has

$$\frac{\mathrm{d}\rho''}{\mathrm{d}\xi''} = \frac{\xi''}{\rho''}, \qquad \frac{\mathrm{d}\rho''}{\mathrm{d}\eta''} = \frac{\eta''}{\rho''}, \qquad \frac{\mathrm{d}\rho''}{\mathrm{d}\zeta''} = \frac{\zeta''}{\rho''},$$

and since

$$\rho''^2_{\prime} = (\xi'' - \xi')^2 + (\eta'' - \eta')^2 + (\zeta'' - \zeta')^2$$

there will result

$$\frac{\mathrm{d}\rho''}{\mathrm{d}\xi''}\xi' + \frac{\mathrm{d}\rho''}{\mathrm{d}\eta''}\eta' + \frac{\mathrm{d}\rho''}{\mathrm{d}\zeta''}\zeta' = \frac{\rho'^2 + \rho''^2 - \rho''^2_{\prime}}{2}$$

and similarly

$$\frac{\mathrm{d}\rho'''}{\mathrm{d}\xi'''}\xi' + \frac{\mathrm{d}\rho'''}{\mathrm{d}\eta'''}\eta' + \frac{\mathrm{d}\rho'''}{\mathrm{d}\zeta'''}\zeta' = \frac{\rho'^2 + \rho'''^2 - \rho'''^2_{\prime}}{2}$$

and so on for the other similar expressions. Therefore, denoting the total integral by V', one will have

$$V' = (m+m')\int R'\,\mathrm{d}\rho' + m''\int R''_{,}\,\mathrm{d}\rho''_{,} + m'''\int R'''_{,}\,\mathrm{d}\rho'''_{,} + \cdots$$
$$+ m''R''\frac{\rho'^2+\rho''^2-\rho''^2_{,}}{2\rho''} + m'''R'''\frac{\rho'^2+\rho'''^2-\rho'''^2_{,}}{2\rho'''} + \cdots$$

Thus after the transformation of coordinates, the terms multiplied by $\delta\xi$ will be reduced to $(\delta V'/\delta\xi)\delta\xi$. Since the new coordinates ξ, ψ, φ are assumed independent, each of them, such as ξ, will give an equation of the form

$$\mathrm{d}\frac{\delta T'}{\delta \mathrm{d}\xi'} - \frac{\delta T'}{\delta\xi} + \frac{\delta V'}{\delta\xi} = 0$$

91. If there were only two bodies, m and m', the expression for V' becomes

$$V' = (m+m')\int R'\,\mathrm{d}\rho'$$

Thus the values of T' and V' are the same as for a body attracted towards a fixed center by a force $(m+m')R'$ proportional to a function of the distance ρ' (Article 4). Therefore, the relative motion of a body m' about a body m will be the same as if one were fixed and the attracting mass were the sum of the two masses, which is known from Newton.

When the mass m of the body about which the others are assumed to move is much larger than the total sum of the masses m', m'', etc., which is the case for the Sun with respect to the other planets, one has, by some very small approximations

$$V' = (m+m')\int R'\,\mathrm{d}\rho'$$

The motion of the body m' about the body m will thus, in this case, be with some very small approximations, the same as if this one were fixed and that the sum of the masses $(m+m')$ were added. By viewing the other forces $m''R'$, $m''R''$, etc., as perturbing forces, the theory of the variation of arbitrary constants can be used to determine the effect of these forces. It will only be necessary to take, in accordance with Article 9 of SECTION V, $-\Omega$ equal to the sum of all the other terms of the expression for V' given above. Thus after marking the letter Ω with a prime to indicate that it refers to the planet m' there results

$$\Omega' = -m''\int R''_{,}\,\mathrm{d}\rho''_{,} - m'''\int R'''_{,}\,\mathrm{d}\rho'''_{,} - \cdots$$
$$- m''R''\frac{\rho'^2+\rho''^2-\rho''^2_{,}}{2\rho''} - m'''R'''\frac{\rho'^2+\rho'''^2-\rho'''^2_{,}}{2\rho'''} + \cdots$$

and by the general formulas of Article 14 of the same section, one will have the variations of the elements of the motion of the body m' about the body m, which is viewed as if it were fixed.

Subsection II
General Formulas For The Long Term Variation Of The Elements Of The Orbits Of Planets About The Sun

92. In order to apply these formulas to the motion of the planets about the Sun, take the mass m for the mass of the Sun and the mass m' for the mass of the planet for which the perturbations have to be found and the masses m'', m''', etc. for the masses of the perturbing planets. Then the following equations will result

$$R' = \frac{1}{\rho'^2}, \qquad R'' = \frac{1}{\rho''^2} \cdots, \qquad R''_{,} = \frac{1}{\rho''^2_{,}} \cdots$$

Then, in the function Ω', the coordinates ξ', η', ζ', ξ'', η'', ζ'', etc., of the different bodies about m will be substituted with their values expressed as a function of t, in accordance with the formulas which were given in Chapter I for the expressions of the coordinates x, y, z, by defining

$$g = m + m'$$

or simply g', to refer it to the body m'. One will have, beginning with Article 69 and what follows, the variation of six elements of the orbit of the planet about the Sun.

It will suffice here to find the **secular** variations of the elements which are the most important and which only depend on the first constant term in the development of Ω.

The expression for Ω' will become

$$\Omega' = m''\left(\frac{1}{\rho''_{,}} - \frac{\rho'^2 + \rho''^2 - \rho''^2_{,}}{2\rho''^3}\right) + m'''\left(\frac{1}{\rho'''_{,}} - \frac{\rho'^2 + \rho'''^2 - \rho'''^2_{,}}{2\rho'''^3}\right) + \cdots$$

93. Let us begin by developing the quantity

$$\rho''_{,} = \sqrt{(\xi'' - \xi')^2 + (\eta'' - \eta')^2 + (\zeta'' - \zeta')^2}$$

At the outset, replace ξ, η, ζ, with the expressions for x, y, z of Article 13 by marking with one or two primes the quantities relative to the masses m' and m''. Thus there will result

$$\begin{aligned}
\rho^2_{,} &= \rho'^2 + \rho''^2 - 2\rho'\rho''(\alpha' \cos\Phi' + \beta' \sin\Phi')(\alpha'' \cos\Phi'' + \beta'' \sin\Phi'') \\
&-2\rho'\rho''(\alpha'_1 \cos\Phi' + \beta'_1 \sin\Phi')(\alpha''_1 \cos\Phi'' + \beta''_1 \sin\Phi'') \\
&-2\rho'\rho''(\alpha'_2 \cos\Phi' + \beta'_2 \sin\Phi')(\alpha''_2 \cos\Phi'' + \beta''_2 \sin\Phi'')
\end{aligned}$$

If the multiplications are carried out, and the products of the sines and cosines are developed, then in order to shorten the expressions one makes the following definitions

$$A = \alpha'\alpha'' + \alpha'_1\alpha''_1 + \alpha'_2\alpha''_2$$

$$B = \alpha'\beta'' + \alpha_1'\beta_1' + \alpha_2'\beta_2'$$
$$A_1 = \alpha''\beta' + \alpha_1''\beta_1' + \alpha_2''\beta_2'$$
$$B_2 = \beta'\beta'' + \beta_1'\beta_1'' + \beta_2'\beta_2''$$

the following equation results

$$\begin{aligned}\rho_\prime''^2 = \rho'^2 + \rho''^2 &- (A + B_1)\rho'\rho''\cos(\Phi' - \Phi'') - (A - B_1)\rho'\rho''\cos(\Phi' + \Phi'') \\ &- (A_1 - B)\rho'\rho''\sin(\Phi' - \Phi'') - (A_1 + B)\rho'\rho''\sin(\Phi' + \Phi'')\end{aligned}$$

The quantities α', β', α_1', etc., are functions of the elements h', i', k' of the orbit of the planet m', given by the formulas of Article 13, by marking all the letters with one prime and the quantities α'', β'', α_1'', etc. are similar functions of the elements h'', i'', k'' of the orbit of the planet m'' by marking the letters with two primes. Thus the quantities A, B, A_1, B_1 are functions of the same elements. By the following consideration, their meaning will be clear.

94. The original coordinates with respect to a given plane are x, y, z. In order to transform them to the coordinates x', y', z', referred to the plane of the orbit of m', one has from the general formulas of Article 14

$$x = \alpha'x' + \beta'y' + \gamma'z'$$
$$y = \alpha_1'x' + \beta_1'y' + \gamma_1'z'$$
$$z = \alpha_2'x' + \beta_2'y' + \gamma_2'z'$$

where the coefficients α', β', etc., are dependent on the constants h', i', k' which determine the position of the new axes with respect to the original coordinate axes. The quantity i' is the inclination of these two planes.

Similarly, if the same coordinates are required to be transformed to the coordinates x'', y'', z'' referred to the plane of the orbit of m'', one would have

$$x = \alpha''x'' + \beta''y'' + \gamma''z''$$
$$y = \alpha_1''x'' + \beta_1''y'' + \gamma_1''z''$$
$$z = \alpha_2''x'' + \beta_2''y'' + \gamma_2''z''$$

where the coefficients α'', β'', etc., would be similar functions of the constants h'', i'', k'' which determine the position of this new plane with respect to the original plane and where i'' would be the inclination of the planes.

Now if these expressions are compared, one has

$$\alpha'x' + \beta'y' + \gamma'z' = \alpha''x'' + \beta''y'' + \gamma''z''$$
$$\alpha_1'x' + \beta_1'y' + \gamma_1'z' = \alpha_1''x'' + \beta_1''y'' + \gamma_1''z''$$
$$\alpha_2'x' + \beta_2'y' + \gamma_2'z' = \alpha_2''x'' + \beta_2''y'' + \gamma_2''z''$$

Since the coefficients α', β', γ', α_1', etc., are subject to the same equations of condition as the coefficients α, β, γ, α_1, of Article 14, if the three preceding equations are added together, after having multiplied each respectively by α', α'_1, α'_2, by β', β'_1, β'_2 and by γ', γ'_1, γ'_2, one will obtain from these equations

$$x' = Ax'' + By'' + Cz''$$
$$y' = A_1x'' + B_1y'' + C_1z''$$
$$z' = A_2x'' + B_2y'' + C_2z''$$

In order to shorten the expressions, one makes the following definitions

$$A = \alpha'\alpha'' + \alpha'_1\alpha''_1 + \alpha'_2\alpha''_2$$
$$B = \alpha'\beta'' + \alpha'_1\beta''_1 + \alpha'_2\beta''_2$$
$$C = \alpha'\gamma'' + \alpha'_1\gamma''_1 + \alpha'_2\gamma''_2$$
$$A_1 = \beta'\alpha'' + \beta'_1\alpha''_1 + \beta'_2\alpha''_2$$
$$B_1 = \beta'\beta'' + \beta'_1\beta''_1 + \beta'_2\beta''_2$$
$$C_1 = \beta'\gamma'' + \beta'_1\gamma''_1 + \beta'_2\gamma''_2$$
$$A_2 = \gamma'\alpha'' + \gamma'_1\alpha''_1 + \gamma'_2\alpha''_2$$
$$B_2 = \gamma'\beta'' + \gamma'_1\beta''_1 + \gamma'_2\beta''_2$$
$$C_2 = \gamma'\gamma'' + \gamma'_1\gamma''_1 + \gamma'_2\gamma''_2$$

It is evident that by using these formulas the coordinates x', y', z' are transformed to the coordinates x'', y'', z''. Thus the coefficients A, B, C, A_1, B_1, etc., will be expressed in a manner similar to the analogous coefficients α', β', γ', α'_1, β'_1, etc. By taking the constants H, I, K, instead of h', i', k', one will have from the general formulas of Article 13

$$A = \cos K \cos H - \sin H \sin I \cos K$$
$$B = -\sin K \cos H - \cos K \sin H \cos I$$
$$C = \sin H \sin I$$
$$A_1 = \cos K \sin H + \sin K \cos H \cos I$$
$$B_1 = -\sin K \sin H + \cos K \cos H \cos I$$
$$C_1 = -\cos H \sin I$$
$$A_2 = -\sin K \sin I$$
$$B_2 = \cos K \sin I$$
$$C_2 = \cos I$$

The constant I will represent the angle of inclination between the two planes which contain the orbits of the planets m' and m'' and we will denote it by $I'_{\prime}$ to indicate that it is with respect to the orbits of m' and m''. If in the expression for C_2 as a function of $\gamma', \gamma'', \gamma'''$, γ'_1, etc., the values of these coefficients were substituted as functions of $h', k', i', h'', k'', i''$ (Article 13), one has

$$\cos I''_{\prime} = \cos i' \cos i'' + \cos(h' - h'') \sin i' \sin i''$$

It is clear that the quantities that we just denoted by A, B, A_1, B_1 are the same functions of α', α'', β', etc., as those we denoted by the same letters in Article 93. Thus one will obtain for the formulas of this article, after substituting for these quantities the expressions that were just found

$$A + B_1 = 2\cos(H + K)(\cos\tfrac{1}{2}I_{\prime}'')^2$$
$$A - B_1 = 2\cos(H - K)(\sin\tfrac{1}{2}I_{\prime}'')^2$$
$$A_1 - B = 2\sin(H + K)(\cos\tfrac{1}{2}I_{\prime}'')^2$$
$$A_1 + B = 2\sin(H - K)(\sin\tfrac{1}{2}I_{\prime}'')^2$$

95. Thus after carrying out these substitutions in the expression for $\rho_{\prime}''^2$ of Article 93, one will have

$$\rho_{\prime}''^2 = \rho'^2 + \rho''^2 - 2\rho'\rho''\cos(\Phi' - \Phi'' - H - K)(\cos\tfrac{1}{2}I_{\prime}'')^2$$
$$-2\rho'\rho''\cos(\Phi' + \Phi'' - H + K)(\sin\tfrac{1}{2}I_{\prime}'')^2$$

Let us define temporarily

$$\Delta = \cos(\Phi' + \Phi'' - H + K) - \cos(\Phi - \Phi' - H - K)$$

so that one will have

$$\frac{1}{\rho_{\prime}''} = \frac{1}{\sqrt{\rho'^2 + \rho''^2 - 2\rho'\rho''\cos(\Phi' - \Phi'' - H - K) - 2\rho'\rho''\Delta(\sin\tfrac{1}{2}I_{\prime}'')^2}}$$

This is the expression which should be substituted in the equation for Ω of Article 92. In the same fashion, the following equation will be obtained

$$\frac{\rho'^2 + \rho''^2 - \rho_{\prime}''^2}{2\rho''^3} = \frac{\rho'\cos(\Phi' - \Phi'' - H - K)}{\rho''^2} + \frac{\rho'\Delta(\sin\tfrac{1}{2}I_{\prime}'')^2}{\rho''^2}$$

By marking the letters which are now marked with only two primes, with three primes, the terms multiplied by m''' in Ω will be obtained and so on.

One should then substitute for ρ', ρ'', etc. and for Φ', Φ'', etc. their expressions which are expressed by the average anomalies u', u'', etc., according to the formulas of Article 21 and 22. In the development, we will only consider the second orders of the eccentricities e', e'', etc. and of the mutual inclinations $I_{\prime}''$, $I_{\prime}'''$ of the orbits of m'' and m''', etc. to the orbit of m' by treating the quantities as very small and all of the same order and by neglecting the terms where they would form products of more than the second order.

One will thus have

$$\frac{1}{\rho_{\prime}''} = \frac{1}{\sqrt{\rho'^2 + \rho''^2 - 2\rho'\rho''\cos(\Phi' - \Phi'' - H - K)}}$$
$$-\frac{\rho\rho'[\cos(\Phi' + \Phi'' - H + K) - \cos(\Phi' - \Phi'' - H - K)]}{[\rho'^2 + \rho''^2 - 2\rho'\rho''\cos(\Phi' - \Phi'' - H - K)]^{3/2}}(\sin\tfrac{1}{2}I_{\prime}'')^2$$

96. It is known that the powers of a function of the form

$$\rho'^2 + \rho''^2 - 2\rho'\rho'' \cos\varphi$$

can be developed in a series of cosines which are multiples of the angle φ. Thus it can be assumed that

$$(\rho'^2 + \rho''^2 - 2\rho'\rho'' \cos\varphi)^{-1/2} = (\rho', \rho'') + (\rho', \rho'')_1 \cos\varphi$$
$$+(\rho', \rho'')_2 \cos 2\varphi + (\rho', \rho'')_3 \cos 3\varphi + \cdots$$
$$(\rho'^2 + \rho''^2 - 2\rho'\rho'' \cos\varphi)^{-3/2} = [\rho', \rho''] + [\rho', \rho'']_1 \cos\varphi$$
$$+[\rho', \rho'']_2 \cos 2\varphi + [\rho', \rho'']_3 \cos 3\varphi + \cdots$$

where (ρ', ρ''), $(\rho', \rho'')_1$, etc., $[\rho', \rho'']$, $[\rho', \rho'']_1$, etc. are functions of ρ', ρ'' expressed in a series or by definite integrals in which the quantities ρ', ρ'' are included in the same manner and form homogeneous functions of dimensions -1 or -3.

Thus after making the following definition

$$\varphi = \Phi' - \Phi'' - H - K$$

one will have

$$\frac{1}{\rho_{\prime}''} = (\rho', \rho'') + (\rho', \rho'')_1 \cos\varphi + (\rho', \rho'')_2 \cos 2\varphi + \cdots$$
$$+\rho'\rho''([\rho', \rho''] + [\rho', \rho'']_1 \cos\varphi + [\rho', \rho'']_2 \cos 2\varphi + \cdots)$$
$$\times[\cos(\varphi + 2\Phi'' + 2K) - \cos\varphi](\sin \tfrac{1}{2} I_{\prime}'')^2$$

where the following definitions must be made (Articles 21 and 22)

$$\rho' = a'(1 - e' \cos u') + \frac{e'^2}{2} - \frac{e'^2}{2} \cos 2u'$$
$$\rho'' = a''(1 - e'' \cos u'') + \frac{e''^2}{2} - \frac{e''^2}{2} \cos 2u''$$
$$\Phi' = u' + 2e' \sin u' + \frac{5e'^2}{4} \sin 2u'$$
$$\Phi'' = u'' + 2e'' \sin u'' + \frac{5e''^2}{4} \sin 2u''$$

Consequently

$$\varphi = u' - u'' - L + 2(e' \sin u' - e'' \sin u'') + \tfrac{5}{4}(e'^2 \sin 2u' - e''^2 \sin 2u'')$$

where L is equal to $H + K$.

Since we neglect the above quantities of the second order in the terms multiplied by $\sin^2(I/2)$, one can immediately put a', a'', in place of ρ', ρ'' and u', u'' in place of Φ', Φ''

and $u' - u'' - L$ in place of φ. By developing the products of the cosines, it is easily seen that there will be only one term independent of the angles u' and u'', that is, the following expression

$$-\tfrac{1}{2}a'a''[a',a'']_1(\sin\tfrac{1}{2}I_{,}'')^2$$

Let us now consider the functions (ρ',ρ''), $(\rho',\rho'')_1$, etc. After carrying out the preceding substitutions in place of ρ' and ρ'' and by keeping the second-order terms of e' and of e'', one will have

$$\begin{aligned}(\rho',\rho'') = (a',a'') &+ \frac{\mathrm{d}(a',a'')}{\mathrm{d}a'}\left(-a'e'\cos u' + \frac{a'e'^2}{2} - \frac{a'e'^2}{2}\cos 2u'\right)\\ &+ \frac{\mathrm{d}^2(a',a'')}{2\,\mathrm{d}a'^2}\frac{a'^2e'^2}{2}(1+\cos 2u')\\ &+ \frac{\mathrm{d}(a',a'')}{\mathrm{d}a''}\left(-a''e''\cos u'' + \frac{a''e''^2}{2} - \frac{a''e''^2}{2}\cos 2u''\right)\\ &+ \frac{\mathrm{d}^2(a',a''')}{2\,\mathrm{d}a''^2}\frac{a''^2e''^2}{2}(1+\cos 2u'')\\ &+ \frac{\mathrm{d}^2(a',a'')}{\mathrm{d}a'\,\mathrm{d}a''}\frac{a'a''e'e''}{2}[\cos(u'-u'') - \cos(u'+u'')]\end{aligned}$$

and it will be the same for the other similar functions. Similarly, one will have

$$\begin{aligned}\cos\varphi = \cos(u'-u''-L) &- \sin(u'-u''-L)\left\{\begin{matrix}2e'\sin u' + \dfrac{5e'^2}{4}\sin 2u'\\ -2e''\sin u'' - \dfrac{5e''^2}{4}\sin 2u''\end{matrix}\right\}\\ &- \cos(u'-u''-L)\left\{\begin{matrix}e'^2 - e'^2\cos 2u' + e''^2 - e''^2\cos 2u''\\ -2e'e''[\cos(u'-u'') - \cos(u'+u'')]\end{matrix}\right\}\end{aligned}$$

In the same manner, the cosines of the angles which are multiples of φ will be developed. The expressions for the cosine functions will be multiplied by the coefficients of (ρ',ρ''), $(\rho',\rho'')_1$, etc.

Since we are only examining the terms independent of any period, all the terms which are multiplied by the cosines of angles which are multiples of u' and u'' and which are the angles of the average motions of m' and m'' should be discarded.

Thus the term (ρ',ρ'') of the expression $1/\rho_{,}''$ will only produce the following expression

$$(a',a'') + \left(\frac{\mathrm{d}(a',a'')}{2\,\mathrm{d}a'}a' + \frac{\mathrm{d}^2(a',a'')}{4\,\mathrm{d}a'^2}a'^2\right)e'^2 + \left(\frac{\mathrm{d}(a',a'')}{2\,\mathrm{d}a''}a'' + \frac{\mathrm{d}^2(a',a'')}{4\,\mathrm{d}a''^2}a''^2\right)e''^2$$

The term $(\rho',\rho'')_1\cos\varphi$ will lead to the following expression

$$\left((a',a'')_1 + \frac{\mathrm{d}(a',a'')_1}{2\,\mathrm{d}a'}a' + \frac{\mathrm{d}(a',a'')_1}{2\,\mathrm{d}a''}a'' + \frac{\mathrm{d}^2(a',a'')_1}{4\,\mathrm{d}a'\,\mathrm{d}a''}a'a''\right)e'e''\cos L$$

The term $(\rho', \rho'')_2 \cos 2\varphi$, and what follows, will give only terms where e', e'', i', i'' will have more than two orders and we will neglect them.

It remains to develop the quantity (Articles 93 and 94)

$$\frac{\rho'^2 + \rho''^2 - \rho_{\prime}''^2}{2\rho''^2} = \frac{\rho' \cos \varphi}{\rho''^2} + \frac{\rho' \Delta (\sin \frac{1}{2} I_{\prime}'')^2}{\rho''^2}$$

The same substitutions will be made here for ρ' and ρ'' as were made above. There will immediately result

$$\begin{aligned}\frac{\rho'}{\rho''^2} &= \frac{a'}{a''^2}(1 - e' \cos u') + 2e'' \cos u'' + \frac{e'^2}{2}(1 - \cos 2u') \\ &+ \frac{e''^2}{2}(1 + 5 \cos 2u'') + e'e''[\cos(u' - u'') + \cos(u' - u'')]\end{aligned}$$

but in the term which contains $\sin(I_{\prime}''/2)$ and which is already of the second order, it will suffice to replace ρ'/ρ''^2 with a'/a''^2 and since

$$\Delta = \cos(\Phi + \Phi'' - H + K) - \cos(\Phi' - \Phi'' - H - K)$$

it is clear that this term will not give a constant quantity.

By multiplying the preceding expression for ρ'/ρ''^2 by $\cos \varphi$ of the preceding article and keeping only the constant terms where e' and e'' do not exceed the second-order

$$\frac{a'}{a''}\left(-\frac{e'e''}{2} - e'e'' + \frac{e'e''}{2} + e'e''\right) \cos L = 0$$

so that in the quantity in question the constant terms cancel one another.

97. The sum of all the terms which we just found after multiplication by m'' will be the constant part of the function Ω', resulting from the action of the planet m''. A similar expression will be obtained for the part resulting from the action of the planet m''' by referring these quantities relative to the planet m''.

Earlier we designated by (Ω) the non-periodic part of the function Ω. Thus if the following definitions are made in order to shorten the expressions

$$((a', a'')) = \frac{\mathrm{d}(a', a'')}{2\, \mathrm{d}a'} a' + \frac{\mathrm{d}^2(a', a'')}{4\, \mathrm{d}a'^2} a'^2$$

and consequently, since a' and a'' are present in the same manner in the function (a', a'')

$$((a'', a')) = \frac{\mathrm{d}(a', a'')}{2\, \mathrm{d}a''} a'' + \frac{\mathrm{d}^2(a', a'')}{4\, \mathrm{d}a''^2} a''^2$$

Moreover

$$[(a',a'')] = (a',a'')_1 + \frac{d(a',a'')_1}{2\,da'}a' + \frac{d(a',a'')_1}{2\,da''}a'' + \frac{d^2(a',a'')_1}{4\,da'\,da''}a'a''$$

one will have

$$-(\Omega') = m'' \left\{ \begin{array}{l} (a',a'') + ((a',a''))e'^2 + ((a'',a'))e''^2 \\ +[(a',a'')]e'e''\cos L - \frac{1}{2}a'a''[a',a'']_1(\sin\frac{1}{2}I_{\prime}'')^2 \end{array} \right\} + \cdots$$

This expression is exact to the quantities of third order, by viewing the eccentricities e' and e'' of the orbits of m' and m'' and their mutual inclination I as very small quantities of the first order, whatever the inclinations of these orbits on the fixed plane of projection.

98. The expressions for the functions $((a',a''))$ and $[(a',a'')]$ can be greatly simplified using the known properties of the coefficients of the series in $\cos\varphi$, $\cos 2\varphi$, etc. Indeed, if one differentiates logarithmically with respect to φ and then with respect to a' the identity results

$$\begin{aligned}(a'^2 - 2a'a''\cos\varphi + a''^2)^{-1/2} &= (a',a'') + (a',a'')_1\cos\varphi \\ &\quad + (a',a'')_2\cos 2\varphi + \cdots\end{aligned}$$

and if after having cross-multiplied, one compares the terms multiplied by the same cosines, the following equation results immediately

$$3a'a''(a',a'')_2 = -2a'a''(a',a'') + 2(a'^2 + a''^2)(a',a'')_1$$

then the differentials relative to a' and a'' will give

$$\begin{aligned}
\frac{d(a',a'')}{da'} &= \frac{a'(a',a'') - a''(a',a'')_1}{(a''^2 - a'^2)} \\
\frac{d(a',a'')_1}{da'} &= \frac{a'a''(a',a'') - a''^2(a',a'')_1}{a'(a''^2 - a'^2)} \\
\frac{d^2(a',a'')}{da'^2} &= \frac{4a'^3(a',a'') + a''(a''^2 - 3a'^2)(a',a'')_1}{a'(a''^2 - a'^2)^2} \\
\frac{d^2(a',a'')}{da'\,da''} &= \frac{-2(a'^2 + a''^2) + 2a'a''(a',a'')_1}{(a''^2 - a'^2)^2}
\end{aligned}$$

After substituting these expressions, there results

$$\begin{aligned}
((a',a'')) &= \frac{4a'^2a''^2(a',a'') - a'a''(a'^2 + a''^2)(a',a'')_1}{8(a''^2 - a'^2)^2} \\
[(a',a'')] &= \frac{-a'a''(a'^2 + a'')(a',a'') + (a'^2 + a''^2 - a'a'')(a',a'')_1}{2(a''^2 - a'^2)^2}
\end{aligned}$$

But simpler expressions can be obtained by using the coefficients of the series

$$(a'^2 - 2a'a''\cos\varphi + a''^2)^{-3/2} = [a', a''] + [a', a'']_1 \cos\varphi + \cdots$$

since after differentiating logarithmically and then cross-multiplying, it is immediately found, as above, that

$$a'a''[a', a'']_2 = 2(a'^2 + a''^2)[a', a'']_1 - 6a'a''[a', a'']$$

After substituting this expression for $[a', a'']_2$ and then comparing the series multiplied by $(a'^2 - 2a'a''\cos\varphi + a''^2)$ with the series $(a', a'') + (a', a'')_1 \cos\varphi +$ etc. with which it must be identical, it is easy to deduce the following relations

$$(a', a'') = (a'^2 + a''^2)[a', a''] - a'a''[a', a'']_1$$
$$(a', a'')_1 = 4a'a''[a', a''] - (a'^2 + a''^2)[a', a'']_1$$

and with the substitution of these expressions, one will obtain the following equation

$$((a', a'')) = \tfrac{1}{8}a'a''[a', a'']_1 = ((a'', a'))$$
$$[(a', a'')] = \tfrac{3}{2}a'a''[a', a''] - \tfrac{1}{2}(a'^2 + a''^2)[a', a'']_1 = -\tfrac{1}{4}a'a''[a', a'']_2$$

which will be substituted in the expression for (Ω') in the preceding article.

With respect to the expressions for the coefficients (a', a''), $(a', a'')_1$, etc., $[a', a'']$, $[a', a'']_1$, etc. as a function of a', a'', they can be found by the development of the radicals in powers of $\cos\varphi$ and by the development of these powers in cosines of multiple angles of φ, as Euler was the first to do in his research on Jupiter and Saturn. But I found long ago that they could be obtained in a simpler manner by decomposing the binomial $a'^2 - 2a'a''\cos\varphi + a''^2$ into its two imaginary factors

$$(a' - a''e^{\varphi\sqrt{-1}})(a' - a''e^{-\varphi\sqrt{-1}})$$

and by using the binomial formula to develop the powers $-1/2$, $-3/2$ of each of the factors.

Let there be defined in order to shorten the expressions

$$n' = \frac{n(n+1)}{2}, \qquad n'' = \frac{n(n+1)(n+2)}{2.3}, \qquad \ldots$$

One will have in general

$$(a' - a''e^{\varphi\sqrt{-1}})^{-n} = a'^{-n} + n(a')^{-n-1}a''e^{\varphi\sqrt{-1}} + n(a')^{-n-2}a''^2e^{2\varphi\sqrt{-1}}$$

and if the two series which correspond to $\sqrt{-1}$ and to $-\sqrt{-1}$ are multiplied together and if the imaginary exponentials are replaced by cosines of multiple angles, one will have

$$(a'^2 - 2a'a'' \cos\varphi + a''^2)^{-n} = A + B\cos\varphi + C\cos 2\varphi + \cdots$$

by defining

$$A = \frac{1}{a'^{2n}}[1 + n^2\left(\frac{a''}{a'}\right)^2 + n'^2\left(\frac{a''}{a'}\right)^4 + n''^2\left(\frac{a''}{a'}\right)^6 + \cdots]$$
$$B = \frac{2}{a'^{2n}}[n\left(\frac{a''}{a'}\right) + nn'\left(\frac{a''}{a'}\right)^3 + n'n''\left(\frac{a''}{a'}\right)^5 + \cdots]$$
$$C = \frac{3}{a'^{2n}}[n'\left(\frac{a''}{a'}\right)^2 + nn''\left(\frac{a''}{a'}\right)^4 + n'n'''\left(\frac{a''}{a'}\right)^6 + \cdots]$$
$$\vdots$$

These series are always convergent when $a' > a''$. But if $a'' > 1$, it will only be necessary to replace a' by a'' and a'' by a' since in the undeveloped function the quantities a' and a'' enter in the same fashion.

One consequence which results from the form of these series is that as long as n is a positive number all of the coefficients A, B, C, etc. are always positive.

If one puts $n = 1/2$, these coefficients will become (a', a''), $(a', a'')_1$, $(a', a'')_2$, etc. and if one puts $n = 3/2$ they will become $[a', a'']$, $[a', a'']_1$, $[a', a'']_2$, etc.

99. It remains to determine the angle L. Since we neglected the quantities of the third order and since in the expression for A, $\cos L$ is already multiplied by the quantity $e'e''$, the very small quantities of the first order can be neglected in the determination of L. Consequently, it can be assumed that $I_{,}''$ is equal to zero. But $L = H + K$ (Article 96) and by equating $I_{,}''$ to zero in the formulas of Article 94, one has

$$A = \cos(H + K), \qquad A_1 = -B = \sin(H + K), \qquad A_2 = 0$$

One also has by the formulas of this article

$$A = \alpha'\alpha'' + \alpha_1'\alpha_1'' + \alpha_2'\alpha_2'' = \cos L$$

Let us differentiate the expression for $\cos L$, varying the quantities α', α'', α_1', etc. and substituting for their differentials the expressions given in Article 67, accenting the respective quantities and replacing the quantities A_1, A_2, etc., by the expressions for α', α'', β', etc. One will easily find

$$-\sin L\, dL = A_1\, d\chi' + A_2\, d\pi' + B\, d\chi'' + C\, d\pi''$$

but

$$A_1 = \sin L, \qquad A_2 = 0, \qquad B = -\sin L, \qquad C = 0$$

thus by dividing by $\sin L$, one will obtain

$$\mathrm{d}L = \mathrm{d}\chi'' - \mathrm{d}\chi'$$

and after integrating

$$L = \chi'' - \chi'$$

where it is not necessary to add the constant of integration because the origin of the angles χ' and χ'' is arbitrary. The angle χ[29] is in general described by the orbit rotating in its plane and that which we have substituted for the longitude k of the perihelion (Article 68).

100. The function (Ω') is now reduced to the simpler and proper form for the calculation of **secular** variations. It will only be necessary to substitute it in the formulas of Article 71 by marking the letters of these formulas with one prime in order to refer them to the planet m' for which the variations are sought. By simply interchanging the letters marked with one and two primes, similar formulas for the variations of the planet m'' will be obtained and similarly for the others.

It is clear that this function is composed of two distinct parts of which only one contains the eccentricities and the locations of the aphelion in the orbits and the other contains only the inclinations of the orbits on a fixed plane with the locations of their nodes. If the first is designated by $(\Omega')_1$ and the second by $(\Omega')_2$, so that one has

$$(\Omega') = (\Omega')_1 + (\Omega')_2$$

the following equations result

$$\begin{aligned}(\Omega')_1 = {} & \tfrac{1}{8}m'' \left\{ \begin{array}{l} 8(a', a'') + a'a''[a', a'']_1(e'^2 + e''^2) \\ -2a'a''[a', a'']_2 e'e'' \cos(\chi' - \chi'') \end{array} \right\} \\ & + \tfrac{1}{8}m''' \left\{ \begin{array}{l} 8(a', a''') + a'a'''[a', a''']_1(e'^2 + e'''^2) \\ -2a'a'''[a', a''']_2 e'e''' \cos(\chi' - \chi''') \end{array} \right\} \\ & \ldots \end{aligned}$$

$$\begin{aligned}(\Omega')_2 = {} & -\tfrac{1}{4}m''a'a''[a', a'']_1(1 - \cos I_{\prime}'') \\ & -\tfrac{1}{4}m'''a'a'''[a', a''']_1(1 - \cos I_{\prime}''') \\ & \ldots \end{aligned}$$

where

$$\cos I_{\prime}'' = \cos i' \cos i'' + \cos(h' - h'') \sin i' \sin i''$$
$$\cos I_{\prime}''' = \cos i' \cos i''' + \cos(h' - h''') \sin i' \sin i'''$$

The angles $I_{\prime}''$, $I_{\prime}'''$, etc. are the inclinations of the orbit of the planet m' to the orbits of the planets m'', m''', etc.

Thus the expression $(\Omega')_1 + (\Omega')_2$ will be substituted for (Ω') in the equations of the **secular** variations (Article 76) and one will mark the letters with primes to refer them to the planet m' for which the variations are sought. By neglecting e'^2 and simply replacing b by a' in the coefficients of the functions $(\Omega)_1$ and $(\Omega)_2$, which are of the second order, the following equations result

$$\frac{de'}{dt} = -\frac{1}{\sqrt{g'a'}}\frac{d(\Omega')_1}{e'\,d\chi'}, \qquad \frac{d\chi'}{dt} = \frac{1}{\sqrt{g'a'}}\frac{d(\Omega')_1}{e'\,de'},$$
$$\frac{di'}{dt} = -\frac{1}{\sqrt{g'a'}}\frac{d(\Omega')_2}{\sin i' dh'}, \qquad \frac{dh'}{dt} = \frac{1}{\sqrt{g'a'}}\frac{d(\Omega')_2}{\sin i' di'},$$

Similar equations will be obtained for the variations of the elements of the planet m'' in its orbit about m. It is only necessary to mark with two primes the letters which are marked with one prime and on the contrary, to mark with one prime those which are marked with two in order to obtain these equations.

Thus if it is noted that the functions a' and a'' which are represented by parentheses, do not change if the quantities a', a'' are interchanged one will have

$$(\Omega')_1 = \tfrac{1}{8}m' \left\{ \begin{array}{l} 8(a', a'') + a'a''[a', a'']_1(e'^2 + e''^2) \\ -2[(a', a'')]_2\, e'e'' \cos(\chi' - \chi'') \end{array} \right\}$$
$$+ \tfrac{1}{8}m''' \left\{ \begin{array}{l} 8(a'', a''') + a''a'''[a'', a''']_1(e''^2 + e'''^2) \\ -2[(a'', a''')]_2\, e''e''' \cos(\chi'' - \chi''') \end{array} \right\}$$

....

$$(\Omega'')_2 = -\tfrac{1}{2}m'a'a''[a', a'']_1(1 - \cos I_{\prime\prime}')$$
$$-\tfrac{1}{4}m'''a''a'''[a'', a'''](1 - \cos I_{\prime\prime}''')$$

....

where

$$\cos I_{\prime\prime}' = \cos I_{\prime}'', \qquad \cos I_{\prime\prime}''' = \cos i'' \cos i''' + \sin(h'' - h''') \sin i'' \sin i'''$$

and the equations for the variations will be

$$\frac{de''}{dt} = -\frac{1}{\sqrt{g''a''}}\frac{d(\Omega'')_1}{e''\,d\chi''}, \qquad \frac{d\chi''}{dt} = \frac{1}{\sqrt{g''a''}}\frac{d(\Omega'')_1}{e''\,de''},$$

$$\frac{di''}{dt} = -\frac{1}{\sqrt{g''a''}}\frac{d(\Omega'')_2}{\sin i''dh''}, \qquad \frac{dh''}{dt} = \frac{1}{\sqrt{g''a''}}\frac{d(\Omega'')_2}{\sin i''di''},$$

and similarly for the variations of the elements of the orbits m''', m^{IV}, etc. about m.

101. But it should be noted that the different functions $(\Omega')_1$, $(\Omega'')_1$, etc., as well as the functions $(\Omega')_2$, $(\Omega'')_2$, etc. can be reduced to only one function which will introduce simplicity and uniformity in the formulas of variation. Indeed, if the following equation is defined

$$\begin{aligned}\Phi = {} & \frac{m'm''a'a''}{8}[8(a',a'') + [a',a'']_1(e'^2 + e''^2) - 2[a',a'']_2 e'e'' \cos(\chi' - \chi'')] \\ & + \frac{m'm'''a'a'''}{8}[8(a',a''') + [a',a''']_1(e'^2 + e'''^2) - 2[a',a''']_2 e'e''' \cos(\chi' - \chi''')] \\ & + \frac{m''m'''a''a'''}{8}[8(a'',a''') + [a'',a''']_1(e''^2 + e'''^2) - 2[a'',a''']_2 e''e''' \cos(\chi'' - \chi''')] \\ & + \cdots\end{aligned}$$

by taking all combinations of the masses m', m'', m''', etc. and of the functions which are related, it is easy to see that in the partial differences of $(\Omega')_1$, $(\Omega'')_1$, etc., one will be able to change these expressions to functions of Φ if the partial differences relative to e' and χ' are divided by m', the partial differences of e'', χ'' by m'' and so on.

Thus the equations for the variations of the eccentricities and aphelions will become

$$\frac{de'}{dt} = -\frac{1}{m'\sqrt{g'a'}}\frac{d\Phi}{e'\,d\chi'}, \qquad \frac{d\chi'}{dt} = \frac{1}{m'\sqrt{g'a'}}\frac{d\Phi}{e'\,de'},$$

$$\frac{de''}{dt} = -\frac{1}{m''\sqrt{g''a''}}\frac{d\Phi}{e''\,d\chi''}, \qquad \frac{d\chi''}{dt} = \frac{1}{m''\sqrt{g''a''}}\frac{d\Phi}{e''\,de''},$$

$$\cdots$$

These equations give

$$\frac{d\Phi}{d\chi'}d\chi' + \frac{d\Phi}{de'}de' = 0, \qquad \frac{d\Phi}{d\chi''}d\chi'' + \frac{d\Phi}{de''}de'' = 0, \qquad \cdots$$

Therefore, since Φ is a function of the variables e', χ', e'', χ'', etc. without t, one will have $d\Phi$ equal to zero and consequently, Φ equal to a constant. This is a general relation between the eccentricities and the locations of the aphelions of planets which must always exist, whatever variations the eccentricities and the locations of the aphelions have ultimately, as long as they are very small.

102. But the nature of the function Φ still leads to another general relation between these same elements. Indeed, it is easy to see that there is the equation

$$\frac{\mathrm{d}\Phi}{\mathrm{d}\chi'} + \frac{\mathrm{d}\Phi}{\mathrm{d}\chi''} + \frac{\mathrm{d}\Phi}{\mathrm{d}\chi'''} + \cdots = 0$$

and if these partial differences are substituted with their values $m'\sqrt{g'a'}e'(\mathrm{d}e'/\mathrm{d}t)$, $m''\sqrt{g''a''}e''(\mathrm{d}e''/\mathrm{d}t)$, etc. given by the equations of the preceding article, one will have the following finite equation after integrating with respect to t

$$m'\sqrt{g'a'}e'^2 + m''\sqrt{g''a''}e''^2 + m'''\sqrt{g'''a'''}e'''^2 + \cdots = K^2$$

where K^2 is a constant equal to the value of the first member of this equation at an arbitrary instant.

This equation shows that the eccentricities e', e'', e''', etc. have limits which they can not exceed. Because they are necessarily real, if the orbits are conic sections, each term such as $m'\sqrt{g'a'}\,e'^2$ will always be positive and its *maximum* will be the constant K^2.

From this result, it follows that if the eccentricities of the orbits which belong to very large masses are once very small they will remain so, which is the case for Jupiter and Saturn. But those which belong to very small masses could increase to unity and above and their veritable limits could only be determined by integration of the differential equations, as will be shown below.[30]

Moreover, because the quantity Φ, viewed as a function of e', e'', e''', etc. is a homogeneous function of second degree, one will have from the known property of these functions

$$\frac{\mathrm{d}\Phi}{\mathrm{d}e'}e' + \frac{\mathrm{d}\Phi}{\mathrm{d}e''}e'' + \frac{\mathrm{d}\Phi}{\mathrm{d}e'''}e''' + \cdots = 2\Phi$$

After substituting in this equation the expressions for the partial differences of Φ relative to e', e'', e''', etc., obtained from the same equations of the preceding article, one will have

$$m'\sqrt{g'a'}\frac{e'^2\,\mathrm{d}\chi'}{\mathrm{d}t} + m''\sqrt{g''a''}\frac{e''^2\,\mathrm{d}\chi''}{\mathrm{d}t} + m'''\sqrt{g'''a''}\frac{e'''^2\,\mathrm{d}\chi'''}{\mathrm{d}t} + \cdots = 2F$$

where F is the value of Φ at an arbitrary instant.

In this equation, the quantities $\mathrm{d}\chi'/\mathrm{d}t$, $\mathrm{d}\chi''/\mathrm{d}t$, etc. express the angular velocities of the motions of the aphelions. Consequently, this equation gives an invariable relation between these velocities from which it is seen that they are also necessarily limits if they are all of the same sign.

103. If the transformations of Article 73 are used, such as

$$\mathbf{m}' = e'\sin\chi', \qquad \mathbf{n}' = e'\cos\chi', \qquad \mathbf{m}'' = e''\sin\chi'', \qquad \mathbf{n}'' = e''\cos\chi'', \qquad \cdots$$

one will have

$$\Phi = \frac{m'm''a'a''}{8}([a',a'']_1(\mathbf{m}'^2 + \mathbf{n}'^2 + \mathbf{m}''^2 + \mathbf{n}''^2) - 2[a',a'']_2(\mathbf{m}'\mathbf{m}'' + \mathbf{n}'\mathbf{n}''))$$
$$+ \frac{m'm'''a'a'''}{8}([a',a''']_1(\mathbf{m}'^2 + \mathbf{n}'^2 + \mathbf{m}'''^2 + \mathbf{n}'''^2) - 2[a',a''']_2(\mathbf{m}'\mathbf{m}''' + \mathbf{n}'\mathbf{n}'''))$$
$$+ \frac{m''m'''a''a'''}{8}([a'',a''']_1(\mathbf{m}''^2 + \mathbf{n}''^2 + \mathbf{m}'''^2 + \mathbf{n}'''^2) - 2[a'',a''']_2(\mathbf{m}''\mathbf{m}''' + \mathbf{n}''\mathbf{n}'''))$$
$$+ \cdots\cdots\cdots\cdots$$

and the equations of the variations will be

$$m'\frac{\mathbf{dm}'}{\mathrm{d}t} = \frac{1}{\sqrt{g'a'}}\frac{\mathrm{d}\Phi}{\mathbf{dn}'}, \qquad m'\frac{\mathbf{dn}'}{\mathrm{d}t} = -\frac{1}{\sqrt{g'a'}}\frac{\mathrm{d}\Phi}{\mathbf{dm}'}$$
$$m''\frac{\mathbf{dm}''}{\mathrm{d}t} = \frac{1}{\sqrt{g''a''}}\frac{\mathrm{d}\Phi}{\mathbf{dn}''}, \qquad m''\frac{\mathbf{dn}''}{\mathrm{d}t} = -\frac{1}{\sqrt{g''a''}}\frac{\mathrm{d}\Phi}{\mathbf{dm}''}$$
$$\cdots$$

If the expression for Φ were substituted in these equations and if the partial differences were obtained, one will have linear equations in $\mathbf{m}'$, $\mathbf{n}'$, $\mathbf{m}''$, $\mathbf{n}''$, etc. which are easy to integrate. These equations will be entirely identical to those I have found by other means in the *Mémoires de Berlin* of 1781, page 262, as it is easy to check by comparing the different denominations of the same quantities.[31]

In the *Mémoires* of 1782, I applied these equations to six major planets using for their masses the most likely values and I obtained by integration general formulas for the variations of their eccentricities and of the locations of their aphelions, which give the values of these elements for the Earth as well as for the other planets after an arbitrary time, either before or after the year of 1700. Since with these formulas the eccentricities always remain very small, as was assumed in the calculations, one is sure of their exactitude for all times, past and future. It is found in the volumes for the same *Mémoires* of 1786–1787, printed in 1792, a supplement to this theory, relative to the new planet of Herschel, where the **secular** variations of the eccentricity and of the location of the aphelion of this planet, produced by the actions of Jupiter and Saturn, are determined in the same manner and also by general formulas. Only the effect of the action of Herschel's planet on these two planets, as well as on the other more distant planets has been neglected because of its small mass and its distant location.

104. The equations of the variations of nodes and inclinations can similarly be reduced to a simpler form. Let

$$\Psi = -\tfrac{1}{4}m'm''a'a''[a',a'']_1(1 - \cos I_{\prime}'')$$
$$- \tfrac{1}{4}m'm'''a'a'''[a',a''']_1(1 - \cos I_{\prime}''')$$
$$- \tfrac{1}{4}m''m'''a''a'''[a'',a''']_1(1 - \cos I_{\prime\prime}''')$$
$$- \cdots\cdots\cdots\cdots$$

by performing as such all the combinations two by two of the masses m', m'', m''', etc., which are assumed acting one on another with the corresponding functions of a', a'', a''', etc., and of the inclinations $I_{\prime}''$, $I_{\prime}'''$, $I_{\prime\prime}'''$, etc. which are determined in general by the formula

$$\cos I_m^n = \cos i^{(m)} \cos i^{(n)} + \cos(h^{(m)} - h^{(n)}) \sin i^{(m)} \sin i^{(n)}$$

One will have after substituting $\mathrm{d}\Psi/m'\,\mathrm{d}i'$, $\mathrm{d}\Psi/m'\,\mathrm{d}h'$, etc. in place of $\mathrm{d}(\Omega)_2/\mathrm{d}i'$, $\mathrm{d}(\Omega)_2/\mathrm{d}h'$ etc., the following equations

$$\frac{\mathrm{d}i'}{\mathrm{d}t} = -\frac{1}{m'\sqrt{g'a'}}\frac{\mathrm{d}\Psi}{\sin i'\,\mathrm{d}h'}, \qquad \frac{\mathrm{d}h'}{\mathrm{d}t} = \frac{1}{m'\sqrt{g'a'}}\frac{\mathrm{d}\Psi}{\sin i'\,\mathrm{d}i'}$$

$$\frac{\mathrm{d}i''}{\mathrm{d}t} = -\frac{1}{m''\sqrt{g''a''}}\frac{\mathrm{d}\Psi}{\sin i''\,\mathrm{d}h''}, \qquad \frac{\mathrm{d}h''}{\mathrm{d}t} = \frac{1}{m''\sqrt{g''a''}}\frac{\mathrm{d}\Psi}{\sin i''\,\mathrm{d}i''}$$

These equations also give

$$\frac{\mathrm{d}\Psi}{\mathrm{d}h'}\mathrm{d}h' + \frac{\mathrm{d}\Psi}{\mathrm{d}i'}\mathrm{d}i' = 0, \qquad \frac{\mathrm{d}\Psi}{\mathrm{d}h''}\mathrm{d}h'' + \frac{\mathrm{d}\Psi}{\mathrm{d}i''}\mathrm{d}i'' = 0$$

Consequently, since Ψ is a function of h', i', h'', i'', etc. without any other variables, $\mathrm{d}\Psi = 0$ and Ψ is equal to a constant.

Moreover, it is obvious from the form of the function Ψ that we have the following equation

$$\frac{\mathrm{d}\Psi}{\mathrm{d}h'} + \frac{\mathrm{d}\Psi}{\mathrm{d}h''} + \frac{\mathrm{d}\Psi}{\mathrm{d}h'''} + \cdots = 0$$

After substituting for the differentials of Ψ relative to h' h'', h''', etc. the expressions given by the preceding equations, one will obtain a differential equation in i', i'', i''', etc. for which the integral will be

$$m'\sqrt{g'a'}\cos i' + m''\sqrt{g''a''}\cos i'' + m'''\sqrt{g'''a'''}\cos i''' + \cdots = \text{const.}$$

and which can be put in the form

$$m'\sqrt{g'a'}\left(\sin\frac{i'}{2}\right)^2 + m''\sqrt{g''a''}\left(\sin\frac{i''}{2}\right)^2 + \cdots = H^2$$

where H^2 is the value of the first member at an arbitrary instant. One can obtain from this equation, relative to the limits of the quantities $\sin(i'/2)$, $\sin(i''/2)$, etc., analogous consequences to those we have deduced from a similar equation in e', e'', etc. in Article 101.

105. In the case where only the action of two planets m' and m'' are considered, the expression for Ψ is reduced to only one term multiplied by $m'm''$ and the mutual inclination $I_{\prime}''$ of the two orbits becomes a constant. This is, at least, the case with Jupiter and Saturn.

In this case, we will also remark that if one assumed that the plane of the perturbing planet m'' coincided at an instant with the fixed plane, one will have i'' equal to zero and consequently $\cos I_{\prime}'' = \cos i'$ which will give

$$\Psi = -\tfrac{1}{4}m''a'a''[a', a'']_1(1 - \cos i')$$

and from this result

$$\frac{\mathrm{d}h'}{\mathrm{d}t} = -\frac{m''a'a''[a', a'']_1}{4\sqrt{g'a'}}$$

This is the expression for the velocity of the retrograde motion of the node of the orbit of m' on the plane of the orbit of m'', as long as the mutual inclination of the planes remains constant. From this result, it is clear that the action of the planet m'' on the planet m' to change the orientation of its orbit is reduced to giving to the node of its orbit on the orbit of the perturbing planet m'' an instantaneous retrograde motion expressed by

$$-\frac{m''a'a''[a', a'']_1}{4\sqrt{g'a'}}\mathrm{d}t$$

without affecting the mutual inclination of the two orbits.

In the same manner, the action of the planet m' on the planet m'' to change the orientation of its orbit will retrograde the node of this planet on the plane of the orbit of m' with an instantaneous motion of

$$-\frac{m'a'a''[a', a'']_1}{4\sqrt{g''a''}}\mathrm{d}t$$

and similarly for the other planets.

By considering the planets two at a time, one can find the variations of their nodes and their mutual inclinations because from the nature of the differential calculus the sum of the particular values of a differential compose the total value. This is how the annual changes of the nodes and of the inclinations of planets, produced by their mutual attractions were found, before a direct and general theory of **secular** variations was available.

106. In order to give an example of this method, let us consider three planets m', m'', m''' for which the orbits intersect with inclinations $I_{\prime}''$, $I_{\prime}'''$ which are the inclinations of the second and third planets with respect to the first, and $I_{\prime\prime}'''$ which is the inclination of the second planet to the third. It is easy to see that they will form on a spheroid a spherical triangle for which the three angles, assuming that the inclinations of m'' and m''' are on

the same side, will be $I_{,}^{\prime\prime}$, $180^\circ - I_{,}^{\prime\prime\prime}$ and $I_{,,}^{\prime\prime\prime}$, which we will denote for more simplicity by α, β, γ.

The planet m' will be retrograde to the node of the planet m'' on its orbit by the single quantity

$$\frac{m'a'a''[a', a'']_1}{4\sqrt{g''a''}} dt$$

and the same planet will be retrograde at the same time in its orbit to the node of the planet m''', by the elementary quantity

$$\frac{m'a'a'''[a', a''']_1}{4\sqrt{g'''a'''}} dt$$

while the inclinations $I_{,}^{\prime\prime}$ and $I_{,}^{\prime\prime\prime}$ remain constant.

Thus, in the triangle formed by the intersection of the three orbits, the portion of the orbit of m' intercepted between the orbits of m'' and m''', that is, the side adjacent to the angles α and β will increase by the quantity $A\,dt$. In order to shorten the expressions, let us define

$$A = \frac{m'}{4}\left(\frac{a'a''[a', a'']_1}{\sqrt{g''a''}} - \frac{a'a'''[a', a''']_1}{\sqrt{g'''a'''}}\right)$$

where the angles α and β remain constant.

But, in a spherical triangle for which the angles are α, β, γ and for which the adjacent side to α and β, and consequently, opposite to γ, is c, one has

$$\cos\gamma = \sin\alpha \sin\beta \cos c - \cos\alpha \cos\beta$$

Therefore, by varying c of $A\,dt$, one will have

$$\mathrm{d}(\cos\gamma) = -\sin\alpha \sin\beta \sin c A\,\mathrm{d}t$$

But the same equation gives

$$\cos c = \frac{\cos\gamma + \cos\alpha\cos\beta}{\sin\alpha\sin\beta}$$

from which the following equation is obtained

$$\sin c = \frac{\sqrt{\sin^2\alpha\sin^2\beta - (\cos\gamma + \cos\alpha\cos\beta)^2}}{\sin\alpha\sin\beta}$$
$$= \frac{\sqrt{1 - \cos^2\alpha - \cos^2\beta - \cos^2\gamma - 2\cos\alpha\cos\beta\cos\gamma}}{\sin\alpha\sin\beta}$$

In order to shorten the expressions, let us define

$$u = \sqrt{1 - \cos^2\alpha - \cos^2\beta - \cos^2\gamma - 2\cos\alpha\cos\beta\cos\gamma} = \sin\alpha\sin\beta\sin c$$

one will have

$$\mathrm{d}(\cos\gamma) = -Au\,\mathrm{d}t$$

In the same manner, it will be found by considering the retrogradation of the orbits of m' and m''' with respect to the orbit of m'', which increases the side adjacent to the angles α, γ and consequently, opposite to the angle β by the elementary quantity $B\,dt$ after making the following definition

$$B = \frac{m''}{4}\left(\frac{a'a''[a',a'']_1}{\sqrt{g'a'}} - \frac{a''a'''[a'',a''']_1}{\sqrt{g'''a'''}}\right)$$

while the angles α and γ remain constant

$$\mathrm{d}(\cos\beta) = -Bu\,\mathrm{d}t$$

because the quantity u is a symmetrical function of three cosines.

Finally, the retrogradation of the orbits of m' and m'' with respect to the orbit of m''' also gives

$$\mathrm{d}(\cos\alpha) = -Cu\,\mathrm{d}t$$

after defining

$$C = \frac{m'''}{4}\left(\frac{a'a'''[a',a''']_1}{\sqrt{g'a'}} - \frac{a''a'''[a'',a''']_1}{\sqrt{g''a''}}\right)$$

In these equations, the three coefficients A, B, C are constants. Therefore, the only variable is the quantity u, which is a function of the three cosines of α, β, γ, that is, the respective inclinations of the orbits $I_{\prime}''$, $I_{\prime}'''$, $I_{\prime\prime}'''$. Thus their expressions can be determined as a function of t.

If these three equations are added together, after having multiplied the first by $m''m'''a''a'''$ $[a'',a''']_1$, the second by $-m'm'''a'a'''[a',a''']$, the third by $m'm''a'a''[a',a'']$, one will have

$$\begin{aligned}
&m''m'''a''a'''[a'',a''']_1\,\mathrm{d}(\cos\gamma) - m'm'''a'a'''[a',a''']\,\mathrm{d}(\cos\beta)\\
&+m'm''a'a''[a',a'']_1\,\mathrm{d}(\cos\alpha)\\
&= -\begin{pmatrix} m''m'''a''a'''[a'',a''']_1A - m'm'''a'a'''[a',a''']_1B \\ +m'm''a'a''[a',a'']_1C \end{pmatrix} u\,\mathrm{d}t
\end{aligned}$$

But, by substituting the expressions of A, B, C, it is clear that the second member is reduced to zero, by the mutual cancellation of all the terms, and since the first member is integrable, one has, after replacing α, β, γ by the expressions of $I_{,}''$, $180° - I_{,}''$, $I_{,,}'''$, the following equation

$$m''m'''a''a'''[a'', a''']_1 \cos I_{,,}''' + m'm'''a'a'''[a', a''']_1 \cos I_{,}'' \\ +m'm''a'a''[a', a'']_1 \cos I_{,}'' = \text{const.}$$

an equation which is in agreement, for the case of three orbits, with the integral Ψ equal to the constant found above (Article 104).

If the following definitions are made for greater simplicity

$$\cos\alpha = x, \qquad \cos\beta = y, \qquad \cos\gamma = z$$

one has the three equations

$$\mathrm{d}x = -Cu\,\mathrm{d}t, \qquad \mathrm{d}y = -Bu\,\mathrm{d}t, \qquad \mathrm{d}z = -Au\,\mathrm{d}t$$
$$\text{where} \quad u = \sqrt{1 - x^2 - y^2 - z^2 - 2xyz}$$

The first equation combined with the second and the third, gives after elimination of u

$$\mathrm{d}y = \frac{B}{C}\mathrm{d}x, \qquad \mathrm{d}z = \frac{A}{C}\mathrm{d}x$$

and from this result

$$y = \frac{Bx + a}{C}, \qquad z = \frac{Ax + b}{C}$$

After substituting these expressions in the equation for u, the variable x will increase to the third degree under the radical sign and the equation $\mathrm{d}x = -Cu\,\mathrm{d}t$ will give $\mathrm{d}t = -(\mathrm{d}x/Cu)$, an equation where the variables are separated. But the second member will only be integrable after the rectification of the conic section.

But since the mutual inclinations of the orbits must be assumed very small, the following definitions can be made

$$x = 1 - \xi, \qquad y = 1 - \eta, \qquad z = 1 - \zeta$$

which lead to the following equations

$$\xi = (\tfrac{1}{2}I_{,}'')^2, \qquad \eta = (\tfrac{1}{2}I_{,}''')^2, \qquad \zeta = (\tfrac{1}{2}I_{,,}''')^2$$

The quantities ξ, η, ζ will be very small and the third order term with respect to the second can be neglected in the expression for u. Therefore, there results

$$u^2 = 2(\zeta\eta + \xi\zeta + \eta\xi) - \xi^2 - \eta^2 - \zeta^2$$

and

$$d\xi = -Cu\,dt, \qquad d\eta = -Bu\,dt, \qquad d\zeta = -Au\,dt$$

If the equation for u^2 is differentiated and after the substitution of the expressions of $d\xi$, $d\eta$, $d\zeta$, one divides the equation by $u\,dt$ and then re-differentiates it and again the same substitutions are made, one will have since dt is constant

$$\frac{d^2u}{dt^2} = [2(-AC - AB - BC) - A^2 - B^2 - C^2]u$$

an equation integrable by exponentials or sines, depending on whether the coefficient of u is positive or negative. But since $u = \sin\alpha \sin\beta \sin c$, it is obvious that the expression for u as a function of t can not contain exponentials. Thus designating by $-\mu^2$ the coefficient of u in the preceding equation, one will have

$$u = K\cos(\mu t + k)$$

where K and k are two arbitrary constants which must be determined from the initial conditions. Since $\sin\alpha$ and $\sin\beta$ are by hypothesis very small quantities, K will also have a very small value.

From this result, one will have the values of ξ, η, ζ by an integration which will only contain t in the expression $\sin(\mu t + k)$ and which since it is again very small, it will necessarily remain small so that the solution will always be valid.

Thus the reciprocal inclinations of the orbits for an arbitrary instant will be known but their absolute location in space which depends on the angles h', h'', etc., i', i'', etc. will still not be known. This is why it is simpler to find these angles directly by integration of the formulas of Article 104.

107. But instead of using these equations in the presented form, it will be more advantageous to transform them by the substitutions of Article 73, after making the following definitions

$$\begin{array}{lll} p' = \sin i' \sin h', & p'' = \sin i'' \sin h'', & \cdots \\ q' = \sin i' \cos h', & q'' = \sin i'' \cos h'', & \cdots \end{array}$$

Thus one will have, after marking with primes the letters p and q to refer them to the planets m', m'', etc., respectively and after putting the function Ψ/m in place of (Ω) (Article 104)

$$\begin{array}{ll} \dfrac{dp'}{dt} = \dfrac{\sqrt{1 - p'^2 - q'^2}}{m'\sqrt{g'a'}}\dfrac{d\Psi}{dq'}, & \dfrac{dq'}{dt} = -\dfrac{\sqrt{1 - p'^2 - q'^2}}{m'\sqrt{g'a'}}\dfrac{d\Psi}{dp'} \\ \dfrac{dp''}{dt} = \dfrac{\sqrt{1 - p''^2 - q''^2}}{m''\sqrt{g''a''}}\dfrac{d\Psi}{dq''}, & \dfrac{dq''}{dt} = -\dfrac{\sqrt{1 - p''^2 - q''^2}}{m''\sqrt{g''a''}}\dfrac{d\Psi}{dp''} \\ \vdots & \end{array}$$

The function Ψ will be, as in the cited article

$$
\begin{aligned}
\Psi = & -\tfrac{1}{4}m'm''a'a''[a',a'']_1(1-\cos I_{\prime}'') \\
& -\tfrac{1}{4}m'm'''a'a'''[a',a''']_1(1-\cos I_{\prime}''') \\
& - \cdots\cdots\cdots\cdots
\end{aligned}
$$

but the expressions for $\cos I_{\prime}$, $\cos I_{\prime}'''$, $\cos I_{\prime\prime}'''$, etc. will become after the same substitutions are made

$$
\begin{aligned}
\cos I_{\prime} &= \sqrt{1-p'^2-q'^2}\sqrt{1-p''^2-q''^2}+p'p''+q'q'' \\
\cos I_{\prime}''' &= \sqrt{1-p'^2-q'^2}\sqrt{1-p'''^2-q'''^2}+p'p'''+q'q''' \\
\cos I_{\prime\prime}''' &= \sqrt{1-p''^2-q''^2}\sqrt{1-p'''^2-q'''^2}+p''p'''+q''q''' \\
& \cdots\cdots\cdots\cdots
\end{aligned}
$$

After these substitutions are made and the differentiations with respect to $p', q', p'', q'', \cdots$ are carried out

$$
\begin{aligned}
\frac{dp'}{dt} = & -\frac{m''a'a''[a',a'']_1}{4\sqrt{g'a'}}\left(q'\sqrt{1-p''^2-q''^2}-q''\sqrt{1-p'^2-q'^2}\right) \\
& -\frac{m'''a'a'''[a',a''']_1}{4\sqrt{g'a'}}\left(q'\sqrt{1-p'''^2-q'''^2}-q'''\sqrt{1-p'^2-q'^2}\right) \\
& - \cdots\cdots\cdots\cdots
\end{aligned}
$$

$$
\begin{aligned}
\frac{dq'}{dt} = & \frac{m''a'a''[a',a'']_1}{4\sqrt{g'a'}}\left(p'\sqrt{1-p''^2-q''^2}-p''\sqrt{1-p'^2-q'^2}\right) \\
& +\frac{m'''a'a'''[a',a''']_1}{4\sqrt{g'a'}}\left(p'\sqrt{1-p'''^2-q'''^2}-p'''\sqrt{1-p'^2-q'^2}\right) \\
& + \cdots\cdots\cdots\cdots
\end{aligned}
$$

$$
\begin{aligned}
\frac{dp''}{dt} = & -\frac{m'a'a''[a',a'']_1}{4\sqrt{g''a''}}\left(q''\sqrt{1-p'^2-q'^2}-q'\sqrt{1-p''^2-q''^2}\right) \\
& -\frac{m'''a'a'''[a',a''']_1}{4\sqrt{g''a''}}\left(q''\sqrt{1-p'''^2-q'''^2}-q'''\sqrt{1-p''^2-q''^2}\right) \\
& - \cdots\cdots\cdots\cdots
\end{aligned}
$$

$$
\begin{aligned}
\frac{dq''}{dt} = & \frac{m'a'a''[a',a'']_1}{4\sqrt{g''a''}}\left(p''\sqrt{1-p'^2-q'^2}-p'\sqrt{1-p''^2-q''^2}\right) \\
& +\frac{m'''a''a'''[a'',a''']_1}{4\sqrt{g''a''}}\left(p''\sqrt{1-p'''^2-q'''^2}-p'''\sqrt{1-p''^2-q''^2}\right)+\cdots \\
& \cdots
\end{aligned}
$$

108. These equations hold, whatever the values of the variables p', q', p'', q'', etc., because our formulas do not assume that the inclinations i', i'', etc. of the orbits on the fixed plane

are very small, as was true heretofore in all the formulas which were given for the motion of the nodes and the variations of the inclinations. The formulas only require that the mutual inclinations of the orbit are very small.

In general, their integration seems very difficult and perhaps it may only be for the case of two orbits where it can be carried out.

In this case, in order to shorten the expressions, let us make the following definitions

$$\frac{m''a'a''[a',a'']_1}{4\sqrt{g'a'}} = M, \qquad \frac{m'a'a''[a',a'']_1}{4\sqrt{g''a''}} = N$$
$$\sqrt{1-p'^2-q'^2} = x, \qquad \sqrt{1-p''^2-q''^2} = y$$

One will have these equations

$$\frac{\mathrm{d}p'}{\mathrm{d}t} = -M(q'y - q''x), \qquad \frac{\mathrm{d}q'}{\mathrm{d}t} = M(p'y - p''x)$$
$$\frac{\mathrm{d}p''}{\mathrm{d}t} = -N(q''x - q'y), \qquad \frac{\mathrm{d}q''}{\mathrm{d}t} = N(p''x - p'y)$$

They give immediately

$$N\frac{\mathrm{d}p'}{\mathrm{d}t} + M\frac{\mathrm{d}p''}{\mathrm{d}t} = 0, \qquad N\frac{\mathrm{d}q'}{\mathrm{d}t} + M\frac{\mathrm{d}q''}{\mathrm{d}t} = 0$$

from which one has

$$Np' + Mp'' = b, \qquad Nq' + Mq'' = c$$

where b and c are constants.

Now, if the equation $x^2 = 1 - p'^2 - q'^2$ is differentiated and the expressions for $\mathrm{d}p'$ and $\mathrm{d}q'$ are substituted, one obtains after division by $x\,\mathrm{d}t$

$$\frac{\mathrm{d}x}{\mathrm{d}t} = -M(p'q'' - q'p'')$$

Similarly, it will be found that

$$\frac{\mathrm{d}y}{\mathrm{d}t} = -N(q'p'' - p'q'')$$

and from this equation

$$N\frac{\mathrm{d}x}{\mathrm{d}t} + M\frac{\mathrm{d}y}{\mathrm{d}t} = 0, \qquad Nx + My = f$$

where f is a constant.

The integrals that we just found give

$$Mp'' = b - Np', \qquad Mq'' = c - Nq', \qquad My = f - Nx$$

After substituting the preceding expressions in these three equations

$$\frac{dp'}{dt} + M(q'y - q''x) = 0, \qquad \frac{dq'}{dt} - M(p'y - p''x) = 0$$
$$\frac{dx}{dt} + M(p'q'' - q'p'') = 0$$

the following equations are obtained

$$\frac{dp'}{dt} + fq' - cx = 0, \qquad \frac{dq'}{dt} - fp' + bx = 0$$
$$\frac{dx}{dt} + cp' - bq' = 0$$

Since these equations are linear with constant coefficients, they will always be integrable.

Similar equations will be obtained by replacing p', q', x with p'', q'', y. But when the first three of these quantities are known, the last three will be obtained from the preceding integrals.

The expressions for p', q', x as functions of t will contain three arbitrary constants and since the constants b, c, f are also arbitrary, six arbitrary constants will result which will reduce to four in order to satisfy the postulated equations $p'^2+q'^2+x^2 = 1$ and $p''^2+q''^2+y^2 = 1$. Thus the complete values of the four variables p', q', p'', q'' will be obtained which give the orientation of the two orbits in space.

But our analysis is founded on the assumption that the mutual inclination of the two orbits is very small. The cosine of this inclination is expressed by the formula (Article 107)

$$xy + p'p'' + q'q''$$

for which the differential is equal to zero, according to the above differential equations. This quantity will thus be equal to a constant as we have already shown (Article 105) and this constant should be very small for the preceding solution to be accurate.

It would be difficult, perhaps even impossible, to solve in the same manner the case of three or more orbits. But we will observe that because the location of the plane of projections is arbitrary, it can always be taken in such a manner that the inclinations of the orbits on this plane are very small since their mutual inclinations must be very small. If the inclinations are very small, the solution founded on this assumption will be acceptable.

109. By assuming the inclinations i', i'', etc. of the orbits to be very small on the fixed plane, the variables p', q', p'', q'', etc., will also be very small and one can, in the

equations of Article 107 between these variables, replace the radicals $\sqrt{1-p'^2-q'^2}$, $\sqrt{1-p''^2-q''^2}$, etc. by unity, which will reduce them to a linear form for which the integration is easy.

Thus one will obtain equations entirely similar to those I have found by another method in the *Mémoires de l'Academie de Berlin* of 1782 and which I also applied to the six principal planets by giving the finite expressions of the variables for an indefinite time. The *Mémoires* of 1787 for the same academy also contain what is related to the orbit of Herschel's planet. It should be noted that in the formulas of these memoirs the tangents of the inclinations account for the sines contained in the values of the variables

$$\begin{aligned} p' &= \sin i' \sin h', & p'' &= \sin i'' \sin h'', & \ldots \\ q' &= \sin i' \cos h', & q'' &= \sin i'' \cos h'', & \ldots \end{aligned}$$

but because the inclinations are small, this difference is of no importance.

When the values of these variables are known, the mutual inclinations of the orbits can immediately be determined by the formulas of Article 107. But these formulas simplify in the case where the quantities p', q', p'', q'', etc. are very small. In this case, by neglecting the third order terms of these quantities

$$\cos I_{,}'' = 1 - \tfrac{1}{2}(p'^2 + q'^2 + p''^2 + q''^2) + pp' + qq'$$

and from this result, since $1 - \cos I_{,}'' = 2(\sin \frac{1}{2} I_{,}'')^2$ there will result

$$\sin \tfrac{1}{2} I_{,}'' = \tfrac{1}{2}\sqrt{(p'-p'')^2 + (q'-q'')^2}$$

Similarly, one will obtain

$$\sin \tfrac{1}{2} I_{,}''' = \tfrac{1}{2}\sqrt{(p'-p''')^2 + (q'-q''')^2}$$

and the same for the others.

110. It still remains, in order to complete the theory of **secular** variations, to consider the variation of the average motion that we have designated by $\mathrm{d}\lambda$ in Article 77 and which becomes, by neglecting the square of the eccentricity e, which is assumed very small compared to unity and by marking the letters with primes to refer them successively to the planets m', m'', etc., the following equation

$$\mathrm{d}\lambda' = -2\sqrt{\frac{a'}{g'}}\frac{\mathrm{d}(\Omega')}{\mathrm{d}a'}\mathrm{d}t + \frac{e'}{\sqrt{g'a'}}\frac{\mathrm{d}(\Omega')}{\mathrm{d}e'}\mathrm{d}t$$

for the planet m'. Similarly, one will obtain for the planet m'' the variation $\mathrm{d}\lambda'$, by adding a prime to the letters which have only one, and similarly for the others.

One will thus substitute in this formula, instead of the function (Ω'), the sum $(\Omega')_1 + (\Omega')_2$, according to Article 100. Because the function $(\Omega')_2$ does not contain the eccentricities e', e'', etc., one will simply obtain

$$\mathrm{d}\lambda' = -2\sqrt{\frac{a'}{g'}}\left(\frac{\mathrm{d}(\Omega')_1}{\mathrm{d}a'} + \frac{\mathrm{d}(\Omega')_2}{\mathrm{d}a'}\right) + \frac{e'}{\sqrt{g'a'}}\frac{\mathrm{d}(\Omega')_1}{\mathrm{d}e'}$$

and in order to obtain a uniform formula for all the planets m', m'', etc., it will only be necessary to replace, following the remarks of Articles 101 and 104, the quantities $\mathrm{d}(\Omega')_1/\mathrm{d}a'$, $\mathrm{d}(\Omega')_1/\mathrm{d}e'$, by $\mathrm{d}\varphi/m'\,\mathrm{d}a'$, $\mathrm{d}\varphi/m'\,\mathrm{d}e'$, and for the quantity $\mathrm{d}(\Omega')_2/\mathrm{d}a'$, put $\mathrm{d}\Psi/m'\,\mathrm{d}a$ which will give

$$\mathrm{d}\lambda' = -2\sqrt{\frac{a'}{g'}}\left(\frac{\mathrm{d}\Phi}{m'\,\mathrm{d}a'} + \frac{\mathrm{d}\Psi}{m'\,\mathrm{d}a'}\right) + \frac{e'}{\sqrt{g'a'}}\frac{\mathrm{d}\Phi}{m'\,\mathrm{d}e'}$$

where the functions Φ and Ψ are given in the same articles and are the same for all planets.

But if instead of these functions expressed in terms of e', ρ', i', h', e'', etc., one wants to use the expressions of Articles 103 and 107 as functions of $\mathbf{m}', \mathbf{n}', p', q', \mathbf{m}''$, etc. in this case, $e'(\mathrm{d}\Phi/\mathrm{d}e')$ shall be changed to $m'\,\mathrm{d}\Phi/\mathrm{d}\mathbf{m}' + n'\,\mathrm{d}\Phi/\mathrm{d}\mathbf{n}'$ following Article 75. Thus one will obtain

$$\mathrm{d}\lambda' = -2\sqrt{\frac{a'}{g'}}\left(\frac{\mathrm{d}\Phi}{m'\,\mathrm{d}a'} + \frac{\mathrm{d}\Psi}{m'\,\mathrm{d}a'}\right) + \frac{1}{\sqrt{g'a'}}\left(\frac{\mathbf{m}'\,\mathrm{d}\Phi}{m'\,\mathrm{d}\mathbf{m}'} + \frac{\mathbf{n}'\,\mathrm{d}\Phi}{m'\,\mathrm{d}\mathbf{n}'}\right)$$

and in order to obtain $\mathrm{d}\lambda''$, $\mathrm{d}'''\lambda$, etc., it will suffice to change $g', a', m', \mathbf{m}', \mathbf{n}'$, etc. to g'', $a'', m'', \mathbf{m}'', \mathbf{n}''$, etc.

In these formulas, the differentials relative to a' only affect the coefficients of $a'a''(a', a'')$, $a'a''[a', a'']_1$, $a'a''[a', a'']_2$, $a'a'''(a', a''')$, etc. and for the functions Φ and Ψ in whose place it suffices to substitute

$$a'(a', a'') + a'a''\frac{\mathrm{d}(a', a'')}{\mathrm{d}a'}$$
$$a'[a', a'']_1 + a'a''\frac{\mathrm{d}[a', a'']_1}{\mathrm{d}a'}, \qquad \ldots$$

to obtain the values of $\mathrm{d}\Phi/\mathrm{d}a'$ and $\mathrm{d}\Psi/\mathrm{d}a'$. From the formulas of Article 98, one will find the values of the partial differences $\mathrm{d}(a', a'')/\mathrm{d}a'$, $\mathrm{d}[a', a'']_1/\mathrm{d}a'$, etc.

Then one shall substitute of the expressions of $\mathbf{m}'$, $\mathbf{n}'$, p', q', $\mathbf{m}''$, etc., those found as a function of t after integration of the differential equations of Articles 103 and 109 and which we have given for all planets in the cited memoirs of the Académie de Berlin. Since these values are expressed by a series of sines and cosines, the variations $\mathrm{d}\lambda'$, $\mathrm{d}\lambda''$, etc., will be integrable. The constant terms will give in the expressions for λ', λ'', etc., terms proportional to t, which will be mixed with the average motions and the terms in sines and cosines will give similar terms, which will express the **secular** variations of these motions.

I had found by another method in the cited memoirs of the Académie de Berlin formulas to determine the **secular** variations of the average motions of planets and they produced for Jupiter and Saturn almost negligible results. But the preceding formulas are perhaps more rigorous and it will be good to apply them to the planets. This is an object which I will treat elsewhere. Here, I have only the intention to show the use of the new theory of the variation of arbitrary constants in the determination of **secular** variations of elements of orbits of planets.

Subsection III
The Secular Equations For The Elements of Planetary Motion Within A Medium Of Very Low Density

111. In order to complete the **secular** variations of planets, we must still consider the effect of a rare resisting medium in which they may possibly move and in which they necessarily would move if light is due to the oscillations of a rare fluid.[32]

We have already seen in Article 79 that with respect to resistance it suffices to add to the value of $\delta\Omega$ the terms

$$-\frac{\Gamma\sqrt{\mathrm{d}r^2 + r^2\,\mathrm{d}\Phi^2}(\mathrm{d}r\,\delta r + r^2\,\mathrm{d}\Phi\,\delta\Phi + r^2\,\mathrm{d}\Phi\,\delta\chi)}{\mathrm{d}t^2}$$

where Γ is the density of the medium which can be a function of r and which must be very small, and to substitute for r and Φ their values as a function of t given by the elliptical motion of the planet, keeping in mind that the operator d is with respect to time t and that the operator δ refers to the elements of the planet.

Since we seek only the **secular** variations, we shall have to discard all the periodical terms and retain only constant terms, as we have done in the past.

112. By denoting, as in Article 21, the average anomaly of the planet by $u = \sqrt{g/a^3}(t-c)$, it has been shown that r and $\Phi - u$ can be expressed by series. The series for r contains only cosines and for the second quantity the series contains only the sines of angles which are multiples of u.

Therefore, the differential $\mathrm{d}r$ will contain only sines without constant terms and the differential $\mathrm{d}\Phi$ cosines of the same angles. Consequently, the quantity $\mathrm{d}r^2 + r^2\,\mathrm{d}\Phi^2$ will only contain cosines. It will be the same for the series which expresses the value of $\sqrt{\mathrm{d}r^2 + r^2\,\mathrm{d}\Phi^2}$. Thus, by making $\Gamma = fr$, the quantity $\Gamma\sqrt{\mathrm{d}r^2 + r^2\,\mathrm{d}\Phi}$ will be expressed by a series of cosines which are multiples of u.

Now, in order to obtain δr and $\delta\Phi$, it remains to vary in the series for r and Φ the coefficients of the cosines and sines which are given in terms of a and e, and moreover, the angle u in terms of a and e. Let us denote by $\delta(r)$ and $\delta(\Phi)$ the parts of δr and $\delta\Phi$ which contain the variations of the coefficients. One will have $\delta r = \delta(r) + (\mathrm{d}r/\mathrm{d}u)\delta u$ and similarly, $\delta\Phi = \delta(\Phi) + (\mathrm{d}\Phi/\mathrm{d}u)\delta u$. Therefore

$$\mathrm{d}r\,\delta r + r^2\,\mathrm{d}\Phi\,\delta\Phi = \delta r\,\delta(r) + r^2\,\mathrm{d}\Phi\,\delta(\Phi) + (\mathrm{d}r^2 + r^2\,\mathrm{d}\Phi^2)\frac{\delta u}{\mathrm{d}u}$$

It is clear that $\delta(r)$ will only contain cosines and since $\mathrm{d}r$ contains only sines, $\mathrm{d}r\,\delta(r)$ will also contain only sines without constant terms. Similarly, $\delta(\Phi)$ will only contain sines and since $\delta\Phi$ contains only cosines, $\mathrm{d}\Phi\,\delta(\Phi)$ will only contain sines. Also, r^2 only contains cosines and consequently, $r^2\,\mathrm{d}\Phi\,\delta(\Phi)$ will contain only sines. Therefore, the quantity $\mathrm{d}r\,\delta(r) + r^2\,\mathrm{d}\Phi\,\delta(\Phi)$ which contains only sines of angles which are multiples of u without any constant terms, shall be neglected.

113. Thus one will simply have for the **secular** equations

$$\mathrm{d}r\,\delta r + r^2\,\mathrm{d}\Phi\,\delta\Phi = (\mathrm{d}r^2 + r^2\,\mathrm{d}\Phi^2)\frac{\delta u}{\mathrm{d}u}$$

Since $\mathrm{d}u = \mathrm{d}t\sqrt{g/a^3}$, after making these substitutions, one will have for the terms added to $\delta\Omega$, because of the resistance of the medium

$$-\frac{\Gamma(\mathrm{d}r^2 + r^2\,\mathrm{d}\Phi^2)^{3/2}}{\mathrm{d}t^2}\sqrt{\frac{a^3}{g}}\,\delta u - \frac{\Gamma(\sqrt{\mathrm{d}r^2 + r^2\,\mathrm{d}\Phi^2})r^2\,\mathrm{d}\Phi^2}{\mathrm{d}t^2}\delta\chi$$

where it will be necessary to substitute for r and Φ their expressions as functions of t or u and to retain only the terms independent of sines and cosines of u in the results.

From the properties of elliptical motion, one has immediately

$$r^2\,\mathrm{d}\Phi = D\,\mathrm{d}t = \mathrm{d}t\sqrt{ga(1-e^2)}$$
$$\mathrm{d}r^2 + r^2\,\mathrm{d}\Phi^2 = \left(\frac{2}{r} - \frac{1}{a}\right)g\,\mathrm{d}t^2$$

thus the terms in question will be reduced to

$$-g\Gamma\left(\frac{2}{r} - \frac{1}{a}\right)^{3/2}\sqrt{a^3}\,\delta u - g\Gamma\sqrt{\frac{2}{r} - \frac{1}{a}}\sqrt{a(1-e^2)}\,\delta\chi$$

where

$$\delta u = -\sqrt{\frac{g}{a^3}}\,\delta c - \tfrac{3}{2}\sqrt{\frac{g}{a^3}}(t-c)\delta a$$

114. These terms must be substituted for $\delta\Omega$, in the general formulas for the variations of the elements of the planets (Article 74). After the substitution is made, one will have

$$da = -2a^2\Gamma\left(\frac{2}{r}-\frac{1}{a}\right)^{3/2}\sqrt{g}\,dt$$

$$dc = 3a\Gamma\left(\frac{2}{r}-\frac{1}{a}\right)^{3/2}(t-c)\sqrt{g}\,dt$$

$$de = \frac{1-e^2}{e}\Gamma\left[\sqrt{\frac{2}{r}-\frac{1}{a}}-a\left(\frac{2}{r}-\frac{1}{a}\right)^{3/2}\right]\sqrt{g}\,dt$$

and the variations of the other elements χ, h, i will be equal to zero. From this result, one can conclude at the outset that the major axis or the line of the apsides, as well as the node and the inclination, will not be subject to any **secular** variation. Consequently, the resistance will not displace the orbit of the planet but it will only change the major axis and the eccentricity over time. It will also produce at the same time a **secular** equation in the average anomaly which is dependent upon the variation of c.

If the first two equations are combined, one has

$$dc = -\frac{3(t-c)da}{2a}$$

therefore

$$dt - dc - \frac{3(t-c)da}{2a} = dt$$

After dividing by $\sqrt{a^3}$ and integrating, one finds that

$$\frac{t-c}{\sqrt{a^3}} = \int\frac{dt}{\sqrt{a^3}}$$

and since, $u = (t-c)\sqrt{g/a^3}$, one will have $u = \sqrt{g}\int dt/\sqrt{a^3}$, assuming that the integral $\int dt/\sqrt{a^3}$ begins at the point where $u = 0$. Thus everything depends upon the variation of the average distance a.

If in the first approximation the eccentricity e, which is assumed to be very small, is neglected,[33] one has $r = a(2/r - 1/a)^{3/2} = 1/\sqrt{a^3}$. Since the density of the medium Γ can only be a function of r, it will then be a function of a and the first equation will give $dt = -da/(2\Gamma\sqrt{ga})$.

Let us assume that Γ is a constant. After integration, the following equation results

$$t = \frac{\sqrt{A}-\sqrt{a}}{\Gamma\sqrt{g}}$$

where A is the value of a when $t = 0$. Therefore, $a = (\sqrt{A} - t\Gamma\sqrt{g})^2$ and the value of u will become

$$u = \sqrt{g}\int \frac{\mathrm{d}t}{(\sqrt{A} - t\Gamma\sqrt{g})^3} = \frac{1}{2\Gamma}\left[\frac{1}{(\sqrt{A} - t\Gamma\sqrt{g})^2} - \frac{1}{A}\right] = \frac{t\sqrt{\frac{g}{A^3}} - \Gamma t^2 \frac{g}{2A^2}}{\left(1 - \Gamma t\sqrt{\frac{g}{A}}\right)^2}$$

where the coefficient Γ must be assumed very small.

115. In order to obtain the **secular** variation of the eccentricity e, it remains to substitute in the irrational expressions $\sqrt{2/r - 1/a}$ and $(2/r - 1/a)^{3/2}$ of the expression for de the expression of r as a function of u, and to only retain the constant terms in the development. But by keeping only the second order terms of e, one has from Article 21

$$r = a(1 - e\cos u + \frac{e^2}{2} - \frac{e^2}{2}\cos 2u)$$

from which

$$\frac{1}{r} = \frac{1}{a}(1 + e\cos u + e^2\cos 2u)$$

$$\sqrt{\frac{2}{r} - \frac{1}{a}} = \frac{1}{\sqrt{a}}\left(1 + e\cos u - \frac{e^2}{4} + \frac{3e^2}{4}\cos 2u\right)$$

$$\left(\frac{2}{r} - \frac{1}{a}\right)^{3/2} = \frac{1}{\sqrt{a^3}}\left(1 + 3e\cos u + \frac{3}{4}e^2 + \frac{15}{4}e^2\cos 2u\right)$$

Consequently, neglecting the $\cos u$ and $\cos 2u$, one will have[34]

$$\mathrm{d}e = -\Gamma\sqrt{g}(1 - e^2)e\frac{\mathrm{d}t}{\sqrt{a}}$$

If the quantity $\sqrt{a}$ is replaced by its equivalent expression $\sqrt{A} - t\Gamma\sqrt{g}$, and if the e^3 terms are neglected, one will have the equation

$$\mathrm{d}e(\sqrt{A} - t\Gamma\sqrt{g}) = \Gamma\sqrt{g}\, e\, \mathrm{d}t$$

whose integral gives

$$e = E\left(1 - t\Gamma\sqrt{\frac{g}{A}}\right)$$

where E is the value of e when t is equal to zero.

Since the existence of a resisting medium and moreover, since the law of density of this medium is only hypothetical, the preceding results must only be viewed as an application of our general formulas for **secular** variations.

Subsection IV
The Motion Of The Center Of Gravity Of Several Bodies Which Attract One Another

116. We demonstrated in Article 6 of SECTION III that in all free systems the equations of motion of the bodies of the system are the same, either with respect to the center of gravity of the system or to an arbitrary fixed point outside the system. Thus in the formulas of Article 8, one can establish the origin of the coordinates x, y, z, x', y', etc. at the center of gravity of all the bodies m, m', m'', etc. Furthermore, from the properties of the center of gravity, the three following equations can be derived

$$mx + m'x' + m''x'' + \cdots = 0$$
$$my + m'y' + m''y'' + \cdots = 0$$
$$mz + m'z' + m''z'' + \cdots = 0$$

which immediately give the values for the coordinates of m, in terms of those of the other bodies m', m'', etc.

Let us consider, in particular, the motion of the body m' about the common center of gravity. Since its coordinates x', y', z', are independent, it is possible to reduce the quantities T and V in the formulas of the cited article, to the terms multiplied by m', which are the only ones which contain the variables x', y', z' and then divide these quantities by m'. Thus, in the general equation, one will be able to substitute T' and V' for T and V by

$$T' = \frac{\mathrm{d}x'^2 + \mathrm{d}y'^2 + \mathrm{d}z'^2}{2\,\mathrm{d}t^2}$$

and

$$V' = m\int R'\,\mathrm{d}\rho' + m''\int R''_{\prime}\,\mathrm{d}\rho''_{\prime} + m'''\int R'''_{\prime}\,\mathrm{d}\rho'''_{\prime} + \cdots$$

and for each of the three coordinates of the orbit of m' about the common center of gravity an equation of the following form will be obtained

$$\mathrm{d}\frac{\delta T'}{\delta \mathrm{d}\xi} - \frac{\delta T'}{\delta \xi} + \frac{\delta V'}{\delta \xi} = 0$$

where ξ can be any one of the coordinates.

117. If there were only two bodies m and m', the value of V' will be reduced to one term $m\int R'\,\mathrm{d}\rho'$, and one would have $\delta V' = mR'\,\delta\rho'$, where R' is assumed to be a function of ρ.

In order to obtain the expression for $\delta V'$ with respect to ξ, one must differentiate the following expression

$$\rho' = \sqrt{(x'-x)^2 + (y'-y)^2 + (z'-z)^2}$$

by varying only x', y', z' and then substituting for x, y, z the expressions

$$x = -\frac{m'x'}{m}, \qquad y = -\frac{m'y'}{m}, \qquad z = -\frac{m'z'}{m}$$

One will thus have

$$\delta\rho' = \frac{m+m'}{m}\frac{x'\,\delta x' + y'\,\delta y' + z,\,\delta z'}{\rho'}$$

But, with the same substitutions, one has

$$\rho' = \frac{m+m'}{m}\sqrt{x'^2 + y'^2 + z'^2}$$

Therefore, by making the following definition

$$r' = \sqrt{x'^2 + y'^2 + z'^2}$$

where r' is the radius vector of the orbit of m', one will have $\rho' = ((m+m')/m)r'$ and consequently

$$\delta\rho' = \frac{x'\,\delta x' + y'\,\delta y' + z'\,\delta z'}{r'} = \delta r'$$

Therefore, one will also have $\mathrm{d}\rho' = \mathrm{d}r'$, such that the value of V' will become $m \int R'\,\mathrm{d}r$, where R' is now a function of $((m+m')/m)r'$ similar to the assumed function of ρ'. In the case of natural phenomena, one has $R' = 1/\rho'^2$. Therefore, the force R', directed toward the common center of gravity, will be represented by $m^2/((m+m')^2r'^2)$, which is already known.

118. Let us now consider the case where the system is composed of more than two bodies. Furthermore, let us assume, in order to simplify the problem, that the mass m is much larger than each of the other masses m', m'', etc., which is the case of the planets with respect to the Sun. The quantities x, y, z will become very small compared to the quantities x', y', x', etc. in the ratio of the masses m', m'', etc., to the mass m from the equations given in the preceding article. It is acceptable in the development, to consider only the first powers of x, y, z, at least, if the squares of the masses are not to be considered.

Since R' is assumed to be a function of ρ', then the integral $\int R'\,\mathrm{d}\rho'$ will also be a function of ρ' which we will denote by $F\rho'$. The quantity R will be expressed by $F'\rho'$, according to the notation of derived functions.[35] Now the following equation results

$$\rho' = \sqrt{(x'-x)^2 + (y'-y)^2 + (z'-z)^2}$$

and

$$r' = \sqrt{x'^2 + y'^2 + z'^2}$$

thus

$$F\rho' = Fr' - \frac{\mathrm{d}Fr'}{\mathrm{d}x'}x - \frac{\mathrm{d}Fr'}{\mathrm{d}y'}y - \frac{\mathrm{d}Fr'}{\mathrm{d}z'}z$$

and after differentiating with respect to δ, the quantities x, y, z remain constant

$$\delta F\rho' = \delta Fr' - x\,\delta\frac{\mathrm{d}Fr'}{\mathrm{d}x'} - y\,\delta\frac{\mathrm{d}Fr'}{\mathrm{d}y'} - z\,\delta\frac{\mathrm{d}Fr'}{\mathrm{d}z'}$$

where x, y, z will be replaced by their expressions

$$x = -\frac{m'x' + m''x'' + m'''x''' + \cdots}{m}$$
$$y = -\frac{m'y' + m''y'' + m'''y''' + \cdots}{m}$$
$$z = -\frac{m'z' + m''z'' + m'''z''' + \cdots}{m}$$

119. Let us assume that the force of attraction R' is as the power $\mu + 1$ of the distance ρ'. Thus we will have $F\rho' = \rho'^{\mu+1}/(\mu + 1)$ and the function $F\rho'$ will be a homogeneous function of x', y', z' of degree $\mu+1$ such that one will have, by a property of these functions

$$x'\frac{\mathrm{d}Fr'}{\mathrm{d}x'} + y'\frac{\mathrm{d}Fr'}{\mathrm{d}y'} + z'\frac{\mathrm{d}Fr'}{\mathrm{d}z'} = (\mu + 1)Fr'$$

Thus after differentiating with respect to δ and noting that

$$\frac{\mathrm{d}Fr'}{\mathrm{d}x'}\delta x' + \frac{\mathrm{d}Fr'}{\mathrm{d}y'}\delta y' + \frac{\mathrm{d}Fr'}{\mathrm{d}z'}\delta z' = \delta Fr'$$

one will have

$$x'\,\delta\frac{\mathrm{d}Fr'}{\mathrm{d}x'} + y'\,\delta\frac{\mathrm{d}Fr'}{\mathrm{d}y'} + z'\,\delta\frac{\mathrm{d}Fr'}{\mathrm{d}z'} = \mu\delta Fr'$$

Therefore, if in order to shorten the expressions, the following definitions are made

$$(R'') = x''\frac{\mathrm{d}Fr'}{\mathrm{d}x'} + y''\frac{\mathrm{d}Fr'}{\mathrm{d}y'} + z''\frac{\mathrm{d}Fr'}{\mathrm{d}z'}$$
$$(R''') = x'''\frac{\mathrm{d}Fr'}{\mathrm{d}x'} + y'''\frac{\mathrm{d}Fr'}{\mathrm{d}y'} + z'''\frac{\mathrm{d}Fr'}{\mathrm{d}z'}$$
$$\vdots$$

one will have after making the substitutions

$$\delta F \rho' = \frac{m + \mu m'}{m} \delta F r' + \frac{m''}{m} \delta (R'') + \frac{m'''}{m} \delta (R''') + \cdots$$

which will be the expression for $\delta \int R' \, \mathrm{d}\rho'$ in the difference equation for $\delta V'$ (Article 116). Thus one will have for the first term of V', the quantity $m \int R' \, \mathrm{d}\rho'$, the following result

$$m \int R' \, \mathrm{d}\rho' = (m + \mu m') F r' + m''(R'') + m'''(R''') + \cdots$$

where $Fr' = r'^{\mu+1}/(\mu + 1)$.

In the system of the world, the attraction of planets is inversely proportional to the square of the distances between them. Thus one has

$$\mu = -2, \qquad Fr' = -\frac{1}{r'}$$

and in addition

$$\begin{aligned} (R'') &= \frac{x'x'' + y'y'' + z'z''}{r'^3} = \frac{r'^2 + r''^2 - \rho_{,}''^2}{2r'^3} \\ (R''') &= \frac{x'x''' + y'y''' + z'z'''}{r'^3} = \frac{r'^2 + r'''^2 - \rho_{,}'''^2}{2r'^3} \\ &\vdots \end{aligned}$$

after putting

$$r'' = \sqrt{x''^2 + y''^2 + z''^2}, \qquad r''' = \sqrt{x'''^2 + y'''^2 + z'''^2} \qquad \cdots$$

Therefore, if these substitutions are made in the expression for V' (Article 116), and if it is also assumed that $R_{,}'' = 1/\rho_{,}''^2$, $R_{,}''' = 1/\rho_{,}'''^2$, etc. there will result, for the motion of the body m' about the common center of gravity

$$\begin{aligned} V' = &-\frac{m - 2m'}{r'} - m'' \left(\frac{1}{\rho_{,}''} - \frac{r'^2 + r''^2 - \rho_{,}''^2}{2r'^3} \right) \\ &- m''' \left(\frac{1}{\rho_{,}'''} - \frac{r'^2 - r'''^2 - \rho_{,}'''^2}{2r'^3} \right) - \cdots \end{aligned}$$

The first term of V' would give, if it were by itself as was shown in Chapter I, an elliptical orbit in which $g = m - 2m'$. Since the other terms are very small with respect to this term, after multiplication by the masses m'', m''', etc. which are assumed to be very small with respect to m, they can be viewed as resulting from perturbing forces whose effect is to vary the elements of the elliptical orbit. Thus by proceeding as in Article 90, we obtain the following equation

$$\Omega' = m''\left(\frac{1}{\rho_{\prime}''} - \frac{r'^2 + r''^2 - \rho_{\prime}''^2}{2r'^3}\right) + m'''\left(\frac{1}{\rho_{\prime}'''} - \frac{r'^2 + r'''^2 - \rho_{\prime}'''^2}{2r'^3}\right) + \cdots$$

It will be possible to determine, from the formulas given in Article 74, the variations of these elements.

120. If the expression for Ω' which was found for the motion of the body m' about the center of gravity of the system, is compared to the expression for Ω' for the motion of the same body about the body m, it is clear that they are the same. The radius vectors are represented in this latter expression by ρ', ρ'', etc. and in the former by r', r'', etc. But since the quantities $\rho_{\prime}'$, $\rho_{\prime}'''$, are the same in both expressions because they represent the rectilinear distances of the body m' to the other bodies m'', m''', etc. the only difference results from replacing ρ', ρ'', etc., by r', r'', etc. The quantities r', r'', as well as the quantities r', r''' must also be replaced and similarly for the others. But if only the **secular** variations of the elements are sought, as was done for the orbit of m' about m, it is easy to see that one will have the same expression for (Ω') for both orbits. Consequently, the same formulas for these variations will be obtained which is rather remarkable.

Also, in the expression for Ω' found above, the terms $-m''r'^2/(2r'^3) - m'''r'^2/(2r'^3) - \cdots$ could be removed because they are of the same form as the first term $-(m - 2m')/r'$ of the quantity V' and consequently, they can be added to this term, which would become

$$-\frac{m - 2m' - m'' - m''' - \cdots}{r'}$$

such that the body m' would describe an orbit about the center of gravity as if there were at this center, a mass equal to

$$m - 2m' - m'' - m''' - \cdots$$

Consequently, the perturbing forces of this orbit will be less than those of the orbit of the same body m' about the largest body m, everything else being equal.

SECTION VIII
THE MOTION OF CONSTRAINED BODIES WHICH INTERACT IN AN ARBITRARY FASHION

1. In the preceding section, we assumed that the bodies were free and consequently, all motions that the perturbing forces could impress upon them were possible. With this

assumption, the coordinates of each of the bodies could be taken as independent variables and each of them would give an equation of the form (SECTION VII, Article 1)

$$\mathrm{d}\frac{\delta T}{\delta\,\mathrm{d}\xi} - \frac{\delta T}{\delta \xi} + \frac{\delta V}{\delta \xi} = 0$$

When the bodies are constrained, either because they are forced to move on given surfaces or lines or because they are attached by strings or rods or because their motions are modified in other arbitrary ways, the conditions, if they are expressed analytically, can always be reduced to equations of condition between the different coordinates of the same bodies. Some of the coordinates will depend upon others and could be expressed as functions of those they depend upon. In this situation, the number of independent variables will be reduced. But each of these variables will still give the same equation as if it belonged to a free body. Thus the same formulas we have given in Articles 1 and 2 of the preceding section will be used as the basis in this case.

In addition, whatever the connections between the bodies, the equation of **forces vives** will be $T + V = H$.

2. If the motion took place in a resisting medium, we have seen in Article 3 of the same section that the resistance R will give for each body m the terms

$$R\frac{\mathrm{d}x\,\delta x + \mathrm{d}y\,\delta y + \mathrm{d}z\,\delta z}{\mathrm{d}s}$$

to add to δV. Thus, it will only be necessary to reduce the differences $\delta x, \delta y, \delta z$ to differences related to the new independent variables.

A general form can be given to this result using the analysis of Article 4 of SECTION IV. Because after denoting the new variables by ξ, ψ, φ, one saw earlier that the quantity $\mathrm{d}x\,\delta x + \mathrm{d}y\,\delta y + \mathrm{d}z\,\delta z$ can be transformed to

$$F\,\mathrm{d}\xi\,\delta\xi + G(\mathrm{d}\xi\,\delta\psi + \mathrm{d}\psi\,\delta\xi) + H\mathrm{d}\psi\,\delta\psi + I(\mathrm{d}\xi\,\delta\varphi + \mathrm{d}\varphi\,\delta\xi) + \cdots$$

where it is seen that the coefficient of $\delta\xi$ is

$$F\,\mathrm{d}\xi + G\,\mathrm{d}\psi + I\,\mathrm{d}\varphi$$

If the operator δ is replaced by d, then the transformation of $\mathrm{d}x^2 + \mathrm{d}y^2 + \mathrm{d}z^2$ becomes

$$F\,\mathrm{d}\xi^2 + 2G\,\mathrm{d}\xi\,\mathrm{d}\psi + H\,\mathrm{d}\psi^2 + 2I\,\mathrm{d}\xi\,\mathrm{d}\varphi + \cdots$$

If this transformation is denoted by Φ, it is clear that one has

$$F\,\mathrm{d}\xi + G\,\mathrm{d}\psi + I\,\mathrm{d}\varphi = \frac{\mathrm{d}\Phi}{2\,\delta\,\mathrm{d}\xi}$$

In general, it follows from this result that

$$\mathrm{d}x\,\delta x + \mathrm{d}y\,\delta y + \mathrm{d}z\,\delta z = \frac{\mathrm{d}\Phi}{2\,\delta\,\mathrm{d}\xi}\delta\xi + \frac{\mathrm{d}\Phi}{2\,\delta\,\mathrm{d}\psi}\delta\psi + \frac{\mathrm{d}\Phi}{2\delta\,\mathrm{d}\varphi}\delta\varphi$$

The resistance of fluids is generally proportional to the square of the velocity $\mathrm{d}s/\mathrm{d}t$ where s is the distance traversed by the body. If the density of the fluid is denoted by Γ, we will have $R = \Gamma(\mathrm{d}s/\mathrm{d}t)^2$ and the terms to be added to δV will be

$$\Gamma\,\mathrm{d}s\frac{\mathrm{d}x\,\delta x + \mathrm{d}y\,\delta y + \mathrm{d}z\,\delta z}{\mathrm{d}t^2}$$

Thus if we retain the meaning of the quantity T of Article 1 of the preceding section, it only remains to add to δV the following terms

$$\frac{\delta T}{\delta\,\mathrm{d}\xi}\delta\xi + \frac{\delta T}{\delta\,\mathrm{d}\psi}\delta\psi + \frac{\delta T}{\delta\,\mathrm{d}\varphi}\delta\varphi + \cdots$$

and then to replace m, m', m'', etc. with $\Gamma\,\mathrm{d}s$, $\Gamma'\,\mathrm{d}s'$, $\Gamma''\,\mathrm{d}s''$, etc. Since the resistance is not proportional to the mass but only to the surface area, it is only necessary to express by Γ, Γ', Γ'', etc. the resistance that the bodies m, m', m'', etc., would be subject to while moving with a velocity equal to unity.

Thus the equation of Article 1, with respect to ξ, will become

$$\mathrm{d}\frac{\delta T}{\delta\,\mathrm{d}\xi} - \frac{\delta T}{\delta\xi} + \frac{\delta V}{\delta\xi} + \frac{\delta T}{\delta\,\mathrm{d}\xi} = 0$$

But the equation of **forces vives** will not hold for this case.

3. Instead of reducing at the outset all of the variables of the problem to a smaller number of independent variables by means of equations of condition given by the nature of the problem, all of the variables can immediately be treated as independent. If $L = 0$, $M = 0$, etc. are the equations of condition between these variables, it will suffice to add to the equation relative to each of these variables, terms of the form

$$\lambda\frac{\delta L}{\delta\xi} + \mu\frac{\delta M}{\delta\xi} + \cdots$$

Thus the following equation, with respect to an arbitrary variable ξ, will be obtained

$$\mathrm{d}\frac{\delta T}{\delta\,\mathrm{d}\xi} - \frac{\delta T}{\delta\xi} + \frac{\delta V}{\delta\xi} + \lambda\frac{\delta L}{\delta\xi} + \mu\frac{\delta M}{\delta\xi} + \cdots = 0$$

where the quantities λ, μ, etc. are undetermined quantities which have to be eliminated by means of the equations of condition.

With respect to these equations, we noted earlier that it is not necessary that they be in a finite form. It suffices that they are differentiable and of the first order. Then by replacing the operator d by δ, the partial differences relative to each variable ξ will also be obtained.

Finally, if the system were composed of an infinite number of particles joined together in an arbitrary fashion, one should follow, with respect to the terms resulting from the equations of condition, the same rules which were given in SECTION VI of PART I (Article 10) because these terms are the same in the general formulas of motion as in those of equilibrium.

4. After the problem is reduced to a fixed number of independent variables, a differential equation of the second order will be obtained for each of these variables whose integration will introduce two arbitrary constants. Consequently, the complete solution will contain twice as many arbitrary constants as there are independent variables, which will have to be determined from the initial state of the system. But if during the motion of the system, it should happen that one or several bodies which are part of the system receive at some instant arbitrary external impulses, these impulses, since they act for only an instant, will not change the form of the equations, but only the value of the arbitrary constants. If the impulses became infinitesimal and continuous, the arbitrary constants would cease to be constant and would become variables themselves.

We have already given in Chapter II of the preceding section the theory of the variation of arbitrary constants for free bodies and we have applied the theory to the elements of the orbits of planets. We will begin this section by further generalizing this theory and by making it applicable to all systems of bodies which act upon one another.

Chapter I
GENERAL FORMULAS FOR THE VARIATION OF ARBITRARY CONSTANTS IN THE MOTION OF AN ARBITRARY SYSTEM OF BODIES, PRODUCED BY EITHER FINITE AND INSTANTANEOUS IMPULSES OR BY INFINITESIMAL AND CONTINUOUS IMPULSES

5. Let us denote by ξ, ψ, φ, etc. the independent variables to which all the coordinates x, y, z of the bodies of the system have been reduced by means of the equations of condition which depend on the connections between the bodies. One will always be able to express each constant by a given function of ξ, ψ, φ, etc., and the derivatives $\mathrm{d}\xi/\mathrm{d}t$, $\mathrm{d}\psi/\mathrm{d}t$, $\mathrm{d}\varphi/\mathrm{d}t$, etc. But the finite variables ξ, ψ, φ, etc. do not depend on the instantaneous position of the bodies in space. Consequently, they cannot express any changes due to external impulses. Thus, only the derivatives $\mathrm{d}\xi/\mathrm{d}t$, $\mathrm{d}\psi/\mathrm{d}t$, $\mathrm{d}\varphi/\mathrm{d}t$, could have their values changed by these impulses.

Let us assume they become $\mathrm{d}\xi/\mathrm{d}t + \dot{\xi}$, $\mathrm{d}\psi/\mathrm{d}t + \dot{\psi}$, $\mathrm{d}\varphi/\mathrm{d}t + \dot{\varphi}$, etc. The increments $\dot{\xi}$, $\dot{\psi}$, $\dot{\varphi}$, etc. are due to the impulses. They are the velocities projected along the coordinates ξ, ψ, φ, etc. which the impulses produce initially and which need to be determined.

Let P, Q, R, etc., be the forces of impulsion applied to each body m of the system, in the directions of the lines p, q, r, etc. and with a tendency to shorten the lines. Let $\dot{x}$, $\dot{y}$, $\dot{z}$ be the initial velocities of this body in the directions of the rectangular coordinates x, y, z and in a direction which increases these coordinates, if the entire system was at rest. From Article 11 of SECTION II, one will have the equation

$$\mathrm{S}(\dot{x}\,\delta x + \dot{y}\,\delta y + \dot{z}\,\delta z)m - \mathrm{S}(P\,\delta p + Q\,\delta q + R\,\delta r + \cdots) = 0$$

which must be verified independent of the variations $\delta\xi$, $\delta\psi$, $\delta\varphi$, etc. of each of the independent variables. Thus it will remain to substitute in this equation the values of x, y, z and p, q, r, etc., as functions of ξ, ψ, φ, etc. by noting that the velocities $\dot{x}$, $\dot{y}$, $\dot{z}$ can be expressed in the same fashion as all the velocities by $\mathrm{d}x/\mathrm{d}t$, $\mathrm{d}y/\mathrm{d}t$, $\mathrm{d}z/\mathrm{d}t$.

With these substitutions, one will have the transformed equation

$$\mathrm{S}(\dot{x}\,\delta x + \dot{y}\,\delta y + \dot{z}\,\delta z)m = \Xi\,\delta\xi + \Psi\,\delta\psi + \Phi\,\delta\varphi + \cdots$$

and if one puts, as in Article 62 of the preceding section

$$\delta\Omega = -\mathrm{S}(P\,\delta p + Q\,\delta q + R\,\delta r + \cdots)$$

the following equations will be obtained

$$\Xi = \frac{\delta\Omega}{\delta\xi}, \qquad \Psi = \frac{\delta\Omega}{\delta\psi}, \qquad \Phi = \frac{\delta\Omega}{\delta\varphi}, \qquad \ldots$$

There will be as many equations as there are variables ξ, ψ, φ, etc.

But it is easy to see that the quantities Ξ, Ψ, Φ, etc. will be functions of ξ, ψ, φ, etc. and of their derivatives $\mathrm{d}\xi/\mathrm{d}t$, $\mathrm{d}\psi/\mathrm{d}t$, $\mathrm{d}\varphi/\mathrm{d}t$, etc. The derivatives will be nothing but the initial velocities we have denoted above by $\dot{\xi}$, $\dot{\psi}$, $\dot{\varphi}$, etc. and which, consequently, can be determined from the preceding equations.

Since the quantities $\dot{x}$, $\dot{y}$, $\dot{z}$ are equivalent to $\mathrm{d}x/\mathrm{d}t$, $\mathrm{d}y/\mathrm{d}t$, $\mathrm{d}z/\mathrm{d}t$, the equation $\dot{x}\,\delta x + \dot{y}\,\delta y + \dot{z}\,\delta z$ can also be expressed by

$$\frac{\mathrm{d}x\,\delta x + \mathrm{d}y\,\delta y + \mathrm{d}z\,\delta z}{\mathrm{d}t}$$

It follows from what we have demonstrated above (Article 2) that if in the formula

$$T = \mathrm{S}\frac{\mathrm{d}x^2 + \mathrm{d}y^2 + \mathrm{d}z^2}{2\,\mathrm{d}t^2}m$$

the variables x, y, z are replaced by ξ, ψ, φ, etc. and if $\mathrm{d}\xi/dt$ is replaced by $\dot{\xi}$, $\mathrm{d}\psi/dt$ by $\dot{\psi}$, $\mathrm{d}\varphi/dt$ by $\dot{\varphi}$, one will have from the partial differences with respect to $\dot{\xi}$, $\dot{\psi}$, $\dot{\varphi}$, etc.

$$\Xi = \frac{\mathrm{d}T}{\mathrm{d}\dot{\xi}}, \qquad \Psi = \frac{\mathrm{d}T}{\mathrm{d}\dot{\psi}}, \qquad \Phi = \frac{\mathrm{d}T}{\mathrm{d}\dot{\varphi}}, \qquad \cdots$$

and the equations to determine $\dot{\xi}, \dot{\psi}, \dot{\varphi}$, etc. will be

$$\frac{\mathrm{d}T}{\mathrm{d}\dot{\xi}} = \frac{\delta\Omega}{\delta\xi}, \qquad \frac{\mathrm{d}T}{\mathrm{d}\dot{\psi}} = \frac{\delta\Omega}{\delta\psi}, \qquad \frac{\mathrm{d}T}{\mathrm{d}\dot{\varphi}} = \frac{\delta\Omega}{\delta\varphi}, \qquad \cdots$$

where it should be noted that the unknowns will only be of the first degree because they can only be of the second degree in the quantity T.

Thus the effect of instantaneous and finite impulses P, Q, R, etc., will be to increment the derivatives $\mathrm{d}\xi/dt$, $\mathrm{d}\psi/dt$, $\mathrm{d}\varphi/dt$, etc. of the quantities $\dot{\xi}, \dot{\psi}, \dot{\varphi}$, etc. in the expressions for the arbitrary constants of the problem.

6. In order to apply this theory to the case of very small and continuous impulses, it is necessary that P, Q, R, etc. are replaced by $P\,\mathrm{d}t, Q\,\mathrm{d}t, R\,\mathrm{d}t$, etc. which will replace $\delta\Omega$ by $\delta\Omega\,\mathrm{d}t$ and the quantities $\dot{\xi}, \dot{\psi}, \dot{\varphi}$, etc. will become very small of the first order. The arbitrary constants will become continuous variables and the quantities $\dot{\xi}$, $\dot{\psi}$, $\dot{\varphi}$, etc. will be the variations of $\mathrm{d}\xi/dt$, $\mathrm{d}\psi/dt$, $\mathrm{d}\varphi/dt$, etc. in the expressions for these constants such that since a is one of the constants which became variable, one will have, after defining

$$\frac{\mathrm{d}\xi}{\mathrm{d}t} = \xi', \qquad \frac{\mathrm{d}\psi}{\mathrm{d}t} = \psi', \qquad \frac{\mathrm{d}\varphi}{\mathrm{d}t} = \varphi', \qquad \cdots$$

the following equation

$$\mathrm{d}a = \frac{\mathrm{d}a}{\mathrm{d}\xi'}\dot{\xi} + \frac{\mathrm{d}a}{\mathrm{d}\psi'}\dot{\psi} + \frac{\mathrm{d}a}{\mathrm{d}\varphi'}\dot{\varphi} + \cdots$$

The finite variables ξ, ψ, φ, etc. are not to be incremented. It is only necessary to substitute for $\dot{\xi}$, $\dot{\psi}$, $\dot{\varphi}$, etc. their expressions obtained from the equations above. But in the present case, these equations can be put in a simple form with the following consideration.

By treating the variables ξ, ψ, φ, etc. as well as the differentials ξ', ψ', φ', etc. as functions of arbitrary constants a, b, c, etc. and of the time t and by denoting by δ the variation of the resulting constants of the variation, it is clear that

$$\delta\xi = 0, \qquad \delta\psi = 0, \qquad \delta\varphi = 0, \qquad \delta\xi' = \dot{\xi}, \qquad \delta\psi' = \dot{\psi}, \qquad \delta\varphi' = \dot{\varphi}, \cdots$$

Since the partial differences $\mathrm{d}T/\mathrm{d}\dot{\xi}$, $\mathrm{d}T/\mathrm{d}\dot{\psi}$ contain only terms of the first degree for $\dot{\xi}$, $\dot{\psi}$, $\dot{\varphi}$, etc., it is easy to see that they can be reduced to $\delta(\mathrm{d}T/\mathrm{d}\xi')$, $\delta(\mathrm{d}T/\mathrm{d}\psi')$, etc. by viewing T as function of ξ', ψ', φ', etc. Thus the equations in question will become

$$\delta\frac{\mathrm{d}T}{\mathrm{d}\xi'} = \frac{\delta\Omega}{\delta\xi}\mathrm{d}t, \qquad \delta\frac{\mathrm{d}T}{\mathrm{d}\psi'} = \frac{\delta\Omega}{\delta\psi}\mathrm{d}t, \qquad \delta\frac{\mathrm{d}T}{\mathrm{d}\varphi'} = \frac{\delta\Omega}{\delta\varphi}\mathrm{d}t, \qquad \cdots$$

and one will have

$$\mathrm{d}a = \frac{\mathrm{d}a}{\mathrm{d}\xi'}\delta\xi' + \frac{\mathrm{d}a}{\mathrm{d}\psi'}\delta\psi' + \frac{\mathrm{d}a}{\mathrm{d}\varphi'}\delta\varphi' + \cdots$$

where the values of $\delta\xi'$, $\delta\psi'$, $\delta\varphi'$, etc. obtained from these equations will have to be substituted.

If the partial differences of Ω relative to ξ, ψ, φ, etc. are changed to partial differences relative to the constants a, b, c, etc., formulas similar to those of Article 60 of the preceding section will be obtained, in which the coefficients of $\mathrm{d}\Omega/\mathrm{d}a$, $\mathrm{d}\Omega/\mathrm{d}b$, etc. will have the property of being independent of time t. The direct demonstration of this singular property is very difficult as shown by the memoir of Poisson on this subject in Volume VIII of the *Journal de l'Ecole Polytechnique*. No effort would have been made to search for this property, if its existence were not known beforehand.[36]

Since I already gave in SECTION V a complete theory of the variations of arbitrary constants, I will not expound on it here. I will add only two remarks on the formulas of this theory.

7. The first remark is with respect to the general formula of Article 1 of the cited section in which, in order to shorten the expression the following definitions are in order

$$\frac{\mathrm{d}T}{\mathrm{d}\xi'} = T', \qquad \frac{\mathrm{d}T}{\mathrm{d}\psi'} = T'', \qquad \frac{\mathrm{d}T}{\mathrm{d}\varphi'} = T''', \qquad \cdots$$

which, in turn, reduces the expression to

$$\begin{aligned} \Delta(\Omega\,\mathrm{d}t) &= \Delta\xi\,\delta T' + \Delta\psi\,\delta T'' + \Delta\varphi\delta T''' + \cdots \\ &\quad - \delta\xi\Delta T' - \delta\psi\Delta T'' - \delta\varphi\Delta T''' - \cdots \end{aligned}$$

The operator δ indicates the variations of all the constants a, b, c, etc., which have become variables, but the operator Δ applies generally to each of these constants. By first applying it to an arbitrary constant, such as a and developing the variations indicated by δ, one has immediately the formula

$$\frac{\mathrm{d}\Omega}{\mathrm{d}a}\mathrm{d}t = [a,b]\mathrm{d}b + [a,c]\mathrm{d}c + [a,k]\mathrm{d}k + \cdots$$

in which

$$\begin{aligned} [a,b] &= \frac{\mathrm{d}\xi}{\mathrm{d}a}\frac{\mathrm{d}T'}{\mathrm{d}b} + \frac{\mathrm{d}\psi}{\mathrm{d}a}\frac{\mathrm{d}T''}{\mathrm{d}b} + \frac{\mathrm{d}\varphi}{\mathrm{d}a}\frac{\mathrm{d}T'''}{\mathrm{d}b} + \cdots \\ &\quad - \frac{\mathrm{d}T'}{\mathrm{d}a}\frac{\mathrm{d}\xi}{\mathrm{d}b} - \frac{\mathrm{d}T''}{\mathrm{d}a}\frac{\mathrm{d}\psi}{\mathrm{d}b} - \frac{\mathrm{d}T'''}{\mathrm{d}a}\frac{\mathrm{d}\varphi}{\mathrm{d}b} - \cdots \\ [a,c] &= \frac{\mathrm{d}\xi}{\mathrm{d}a}\frac{\mathrm{d}T'}{\mathrm{d}c} + \frac{\mathrm{d}\psi}{\mathrm{d}a}\frac{\mathrm{d}T''}{\mathrm{d}c} + \frac{\mathrm{d}\varphi}{\mathrm{d}a}\frac{\mathrm{d}T'''}{\mathrm{d}c} + \cdots \end{aligned}$$

$$- \frac{dT'}{da}\frac{d\xi}{dc} - \frac{dT''}{da}\frac{d\psi}{dc} - \frac{dT'''}{da}\frac{d\varphi}{dc} - \cdots$$

$$\vdots$$

and where the values of the coefficients $[a, b]$, $[a, c]$, etc. become independent of t, after substitution of the expressions for ξ, ψ, φ, etc. in place of a, b, c, etc. and t.

Thus the formulas are obtained which I found originally in the memoir on the variation of arbitrary constants in the problems of mechanics. Poisson later found more direct formulas, which are the same as those I gave in Article 18 of SECTION V. But, although these latter equations seem simpler because they immediately give the values of the variations da, db, etc., instead of having to deduce them from the others by elimination, this advantage is only apparent as we have already noted earlier (SECTION VII, Article 66). It can still be said that on several occasions the advantage will lie entirely with the preceding formulas because they do not require any preliminary simplifications and they can be applied immediately, every time that the expression of each variable occurs as a function of time, in which the arbitrary constants are included in some manner. This is the reason which led me to present them again.

8. The second remark is with regard to the lessons that can be derived from these formulas, relative to the nature of perturbing forces. We have always assumed that these forces were such that when multiplied by the elements of their direction, the sum would become integrable and could be expressed by a function of independent variables which we have denoted by $-\Omega$.

But we have already noted in Article 62 of the preceding section that whatever the perturbing forces R, Q, P, etc., it suffices to define

$$-\delta\Omega = R\,\delta r + Q\,\delta q + P\,\delta p + \cdots$$

by referring the partial differences relative to the operator δ only to the variables r, q, p, etc.

In general, it is not necessary for the accuracy of the formulas of variations that the perturbing forces, which we have denoted by partial differences $d\Omega/d\xi$, $d\Omega/d\psi$, $d\Omega/d\varphi$, etc. be indeed partial differences of the same quantity. One can assume that these forces are expressed by arbitrary quantities, which we will denote by Ω', Ω'', etc. Then, instead of $\Delta\Omega$ in the formulas of Article 11 of SECTION I, we will have

$$\Omega'\,\Delta\xi + \Omega''\,\Delta\psi + \Omega'''\,\Delta\varphi + \cdots$$

and the equation of the preceding article will become

$$\begin{aligned}&(\Omega'\,\Delta\xi + \Omega''\,\Delta\psi + \Omega'''\,\Delta\varphi + \cdots)dt\\ &= \Delta\xi\,\delta T' + \Delta\psi\,\delta T'' + \Delta\varphi\delta T''' + \cdots\\ &-\delta\xi\,\Delta T' - \delta\psi\,\Delta T' - \delta\varphi\,\Delta T''' - \cdots\end{aligned}$$

from which, referring the operator Δ to the arbitrary constant a, one also has

$$\left(\Omega'\frac{\mathrm{d}\xi}{\mathrm{d}a}+\Omega''\frac{\mathrm{d}\psi}{\mathrm{d}a}+\Omega'''\frac{\mathrm{d}\varphi}{\mathrm{d}a}+\cdots\right)\mathrm{d}t$$
$$=[a,b]\mathrm{d}b+[a,c]\mathrm{d}c+[a,k]\mathrm{d}k+\cdots$$

after viewing the variables ξ, ψ, φ, etc. as functions of a, b, c, k, etc.

The same thing will hold for the formulas of Articles 14 and 18 of SECTION V, by putting everywhere

$$\Omega'\,\mathrm{d}\xi+\Omega''\,\mathrm{d}\psi+\Omega'''\,\mathrm{d}\varphi+\cdots$$

in place of $\mathrm{d}\Omega$ and referring to the variables ξ, ψ, φ, etc., the partial differences of Ω relative to the constants α, β, γ, etc., λ, μ, ν, etc. or a, b, c, k, etc.

9. Finally, the perturbing forces can be neglected and only the function $-\Omega$ need be considered as a quantity which, being added to the function V resulting from the principal forces, produces the variations of the arbitrary constants in the motions resulting from these forces. Since in the calculus of these variations there are only partial differences of Ω relative to the independent variables ξ, ψ, φ, etc., it is not necessary that the differential $\mathrm{d}\Omega$ be an exact differential. It suffices that the differentials it contains are themselves exact differentials for which partial differences with respect to the variables ξ, ψ, φ, etc. can be obtained.

This extension of our formulas which was mentioned in the preface to Volume I can be useful for several problems where the perturbing forces will not only be functions of the independent variables ξ, ψ, φ, etc., but also of their derivatives $\mathrm{d}\xi/\mathrm{d}t$, $\mathrm{d}\psi/\mathrm{d}t$, $\mathrm{d}\varphi/\mathrm{d}t$, etc. and of the time t. For example, if after having solved a problem of mechanics in a vacuum, one also wanted to consider the resistance of the medium, as we have done, for the planets in the preceding section.

But the same extension can not hold for the main forces which are in the differential equations for which the integration introduces arbitrary constants. These forces, each multiplied by its element of direction, must always form an integrable quantity that we have designated by V (SECTION IV, Article 9) and which must be a function of independent variables without their derivatives. Otherwise, the simplification of these equations to the form of Article 2 of SECTION V will not hold and the analysis of Subsection I of the same section will cease to be exact. However, nothing prevents the expressions of these forces from containing the time t because as the quantity V disappears in the partial differences of $Z = T - V$, relative to ξ', ψ', φ', etc. the final result of Article 7 will always hold, because it is independent of V. But it will cease to exist if this quantity were a function of ξ, ψ, φ, etc. and of ξ', ψ', φ', etc.

Now we will solve some particular problems.

Chapter II
THE MOTION OF A BODY ON A GIVEN SURFACE OR LINE

10. If only an isolated body is considered, its mass can be neglected or it can be assumed to be equal to unity. Then there results, as in Article 3 of the preceding section

$$T = \frac{dx^2 + dy^2 + dz^2}{2\,dt^2}, \qquad \delta V = R\,\delta r + Q\,\delta q + P\,\delta p + \cdots$$

The equation of the surface will give z as a function of x and y. One will thus have

$$dz = \frac{dz}{dx}dx + \frac{dz}{dy}dy$$

and if the variables x and y are treated as independent, each of them will give an equation of the form

$$d\frac{\delta T}{\delta dx} \quad \frac{\delta T}{\delta x} + \frac{\delta V}{\delta x} - 0$$

The term $dx^2/2\,dt^2$ in the expression for T immediately gives d^2x/dt^2. The term $dz^2/2\,dt^2$ which is supposed to be a function of x, y, and dx, dy, will at first give these two terms

$$\frac{d\left(dz\dfrac{dz}{dx}\right)}{dt^2} - \frac{\delta dz^2}{2\,dx\,dt^2}$$

Now dz/dx is equivalent to $\delta z/\delta x$ and $\delta dz^2/2\,\delta x$ is equivalent to $dz\,\delta dz/\delta x$ or $dz\,d\delta z/\delta x$. Therefore, the two terms in question will be reduced to $d^2z\,\delta z/dt^2\,\delta x$. Thus the equation relative to x will be

$$\frac{d^2x}{dt^2} + \frac{d^2z}{dt^2}\frac{\delta z}{\delta x} + \frac{\delta V}{\delta x} = 0$$

and similarly, with respect to y, we will have

$$\frac{d^2y}{dt^2} + \frac{d^2z}{dt^2}\frac{\delta z}{\delta y} + \frac{\delta V}{\delta y} = 0$$

If the body were constrained to move on a given line, then y would be a function of x. The term $dy^2/2\,dt^2$ in the expression for T would give the terms

$$\frac{d\left(dy\dfrac{dy}{dx}\right)}{dt^2} - \frac{\delta(dy^2)}{2\,dt^2}$$

which would be reduced in the same fashion to $(\mathrm{d}^2y/\mathrm{d}t^2)(\delta y/\delta x)$. Similarly, the term $\mathrm{d}z^2/2\,\mathrm{d}t^2$ would give $(\mathrm{d}^2z/\mathrm{d}t^2)(\delta z/\delta x)$ and one would obtain, with respect to x, which is the only variable, the equation

$$\frac{\mathrm{d}^2x}{\mathrm{d}t^2}+\frac{\mathrm{d}^2y}{\mathrm{d}t^2}\frac{\delta y}{\delta x}+\frac{\mathrm{d}^2z}{\mathrm{d}t^2}\frac{\delta z}{\delta x}+\frac{\delta V}{\delta x}=0$$

It is clear from the preceding analysis that each term of the quantity T which will be of the form $k(\mathrm{d}z^2/\mathrm{d}t^2)$, where z is a given function of the two other variables x and y, will give

$$\mathrm{d}\frac{\delta T}{\delta \mathrm{d}x}-\frac{\delta T}{\delta x}=2k\frac{\mathrm{d}^2z}{\mathrm{d}t^2}\frac{\delta z}{\delta x}$$
$$\mathrm{d}\frac{\delta T}{\delta \mathrm{d}y}-\frac{\delta T}{\delta y}=2k\frac{\mathrm{d}^2z}{\mathrm{d}t^2}\frac{\delta z}{\delta y}$$

simplifications, which can be useful at times.

11. If instead of the rectangular coordinates x, y, z, one wanted to use, for the surface, a radius r with two angles φ and ψ, as in Article 4 of the preceding section, one would have

$$T=\frac{r^2(\mathrm{d}t^2+\cos^2\psi\,\mathrm{d}\varphi^2)+\mathrm{d}r^2}{2\,\mathrm{d}t^2}$$

where r would be a function of φ and ψ, depending on the nature of the surface, and there would result, relative to φ and ψ, two equations of the form

$$\mathrm{d}\frac{\delta T}{\delta \mathrm{d}\psi}-\frac{\delta T}{\delta \psi}+\frac{\delta V}{\delta \psi}=0$$

The term $\mathrm{d}r^2/(2\,\mathrm{d}t^2)$ in the expression for T would give $(\mathrm{d}^2r/\mathrm{d}t^2)(\delta r/\delta\psi)$ with respect to ψ and $(\mathrm{d}^2r/\mathrm{d}t^2)(\delta r/\delta\varphi)$ with respect to φ, and the following two equations would be obtained

$$\mathrm{d}\frac{r^2\,\mathrm{d}\psi}{\mathrm{d}t^2}+\frac{r^2\sin\psi\cos\psi\,\mathrm{d}\varphi^2}{\mathrm{d}t^2}+\frac{\mathrm{d}^2r}{\mathrm{d}t^2}\frac{\delta r}{\delta\psi}+\frac{\delta V}{\delta\psi}=0$$
$$\mathrm{d}\frac{r^2\cos^2\psi\,\mathrm{d}\varphi}{\mathrm{d}t^2}+\frac{\mathrm{d}^2r}{\mathrm{d}t^2}\frac{\delta r}{\delta\varphi}+\frac{\delta V}{\delta\varphi}=0$$

which ordinary methods would only give after several simplifications.

12. It is useful to note that the equation $T+V=H$, which always holds when a body is only acted upon by forces proportional to functions of their distances to given centers, immediately gives the velocity of the body at an arbitrary point of the curve it describes. Since u is the velocity and s is the described space, one has

$$u=\frac{\mathrm{d}s}{\mathrm{d}t}=\frac{\sqrt{\mathrm{d}x^2+\mathrm{d}y^2+\mathrm{d}z^2}}{\mathrm{d}t}$$

thus $T = u^2/2$ and $u = \sqrt{2(H - V)}$ consequently such that since V is a finite function of the coordinates, the velocity will only depend upon the position of the body in space.

If the body is not acted upon by any accelerating forces, one has $V = 0$ and the velocity is constant. In this case, as we have demonstrated in general, the formula $\int u\,\mathrm{d}s$ is always a maximum or a minimum for the given limits (SECTION III, Article 39). The quantity $\int \mathrm{d}s$ or s, that is, the length of the curve described by the body, will also be a maximum or a minimum but it is obvious that it can only be a minimum[37] because the maximum does not exist. From this result, the known theorem is obtained that a body projected on an arbitrary surface will always describe the shortest line between given points.

13. But, in the solution of these problems, it is often simpler to view all the coordinates as independent variables, to use equations of the given surface or line as equations of condition which, being represented by $L = 0$, $M = 0$, will simply give for each variable the terms $\lambda\,\delta L$, $\mu\,\delta M$ to add to δV. The coefficients λ, μ are indeterminate and have to be eliminated.

But, from what we have demonstrated in Article 5 of SECTION IV of PART I, it follows that each term, such as $\lambda\,\delta L$, can represent a moment of a force equal to

$$\lambda\sqrt{\left(\frac{\mathrm{d}L}{\mathrm{d}x}\right)^2 + \left(\frac{\mathrm{d}L}{\mathrm{d}y}\right)^2 + \left(\frac{\mathrm{d}L}{\mathrm{d}z}\right)^2}$$

and which acts perpendicular to the surface for which the equation is $\mathrm{d}L = 0$. Consequently, this force can not be the one which derives from the resistance that the surface exerts on the body but is equal to the pressure that the body exerts on the surface.

Thus the coefficient λ will be used to determine with the equation $L = 0$ the pressure of the body on the given surface. If the body were to move on a given line, by viewing the line as produced by the intersection of two surfaces represented by the equations $L = 0$, $M = 0$, the two coefficients λ and μ will be used to determine the pressures that the body exerts on this line, perpendicular to the two surfaces.

14. In general, one can compare the term $\lambda\,\delta L$ to the term δV. Since $\delta V = R\,\delta r + Q\,\delta q +$ etc., where R, Q, etc. are forces which act along the lines r, q, etc. and which have a tendency to shorten them and if L is a function of the coordinates ξ, ψ, φ, one will have

$$\delta L = \frac{\mathrm{d}L}{\mathrm{d}\xi}\delta\xi + \frac{\mathrm{d}L}{\mathrm{d}\psi}\delta\psi + \frac{\mathrm{d}L}{\mathrm{d}\varphi}\delta\varphi$$

and the terms $\lambda(\mathrm{d}L/\mathrm{d}\xi)$, $\lambda(\mathrm{d}L/\mathrm{d}\psi)$, $\lambda(\mathrm{d}L/\mathrm{d}\varphi)$ will express the forces which result from the resistance of the surface for which the equation is $L = 0$, along the direction of the coordinates[38] ξ, ψ, φ, and which have a tendency to decrease these coordinates.

If the equation of the surface were $\xi = a$, where a is a constant, which can always be obtained by a judicious selection of coordinates, one would have

$$L = \xi - a, \qquad \frac{\delta L}{\delta \xi} = 1, \qquad \frac{\delta L}{\delta \psi} = 0, \qquad \frac{\delta L}{\delta \varphi} = 0$$

and the equation relative to ξ (Article 3) would be $\mathrm{d}(\delta T/\delta \mathrm{d}\xi) - (\delta T/\delta \xi) + \lambda + (\delta V/\delta \xi) = 0$. The equations for the two other variables are unchanged. Thus one will immediately obtain the pressure λ of the body on the surface, after substituting in the expression for λ

$$\lambda = \frac{\delta T}{\delta \xi} - \mathrm{d}\frac{\delta T}{\delta \mathrm{d}\xi} - \frac{\delta V}{\delta \xi}$$

$$\xi = a, \qquad \mathrm{d}\xi = 0$$

Since the application of our general formula does not present any difficulty, only one or two examples of the formula will be given.

Subsection I
The Oscillation Of A Simple Pendulum Of Given Length

15. We will place the origin of the coordinate axes at the point of suspension of the pendulum with the z-coordinate, vertical and directed upwards. But, instead of the rectangular coordinates x, y, z we will use a radius r which will be the length of the pendulum with two angles ψ and φ, where the first angle will be the inclination of the pendulum to the vertical and the second angle will give the position of the pendulum as it rotates about the vertical. Thus one will have

$$x = r \sin\psi \cos\varphi, \qquad y = r \sin\psi \sin\varphi, \qquad z = r \cos\psi$$

and the quantity T will become, since r is constant

$$T = \frac{r^2(\sin^2\psi \, \mathrm{d}\varphi^2 + \mathrm{d}\psi^2)}{2\,\mathrm{d}t^2}$$

It is noteworthy to observe that the angle ψ which is used here is the complement of the angle ψ that we used earlier and which represented the inclination of the radius r to the horizontal plane, instead of representing its inclination to the vertical.

The force R directed towards the origin of the radius r will be equal to zero. The force Q can be taken to represent gravity, which will be designated by g. Since it acts parallel to the z-coordinate and tends to increase this coordinate, instead of having the force Q assumed acting to shorten the distance q, one should have $\mathrm{d}q = -\mathrm{d}z = -\mathrm{d}(r\cos\psi)$ assuming that the center of this force is at an infinite distance. Thus one will simply have

$$\delta V = -g\,\delta(r\cos\psi) = gr\sin\psi\,\delta\psi$$

The equations relative to ψ and φ will then become, after dividing by r^2

$$\frac{\mathrm{d}^2\psi}{\mathrm{d}t^2} - \frac{\sin\psi\cos\psi\,\mathrm{d}\varphi^2}{\mathrm{d}t^2} + \frac{g}{r}\sin\psi = 0$$
$$\frac{\mathrm{d}^2(\sin^2\psi\,\mathrm{d}\varphi)}{\mathrm{d}t^2} = 0$$

The integral of the second of these equations is $\sin^2\psi\,\mathrm{d}\varphi/\mathrm{d}t = C$ and the expression for $\mathrm{d}\varphi$ obtained from this equation can be substituted in the first equation in order to obtain

$$\frac{\mathrm{d}^2\psi}{\mathrm{d}t^2} - \frac{C^2\cos\psi}{\sin^3\psi} + \frac{g}{r}\sin\psi = 0$$

After multiplying this equation by $2\,\mathrm{d}\psi$ and integrating, the following equation results

$$\frac{\mathrm{d}\psi^2}{\mathrm{d}t^2} + \frac{C^2}{\sin^2\psi} - 2\frac{g}{r}\cos\psi = E$$

where C and E are two constants which depend upon the initial conditions.

This last integral immediately gives

$$\mathrm{d}t = \frac{\sin\psi\,\mathrm{d}\psi}{\sqrt{\left(E + 2\frac{g}{r}\cos\psi\right)\sin^2\psi - C^2}}$$

and since there results from the first equation, $\mathrm{d}\varphi = C\,\mathrm{d}t/\sin^2\psi$, one will have

$$\mathrm{d}\varphi = \frac{C\,\mathrm{d}\psi}{\sin\psi\sqrt{\left(E + \frac{2g}{r}\cos\psi\right)\sin^2\psi - C^2}}$$

an equation whose variables have been separated but whose second members are only integrable by the rectification of conic sections.

The equation in t and ψ will give the time required by the pendulum to traverse vertically the angle ψ, and the equation in φ and ψ will give the curve described by the body which composes the pendulum, which will be a sort of spherical spiral. If the equation $r\sin\psi = \rho$ is substituted, an equation will be obtained for the projection of this spiral on the horizontal plane between the radius vector ρ and the angle ψ described by this radius about the vertical.

16. If the quantity under the radical sign is equated to zero the following equation results

$$\left(E + \frac{2g}{r}\cos\psi\right)\sin^2\psi - C^2 = 0$$

which will give the largest and smallest values of the angle of inclination ψ. This equation, because of the identity $\sin^2\psi = 1 - \cos^2\psi$, is of the third degree, relative to the unknown $\cos\psi$. It will thus have a real root. But it is easy to see, from the nature of the problem, that there can be a maximum for ψ, without simultaneously having a minimum and vice versa. From this result, it follows that the three roots will necessarily be real[39] and two will give a maximum value and the third a minimum.

Let us designate by α and β the largest and smallest values of ψ. Then these two equations will result

$$\left(E + \frac{2g}{r}\cos\alpha\right)\sin^2\alpha - C^2 = 0$$

$$\left(E + \frac{2g}{r}\cos\beta\right)\sin^2\beta - C^2 = 0$$

which give

$$E = \frac{2g(\cos\alpha\sin^2\alpha - \cos\beta\sin^2\beta)}{r(\sin^2\beta - \sin^2\alpha)}$$

$$C^2 = \frac{2g\sin^2\alpha\sin^2\beta(\cos\alpha - \cos\beta)}{r(\sin^2\beta - \sin^2\alpha)}$$

These expressions can be simplfied to the following form

$$E = \frac{2g(1 - \cos^2\alpha - \cos^2\beta - \cos\alpha\cos\beta)}{r(\cos\alpha + \cos\beta)}$$

$$C^2 = \frac{2g\sin^2\alpha\sin^2\beta}{r(\cos\alpha + \cos\beta)}$$

The expressions are to be substituted in the equation for $\cos\psi$, which after changing the signs, has the form

$$\frac{2g}{r}\cos^3\psi + E\cos^2\psi - \frac{2g}{r}\cos\psi + C^2 - E = 0$$

and from the nature of the equations, its first member will become

$$\frac{2g}{r}(\cos\psi - \cos\alpha)(\cos\psi - \cos\beta)\left(\cos\psi + \cos\alpha + \cos\beta + \frac{Er}{2g}\right)$$

This quantity, taken with the minus sign, will be identical to the quantity under the radical sign in the last two equations of the preceding article.

But after reducing the expression

$$\cos\alpha + \cos\beta + \frac{Er}{2g} = \frac{1 + \cos\alpha\cos\beta}{\cos\alpha + \cos\beta}$$

for which the preceding quantity becomes

$$-\frac{2g}{r}(\cos\psi - \cos\alpha)(\cos\psi - \cos\beta)\left(\cos\psi + \frac{1+\cos\alpha\cos\beta}{\cos\alpha + \cos\beta}\right)$$

17. Let us now assume

$$\cos\psi - \cos\alpha\sin^2\sigma + \cos\beta\cos^2\sigma$$

It is clear that the value of β for ψ, which is assumed to be the smallest, will correspond to $\sigma = 0, 2\pi, 4\pi$, etc. and that the value α, which is assumed to be the largest, will correspond to $\sigma = \pi/2, 3\pi/2, 5\pi/2$, etc. where π is 180 degrees. Thus one will have

$$\cos\psi - \cos\alpha = (\cos\beta - \cos\alpha)\cos^2\sigma$$
$$\cos\psi - \cos\beta = (\cos\alpha - \cos\beta)\sin^2\sigma$$
$$\cos\psi + \frac{1+\cos\alpha\cos\beta}{\cos\alpha+\cos\beta} = \frac{1 + 2\cos\alpha\cos\beta + \cos^2\alpha\sin^2\sigma + \cos^2\beta\cos^2\sigma}{\cos\alpha+\cos\beta}$$

In addition, there is the following equation

$$\sin\psi\,\mathrm{d}\psi = -\mathrm{d}\cos\psi = 2(\cos\beta - \cos\alpha)\sin\sigma\cos\sigma\,\mathrm{d}\sigma$$

Therefore, after making these substitutions in the differential equations in t and ψ of the preceding article, it will become

$$\mathrm{d}t = \frac{2\,\mathrm{d}\sigma\sqrt{\cos\alpha+\cos\beta}}{\sqrt{\frac{2g}{r}(1 + 2\cos\alpha\cos\beta + cos^2\alpha\sin^2\sigma + \cos^2\beta\cos^2\sigma)}}$$

and if the following definitions are made in order to shorten the expressions

$$\kappa^2 = \frac{\cos\alpha+\cos\beta}{2 + 4\cos\alpha\cos\beta + \cos^2\alpha + \cos^2\beta}$$
$$\Sigma = \sqrt{1 + k^2(\cos\beta - \cos\alpha)\cos 2\sigma}$$

it will reduce to

$$\mathrm{d}t = \sqrt{\frac{r}{g}}\,\frac{2\kappa\,\mathrm{d}\sigma}{\Sigma}$$

Then the following equation results

$$\mathrm{d}\varphi = \frac{C\,\mathrm{d}t}{\sin^2\psi} = \sqrt{\frac{2g}{r}}\,\frac{\sin\alpha\sin\beta}{\sqrt{\cos\alpha+\cos\beta}}\,\frac{\mathrm{d}t}{\sin^2\psi}$$
$$= \frac{\kappa\sqrt{2}\sin\alpha\sin\beta}{\sqrt{\cos\alpha+\cos\beta}}\left[\frac{\mathrm{d}\sigma}{(1+\cos\psi)\Sigma} + \frac{\mathrm{d}\sigma}{(1-\cos\psi)\Sigma}\right]$$

where one shall substitute for $\cos\psi$ its equivalent $\cos 2\sigma$

$$\cos\psi = \tfrac{1}{2}(\cos\alpha + \cos\beta) + \tfrac{1}{2}(\cos\beta - \cos^2\alpha)\cos 2\sigma$$

By integrating these equations from $\sigma = 0$ to $\sigma = \pi/2$, one will have the time and the angle of rotation included between the lowest point where the inclination of the pendulum with the vertical is β and the highest point where the inclination is α but these integrations generally depend upon the rectification of conic sections. If the value of φ between these two limits of s is a multiple of π, the spiral described by the pendulum will repeat itself after executing a given number of spirals. But if this value is not a multiple of π, the spiral will describe an infinity of different revolutions.

18. When the pendulum executes only rather small excursions in height, in such a manner that the angles α and β are not very different, the difference $(\cos\beta - \cos\alpha)$ will also be rather small so that the radical Σ can be reduced to a converging series.

Let us assume $\kappa(\cos\beta - \cos\alpha) = \sin 2\gamma = 2\tan\gamma/(1+\tan^2\gamma)$. The function Σ will become

$$\Sigma = \cos\gamma\sqrt{1+\tan^2\gamma + 2\tan\gamma\cos 2\sigma}$$

The irrational function $(1+\tan^2\gamma + 2\tan\gamma\cos 2\sigma)^{-1/2}$ can be reduced to a series of the form

$$A + B\cos 2\sigma + C\cos 4\sigma + D\cos 6\sigma + \cdots$$

in which one will have, after putting in the last formulas of Article 98 of the preceding section

$$\begin{aligned}
&\varphi = 2\sigma, \quad a' = 1, \quad a'' = -\tan\gamma, \quad n = \tfrac{1}{2}, \quad n' = \frac{1.3}{2.4}, \quad n'' = \frac{1.3.5}{2.4.6}, \quad \ldots\\
&A = 1 + n^2\tan^2\gamma + n'^2\tan^4\gamma + n''^2\tan^6\gamma + \cdots\\
&B = -2(\tan\gamma + nn'\tan^3\gamma + n'n''\tan^5\gamma + \cdots\\
&C = 2(n'\tan^2\gamma + nn''\tan^4\gamma + n''n'''\tan^6\gamma + \cdots\\
&\vdots
\end{aligned}$$

Thus one will have after substituting

$$dt = \sqrt{\frac{r}{g}}\frac{2\kappa}{\cos\gamma}(A + B\cos 2\sigma + C\cos 4\sigma + \cdots)d\sigma$$

and after integrating with the integral's lower limit beginning at $\sigma = 0$

$$t = \sqrt{\frac{r}{g}}\frac{2\kappa}{\cos\gamma}(A\sigma + \tfrac{1}{2}B\sin 2\sigma + \tfrac{1}{4}C\sin 4\sigma + \cdots)$$

By putting $\sigma = \pi/2$, the time for the pendulum to move from the highest point to the lowest point will be obtained. If this time is denoted by T, one will have

$$T = A\pi\sqrt{\frac{r}{g}\frac{\kappa}{\cos\gamma}}$$

If the values of t which correspond to

$$s = \frac{3\pi}{2}, \frac{5\pi}{2} \quad \cdots$$

are denoted by T', T'', etc. we will have that

$$T' = 3T, \qquad T'' = 5T, \qquad \cdots$$

It is clear from the preceding equation that the pendulum will always return to the same height after a time equal to $2T$, which consequently, is the duration of onc complete oscillation.

19. In the same manner, one can obtain the corresponding angle φ. In order to obtain φ, make the following definitions

$$\frac{\cos\beta - \cos\alpha}{2 + \cos\alpha + \cos\beta} = \sin 2\mu$$
$$\frac{\cos\beta - \cos\alpha}{2 - \cos\alpha - \cos\beta} = \sin 2\nu$$

and one will have

$$\frac{1}{1 + \cos\psi} = \frac{2}{(2 + \cos\alpha + \cos\beta)\cos^2\mu(1 + \tan^2\mu + 2\tan\mu\cos 2\sigma)}$$
$$\frac{1}{1 - \cos\psi} = \frac{2}{(2 - \cos\alpha - \cos\beta)\cos^2\nu(1 + \tan^2\nu - 2\tan\nu\cos 2\sigma)}$$

If in the same formulas of Article 98 of SECTION VII, the quantity n is put equal to unity, there will result

$$n' = 1, \qquad n'' = 1, \qquad \cdots$$

then

$$(1 + \tan^2\mu + 2\tan\mu\cos 2\sigma)^{-1} = (\mathrm{A}) + (\mathrm{B})\cos 2\sigma$$
$$+(\mathrm{C})\cos 4\sigma + (\mathrm{D})\cos 6\sigma + \cdots$$

where

$$(A) = 1 + \tan^2\mu + \tan^4\mu + \tan^6\mu + \cdots = \frac{1}{1 - \tan^2\mu}$$

$$(B) = -2\tan\mu(1 + \tan^2\mu + \tan^4\mu + \cdots) = \frac{-2\tan\mu}{1 - \tan^2\mu}$$

$$(C) = 2\tan^2\mu(1 + \tan^2\mu + \tan^4\mu + \cdots) = \frac{2\tan^2\mu}{1 - \tan^2\mu}$$

$$\vdots$$

Thus one will have

$$(1 + \tan^2\mu + 2\tan\mu\cos 2\sigma)^{-1}$$
$$= \frac{1}{1 - \tan^2\mu}(1 - 2\tan\mu\cos 2\sigma + 2\tan^2\mu\cos 4\sigma - 2\tan^3\mu\cos 6\sigma + \cdots$$

If this series is multiplied by the following series

$$A + B\cos 2\sigma + C\cos 4\sigma + \cdots$$

the product will again be of the form

$$A' + B'\cos 2\sigma + C'\cos 4\sigma + \cdots$$

and the following equation will be obtained

$$A' = \frac{A - B\tan\mu + C\tan^2\mu + D\tan^3\mu + \cdots}{1 - \tan^2\mu}$$

One will also have

$$\frac{1}{(1 + \cos\psi)\Sigma} = \frac{2}{(2 + \cos\alpha + \cos\beta)\cos^2\mu\cos\gamma}(A' + B'\cos 2\sigma + C'\cos 4\sigma + \cdots)$$

Similarly, the following equation will be obtained

$$\frac{1}{(1 - \cos\psi)\Sigma} = \frac{2}{(2 - \cos\alpha - cos\beta)\cos^2\nu\cos\gamma}(A'' + B''\cos 2\sigma + C''\cos 4\sigma + \cdots)$$

where, after replacing μ by $-\nu$, the following equation is obtained

$$A'' = \frac{A + B\tan\nu + C\tan^2\nu + D\tan^3\nu + \cdots}{1 - \tan^2\nu}$$

After making these substitutions in the expression for $d\varphi$ of Article 17 and integrating such that φ is equal to zero when $\sigma = 0$, one will have

$$\varphi = \frac{k\sqrt{2\sin\alpha\sin\beta}}{\sqrt{\cos\alpha+\cos\beta}}\left\{\begin{array}{l}\dfrac{2A's + B'\cos 2\sigma + \frac{1}{2}C'\cos 4\sigma + \cdots}{(2+\cos\alpha+\cos\beta)\cos^2\mu\cos\gamma} \\ +\dfrac{2A''s + B''\cos 2\sigma + \frac{1}{2}C''\cos 4\sigma + \cdots}{(2-\cos\alpha-\cos\beta)\cos^2\nu\cos\gamma}\end{array}\right\}$$

By putting $\sigma = \pi/2$, one will obtain the angle between the planes passing through the vertical and through the highest and lowest points of the curve described by the pendulum. This angle, which is denoted by Φ, will be given by the following equation

$$\Phi = \frac{\pi A'k\sqrt{2\sin\alpha\sin\beta}}{\sqrt{\cos\alpha+\cos\beta}(2+\cos\alpha+\cos\beta)\cos^2\mu\cos\gamma} + \frac{\pi A''k\sqrt{2\sin\alpha\sin\beta}}{\sqrt{\cos\alpha\cos\beta}(2-\cos\alpha-\cos\beta)\cos^2\nu\cos\gamma}$$

Because all the highest points or the peaks of the curve, correspond to $s = \pi/2,\ 3\pi/2,\ 5\pi/2$, etc., if one denotes by Φ', Φ'', etc. the values of φ for $s = 3\pi/2,\ 5\pi/2$, etc., one will have

$$\Phi' = 3\Phi, \qquad \Phi'' = 5\Phi, \qquad \cdots$$

Thus the angle included between two consecutive peaks, and corresponding to a complete oscillation of the pendulum, will be equal to 2Φ.

20. If it is assumed that the angles α and β are very small of the first order, the quantity $(\cos\beta - \cos\alpha)$ will be very small of the second order. Consequently, the angle γ will also be very small of the second order. Thus by neglecting only the very small quantities of the fourth order, one will have

$$A = 1, \qquad \cos\gamma = 1$$

thus

$$T = 2\pi\sqrt{\frac{r}{g}}\sqrt{\frac{\cos\alpha+\cos\beta}{2+4\cos\alpha\cos\beta+\cos^2\alpha+\cos^2\beta}}$$

and $2T$ will be of the fourth order of magnitude. It represents the time required for a complete oscillation.

If the quantities of the second order were neglected, this expression will be $\pi\sqrt{r/g}$. This is the known expression for the duration of very small oscillations of a pendulum for which the length is r and where g can be set equal to unity. But the preceding analysis shows that the duration is the same, whatever the magnitude of the oscillations, whether they take place in a vertical plane or whether the pendulum has simultaneously a rotational motion about the vertical.

By retaining the quantities of the second order, one can simplify the preceding formula by replacing $\cos\alpha$ and $\cos\beta$ with their approximative values to the fourth order $1-\alpha^2/2$, $1-\beta^2/2$. By always neglecting the terms of the fourth order, one will have the duration of very small oscillations to the fourth order, expressed by

$$\pi\sqrt{\frac{r}{g}}\left(1+\frac{\alpha^2+\beta^2}{16}\right)$$

21. When the angle β, which corresponds to the lowest point of the pendulum is zero, the pendulum always assumes the vertical position and the oscillations are in the vertical plane. By putting $\beta=0$, it is clear from the formula of Article 5 that the angle φ is zero. This is the case which is ordinarily considered and which occurs each time the pendulum is displaced from the vertical by an angle α, and it is left to fall free without any initial impulse. But if the pendulum receives any impulse in a direction different from the vertical, it will perform conical oscillations and the angle β will not be zero.

In this case, if it is also assumed that the angles α and β are very small and if, as a first approximation, the very small quantities of second order are neglected, one will have

$$k=\tfrac{1}{2},\qquad \gamma=0,\qquad A=1,\qquad B=0,\qquad C=0,\qquad \cdots$$
$$\mu=0,\qquad \sin 2\nu=\frac{\alpha^2-\beta^2}{\alpha^2+\beta^2},\qquad A'=1,\qquad A''=\frac{1}{1+\tan^2\nu}=\cos^2\nu$$

therefore[40] $\Phi=\pi\alpha\beta/(\alpha^2+\beta^2)$ and 2Φ will be the angle to the vertical made by two consecutive peaks of the curve. Thus, if the ratio of $\alpha\beta$ to $\alpha^2+\beta^2$ is rational, the angle 2Φ will be a multiple of π and the curve described by the pendulum will only be formed by a given number of repeating peaks. Otherwise, the curve will be a sort of continuous spiral. But these conclusions are only approximate and to obtain more precise results, one shall extend the approximation by means of a series which was given earlier.

This problem was solved long ago by Clairaut in the *Mémoires de l'Académie des Sciences* for 1735 but in a less complete manner. The approximate results which were just found are in agreement with his result if it is assumed that $\beta=0$ in the expression for T and $\beta=\alpha$ in the expression for Φ.

22. The preceding formulas hold as long as the angle α is different from β because, however small their difference, there is always a maximum and minimum in the vertical motions of the pendulum. But, if one has rigorously that $\alpha=\beta$, there is no longer a maximum or minimum. The pendulum always makes the angle α with the vertical and consequently, it will describe a cone with a circular base.

This assumption is possible because then (Article 16 and 17) the quantity under the radical sign in the expression for $\mathrm{d}t$ has two equal factors $(\cos\psi - \cos\alpha)$ such that, from the theory given in Article 83 of the preceding section, one must[41] always put $\cos\psi = \cos\alpha$. This is the case of conical oscillations that Huygens was the first to consider.

In this case, the equation (Article 4)

$$\mathrm{d}\varphi = \frac{C\,\mathrm{d}t}{\sin^2\psi} = \sqrt{\frac{2g}{r\cos\alpha}}\,\mathrm{d}t$$

will give $\varphi = t\sqrt{2g/r\cos\alpha}$ such that the time for a complete revolution will be expressed by $2\pi\sqrt{r/2g}$. For this case to exist, the pendulum must receive an angular velocity of rotation about the vertical, expressed by $\mathrm{d}\varphi/\mathrm{d}t = \sqrt{2g/r\cos\alpha}$ which only depends upon the height of the cone which it describes.

23. If the pendulum moves in a medium whose resistance varies as the square of the velocity and whose density is expressed by Γ, the following terms should be added to δV in order to obtain the equations of motion (Article 2)

$$\Gamma\,\mathrm{d}s\left(\frac{\delta T}{\delta \mathrm{d}\psi}\delta\psi + \frac{\delta T}{\delta \mathrm{d}\varphi}\delta\varphi\right)$$

keeping the expression for the quantity T of Article 11 in which r is constant.

Thus, one will have to add the term $\Gamma\,\mathrm{d}s\,\mathrm{d}\psi/\mathrm{d}t^2$ to the first member of the first differential equation of this article and the term $\Gamma\sin^2\psi\,\mathrm{d}s\,\mathrm{d}\varphi/\mathrm{d}t^2$ to the first member of the second differential equation.

With the addition of these terms, the equations which were non-integrable, become integrable. But when the resistance is very small with respect to the force of gravity, which holds for bodies moving slowly in air, these equations can be solved by an approximation with the substitution of the terms owing to the resistance of the values of ψ and φ as functions of t, which occur in a vacuum, and by finding the small quantities that these known terms will add to these same expressions.

The two equations in question will be

$$\frac{\mathrm{d}^2\psi}{\mathrm{d}t^2} - \frac{\sin\psi\cos\psi\,\mathrm{d}\varphi^2}{\mathrm{d}t^2} + \frac{g}{r}\sin\psi + \frac{\Gamma\,\mathrm{d}s\,\mathrm{d}\psi}{\mathrm{d}t^2} = 0$$

$$\frac{\mathrm{d}(\sin^2\psi\,\mathrm{d}\varphi)}{\mathrm{d}t^2} + \frac{\Gamma\sin^2\psi\,\mathrm{d}s\,\mathrm{d}\varphi}{\mathrm{d}t^2} = 0$$

The second equation, after division by $\sin^2\psi\,\mathrm{d}\varphi/\mathrm{d}t^2$ and then integrated, gives $\sin^2\psi\,\mathrm{d}\varphi/\mathrm{d}t = Ci^{-\Gamma s}$, where i is the number for which the hyperbolic logarithm is unity.

Then, after the first equation is multiplied by $2\,\mathrm{d}\psi$ and added to the second equation which is multiplied by $2\,\mathrm{d}\varphi$, the following integral equation is obtained

$$\frac{\mathrm{d}\psi^2 + \sin^2\psi\,\mathrm{d}\varphi^2}{\mathrm{d}t^2} - \frac{2g\cos\psi}{r} + \frac{2\Gamma}{r^2}\int\frac{\mathrm{d}s^2}{\mathrm{d}t^2}\mathrm{d}s = E$$

because $r^2(\mathrm{d}\psi^2 + \sin^2\psi\,\mathrm{d}\varphi^2) = \mathrm{d}s^2$.

Thus with the substitution of $Ci^{-\Gamma s}$ in place of C and $E - (2\gamma/r^2)\int(\mathrm{d}s^2/\mathrm{d}t^2)\mathrm{d}s$ in place of E the same differential equations in terms of t, φ and ψ are obtained which were found in Article 41. So, the effect of resistance will be reduced to varying these constants in the general solution given above in Article 13 where we did not consider the resistance and where the relations between the variables ψ, φ and t must be deduced from the equations

$$\frac{\sin^2\psi\,\mathrm{d}\varphi}{\mathrm{d}t} = C, \qquad \frac{\mathrm{d}\psi^2}{\mathrm{d}t^2} + \frac{C^2}{\sin^2\psi} - 2\frac{g}{r}\cos\psi = E$$

Therefore, if the quantities C and E are viewed as variables, the following equations will result

$$\mathrm{d}C = \Gamma C\,\mathrm{d}s, \qquad \mathrm{d}E = -\frac{2\Gamma}{r^2}\frac{\mathrm{d}s^2}{\mathrm{d}t^2}\mathrm{d}s = -\frac{2\Gamma}{r^2}\left(E + \frac{2g\cos\psi}{r}\right)\mathrm{d}s$$

and

$$\mathrm{d}s = \frac{\sin\psi\sqrt{\left(E + \frac{2g}{r}\cos\psi\right)}}{\sqrt{\left(E + \frac{2g\cos\psi}{r}\right)\sin^2\psi - C^2}}\mathrm{d}\psi$$

When the pendulum performs only vertical oscillations, one has $C = 0$, and consequently, $\mathrm{d}s = \mathrm{d}\psi$. The equation in E becomes integrable after it is multiplied by $i^{2\Gamma\psi/r^2}$ and the integral is

$$Ei^{2\Gamma\psi/r^2} = (E) - \frac{2\Gamma}{r^2}\int i^{2\Gamma\psi/r^2}\cos\psi\,\mathrm{d}\psi$$

where (E) is an arbitrary constant which replaces the constant E which became variable. After integration by parts the following equation will be found

$$\int i^{2\Gamma\psi/r^2}\cos\psi\,\mathrm{d}\psi = \frac{i^{2\Gamma\psi/r^2}\left(\sin\psi - \frac{2\Gamma}{r^2}\cos\psi\right)}{1 + \frac{4\Gamma^2}{r^2}}$$

thus one will have

$$E = (E)i^{-2\Gamma\psi/r^2} - \frac{2\Gamma}{r^2 + 4\Gamma^2}\left(\sin\psi - \frac{2\Gamma}{r^2}\cos\psi\right)$$

This is the expression which has to be substituted for E in the differential equation which will give the value of t as a function of ψ. By assuming the coefficient Γ to be small, one will easily obtain the alteration produced in the expression for the time t by the resistance of the medium.

24. In the case of the pendulum, by taking, as we just have done r, φ, ψ for the three coordinates, one has the equation $r = a$ where a is the given length of the pendulum. Thus from Article 14, by replacing ξ by r, one will immediately have the value of λ, which will express the tensile force in the string which holds the body on the spherical surface.

This force is thus expressed by

$$\frac{\delta T}{\delta r} - \mathrm{d}\frac{\delta T}{\delta \mathrm{d}r} - \frac{\delta V}{\delta r}$$

and after substituting for T and V the following complete expressions

$$T = \frac{r^2(\mathrm{d}\psi^2 + \sin^2\psi\,\mathrm{d}\varphi^2) + \mathrm{d}r^2}{2\,\mathrm{d}t^2}, \qquad V = -gr\cos\psi$$

and then setting r equal to a constant, the following equation is obtained

$$\frac{\delta T}{\delta \mathrm{d}r} = 0, \qquad \frac{\delta T}{\delta r} = \frac{r(\mathrm{d}\psi^2 + \sin^2\psi\,\mathrm{d}\varphi^2)}{\mathrm{d}t^2}, \qquad \frac{\delta V}{\delta r} = -g\cos\psi$$

Consequently

$$\lambda = \frac{r(\mathrm{d}\psi^2 + \sin^2\psi\,\mathrm{d}\varphi^2)}{\mathrm{d}t^2} + g\cos\psi = \frac{2T}{r} - \frac{V}{r}$$

where it is clear that $2T = u^2$ (Article 12) such that the tension in the cord which is part of the pendulum will be expressed by $u^2/r + g\cos\psi$.

When the pendulum moves in a vacuum, one has from the same article, where c is the velocity when $\psi = 0$, the following equation

$$u^2 = 2(H - V) = c^2 - 2gr(1 - \cos\psi)$$

The tension, denoted by λ, will be expressed by

$$\lambda = \frac{c^2}{r} - g(2 - 3\cos\psi)$$

25. We have assumed until now that the length of the pendulum is invariable. But, if this length changed from one instant to the next, according to a known law where r is a given function of t, then r should be assumed variable in the differential equations. But one would equally have $\delta r = 0$, as in the case where r is a constant. Thus we will begin with the equations

$$T = \frac{r^2(\sin^2\psi\, \mathrm{d}\varphi^2 + \mathrm{d}\psi^2) + \mathrm{d}r^2}{2\,\mathrm{d}t^2}, \qquad V = -gr\cos\psi$$

The equation relative to r would not hold but the other two would become

$$\frac{\mathrm{d}(r^2\,\mathrm{d}\psi)}{\mathrm{d}t} - \frac{r^2\sin\psi\cos\psi\,\mathrm{d}\varphi^2}{\mathrm{d}t^2} + gr\sin\psi = 0, \qquad \mathrm{d}\frac{r^2\sin^2\psi\,\mathrm{d}\varphi}{\mathrm{d}t^2} = 0$$

Finally, if the cord which holds the body were elastic and extensible, and if the tensile force in the cord is denoted by F which can only be a function of r, it will only be necessary to add $F\,\delta r$ to δV. Then for the equation relative to r there will result

$$\frac{\mathrm{d}^2 r}{\mathrm{d}t^2} - \frac{r(\sin^2\psi\,\mathrm{d}\varphi^2 + \mathrm{d}\psi^2)}{\mathrm{d}t^2} + F - g\cos\psi = 0$$

and the other two equations will remain the same. In this case, one will always have the integral $T + V = H$, or $V = \int F\,\mathrm{d}r - gr\cos\psi$.

Subsection II
The Motion Of A Heavy Body On An Arbitrary Surface Of Revolution

26. The axis of revolution is taken to be the z-axis. If one puts $x = \rho\cos\varphi$ and $y = \rho\sin\varphi$, then z will be the abscissa and ρ will be the ordinate of the curve which when revolved about the axis of the abscissas, generates the proposed solid. Thus an equation between z and ρ will be obtained for which z will be a given function of ρ.

If it is assumed that the z-axis is vertical and that the z-ordinates are directed upward, there results

$$T = \frac{\rho^2\,\mathrm{d}\varphi^2 + \mathrm{d}\rho^2 + \mathrm{d}z^2}{2\,\mathrm{d}t^2}, \qquad V = -gz$$

If ρ and φ are taken for the two independent variables, the two equations relative to these variables will immediately (Article 11) be obtained

$$\frac{\mathrm{d}^2\rho}{\mathrm{d}t^2} - \frac{\rho\,\mathrm{d}\varphi^2}{\mathrm{d}t^2} + \left(\frac{\mathrm{d}^2 z}{\mathrm{d}t^2} - g\right)\frac{\delta z}{\delta\rho} = 0, \qquad \mathrm{d}\frac{\rho^2\,\mathrm{d}\varphi}{\mathrm{d}t^2} = 0$$

If the z-axis were not vertical, but inclined to the vertical by an angle α, the expression for T will remain the same, but the expression for V will become $-g(z\cos\alpha - x\sin\alpha)$ such that it will only be necessary to change g in the first equation to $g\cos\alpha$ and add to its first member the term $g\sin\alpha\cos\varphi$ and then to add to the first member of the second equation, the term $g\sin\alpha\sin\varphi$.

In general, whatever the change given to the location of the surface or the line on which the body moves, the expression for T from which are derived the differential terms of the equation does not change. Only the expression for V depends upon the position of the surface or the line.

SECTION IX
ROTATIONAL MOTION

The importance and difficulty of this problem requires a separate section in order to treat it fully. At the outset, the most general formulas will be given and at the same time the simplest approach to representing the motion of rotation of a body or a system of bodies about a point will be developed. Then from these formulas using the methods of SECTION IV, the equations necessary to determine the motion of rotation of a system of bodies acted upon by arbitrary forces will be deduced. Finally, various applications of these equations will be presented.

Although this topic has already been treated by several mathematicians, the theory presented here will still prove useful. On the one hand, it will give a new way of solving the well-known problem of the rotation of bodies with an arbitrary configuration. On the other hand, it will be used to bring together and unite under the same point of view the solutions which have already been given for this problem and which are all founded on different principles and presented in diverse forms. This sort of harmonization is always instructive and can only prove to be very useful to the progress of analysis. It can even be said that it is necessary at the stage where mechanics is today because as this science progresses and includes new methods, it also becomes more complex and it can only be simplified if the applicable methods can be generalized and reduced.

Chapter I
The Rotation Of An Arbitrary System Of Bodies

Subsection I
General Formulas For Rotational Motion

The differential formulas found in PART I to express the variations given to the coordinates of an arbitrary system of points, for which the distances are assumed invariable, are a natural application of the investigation in question because this assumption only cancels the terms which would result from the variations of the distances between different points. So, the remaining terms express what is general and common to all points in the motion of the system, not accounting for their relative motion. This is precisely the common and absolute motion that we propose to examine here.

1. Let us reconsider the formulas of Article 55 of SECTION V, which were found by a direct analysis uniquely founded on the assumption that invariable distances are maintained between the points of the system. By changing the operator δ to d, one will have for the absolute motion of the system the following three equations

$$\begin{aligned}
dx &= d\lambda + z\,dM - y\,dN \\
dy &= d\mu + x\,dN - z\,dL \\
dz &= d\nu + y\,dL - x\,dM
\end{aligned}$$

where x, y, z represent, as usual, the coordinates of each point of the system with respect to three fixed and mutually perpendicular axes. The differentials $d\lambda$, $d\mu$, $d\nu$, dL, dM, dN are indeterminate quantities, the same for all points, and which depend only upon the motion of the system in general.

Now let x', y', z' be the coordinates of a given point of the system. Thus we will have

$$\begin{aligned}
dx' &= d\lambda + z'\,dM - y'\,dN \\
dy' &= d\mu + x'\,dN - z'\,dL \\
dz' &= d\nu + y'\,dL - x'\,dM
\end{aligned}$$

Consequently, if these formulas are subtracted from the preceding formulas and if for additional simplicity the following definitions are made

$$x = x' + \xi, \qquad y = y' + \eta, \qquad z = z' + \zeta$$

one will have these differential equations

$$d\xi = \zeta\,dM - \eta\,dN, \qquad d\eta = \xi\,dN - \zeta\,dL, \qquad d\zeta = \eta\,dL - \xi\,dM$$

in which the variables ξ, η, ζ will represent the coordinates of different points of the system, measured from a given point of the system. This point will be called from now on the *center of the system*. Since these equations are linear and solely of the first order, it follows from the known theory for this type of equation that if one designates by ξ', ξ'', ξ''' three particular values of ξ and by η', η'', η''', and ζ', ζ'', ζ''' the corresponding values of η and ζ, the following complete integrals will be obtained

$$\begin{aligned}
\xi &= a\xi' + b\xi'' + c\xi''' \\
\eta &= a\eta' + b\eta'' + c\eta''' \\
\zeta &= a\zeta' + b\zeta'' + c\zeta'''
\end{aligned}$$

where a, b, c are arbitrary constants.

It is clear that ξ', η', ζ' are nothing more than the coordinates of an arbitrary point of the system and that similarly, ξ'', η'', ζ'', and ξ''', η''', ζ''' are the coordinates of two other arbitrary points of the system.

Thus, by knowing the coordinates of three given points, one will have with the preceding equations, the values of the coordinates for any other points dependent upon the constants a, b, c. But it remains to determine the values of these constants.

2. If it is assumed, which is permissible, that in the initial state of the system the three given points lie on the three coordinate axes and that their distance from the origin is one unit, it is clear that one will have

$$\begin{array}{lll} \xi' = 1, & \eta' = 0, & \zeta' = 0 \\ \xi'' = 0, & \eta'' = 1, & \zeta'' = 0 \\ \xi''' = 0, & \eta''' = 0, & \zeta''' = 1 \end{array}$$

which will give $\xi = a, \eta = b, \zeta = c$.

Thus the quantities a, b, c, will be nothing more than the coordinates of an arbitrary point of the system referred to the same axes. But because of the motion of the system, the coordinate axes change their positions in space but remain fixed in the system because these coordinates are constants for the same point and only vary from one point to the next. The position of their axes at an arbitrary instant, with respect to the immobile axes ξ, η, ζ will only depend upon the coefficients $\xi', \xi'', \xi''', \eta', \eta''$, etc. Indeed if one puts $b = 0, c = 0$, one obtains $\xi = a\xi'$, $\eta = a\eta'$, $\zeta = a\zeta'$. Consequently, there results $a = \sqrt{\xi^2 + \eta^2 + \zeta^2}$.

It is easy to see that the coefficients ξ', η', ζ' are the cosines of the angles between the a-axis and the ξ-, η-, ζ-axes. It is even clear, by assuming a and c equal to zero, then a and b equal to zero, that the coefficients ξ'', η'', ζ'' are the cosines of the angles between the b-axis and the ξ-, η-, ζ-axes and the coefficients ξ''', η''', ζ''' are the cosines of the angles between the c-axis and the ξ-, η-, ζ-axes.

3. Since these coefficients represent, in general, the coordinates of three given points of the system assumed to be a unit distance from the origin, and placed at the origin of the axes of the rectangular coordinates a, b, c, one will have at the outset these three equations

$$\xi'^2 + \eta'^2 + \zeta'^2 = 1, \qquad \xi''^2 + \eta''^2 + \zeta''^2 = 1, \qquad \xi'''^2 + \eta'''^2 + \zeta'''^2 = 1$$

Then, because the mutual distances of these points are the hypotenuses of rectilinear triangles for which the length of the sides are equal to unity, one will have

$$(\xi' - \xi'')^2 + (\eta' - \eta'')^2 + (\zeta' - \zeta'')^2 = 2$$
$$(\xi' - \xi''')^2 + (\eta' - \eta''')^2 + (\zeta' - \zeta''')^2 = 2$$
$$(\xi'' - \xi''')^2 + (\eta'' - \eta''')^2 + (\zeta'' - \zeta''')^2 = 2$$

from which these three equations are obtained

$$\xi'\xi'' + \eta'\eta'' + \zeta'\zeta'' = 0, \quad \xi'\xi''' + \eta'\eta''' + \zeta'\zeta''' = 0, \quad \xi''\xi''' + \eta''\eta''' + \zeta''\zeta''' = 0,$$

Thus, there are six equations of condition between the nine coefficients $\xi', \xi'', \xi''', \eta', \eta''$, etc., which reduces the nine coefficients to three indeterminates.

4. By means of these equations, the general expressions of the coordinates ξ, η, ζ of Article 1 satisfy the original condition that the distance between two arbitrary points of the system remains invariable. Indeed, if ξ, η, ζ are the coordinates of one of these points, and $\xi 1, \eta 1, \zeta 1$ are the coordinates of another point, the square of their distance apart will be expressed by

$$(\xi - \xi 1)^2 + (\eta - \eta 1)^2 + (\zeta - \zeta 1)^2$$

and if the coordinates relative to the axes of a, b, c for the second point are designated by $a1, b1, c1$, one will have the values of $\xi 1, \eta 1, \zeta 1$ by changing a, b, c to $a1, b1, c1$ in the expressions for ξ, η, ζ.

After making these substitutions in the preceding expression and taking into account the six equations of condition, it will be reduced to

$$(a - a1)^2 + (b - b1)^2 + (c - c1)^2$$

Consequently, it will remain constant during the motion. From this result, it can be concluded that the six equations of condition are solely necessary for the respective positions of the different points of the system to depend only upon the constants a, b, c, and not upon the variables ξ', η', ζ', etc.

Also, it is clear that the coordinates ξ, η, ζ are only transformations of the coordinates a, b, c and that the six equations of condition are the result of the general condition

$$\xi^2 + \eta^2 + \zeta^2 = a^2 + b^2 + c^2$$

This conclusion is clear from the comparison of these formulas with those of Article 15 of SECTION III of PART I, in which the coordinates x, y, z, x', y', z', correspond to ξ, η, ζ, a, b, c and the coefficients α, β, γ, α', β', γ', $\alpha'', \beta'', \gamma''$ correspond to ξ', ξ'', ξ''', η', η'', η''', $\zeta', \zeta'', \zeta'''$.

5. If the expressions for ξ, η, ζ of Article 1 are added together after they are multiplied by ξ', η', ζ', respectively, then with ξ'', η'', ζ'', and finally with ξ''', η''', ζ''', one will immediately have, from the equations of condition of Article 3, the following inverse formulas

$$a = \xi\xi' + \eta\eta' + \zeta\zeta'$$
$$b = \xi\xi'' + \eta\eta'' + \zeta\zeta''$$
$$c = \xi\xi''' + \eta\eta''' + \zeta\zeta'''$$

and if these expressions for a, b, c, are substituted in the equation

$$\xi^2 + \eta^2 + \zeta^2 = a^2 + b^2 + c^2$$

which must hold whatever the values of ξ, η, ζ, it will give by a comparison of terms these new equations of condition

$$\begin{array}{lll} \xi'^2 + \xi''^2 + \xi'''^2 = 1, & \eta'^2 + \eta''^2 + \eta'''^2 = 1, & \zeta'^2 + \zeta''^2 + \zeta'''^2 = 1, \\ \xi'\eta' + \xi''\eta'' + \xi'''\eta''' = 0, & \xi'\zeta' + \xi''\zeta'' + \xi'''\zeta''' = 0, & \eta'\zeta' + \eta''\zeta'' + \eta'''\zeta''' = 0 \end{array}$$

which necessarily follow from those of Article 3 because these and the others also result from the general condition

$$\xi^2 + \eta^2 + \zeta^2 = a^2 + b^2 + c^2$$

6. But, if the expressions for a, b, c are directly investigated by the resolution of the equations of Article 1, one will have, from the known formulas

$$a = \frac{\xi(\eta''\zeta''' - \eta'''\zeta'') + \eta(\zeta''\xi''' - \zeta'''\xi'') + \zeta(\xi''\eta''' - \xi'''\eta'')}{k}$$
$$b = \frac{\xi(\zeta'\eta''' - \zeta'''\eta') + \eta(\xi'\zeta''' - \xi'''\zeta') + \zeta(\eta'\xi''' - \eta'''\xi')}{k}$$
$$c = \frac{\xi(\eta'\zeta'' - \eta''\zeta') + \eta(\zeta'\xi'' - \zeta''\xi') + \zeta(\xi'\eta'' - \xi''\eta')}{k}$$

assuming

$$k = \xi'\eta''\zeta''' - \eta'\xi''\zeta''' + \zeta'\xi''\eta''' - \xi'\zeta''\eta''' + \eta'\zeta''\xi''' - \zeta'\eta''\xi'''$$

These expressions must then be identical with those of the preceding article. Thus by comparing the coefficients of the quantities ξ, η, ζ, the following equations will be obtained

$$\begin{array}{lll} \eta''\zeta''' - \eta'''\zeta'' = k\xi', & \zeta''\xi''' - \zeta'''\xi'' = k\eta', & \xi''\eta''' - \xi'''\eta'' = k\zeta' \\ \zeta'\eta''' - \zeta'''\eta' = k\xi'', & \xi'\zeta''' - \xi'''\zeta' = k\eta'', & \eta'\xi''' - \eta'''\xi' = k\zeta'' \\ \eta'\zeta'' - \eta''\zeta' = k\xi''', & \zeta'\xi'' - \zeta''\xi' = k\eta''', & \xi'\eta'' - \xi''\eta' = k\zeta'' \end{array}$$

But if the squares of the first three are added together, one has

$$(\eta''\zeta''' - \eta'''\zeta'')^2 + (\zeta''\xi''' - \zeta'''\xi'')^2 + (\xi''\eta''' - \xi'''\eta'')^2 = k^2(\xi'^2 + \eta'^2 + \zeta'^2)$$

The first member of this equation can be put in the form

$$(\xi''^2 + \eta''^2 + \zeta''^2)(\xi'''^2 + \eta'''^2 + \zeta'''^2) - (\xi''\xi''' + \eta''\eta''' + \zeta''\zeta''')^2$$

Thus with the equations of condition of Article 3, this equation is reduced to $k^2 = 1$ from which $k = \pm 1$. In order to decide which of the two signs should be taken, one has only to consider the value of k for a particular case. The simplest case is the one where the three axes of the coordinates a, b, c coincide with the three axes of the coordinates ξ, η, ζ in which case one would have $\xi = a, \eta = b, \zeta = c$. Consequently, with the formulas of Article 1, we obtain $\xi' = 1, \eta'' = 1, \zeta''' = 1$. and the other quantities ξ'', ξ''', etc., are equal to zero. By making these substitutions in the general expression for k, it becomes equal to unity. Therefore, one will always have k positive and equal to unity.

7. Between the nine indeterminates $\xi', \xi'', \xi''', \eta', \eta'', \eta''', \zeta', \zeta'', \zeta'''$, there are essentially six equations of condition. These indeterminates can be reduced to three. It will suffice to reduce the six indeterminates $\xi', \xi'', \eta', \eta'', \zeta', \zeta''$ by means of the three equations of condition

$$\xi'^2 + \eta'^2 + \zeta'^2 = 1, \qquad \xi''^2 + \eta''^2 + \zeta''^2 = 1, \qquad \xi'\xi'' + \eta'\eta'' + \zeta'\zeta'' = 0$$

because the three other indeterminates $\xi''', \eta''', \zeta'''$ are already known as functions of those from the preceding formulas.

But this simplification is made much easier by using the sines or cosines of angles. One can even directly arrive at this result by the known transformations of coordinates.

Indeed, since ξ, η, ζ are the rectangular coordinates of an arbitrary point of the body with respect to three axes drawn from its center parallel to the fixed axes with coordinates x, y, z and since a, b, c are the rectangular coordinates of the same point with respect to three different axes passing through the same center, but fixed inside the body, and consequently, in variable locations with respect to the ξ-, η-, ζ-axes, it follows that to obtain the expressions for ξ, η, ζ as functions of a, b, c, it will suffice to transform, in the same general manner, these coordinates to the others.

In order to achieve this result, we will denote by ω the angle between the plane of the coordinates a and b with the plane of the coordinates ξ and η. We will also denote by ψ the angle made by the intersection of these two planes and the ξ-axis, and finally by φ the angle made by the a-axis and the same line of intersection. These three quantities ω, ψ, φ will be used, as is seen, to determine the position of the axes of the coordinates a, b, c, relative to the axes of the coordinates ξ, η, ζ. Consequently, these latter coordinates can be expressed as functions of the former by this means.

In order to fix these ideas, let us assume that the body under consideration is the Earth, that the plane of a, b is the plane of the equator, that the a-axis passes through the meridian, that the plane of ξ, η is the plane of the ecliptic and that the ξ-axis is directed toward the first point of Aries. It is clear that the angle ω will become the obliquity of the ecliptic and that the angle ψ will be the longitude of the equinox of autumn or the ascending node of the equator on the ecliptic and further, that φ will be the distance from the given meridian to the equinox.

In general, φ will be the angle that the body describes by rotating about the axis of coordinates c, which can be simply called because of this fact *the axis of the body*. The quantity $(90° - \omega)$ will be the angle of inclination of this axis on the plane of the coordinates ξ, η and the quantity $(\omega - 90°)$ will be the angle between the projection of this same axis and the axis of the ξ coordinates.

After having stated these facts, let us assume at the outset that the two coordinates a, b are changed to a', b', which are located in the same plane in such a manner that the axis of a' is at the intersection of the two planes and the axis of b' is perpendicular to this intersection. One will have

$$a' = a\cos\varphi - b\sin\varphi, \qquad b' = b\cos\varphi + a\sin\varphi$$

Let us then assume that the two coordinates b', c are replaced by b'', c' of which b'' is always perpendicular to the intersection of the planes, but is located in the plane of ξ, η and for which c' is perpendicular to this last plane. Similarly, one will find

$$b'' = b'\cos\omega - c\sin\omega, \qquad c'' = c\cos\omega + b'\sin\omega$$

Finally, let us also assume that the coordinates a', b'' which are already in the plane of ξ, η are replaced by a'', b''' located in this same plane but such that the a''-axis coincides with the ξ-axis. In the same manner, one will find

$$a'' = a'\cos\psi - b''\sin\psi, \qquad b''' = b''\cos\psi + a'\sin\psi$$

It is obvious that the three coordinates a'', b''', c' will be the same as the coordinates ξ, η, ζ, because they are referred to the same axes so that by successively substituting the values of a', b'', b', one will obtain the expressions for ξ, η, ζ as functions of a, b, c which will be of the same form as those of Article 1, assuming that

$$\begin{aligned}
\xi' &= \cos\varphi\cos\psi - \sin\varphi\sin\psi\sin\omega \\
\xi'' &= -\sin\varphi\cos\psi - \cos\varphi\sin\psi\cos\omega \\
\xi''' &= \sin\psi\sin\omega \\
\eta' &= \cos\varphi\sin\psi + \sin\varphi\cos\psi\cos\omega \\
\eta'' &= -\sin\varphi\sin\psi + \cos\varphi\cos\psi\cos\omega \\
\eta''' &= -\cos\psi\sin\omega \\
\zeta' &= \sin\varphi\sin\omega \\
\zeta'' &= \cos\varphi\sin\omega \\
\zeta''' &= \cos\omega
\end{aligned}$$

These expressions also satisfy the six equations of condition of Article 3 as well as those of Article 5 and they solve these equations entirely because they contain three indeterminate variables φ, ψ, ω.

By substituting these expressions, the equations for the coordinates ξ, η, ζ become simpler but it is useful to keep the coefficients ξ', η', ζ', etc. to maintain the symmetry in the formulas and to facilitate their simplification.

8. Since the quantities ξ', η', ζ', are particular values of ξ, η, ζ, they must also satisfy the differential equations of Article 1 for the latter variables. Thus one will have

$$\mathrm{d}\xi' = \zeta'\,\mathrm{d}M - \eta'\mathrm{d}N, \qquad \mathrm{d}\eta' = \xi'\,\mathrm{d}N - \zeta'\,\mathrm{d}L, \qquad \mathrm{d}\zeta' = \eta'\,\mathrm{d}L - \xi'\,\mathrm{d}M$$

and similarly

$$\begin{aligned} \mathrm{d}\xi'' &= \zeta''\,\mathrm{d}M - \eta''\,\mathrm{d}N, \qquad \mathrm{d}\eta'' = \xi''\,\mathrm{d}N - \zeta''\,\mathrm{d}L, \qquad \mathrm{d}\zeta'' = \eta''\,\mathrm{d}L - \xi''\,\mathrm{d}M \\ \mathrm{d}\xi''' &= \zeta'''\,\mathrm{d}M - \eta'''\,\mathrm{d}N, \quad \mathrm{d}\eta''' = \xi'''\,\mathrm{d}N - \zeta'''\,\mathrm{d}L, \quad \mathrm{d}\zeta''' = \eta'''\,\mathrm{d}L - \xi'''\,\mathrm{d}M \end{aligned}$$

From this result, the values of the quantities $\mathrm{d}L$, $\mathrm{d}M$, $\mathrm{d}N$, can easily be obtained as functions of $\xi', \eta', \zeta', \xi''$, etc. Indeed, if the values of $\mathrm{d}\zeta'$, $\mathrm{d}\zeta''$, $\mathrm{d}\zeta'''$ are added together after having multiplied them by η', η'', η''', one will obtain from the equations of condition

$$\mathrm{d}L = \eta'\,\mathrm{d}\zeta' + \eta''\,\mathrm{d}\zeta'' + \eta'''\,\mathrm{d}\zeta'''$$

Similarly, one will find by multiplying $\mathrm{d}\xi', \mathrm{d}\xi'', \mathrm{d}\xi'''$ by $\zeta', \zeta'', \zeta'''$ and $\mathrm{d}\eta', \mathrm{d}\eta'', \mathrm{d}\eta'''$ by ξ', ξ'', ξ''' the following equations

$$\begin{aligned} \mathrm{d}M &= \zeta'\,\mathrm{d}\xi' + \zeta''\,\mathrm{d}\xi'' + \zeta'''\,\mathrm{d}\xi''' \\ \mathrm{d}N &= \xi'\,\mathrm{d}\eta' + \xi''\,\mathrm{d}\eta'' + \xi'''\,\mathrm{d}\eta''' \end{aligned}$$

Thus with the values of $\mathrm{d}L, \mathrm{d}M, \mathrm{d}N$, as functions of $\zeta', \zeta'', \zeta'''$, etc. we can substitute them as functions of the angles φ, ψ, ω (Article 7) to obtain after simplification, the following rather simple expressions

$$\begin{aligned} \mathrm{d}L &= \sin\psi\sin\omega\,\mathrm{d}\varphi + \cos\psi\,\mathrm{d}\omega \\ \mathrm{d}M &= -\cos\psi\sin\omega\,\mathrm{d}\varphi + \sin\psi\,\mathrm{d}\omega \\ \mathrm{d}N &= \cos\omega\,\mathrm{d}\varphi + \mathrm{d}\psi \end{aligned}$$

9. The axis about which the system can rotate and describe the angle φ and for which its position is dependent upon the two angles ψ and ω is assumed to be fixed in the system and mobile in space. But we have seen, in SECTION III of PART I (Articles 11, 12) that there is always an axis about which the system actually rotates at any instant. We called this axis the *instantaneous axis of rotation*. Therefore, the instantaneous position of this axis can be determined as well as the elementary angle of rotation with angles $\bar{\psi}, \bar{\omega}, \bar{\varphi}$ analogous to the angles ψ, ω, φ. Since the expressions for $\mathrm{d}L, \mathrm{d}M, \mathrm{d}N$ are general for any arbitrary positions of the axis of rotation φ, they will also hold for the instantaneous axis of rotation by changing ψ, ω, φ to $\bar{\psi}, \bar{\omega}, \bar{\varphi}$. Since a property of this last axis is to remain immobile for

an instant, the differentials $\mathrm{d}\bar{\psi}$ and $\mathrm{d}\bar{\omega}$ resulting from a change in the position of this axis must be equal to zero. Thus one will obtain for the axis in question

$$\sin\bar{\psi}\sin\bar{\omega}\,\mathrm{d}\bar{\varphi} = \mathrm{d}L$$
$$\cos\bar{\psi}\sin\bar{\omega}\,\mathrm{d}\bar{\varphi} = -\mathrm{d}M$$
$$\cos\bar{\omega}\,\mathrm{d}\bar{\varphi} = \mathrm{d}N$$

from which one obtains the equation

$$\mathrm{d}\bar{\varphi} = \sqrt{\mathrm{d}L^2 + \mathrm{d}M^2 + \mathrm{d}N^2}$$

which is the angle of instantaneous rotation, which was denoted by $\mathrm{d}\theta$ in the cited articles of PART I.

Then, the orientation of this axis will be obtained from the two angles $\bar{\omega}$ and $\bar{\psi}$. But, in order to refer it to the fixed axes ξ, η, ζ, it suffices to consider that after having taken the c-axis for the axis of rotation, one has for all the points of this axis, $a = 0$, $b = 0$. Thereforc, if one denotes by $\bar{\xi}$, η, $\bar{\zeta}$ the coordinates which correspond to the point where $c = 1$, and which are at the same time the cosines of the angles that the axis of rotation makes with the ξ-, η-, ζ-axes, one has with the formulas of Article 8

$$\bar{\xi} = \frac{\mathrm{d}L}{\mathrm{d}\bar{\varphi}}, \qquad \bar{\eta} = \frac{\mathrm{d}M}{\mathrm{d}\bar{\varphi}}, \qquad \bar{\zeta} = \frac{\mathrm{d}N}{\mathrm{d}\bar{\varphi}}$$

Indeed, the values of $\bar{\xi}, \bar{\eta}, \bar{\zeta}$ render those of their differentials equal to zero as is seen from the formulas of Article 1, which is the property of all the points of the instantaneous axis of rotation and with which we have determined this axis in SECTION III of PART I.

From this result, it is clear that the quantities $\mathrm{d}L, \mathrm{d}M, \mathrm{d}N$ correspond exactly to the angles of rotation which were denoted by $\mathrm{d}\psi$, $\mathrm{d}\omega$, $\mathrm{d}\varphi$ in the section that we just cited, and that we have retained in the cited SECTION III.

10. Now, if the same expressions for $\bar{\xi}$, $\bar{\eta}$, $\bar{\zeta}$ were substituted in the general expressions for a, b, c of Article 5 in place of ξ, η, ζ, one will obtain the coordinates of a, b, c, which correspond to the instantaneous axis of rotation which we will denote by $\bar{a}, \bar{b}, \bar{c}$. Thus in order to shorten the expressions, the following definitions are necessary

$$\mathrm{d}P = \xi'\,\mathrm{d}L + \eta'\,\mathrm{d}M + \zeta'\,\mathrm{d}N$$
$$\mathrm{d}Q = \xi''\,\mathrm{d}L + \eta''\,\mathrm{d}M + \zeta''\,\mathrm{d}N$$
$$\mathrm{d}R = \xi'''\,\mathrm{d}L + \eta'''\,\mathrm{d}M + \zeta'''\,\mathrm{d}N$$

which gives with the equations of condition of Article 5

$$\mathrm{d}P^2 + \mathrm{d}Q^2 + \mathrm{d}R^2 = \mathrm{d}L^2 + \mathrm{d}M^2 + \mathrm{d}N^2 = \mathrm{d}\bar{\varphi}^2$$

Thus with the following equations

$$\bar{a} = \frac{\mathrm{d}P}{\mathrm{d}\bar{\varphi}}, \qquad \bar{b} = \frac{\mathrm{d}Q}{\mathrm{d}\bar{\varphi}}, \qquad \bar{c} = \frac{\mathrm{d}R}{\mathrm{d}\bar{\varphi}}$$

which are expressions entirely similar to those of ξ, η, ζ, and in which it is clear that the quantities[42] $\mathrm{d}P$, $\mathrm{d}Q$, $\mathrm{d}R$ correspond to the quantities $\mathrm{d}L$, $\mathrm{d}M$, $\mathrm{d}N$. The values of $\bar{a}, \bar{b}, \bar{c}$ will similarly be the cosines of the angles between the axis of rotation and the axes of coordinates a, b, c.

11. In order to obtain the expressions for $\mathrm{d}P$, $\mathrm{d}Q$, $\mathrm{d}R$ as functions of the variables ξ', η', ζ', ξ'', etc., it will suffice to substitute in place of $\mathrm{d}L$, $\mathrm{d}M$, $\mathrm{d}N$ the expressions given in Article 8. But, in order to obtain the simplest formulas, it will be convenient to put these latter expressions in the following form, which are equivalent to the expressions in the cited article because of the equations of condition given in Article 5

$$\begin{aligned}
2\,\mathrm{d}L &= \eta'\,\mathrm{d}\zeta' + \eta''\,\mathrm{d}\zeta'' + \eta'''\,\mathrm{d}\zeta''' - \zeta'\,\mathrm{d}\eta' - \zeta''\,\mathrm{d}\eta'' - \zeta'''\,\mathrm{d}\eta''' \\
2\,\mathrm{d}M &= \zeta'\,\mathrm{d}\xi' + \zeta''\,\mathrm{d}\xi'' + \zeta'''\,\mathrm{d}\xi''' - \xi'\,\mathrm{d}\zeta' - \xi''\,\mathrm{d}\zeta'' - \xi'''\,\mathrm{d}\zeta''' \\
2\,\mathrm{d}N &= \xi'\,\mathrm{d}\eta' + \xi''\,\mathrm{d}\eta'' + \xi'''\,\mathrm{d}\eta''' - \eta'\,\mathrm{d}\xi' - \eta''\,\mathrm{d}\xi'' - \eta'''\,\mathrm{d}\xi'''
\end{aligned}$$

Thus one will have after substitution and ordering of the terms

$$\begin{aligned}
2\,\mathrm{d}P = {} & (\xi'\eta'' - \eta'\xi'')\mathrm{d}\zeta'' + (\xi'\eta''' - \eta'\xi''')\mathrm{d}\zeta''' \\
& + (\zeta'\xi'' - \xi'\zeta'')\mathrm{d}\eta'' + (\zeta'\xi''' - \xi'\zeta''')\mathrm{d}\eta''' \\
& + (\eta'\zeta'' - \zeta'\eta'')\mathrm{d}\xi'' + (\eta'\zeta''' - \zeta'\eta''')\mathrm{d}\xi'''
\end{aligned}$$

which is reduced by the formulas of Article 6 to

$$2\,\mathrm{d}P = \zeta'''\,\mathrm{d}\zeta'' - \zeta''\,\mathrm{d}\zeta''' + \eta'''\,\mathrm{d}\eta'' - \eta''\,\mathrm{d}\eta''' + \xi'''\,\mathrm{d}\xi'' - \xi''\,\mathrm{d}\xi'''$$

Finally, after differentiation and with the three equations of condition of Article 5, this simple expression will be obtained

$$\mathrm{d}P = \xi'''\,\mathrm{d}\xi'' + \eta'''\,\mathrm{d}\eta'' + \zeta'''\,\mathrm{d}\zeta''$$

and similarly, one will obtain

$$\begin{aligned}
\mathrm{d}Q &= \xi'\,\mathrm{d}\xi''' + \eta'\,\mathrm{d}\eta''' + \zeta'\,\mathrm{d}\zeta''' \\
\mathrm{d}R &= \xi''\,\mathrm{d}\xi' + \eta''\,\mathrm{d}\eta' + \zeta''\,\mathrm{d}\zeta'
\end{aligned}$$

If in place of the quantities ξ', ξ'', ξ''', etc., the expressions for ψ, ω, φ given in Article 7 are substituted, one has after some simplification

$$\begin{aligned}
\mathrm{d}P &= \sin\varphi \sin\omega\,\mathrm{d}\psi + \cos\varphi\,\mathrm{d}\omega \\
\mathrm{d}Q &= \cos\varphi \sin\omega\,\mathrm{d}\psi - \sin\varphi\,\mathrm{d}\omega \\
\mathrm{d}R &= \mathrm{d}\varphi + \cos\omega\,\mathrm{d}\psi
\end{aligned}$$

12. It is easy to convince oneself that the expressions for $\bar{a}, \bar{b}, \bar{c}$ make the differentials of the coordinates ξ, η, ζ equal to zero. Because after differentiating and setting $d\xi = 0$, $d\eta = 0$, $d\zeta = 0$ in the formulas of Article 1 and then changing a, b, c to $\bar{a}, \bar{b}, \bar{c}$ to refer them to the instantaneous axis of rotation, the three following equations are obtained

$$\bar{a}\,d\xi' + \bar{b}\,d\xi'' + \bar{c}\,d\xi''' = 0$$
$$\bar{a}\,d\eta' + \bar{b}\,d\eta'' + \bar{c}\,d\eta''' = 0$$
$$\bar{a}\,d\zeta' + \bar{b}\,d\zeta'' + \bar{c}\,d\zeta''' = 0$$

By adding these equations together, after having multiplied them successively by ξ', η', ζ', by ξ'', η'', ζ'' and by $\xi''', \eta''', \zeta'''$ and after taking into account the equations of condition of Article 2, one has the following equations

$$\bar{b}(\xi'\,d\xi'' + \eta'\,d\eta'' + \zeta'\,d\zeta'') + \bar{c}(\xi'\,d\xi''' + \eta'\,d\eta''' + \zeta'\,d\zeta''') = 0$$
$$\bar{a}(\xi''\,d\xi' + \eta''\,d\eta' + \zeta''\,d\zeta') + \bar{c}(\xi''\,d\xi''' + \eta''\,d\eta''' + \zeta''\,d\zeta''') = 0$$
$$\bar{a}(\xi'''\,d\xi' + \eta'''\,d\eta' + \zeta'''\,d\zeta') + \bar{b}(\xi'''\,d\xi'' + \eta'''\,d\eta'' + \zeta'''\,d\zeta'') = 0$$

Then, after taking into account the three other equations of condition of Article 5 and assuming for dP, dQ, dR the expressions given above, the three equations become

$$\bar{c}\,dQ - \bar{b}\,dR = 0, \qquad \bar{a}\,dR - \bar{c}\,dP = 0, \qquad \bar{b}\,dP - \bar{a}\,dQ = 0$$

which obviously satisfy the expressions for $\bar{a}, \bar{b}, \bar{c}$ given above.

13. In the same fashion, the quantities dL, dM, dN are used to express in an uniform manner the differentials of the quantities ξ', ξ'', ξ''', etc. As was shown in Article 8, these differentials can also be expressed by the quantities dP, dQ, dR.

Indeed, if these three equations are considered

$$\xi'\,d\xi' + \eta'\,d\eta' + \zeta'\,d\zeta' = 0$$
$$\xi''\,d\xi' + \eta''\,d\eta' + \zeta''\,d\zeta' = dR$$
$$\xi'''\,d\xi' + \eta'''\,d\eta' + \zeta'''\,d\zeta' = -dQ$$

and if they are added together after having multiplied them successively by ξ', ξ'', ξ''', by η', η'', η''', and by $\zeta', \zeta'', \zeta'''$, one will immediately obtain with the equations of condition of Article 5

$$d\xi' = \xi''\,dR - \xi'''\,dQ$$
$$d\eta' = \eta''\,dR - \eta'''\,dQ$$
$$d\zeta' = \zeta''\,dR - \zeta'''\,dQ$$

Similarly, if the three equations

$$\xi'\,d\xi'' + \eta'\,d\eta'' + \zeta'\,d\zeta'' = -dR$$
$$\xi''\,d\xi'' + \eta''\,d\eta'' + \zeta''\,d\zeta'' = 0$$
$$\xi'''\,d\xi'' + \eta'''\,d\eta'' + \zeta'''\,d\zeta'' = dP$$

after being successively multiplied by ξ', ξ'', ξ''', by η', η'', η''', and by $\zeta', \zeta'', \zeta'''$ are then added together, using the same equations of condition, the following equations will result

$$\begin{aligned} d\xi'' &= \xi''' \, dP - \xi' \, dR \\ d\eta'' &= \eta''' \, dP - \eta' \, dR \\ d\zeta'' &= \zeta''' \, dP - \zeta' \, dR \end{aligned}$$

Finally, the equations

$$\begin{aligned} \xi' \, d\xi''' + \eta' \, d\eta''' + \zeta' \, d\zeta''' &= dQ \\ \xi'' \, d\xi''' + \eta'' \, d\eta''' + \zeta'' \, d\zeta''' &= -dP \\ \xi''' \, d\xi''' + \eta''' \, d\eta''' + \zeta''' \, d\zeta''' &= 0 \end{aligned}$$

will give in the same manner

$$\begin{aligned} d\xi''' &= \xi' \, dQ - \xi'' \, dP \\ d\eta''' &= \eta' \, dQ - \eta'' \, dP \\ d\zeta''' &= \zeta' \, dQ - \zeta'' \, dP \end{aligned}$$

14. By means of these formulas, the variations of the coordinates ξ, η, ζ can be represented in a very simple manner when it is necessary to consider simultaneously the change in the orientation of the system about its center and the change in the relative distances between the points of the system. For that reason, it is clear that one must differentiate the expressions of ξ, η, ζ regarding as variables all the quantities $\xi', \eta', \zeta', \eta''$, etc. simultaneously as well as a, b, c which gives

$$\begin{aligned} d\xi &= a \, d\xi' + b \, d\xi'' + c \, d\xi''' + \xi' \, da + \xi'' \, db + \xi''' \, dc \\ d\eta &= a \, d\eta' + b \, d\eta'' + c \, d\eta''' + \eta' \, da + \eta'' \, db + \eta''' \, dc \\ d\zeta &= a \, d\zeta' + b \, d\zeta'' + c \, d\zeta''' + \zeta' \, da + \zeta'' \, db + \zeta''' \, dc \end{aligned}$$

After substituting the expressions for $d\xi'$, $d\eta'$, $d\zeta'$, $d\xi''$, etc., which were just found and defining for the purpose of shortening the expressions

$$\begin{aligned} da' &= da + c \, dQ - b \, dR \\ db' &= db + a \, dR - c \, dP \\ dc' &= dc + b \, dP - a \, dQ \end{aligned}$$

one will obtain these very simple differentiated formulas

$$\begin{aligned} d\xi &= \xi' \, da' + \xi'' \, db' + \xi''' \, dc' \\ d\eta &= \eta' \, da' + \eta'' \, db' + \eta''' \, dc' \\ d\zeta &= \zeta' \, da' + \zeta'' \, db' + \zeta''' \, dc' \end{aligned}$$

If these expressions were differentiated and the expressions for $d\xi', d\eta', d\zeta'$, $d\xi''$, etc., were again substituted in the above equations and if in order to shorten the expressions, the following definitions are made

$$d^2a'' = d^2a' + dc'\,dQ - db'\,dR$$
$$d^2b'' = d^2b' + da'\,dR - dc'\,dP$$
$$d^2c'' = d^2c' + db'\,dP - da'\,dQ$$

one will obtain the second differentials

$$d^2\xi = \xi'\,d^2a'' + \xi''d^2b'' + \xi'''d^2c''$$
$$d^2\eta = \eta'\,d^2a'' + \eta''\,d^2b'' + \eta'''\,d^2c''$$
$$d^2\zeta = \zeta'\,d^2a'' + \zeta''\,d^2b'' + \zeta'''\,d^2c''$$

It is clear that the first and second differentials are similar to the finite expressions for ξ, η, ζ (Article 1) and that the quantities $\xi', \eta', \zeta', \xi''$, etc. are represented in the same manner. It would be the same with the differentials of all the other orders, which makes the use of the quantities dP, dQ, dR very advantageous in calculations involving rotational motion.

15. But there is an important remark to be made on the use of these quantities. It is that although they are presented in differential form, it would be a mistake to treat them as such in the differentiations relative to the operator δ. Therefore, in this case it is not permitted to simply replace δdP with $d\delta P$, etc., in the expression for δT.

It is clear at the outset that nothing prevents us from replacing in the differential formulas of Article 13 the operator d by δ, which will introduce in the expressions for the variations $\delta\xi', \delta\eta', \delta\zeta', \delta\xi''$, etc., the three indeterminates δP, δQ, δR, which will be used to reduce all of the variations to three arbitrary ones.

Thus after having found (Article 13) that

$$dP = \xi'''\,d\xi'' + \eta'''\,d\eta'' + \zeta'''\,d\zeta''$$

by replacing d by δ, the following equation will be obtained

$$\delta P = \xi'''\,\delta\xi'' + \eta'''\,\delta\eta'' + \zeta'''\,\delta\zeta''$$

and similarly for the quantities dQ, dR, which will become δQ and δR.

Now, after differentiating dP with respect to δ, there will result

$$\delta dP = \xi'''\,\delta d\xi'' + \eta'''\,\delta d\eta'' + \zeta'''\,\delta d\zeta'' + \delta\xi'''\,d\xi'' + \delta\eta'''\,d\eta'' + \delta\zeta'''\,d\zeta''$$

and after differentiating δP with respect to d

$$d\delta P = \xi'''\,d\delta\xi'' + \eta'''\,d\delta\eta'' + \zeta'''\,d\delta\zeta'' + d\xi'''\,\delta\xi'' + d\eta'''\,\delta\eta'' + d\zeta'''\,\delta\zeta''$$

but $\delta \mathrm{d}\xi''$, $\delta \mathrm{d}\eta''$, $\delta \mathrm{d}\zeta''$ are the same as $\mathrm{d}\delta\xi''$, $\mathrm{d}\delta\eta''$, $\mathrm{d}\delta\zeta''$ because the quantities ξ'', η'', ζ'' are finite variables. Therefore, one will have

$$\delta \mathrm{d}P - \mathrm{d}\delta P = \delta\xi'''\,\mathrm{d}\xi'' + \delta\eta'''\,\mathrm{d}\eta'' + \delta\zeta'''\,\mathrm{d}\zeta'' - \mathrm{d}\xi'''\,\delta\xi'' - \mathrm{d}\eta'''\,\delta\eta'' - \mathrm{d}\zeta'''\,\delta\zeta''$$

Let us substitute for $\mathrm{d}\xi''$, $\mathrm{d}\eta''$, $\mathrm{d}\zeta''$ and $\mathrm{d}\xi'''$, $\mathrm{d}\eta'''$, $\mathrm{d}\zeta'''$ the expressions in terms of $\mathrm{d}P$, $\mathrm{d}Q$, $\mathrm{d}R$ (Article 21), and for $\delta\xi''$, $\delta\eta''$, $\delta\zeta''$ and $\delta\xi'''$, $\delta\eta'''$, $\delta\zeta'''$ the analogous expressions which are the result of replacing the operator d by δ. One will obtain with the equations of condition of Article 2

$$\delta\xi'''\,\mathrm{d}\xi'' + \delta\eta'''\,\mathrm{d}\eta'' + \delta\zeta'''\,\mathrm{d}\zeta'' = -\delta Q\,\mathrm{d}R$$
$$\mathrm{d}\xi'''\,\delta\xi'' + \mathrm{d}\eta'''\,\delta\eta'' + \mathrm{d}\zeta'''\,\delta\zeta'' = -\mathrm{d}Q\,\delta R$$

Thus

$$\delta \mathrm{d}P = \mathrm{d}\delta P + \mathrm{d}Q\,\delta R - \mathrm{d}R\,\delta Q$$

and by a similar calculation, one will find that

$$\delta\,\mathrm{d}Q = \mathrm{d}\delta Q + \mathrm{d}R\,\delta P - \mathrm{d}P\,\delta R$$
$$\delta\,\mathrm{d}R = \mathrm{d}\delta R + \mathrm{d}P\,\delta Q - \mathrm{d}Q\,\delta P$$

(Note of the editors of the Second Edition) This is the end of what could be found of finished works on rotational motion in Lagrange's manuscripts. We propose to continue this chapter with the sections of the first edition, taking advantage of several changes noted in Lagrange's copy. We will include in a note at the end of the volume some passages relative to this topic which were going to be used as the basis for a section on the general equations of the rotational motion of an arbitrary system of bodies. They are not in a finished form to be included in the text. However, geometers would be left wanting if they were not included.

Subsection II
The Equations of Rotational Motion for a Solid Body Acted Upon by Arbitrary Forces

16. We have just seen in the preceding section that whatever the motion a solid body acquires, this motion only depends upon six variables, of which three are related to the translational motion of a unique point of the body which is called *the center of the system*[43] and of which the three remaining variables are used to specify the rotational motion of the body about this center. From this result, it follows that the equations to be developed can only be six at the most. Consequently, it is clear that these equations can be deduced from those we have already given in SECTION III, Subsections I and II, and which are

applicable to all systems of bodies. But two cases must be distinguished; to wit, one when the body is entirely free and the other when it has to move about a fixed point.

17. Let us consider at the outset an absolutely free solid body. Let us take for the center of the body its center of gravity and denote the three rectangular coordinates of this center by x', y', z'. Let the entire mass of the body be m, with $\mathrm{D}m$ one of its elements and X, Y, Z the accelerating forces which act upon this element in the same directions as the coordinates. At the outset, the following three equations (SECTION III, Article 3) will be obtained

$$\frac{\mathrm{d}^2 x'}{\mathrm{d}t^2} m + \mathrm{S}\, X\, \mathrm{D}m = 0$$
$$\frac{\mathrm{d}^2 y'}{\mathrm{d}t^2} m + \mathrm{S}\, Y\, \mathrm{D}m = 0$$
$$\frac{\mathrm{d}^2 z'}{\mathrm{d}t^2} m + \mathrm{S}\, Z\, \mathrm{D}m = 0$$

where the operator S refers to the total integral relative to the entire mass of the body. These equations will be used, as will be shown, to determine the motion of the center of gravity of the body.

Secondly, if the rectangular coordinates of each element $\mathrm{D}m$, measured from the center of gravity and parallel to the same axes of coordinates x', y', z' of this center, are denoted by ξ, η, ζ, the three following equations will be obtained (cited section, Article 12)

$$\mathrm{S}(\xi \frac{\mathrm{d}^2\eta}{\mathrm{d}t^2} - \eta \frac{\mathrm{d}^2\xi}{\mathrm{d}t^2} + \xi Y - \eta X)\mathrm{D}m = 0$$
$$\mathrm{S}(\xi \frac{\mathrm{d}^2\zeta}{\mathrm{d}t^2} - \zeta \frac{\mathrm{d}^2\xi}{\mathrm{d}t^2} + \xi Z - \zeta X)\mathrm{D}m = 0$$
$$\mathrm{S}(\eta \frac{\mathrm{d}^2\zeta}{\mathrm{d}t^2} - \zeta \frac{\mathrm{d}^2\eta}{\mathrm{d}t^2} + \eta Z - \zeta Y)\mathrm{D}m = 0$$

But we demonstrated in the preceding paragraph that the expressions for the quantities ξ, η, ζ are always of the form

$$\xi = a\xi' + b\xi'' + c\xi'''$$
$$\eta = a\eta' + b\eta'' + c\eta'''$$
$$\zeta = a\zeta' + b\zeta'' + c\zeta'''$$

It was shown that for solid bodies the quantities a, b, c are necessarily constants with respect to time and only variables with respect to the difference elements $\mathrm{D}m$, because these quantities represent the rectangular coordinates of each of these elements, referred to three axes which intersect at the center of the body and which are fixed in the body.

On the other hand, the quantities ξ', ξ'', etc., are variables with respect to time, and constants with respect to all the elements of the body because these quantities are all functions of three angles φ, ψ, ω, which determine the various rotational motions the body has about its center. Thus, if these different substitutions are made in the preceding equations after putting outside the operator S the variables φ, ψ, ω and their differences, one will obtain three second order differential equations between the same variables and the time t, which will be used to determine all three as a function of t.

These equations will be similar to those which d'Alembert was the first to find for the rotational motion of a body of arbitrary shape and which were so useful in his research on the precession of the equinoxes.

For this reason and because the form of these equations does not have all the simplicity they are capable of, we will not develop them here but we will rather directly solve the problem using the general method of SECTION IV, which will immediately give simpler and easier equations for the calculation.

18. In order to use this method in the most general and simple manner, it will be assumed, which is natural, that each particle Dm of the body is attracted by the forces $\bar{P}, \bar{Q}, \bar{R}$, etc., proportional to arbitrary functions of the distances $\bar{p}, \bar{q}, \bar{r}$, etc. of the same particle to the centers of forces and one will have from this the algebraic equation

$$\Pi = \int (\bar{P}\,\mathrm{d}\bar{p} + \bar{Q}\,\mathrm{d}\bar{q} + \bar{R}\,\mathrm{d}\bar{r} + \cdots)$$

Then the following two expressions will be considered

$$T = \mathrm{S}\left(\frac{\mathrm{d}x^2 + \mathrm{d}y^2 + \mathrm{d}z^2}{2\,\mathrm{d}t^2}\right)\mathrm{D}m, \qquad V = \mathrm{S}\,\Pi\,\mathrm{D}m$$

where the integral operator S operates only on the elements $\mathrm{D}m$ of the body and on the quantities relative to the position of these elements in the body.

These two equations will be expanded as functions of arbitrary variables ξ, ψ, φ etc. relative to the various motions of the body and one will develop from this result the following general formula (SECTION IV, Article 10)

$$0 = \left(\mathrm{d}\frac{\delta T}{\delta \mathrm{d}\xi} - \frac{\delta T}{\delta \xi} + \frac{\delta V}{\delta \xi}\right)\delta\xi + \left(\mathrm{d}\frac{\delta T}{\delta \mathrm{d}\psi} - \frac{\delta T}{\delta \psi} + \frac{\delta V}{\delta \psi}\right)\delta\psi + \left(\mathrm{d}\frac{\delta T}{\delta \mathrm{d}\varphi} - \frac{\delta T}{\delta \varphi} + \frac{\delta V}{\delta \varphi}\right)\delta\varphi + \cdots\cdots$$

If the variables ξ, ψ, φ, etc. are, because of the nature of the problem, independent of each other (and they can always be defined so that they are), one will equate separately to zero the quantities multiplied by each of the indeterminate variables $\delta\xi$, $\delta\psi$, $\delta\varphi$, etc. Then there will be as many equations between the variables ξ, ψ, φ, etc. as there are variables.

If these variables are not entirely independent so that there are one or several equations of condition between them, the differentiation of these equations will provide an equal number of equations of condition between the variations $\delta\xi$, $\delta\psi$, $\delta\varphi$, etc. by means of which these variations can be reduced to a smaller number.

After having performed this simplification of the general formula, one will similarly equate to zero each of the remaining coefficients of the variation. The equations which result, added to those of the given conditions, will suffice to solve the problem.

In the problem in question, it will only be necessary to use the transformations described in the preceding paragraph. Thus, one will first substitute $x' + \xi, y' + \eta, z' + \zeta$, in place of x, y, z. Then $a\xi' + b\xi'' + c\xi'''$, $a\eta' + b\eta'' + c\eta'''$, $a\zeta' + b\zeta'' + c\zeta'''$, will be substituted for ξ, η, ζ (Article 1). Finally, the quantities ξ', η', etc. should be replaced with the expressions for φ, ψ, ω of Article 7. Thus the quantities T, V will be expressed as functions of the six variables $x', y', z', \varphi, \psi, \omega$. Instead of the latter quantities, one could also, if it is judged to be appropriate, introduce other equivalent ones. Each of them will provide, for the determination of the motion of the body, an equation of the form

$$\mathrm{d}\frac{\delta T}{\delta \mathrm{d}\alpha} - \frac{\delta T}{\delta \alpha} + \frac{\delta V}{\delta \alpha} = 0$$

where α is one of these variables.

19. Let us begin by introducing in the expression for T these new variables $x' + \xi, y' + \eta, z' + \zeta$, in place of x, y, z, and placing outside the integral sign S the quantities x', y', z', which are the same for all points of the body because these are the coordinates of the center of the body. The function T becomes

$$\frac{\mathrm{d}x'^2 + \mathrm{d}y'^2 + \mathrm{d}z'^2}{2\,\mathrm{d}t^2}\mathrm{S}\,\mathrm{D}m + \mathrm{S}\left(\frac{\mathrm{d}\xi^2 + \mathrm{d}\eta^2 + \mathrm{d}\zeta^2}{2\,\mathrm{d}t^2}\right)\mathrm{D}m$$
$$+\frac{\mathrm{d}x'\,\mathrm{S}\,\mathrm{d}\xi\,\mathrm{D}m + \mathrm{d}y'\,\mathrm{S}\,\mathrm{d}\eta\,\mathrm{D}m + \mathrm{d}z'\,\mathrm{S}\,\mathrm{d}\zeta\,\mathrm{D}m}{\mathrm{d}t^2}$$

This expression is clearly composed of three parts. The first part contains only the variables x', y', z' and expresses the value of T for the case where the body could be viewed as a point. Therefore, if these variables are independent of the other variables ξ, η, ζ which holds if the body is free to rotate in any direction about its center, the formula in question should be treated separately and will give for the motion of this center the same equations as if the body were concentrated there. Consequently, this part of the problem is part of the one which has been solved in the preceding sections and to which we refer.

The third part of the preceding expression, the one which contains the differences $\mathrm{d}x', \mathrm{d}y'$, $\mathrm{d}z'$, multiplied by the differences $\mathrm{d}\xi$, $\mathrm{d}\eta$, $\mathrm{d}\zeta$ will not appear in two cases: when the center of the body is fixed, which is obvious because then the differences $\mathrm{d}x', \mathrm{d}y', \mathrm{d}z'$, of the coordinates of the center are zero and secondly, when the center is assumed located at the center of gravity of the body because then the integrals $\mathrm{S}\,\mathrm{d}\xi\,\mathrm{D}m$, $\mathrm{S}\,\mathrm{d}\eta\,\mathrm{D}m$, $\mathrm{S}\,\mathrm{d}\zeta\,\mathrm{D}m$ are equal to zero. Indeed, by substituting for $\mathrm{d}\xi, \mathrm{d}\eta, \mathrm{d}\zeta$ the expressions $a\,\mathrm{d}\xi' + b\,\mathrm{d}\xi'' + c\,\mathrm{d}\xi'''$, $a\,\mathrm{d}\eta' + b\,\mathrm{d}\eta'' + c\,\mathrm{d}\eta'''$, $a\,\mathrm{d}\zeta' + b\,\mathrm{d}\zeta'' + c\,\mathrm{d}\zeta'''$ (preceding article), and putting outside the integral sign S the quantities $\mathrm{d}\xi''$, etc., which are independent of the position of the particles $\mathrm{d}m$ of the body, each term of these integrals will be multiplied by one of the three quantities $\mathrm{S}\,a\,\mathrm{D}m$, $\mathrm{S}\,b\,\mathrm{D}m$, $\mathrm{S}\,c\,\mathrm{D}m$. But these quantities are precisely the sums of the products of each element dm, multiplied by the distance to three planes passing through the center of the body and perpendicular to the axes of coordinates a, b, c. Thus they are zero when this center coincides with the center of gravity of all the bodies, because of the known properties of this latter center. Therefore, the three integrals S dξ Dm, S dη Dm, S dζ Dm will also be equal to zero in this case.

In both cases, it only remains to consider in the expression for T the formula $\mathrm{S}((\mathrm{d}\xi^2+\mathrm{d}\eta^2+\mathrm{d}\zeta^2)/2\,\mathrm{d}t^2)\mathrm{D}m$ which only applies to the motion of rotation of the system about its center and which will, consequently, be used to determine the laws of this motion independent of the motion of this center in space.

20. In order to make the solution as simple as possible, it is appropriate to use the expressions for dξ, dη, dζ of Article 14, which give, by putting d$a = 0$, d$b = 0$, d$c = 0$

$$\begin{aligned}&\mathrm{d}\xi^2 + \mathrm{d}\eta^2 + \mathrm{d}\zeta^2 = (c\,\mathrm{d}Q - b\,\mathrm{d}R)^2 + (a\,\mathrm{d}R - c\,\mathrm{d}P)^2 + (b\,\mathrm{d}P - a\,\mathrm{d}Q)^2\\ &= (b^2 + c^2)\mathrm{d}P^2 + (a^2 + c^2)\mathrm{d}Q^2 + (a^2 + b^2)\mathrm{d}R^2 - 2bc\,\mathrm{d}Q\,\mathrm{d}R - 2ac\,\mathrm{d}P\,\mathrm{d}R\\ &-2ab\,\mathrm{d}P\,\mathrm{d}Q\end{aligned}$$

But since the quantities a, b, c are the only variables relative to the position of the particles Dm in the body, it follows that to obtain the value of $\mathrm{S}(\mathrm{d}\xi^2+\mathrm{d}\eta^2+\mathrm{d}\zeta^2)^2\mathrm{D}m$, it will suffice to multiply each term of the preceding quantity by Dm and then to integrate relative to the operator S, putting outside this sign the quantities dP, dQ, dR, which are independent. Thus the quantity $\mathrm{S}\left((\mathrm{d}\xi^2 + \mathrm{d}\eta^2 + \mathrm{d}\zeta^2)/\,\mathrm{d}t^2\right)\mathrm{D}m$ will become

$$\frac{A\,\mathrm{d}P^2 + B\,\mathrm{d}Q^2 + C\,\mathrm{d}R^2}{2\,\mathrm{d}t^2} - \frac{F\,\mathrm{d}Q\,\mathrm{d}R + G\,\mathrm{d}P\,\mathrm{d}R + H\,\mathrm{d}P\,\mathrm{d}Q}{\mathrm{d}t^2}$$

after setting, in order to shorten the expressions

$$\begin{aligned}&A = \mathrm{S}(b^2 + c^2)\mathrm{D}m, \qquad B = \mathrm{S}(a^2 + c^2)\mathrm{D}m, \qquad C = \mathrm{S}(a^2 + b^2)\mathrm{D}m\\ &F = \mathrm{S}\,bc\,\mathrm{D}m, \qquad G = \mathrm{S}\,ac\,\mathrm{D}m, \qquad H = \mathrm{S}\,ab\,\mathrm{D}m\end{aligned}$$

These integrations are relative to the whole mass of the body so that A, B, C, F, G, H must now be viewed and treated as constants given by the shape of the body.

21. If the following definitions are made for greater simplicity

$$\frac{\mathrm{d}P}{\mathrm{d}t} = p, \qquad \frac{\mathrm{d}Q}{\mathrm{d}t} = q, \qquad \frac{\mathrm{d}R}{\mathrm{d}t} = r$$

one will have considering only the terms relative to the motion of rotation in the function T

$$T = \tfrac{1}{2}(Ap^2 + Bq^2 + Cr^2) - Fqr - Gpr - Hpq$$

Thus the quantity T is solely a function of p, q, r. After differentiating by δ, there results

$$\delta T = \frac{\mathrm{d}T}{\mathrm{d}p}\delta p + \frac{\mathrm{d}T}{\mathrm{d}q}\delta q + \frac{\mathrm{d}T}{\mathrm{d}r}\delta r$$

But, by the formulas of Article 11, we have that

$$p = \frac{\sin\varphi \sin\omega\, \mathrm{d}\psi + \cos\varphi\, \mathrm{d}\omega}{\mathrm{d}t}$$
$$q = \frac{\cos\varphi \sin\omega\, \mathrm{d}\psi - \sin\varphi\, \mathrm{d}\omega}{\mathrm{d}t}$$
$$r = \frac{\mathrm{d}\varphi + \cos\omega\, \mathrm{d}\psi}{\mathrm{d}t}$$

Since $\mathrm{d}t$ is always constant

$$\begin{aligned}\delta T &= \left(\frac{\mathrm{d}T}{\mathrm{d}p}q - \frac{\mathrm{d}T}{\mathrm{d}q}p\right)\delta\varphi + \frac{\mathrm{d}T}{\mathrm{d}r}\frac{\delta \mathrm{d}\varphi}{\mathrm{d}t} \\ &+ \left(\frac{\mathrm{d}T}{\mathrm{d}p}\sin\varphi\sin\omega + \frac{\mathrm{d}T}{\mathrm{d}q}\cos\varphi\sin\omega + \frac{\mathrm{d}T}{\mathrm{d}r}\cos\omega\right)\frac{\delta\mathrm{d}\psi}{\mathrm{d}t} \\ &+ \left(\frac{\mathrm{d}T}{\mathrm{d}p}\sin\varphi\cos\omega + \frac{\mathrm{d}T}{\mathrm{d}q}\cos\varphi\cos\omega - \frac{\mathrm{d}T}{\mathrm{d}r}\sin\omega\right)\frac{\mathrm{d}\psi\,\delta\varphi}{\mathrm{d}t} \\ &+ \left(\frac{\mathrm{d}T}{\mathrm{d}p}\cos\varphi - \frac{\mathrm{d}T}{\mathrm{d}q}\sin\varphi\right)\frac{\delta\mathrm{d}\omega}{\mathrm{d}t}\end{aligned}$$

from which one obtains immediately for the rotational motion of the body, the three following equations of second order

$$\frac{\mathrm{d}\frac{\mathrm{d}T}{\mathrm{d}r}}{\mathrm{d}t} - \frac{\mathrm{d}T}{\mathrm{d}p}q + \frac{\mathrm{d}T}{\mathrm{d}q}p + \frac{\delta V}{\delta\varphi} = 0$$

$$\frac{\mathrm{d}\left(\frac{\mathrm{d}T}{\mathrm{d}p}\sin\varphi\sin\omega + \frac{\mathrm{d}T}{\mathrm{d}q}\cos\varphi\sin\omega + \frac{\mathrm{d}T}{\mathrm{d}r}\cos\omega\right)}{\mathrm{d}t} + \frac{\delta V}{\delta\psi} = 0$$

$$\frac{\mathrm{d}\left(\frac{\mathrm{d}T}{\mathrm{d}p}\cos\varphi - \frac{\mathrm{d}T}{\mathrm{d}q}\sin\varphi\right)}{\mathrm{d}t} - \left(\frac{\mathrm{d}T}{\mathrm{d}p}\sin\varphi\cos\omega + \frac{\mathrm{d}T}{\mathrm{d}q}\cos\varphi\cos\omega - \frac{\mathrm{d}T}{\mathrm{d}r}\sin\omega\right)\frac{\mathrm{d}\psi}{\mathrm{d}t}$$
$$+ \frac{\delta V}{\delta\omega} = 0$$

With respect to the quantity V, since it depends upon the forces which act on the body, it will be equal to zero if the body is not acted upon by any forces. Thus in this case, the three quantities $\delta V/\delta\varphi$, $\delta V/\delta\psi$, $\delta V/\delta\omega$ will also be equal to zero and the second of the three preceding equations will be integrable separately. But the general integration of all these equations will remain very difficult.

In general, since $V = \mathrm{S}\,\Pi\,\mathrm{D}m$ and since Π is an algebraic function of the distances $\bar{p}$, $\bar{q}$, etc. (Article 18) of which each is expressed by

$$\sqrt{(x-f)^2+(y-g)^2+(z-h)^2}$$

after denoting by f, g, h the coordinates of the fixed center of the forces, it remains to make the same substitutions in the function Π as those made above. Then after having integrated with respect to the entire mass of the body, the expression for V in terms of φ, ψ, ω will be obtained from which the expressions $\delta V/\delta\varphi$, $\delta V/\delta\psi$, $\delta V/\delta\omega$, which are the same as $\mathrm{d}V/\mathrm{d}\varphi$, $\mathrm{d}V/\mathrm{d}\psi$, $\mathrm{d}V/\mathrm{d}\omega$, will be obtained by means of ordinary differentiation. Because this calculation presents no difficulty, we will not dwell on it. We will only note that the preceding equations are reduced to those used in my first research on the *libration of the moon*.

22. Although the use of the angles φ, ψ, ω, in our method seems to be the simplest approach to obtaining the equations of rotation for the body, nevertheless, these equations can be obtained more directly and even more elegantly. More convenient formulas can be obtained for the calculation in several cases by considering at the outset the variations of the quantities $\mathrm{d}P$, $\mathrm{d}Q$, $\mathrm{d}R$ given by the formulas of Article 15, that is

$$\begin{aligned}
\delta \mathrm{d}P &= \mathrm{d}\delta P + \mathrm{d}Q\,\delta R - \mathrm{d}R\,\delta Q\\
\delta \mathrm{d}Q &= \mathrm{d}\delta Q + \mathrm{d}R\,\delta P - \mathrm{d}P\,\delta R\\
\delta \mathrm{d}R &= \mathrm{d}\delta R + \mathrm{d}P\,\delta Q - \mathrm{d}Q\,\delta P
\end{aligned}$$

and after substituting these expressions in the equation for δT and replacing $\mathrm{d}P/\mathrm{d}t$, $\mathrm{d}Q/\mathrm{d}t$, $\mathrm{d}R/\mathrm{d}t$ by p, q, r, one will have

$$\begin{aligned}
\delta T = {}& \frac{\mathrm{d}T}{\mathrm{d}p}\left(\frac{\mathrm{d}\delta P}{\mathrm{d}t} - q\,\delta R - r\,\delta Q\right)\\
&+\frac{\mathrm{d}T}{\mathrm{d}q}\left(\frac{\mathrm{d}\delta Q}{\mathrm{d}t} + r\,\delta P - p\,\delta R\right)\\
&+\frac{\mathrm{d}T}{\mathrm{d}r}\left(\frac{\mathrm{d}\delta R}{\mathrm{d}t} + p\,\delta Q - q\,\delta P\right)
\end{aligned}$$

With respect to the terms relative to the variation of V, since V becomes an algebraic function of $\xi', \xi'', \xi''', \eta'$, etc. after substitution of $x'+a\xi'+b\xi''+c\xi'''$, $y'+a\eta'+b\eta''+c\eta'''$, $z'+a\zeta'+b\zeta''+c\zeta'''$ in place of x, y, z, and since the integral operator S has no relation to the quantities a, b, c, it will only be necessary to differentiate with respect to δ and then to replace $\delta\xi'$, $\delta\xi''$, etc., by their expressions of δP, δQ, δR. Thus, since $\delta V/\delta\xi' = \mathrm{d}V/\mathrm{d}\xi'$, $\delta V/\delta\xi'' = \mathrm{d}V/\mathrm{d}\xi''$ etc. there will exist in the same equation the following terms for δV

$$\begin{aligned}
&\frac{\mathrm{d}V}{\mathrm{d}\xi'}(\xi''\,\delta R - \xi'''\,\delta Q) + \frac{\mathrm{d}V}{\mathrm{d}\xi''}(\xi'''\,\delta P - \xi'\,\delta R)\\
&+\frac{\mathrm{d}V}{\mathrm{d}\xi'''}(\xi'\,\delta Q - \xi''\,\delta P) + \frac{\mathrm{d}V}{\mathrm{d}\eta'}(\eta''\,\delta R - \eta'''\,\delta Q) + \cdots
\end{aligned}$$

Finally, gathering all the terms multiplied by each of the three quantities $\delta P, \delta Q, \delta R$, one will have a general equation of this form

$$0 = (P)\delta P + (Q)\delta Q + (R)\delta R$$

where

$$(P) = \frac{\mathrm{d}\frac{\mathrm{d}T}{\mathrm{d}p}}{\mathrm{d}t} + q\frac{\mathrm{d}T}{\mathrm{d}r} - r\frac{\mathrm{d}T}{\mathrm{d}q} + \xi'''\frac{\mathrm{d}V}{\mathrm{d}\xi''} + \eta'''\frac{\mathrm{d}V}{\mathrm{d}\eta''}$$
$$+\zeta'''\frac{\mathrm{d}V}{\mathrm{d}\zeta''} - \xi''\frac{\mathrm{d}V}{\mathrm{d}\xi'''} - \eta''\frac{\mathrm{d}V}{\mathrm{d}\eta'''} - \zeta''\frac{\mathrm{d}V}{\mathrm{d}\zeta'''}$$
$$(Q) = \frac{\mathrm{d}\frac{\mathrm{d}T}{\mathrm{d}q}}{\mathrm{d}t} + r\frac{\mathrm{d}T}{\mathrm{d}p} - p\frac{\mathrm{d}T}{\mathrm{d}r} + \xi'\frac{\mathrm{d}V}{\mathrm{d}\xi'''} + \eta'\frac{\mathrm{d}V}{\mathrm{d}\eta'''}$$
$$+\zeta'\frac{\mathrm{d}V}{\mathrm{d}\zeta'''} - \xi'''\frac{\mathrm{d}V}{\mathrm{d}\xi'} - \eta'''\frac{\mathrm{d}V}{\mathrm{d}\eta'} - \zeta'''\frac{\mathrm{d}V}{\mathrm{d}\zeta'}$$
$$(R) = \frac{\mathrm{d}\frac{\mathrm{d}T}{\mathrm{d}r}}{\mathrm{d}t} + p\frac{\mathrm{d}T}{\mathrm{d}q} - q\frac{\mathrm{d}T}{\mathrm{d}p} + \xi''\frac{\mathrm{d}V}{\mathrm{d}\xi'} + \eta''\frac{\mathrm{d}V}{\mathrm{d}\eta'}$$
$$+\zeta''\frac{\mathrm{d}V}{\mathrm{d}\zeta'} - \xi'\frac{\mathrm{d}V}{\mathrm{d}\xi'''} - \zeta'\frac{\mathrm{d}V}{\mathrm{d}\eta''} - \zeta'\frac{\mathrm{d}V}{\mathrm{d}\zeta''}$$

Since the three quantities $\delta P, \delta Q, \delta R$ are independent of each other and at the same time arbitrary, the following three particular equations result

$$(P) = 0, \qquad (Q) = 0, \qquad (R) = 0$$

which after being combined with the six equations of condition between the nine variables ξ', ξ'', etc. (Article 5) will be used to determine each of these variables.

The terms of these equations which depend upon the quantity V can be reduced, if desired, to a simpler form. Because $V = \mathrm{S}\,\Pi\,\mathrm{D}m$, one will have (since the integral operator S does not apply to the variables ξ', ξ'', etc.)

$$\xi''\frac{\mathrm{d}V}{\mathrm{d}\xi'} = \mathrm{S}\,\xi''\frac{\mathrm{d}\Pi}{\mathrm{d}\xi'}\mathrm{D}m, \qquad \eta''\frac{\mathrm{d}V}{\mathrm{d}\eta'} = \mathrm{S}\,\eta''\frac{\mathrm{d}\Pi}{\mathrm{d}\eta'}\mathrm{D}m, \qquad \cdots$$

Since Π is an algebraic function of

$$a\xi' + b\xi'' + c\xi''', \qquad a\eta' + b\eta'' + c\eta''', \qquad a\zeta' + b\zeta'' + c\zeta'''$$

it is easy to see that by varying separately a, b, c, the following equations will be obtained

$$\xi'''\frac{\mathrm{d}\Pi}{\mathrm{d}\xi''} + \eta'''\frac{\mathrm{d}\Pi}{\mathrm{d}\eta''} + \zeta'''\frac{\mathrm{d}\Pi}{\mathrm{d}\zeta''} = b\frac{\mathrm{d}\Pi}{\mathrm{d}c}, \qquad \xi''\frac{\mathrm{d}\Pi}{\mathrm{d}\xi'''} + \eta''\frac{\mathrm{d}\Pi}{\mathrm{d}\eta'''} + \zeta''\frac{\mathrm{d}\Pi}{\mathrm{d}\zeta'''} = c\frac{\mathrm{d}\Pi}{\mathrm{d}b}$$

and so on such that we will have in this manner

$$\xi'''\frac{dV}{d\xi''}+\eta'''\frac{dV}{d\eta''}+\zeta'''\frac{dV}{d\zeta''}-\xi''\frac{dV}{d\xi'''}-\eta''\frac{dV}{d\eta'''}-\zeta''\frac{dV}{d\zeta'''}$$
$$=\mathrm{S}\left(b\frac{d\Pi}{dc}-c\frac{d\Pi}{db}\right)\mathrm{D}m$$
$$\xi'\frac{dV}{d\xi'''}+\eta'\frac{dV}{d\eta'''}+\zeta'\frac{dV}{d\zeta'''}-\xi'''\frac{dV}{d\xi'}-\eta'''\frac{dV}{d\eta'}-\zeta'''\frac{dV}{d\zeta'}$$
$$=\mathrm{S}\left(c\frac{d\Pi}{da}-a\frac{d\Pi}{dc}\right)\mathrm{D}m$$
$$\xi''\frac{dV}{d\xi'}+\eta''\frac{dV}{d\eta'}+\zeta''\frac{dV}{d\zeta'}-\xi'\frac{dV}{d\xi''}-\eta'\frac{dV}{d\eta''}-\zeta'\frac{dV}{d\zeta''}$$
$$=\mathrm{S}\left(a\frac{d\Pi}{db}-b\frac{d\Pi}{da}\right)\mathrm{D}m$$

But if this transformation simplifies the formulas, it does not simplify the calculation, because instead of the single integration required for V, there are now three integrations to perform.

23. When the distances from the centers of the forces to the center of the body are very large with respect to the dimensions of this body, one can then reduce the quantity Π to a rapidly converging series with terms proportional to the powers of a, b, c, such that the integration $\mathrm{S}\,\Pi\,\mathrm{D}m$ will pose no difficulty. This is true for the case of the planets since they attract each other.

If the attractive force $\bar{P}$ is simply proportional to the distance $\bar{p}$ such that $\bar{P}=k\bar{p}$, where k is a constant coefficient, the term $\int \bar{P}\,d\bar{p}$ of the function Π (Article 18) becomes $k\bar{p}^2/2$ and since $\bar{p}$ is expressed in general by $\sqrt{(x-f)^2+(y-g)^2+(z-h)^2}$ where the coordinates of the center of forces are expressed by f, g, h, the term in question will give the following expression

$$\frac{k}{2}[(x-f)^2+(y-g)^2+(z-h)^2]$$

Thus after substituting for x, y, z the expressions $x'+\xi$, $y'+\eta$, $z'+\zeta$, multiplying by $\mathrm{D}m$ and integrating according to S, one will have in the expression for $V=\mathrm{S}\,\Pi\,\mathrm{D}m$, the following terms

$$\frac{k}{2}[(x'-f)^2+(y'-g)^2+(z'-h)^2]\mathrm{S}\,\mathrm{D}m+k(x'-f)\mathrm{S}\,\xi\,\mathrm{D}m$$
$$+k(y'-g)\mathrm{S}\,\eta\,\mathrm{D}m+k(z'-h)\mathrm{S}\,\zeta\,\mathrm{D}m+\frac{k}{2}\mathrm{S}(\xi^2+\eta^2+\zeta^2)\mathrm{D}m$$

But

$$\xi=a\xi'+b\xi''+c\xi''',\qquad \eta=a\eta'+b\eta''+c\eta''',\qquad \zeta=a\zeta'+b\zeta''+c\zeta'''$$

thus

$$\mathrm{S}\,\xi\,\mathrm{D}m = \xi'\,\mathrm{S}\,a\,\mathrm{D}m + \xi''\mathrm{S}\,b\,\mathrm{D}m + \xi'''\mathrm{S}\,c\,\mathrm{D}m$$

and similarly for the others. The equation (Article 5)

$$\mathrm{S}(\xi^2 + \eta^2 + \zeta^2)\mathrm{D}m = \mathrm{S}(a^2 + b^2 + c^2)\mathrm{D}m$$

will be equal to a constant which we will call E.

But, if the arbitrary center of the body is taken at its center of gravity, there will result

$$\mathrm{S}\,a\,\mathrm{D}m = 0, \qquad \mathrm{S}\,b\,\mathrm{D}m = 0, \qquad \mathrm{S}\,c\,\mathrm{D}m = 0$$

as we have already shown above (Article 19). Therefore, in this case, the quantity V will only contain, relative to the force in question, the following terms

$$\frac{k}{2}[(x' - f)^2 + (y' - g)^2 + (z' - h)^2] + \frac{k}{2}E$$

so that all the partial differences $\mathrm{d}V/\mathrm{d}\xi'$, $\mathrm{d}V/\mathrm{d}\xi''$, etc. will be equal to zero.

From which it follows that the effect of this force will be zero with respect to the rotational motion about the center of gravity.

Since the preceding expression for V, with the exception of the constant term $kE/2$, is the same as if the body were concentrated at its center, in which case $x = x'$, $y = y'$, $z = z'$, one will have for the progressive motion of this center the same equations as if the body were reduced to a point because the partial differences of V, relative to the variables x', y', z', will be the same as in this assumption.

If the body is considered to be heavy and the accelerating force of gravity is assumed to be equal to unity and also, if the axis of the z-coordinate is oriented vertically upwards, one will have

$$\bar{P} = 1, \qquad \bar{p} = h - z$$

Thus

$$\int P\,\mathrm{d}p = h - z = h - z' - a\zeta' - b\zeta'' - c\zeta'''$$

so that the quantity V will contain, because of the gravity of the body, the terms

$$(h - z')\mathrm{S}\,\mathrm{D}m - \zeta'\mathrm{S}\,a\,\mathrm{D}m - \zeta''\mathrm{S}\,b\,\mathrm{D}m - \zeta'''\mathrm{S}\,c\,\mathrm{D}m$$

Therefore, if the center of the body is taken at its center of gravity, the terms which contain the variables ζ', ζ'', etc. will disappear and consequently, the effect of gravity on the rotational motion will be non-existent, as in the preceding case. The expression for V, to the extent that it is a result of gravity, will then be reduced to $(h - z')\mathrm{S}\,\mathrm{D}m$, that is, to what it would be if the body were reduced to a point, keeping its mass equal to $\mathrm{S}\,\mathrm{D}m$. Therefore, the motion of translation of the body will be the same as in this case.

Subsection III
The Determination of the Motion of a Heavy Body of Arbitrary Shape

24. This problem, as difficult as it is, is, nevertheless, one of the simplest in mechanics when the nature of the problem is considered in its entirety. Since all the bodies are essentially heavy and extended they cannot be stripped of any of these characteristics without changing their nature and the problems where simplifications are not accounted for, will only be for pure curiosity.

We will begin by investigating the motion of free bodies, such as projectiles. Then we will examine the motion of bodies restrained by a fixed point such as pendulums.

In the first case, the center of the body will be taken at its center of gravity. Then because the effect of gravity on the rotational motion is non-existent, as has been shown, the laws of this rotation will be determined by the following three equations (Article 22)

$$(\mathrm{A})\quad\begin{cases} \dfrac{\mathrm{d}\dfrac{\mathrm{d}T}{\mathrm{d}p}}{\mathrm{d}t} + q\dfrac{\mathrm{d}T}{\mathrm{d}r} - r\dfrac{\mathrm{d}T}{\mathrm{d}q} = 0 \\[2ex] \dfrac{\mathrm{d}\dfrac{\mathrm{d}T}{\mathrm{d}q}}{\mathrm{d}t} + r\dfrac{\mathrm{d}T}{\mathrm{d}p} - p\dfrac{\mathrm{d}T}{\mathrm{d}r} = 0 \\[2ex] \dfrac{\mathrm{d}\dfrac{\mathrm{d}T}{\mathrm{d}r}}{\mathrm{d}t} + p\dfrac{\mathrm{d}T}{\mathrm{d}q} - q\dfrac{\mathrm{d}T}{\mathrm{d}p} = 0 \end{cases}$$

assuming that (Article 21)

$$p = \frac{\mathrm{d}P}{\mathrm{d}t}, \qquad q = \frac{\mathrm{d}Q}{\mathrm{d}t}, \qquad r = \frac{\mathrm{d}R}{\mathrm{d}t}$$

and

$$T = \tfrac{1}{2}(Ap^2 + Bq^2 + Cr^2) - Fqr - Gpr - Hpq$$

The motion of the center of the body follows from the known laws for the motion of projectiles treated as particles or points. Thus the determination of this motion presents no difficulty and we will not dwell on it.

In the second case, the fixed point of suspension is taken at the center of the body and assuming the z-coordinate vertical and positive in the upward direction, one will have (Article 23)

$$V = (a - z')\mathrm{S}\,\mathrm{D}m - \zeta'\mathrm{S}\,a\,\mathrm{D}m - \zeta''\mathrm{S}\,b\,\mathrm{D}m - \zeta'''\mathrm{S}\,c\,\mathrm{D}m$$

from which the following equation is obtained

$$\frac{\mathrm{d}V}{\mathrm{d}\zeta'} = -\mathrm{S}\,a\,\mathrm{D}m, \qquad \frac{\mathrm{d}V}{\mathrm{d}\zeta''} = -\mathrm{S}\,b\,\mathrm{D}m, \qquad \frac{\mathrm{d}V}{\mathrm{d}\zeta'''} = -\mathrm{S}\,c\,\mathrm{D}m$$

and all of the other partial differences of V are zero so that the equations of motion for the motion of rotation will be (Article 22)

$$(\mathrm{B}) \begin{cases} \dfrac{\mathrm{d}\dfrac{\mathrm{d}T}{\mathrm{d}p}}{\mathrm{d}t} + q\dfrac{\mathrm{d}T}{\mathrm{d}r} - r\dfrac{\mathrm{d}T}{\mathrm{d}q} - \zeta'''\mathrm{S}\,b\,\mathrm{D}m + \zeta''\mathrm{S}\,c\,\mathrm{D}m = 0 \\ \dfrac{\mathrm{d}\dfrac{\mathrm{d}T}{\mathrm{d}q}}{\mathrm{d}t} + r\dfrac{\mathrm{d}T}{\mathrm{d}p} - p\dfrac{\mathrm{d}T}{\mathrm{d}r} - \zeta'\mathrm{S}\,c\,\mathrm{D}m + \zeta'''\mathrm{S}\,a\,\mathrm{D}m = 0 \\ \dfrac{\mathrm{d}\dfrac{\mathrm{d}T}{\mathrm{d}r}}{\mathrm{d}t} + p\dfrac{\mathrm{d}T}{\mathrm{d}q} - q\dfrac{\mathrm{d}T}{\mathrm{d}p} - \zeta''\mathrm{S}\,a\,\mathrm{D}m + \zeta'\mathrm{S}\,b\,\mathrm{D}m = 0 \end{cases}$$

where the quantities $\mathrm{S}\,a\,\mathrm{D}m$, $\mathrm{S}\,b\,\mathrm{D}m$, $\mathrm{S}\,c\,\mathrm{D}m$ must be viewed as constants given by the configuration of the body and by the location of the point of suspension.

25. The solution of the first case, where the body is assumed to be entirely free and where only the rotation about the center of gravity is considered, uniquely depends upon the integration of the three equations (A).

Now it is easy to find two integrals of these equations because

1. If they are multiplied respectively by $\mathrm{d}T/\mathrm{d}p, \mathrm{d}T/\mathrm{d}q, \mathrm{d}T/\mathrm{d}r$ and then added together, an integrable equation results for which the integral will be

$$\left(\frac{\mathrm{d}T}{\mathrm{d}p}\right)^2 + \left(\frac{\mathrm{d}T}{\mathrm{d}q}\right)^2 + \left(\frac{\mathrm{d}T}{\mathrm{d}r}\right)^2 = f^2$$

   where f^2 is an arbitrary constant.

2. If these equations are multiplied by p, q, r, and added together, one will obtain the following expression

$$p\,\mathrm{d}\frac{\mathrm{d}T}{\mathrm{d}p} + q\,\mathrm{d}\frac{\mathrm{d}T}{\mathrm{d}q} + r\,\mathrm{d}\frac{\mathrm{d}T}{\mathrm{d}r} = 0$$

Since T is only a function of p, q, r, we have that

$$dT = \frac{dT}{dp}dp + \frac{dT}{dq}dq + \frac{dT}{dr}dr$$

This equation is also integrable and its integral is

$$p\frac{dT}{dp} + q\frac{dT}{dq} + r\frac{dT}{dr} - T = h^2$$

where h^2 is a new arbitrary constant.

By replacing the expressions for T, dT/dp, dT/dq, dT/dr by the values given by these equations, two equations of the second order between p, q, r, will be obtained with which the value of two of these variables can be determined as a function of the third. If the values are then substituted in any of the three equations denoted by (A), an equation of the first order between t and the variable in question will be obtained. Thus the values of p, q, r as a function of t can be obtained by this means. This is the result of our development.

It should be noted at the outset that the second of the two integrals can be reduced to a simpler form by noting that since T is a homogeneous function of two dimensions of p, q, r, one has from the properties of these types of functions

$$p\frac{dT}{dp} + q\frac{dT}{dq} + r\frac{dT}{dr} = 2T$$

which reduces the integral equation in question to $T = h^2$, which expresses the **Conservation des Forces Vives** for the motion of rotation.

Secondly, it should be noted that since the quantity

$$\left(r\frac{dT}{dq} - q\frac{dT}{dr}\right)^2 + \left(p\frac{dT}{dr} - r\frac{dT}{dp}\right)^2 + \left(q\frac{dT}{dp} - p\frac{dT}{dq}\right)^2$$

is equivalent to

$$(p^2 + q^2 + r^2)\left[\left(\frac{dT}{dp}\right)^2 + \left(\frac{dT}{dq}\right)^2 + \left(\frac{dT}{dr}\right)^2\right] - \left(p\frac{dT}{dp} + q\frac{dT}{dq} + r\frac{dT}{dr}\right)^2$$

which becomes

$$f^2(p^2 + q^2 + r^2) - 4h^4$$

Because of the two preceding integrals, a simpler differential equation will be obtained by adding together the squares of the values of $d(dT/dp)$, $d(dT/dq)$, $d(dT/dr)$ in the three differential equations (A). This equation can also be used in place of any of the equations of the set (A).

In this manner, the determination of the quantities p, q, r, as a function of t will only depend upon these three equations

$$T = h^2$$
$$\left(\frac{dT}{dp}\right)^2 + \left(\frac{dT}{dq}\right)^2 + \left(\frac{dT}{dr}\right)^2 = f^2$$
$$\left(d\frac{dT}{dp}\right)^2 + \left(d\frac{dT}{dq}\right)^2 + \left(d\frac{dT}{dr}\right)^2 = [f^2(p^2 + q^2 + r^2) - 4h^4]dt^2$$

where

$$T = \tfrac{1}{2}(Ap^2 + Bq^2 + Cr^2) - Fqr - Gpr - Hpq$$

26. The determination of the quantities p, q, r is rather simple when the three constants F, G, H are zero because, one has simply

$$T = \tfrac{1}{2}(Ap^2 + Bq^2 + Cr^2)$$

Therefore

$$\frac{dT}{dp} = Ap, \qquad \frac{dT}{dq} = Bq, \qquad \frac{dT}{dr} = Cr$$

such that the three equations to be solved will be of the following form

$$Ap^2 + Bq^2 + Cr^2 = 2h^2$$
$$A^2p^2 + B^2q^2 + C^2r^2 = f^2$$
$$\frac{A^2\, dp^2 + B^2\, dq^2 + C^2\, dr^2}{dt^2} = f^2(p^2 + q^2 + r^2) - 4h^4$$

Therefore, if we define $p^2 + q^2 + r^2 = u$ and if the expressions for p, q, r, are obtained from the following three equations

$$p^2 + q^2 + r^2 = u$$
$$Ap^2 + Bq^2 + Cr^2 = 2h^2$$
$$A^2p^2 + B^2q^2 + C^2r^2 = f^2$$

one will have

$$p^2 = \frac{BCu - 2h^2(B + C) + f^2}{(A - B)(A - C)}$$
$$q^2 = \frac{ACu - 2h^2(A + C) + f^2}{(B - A)(B - C)}$$
$$r^2 = \frac{ABu - 2h^2(A + B) + f^2}{(C - A)(C - B)}$$

After substituting these expressions in the above differential equations, the first member of this equation will become after reduction of the terms

$$\frac{A^2B^2C^2(4h^4 - f^2u)\mathrm{d}u^2}{4[BCu - 2h^2(B+C) + f^2][ACu - 2h^2(A+C) + f^2][ABu - 2h^2(A+B) + f^2]\mathrm{d}t^2}$$

and the second member will become $(f^2u - 4h^4)$ such that by dividing the entire equation by $(f^2u - 4h^4)$ and taking the square root, one will finally obtain

$$\mathrm{d}t = \frac{ABC\,\mathrm{d}u}{2\sqrt{-[BCu - 2h^2(B+C) + f^2][ACu - 2h^2(A+C) + f^2][ABu - 2h^2(A+B) + f^2]}}$$

from which, by integration, t will be obtained as a function of u and reciprocally.

27. Let us now assume that the constants F, G, H are not zero and let us see how this case can be reduced to the preceding one by means of some substitutions.

For this purpose, substitute in place of the variables, p, q, r functions of other variables x, y, z which should not be confused with those used earlier to represent the coordinates of different points of the body. It must be assumed at the outset that these functions are such that

$$p^2 + q^2 + r^2 = x^2 + y^2 + z^2$$

It is evident that in order to satisfy this condition, these variables can only be linear. Consequently, they must have the following form

$$p = p'x + p''y + p'''z, \qquad q = q'x + q''y + q'''z, \qquad r = r'x + r''y + r'''z$$

The quantities p', p'', p''', q', etc. will be arbitrary constants and among them in virtue of the equation

$$p^2 + q^2 + r^2 = x^2 + y^2 + z^2$$

there must be the six equations of condition below

$$p'^2 + q'^2 + r'^2 = 1, \qquad p''^2 + q''^2 + r''^2 = 1, \qquad p'''^2 + q'''^2 + r'''^2 = 1$$

$$p'p'' + q'q'' + r'r'' = 0, \quad p'p''' + q'q''' + r'r''' = 0, \quad p''p''' + q''q''' + r''r''' = 0$$

such that since the unknowns in question are nine, three will remain arbitrary after the six equations are satisfied.

Now substitute the expressions for p, q, r in the equation for T and by means of the three arbitrary quantities, which I will discuss shortly, the three terms which contain the products

xy, xz, yz can be eliminated from the expression for T such that this quantity is reduced to this form

$$\frac{\alpha x^2 + \beta y^2 + \gamma z^2}{2}$$

But, in order to simplify the calculation, substitute in this equation for the values of x, y, z those of p, q, r and then compare the result with the expression for T. The arbitrary quantities in question will not only be determined, but also the unknowns α, β, γ. If the above expressions for p, q, r are multiplied respectively by p', q', r' and p'', q'', r'' and finally, by p''', q''', r''' and then added together, they immediately give in virtue of the equation of condition between the coefficients p', p'', etc.

$$x = p'p + q'q + r'r, \qquad y = p''p + q''q + r''r, \qquad z = p'''p + q'''q + r'''r$$

The substitution of these expressions in the equation $(\alpha x^2 + \beta y^2 + \gamma z^2)/2$ and its comparison with the expression for T in Article 25 will also give the six following equations

$$\begin{aligned}
&\alpha p'^2 + \beta p''^2 + \gamma p'''^2 = A\\
&\alpha q'^2 + \beta q''^2 + \gamma q'''^2 = B\\
&\alpha r'^2 + \beta r''^2 + \gamma r'''^2 = C\\
&\alpha q'r' + \beta q''r'' + \gamma q'''r''' = -F\\
&\alpha p'r' + \beta p''r'' + \gamma p'''r''' = -G\\
&\alpha p'q' + \beta p''q'' + \gamma p'''q''' = -H
\end{aligned}$$

which will be used to determine the six unknowns.

This determination presents no difficulty because if the first equation is multiplied by p' and then added to the sixth equation multiplied by q' and to the fifth equation multiplied by r', one will obtain by means of the equations of condition presented earlier

$$\alpha p' = Ap' - Hq' - Gr'$$

by adding the second, fourth and the sixth equations, multiplied respectively by q', r', p', one will similarly obtain

$$\alpha q' = Bq' - Fr' - Hp'$$

Finally, by adding together the third, fifth and the fourth equations multiplied respectively by r', p', q', one will obtain

$$\alpha r' = Cr' - Gp' - Fq'$$

After these three equations are combined with the equation of condition

$$p'^2 + q'^2 + r'^2 = 1$$

they will be used to determine the four unknowns α, p', q', r'.

The first two equations give

$$q' = \frac{HG + F(A-\alpha)}{FH + G(B-\alpha)}p', \qquad r' = \frac{(A-\alpha)(B-\alpha) - H^2}{FH + G(B-\alpha)}p'$$

After substituting these expressions in the third equation, the following equation is obtained as a function of α after dividing by p'

$$\begin{aligned}&(\alpha - A)(\alpha - B)(\alpha - C) - F^2(\alpha - A)\\ &-G^2(\alpha - B) - H^2(\alpha - C) + 2FGH = 0\end{aligned}$$

which, since it is of the third-order, will necessarily have a real root.

After substituting the same expressions in the fourth equation, one will obtain the equations for p', q', r' as a function of α. In order to shorten the expressions, we will first make the following definition

$$(\alpha) = \sqrt{[(A-\alpha)(B-\alpha) - H^2]^2 + [HG + F(A-\alpha)]^2 + [FH + G(B-\alpha)]^2}$$

Then, the equations can be expressed as

$$p' = \frac{FH + G(B-\alpha)}{(\alpha)}, \qquad q' = \frac{HG + F(A-\alpha)}{(\alpha)}, \qquad r' = \frac{(A-\alpha)(B-\alpha) - H^2}{(\alpha)}$$

If the same combinations of the above equations are made again but in this case, by taking for multipliers the quantities p'', q'', r'' instead of p', q', r', the following equations will be obtained

$$\begin{aligned}\beta p'' &= Ap'' - Hq'' - Gr''\\ \beta q'' &= Bq'' - Fr'' - Hp''\\ \beta r'' &= Cr'' - Gp'' - Fq''\end{aligned}$$

which in conjunction with the equation of condition $p''^2 + q''^2 + r''^2 = 1$, will be used to determine the four unknowns β, p'', q'', r''. Since these equations differ from the earlier equations in that the unknowns are β, p'', q'', r'' rather than α, p', q', r', one will immediately conclude that the equation in terms of β as well as the expressions for p'', q'', r'' as a function of β will be the same as those just found as a function of α.

Finally, if the same operations are repeated but in this case taking p''', q''', r''' for multipliers, the following three equations will be found in a similar fashion

$$\gamma p''' = Ap''' - Hq''' - Gr'''$$
$$\gamma q''' = Bq''' - Fr''' - Hp'''$$
$$\gamma r''' = Cr''' - Gp''' - Fq'''$$

to which the equation $p'''^2 + q'''^2 + r'''^2 = 1$ must be appended. Since these equations are in every way similar to the preceding ones, similar conclusions will be drawn.

In general, it can be concluded that the equation found above as a function of α will have for roots the values of the three quantities α, β, γ and if these roots are successively substituted in the expressions for p', q', r' as functions of α one will immediately have the values of $p', q', r', p'', q'', r''$ and p''', q''', r''' so that all the quantities will be known after resolution of this equation.

Also, since this equation is of the third degree it will always have a real root which when taken for α will make p', q', r', real quantities. With respect to the other roots β and γ, if they are imaginary they will, as is known, be of the form

$$b + c\sqrt{-1}, \qquad b - c\sqrt{-1}$$

such that the quantities p'', q'', r'' which are rational functions of β will also be of the form

$$m + n\sqrt{-1}, \qquad m' + n'\sqrt{-1}, \qquad m'' + n''\sqrt{-1}$$

The quantities p''', q''', r''' which are conjugate functions of γ will be of the form

$$m - n\sqrt{-1}, \qquad m' - n'\sqrt{-1}, \qquad m'' - n''\sqrt{-1}$$

Therefore, the equation of condition $p''p''' + q''q''' + r''r''' = 0$ will become

$$m^2 + n^2 + m'^2 + n'^2 + m''^2 + n''^2 = 0$$

Consequently, it is possible that m, n, m', n', m'', n'' are real quantities. From this result, it follows that β and γ can not be imaginary numbers.[44]

In order to assure oneself of the veracity of this statement, consider again the equation discussed earlier, but never formulated. This equation can be put in the following form

$$\alpha - C = \frac{F^2(\alpha - A) + G^2(\alpha - B) - 2FGH}{(\alpha - A)(\alpha - B) - H^2}$$

Now substitute successively in this equation the two other roots β and γ for α and then substitute the two resulting equations. After the simplifications and the division by $(\beta-\gamma)$ are made, the following transformed equation is the result

$$\begin{aligned}&[(\beta-A)(\beta-B)-H^2][(\gamma-A)(\gamma-B)-H^2]\\&+(F^2+G^2)\beta\gamma-(AF^2+BG^2+2FGH)(\beta+\gamma)\\&+(F^2+G^2)H^2+A^2F^2+B^2G^2+2FGH(A+B)=0\end{aligned}$$

which is reducible to this form

$$\begin{aligned}&[(\beta-A)(\beta-B)-H^2][(\gamma-A)(\gamma-B)-H^2]\\&+[F(\beta-A)-GH][F(\gamma-A)-GH]\\&+[G(\beta-B)-HF][G(\gamma-B)-HF]=0\end{aligned}$$

This expression is equivalent to the equation[45]

$$p''p'''+q''q'''+r''r'''=0$$

which consequently provides similar conclusions.

Therefore, the three roots α, β, γ will necessarily be real numbers and the nine coefficients p', q', r', p'', etc., which are rational functions of these roots will also be real.

28. The values of these coefficients were determined so that one has

$$p^2+q^2+r^2=x^2+y^2+z^2, \qquad T=\frac{\alpha x^2+\beta y^2+\gamma z^2}{2}$$

but by successively varying p, q, r one will obtain

$$\begin{aligned}\frac{dT}{dp}&=\alpha x\frac{dx}{dp}+\beta y\frac{dy}{dp}+\gamma z\frac{dz}{dp}\\\frac{dT}{dq}&=\alpha x\frac{dx}{dq}+\beta y\frac{dy}{dq}+\gamma z\frac{dz}{dq}\\\frac{dT}{dr}&=\alpha x\frac{dx}{dr}+\beta y\frac{dy}{dr}+\gamma z\frac{dz}{dr}\end{aligned}$$

since x, y, z are functions of these variables. But it was shown earlier that

$$x=p'p+q'q+r'r, \qquad y=p''p+q''q+r''r, \qquad z=p'''p+q'''q+r'''r$$

Therefore, as one saw earlier,

$$\frac{dx}{dp}=p', \qquad \frac{dx}{dq}=q', \qquad \frac{dx}{dr}=r', \qquad \frac{dy}{dp}=p'', \qquad \frac{dy}{dq}=q'', \qquad \dots$$

After substituting these expressions, one obtains

$$\frac{dT}{dp} = p'\alpha x + p''\beta y + p'''\gamma z$$
$$\frac{dT}{dq} = q'\alpha x + q''\beta y + q'''\gamma z$$
$$\frac{dT}{dr} = r'\alpha x + r''\beta y + r'''\gamma z$$

Because of the equations of condition between the coefficients p', q', r', p'', etc., the following equation is obtained

$$\left(\frac{dT}{dp}\right)^2 + \left(\frac{dT}{dq}\right)^2 + \left(\frac{dT}{dr}\right)^2 = \alpha^2 x^2 + \beta^2 y^2 + \gamma^2 z^2$$

and

$$\left(d\frac{dT}{dp}\right)^2 + \left(d\frac{dT}{dq}\right)^2 + \left(d\frac{dT}{dr}\right)^2 = \alpha^2\, dx^2 + \beta^2\, dy^2 + \gamma^2\, dz^2$$

Consequently, the three final equations of Article 24 will be reduced to the following three

$$\alpha x^2 + \beta y^2 + \gamma z^2 = 2h^2$$
$$\alpha^2 x^2 + \beta^2 y^2 + \gamma^2 z^2 = f^2$$
$$\frac{\alpha^2\, dx^2 + \beta^2\, dy^2 + \gamma^2\, dz^2}{dt^2} = f^2(x^2 + y^2 + z^2) - 4h^4$$

which, it is clear, are entirely similar to those of Article 25 where the quantities x, y, z, α, β, γ correspond to the quantities p, q, r, A, B, C.

From this result, it follows that if one puts, as in the cited article

$$u = p^2 + q^2 + r^2 = x^2 + y^2 + z^2$$

the same formulas will be obtained between the variables x, y, z, u, t, as were found between p, q, r, u, t, after replacing A, B, C, with α, β, γ.

Thus with the expressions for x, y, z, as functions of u or t, one will obtain the complete expressions for p, q, r, from the formulas of Article 27.

29. The quantities p, q, r do not suffice to define all cases of rotational motion of the body. They are only useful to obtain its instantaneous rotation. Indeed, since

$$p = \frac{dP}{dt}, \qquad q = \frac{dQ}{dt}, \qquad r = \frac{dR}{dt}$$

it follows from what has been shown in Article 10 that the instantaneous axis of rotation of the body at each instant will make with the coordinate axes a, b, c, the angles for which the cosines will be respectively

$$\frac{p}{\sqrt{p^2+q^2+r^2}}, \qquad \frac{q}{\sqrt{p^2+q^2+r^2}}, \qquad \frac{r}{\sqrt{p^2+q^2+r^2}}$$

and the angular velocity about this axis will be represented by

$$\sqrt{p^2+q^2+r^2}$$

In order to completely define the rotation of the body, one must still determine the values of the nine quantities $\xi', \eta', \zeta', \xi''$ etc. on which depend the values of the coordinates ξ, η, ζ which give the absolute location of each point of the body in space relative to a center of gravity viewed as an immobile point (Article 17). Three more integrations are still required.

To this end, the differential formulas of Article 13 will be reconsidered and after replacing $\mathrm{d}P, \mathrm{d}Q, \mathrm{d}R$ with $p\,\mathrm{d}t$, $q\,\mathrm{d}t$, $r\,\mathrm{d}t$, the following equations will be obtained

$$(\mathrm{C})\begin{cases} \mathrm{d}\xi' + (q\xi''' - r\xi'')\mathrm{d}t = 0 \\ \mathrm{d}\xi'' + (r\xi' - p\xi''')\mathrm{d}t = 0 \\ \mathrm{d}\xi''' + (p\xi'' - q\xi')\mathrm{d}t = 0 \end{cases}$$

and as many similar equations in η', η'', η''' and ζ', ζ'', ζ''' will be obtained by replacing only ξ with η and then with ζ.

If these equations are compared with the differential equations (A) for the quantities $\mathrm{d}T/\mathrm{d}p$, $\mathrm{d}T/\mathrm{d}q$, $\mathrm{d}T/\mathrm{d}r$ of Article 24, it is obvious that they are entirely similar. Therefore, these quantities correspond to the quantities ξ', ξ'', ξ''' as well as to the quantities η', η'', η''' and $\zeta', \zeta'', \zeta'''$.

From this result, it can be concluded that the last set of variables can be viewed as particular values of the variables $\mathrm{d}T/\mathrm{d}p$, $\mathrm{d}T/\mathrm{d}q$, $\mathrm{d}T/\mathrm{d}r$. Also, since the equations between these variables are linear, one will have, with the three arbitrary constants ℓ, m, n, the three complete integral equations

$$(\mathrm{D})\begin{cases} \dfrac{\mathrm{d}T}{\mathrm{d}p} = \ell\xi' + m\eta' + n\zeta' \\ \dfrac{\mathrm{d}T}{\mathrm{d}q} = \ell\xi'' + m\eta'' + n\zeta'' \\ \dfrac{\mathrm{d}T}{\mathrm{d}r} = \ell\xi''' + m\eta''' + n\zeta''' \end{cases}$$

By combining these three equations with the six equations of condition between the same variables ξ', η', etc., it appears that these nine variables could be determined. By taking a closer look at the preceding equations, it is clear that they really constitute only two equations because, by adding together their squares all the unknowns ξ', η', ζ', etc., disappear simultaneously by reason of the equations of condition (Article 5). Therefore, only the following equation is obtained

$$\left(\frac{\mathrm{d}T}{\mathrm{d}p}\right)^2 + \left(\frac{\mathrm{d}T}{\mathrm{d}q}\right)^2 + \left(\frac{\mathrm{d}T}{\mathrm{d}r}\right)^2 = \ell^2 + m^2 + n^2$$

which, it is clear, is comparable to the first of the two integrals found above (Article 25). The comparison of these equations gives

$$f^2 = \ell^2 + m^2 + n^2$$

so that three of the four constants f, ℓ, m, n remain arbitrary.

From this result, it must be concluded that the complete solution requires a new integration for which any one of the differential equations above or a combination of these differential equations could be used.

30. But the calculation can be made more general and simpler by directly examining the values of the coordinates ξ, η, ζ, which immediately determine the absolute position of an arbitrary point of the body for which the relative coordinates with respect to the axes of the body are a, b, c.

In order to demonstrate the procedure, begin by adding together the three integral equations (D) found above, after multiplying the first by a, the second by b and the third by c. This operation gives (Article 1) the following equation

$$\ell\xi + m\eta + n\zeta = a\frac{\mathrm{d}T}{\mathrm{d}p} + b\frac{\mathrm{d}T}{\mathrm{d}q} + c\frac{\mathrm{d}T}{\mathrm{d}r}$$

But it has already been shown from the nature of the quantities ξ, η, ζ (Article 5) that

$$\xi^2 + \eta^2 + \zeta^2 = a^2 + b^2 + c^2$$

Finally, by replacing $\mathrm{d}P$, $\mathrm{d}Q$, $\mathrm{d}R$ with $p\,\mathrm{d}t$, $q\,\mathrm{d}t$, $r\,\mathrm{d}t$ and with a, b, c as constants (Article 14), there will result

$$\frac{\mathrm{d}\xi^2 + \mathrm{d}\eta^2 + \mathrm{d}\zeta^2}{\mathrm{d}t^2} = (cq - br)^2 + (ar - cp)^2 + (bp - aq)^2$$

Thus from these three equations the values of ξ, η, ζ can be obtained with only one integration.

Then, if the values of $\xi', \eta', \zeta', \xi''$, etc. are individually required it will only be necessary to assume in the general expressions for ξ, η, ζ the following values for the constants

$$a = 1, \quad b = 0, \quad c = 0, \quad \text{or} \quad a = 0, \quad b = 1, \quad c = 0, \quad \text{or} \quad a = 0, \quad b = 0, \quad c = 1$$

Let us assume in order to shorten the expressions

$$L = a\frac{\mathrm{d}T}{\mathrm{d}p} + b\frac{\mathrm{d}T}{\mathrm{d}q} + c\frac{\mathrm{d}T}{\mathrm{d}r}$$
$$M = a^2 + b^2 + c^2$$
$$N = (cq - br)^2 + (ar - cp)^2 + (bp - aq)^2$$

Thus it will be necessary to solve these three equations

$$\ell\xi + m\eta + n\zeta = L$$
$$\xi^2 + \eta^2 + \zeta^2 = M$$
$$\frac{\mathrm{d}\xi^2 + \mathrm{d}\eta^2 + \mathrm{d}\zeta^2}{\mathrm{d}t^2} = N$$

in which M is a given constant, L and N are assumed known as functions of t, and ℓ, m, n are arbitrary constants.

It is clear at the outset that if ℓ and m were simultaneously equal to zero the first equation would give $\zeta = L/n$ and if this equation were substituted in the two remaining equations, one would have

$$\xi^2 + \eta^2 = M - \frac{L^2}{n^2}, \qquad \frac{\mathrm{d}\xi^2 + \mathrm{d}\eta^2}{\mathrm{d}t^2} = N - \frac{\mathrm{d}L^2}{n^2\,\mathrm{d}t^2}$$

These equations are very easy to integrate if the expressions, $\xi = \rho\cos\theta$, $\eta = \rho\sin\theta$, are substituted in these equations. The following equations result after making these substitutions

$$\rho^2 = M - \frac{L^2}{n^2}, \qquad \frac{\rho^2\,\mathrm{d}\theta^2 + \mathrm{d}\rho^2}{\mathrm{d}t^2} = N - \frac{\mathrm{d}L^2}{n^2\,\mathrm{d}t^2}$$

The first equation will give the value of ρ and the second equation will give the angle θ by integration of the expression

$$\mathrm{d}\theta = \frac{\mathrm{d}t}{\rho}\sqrt{N - \frac{\mathrm{d}L^2}{n^2\,\mathrm{d}t^2} - \frac{\mathrm{d}\rho^2}{\mathrm{d}t^2}}$$

Let us now assume that ℓ and m are not equal to zero and we will show how this case can be reduced to the earlier one. It is clear that if the following definitions are made

$$\ell\xi + m\eta = x\sqrt{\ell^2 + m^2}, \qquad m\xi - \ell\eta = y\sqrt{\ell^2 + m^2}$$

one will also have

$$\xi^2 + \eta^2 = x^2 + y^2, \qquad d\xi^2 + d\eta^2 = dx^2 + dy^2$$

Thus at the outset the developed equations will be reduced to this form

$$x\sqrt{\ell^2 + m^2} + n\zeta = L$$
$$x^2 + y^2 + \zeta^2 = M$$
$$\frac{dx^2 + dy^2 + d\zeta^2}{dt^2} = N$$

If it is assumed that

$$x\sqrt{\ell^2 + m^2} + n\zeta = z\sqrt{\ell^2 + m^2 + n^2}$$
$$nx - \zeta\sqrt{\ell^2 + m^2} = u\sqrt{\ell^2 + m^2 + n^2}$$

and in addition, that we have

$$x^2 + \zeta^2 = z^2 + u^2, \qquad dx^2 + d\zeta^2 = dz^2 + du^2$$

then the following transformed equations will be obtained

$$z\sqrt{\ell^2 + m^2} + n^2 = L$$
$$u^2 + y^2 + z^2 = M$$
$$\frac{du^2 + dy^2 + dz^2}{dt^2} = N$$

which are, it is clear, entirely similar to those which were just solved above so that u, y, z will have the same expressions as those found for ξ, η, ζ by only replacing n with $\sqrt{\ell^2 + m^2 + n^2}$.

Since these expressions are known, the general values of ξ, η, ζ will be obtained with the following equations

$$\xi = \frac{\ell x + my}{\sqrt{\ell^2 + m^2}}, \qquad \eta = \frac{mx - \ell y}{\sqrt{\ell^2 + m^2}}, \qquad \zeta = \frac{nz - u\sqrt{\ell^2 + m^2}}{\sqrt{\ell^2 + m^2 + n^2}}$$

31. If I am not mistaken, this is the most general solution and at the same time the simplest which can be given for the famous problem of rotational motion of free bodies. It is analogous to the solution which I gave in the mémoires of the Académie de Berlin in 1773, but it is at the same time more direct and simpler in some respects. Here, I started from three integrable equations which correspond to equations (D) of Article 29. These equations were obtained directly from the known principles of areas and moments and to which I had added the equation of **force vives** $T = h^2$ (Article 24). Here, the entire solution is deduced from the three original differential equations and I believe that the

solution has all the clarity and (if I dare say so) all the elegance of which it is capable. For this reason, I flatter myself that no one will object to my having treated this problem again. Although this problem presents mere curiosity, I certainly have no doubt that it can be of some use to the progress of analysis.

It seems to me that there is something more remarkable in the preceding solution. This is the use which has been made of the quantities ξ', η', ζ', ξ'', etc. without knowing their values but only the equations of condition by which they are constrained. These quantities disappear at the end of the calculation. I have no doubt that this type of analysis can also be of some use in other instances.

By the way, if this solution was somewhat lengthy, it is because of the great generality which we wanted to retain. It has been noted that there are two approaches to simplification. One is to assume that the constants F, G, H are equal to zero (Article 25) and the other is to equate to zero the constants ℓ and m (Article 30).

The first of these two assumptions had always been viewed as indispensable to obtain a complete solution of the problem until I gave in my memoir of 1773 the means to avoid it. Indeed, this assumption consists of taking for coordinate axes a, b, c, straight lines such that the sums $\mathrm{S}\, ab\, \mathrm{D}m$, $\mathrm{S}\, ac\, \mathrm{D}m$, $\mathrm{S}\, bc\, \mathrm{D}m$ are equal to zero (Article 19). Euler was the first to demonstrate that this is always possible, whatever the shape of the body. The axes which have been determined are the natural axes of rotation, that is, the bodies can freely rotate about each of these axes. Although it is always possible to find axes which have this property and in addition, the property that the location of the axes of the body are arbitrary, it is not without interest to obtain a solution at the same time direct and independent of these particular considerations.

The second of these two assumptions depends upon the location of the coordinate axes ξ, η, ζ in space. Since the location of the system of coordinates is also arbitrary, it can always be assumed that the constants ℓ and m can be equal to zero. This assertion can be demonstrated from the general expressions of ξ, η, ζ that we have found.

32. By assuming F, G, H equal to zero, one has, as shown in Article 26

$$\frac{\mathrm{d}T}{\mathrm{d}p} = Ap, \qquad \frac{\mathrm{d}T}{\mathrm{d}q} = Bq, \qquad \frac{\mathrm{d}T}{\mathrm{d}r} = Cr$$

If these expressions are substituted in the three differential equations denoted by (A), the following equations are obtained

$$\mathrm{d}p + \frac{C-B}{A} qr\,\mathrm{d}t = 0, \qquad \mathrm{d}q + \frac{A-C}{B} pr\,\mathrm{d}t = 0, \qquad \mathrm{d}r + \frac{B-A}{C} pq\,\mathrm{d}t = 0$$

These equations are consistent with those given by Euler. He was the first to present a solution for this problem. (The reader should refer to the Mémoires of the Académie de Berlin in 1758 for additional information.) In order to be convinced that the first assertion is accurate, it is sufficient to observe that the constants A, B, C (Article 20) are identical to those Euler calls the **moments of inertia** of the body about the coordinate axes a, b, c and that the variables p, q, r depend upon the free and instantaneous rotational motion in such a way that if α, β, γ are the angles between the axis about which the body rotates freely at each instant and the coordinate axes a, b, c, and if ρ is the angular velocity about this axis, one has (Article 29)

$$p = \rho \cos \alpha, \qquad q = \rho \cos \beta, \qquad r = \rho \cos \gamma$$

With respect to the other equations derived by Euler which are used to determine the location of these axes of the body in space, they are related to our equations (C) of Article 29. Indeed, since the nine quantities ξ', η', ζ', ξ'', etc. are simply the rectangular coordinates of the three points of the body measured with respect to these three axes at a distance i from the center (which evidently comes from the fact that these quantities derive from ξ, η, ζ, by successively putting $a = 1, b = 0, c = 0$, then $a = 0, b = 1, c = 0$, and finally $a = 0, b = 0, c = 1$), it is clear that if one designates, as Euler did, by ℓ, m, n, the complements of the angles of inclination of these axes on the fixed plane of ξ and η, and by λ, μ, ν the angles that the projections of the same axes make with the fixed ξ-axis, one will obtain these three expressions

$$\begin{aligned}
\zeta' &= \cos \ell, & \eta' &= \sin \ell \sin \lambda, & \xi' &= \sin \ell \cos \lambda \\
\zeta'' &= \cos m, & \eta'' &= \sin m \sin \mu, & \xi'' &= \sin m \cos \mu \\
\zeta''' &= \cos n, & \eta''' &= \sin n \sin \nu, & \xi''' &= \sin n \cos \nu
\end{aligned}$$

By means of these substitutions, it is easy to obtain the equations which Euler derived by geometrical and trigonometrical considerations.

33. Also, by adopting simultaneously the two assumptions that F, G, H are equal to zero, and that ℓ and m are both equal to zero, one will have the simplest solution with the three equations (D) of Article 29, by substituting the values of $\zeta', \zeta'', \zeta'''$, and of p, q, r in φ, ψ, ω (Articles 7 and 20) because in this manner the following three equations of the first order will be obtained

$$A \frac{\sin \varphi \sin \omega \,\mathrm{d}\psi + \cos \varphi \,\mathrm{d}\omega}{\mathrm{d}t} = n \sin \varphi \sin \omega$$

$$B \frac{\cos \varphi \sin \omega \,\mathrm{d}\psi - \sin \varphi \,\mathrm{d}\omega}{\mathrm{d}t} = n \cos \varphi \sin \omega$$

$$C \frac{\mathrm{d}\varphi + \cos \omega \,\mathrm{d}\psi}{\mathrm{d}t} = n \cos \omega$$

which obviously reduce to these equations

$$n \,\mathrm{d}t - A \,\mathrm{d}\psi = \frac{A \,\mathrm{d}\omega}{\tan \varphi \sin \omega}$$

$$n \,\mathrm{d}t - B \,\mathrm{d}\psi = -\frac{B \tan \varphi \,\mathrm{d}\omega}{\sin \omega}$$

$$n \,\mathrm{d}t - C \,\mathrm{d}\psi = \frac{C \,\mathrm{d}\varphi}{\cos \omega}$$

But, if $\mathrm{d}t$ and $\mathrm{d}\psi$ were eliminated by adding these three equations together after having multiplied them separately by $C-B$, $A-C$, $B-A$, one will obtain the following equation

$$A(C-B)\frac{\mathrm{d}\omega}{\tan\varphi\sin\omega} - B(A-C)\frac{\tan\varphi\,\mathrm{d}\omega}{\sin\omega} + C(B-A)\frac{\mathrm{d}\varphi}{\cos\omega} = 0$$

which reduces to

$$\frac{\cos\omega\,\mathrm{d}\omega}{\sin\omega} = \frac{C(B-A)\,\mathrm{d}\varphi}{B(A-C)\tan\varphi - \dfrac{A(C-B)}{\tan\varphi}}$$

where the variables are separated.

The second member of this equation can be transformed to

$$\frac{C(B-A)\sin\varphi\cos\varphi\,\mathrm{d}\varphi}{B(A-C)\sin^2\varphi - A(C-B)\cos^2\varphi}$$

or yet to

$$\frac{C(B-A)\sin 2\varphi\,\mathrm{d}\varphi}{2AB - C(A+B) + C(B-A)\cos 2\varphi}$$

By integrating logarithmically and then by eliminating the logarithms, one will have

$$2AB - C(A+B) + C(B-A)\cos 2\varphi = \frac{K}{\sin^2\omega}$$

where K is an arbitrary constant. But $\tan\varphi = \sqrt{(1-\cos 2\varphi)/(1+\cos 2\varphi)}$, therefore, after substituting the preceding expression, one will obtain

$$\tan\varphi = \sqrt{\frac{2A(B-C)\sin^2\omega - K}{2B(C-A)\sin^2\omega + K}}$$

After substituting this equation for $\tan\varphi$ in the first two differential equations, the following equations will be obtained

$$n\,\mathrm{d}t - A\,\mathrm{d}\psi = \frac{A\,\mathrm{d}\omega}{\sin\omega}\sqrt{\frac{2B(C-A)\sin^2\omega + K}{2A(B-C)\sin^2\omega - K}}$$

$$n\,\mathrm{d}t - B\,\mathrm{d}\psi = -\frac{B\,\mathrm{d}\omega}{\sin\omega}\sqrt{\frac{2A(B-C)\sin^2\omega - K}{2B(C-A)\sin^2\omega + K}}$$

where the variables are separated and which, after being integrated, will give t and ψ as functions of ω.

This solution is identical to the solution given by d'Alembert in the fourth volume of his *Opuscules*.

34. Let us now consider the second case, where it is assumed that the heavy body is suspended from a fixed point, about which it can rotate in any direction. By taking this point as the center of the body, that is, for the common origin of the coordinates ξ, η, ζ and a, b, c, and furthermore, assuming that the ordinates ζ are vertical and directed upward, one will have for the rotational motion of the body equations (B) of Article 23. These equations are more complex than those of the preceding case because of the terms multiplied by the quantities $\mathrm{S}\, a\, \mathrm{D}m$, $\mathrm{S}\, b\, \mathrm{D}m$, $\mathrm{S}\, c\, \mathrm{D}m$, which are not equal to zero when the center of the body, for which the location is given here, falls outside its center of gravity. Nevertheless, it is still possible to eliminate two of these quantities by having one of the axes of coordinates a, b, c pass through the center of gravity. This axis can be oriented arbitrarily in the body which will simplify the equations somewhat.

If it is assumed that the axis of coordinates c passes through the center of gravity of the body, then, one will obtain from the properties of this center that

$$\mathrm{S}\, a\, \mathrm{D}m = 0, \qquad \mathrm{S}\, b\, \mathrm{D}m = 0$$

If the letter k designates the distance between the center of the body, which is the point of suspension, and its center of gravity, it is clear that we will also have

$$\mathrm{S}(c - k)\mathrm{D}m = 0$$

Therefore

$$\mathrm{S}\, c\, \mathrm{D}m = \mathrm{S}\, k\, \mathrm{D}m = k\, \mathrm{S}\, \mathrm{D}m = km$$

where m denotes the mass of the body.

After making these substitutions and replacing K by km, the three following equations will be obtained

$$(\mathrm{E}) \begin{cases} \dfrac{\mathrm{d}\dfrac{\mathrm{d}T}{\mathrm{d}p}}{\mathrm{d}t} + q\dfrac{\mathrm{d}T}{\mathrm{d}r} - r\dfrac{\mathrm{d}T}{\mathrm{d}q} + K\zeta'' = 0 \\ \dfrac{\mathrm{d}\dfrac{\mathrm{d}T}{\mathrm{d}q}}{\mathrm{d}t} + r\dfrac{\mathrm{d}T}{\mathrm{d}p} - p\dfrac{\mathrm{d}T}{\mathrm{d}r} - K\zeta' = 0 \\ \dfrac{\mathrm{d}\dfrac{\mathrm{d}T}{\mathrm{d}r}}{\mathrm{d}t} + p\dfrac{\mathrm{d}T}{\mathrm{d}q} - q\dfrac{\mathrm{d}T}{\mathrm{d}p} = 0 \end{cases}$$

where

$$T = \tfrac{1}{2}(Ap^2 + Bq^2 + Cr^2) - Fqr - Gpr - Hpq$$

35. The two integrals of these equations can be found at the outset by adding the equations together, after having multiplied them respectively by p, q, r or by ζ', ζ'', ζ''', because (Article 15)

$$\mathrm{d}\zeta' = (\zeta''r - \zeta'''q)\mathrm{d}t, \qquad \mathrm{d}\zeta'' = (\zeta'''p - \zeta'r)\mathrm{d}t, \qquad \mathrm{d}\zeta''' = (\zeta'q - \zeta''p)\mathrm{d}t$$

Thus the two following equations will be obtained

$$p\,\mathrm{d}\frac{\mathrm{d}T}{\mathrm{d}p} + q\,\mathrm{d}\frac{\mathrm{d}T}{\mathrm{d}q} + r\,\mathrm{d}\frac{\mathrm{d}T}{\mathrm{d}r} - K\,\mathrm{d}\zeta''' = 0$$
$$\zeta'\,\mathrm{d}\frac{\mathrm{d}T}{\mathrm{d}p} + \zeta''\,\mathrm{d}\frac{\mathrm{d}T}{\mathrm{d}q} + \zeta'''\,\mathrm{d}\frac{\mathrm{d}T}{\mathrm{d}r} + \frac{\mathrm{d}T}{\mathrm{d}p}\mathrm{d}\zeta' + \frac{\mathrm{d}T}{\mathrm{d}q}\mathrm{d}\zeta'' + \frac{\mathrm{d}T}{\mathrm{d}r}\mathrm{d}\zeta''' = 0$$

for which the integrals are

$$p\frac{\mathrm{d}T}{\mathrm{d}p} + q\frac{\mathrm{d}T}{\mathrm{d}q} + r\frac{\mathrm{d}T}{\mathrm{d}r} - T - K\zeta''' = f$$
$$\zeta'\frac{\mathrm{d}T}{\mathrm{d}p} + \zeta''\frac{\mathrm{d}T}{\mathrm{d}q} + \zeta'''\frac{\mathrm{d}T}{\mathrm{d}r} = h$$

where f and h are arbitrary constants.

It appears difficult to integrate these equations again and consequently, to obtain the general solution. But a solution can be found by assuming that the shape of the body is constrained by particular conditions.

Thus, by assuming $F = 0$, $G = 0$, $H = 0$ and moreover, that $A = B$, one will obtain

$$\frac{\mathrm{d}T}{\mathrm{d}p} = Ap, \qquad \frac{\mathrm{d}T}{\mathrm{d}q} = Aq$$

and the third of equations (E) will become $\mathrm{d}(\mathrm{d}T/\mathrm{d}r)$, for which the integral is $\mathrm{d}T/\mathrm{d}r =$ constant.

This is the case where the c-coordinate axis, that is, the straight line passing through the point of suspension and through the center of gravity is a natural axis of rotation and where the **moments of inertia** about the two other axes are equal (Article 32), which occurs in general for all solids of revolution when the fixed point is taken on the axis of revolution. The solution of this case, using the three integrals which have been found, is easy.[46]

Indeed, since $T = A(p^2 + q^2)/2 + Cr^2/2$, it is obvious that these three integrals will be reduced to this form

$$A(p^2 + q^2) + Cr^2 - 2K\zeta''' = 2f$$
$$A(\zeta'p + \zeta''q) + C\zeta'''r = h, \qquad r = n$$

where f, h, n are arbitrary constants.

Therefore, if for the quantities ζ', ζ'', ζ''' and p, q, r their expressions as functions of φ, ψ, ω (Articles 7 and 20) are substituted, one will obtain the following three equations

$$A\frac{\sin^2\omega\,\mathrm{d}\psi^2+\mathrm{d}\omega^2}{\mathrm{d}t^2}+Cn^2-2C\cos\omega=2f$$
$$A\frac{\sin^2\omega\,\mathrm{d}\psi}{\mathrm{d}t}+Cn\cos\omega=h,\qquad \frac{\mathrm{d}\varphi+\cos\omega\,\mathrm{d}\psi}{\mathrm{d}t}=n$$

which have the advantage that they do not contain the finite angles ψ and φ.

The second equation gives at the outset

$$\frac{\mathrm{d}\psi}{\mathrm{d}t}=\frac{h-Cn\cos\omega}{A\sin^2\omega}$$

and after this latter equation is substituted in the first equation, one will have

$$\mathrm{d}t=\frac{A\sin\omega\,\mathrm{d}\omega}{\sqrt{A\sin^2\omega(2f-Cn^2+2K\cos\omega)-(h-Cn\cos\omega)^2}}$$

then the second and third equations will give

$$\mathrm{d}\psi=\frac{(h-Cn\cos\omega)\mathrm{d}\omega}{\sin\omega\sqrt{A\sin^2\omega(2f-Cn^2+2K\cos\omega)-(h-Cn\cos\omega)^2}}$$
$$\mathrm{d}\varphi=\frac{(An-h\cos\omega+(C-A)n\cos^2\omega)\mathrm{d}\omega}{\sin\omega\sqrt{A\sin^2\omega(2f-Cn^2+2K\cos\omega)-(h-Cn\cos\omega)^2}}$$

where the variables are separated, but for which the integration depends in general upon the rectification of conic sections.

36. After substituting the expressions of $\mathrm{d}T/\mathrm{d}p$, $\mathrm{d}T/\mathrm{d}q$, $\mathrm{d}T/\mathrm{d}r$ in terms of p, q, r, in equations (E), they will become

$$\frac{A\,\mathrm{d}p-G\,\mathrm{d}r-H\,\mathrm{d}q}{\mathrm{d}t}+(C-B)qr+F(r^2-q^2)-Gpq+Hpr+K\zeta''=0$$
$$\frac{B\,\mathrm{d}q-F\,\mathrm{d}r-H\,\mathrm{d}p}{\mathrm{d}t}+(A-C)pr+G(p^2-r^2)-Hqr+Fpq-K\zeta'=0$$
$$\frac{C\,\mathrm{d}r-F\,\mathrm{d}q-G\,\mathrm{d}p}{\mathrm{d}t}+(B-A)pq+H(q^2-p^2)-Fpr+Gqr=0$$

When the body is in a state of equilibrium, the three quantities p, q, r, are equal to zero because $\sqrt{p^2+q^2+r^2}$ is the instantaneous rotational velocity (Article 29). Therefore, one will have $\zeta'=0$ and $\zeta''=0$ because $\zeta'^2+\zeta''^2+\zeta'''^2=1$ and as a consequence of $\zeta'''=1$, the ζ-coordinate axis will coincide with the c-coordinate axis. This means that this latter axis which passes through the center of gravity of the body and which we will call from now on, the *axis of the body*, will be vertical, which is the state of equilibrium of the body. This result is easier shown from the formulas of Article 7, which give $\sin\varphi \sin\omega = 0$, $\cos\varphi \sin\omega = 0$ and as a consequence $\omega = 0$, where ω is the angle between the c and ζ coordinate axes.

Therefore, if it is assumed that the body is in motion, it must also be assumed simultaneously that its axis departs very little from the vertical such that the angle of deviation ω will always remain very small. Then the quantities ζ' and ζ'' will be very small and one will have the case where the body makes only very small oscillations about the vertical and has at the same time an arbitrary rotational motion about its axis.

This case, which has not yet been resolved, can be easily and completely solved with our formulas. By viewing ζ'', ζ''' as very small of the first order and neglecting the very small quantities of the second and higher orders, one finds by the equations of condition of Article 5

$$\zeta''' = 1, \qquad \xi''' = -\xi'\zeta' - \xi''\zeta'', \qquad \eta''' = -\eta'\zeta' - \eta''\zeta''$$

and

$$\xi'^2 + \xi''^2 = 1, \qquad \eta'^2 + \eta''^2 = 1, \qquad \xi'\eta' + \xi''\eta'' = 0$$

therefore

$$\xi' = \sin\pi, \qquad \xi'' = \cos\pi, \qquad \eta' = \sin\theta, \qquad \eta'' = \cos\theta, \qquad \cos(\pi - \theta) = 0$$

from which

$$\pi = 90^\circ + \theta$$

Consequently

$$\xi' = \cos\theta, \qquad \xi'' = -\sin\theta$$

After substituting these quantities in the expressions for $\mathrm{d}P$, $\mathrm{d}Q$, $\mathrm{d}R$ of Article 11, one will have

$$\mathrm{d}P = \zeta'\,\mathrm{d}\theta + \mathrm{d}\zeta'', \qquad \mathrm{d}Q = \zeta''\,\mathrm{d}\theta - \mathrm{d}\zeta'$$

If the quantities of the second order are always neglected, we will also have $\mathrm{d}R = \mathrm{d}\theta$.

Thus, the following equations result

$$p = \frac{\mathrm{d}P}{\mathrm{d}t} = \frac{\zeta'\,\mathrm{d}\theta + \mathrm{d}\zeta''}{\mathrm{d}t}$$
$$q = \frac{\mathrm{d}Q}{\mathrm{d}t} = \frac{\zeta''\,\mathrm{d}\theta - \mathrm{d}\zeta'}{\mathrm{d}t}$$
$$r = \frac{\mathrm{d}R}{\mathrm{d}t} = \frac{\mathrm{d}\theta}{\mathrm{d}t}$$

After substituting these expressions in the differential equations above and neglecting the powers and products of ζ' and ζ'', linear equations will result for the determination of these variables.

But before making these substitutions, one notes, assuming ζ' and ζ'' are equal to zero, that the equations in question will give

$$-G\frac{\mathrm{d}^2\theta}{\mathrm{d}t^2}+F\left(\frac{\mathrm{d}\theta}{\mathrm{d}t}\right)^2=0,\qquad -F\frac{\mathrm{d}^2\theta}{\mathrm{d}t^2}-G\frac{\mathrm{d}\theta^2}{\mathrm{d}t^2}=0,\qquad C\frac{\mathrm{d}^2\theta}{\mathrm{d}t^2}=0$$

Therefore, since C could not be equal to zero unless the body is reduced to a line because C is equal to $\mathrm{S}(a^2+b^2)\mathrm{D}m$, it follows that these equations can only be satisfied if $\mathrm{d}^2\theta/\mathrm{d}t^2=0$ and then, $\mathrm{d}\theta/\mathrm{d}t=0$, or $F=0$ and $G=0$.

From this result, it is easy to conclude that when ζ' and ζ'' are not equal to zero, but only very small, the values of $\mathrm{d}\theta/\mathrm{d}t$ or of F and G should also be very small, which are two cases which need to be examined separately.

37. Let us first assume that $\mathrm{d}\theta/\mathrm{d}t$ is a very small quantity of the same order as ζ' and ζ''. One will have, considering only quantities less than second order

$$p=\frac{\mathrm{d}\zeta''}{\mathrm{d}t},\qquad q=-\frac{\mathrm{d}\zeta'}{\mathrm{d}t}$$

With these substitutions and neglecting the quantities of second order and also replacing for greater simplicity the symbols ζ' and ζ'' with s and u, the differential equations of the preceding article will become

$$\frac{A\,\mathrm{d}^2u-G\,\mathrm{d}^2\theta+H\,\mathrm{d}^2s}{\mathrm{d}t^2}+Ku=0$$
$$\frac{-B\,\mathrm{d}^2s-F\,\mathrm{d}^2\theta-H\,\mathrm{d}^2u}{\mathrm{d}t^2}-Ks=0$$
$$\frac{C\,\mathrm{d}^2\theta+F\,\mathrm{d}^2s-G\,\mathrm{d}^2u}{\mathrm{d}t^2}=0$$

The last equation will give $\mathrm{d}^2\theta/\mathrm{d}t^2=(G\,\mathrm{d}^2u-F\,\mathrm{d}^2s)/(C\,\mathrm{d}t^2)$. If this last expression is substituted in the first two, the following two equations result

$$\frac{(AC-G^2)\mathrm{d}^2u+(CH+GF)\mathrm{d}^2s}{\mathrm{d}t^2}+CKu=0$$
$$\frac{(BC-F^2)\mathrm{d}^2s+(CH+GF)\mathrm{d}^2u}{\mathrm{d}t^2}+CKs=0$$

for which the integration is easy by known methods.

For this reason, let us assume that

$$s=\alpha\sin(\rho t+\beta),\qquad u=\gamma\sin(\rho t+\beta)$$

where α, β, γ, ρ are undetermined quantities. After substitution, the following two equations of condition will be obtained

$$(AC - G^2)\gamma\rho^2 + (CH + GF)\alpha\rho^2 - CK\gamma = 0$$
$$(BC - F^2)\alpha\rho^2 + (CH + GF)\gamma\rho^2 - CK\alpha = 0$$

which give

$$\frac{\gamma}{\alpha} = \frac{(CH + GF)\rho^2}{CK - (AC - G^2)\rho^2} = \frac{CK - (BC - F^2)\rho^2}{(CH + GF)\rho^2}$$

from which results this equation for ρ

$$\frac{C^2K^2}{\rho^4} - [BC - F^2 + AC - G^2]\frac{CK}{\rho^2}$$
$$+(BC - F^2)(AC - G^2) - (CH + GF)^2 = 0$$

The four roots of this equation are divided into two equal sets and of opposite sign.

Therefore, if one denotes in general by ρ and ρ' the integral roots of this equation, not considering their sign, and if four arbitrary values α, α', β, β' are taken, the following general equation results

$$s = \alpha \sin(\rho t + \beta) + \alpha' \sin(\rho' t + \beta')$$

Consequently

$$u = \frac{(CH + GF)\rho^2\alpha \sin(\rho t + \beta)}{CK - (AC - G^2)\rho^2} + \frac{(CH + GF)\rho'^2\alpha' \sin(\rho' t + \beta')}{CK - (AC - G^2)\rho'^2}$$

Finally, by integrating the expression for $\mathrm{d}^2\theta/\mathrm{d}t^2$, we obtain

$$\theta = f + ht + \frac{Gu - Fs}{C}$$

In this way, all the variables will be known as functions of t and the problem will be solved.

Also, since this solution is founded on the hypothesis that s, u, and $\mathrm{d}\theta/\mathrm{d}t$ are very small quantities, the following two conditions must be met to insure its validity: 1. the constants α, α', h are also very small and 2. the roots ρ and ρ' are real and unequal so that the angle t is always under the operation of the sine function. But this last requirement requires two additional conditions

$$BC - F^2 + AC - G^2 > 0$$
$$[BC - F^2 + AC - G^2]^2 > 4[(BC - F^2)(AC - G^2) - (CH + GF)^2]$$

which depend uniquely upon the shape of the body and the location of the point of suspension.

38. Secondly, let us assume that the constants F and G are also very small of the same order as ζ' and ζ''. Then, neglecting the quantities of second order and replacing ζ' and ζ'' by s and u, the differential equations of Article 36 will become

$$\frac{A(\mathrm{d}(s\,\mathrm{d}\theta)+\mathrm{d}^2u)}{\mathrm{d}t^2}-\frac{G\,\mathrm{d}^2\theta}{\mathrm{d}t^2}-\frac{H(\mathrm{d}(u\,\mathrm{d}\theta)-\mathrm{d}^2s)}{\mathrm{d}t^2}$$
$$+\frac{(C-B)(u\,\mathrm{d}\theta-\mathrm{d}s)\mathrm{d}\theta}{\mathrm{d}t^2}+\frac{F\,\mathrm{d}\theta^2}{\mathrm{d}t^2}+\frac{H(s\,\mathrm{d}\theta+\mathrm{d}u)\mathrm{d}\theta}{\mathrm{d}t^2}+Ku=0$$
$$\frac{B(\mathrm{d}(u\,\mathrm{d}\theta)-\mathrm{d}^2s)}{\mathrm{d}t^2}-\frac{F\,\mathrm{d}^2\theta}{\mathrm{d}t^2}-\frac{H(\mathrm{d}(s\,\mathrm{d}\theta)+\mathrm{d}^2u)}{\mathrm{d}t^2}$$
$$+\frac{(A-C)(s\,\mathrm{d}\theta+\mathrm{d}u)\mathrm{d}\theta}{\mathrm{d}t^2}-\frac{G\,\mathrm{d}\theta^2}{\mathrm{d}t^2}-\frac{H(u\,\mathrm{d}\theta-\mathrm{d}s)\mathrm{d}\theta}{\mathrm{d}t^2}-Ks=0$$
$$\frac{C\,\mathrm{d}^2\theta}{\mathrm{d}t^2}=0$$

The last equation gives $\mathrm{d}^2\theta/\mathrm{d}t^2=0$, and after integration we obtain $\mathrm{d}\theta/\mathrm{d}t=n$, where n is an arbitrary constant.

After substituting the expression for $\mathrm{d}\theta/\mathrm{d}t$ in the two preceding equations, the following equations will be obtained

$$A\frac{\mathrm{d}^2u}{\mathrm{d}t^2}+H\frac{\mathrm{d}^2s}{\mathrm{d}t^2}+(A+B-C)n\frac{\mathrm{d}s}{\mathrm{d}t}$$
$$+(C-B)n^2u+Fn^2+Hn^2s+Ku=0$$
$$B\frac{\mathrm{d}^2s}{\mathrm{d}t^2}+H\frac{\mathrm{d}^2u}{\mathrm{d}t^2}-(A+B-C)n\frac{\mathrm{d}u}{\mathrm{d}t}$$
$$+(C-A)n^2s+Gn^2+Hn^2u+Ks=0$$

The integration of these equations is not difficult.

After dividing each equation by n^2 and for added simplicity, after replacing $n\,\mathrm{d}t$ with $\mathrm{d}\theta$ where $\mathrm{d}\theta$ is now a constant, one will have after ordering the terms and setting $L=K/n^2=Km/n^2$ (Article 34)

$$(C-A+L)s+B\frac{\mathrm{d}^2s}{\mathrm{d}\theta^2}+(C-A-B)\frac{\mathrm{d}u}{\mathrm{d}\theta}+H\left(u+\frac{\mathrm{d}^2u}{\mathrm{d}\theta^2}\right)+G=0$$
$$(C-B+L)u+A\frac{\mathrm{d}^2u}{\mathrm{d}\theta^2}-(C-A-B)\frac{\mathrm{d}s}{\mathrm{d}\theta}+H\left(s+\frac{\mathrm{d}^2s}{\mathrm{d}\theta^2}\right)+F=0$$

In order to integrate these equations, it is necessary to eliminate the constant terms assuming that $s=x+f$, $u=y+h$, and to determine the constants f and h so that the terms F and G disappear, which will give the two following equations of condition

$$(C-A+L)f+Hh+G=0,\qquad (C-B+L)h+Hf+F=0$$

which can be solved for the quantities f and h.

$$f = \frac{FH - G(C - B + L)}{(C - B + L)(C - A + L) - H^2}$$
$$h = \frac{GH - F(C - A + L)}{(C - B + L)(C - A + L) - H^2}$$

The same equations in terms of x, y, θ will be obtained as in terms of s, u, θ with the only difference that the constant terms G and F will no longer appear.

Let us now assume that $x = \alpha e^{i\theta}, y = \beta e^{i\theta}$, where α, β, i are indeterminate quantities and e is the base for which the hyperbolic logarithm is 1.[47] Since all the terms of the equations requiring integration include x and y to the first order, it follows that they will be, after substitution, all divisible by $e^{i\theta}$ and the following two conditions of condition will remain

$$[C - A + L + Bi^2]\alpha + [(C - A - B)i - H(1 + i^2)]\beta = 0$$
$$[C - B + L + Ai^2]\beta - [(C - A - B)i - H(1 + i^2)]\alpha = 0$$

which give

$$\frac{\beta}{\alpha} = -\frac{C - A + L + Bi^2}{(C - A - B)i + H(1 + i^2)} = \frac{(C - A - B) + H(1 + i^2)}{C - B + L + Ai^2}$$

so that the following equation in terms of i results

$$[C - B + L + Ai^2][C - A + L + Bi^2]$$
$$+(C - A - B)^2 i^2 - H^2(1 + i^2)^2 = 0$$

After putting $1 + i^2 = \rho$, it will be reduced to this form

$$(AB - H^2)\rho^2 + [(A - B)(L - C) + C^2]\rho + L^2 - 2L(A + B - C) = 0$$

Since the quantity ρ will be found from this equation, one will have

$$x = \alpha e^{\theta\sqrt{\rho - 1}}, \qquad y = \alpha \frac{(A + B - C)\sqrt{\rho - 1} + H\rho}{A + B - C - L - C\rho} e^{\theta\sqrt{\rho - 1}}$$

and the quantity α will remain indeterminate. But since the equation in terms of ρ has two roots, and because the radical $\sqrt{\rho - 1}$ can also be taken with a plus or minus sign, one will thus obtain four different values of z and y, which being taken together will also satisfy the derived equations because the variables x and y are included only in a linear form. If four different values are taken for α, one will have in this manner the complete values of x and y since these values only depend upon two differential equations of the second order and could not contain more than four arbitrary constants.

39. In order that the expressions for x and y do not contain arcs of a circle, the radical $\sqrt{\rho - 1}$ should be imaginary and thus ρ should be a real quantity less than unity.

Let us denote by ρ and σ the two roots of the equation in terms of ρ, assumed real and less than unity, and let us give to the four constants the following imaginary form

$$\frac{\alpha e^{\beta\sqrt{-1}}}{2\sqrt{-1}}, \qquad -\frac{\alpha e^{-\beta\sqrt{-1}}}{2\sqrt{-1}}, \qquad \frac{\gamma e^{\epsilon\sqrt{-1}}}{2\sqrt{-1}}, \qquad -\frac{\gamma e^{-\epsilon\sqrt{-1}}}{2\sqrt{-1}}$$

one will have, after substitution and after replacing the exponential terms with sine and cosine functions, these complete real expressions for x and y

$$\begin{aligned}
x &= \alpha\sin(\theta\sqrt{1-\rho}+\beta)+\gamma\sin(\theta\sqrt{1-\sigma}+\epsilon)\\
y &= \frac{\alpha(A+B-C)\sqrt{1-\rho}}{B-C+A(1-\rho)-L}\cos(\theta\sqrt{1-\rho}+\beta)\\
&+\frac{\alpha H\rho}{B-C+A(1-\rho)-L}\sin(\theta\sqrt{1-\rho}+\beta)\\
&+\frac{\gamma(A+B-C)\sqrt{(1-\sigma)}}{B-C+A(1-\sigma)-L}\cos(\theta\sqrt{1-\sigma}+\epsilon)\\
&+\frac{\gamma H\sigma}{B-C+A(1-\sigma)-L}\sin(\theta\sqrt{1-\sigma}+\epsilon)
\end{aligned}$$

where α, β, γ, ϵ are arbitrary quantities dependent upon the initial state of the body.

Thus after having obtained x and y, one will have

$$\begin{aligned}
s &= x+\frac{FH+G(B-C-L)}{(A-C-L)(B-C-L)-H^2}\\
u &= y+\frac{GH+F(A-C-L)}{(A-C-L)(B-C-L)-H^2}
\end{aligned}$$

Therefore, taking for θ an arbitrary angle proportional to time, one will have (Article 36) these values for the nine variables ξ', η', ζ', ξ'', η'', etc.

$$\begin{array}{lll}
\xi' = \cos\theta, & \eta' = \sin\theta, & \zeta' = s\\
\xi'' = -\sin\theta, & \eta'' = \cos\theta, & \zeta'' = u\\
\xi''' = -s\cos\theta + u\sin\theta, & \eta''' = -s\sin\theta - u\cos\theta, & \zeta''' = 1
\end{array}$$

so that the coordinates ξ, η, ζ of each point of the body will be known at an arbitrary time (Article 1).

If the preceding expressions for ξ', η', etc. are compared with those of Article 7, it is easy to deduce the equations for the angles of rotation φ, ψ, ω which are

$$\varphi+\psi=\theta, \qquad \sin\varphi\sin\omega = s, \qquad \cos\varphi\sin\omega = u$$

from which one has

$$\tan\omega = \sqrt{s^2+u^2}, \qquad \tan\varphi = \frac{s}{u}, \qquad \psi = \theta-\varphi$$

It is easy to see, from the definitions of Article 7, that ω will be the very small inclination of the axis of the body with the vertical and that ψ will be the angle described by a rotation about the vertical and that φ will be the angle that the same body describes by rotating about the same axis. The two angles φ and ψ have arbitrary magnitudes.

40. For an accurate the solution, the variables s and u should always be very small. Thus not only should the constants α and γ which depend upon the initial state of the body always be very small, but the values of the constants F and G, given by the shape of the body, should also be very small. Moreover, the roots ρ and σ should be real and positive, so that the angle θ is always the argument in the sine or cosine functions.

If it is assumed that $F = 0$, $G = 0$, that is, $\mathrm{S}\, bc\, \mathrm{D}m = 0$, $\mathrm{S}\, ac\, \mathrm{D}m = 0$, one will obtain the necessary conditions for the moments of the centrifugal forces about the axis of the body, which is at the same time the c-coordinate axis, to cancel each other so that the body can rotate uniformly and freely about this axis. But it is known that in each body there are three orthogonal axes passing through the center of gravity, which have this property and which are commonly called, after Euler, the **principal axes of the body**. Therefore, since we have assumed that the axis of the body passes at the same time through the center of gravity and the point of suspension, it follows that the quantities F and G will be zero when the body is suspended from an arbitrary point taken on one of the principal axes.

Therefore, in order that these quantities be very small without being absolutely equal to zero, the point of suspension should be very close to one of its principal axes. This is the first necessary condition for the axis of the body to perform only very small oscillations about the vertical. The body itself will have an arbitrary motion of rotation about this axis.

The other necessary condition for the oscillations to be always very small, depend upon the equation in terms of ρ and can be reduced to the following equations

$$[(A+B)(L-C)+C^2]^2 > 4(AB-H^2)[L^2-2L(A+B-C)]$$

$$\frac{2(AB-H^2)+(A+B)(L-C)+C^2}{AB-H^2} > 0$$

$$\frac{[(A-C-L)(B-C-L)-H^2}{AB-H^2} > 0$$

which depend simultaneously on the location of the point of suspension and on the shape of the body.

41. The solution we just presented utilizes the theory of small oscillations of pendulums in all its generality. It is known that Huygens was the first to present a theory of circular oscillations. Then Clairaut expanded the theory to include conical oscillations, which occur when the pendulum is displaced from its line of rest and receives an impulse whose direction does not pass through this line. But if the pendulum receives at the same time a motion of rotation about its axis, the centrifugal force produced by this motion can greatly disturb the circular or conical oscillations. The determination of these new oscillations is

a problem which was not yet entirely resolved for a pendulum of arbitrary shape. It is for this reason that we have treated it here.

SECTION X
THE PRINCIPLES OF HYDRODYNAMICS

The understanding of the motion of fluids is the object of hydrodynamics. The object of ordinary hydraulics is reduced to the art of piping water and to making it serve for the motion of machines. This art has been nurtured through all time because of its ever present need and the Ancients may have been as capable as we are, based on an evaluation of the works they left behind in this area.

Hydrodynamics is a science created in the last century. Newton was the first to attempt to calculate the motion of fluids from the principles of mechanics and d'Alembert was the first to have reduced the true laws of their motion to analytical equations. Archimedes and Galileo (since the time interval separating these two geniuses is unimportant in the history of mechanics) considered only the equilibrium of fluids.

Torricelli began by examining the motion of water which flows out of a vessel through a very small orifice and he searched for a governing law. He found that by giving a water jet a vertical direction it always reaches, with some minor variations, the level of the water in the vessel. Since he assumed that it would reach this level were it not for the resistance of air and friction, Torricelli concluded that the velocity of the water jet is the same as the velocity which it would have obtained by falling freely from the height of the water level in the vessel and that this velocity, consequently, is proportional to the square root of this height.

However, since he could not develop a rigorous demonstration of this proposition, he had to be satisfied with presenting it as an empirical result at the end of his treatise *de Motu naturaliter accelerato* published in 1643. Newton undertook to demonstrate it in the second book of the *Principia*, which was published in 1687. But one must admit that it is the part of this work which is the least satisfying.

If a column of water is falling freely in a vacuum, it is easy to be convinced that it should take the shape of a conoid formed by the revolution of a hyperbola of the fourth order about the vertical axis because the velocity of each horizontal slice is proportional to the square root of the height from which it falls and also, it must be inversely proportional to the width of this slice because of the continuity of water, and consequently, it is inversely proportional to the square of its radius. From this observation it is clear that the segment of the axis, or the abscissa which represents the height, is inversely proportional to the fourth order of the ordinate of the generating hyperbola. If a vessel is assumed to have the shape of a conoid and if it is always full of water and further, if it is assumed that the motion of the water is in a steady state, it is clear that each particle of water will fall as if it were free and that it will have, consequently, at the orifice of the vessel, the velocity resulting from the height through which it fell.

Now Newton assumes that the water which fills a vertical cylinder with an orifice in its bottom from which the water can flow out is naturally composed of two parts, of which only one contributes to the motion and to the shape of the conoid in question. It is what he calls the *cataract*. The other part is at rest, as if it were frozen. In this manner, it is clear that the water must flow out with a velocity equal to the one it would have acquired by falling from the height of the cylinder as Torricelli found empirically. However, Newton measured the quantity of water which exited the cylinder in a given time and compared it to the size of the orifice and concluded in the first edition of his *Principia*, that the velocity of the water, at the orifice in the cylinder, was only half the value of the height of the water in the cylinder. This error resulted from the failure to account for the constriction of the water jet. He accounted for this phenomenon in the second edition published in 1714, and he recognized that the smaller section of the water jet was, at the orifice, in the ratio of 1 to the square root of 2. So by taking this section for the true section of the orifice, the velocity must be increased in the same ratio and almost corresponds to the height of the water. In this manner, his theory more closely approximates the results of experience but it did not become more exact because the formation of a cataract or a fictitious vessel in which the water is assumed to move, while the water on the lateral sides remains at rest, is obviously contrary to the known laws of the equilibrium of fluids. Indeed, the water which would fall in this cataract with all the force of gravity and cause no lateral pressure, could not possibly resist the pressure of the fluid at rest which surrounds it.

Twenty years earlier, Varignon had given to the Académie des Sciences in Paris a more natural and more plausible explanation of this phenomenon. He observed that when water flows from the cylinder through a small orifice made in the bottom, the water has only a very small and apparently uniform motion for all the particles. He concluded from this observation that no acceleration existed and that the fluid which flows out at each instant was receiving all its motion from the pressure produced by the weight of the column of fluid on its base. Thus this weight which is proportional to the product of the diameter of the opening with the height of the water column must be proportional to the quantity of motion given to the particles exiting at every instant from the same orifice. This quantity of motion is, as one knows, proportional to the velocity and to the mass. The mass is here proportional to the product of the diameter of the opening and of the small distance the particle traverses in the given time, a distance which is evidently proportional to the velocity of the particle. Consequently, the quantity of the motion is proportional to the diameter of the opening multiplied by the square of the velocity. Therefore, the height of the water in the vessel is proportional to the square of the velocity with which it flows out from the vessel, which is Torricelli's theorem.

This reasoning, nevertheless, is still somewhat vague because it is implicitly assumed that the small mass which flows out of the vessel at each instant suddenly receives all its velocity from the pressure of the column of water above the orifice. However, it is known that a pressure can not suddenly create a finite velocity. But by assuming, which is natural, that the weight of the column of water acts on the particle during the entire time that it takes to exit the vessel, it is clear that this particle will receive an accelerating motion for which the magnitude, after an arbitrary time, will be proportional to the pressure multiplied by the time. Thus the product of the weight of the column with the time taken by the particle to exit the vessel will be equal to the product of the mass of this particle with its velocity. Since the mass is the product of the width of the orifice by the small distance that the particle traverses during its exit from the vessel, a distance which is proportional to the product of the velocity with the time because of the nature of uniformly accelerating motion, it follows that the height of the column will again be proportional to the square of the velocity of the particle. Therefore, this conclusion is rigorous if it is understood that each particle, when exiting the vessel, is under a pressure exerted by the entire weight of the column of fluid directly above the particle. Indeed, this is what would take place if the fluid in the vessel was immobile because then the pressure exerted on the bottom where the orifice is located would be equal to the weight of the column. But this pressure should be different when the fluid is in motion. However, it is clear that the closer the fluid is to its state of rest, the closer its pressure is to the total weight of the vertical column. Also, experience shows that the motion of the fluid in the vessel becomes smaller as the opening also becomes smaller. Therefore, the preceding theory will be closer to the truth as the dimensions of the vessel are relatively larger than the opening from which the fluid flows. This statement is confirmed by experience.

By an inverse reasoning, the same theory is inadequate to determine the motion of fluids flowing in pipes with relatively small and nearly constant diameters. In this case, the total motion of the particles of the fluid should simultaneously be considered and examined to determine how they must be changed and modified because of the shape of the pipe. But experience shows that when the pipe has a nearly vertical direction the various horizontal layers of the fluid are almost parallel to one another so that a layer always replaces the preceding layer because of the incompressibility of the fluid and thus the velocity of each horizontal layer, found in the vertical direction should be inversely proportional to the width of this layer, a width which is determined from the shape of the conduit.

Thus it is enough to determine the motion of only one layer of fluid and the problem is somewhat similar to the motion of the compound pendulum. According to James Bernoulli's theory, the motions gained or lost at each instant by the different weights which compose the pendulum are in equilibrium with respect to one another from the standpoint of the lever. Therefore, there must be a state of equilibrium in the conduit between the different layers of the fluid, each one moving with the velocity gained or lost at each instant. From this result, applying the already known principles for the equilibrium of fluids, it would have been possible to first determine the motion of a fluid in a conduit, as the motion of the compound pendulum was determined. But it is never through the most direct and simplest paths that the human mind reaches the truth, whatever it may be, and the topic we are treating provides a striking example.

We showed the various steps in SECTION I which were made to obtain the solution of the problem of the center of oscillation and we saw there that the true theory behind this problem was discovered by James Bernoulli long after Huygens had solved the problem using the indirect principle of the **Conservation des Forces Vives**. It was the same for the problem of fluid motion in vessels and it is surprising that he was not the first to take advantage of the knowledge already gained in the other problem.

The principle of **Conservation des Forces Vives** again provides the first solution to this last problem and was used as a basis for Daniel Bernoulli's *Hydrodynamics*, published in 1738. This work also stands out because of an analysis as elegant in its development as it is simple in its results. But the inaccuracy of this principle, which had not yet been demonstrated in a general manner, was also referred to in the resulting statements and led to a need for a more certain theory uniquely based on the fundamental laws of mechanics. Maclaurin and John Bernoulli undertook the task of filling this need, the former in his *Treatise des Fluxions* and the latter in his *Nouvelle Hydraulique*,[48] published after the publication of his *Oeuvres*. Their methods, although very different, lead to the same results as the principle of the **Conservation des Forces Vives**. But it must be said that Maclaurin's method is not rigorous enough and it appears that he attempted to force it to fit the results which he wanted to obtain. With respect to John Bernoulli's method and without agreeing with all the counter arguments d'Alembert has used against it, it must be admitted that his method lacks clarity and precision.

It was seen in SECTION I how d'Alembert by generalizing James Bernoulli's theory on the pendulum arrived at a simple and general principle of dynamics, which reduces the laws of the motion of bodies to those of their equilibrium. The application of this principle to the motion of fluids was obvious and the author first gave a description of the principle at the end of his dynamics, published in 1743. He then developed it with all the necessary details in his *Traité des Fluides* which was published the following year and which contains solutions as straight forward as they are elegant for the major problems which could be posed for the motion of fluids in vessels.

But these solutions, like those of Daniel Bernoulli, were based on two assumptions which are not true in general: 1. that the different layers of the fluid retain exactly their parallelism to one another so that a layer always replaces the one which precedes it, 2. that the velocity of each layer changes direction, that is, that all the points in the same layer are assumed to have an equal and parallel velocity. When a fluid flows in a very narrow vessel or conduit, the assumptions in question are very plausible and seem to be confirmed by experience. But apart from this case, they are far from the truth and there is then no other means to determine the motion of the fluid but to examine the motion of each particle.

Clairaut had given in his *Théorie de la figure de la Terre*, published in 1743, the general laws for the equilibrium of fluids for which all the particles are acted upon by arbitrary forces. It was only a question of going from these laws to those of their motion by means of the principle to which d'Alembert at this time reduced all of dynamics. The latter, some years later, undertook this important step on the occasion of the prize proposed in 1750 by the Académie de Berlin on the theory of the resistance of fluids and he was the first to give rigorous equations for the motion of either incompressible or compressible and elastic fluids in 1752 in his *Essai d'une nouvelle théorie sur la résistance des fluides*. These equations belong to the class which we call *partial differences* and they are between the different parts of differences with respect to several variables. But these equations did not yet possess all the generality and simplicity they are capable of.[49] Euler was the first to give the general formulas for the motion of fluids, based on the laws of their equilibrium and presented in the simple and clear notation of partial differences (The reader should refer to the volume of the Académie de Berlin, General Principles of the Equilibrium of Fluids, Vol. XI for 1755 for additional information). With this discovery all of fluid mechanics was reduced to a single approach to analysis and if the equations which comprise it were integrable it would be possible in any case to completely determine the parameters of the motion and of the action of a fluid moved by arbitrary forces. Unfortunately, these equations are so difficult to integrate that it was not possible, until now, to solve them with the exception of a few limited cases.

Thus the entire theory of hydrodynamics lies in these equations and in their integration. D'Alembert at the outset used a somewhat complex method to solve them. He later gave a simpler method. But because this method is based on laws of equilibrium peculiar to fluids, hydrodynamics is a science separate from the dynamics of solid bodies. The unification we accomplished in PART I of this work where all the laws for the equilibrium of bodies, either solid or fluid, were brought under the same formula and the application we just made of this formula to the laws of motion leads us naturally to reunite dynamics and hydrodynamics and to consider them as segments of a unique principle and the result of a single general formula.

This is the task which remains to be carried out in order to complete our work on mechanics and to fulfill the promise made in the title of this work.

SECTION XI
THE MOTION OF INCOMPRESSIBLE FLUIDS

1. It should be possible to immediately deduce the laws of motion for these fluids from those of their equilibrium which were found in SECTION VII of PART I. Indeed, from the general principle given in SECTION II, it only remains to add to the actual accelerating forces the new accelerating forces $\mathrm{d}^2x/\mathrm{d}t^2$, $\mathrm{d}^2y/\mathrm{d}t^2$, $\mathrm{d}^2z/\mathrm{d}t^2$, acting in the direction of the rectangular coordinates x, y, z.

Thus since in the formulas of Article 10 and in the formulas of SECTION VII cited above, it was assumed that all the accelerating forces of the fluid were already reduced to three, X, Y, Z, in the direction of the coordinates x, y, z, it will suffice to substitute $X+(\mathrm{d}^2x/\mathrm{d}t^2)$, $Y+(\mathrm{d}^2y/\mathrm{d}t^2)$, $Z+(\mathrm{d}^2z/\mathrm{d}t^2)$, for X, Y, Z in order to apply these formulas to the motion of fluids. But we believe it conforms more to the object of this work which is to directly apply to fluids the general equations given in SECTION IV for the motion of an arbitrary system of bodies.

Subsection I
General Equations For The Motion Of Incompressible Fluids

2. An incompressible fluid can be considered to consist of an infinite number of particles which are free to move independently of one another, without a change in volume. Therefore, the problem fits into those discussed in Article 17 of the section cited above.

Let $\mathrm{D}m$ be the mass of an arbitrary particle or element of fluid and let X, Y, Z represent the accelerating forces which act on this element, reduced for simplicity to the directions of the rectangular coordinates x, y, z and with the tendency to reduce the magnitude of these coordinates. Also, let $L = 0$ be the equation of condition resulting from the incompressibility or invariability of the volume $\mathrm{D}m$, let λ be an undetermined quantity and let S be an integral operator corresponding to the differential operator D and extending over the entire mass of fluid. The motion of the fluid will then be represented by this general equation (SECTION IV)

$$\mathrm{S}\left[\left(\frac{\mathrm{d}^2x}{\mathrm{d}t^2}+X\right)\delta x+\left(\frac{\mathrm{d}^2y}{\mathrm{d}t^2}+Y\right)\delta y+\left(\frac{\mathrm{d}^2z}{\mathrm{d}t^2}+Z\right)\delta z\right]\mathrm{D}m+\mathrm{S}\,\lambda\,\delta L=0$$

Now the expressions for $\mathrm{D}m$ and δL should be substituted in this equation and then, after having eliminated the differences of the variations, if there are any, equate separately to zero the coefficients of the indeterminate variations δx, δy, δz.

Let us retain the operator D to represent the differences relative to the instantaneous position of contiguous particles, while the operator d will uniquely refer to the change of location of the same particle in space. It is clear that the volume of the particle $\mathrm{D}m$ can be represented by the parallelepiped $\mathrm{D}x\,\mathrm{D}y\,\mathrm{D}z$. Thus by denoting by Δ the density of this particle, one will have

$$\mathrm{D}m = \Delta\mathrm{D}x\,\mathrm{D}y\,\mathrm{D}z$$

Moreover, it is obvious that the condition of incompressibility will be included in the equation

$$\mathrm{D}x\,\mathrm{D}y\,\mathrm{D}z = \text{constant}$$

in such a way that

$$L = \mathrm{D}x\,\mathrm{D}y\,\mathrm{D}z = \text{constant}$$

and as a consequence

$$\delta L = \delta(\mathrm{D}x\,\mathrm{D}y\,\mathrm{D}z)$$

In order to determine this differential, the same considerations should be applied as in Article 11 of SECTION VII of PART I. Thus after replacing d by D in the formulas of this section, one will obtain

$$\delta(\mathrm{D}x\,\mathrm{D}y\,\mathrm{D}z) = \mathrm{D}x\,\mathrm{D}y\,\mathrm{D}z\left(\frac{\mathrm{D}\delta x}{\mathrm{D}x} + \frac{\mathrm{D}\delta y}{\mathrm{D}y} + \frac{\mathrm{D}\delta z}{\mathrm{D}z}\right)$$

After being multiplied by λ and integrated with respect to the entire mass of the fluid this expression will give the value of $\mathrm{S}\,\lambda\,\delta L$, in which the operators $\mathrm{D}\delta$ should be eliminated by the means used earlier in Article 17 of the cited section. Thus one will obtain

$$\mathrm{S}\,\lambda\,\delta L = -\mathrm{S}\left(\frac{\mathrm{D}\lambda}{\mathrm{D}x}\delta x + \frac{\mathrm{D}\lambda}{\mathrm{D}y}\delta y + \frac{\mathrm{D}\lambda}{\mathrm{D}z}\delta z\right)\mathrm{D}x\,\mathrm{D}y\,\mathrm{D}z + \mathrm{S}(\lambda''\,\delta x'' - \lambda'\,\delta x')\mathrm{D}y\,\mathrm{D}z$$
$$+\mathrm{S}(\lambda''\,\delta y'' - \lambda'\,\delta y')\mathrm{D}x\,\mathrm{D}z + \mathrm{S}(\lambda''\,\delta z'' - \lambda'\,\delta z')\mathrm{D}x\,\mathrm{D}y$$

Then after making these substitutions in the first member of the general equation, it will contain this total integral formula

$$\text{(a)}\quad \mathrm{S}\left\{\begin{array}{l} \left(\Delta\dfrac{\mathrm{d}^2x}{\mathrm{d}t^2} + \Delta X - \dfrac{\mathrm{D}\lambda}{\mathrm{D}x}\right)\delta x \\ + \left(\Delta\dfrac{\mathrm{d}^2y}{\mathrm{d}t^2} + \Delta Y - \dfrac{\mathrm{D}\lambda}{\mathrm{D}y}\right)\delta y \\ + \left(\Delta\dfrac{\mathrm{d}^2z}{\mathrm{d}t^2} + \Delta Z - \dfrac{\mathrm{D}\lambda}{\mathrm{D}z}\right)\delta z \end{array}\right\}\mathrm{D}x\,\mathrm{D}y\,\mathrm{D}z$$

in which the coefficients of the variations δx, δy, δz should be individually equated to zero, which will provide three indefinite equations for all the points of the fluid mass

$$\text{(A)}\quad \left\{\begin{array}{l} \Delta\left(\dfrac{\mathrm{d}^2x}{\mathrm{d}t^2} + X\right) - \dfrac{\mathrm{D}\lambda}{\mathrm{D}x} = 0 \\ \Delta\left(\dfrac{\mathrm{d}^2y}{\mathrm{d}t^2} + Y\right) - \dfrac{\mathrm{D}\lambda}{\mathrm{D}y} = 0 \\ \Delta\left(\dfrac{\mathrm{d}^2z}{\mathrm{d}t^2} + Z\right) - \dfrac{\mathrm{D}\lambda}{\mathrm{D}z} = 0 \end{array}\right.$$

Then, it will remain to eliminate the partial integrals

$$\mathrm{S}(\lambda''\,\delta x'' - \lambda'\,\delta x')\mathrm{D}y\,\mathrm{D}z$$
$$+\mathrm{S}(\lambda''\,\delta y'' - \lambda'\,\delta y')\mathrm{D}x\,\mathrm{D}z + \mathrm{S}(\lambda''\,\delta z'' - \lambda'\,\delta z')\mathrm{D}x\,\mathrm{D}y$$

which only apply to the external surface of the fluid. It will be accepted, as in Article 18 of SECTION VII cited above, that the value of λ will have to be equal to zero at all points of the surface where the fluid is free. Also, it will be proved, as in Article 31 of the same section, that with respect to the locations where the fluid is contained by fixed walls, the terms of the preceding integrals will cancel mutually so that no equation will remain. In general, by a reasoning similar to those of Articles 32, 38 and 39, the quantity λ, relative to the surface of the fluid, will be the pressure that the fluid exerts and which, when it is not zero, must be equilibrated by the resistance or the action of the walls.

3. Thus the equations which were just found contain the general laws for the motion of incompressible fluids. But the equation which results from the condition of incompressibility of the volume $\mathrm{D}x\,\mathrm{D}y\,\mathrm{D}z$ must still be added when the fluid moves. This equation will be represented by $\mathrm{d}(\mathrm{D}x\,\mathrm{D}y\,\mathrm{D}z) = 0$, so that by replacing δ with d in the expression found above and equating it to zero, one will have

$$\text{(B)} \qquad \frac{\mathrm{Dd}x}{\mathrm{D}x} + \frac{\mathrm{Dd}y}{\mathrm{D}y} + \frac{\mathrm{Dd}z}{\mathrm{D}z} = 0$$

This equation, combined with the three equations (A) of the preceding article, will serve to determine the four unknowns x, y, z and λ.

4. In order to have a clear idea of the nature of these equations, it should be noted that the variables x, y, z, which determine the position of a particle at an arbitrary instant, must simultaneously belong to all the particles which compose the fluid mass. Therefore, they must be functions of time t and of the values that these same variables had at the beginning of the motion or at any other given time. Thus, by calling a, b, c, the values of x, y, z, when t is equal to zero, it should be that the total values of x, y, z are functions of a, b, c and t. In this manner, the differences marked by the operator D will only refer to the variability of a, b, c. The differences marked with the other operator d will simply refer to the variability of t. But since, in the developed equations there are differences relative to the same variables x, y, z, one should reduce those to the differences relative to the same variables a, b, c, which is always possible because it suffices to imagine that one has substituted in the functions before differentiation, the expressions for x, y, z, as functions of a, b, c.

5. By viewing the variables x, y, z, as functions of a, b, c and t, and by representing the differentials using the ordinary notation of partial differences,[50] one will obtain

$$\begin{aligned}
\mathrm{D}x &= \frac{\mathrm{d}x}{\mathrm{d}a}\mathrm{d}a + \frac{\mathrm{d}x}{\mathrm{d}b}\mathrm{d}b + \frac{\mathrm{d}x}{\mathrm{d}c}\mathrm{d}c \\
\mathrm{D}y &= \frac{\mathrm{d}y}{\mathrm{d}a}\mathrm{d}a + \frac{\mathrm{d}y}{\mathrm{d}b}\mathrm{d}b + \frac{\mathrm{d}y}{\mathrm{d}c}\mathrm{d}c \\
\mathrm{D}z &= \frac{\mathrm{d}z}{\mathrm{d}a}\mathrm{d}a + \frac{\mathrm{d}z}{\mathrm{d}b}\mathrm{d}b + \frac{\mathrm{d}z}{\mathrm{d}c}\mathrm{d}c
\end{aligned}$$

and by viewing at the same time the function λ as a function of x, y, z, and as a function of a, b, c, there will result

$$\begin{aligned} \mathrm{D}\lambda &= \frac{\mathrm{D}\lambda}{\mathrm{D}x}\mathrm{D}x + \frac{\mathrm{D}\lambda}{\mathrm{D}y}\mathrm{D}y + \frac{\mathrm{D}\lambda}{\mathrm{D}z}\mathrm{D}z \\ &= \frac{\mathrm{d}\lambda}{\mathrm{d}a}\mathrm{d}a + \frac{\mathrm{d}\lambda}{\mathrm{d}b}\mathrm{d}b + \frac{\mathrm{d}\lambda}{\mathrm{d}c}\mathrm{d}c \end{aligned}$$

The two expressions for $\mathrm{D}\lambda$ should be identical and if the expressions for $\mathrm{D}x$, $\mathrm{D}y$, $\mathrm{D}z$ are substituted in the first expression in place of $\mathrm{d}a$, $\mathrm{d}b$, $\mathrm{d}c$, the coefficients of $\mathrm{d}a$, $\mathrm{d}b$, $\mathrm{d}c$, should be the same in both cases. This result will give three equations which will be used to determine the values of $\mathrm{D}\lambda/\mathrm{d}x$, $\mathrm{D}\lambda/\mathrm{d}y$, $\mathrm{D}\lambda/\mathrm{d}z$ as functions of $\mathrm{d}\lambda/\mathrm{d}a$, $\mathrm{d}\lambda/\mathrm{d}b$, $\mathrm{d}\lambda/\mathrm{d}c$. It will be the same, if in the second expression for $\mathrm{D}\lambda$, the values of $\mathrm{d}a$, $\mathrm{d}b$, $\mathrm{d}c$, were substituted into the expressions for $\mathrm{D}x$, $\mathrm{D}y$, $\mathrm{D}z$ obtained from these last quantities. Then the comparison of the terms dependent upon $\mathrm{D}x$, $\mathrm{D}y$, $\mathrm{D}z$ will immediately give the values of $\mathrm{D}\lambda/\mathrm{D}x$, etc.

From the ordinary rules of elimination, one has

$$\begin{aligned} \mathrm{d}a &= \frac{\alpha\,\mathrm{D}x + \alpha'\,\mathrm{D}y + \alpha''\,\mathrm{D}z}{\theta} \\ \mathrm{d}b &= \frac{\beta\,\mathrm{D}x + \beta'\,\mathrm{D}y + \beta''\,\mathrm{D}z}{\theta} \\ \mathrm{d}c &= \frac{\gamma\,\mathrm{D}x + \gamma'\,\mathrm{D}y + \gamma''\,\mathrm{D}z}{\theta} \end{aligned}$$

assuming that

$$\begin{aligned} \alpha &= \frac{\mathrm{d}y}{\mathrm{d}b}\frac{\mathrm{d}z}{\mathrm{d}c} - \frac{\mathrm{d}y}{\mathrm{d}c}\frac{\mathrm{d}z}{\mathrm{d}b}, & \alpha' &= \frac{\mathrm{d}x}{\mathrm{d}c}\frac{\mathrm{d}z}{\mathrm{d}b} - \frac{\mathrm{d}x}{\mathrm{d}b}\frac{\mathrm{d}z}{\mathrm{d}c}, & \alpha'' &= \frac{\mathrm{d}x}{\mathrm{d}b}\frac{\mathrm{d}y}{\mathrm{d}c} - \frac{\mathrm{d}x}{\mathrm{d}c}\frac{\mathrm{d}y}{\mathrm{d}b} \\ \beta &= \frac{\mathrm{d}y}{\mathrm{d}c}\frac{\mathrm{d}z}{\mathrm{d}a} - \frac{\mathrm{d}y}{\mathrm{d}a}\frac{\mathrm{d}z}{\mathrm{d}c}, & \beta' &= \frac{\mathrm{d}x}{\mathrm{d}a}\frac{\mathrm{d}z}{\mathrm{d}c} - \frac{\mathrm{d}x}{\mathrm{d}c}\frac{\mathrm{d}z}{\mathrm{d}a}, & \beta'' &= \frac{\mathrm{d}x}{\mathrm{d}c}\frac{\mathrm{d}y}{\mathrm{d}a} - \frac{\mathrm{d}x}{\mathrm{d}a}\frac{\mathrm{d}y}{\mathrm{d}c} \\ \gamma &= \frac{\mathrm{d}y}{\mathrm{d}a}\frac{\mathrm{d}z}{\mathrm{d}b} - \frac{\mathrm{d}y}{\mathrm{d}b}\frac{\mathrm{d}z}{\mathrm{d}a}, & \gamma' &= \frac{\mathrm{d}x}{\mathrm{d}b}\frac{\mathrm{d}z}{\mathrm{d}a} - \frac{\mathrm{d}x}{\mathrm{d}a}\frac{\mathrm{d}z}{\mathrm{d}b}, & \gamma'' &= \frac{\mathrm{d}x}{\mathrm{d}a}\frac{\mathrm{d}y}{\mathrm{d}b} - \frac{\mathrm{d}x}{\mathrm{d}b}\frac{\mathrm{d}y}{\mathrm{d}a} \end{aligned}$$

$$\theta = \frac{\mathrm{d}x}{\mathrm{d}a}\frac{\mathrm{d}y}{\mathrm{d}b}\frac{\mathrm{d}z}{\mathrm{d}c} - \frac{\mathrm{d}x}{\mathrm{d}b}\frac{\mathrm{d}y}{\mathrm{d}a}\frac{\mathrm{d}z}{\mathrm{d}c} + \frac{\mathrm{d}x}{\mathrm{d}b}\frac{\mathrm{d}y}{\mathrm{d}c}\frac{\mathrm{d}z}{\mathrm{d}a} - \frac{\mathrm{d}x}{\mathrm{d}c}\frac{\mathrm{d}y}{\mathrm{d}b}\frac{\mathrm{d}z}{\mathrm{d}a} + \frac{\mathrm{d}x}{\mathrm{d}c}\frac{\mathrm{d}y}{\mathrm{d}a}\frac{\mathrm{d}z}{\mathrm{d}b} - \frac{\mathrm{d}x}{\mathrm{d}a}\frac{\mathrm{d}y}{\mathrm{d}c}\frac{\mathrm{d}z}{\mathrm{d}b}$$

After making these substitutions in the following expression

$$\frac{\mathrm{d}\lambda}{\mathrm{d}a}\mathrm{d}a + \frac{\mathrm{d}\lambda}{\mathrm{d}b}\mathrm{d}b + \frac{\mathrm{d}\lambda}{\mathrm{d}c}\mathrm{d}c$$

and after comparing the results from the identical expression

$$\frac{\mathrm{D}\lambda}{\mathrm{D}x}\mathrm{D}x + \frac{\mathrm{D}\lambda}{\mathrm{D}y}\mathrm{D}y + \frac{\mathrm{D}\lambda}{\mathrm{D}z}\mathrm{D}z$$

one will obtain

$$\frac{D\lambda}{Dx} = \frac{\alpha}{\theta}\frac{d\lambda}{da} + \frac{\beta}{\theta}\frac{d\lambda}{db} + \frac{\gamma}{\theta}\frac{d\lambda}{dc}$$
$$\frac{D\lambda}{Dy} = \frac{\alpha'}{\theta}\frac{d\lambda}{da} + \frac{\beta'}{\theta}\frac{d\lambda}{db} + \frac{\gamma'}{\theta}\frac{d\lambda}{dc}$$
$$\frac{D\lambda}{Dz} = \frac{\alpha''}{\theta}\frac{d\lambda}{da} + \frac{\beta''}{\theta}\frac{d\lambda}{db} + \frac{\gamma''}{\theta}\frac{d\lambda}{dc}$$

Thus after substituting these expressions in the three equations (A) of Article 2 and after having multiplied by θ, they will assume the following form

$$\text{(C)}\quad \begin{cases} \theta\,\Delta\left(\dfrac{d^2x}{dt^2} + X\right) - \alpha\dfrac{d\lambda}{da} - \beta\dfrac{d\lambda}{db} - \gamma\dfrac{d\lambda}{dc} = 0 \\ \theta\,\Delta\left(\dfrac{d^2y}{dt^2} + Y\right) - \alpha'\dfrac{d\lambda}{da} - \beta'\dfrac{d\lambda}{db} - \gamma'\dfrac{d\lambda}{dc} = 0 \\ \theta\,\Delta\left(\dfrac{d^2z}{dt^2} + Z\right) - \alpha''\dfrac{d\lambda}{da} - \beta''\dfrac{d\lambda}{db} - \gamma''\dfrac{d\lambda}{dc} = 0 \end{cases}$$

where it can be seen that there are only partial differences relative to a, b, c and t.

In these equations, the quantity Δ, which represents the density, is a given function of a, b, c, without t, because it must remain invariant for each particle. If the fluid were homogeneous, Δ will be a constant independent of a, b, c and t. With respect to the quantities X, Y, Z, which represent the accelerating forces, they will often be given as functions of x, y, z and t.

6. But the preceding equations can be reduced to a simpler expression by adding them together, after having multiplied, respectively and successively by dx/da, dy/da, dz/da, by dx/db, dy/db, dz/db, and by dx/dc, dy/dc, dz/dc, because it is clear that from the expressions for θ, α, β, γ, α', β', etc. given above, one will have

$$\theta = \alpha\frac{dx}{da} + \alpha'\frac{dy}{da} + \alpha''\frac{dz}{da} = \beta\frac{dx}{db} + \beta'\frac{dy}{db} + \beta''\frac{dz}{db}$$
$$= \gamma\frac{dx}{dc} + \gamma'\frac{dy}{dc} + \gamma''\frac{dz}{dc}$$

and then

$$\beta\frac{dx}{da} + \beta'\frac{dy}{da} + \beta''\frac{dz}{da} = 0, \quad \gamma\frac{dx}{da} + \gamma'\frac{dy}{da} + \gamma''\frac{dz}{da} = 0, \quad \alpha\frac{dx}{db} + \alpha'\frac{dy}{db} + \alpha''\frac{dz}{db} = 0$$

and so on. Therefore, with these operations and simplifications, the following transformed equations are obtained

$$
\text{(D)}\quad\begin{cases}
\Delta\left[\left(\dfrac{\mathrm{d}^2x}{\mathrm{d}t^2}+X\right)\dfrac{\mathrm{d}x}{\mathrm{d}a}+\left(\dfrac{\mathrm{d}^2y}{\mathrm{d}t^2}+Y\right)\dfrac{\mathrm{d}y}{\mathrm{d}a}+\left(\dfrac{\mathrm{d}^2z}{\mathrm{d}t^2}+Z\right)\dfrac{\mathrm{d}z}{\mathrm{d}a}\right]-\dfrac{\mathrm{d}\lambda}{\mathrm{d}a}=0\\
\Delta\left[\left(\dfrac{\mathrm{d}^2x}{\mathrm{d}t^2}+X\right)\dfrac{\mathrm{d}x}{\mathrm{d}b}+\left(\dfrac{\mathrm{d}^2y}{\mathrm{d}t^2}+Y\right)\dfrac{\mathrm{d}y}{\mathrm{d}b}+\left(\dfrac{\mathrm{d}^2z}{\mathrm{d}t^2}+Z\right)\dfrac{\mathrm{d}z}{\mathrm{d}b}\right]-\dfrac{\mathrm{d}\lambda}{\mathrm{d}b}=0\\
\Delta\left[\left(\dfrac{\mathrm{d}^2x}{\mathrm{d}t^2}+X\right)\dfrac{\mathrm{d}x}{\mathrm{d}c}+\left(\dfrac{\mathrm{d}^2y}{\mathrm{d}t^2}+Y\right)\dfrac{\mathrm{d}y}{\mathrm{d}c}+\left(\dfrac{\mathrm{d}^2z}{\mathrm{d}t^2}+Z\right)\dfrac{\mathrm{d}z}{\mathrm{d}c}\right]-\dfrac{\mathrm{d}\lambda}{\mathrm{d}c}=0
\end{cases}
$$

These last equations could have been obtained directly by introducing in the formulas of Article 2 the variations of the coordinates in the initial state, δa, δb, δc, instead of the variations δx, δy, δz because by considering x, y, z as functions of a, b, c, one will have[51]

$$
\begin{aligned}
\delta x &= \frac{\mathrm{d}x}{\mathrm{d}a}\delta a+\frac{\mathrm{d}x}{\mathrm{d}b}\delta b+\frac{\mathrm{d}x}{\mathrm{d}c}\delta c\\
\delta y &= \frac{\mathrm{d}y}{\mathrm{d}a}\delta a+\frac{\mathrm{d}y}{\mathrm{d}b}\delta b+\frac{\mathrm{d}y}{\mathrm{d}c}\delta c\\
\delta z &= \frac{\mathrm{d}z}{\mathrm{d}a}\delta a+\frac{\mathrm{d}z}{\mathrm{d}b}\delta b+\frac{\mathrm{d}z}{\mathrm{d}c}\delta c
\end{aligned}
$$

After making these substitutions in formula (a) of Article 2 and equating to zero the quantities multiplied by δa, δb, δc while noting that λ is a function of x, y, z, there results with respect to a, b, c

$$
\begin{aligned}
\frac{\mathrm{d}\lambda}{\mathrm{d}a} &= \frac{\mathrm{D}\lambda}{\mathrm{D}x}\frac{\mathrm{d}x}{\mathrm{d}a}+\frac{\mathrm{D}\lambda}{\mathrm{D}y}\frac{\mathrm{d}y}{\mathrm{d}a}+\frac{\mathrm{D}\lambda}{\mathrm{D}z}\frac{\mathrm{d}z}{\mathrm{d}a}\\
\frac{\mathrm{d}\lambda}{\mathrm{d}b} &= \frac{\mathrm{D}\lambda}{\mathrm{D}x}\frac{\mathrm{d}x}{\mathrm{d}b}+\frac{\mathrm{D}\lambda}{\mathrm{D}y}\frac{\mathrm{d}y}{\mathrm{d}b}+\frac{\mathrm{D}\lambda}{\mathrm{D}z}\frac{\mathrm{d}z}{\mathrm{d}b}\\
\frac{\mathrm{d}\lambda}{\mathrm{d}c} &= \frac{\mathrm{D}\lambda}{\mathrm{D}x}\frac{\mathrm{d}x}{\mathrm{d}c}+\frac{\mathrm{D}\lambda}{\mathrm{D}y}\frac{\mathrm{d}y}{\mathrm{d}c}+\frac{\mathrm{D}\lambda}{\mathrm{D}z}\frac{\mathrm{d}z}{\mathrm{d}c}
\end{aligned}
$$

The equations in question will be immediately obtained, which in the case where $X\,\mathrm{d}x + Y\,\mathrm{d}y + Z\,\mathrm{d}z$ is a complete differential represented by $\mathrm{d}V$, they can be put in the following simpler form

$$
\begin{aligned}
&\Delta\left(\frac{\mathrm{d}^2x}{\mathrm{d}t^2}\frac{\mathrm{d}x}{\mathrm{d}a}+\frac{\mathrm{d}^2y}{\mathrm{d}t^2}\frac{\mathrm{d}y}{\mathrm{d}a}+\frac{\mathrm{d}^2z}{\mathrm{d}t^2}\frac{\mathrm{d}z}{\mathrm{d}a}+\frac{\mathrm{d}V}{\mathrm{d}a}\right)-\frac{\mathrm{d}\lambda}{\mathrm{d}a}=0\\
&\Delta\left(\frac{\mathrm{d}^2x}{\mathrm{d}t^2}\frac{\mathrm{d}x}{\mathrm{d}b}+\frac{\mathrm{d}^2y}{\mathrm{d}t^2}\frac{\mathrm{d}y}{\mathrm{d}b}+\frac{\mathrm{d}^2z}{\mathrm{d}t^2}\frac{\mathrm{d}z}{\mathrm{d}b}+\frac{\mathrm{d}V}{\mathrm{d}b}\right)-\frac{\mathrm{d}\lambda}{\mathrm{d}b}=0\\
&\Delta\left(\frac{\mathrm{d}^2x}{\mathrm{d}t^2}\frac{\mathrm{d}x}{\mathrm{d}c}+\frac{\mathrm{d}^2y}{\mathrm{d}t^2}\frac{\mathrm{d}y}{\mathrm{d}c}+\frac{\mathrm{d}^2z}{\mathrm{d}t^2}\frac{\mathrm{d}z}{\mathrm{d}c}+\frac{\mathrm{d}V}{\mathrm{d}c}\right)-\frac{\mathrm{d}\lambda}{\mathrm{d}c}=0
\end{aligned}
$$

7. In a similar fashion, equation (B) of Article 3 can be transformed. In view of the remark in Article 4, the differentials $\mathrm{d}x$, $\mathrm{d}y$, $\mathrm{d}z$ are only variables with respect to the variable t,

they will first be reduced to the partial differences $(\mathrm{d}x/\mathrm{d}t)/\mathrm{d}t$, $(\mathrm{d}y/\mathrm{d}t)/\mathrm{d}t$, $(\mathrm{d}z/\mathrm{d}t)/\mathrm{d}t$ so that the equation in question after being divided by $\mathrm{d}t$ will be

$$\frac{\mathrm{D}\dfrac{\mathrm{d}x}{\mathrm{d}t}}{\mathrm{D}x}+\frac{\mathrm{D}\dfrac{\mathrm{d}y}{\mathrm{d}t}}{\mathrm{D}y}+\frac{\mathrm{D}\dfrac{\mathrm{d}z}{\mathrm{d}t}}{\mathrm{D}z}=0$$

But, with the formulas found above for the expressions of $\mathrm{D}\lambda/\mathrm{D}x$, $\mathrm{D}\lambda/\mathrm{D}y$, etc., it will similarly be obtained after replacing λ by $\mathrm{d}x/\mathrm{d}t$, $\mathrm{d}y/\mathrm{d}t$, etc.

$$\frac{\mathrm{D}\dfrac{\mathrm{d}x}{\mathrm{d}t}}{\mathrm{D}x}=\frac{\alpha}{\theta}\frac{\mathrm{d}\dfrac{\mathrm{d}x}{\mathrm{d}t}}{\mathrm{d}a}+\frac{\beta}{\theta}\frac{\mathrm{d}\dfrac{\mathrm{d}x}{\mathrm{d}t}}{\mathrm{d}b}+\frac{\gamma}{\theta}\frac{\mathrm{d}\dfrac{\mathrm{d}x}{\mathrm{d}t}}{\mathrm{d}c}$$

Since in the second member of this equation the quantity x is treated as a function of a, b, c, and t, one will obtain

$$\frac{\mathrm{d}\dfrac{\mathrm{d}x}{\mathrm{d}t}}{\mathrm{d}a}=\frac{\mathrm{d}^2x}{\mathrm{d}a\,\mathrm{d}t}$$

and similarly for the other partial differences in x. Thus, it will simply result that

$$\frac{\mathrm{D}\dfrac{\mathrm{d}x}{\mathrm{d}t}}{\mathrm{D}x}=\frac{\alpha}{\theta}\frac{\mathrm{d}^2x}{\mathrm{d}a\,\mathrm{d}t}+\frac{\beta}{\theta}\frac{\mathrm{d}^2x}{\mathrm{d}b\,\mathrm{d}t}+\frac{\gamma}{\theta}\frac{\mathrm{d}^2x}{\mathrm{d}c\,\mathrm{d}t}$$

Similar expressions will be obtained for the values of $\mathrm{D}(\mathrm{d}y/\mathrm{d}t)/\mathrm{D}y$, and $\mathrm{D}(\mathrm{d}z/\mathrm{d}t)/\mathrm{D}z$, and it will only be necessary to replace x by y and z in the preceding formula.

After making these substitutions in the above equation and eliminating the common denominator θ, there will result

$$\alpha\frac{\mathrm{d}^2x}{\mathrm{d}a\,\mathrm{d}t}+\beta\frac{\mathrm{d}^2x}{\mathrm{d}b\,\mathrm{d}t}+\gamma\frac{\mathrm{d}^2x}{\mathrm{d}c\,\mathrm{d}t}+\alpha'\frac{\mathrm{d}^2y}{\mathrm{d}a\,\mathrm{d}t}+\beta'\frac{\mathrm{d}^2y}{\mathrm{d}b\,\mathrm{d}t}$$
$$+\gamma'\frac{\mathrm{d}^2y}{\mathrm{d}c\,\mathrm{d}t}+\alpha''\frac{\mathrm{d}^2z}{\mathrm{d}a\,\mathrm{d}t}+\beta''\frac{\mathrm{d}^2z}{\mathrm{d}b\,\mathrm{d}t}+\gamma''\frac{\mathrm{d}^2z}{\mathrm{d}c\,\mathrm{d}t}=0$$

The first member of this equation is nothing but the expression for $\mathrm{d}\theta/\mathrm{d}t$ as can be verified by differentiation of the expression for θ in Article 5.

Therefore, the equation becomes $\mathrm{d}\theta/\mathrm{d}t = 0$ for which the integral is θ and equal to a function of a, b, c. Let us assume that $t = 0$ in this equation and that K is what becomes of the quantity θ, then one will have the quantity K equal to a function of a, b, c. Consequently, the equation will become $\theta = K$.

But it has been assumed that when $t = 0$, there results

$$x = a, \qquad y = b, \qquad z = c$$

Therefore, there will also result

$$\frac{dx}{da} = 1, \quad \frac{dx}{db} = 0, \quad \frac{dx}{dc} = 0, \quad \frac{dy}{da} = 0, \quad \frac{dy}{db} = 1,$$
$$\frac{dy}{dc} = 0, \quad \frac{dz}{da} = 0, \quad \frac{dz}{db} = 0, \quad \frac{dz}{dc} = 0$$

After these expressions are substituted in the expression for θ in Article 5, one obtains $\theta = 1$ and thus $K = 1$.

Therefore, after replacing θ by its expression in the equation in question, it will become

$$\text{(E)} \quad \left\{ \begin{array}{l} \dfrac{dx}{da}\dfrac{dy}{db}\dfrac{dz}{dc} - \dfrac{dx}{db}\dfrac{dy}{da}\dfrac{dz}{dc} + \dfrac{dz}{db}\dfrac{dy}{dc}\dfrac{dz}{da} \\ -\dfrac{dx}{dc}\dfrac{dy}{db}\dfrac{dz}{da} + \dfrac{dx}{dc}\dfrac{dy}{da}\dfrac{dz}{db} - \dfrac{dx}{da}\dfrac{dy}{dc}\dfrac{dz}{db} \end{array} \right\} = 1$$

Thus this equation, combined with the three equations of (C) or (D) of Articles 5 and 6, will be used to determine the values of λ, x, y, z as functions of a, b, c and t. This equation can also be found in a simpler way, without using the differential equation (B) of Article 3. Indeed, equation (B) only expresses the fact that the variation of the volume $\mathrm{D}x\,\mathrm{D}y\,\mathrm{D}z$ of the particle $\mathrm{D}m$ is equal to zero when the time t varies. Therefore, the quantity $\mathrm{D}x\,\mathrm{D}y\,\mathrm{D}z$ must be a constant equal to the original expression $\mathrm{d}a\,\mathrm{d}b\,\mathrm{d}c$. The expressions for $\mathrm{D}x\,\mathrm{D}y\,\mathrm{D}z$ as a function of $\mathrm{d}a\,\mathrm{d}b\,\mathrm{d}c$ were given in Article 5. However, it must be noted that in the formula $\mathrm{D}x\,\mathrm{D}y\,\mathrm{D}z$, the difference $\mathrm{D}z$ must be used with x and y constant. Also, the difference $\mathrm{D}y$ must be used with x and z constant. Finally, the difference $\mathrm{D}x$ assumes that y and z are constant, which is obvious when considering the rectangular parallelepiped represented by $\mathrm{D}x\,\mathrm{D}y\,\mathrm{D}z$.

Let us first assume that x and y are constants and consequently, the quantities $\mathrm{D}x$ and $\mathrm{D}y$ are zero. The two following equations will be obtained

$$\frac{dx}{da}da + \frac{dx}{db}db + \frac{dx}{dc}dc = 0$$
$$\frac{dy}{da}da + \frac{dy}{db}db + \frac{dy}{dc}dc = 0$$

from which is obtained

$$da = \frac{\dfrac{dx}{db}\dfrac{dy}{dc} - \dfrac{dx}{dc}\dfrac{dy}{db}}{\dfrac{dx}{da}\dfrac{dy}{db} - \dfrac{dy}{da}\dfrac{dx}{db}}dc$$

$$db = \frac{\dfrac{dx}{dc}\dfrac{dy}{da} - \dfrac{dx}{da}\dfrac{dy}{dc}}{\dfrac{dx}{da}\dfrac{dy}{db} - \dfrac{dy}{da}\dfrac{dx}{db}}dc$$

After these expressions are substituted in the expression for Dz there results

$$\mathrm{D}z = \frac{\frac{\mathrm{d}z}{\mathrm{d}a}\left(\frac{\mathrm{d}x}{\mathrm{d}b}\frac{\mathrm{d}y}{\mathrm{d}c} - \frac{\mathrm{d}x}{\mathrm{d}c}\frac{\mathrm{d}y}{\mathrm{d}b}\right) + \frac{\mathrm{d}z}{\mathrm{d}b}\left(\frac{\mathrm{d}x}{\mathrm{d}c}\frac{\mathrm{d}y}{\mathrm{d}a} - \frac{\mathrm{d}x}{\mathrm{d}a}\frac{\mathrm{d}y}{\mathrm{d}c}\right) + \frac{\mathrm{d}z}{\mathrm{d}c}\left(\frac{\mathrm{d}x}{\mathrm{d}a}\frac{\mathrm{d}y}{\mathrm{d}b} - \frac{\mathrm{d}y}{\mathrm{d}a}\frac{\mathrm{d}x}{\mathrm{d}b}\right)}{\frac{\mathrm{d}x}{\mathrm{d}a}\frac{\mathrm{d}y}{\mathrm{d}b} - \frac{\mathrm{d}y}{\mathrm{d}a}\frac{\mathrm{d}x}{\mathrm{d}b}}\mathrm{d}c$$

In order to obtain the expression for Dy, it will be assumed that D$x = 0$ and D$z = 0$ which gives the two following equations

$$\mathrm{d}c = 0, \qquad \frac{\mathrm{d}x}{\mathrm{d}a}\mathrm{d}a + \frac{\mathrm{d}x}{\mathrm{d}b}\mathrm{d}b = 0$$

from which

$$\mathrm{d}a = -\frac{\frac{\mathrm{d}x}{\mathrm{d}b}}{\frac{\mathrm{d}x}{\mathrm{d}a}}\mathrm{d}b$$

If this expression, along with d$c = 0$, is substituted in the expression for Dy there results

$$\mathrm{D}y = \frac{\frac{\mathrm{d}x}{\mathrm{d}a}\frac{\mathrm{d}y}{\mathrm{d}b} - \frac{\mathrm{d}x}{\mathrm{d}b}\frac{\mathrm{d}y}{\mathrm{d}a}}{\frac{\mathrm{d}x}{\mathrm{d}a}}\mathrm{d}b$$

Finally, to obtain the expression for Dx, the expressions for Dy and Dz are set equal to zero which gives

$$\mathrm{d}b = 0, \qquad \mathrm{d}c = 0$$

and consequently

$$\mathrm{D}x = \frac{\mathrm{d}x}{\mathrm{d}a}\mathrm{d}a$$

If the expressions for Dx, Dy and Dz are multiplied together, one has

$$\mathrm{D}x\,\mathrm{D}y\,\mathrm{D}z = \left\{\begin{aligned}&\frac{\mathrm{d}z}{\mathrm{d}a}\left(\frac{\mathrm{d}x}{\mathrm{d}b}\frac{\mathrm{d}y}{\mathrm{d}c} - \frac{\mathrm{d}x}{\mathrm{d}c}\frac{\mathrm{d}y}{\mathrm{d}b}\right) + \frac{\mathrm{d}z}{\mathrm{d}b}\left(\frac{\mathrm{d}x}{\mathrm{d}c}\frac{\mathrm{d}y}{\mathrm{d}a} - \frac{\mathrm{d}x}{\mathrm{d}a}\frac{\mathrm{d}y}{\mathrm{d}c}\right)\\ &+\frac{\mathrm{d}z}{\mathrm{d}c}\left(\frac{\mathrm{d}x}{\mathrm{d}a}\frac{\mathrm{d}y}{\mathrm{d}b} - \frac{\mathrm{d}y}{\mathrm{d}a}\frac{\mathrm{d}x}{\mathrm{d}b}\right)\end{aligned}\right\}\mathrm{d}a\,\mathrm{d}b\,\mathrm{d}c$$

Then assuming that Dx Dy Dz = da db dc, one will immediately obtain equation (E).

It is worthwhile to note that this expression for $\mathrm{D}x\,\mathrm{D}y\,\mathrm{D}z$ is the one which must be used in the triple integrals relative to x, y, z, when one wants to substitute for the variables x, y, z, given functions of other variables a, b, c.

8. Since these equations contain partial differences, the integration process will necessarily introduce different arbitrary functions. The determination of these functions should be reduced, in part, to the initial state of the fluid which must be assumed given and in part, to the consideration of the exterior surface of the fluid, which is also given if the fluid is contained in a vessel and which must be represented by the equation $\lambda = 0$ when the fluid is free (Article 2).

Indeed, in the first case, let the equation $A = 0$ represent the equation of the walls of the vessel where the equation denoted by A is a given function of the coordinates x, y, z of these walls and the time t, if these walls are mobile or of any variable configuration. If we substitute for these variables in terms of the quantities a, b, c and t, an equation between the initial coordinates and the time t is obtained. This equation represents the surface formed in the initial state by the same particles which, after a time t, form the surface defined by the equation $A = 0$. Therefore, if it is required that the same particles which arrive at the surface always remain there and only move along this surface, a condition which seems necessary for the fluid not to separate, and which is generally accepted in the theory of fluids, then the equation in question should not contain the time t. Consequently, the function A of x, y, z should be such that t is eliminated after the substitution of the expressions for x, y, z in terms of a, b, c and t.

For the same reason, the equation $\lambda = 0$ of the free surface should not contain the variable t. Thus the expression for λ should only be a simple function of a, b, c.

Also, there is the case in the motion of a fluid which flows out of a vessel where this condition should not hold. Then the results deriving from this condition are no longer necessary.

9. Such are the equations from which the motion of an arbitrary incompressible fluid can be determined directly. But these equations are in a somewhat complex form and it is possible to reduce them to a simpler form by using for unknowns the velocities $\mathrm{d}x/\mathrm{d}t$, $\mathrm{d}y/\mathrm{d}t$, $\mathrm{d}z/\mathrm{d}t$, positive in the directions of the coordinates, instead of the coordinates x, y, z and to consider these velocities as a function of x, y, z and t.

Indeed, on the one hand, it is clear that since x, y, z are functions of a, b, c and t, the quantities $\mathrm{d}x/\mathrm{d}t$, $\mathrm{d}y/\mathrm{d}t$, $\mathrm{d}z/\mathrm{d}t$, will also be functions of the same variables a, b, c and t. Therefore, it is clear that if the expressions for a, b, c which are obtained from the equations expressing x, y, z as functions of a, b, c, are substituted for x, y, z, one will have $\mathrm{d}x/\mathrm{d}t$, $\mathrm{d}y/\mathrm{d}t$, $\mathrm{d}z/\mathrm{d}t$, expressed as functions of x, y, z and t.

On the other hand, it is clear that to determine the motion of a fluid, it suffices to know at each instant the motion of an arbitrary particle at a given location in space without having to know the prior locations of this particle.

Consequently, it suffices to have the expressions of the velocities $\mathrm{d}x/\mathrm{d}t$, $\mathrm{d}y/\mathrm{d}t$, $\mathrm{d}z/\mathrm{d}t$, as functions of x, y, z and t.

Also, if these expressions are known and if they are denoted by p, q, r, the following equations will be obtained

$$\mathrm{d}x = p\,\mathrm{d}t, \qquad \mathrm{d}y = q\,\mathrm{d}t, \qquad \mathrm{d}z = r\,\mathrm{d}t$$

between x, y, z and t. Then, if these expressions are integrated in such a manner that x, y, z become a, b, c when $t = 0$, they will give the values of x, y, z in terms of a, b, c, and t.

Also, if the quantity dt is eliminated from these differential equations, one will obtain the two following equations

$$p\,\mathrm{d}y = q\,\mathrm{d}x, \qquad p\,\mathrm{d}z = r\,\mathrm{d}x$$

which express the nature of the diverse trajectories of the motion of the fluid at each instant, curves which change location and shape from one instant to the next.

10. Let us reconsider the fundamental equations (A) and (B) of Articles 2 and 3, and let us introduce the variables $p = \mathrm{d}x/\mathrm{d}t$, $q = \mathrm{d}y/\mathrm{d}t$, $r = \mathrm{d}z/\mathrm{d}t$, viewed as functions of x, y, z and t.

It is clear that the quantities $\mathrm{d}^2x/\mathrm{d}t^2$, $\mathrm{d}^2y/\mathrm{d}t^2$, $\mathrm{d}^2z/\mathrm{d}t^2$, can be written in the form $\mathrm{d}(\mathrm{d}x/\mathrm{d}t)/\mathrm{d}t$, $\mathrm{d}(\mathrm{d}y/\mathrm{d}t)/\mathrm{d}t$, $\mathrm{d}(\mathrm{d}z/\mathrm{d}t)/\mathrm{d}t$, where the quantities $\mathrm{d}x/\mathrm{d}t$, $\mathrm{d}y/\mathrm{d}t$, $\mathrm{d}z/\mathrm{d}t$, are assumed to be functions of a, b, c and t.

By viewing the velocities in this fashion, one will have for the difference of $\mathrm{d}x/\mathrm{d}t$

$$\frac{\mathrm{d}\dfrac{\mathrm{d}x}{\mathrm{d}t}}{\mathrm{d}t}\mathrm{d}t + \frac{\mathrm{d}\dfrac{\mathrm{d}x}{\mathrm{d}t}}{\mathrm{d}a}\mathrm{d}a + \frac{\mathrm{d}\dfrac{\mathrm{d}x}{\mathrm{d}t}}{\mathrm{d}b}\mathrm{d}b + \frac{\mathrm{d}\dfrac{\mathrm{d}x}{\mathrm{d}t}}{\mathrm{d}c}\mathrm{d}c$$

and similarly for the others. But by viewing them as functions of x, y, z and t, and after designating them by p, q, t, their total differences are

$$\frac{\mathrm{d}p}{\mathrm{d}t}\mathrm{d}t + \frac{\mathrm{d}p}{\mathrm{d}x}\mathrm{d}x + \frac{\mathrm{d}p}{\mathrm{d}y}\mathrm{d}y + \frac{\mathrm{d}p}{\mathrm{d}z}\mathrm{d}z$$

and similarly for the others. Thus if in these latter expressions one replaces $\mathrm{d}x$, $\mathrm{d}y$, $\mathrm{d}z$ by their expressions in terms of a, b, c, t, they should become identical to the former. But, if the quantity x is viewed as a function of a, b, c and t, one has

$$\mathrm{d}x = \frac{\mathrm{d}x}{\mathrm{d}t}\mathrm{d}t + \frac{\mathrm{d}x}{\mathrm{d}a}\mathrm{d}a + \frac{\mathrm{d}x}{\mathrm{d}b}\mathrm{d}b + \frac{\mathrm{d}x}{\mathrm{d}c}\mathrm{d}c$$

where $\mathrm{d}x/\mathrm{d}t$ is evidently equal to p, after it is assumed that the quantities x, y, z are replaced by a, b, c and t. Thus there results

$$\mathrm{d}x = p\,\mathrm{d}t + \frac{\mathrm{d}x}{\mathrm{d}a}\mathrm{d}a + \cdots$$

and in the same fashion

$$\mathrm{d}y = q\,\mathrm{d}t + \frac{\mathrm{d}y}{\mathrm{d}a}\mathrm{d}a + \cdots, \qquad \mathrm{d}z = r\,\mathrm{d}t + \frac{\mathrm{d}z}{\mathrm{d}a}\mathrm{d}a + \cdots$$

After substituting these expressions in the equations for the complete difference of $\mathrm{d}x/\mathrm{d}t$, the terms with $\mathrm{d}t$ are

$$\left(\frac{\mathrm{d}p}{\mathrm{d}t} + \frac{\mathrm{d}p}{\mathrm{d}x}p + \frac{\mathrm{d}p}{\mathrm{d}y}q + \frac{\mathrm{d}p}{\mathrm{d}z}r\right)\mathrm{d}t$$

which, because they must be identical to the term corresponding to $(\mathrm{d}(\mathrm{d}x/\mathrm{d}t)/\mathrm{d}t)\mathrm{d}t$ or to $(\mathrm{d}^2x/\mathrm{d}t^2)\mathrm{d}t$, one will have

$$\frac{\mathrm{d}^2x}{\mathrm{d}t^2} = \frac{\mathrm{d}p}{\mathrm{d}t} + p\frac{\mathrm{d}p}{\mathrm{d}x} + q\frac{\mathrm{d}p}{\mathrm{d}y} + r\frac{\mathrm{d}p}{\mathrm{d}z}$$

and similarly, it will be found that

$$\frac{\mathrm{d}^2y}{\mathrm{d}t^2} = \frac{\mathrm{d}q}{\mathrm{d}t} + p\frac{\mathrm{d}q}{\mathrm{d}x} + q\frac{\mathrm{d}q}{\mathrm{d}y} + r\frac{\mathrm{d}q}{\mathrm{d}z}$$

$$\frac{\mathrm{d}^2z}{\mathrm{d}t^2} = \frac{\mathrm{d}r}{\mathrm{d}t} + p\frac{\mathrm{d}r}{\mathrm{d}x} + q\frac{\mathrm{d}r}{\mathrm{d}y} + r\frac{\mathrm{d}r}{\mathrm{d}z}$$

These equations should be substituted in equations (A). Since in these same equations the terms $\mathrm{D}\lambda/\mathrm{D}x$, $\mathrm{D}\lambda/\mathrm{D}y$, $\mathrm{D}\lambda/\mathrm{D}z$ represent the partial differences of λ, with respect to x, y, z, assuming t constant, the operator can be changed from D to d.

Thus the following transformed equations result

$$(\mathrm{F}) \quad \begin{cases} \Delta\left(\dfrac{\mathrm{d}p}{\mathrm{d}t} + p\dfrac{\mathrm{d}p}{\mathrm{d}x} + q\dfrac{\mathrm{d}p}{\mathrm{d}y} + r\dfrac{\mathrm{d}p}{\mathrm{d}z} + X\right) - \dfrac{\mathrm{d}\lambda}{\mathrm{d}x} = 0 \\ \Delta\left(\dfrac{\mathrm{d}q}{\mathrm{d}t} + p\dfrac{\mathrm{d}q}{\mathrm{d}x} + q\dfrac{\mathrm{d}q}{\mathrm{d}y} + r\dfrac{\mathrm{d}q}{\mathrm{d}z} + Y\right) - \dfrac{\mathrm{d}\lambda}{\mathrm{d}y} = 0 \\ \Delta\left(\dfrac{\mathrm{d}r}{\mathrm{d}t} + p\dfrac{\mathrm{d}r}{\mathrm{d}x} + q\dfrac{\mathrm{d}r}{\mathrm{d}y} + r\dfrac{\mathrm{d}r}{\mathrm{d}z} + Z\right) - \dfrac{\mathrm{d}\lambda}{\mathrm{d}z} = 0 \end{cases}$$

With respect to equations (B) of Article 3, in which the differences characterized by d are relative to t, and those which are characterized by D are relative to x, y, z, it is only necessary to replace $\mathrm{d}x$, $\mathrm{d}y$, $\mathrm{d}z$ by the quantities $p\,\mathrm{d}t$, $q\,\mathrm{d}t$, $r\,\mathrm{d}t$, and by changing the operator D to d. Since the characteristic has no bearing on the partial differences and since $\mathrm{d}t$ is constant, one will immediately obtain

$$\text{(G)} \quad \frac{\mathrm{d}p}{\mathrm{d}x} + \frac{\mathrm{d}q}{\mathrm{d}y} + \frac{\mathrm{d}r}{\mathrm{d}z} = 0$$

It is clear that these equations are much simpler than equations (C) or (D) and (E) to which they correspond. Thus it is preferable to use them in the theory of fluids.

The two sets of equations denoted by (F) and (G) will give p, q, r and λ as functions of x, y, z and t, viewed as constants during their integration. If the values of x, y, z were then required as a function of t and of the original coordinates a, b, c, as in the first solution, one should only have to integrate the equations

$$\mathrm{d}x = p\,\mathrm{d}t, \qquad \mathrm{d}y = q\,\mathrm{d}t, \qquad \mathrm{d}z = r\,\mathrm{d}t$$

by introducing in them the initial values a, b, c of x, y, z as arbitrary constants.

11. In homogeneous fluids of uniform density, the quantity Δ, which represents the density is a constant. This is the most common case and the only one which we will examine in what follows. But for the case of heterogeneous fluids, this quantity must be a function which is constant with time for the same particle, but variable from one particle to the next, according to a given law. Thus, by considering the fluid in its initial state, where the coordinates x, y, z are a, b, c, the quantity Δ will be a given and known function of a, b, c. Therefore, if the quantity Δ is considered as a function of x, y, z and t, the variable t should cancel out after substitution of the values of x, y, z as functions of a, b, c and t. Consequently, the differential of Δ with respect to t must be equal to zero. Thus the following equations will be obtained where x, y, z are functions of t.

$$\frac{\mathrm{d}\Delta}{\mathrm{d}t} + \frac{\mathrm{d}\Delta}{\mathrm{d}x}\frac{\mathrm{d}x}{\mathrm{d}t} + \frac{\mathrm{d}\Delta}{\mathrm{d}y}\frac{\mathrm{d}y}{\mathrm{d}t} + \frac{\mathrm{d}\Delta}{\mathrm{d}z}\frac{\mathrm{d}z}{\mathrm{d}t} = 0$$

where the values of $\mathrm{d}x/\mathrm{d}t$, $\mathrm{d}y/\mathrm{d}t$, $\mathrm{d}z/\mathrm{d}t$, should be replaced by p, q, r.

Thus, one will have the following equation

$$\text{(H)} \quad \frac{\mathrm{d}\Delta}{\mathrm{d}t} + p\frac{\mathrm{d}\Delta}{\mathrm{d}x} + q\frac{\mathrm{d}\Delta}{\mathrm{d}y} + r\frac{\mathrm{d}\Delta}{\mathrm{d}z} = 0$$

which will serve to determine the unknown Δ in the equations (F) because in these equations the quantity Δ must be considered as a function of x, y, z.

In this regard, they are less useful than the equations (C) or (D), in which the quantity Δ can be viewed as a known function of a, b, c.

12. What we just said with respect to the function Δ, should also be applied to the function A, where $A = 0$ is the equation of the walls of the vessel and where it is assumed that the fluid contiguous to the walls can only move by flowing along the walls in such a manner that the same particles always remain in the surface. This condition requires, as seen in Article 8, that the quantity A is a function of a, b, c without t so that by viewing this quantity as a function of x, y, z and t, one will also have the equation

$$\text{(I)}\quad \frac{\mathrm{d}A}{\mathrm{d}t} + p\frac{\mathrm{d}A}{\mathrm{d}x} + q\frac{\mathrm{d}A}{\mathrm{d}y} + r\frac{\mathrm{d}A}{\mathrm{d}z} = 0$$

For the parts of the surface where the fluid is free, one will have the equations $\lambda = 0$ (Article 2). Consequently, one should have to satisfy the same condition relative to this surface, which is

$$\text{(K)}\quad \frac{\mathrm{d}\lambda}{\mathrm{d}t} + p\frac{\mathrm{d}\lambda}{\mathrm{d}x} + q\frac{\mathrm{d}\lambda}{\mathrm{d}y} + r\frac{\mathrm{d}\lambda}{\mathrm{d}z} = 0$$

13. Here are the most general and simplest formulas for the rigorous determination of the motion of fluids. The only difficulty consists of integrating them. But it is so great that until now we were obliged to content ourselves, even in the simplest problems, with particular methods based on more or less limited hypotheses. In order to reduce this difficulty as much as possible, we will examine how and in which cases these formulas can still be simplified. Then they will be applied to some problems on the motion of fluids in vessels or conduits.

14. Equation (G) of Article 10 is easily satisfied. Indeed, by making the following definitions

$$p = \frac{\mathrm{d}\alpha}{\mathrm{d}z}, \qquad q = \frac{\mathrm{d}\beta}{\mathrm{d}z}$$

equation (G) becomes

$$\frac{\mathrm{d}^2\alpha}{\mathrm{d}x\,\mathrm{d}z} + \frac{\mathrm{d}^2\beta}{\mathrm{d}y\,\mathrm{d}z} + \frac{\mathrm{d}r}{\mathrm{d}z} = 0$$

which is integrable with respect to z. The integration gives

$$r = -\frac{\mathrm{d}\alpha}{\mathrm{d}x} - \frac{\mathrm{d}\beta}{\mathrm{d}y}$$

It is not necessary to include here an arbitrary function because of the indeterminate quantities α and β.

Thus the equation in question will be satisfied by these values

$$p = \frac{d\alpha}{dz}, \qquad q = \frac{d\beta}{dz}, \qquad r = -\frac{d\alpha}{dx} - \frac{d\beta}{dy}$$

After substituting these expressions in the three equations (F) of the same article, there will remain three unknowns α, β, λ. Again, it will be very easy to eliminate λ from the partial differences. Thus in this manner if the density Δ is constant, the problem will be reduced to two unique equations between the unknowns α and β. If the density Δ is variable, it is only necessary to include equation (H) of Article 11. But the integration of the resulting equations is beyond known means of analysis.

15. Let us consider whether equations (F), considered separately, are capable of some simplification.

By only considering the variability of x, y, z, in the function λ, there results

$$d\lambda = \frac{d\lambda}{dx}dx + \frac{d\lambda}{dy}dy + \frac{d\lambda}{dz}dz$$

Therefore, by substituting for $d\lambda/dx$, $d\lambda/dy$, $d\lambda/dz$, the expressions obtained from these equations, one obtains

$$\begin{aligned} d\lambda = &\left(\frac{dp}{dt} + p\frac{dp}{dx} + q\frac{dp}{dy} + r\frac{dp}{dz} + X\right)\Delta dx \\ &+ \left(\frac{dq}{dt} + p\frac{dq}{dx} + q\frac{dq}{dy} + r\frac{dq}{dz} + Y\right)\Delta dy \\ &+ \left(\frac{dr}{dt} + p\frac{dr}{dx} + q\frac{dr}{dy} + r\frac{dr}{dz} + Z\right)\Delta dz \end{aligned}$$

Since the first member of this equation is a complete differential, the second should also be a complete differential with respect to x, y, z. The value of λ which will be obtained will satisfy equations (F).

Let us now assume that the fluid is homogeneous such that the density Δ is constant. Also, for greater simplicity, let us assume that it is equal to unity.

Moreover, let us assume that the accelerating forces X, Y, Z are such that the quantity $X\,dx + Y\,dy + Z\,dz$ is a complete differential. This condition is necessary for the fluid to be in equilibrium under the action of these forces, as was shown in Article 19 of SECTION VII of PART I. It also always holds when the forces are due to one or several attractions proportional to arbitrary functions of the distance to the centers, which is the natural case because by calling the attractions P, Q, R, etc. and the distances p, q, r, etc., one has generally (PART I, SECTION V, Article 7)

$$X\,dx + Y\,dy + Z\,dz = P\,dp + Q\,dq + R\,dr + \cdots$$

Let us define the quantity Δ equal to unity, and

$$X\,dx + Y\,dy + Z\,dz = P\,dp + Q\,dq + R\,dr + \cdots = dV$$

The preceding equation becomes

$$\text{(L)}\quad d\lambda - dV = \begin{cases} \left(\frac{dp}{dt} + p\frac{dp}{dx} + q\frac{dp}{dy} + r\frac{dp}{dz}\right) dx \\ + \left(\frac{dq}{dt} + p\frac{dq}{dx} + q\frac{dq}{dy} + r\frac{dq}{dz}\right) dy \\ + \left(\frac{dr}{dt} + p\frac{dr}{dx} + q\frac{dr}{dy} + r\frac{dr}{dz}\right) dz \end{cases}$$

where the second member of this equation should be a complete differential since the first member is. This equation is also equivalent to equations (F) of Article 10.

But, by considering the differential of $(p^2 + q^2 + r^2)/2$ taken relative to x, y, z, it is not difficult to see that the second member of the equation in question can be given this form

$$\frac{d(p^2 + q^2 + r^2)}{2} + \frac{dp}{dt}dx + \frac{dq}{dt}dy + \frac{dr}{dt}dz$$
$$+ \left(\frac{dp}{dy} - \frac{dq}{dx}\right)(q\,dx - p\,dy) + \left(\frac{dp}{dz} - \frac{dr}{dx}\right)(r\,dx - p\,dz)$$
$$+ \left(\frac{dq}{dz} - \frac{dr}{dy}\right)(r\,dy - q\,dz)$$

It is clear at the outset that this quantity will be a complete differential if the quantity $(p\,dx + q\,dy + r\,dz)$ is a complete differential, since then its differential relative to t, that is, $(dp/dt)dx + (dq/dt)dy + (dr/dt)dz$, will also be one. Moreover, the known conditions of integrability will give

$$\frac{dp}{dy} - \frac{dq}{dx} = 0, \qquad \frac{dp}{dz} - \frac{dr}{dx} = 0, \qquad \frac{dq}{dz} - \frac{dr}{dy} = 0$$

From this result, it follows that equation (L) can be satisfied with the simple assumption that $p\,dx + q\,dy + r\,dz$ is a complete differential and the calculation of the motion of the fluid will then be simplified. But since this is a particular assumption, it is important to examine above all in which cases it can and should hold.

16. In order to shorten the expressions, let us define

$$\alpha = \frac{dp}{dy} - \frac{dq}{dx}, \qquad \beta = \frac{dp}{dz} - \frac{dr}{dx}, \qquad \gamma = \frac{dq}{dz} - \frac{dr}{dy}$$

It will only be a question of making an exact differential of the equation[52]

$$\frac{dp}{dt}dx + \frac{dq}{dt}dy + \frac{dr}{dt}dz + \alpha(q\,dx - p\,dy) + \beta(r\,dx - p\,dz) + \gamma(r\,dy - q\,dz)$$

By viewing p, q, r as functions of t, one can assume

$$\begin{aligned}
p &= p' + p''t + p'''t^2 + p^{\text{IV}}t^3 + \cdots \\
q &= q' + q''t + q'''t^2 + q^{\text{IV}}t^3 + \cdots \\
r &= r' + r''t + r'''t^2 + r^{\text{IV}}t^3 + \cdots
\end{aligned}$$

in which the quantities p', p'', p''', etc., q', q'', q''', etc., r', r'', r''', etc. are functions of x, y, z but without t.

If these expressions are substituted in the three expressions for α, β, γ, they will become

$$\begin{aligned}
\alpha &= \alpha' + \alpha''t + \alpha'''t^2 + \alpha^{\text{IV}}t^3 + \cdots \\
\beta &= \beta' + \beta''t + \beta'''t^2 + \beta^{\text{IV}}t^3 + \cdots \\
\gamma &= \gamma' + \gamma''t + \gamma'''t^2 + \gamma^{\text{IV}}t^3 + \cdots
\end{aligned}$$

assuming

$$\begin{aligned}
\alpha' &= \frac{dp'}{dy} - \frac{dq'}{dx}, & \alpha'' &= \frac{dp''}{dy} - \frac{dq''}{dx}, & &\cdots \\
\beta' &= \frac{dp'}{dz} - \frac{dr'}{dx}, & \beta'' &= \frac{dp''}{dz} - \frac{dr''}{dx}, & &\cdots \\
\gamma' &= \frac{dq'}{dz} - \frac{dr'}{dy}, & \gamma'' &= \frac{dq''}{dz} - \frac{dr''}{dy}, & &\cdots
\end{aligned}$$

Thus the quantity

$$\frac{dp}{dt}dx + \frac{dq}{dt}dy + \frac{dr}{dt}dz + \alpha(q\,dx - p\,dy) + \beta(r\,dx - p\,dz) + \gamma(r\,dy - q\,dz)$$

will become, after these different substitutions, and after ordering the terms according to the powers of t

$$\begin{aligned}
&p''\,dx + q''\,dy + r''\,dz + \alpha'(q'\,dx - p'\,dy) + \beta'(r'\,dx - p'\,dz) + \gamma'(r'\,dy - q'\,dz) \\
&+t\left\{\begin{array}{l} 2(p'''\,dx + q'''\,dy + r'''\,dz) + \alpha'(q''\,dx - p''\,dy) \\ +\beta'(r''\,dx - p''\,dz) + \gamma'(r''\,dy - q''\,dz) + \alpha''(q'\,dx - p'\,dy) \\ +\beta''(r'\,dx - p'\,dz) + \gamma''(r'\,dy - q'\,dz) \end{array}\right\} \\
&+t^2\left\{\begin{array}{l} 3(p^{\text{IV}}\,dx + q^{\text{IV}}\,dy + r^{\text{IV}}\,dz) + \alpha'(q'''\,dx - p'''\,dy) \\ +\beta'(r'''\,dx - p'''\,dz) + \gamma'(r'''\,dy - q'''\,dz) + \alpha''(q''\,dx - p''\,dy) \\ +\beta''(r''\,dx - p''\,dz) + \gamma''(r''\,dy - q''\,dz) + \alpha'''(q'\,dx - p'\,dy) \\ +\beta'''(r'\,dx - p'\,dz) + \gamma'''(r'\,dy - q'\,dz) \end{array}\right\} \\
&+\cdots\cdots\cdots\cdots
\end{aligned}$$

Since this quantity must be an exact differential, independent of the variable t, the quantities multiplied by a power of t should be each an exact differential.

In view of this assumption, let us assume that $(p'\,\mathrm{d}x + q'\,\mathrm{d}y + r'\,\mathrm{d}z)$ is an exact differential. Thus, one will obtain from known theorems

$$\frac{\mathrm{d}p'}{\mathrm{d}y} = \frac{\mathrm{d}q'}{\mathrm{d}x}, \qquad \frac{\mathrm{d}p'}{\mathrm{d}z} = \frac{\mathrm{d}r'}{\mathrm{d}x}, \qquad \frac{\mathrm{d}q'}{\mathrm{d}z} = \frac{\mathrm{d}r'}{\mathrm{d}y}$$

therefore $\alpha' = 0$, $\beta' = 0$, $\gamma' = 0$ and the first quantity, which must be an exact differential, will be reduced to $(p''\,\mathrm{d}x + q''\,\mathrm{d}y + r''\,\mathrm{d}z)$. Consequently, these equations of condition will be obtained

$$\alpha'' = 0, \qquad \beta'' = 0, \qquad \gamma'' = 0$$

Then the second quantity, which must be an exact differential, will become $2(p'''\,\mathrm{d}x + q'''\,\mathrm{d}y + r'''\,\mathrm{d}z)$ and from this result, the following equations are obtained

$$\alpha''' = 0, \qquad \beta''' = 0, \qquad \gamma''' = 0$$

so that the third quantity, which must also be an exact differential will become $3(p^{\mathrm{IV}}\,\mathrm{d}x + q^{\mathrm{IV}}\,\mathrm{d}y + r^{\mathrm{IV}}\,\mathrm{d}z)$. From this result, one will similarly obtain the equations

$$\alpha^{\mathrm{IV}} = 0, \qquad \beta^{\mathrm{IV}} = 0, \qquad \gamma^{\mathrm{IV}} = 0$$

and so on. Therefore, it follows that if the quantity $(p'\,\mathrm{d}x + q'\,\mathrm{d}y + r'\,\mathrm{d}z)$ is an exact differential, it will result that

$$p''\,\mathrm{d}x + q''\,\mathrm{d}y + r''\,\mathrm{d}z, \quad p'''\,\mathrm{d}x + q'''\,\mathrm{d}y + r'''\,\mathrm{d}z, \quad p^{\mathrm{IV}}\,\mathrm{d}x + q^{\mathrm{IV}}\,\mathrm{d}y + r^{\mathrm{IV}}\,\mathrm{d}z, \quad \cdots$$

where each equation taken individually is an exact differential. Consequently, the entire quantity $(p\,\mathrm{d}x + q\,\mathrm{d}y + r\,\mathrm{d}z)$ will be in this case an exact differential with the time t assumed to be very small.

17. It follows from this result that if the quantity $(p\,\mathrm{d}x + q\,\mathrm{d}y + r\,\mathrm{d}z)$ is an exact differential when $t = 0$, it will also be an exact differential when t has any other value. In general, since the origin of t is arbitrary and since t can be taken either positive or negative, it follows that if the quantity $(p\,\mathrm{d}x + q\,\mathrm{d}y + r\,\mathrm{d}z)$ is an exact differential at an arbitrary instant, it should be such for all other instants. Consequently, if there is one instant during the motion for which it is not an exact differential, it cannot be exact for the entire period of motion. If it were exact at another arbitrary instant, it should also be exact at the first instant.

18. When the motion starts form rest, it follows that $p = 0$, $q = 0$, $r = 0$ when $t = 0$. Thus $(p\,\mathrm{d}x + q\,\mathrm{d}y + r\,\mathrm{d}z)$ will be integrable at this time and thus for the entire duration of the motion.

But if there were impressed velocities on the fluid at the beginning of the motion, it is the nature of these velocities which determines whether the quantity $(p\,\mathrm{d}x + q\,\mathrm{d}y + r\,\mathrm{d}z)$ is an integrable quantity or not. In the first case, the quantity $(p\,\mathrm{d}x + q\,\mathrm{d}y + r\,\mathrm{d}z)$ will always be integrable. In the second case, it will never be integrable.

When the initial velocities are produced by an arbitrary impulse applied to the surface of the fluid, such as by the action of a piston, it can be demonstrated that $(p\,\mathrm{d}x + q\,\mathrm{d}y + r\,\mathrm{d}z)$ must be integrable in the first instant.

Since the velocities p, q, r impressed upon each point of the fluid by the impulse at the surface, should be such that if they were cancelled by impressing velocities at each point of the fluid with equal magnitude but of opposite direction, the entire mass of the fluid would remain at rest or in the equilibrium state. Thus equilibrium should hold within this mass because of the impulse applied at the surface and because of the velocities or forces $-p$, $-q$, $-r$, applied to each of the interior points. Consequently, from the general law of equilibrium of fluids (PART I, SECTION VII, Article 19), the quantities p, q, r should be such that $p\,\mathrm{d}x + q\,\mathrm{d}y + r\,\mathrm{d}z$ is an exact differential. Thus, in this case, the same quantity should always be an exact differential at each instant of the motion.

19. Some skepticism may surround the assertion that it is possible to have a motion in the fluid for which $p\,\mathrm{d}x + q\,\mathrm{d}y + r\,\mathrm{d}z$ is not an exact differential. This skepticism can be removed with a very simple example. It is only necessary to consider the case where one would have

$$p = gy, \qquad q = -gx, \qquad r = 0$$

where g is an arbitrary constant. At the outset, it is clear that in this case $p\,\mathrm{d}x + q\,\mathrm{d}y + r\,\mathrm{d}z$ will not be an exact differential because it becomes $g(y\,\mathrm{d}x - x\,\mathrm{d}y)$, which is not integrable. However, equation (L) of Article 15 will be integrable by itself because one will have

$$\frac{\mathrm{d}p}{\mathrm{d}y} = g, \qquad \frac{\mathrm{d}q}{\mathrm{d}x} = -g$$

and all the other partial differences of p and q will be equal to zero so that the equation in question becomes

$$\mathrm{d}\lambda - \mathrm{d}V = -g^2(x\,\mathrm{d}x + y\,\mathrm{d}y)$$

for which the integral is

$$\lambda = V - \frac{g^2}{2}(x^2 + y^2) + f(t)$$

This expression will satisfy the three equations denoted (F) of Article 10.

With respect to equation (G) of the same article, it will also hold since the assumed values give

$$\frac{\mathrm{d}p}{\mathrm{d}x} = 0, \qquad \frac{\mathrm{d}q}{\mathrm{d}y} = 0, \qquad \frac{\mathrm{d}r}{\mathrm{d}z} = 0$$

Also, it is clear that these expressions for p, q, r, represent the motion of a fluid rotating about the fixed z-coordinate axis with a constant angular rotation equal to g. It is further recognized that such a motion is always possible in a fluid.

It can be concluded form this result that in the calculation of the oscillations of the seas due to the attraction of the Sun and Moon, it cannot be assumed that the quantity $p\,\mathrm{d}x + q\,\mathrm{d}y + r\,\mathrm{d}z$ is integrable because it is not when the fluid is at rest with respect to the Earth that only the rotational motion common with it is present.

20. After having resolved the case where it is certain that the quantity $p\,\mathrm{d}x + q\,\mathrm{d}y + r\,\mathrm{d}z$ should be an exact differential, let us see how, besides this case, this condition can be applied to resolve the equations of fluid motion.

Let $p\,\mathrm{d}x + q\,\mathrm{d}y + r\,\mathrm{d}z = \mathrm{d}\varphi$ where φ is an arbitrary function of x, y, z and the variable t. If the variable t is treated as a constant in the expression for $\mathrm{d}\varphi$, there will result

$$p = \frac{\mathrm{d}\varphi}{\mathrm{d}x}, \qquad q = \frac{\mathrm{d}\varphi}{\mathrm{d}y}, \qquad r = \frac{\mathrm{d}\varphi}{\mathrm{d}z}$$

and after substituting these expressions in equation (L) of Article 15, it becomes

$$\begin{aligned}
\mathrm{d}\lambda - \mathrm{d}V &= \left(\frac{\mathrm{d}^2\varphi}{\mathrm{d}t\,\mathrm{d}x} + \frac{\mathrm{d}\varphi}{\mathrm{d}x}\frac{\mathrm{d}^2\varphi}{\mathrm{d}x^2} + \frac{\mathrm{d}\varphi}{\mathrm{d}y}\frac{\mathrm{d}^2\varphi}{\mathrm{d}x\,\mathrm{d}y} + \frac{\mathrm{d}\varphi}{\mathrm{d}z}\frac{\mathrm{d}^2\varphi}{\mathrm{d}x\,\mathrm{d}z}\right)\mathrm{d}x \\
&+ \left(\frac{\mathrm{d}^2\varphi}{\mathrm{d}t\,\mathrm{d}y} + \frac{\mathrm{d}\varphi}{\mathrm{d}x}\frac{\mathrm{d}^2\varphi}{\mathrm{d}x\,\mathrm{d}y} + \frac{\mathrm{d}\varphi}{\mathrm{d}y}\frac{\mathrm{d}^2\varphi}{\mathrm{d}y^2} + \frac{\mathrm{d}\varphi}{\mathrm{d}z}\frac{\mathrm{d}^2\varphi}{\mathrm{d}y\,\mathrm{d}z}\right)\mathrm{d}y \\
&+ \left(\frac{\mathrm{d}^2\varphi}{\mathrm{d}t\,\mathrm{d}z} + \frac{\mathrm{d}\varphi}{\mathrm{d}x}\frac{\mathrm{d}^2\varphi}{\mathrm{d}x\,\mathrm{d}z} + \frac{\mathrm{d}\varphi}{\mathrm{d}y}\frac{\mathrm{d}^2\varphi}{\mathrm{d}y\,\mathrm{d}z} + \frac{\mathrm{d}\varphi}{\mathrm{d}z}\frac{\mathrm{d}^2\varphi}{\mathrm{d}z^2}\right)\mathrm{d}z
\end{aligned}$$

for which the integral with respect to x, y, z is obviously

$$\lambda - V = \frac{\mathrm{d}\varphi}{\mathrm{d}t} + \frac{1}{2}\left(\frac{\mathrm{d}\varphi}{\mathrm{d}x}\right)^2 + \frac{1}{2}\left(\frac{\mathrm{d}\varphi}{\mathrm{d}y}\right)^2 + \frac{1}{2}\left(\frac{\mathrm{d}\varphi}{\mathrm{d}z}\right)^2$$

An arbitrary function of t could be added here because this variable is treated in the integration as a constant. But I observe that this function is supposed implicitly contained in the expression for φ. Indeed, by augmenting φ by an arbitrary function T of t, the expressions for p, q, r remain the same as before and the second member of the preceding equation will be augmented by the function $\mathrm{d}T/\mathrm{d}t$ which is arbitrary. Therefore, it is possible to dispense with the addition of an arbitrary function of t without reducing the generality of this equation.

With this equation, one will obtain

$$\lambda = V + \frac{d\varphi}{dt} + \frac{1}{2}\left(\frac{d\varphi}{dx}\right)^2 + \frac{1}{2}\left(\frac{d\varphi}{dy}\right)^2 + \frac{1}{2}\left(\frac{d\varphi}{dz}\right)^2$$

This equation satisfies the three equations denoted (F) of Article 10 simultaneously. The determination of φ will depend upon equation (G) of the same article, in which after substitution of the expressions $d\varphi/dx$, $d\varphi/dy$, $d\varphi/dz$, for p, q, r becomes

$$\frac{d^2\varphi}{dx^2} + \frac{d^2\varphi}{dy^2} + \frac{d^2\varphi}{dz^2} = 0$$

Thus all the remaining difficulty will now lie in the integration of this last equation.

21. There is still a very general case in which the quantity $p\,dx + q\,dy + r\,dz$ must be an exact differential. This is the case in which it is assumed that the velocities p, q, r are very small and in which the very small quantities of the second order are neglected. Since it is clear that in this assumption the same equation (L) will be reduced to

$$d\lambda - dV = \frac{dp}{dt}dx + \frac{dq}{dt}dy + \frac{dr}{dt}dz$$

from which it is seen that $(dp/dt)/dx + (dq/dt)/dy + (dr/dt)/dz$ must be integrable with respect to x, y, z and the quantity $p\,dx + q\,dy + r\,dz$ should also be integrable. Thus the same formulas as in the preceding article will be obtained by assuming φ a very small function and neglecting the second order terms in φ and in its differentials.

Moreover, it is possible in this case to determine the expression for x, y, z for any arbitrary instant because it will only remain to integrate the following equations (Article 9)

$$dx = p\,dt, \qquad dy = q\,dt, \qquad dz = r\,dt$$

in which since p, q, r are very small and consequently, dx, dy, dz are also very small with respect to dt, the variables x, y, z can be considered constants with respect to t. Thus by treating t as the only variable in the functions p, q, r and by adding the constants a, b, c, one will immediately obtain

$$x = a + \int p\,dt, \qquad y = b + \int q\,dt, \qquad z = c + \int r\,dt$$

Then in order to shorten the expressions, make the following definition, $\Phi = \int \varphi\,dt$ and replace the variables x, y, z with a, b, c in the function Φ. Now one simply obtains

$$x = a + \frac{d\Phi}{da}, \qquad y = b + \frac{d\Phi}{db}, \qquad z = r + \frac{d\Phi}{dc}$$

in which the function Φ should be considered equal to zero when t is equal to zero so that a, b, c are the initial values of x, y, z.

This case has application in the theory of waves and of very small oscillations.

22. In general, when the mass of a fluid has one of its dimensions much smaller than the other two, the z coordinate can be considered very small with respect to the x and y coordinates. This consideration will be applied in every case to facilitate the solution of the general equations.

Since it is clear that the unknowns p, q, r and Δ could take the following form

$$\begin{aligned}
p &= p' + p''z + p'''z^2 + \cdots \\
q &= q' + q''z + q'''z^2 + \cdots \\
r &= r' + r''z + r'''z^2 + \cdots \\
\Delta &= \Delta' + \Delta''z + \Delta'''z^2 + \cdots
\end{aligned}$$

in which p', p'', etc., q', q'', etc., r', r'', etc., Δ', Δ'', etc. would be functions of x, y, t without z. Therefore, after making these substitutions, equations in series would be obtained which would only contain partial differences relative to x, y and t.

In order to provide an example on this point, let us again assume that the fluid is homogeneous, in which $\Delta = 1$. Let us substitute at the outset the preceding values in equation (G) of Article 10 and then order the terms with respect to z. One has

$$\begin{aligned}
0 = \frac{\mathrm{d}p'}{\mathrm{d}x} + \frac{\mathrm{d}q'}{\mathrm{d}y} + r'' + z\left(\frac{\mathrm{d}p''}{\mathrm{d}x} + \frac{\mathrm{d}q''}{\mathrm{d}y} + 2r'''\right) \\
+ z^2\left(\frac{\mathrm{d}p'''}{\mathrm{d}x} + \frac{\mathrm{d}q'''}{\mathrm{d}y} + 3r^{\mathrm{IV}}\right) + \cdots
\end{aligned}$$

so that since p', p'', etc., q', q'', etc., should not contain z, the following particular equations are obtained

$$\begin{aligned}
&\frac{\mathrm{d}p'}{\mathrm{d}x} + \frac{\mathrm{d}q'}{\mathrm{d}y} + r'' = 0 \\
&\frac{\mathrm{d}p''}{\mathrm{d}x} + \frac{\mathrm{d}q''}{\mathrm{d}y} + 2r''' = 0 \\
&\frac{\mathrm{d}p'''}{\mathrm{d}x} + \frac{\mathrm{d}q'''}{\mathrm{d}y} + 3r^{\mathrm{IV}} = 0 \\
&\cdots\cdots\cdots\cdots
\end{aligned}$$

from which the quantities r'', r''', r^{IV}, etc. are first determined and the other quantities r', p', p'', etc., q', q'', etc., remain indeterminate.

If the same substitutions are made in equation (L) of Article 15, which is similar to the three equations (F) of Article 10, it is easy to see that it is reduced to the following form

$$\begin{aligned}
\mathrm{d}\lambda - \mathrm{d}V &= \alpha\,\mathrm{d}x + \beta\,\mathrm{d}y + \gamma\,\mathrm{d}z + z(\alpha'\,\mathrm{d}x + \beta'\,\mathrm{d}y + \gamma'\,\mathrm{d}z) \\
&\quad + z^2(\alpha''\,\mathrm{d}x + \beta''\,\mathrm{d}y + \gamma''\,dz) + \cdots
\end{aligned}$$

If the following definitions are made in order to shorten the expressions

$$\alpha = \frac{dp'}{dt} + p'\frac{dp'}{dx} + q'\frac{dp'}{dy} + r'p''$$
$$\beta = \frac{dq'}{dt} + p'\frac{dq'}{dx} + q'\frac{dq'}{dy} + r'q''$$
$$\gamma = \frac{dr'}{dt} + p'\frac{dr'}{dx} + q'\frac{dr'}{dy} + r'r''$$
$$\alpha' = \frac{dp''}{dt} + p'\frac{dp''}{dx} + p''\frac{dp'}{dx} + q'\frac{dp''}{dy} + q''\frac{dp'}{dy} + 2r'p''' + r''p''$$
$$\beta' = \frac{dq''}{dt} + p'\frac{dq''}{dx} + p''\frac{dq'}{dx} + q'\frac{dq''}{dy} + q''\frac{dq'}{dy} + 2r'q''' + r''q''$$
$$\gamma' = \frac{dr''}{dt} + p'\frac{dr''}{dx} + p''\frac{dr'}{dx} + q'\frac{dr''}{dy} + q''\frac{dr'}{dy} + 2r'r''' + r''r''$$

and so on.

Therefore, in order that the second member of this equation is integrable, the following quantities

$$\alpha\, dx + \beta\, dy, \qquad \gamma\, dz + z(\alpha'\, dx + \beta'\, dy), \qquad \gamma' z\, dz + z^2(\alpha''\, dx + \beta''\, dy) \cdots$$

should each be integrable.

If ω represents a function of x, y, t without z, the following conditions will be obtained

$$\alpha = \frac{d\omega}{dx}, \qquad \beta = \frac{d\omega}{dy}, \qquad \alpha' = \frac{d\gamma}{dx},$$
$$\beta' = \frac{d\gamma}{dy}, \qquad \alpha'' = \frac{d\gamma'}{2\, dx}, \qquad \beta'' = \frac{d\gamma'}{2\, dy} \cdots$$

Then the integration of this equation gives

$$\lambda = V + \omega + \gamma z + \tfrac{1}{2}\gamma' z^2 + \cdots$$

and it will only be a question of satisfying the preceding conditions by means of the indeterminate functions ω, r', p', p'', etc., q', q'', etc.

The calculation would be even easier if the two variables y and z were very small simultaneously with respect to x because it could be assumed that

$$p = p' + p''y + p'''z + p^{\text{IV}}y^2 + p^{\text{V}}yz + \cdots$$
$$q = q' + q''y + q'''z + q^{\text{IV}}y^2 + q^{\text{V}}yz + \cdots$$
$$r = r' + r''y + r'''z + r^{\text{IV}}y^2 + r^{\text{V}}yz + \cdots$$

in which the quantities p', p'', etc., q', q'', etc., r', r'', etc., are simple functions of x.

Thus after making these substitutions in equation (G) and equating separately to zero the terms containing y, z and their products, one would obtain

$$\frac{\mathrm{d}p'}{\mathrm{d}x} + q'' + r''' = 0, \qquad \frac{\mathrm{d}p''}{\mathrm{d}x} + 2q^{\mathrm{IV}} + r^{\mathrm{V}} = 0$$
$$\cdots\cdots\cdots$$

and the equation (L) would assume the form

$$\begin{aligned} \mathrm{d}\lambda - \mathrm{d}V &= \alpha\,\mathrm{d}x + \beta\,\mathrm{d}y + \gamma\,\mathrm{d}z + y(\alpha'\,\mathrm{d}x + \beta'\,\mathrm{d}y + \gamma'\,\mathrm{d}z) \\ &\quad + z(\alpha''\,\mathrm{d}x + \beta''\,\mathrm{d}y + \gamma''\,\mathrm{d}z) + \cdots \end{aligned}$$

assuming that

$$\begin{aligned} \alpha &= \frac{\mathrm{d}p'}{\mathrm{d}t} + p'\frac{\mathrm{d}p'}{\mathrm{d}x} + q'p'' + r'p''' \\ \beta &= \frac{\mathrm{d}q'}{\mathrm{d}t} + p'\frac{\mathrm{d}q'}{\mathrm{d}x} + q'q'' + r'q''' \\ \gamma &= \frac{\mathrm{d}r'}{\mathrm{d}t} + p'\frac{\mathrm{d}r'}{\mathrm{d}x} + q'r'' + r'r''' \\ \alpha' &= \frac{\mathrm{d}p''}{\mathrm{d}t} + p'\frac{\mathrm{d}p''}{\mathrm{d}x} + p''\frac{\mathrm{d}p'}{\mathrm{d}x} + 2q'p^{\mathrm{IV}} + q''p'' + r'p^{\mathrm{V}} + r''p''' \\ &\cdots\cdots\cdots \end{aligned}$$

The following equations of condition are available to facilitate the integration of this equation

$$\alpha' = \frac{\mathrm{d}\beta}{\mathrm{d}x}, \qquad \alpha'' = \frac{\mathrm{d}\gamma}{\mathrm{d}x} \quad \cdots$$

with which this equation will give

$$\lambda = V + \int \alpha\,\mathrm{d}x + \beta y + \gamma z + \cdots$$

Finally, the calculation could at times be simplified by means of substitutions by introducing in place of the coordinates x, y, z other variables ξ, η, ζ, which are given functions of the former. If, depending on the problem treated, the variable ζ, for example, or the two variables η and ζ are very small with respect to ξ, simplifications similar to those demonstrated above could be used.

Subsection II
The Motion Of Dense And Homogeneous Fluids in Vessels or Conduits Of Arbitrary Shape

23. In order to demonstrate the use of the principles and formulas presented above, they will be applied to fluids which move in vessels or conduits of given shape.

It will be assumed that the fluid is dense and homogeneous and that its initial state is one of rest or that its motion is begun by an impulse due to a piston applied to its surface. Thus the velocities p, q, r of each particle should be such that the quantity $p\,\mathrm{d}x + q\,\mathrm{d}y + r\,\mathrm{d}z$ is integrable (Article 18). Consequently, the formulas of Article 20 can be used.

Thus let φ be a function[53] of x, y, z and t, determined from the equation

$$\frac{\mathrm{d}^2\varphi}{\mathrm{d}x^2} + \frac{\mathrm{d}^2\varphi}{\mathrm{d}y^2} + \frac{\mathrm{d}^2\varphi}{\mathrm{d}z^2} = 0$$

First the velocities of each particle, in the directions of the coordinates x, y, z are expressed by

$$p = \frac{\mathrm{d}\varphi}{\mathrm{d}x}, \qquad q = \frac{\mathrm{d}\varphi}{\mathrm{d}y}, \qquad r = \frac{\mathrm{d}\varphi}{\mathrm{d}z}$$

Then it follows that

$$\lambda = V + \frac{\mathrm{d}\varphi}{\mathrm{d}t} + \frac{1}{2}\left(\frac{\mathrm{d}\varphi}{\mathrm{d}x}\right)^2 + \frac{1}{2}\left(\frac{\mathrm{d}\varphi}{\mathrm{d}y}\right)^2 + \frac{1}{2}\left(\frac{\mathrm{d}\varphi}{\mathrm{d}z}\right)^2$$

a quantity which should be zero at the outer surface of the fluid (Article 2).

Now consider the velocity V which depends upon the accelerating forces of the fluid (Article 15). If the letter g denotes the accelerating force of gravity and if ξ, η, ζ are the angles that the axes of the x-, y-, z-coordinates make with the vertical drawn from the point of intersection of these axes, directed upward, one will have

$$X = -g\cos\xi, \qquad Y = -g\cos\eta, \qquad Z = -g\cos\zeta$$

where the minus sign is given to the values of the forces X, Y, Z because these forces are oriented in the opposite directions of the coordinate axes x, y, z. Therefore, since

$$\mathrm{d}V = X\,\mathrm{d}x + Y\,\mathrm{d}y + Z\,\mathrm{d}z$$

one will obtain after integration

$$V = -gx\cos\xi - gy\cos\eta - gz\cos\zeta$$

24. Now let $z = \alpha$ or $z - \alpha = 0$ be the equation of one of the walls of the conduit, in which α is a given function of x, y without z or t. In order that the particles of the fluid are always contiguous to this wall, equation (I) of Article 12 must always be satisfied by assuming that $A = z - \alpha$. Thus one has

$$\frac{d\varphi}{dz} - \frac{d\varphi}{dx}\frac{d\alpha}{dx} - \frac{d\varphi}{dy}\frac{d\alpha}{dy} = 0$$

an equation which the expression $z = \alpha$ should satisfy. Each wall will give a similar equation.

Similarly, since $\lambda = 0$ is the equation of the outer surface of the fluid, the following equation should hold so that the same particles are always contiguous to this surface

$$\frac{d\lambda}{dt} + \frac{d\varphi}{dx}\frac{d\lambda}{dx} + \frac{d\varphi}{dy}\frac{d\lambda}{dy} + \frac{d\varphi}{dz}\frac{d\lambda}{dz} = 0$$

This equation should consequently give the same result for z as the equation $\lambda = 0$. But this equation will not be needed when the condition in question ceases to hold.

25. Having made these remarks, let us begin to determine the function φ. But the equation which it depends upon is not integrable in general by known methods. Thus it will be assumed that one of the dimensions of the fluid mass is very small with respect to the other two. For example, the z-coordinate is very small with respect to x and y. By means of this assumption, it is always possible to represent the quantity φ by a series of this form

$$\varphi = \varphi' + z\varphi'' + z^2\varphi''' + z^3\varphi^{\text{IV}} + \cdots$$

in which φ', φ'', φ''', etc. are functions of x, y and t but do not contain the coordinate z.

Thus after making this substitution in the preceding equation, it becomes

$$\frac{d^2\varphi'}{dx^2} + \frac{d^2\varphi'}{dy^2} + 2\varphi''' + z\left(\frac{d^2\varphi''}{dx^2} + \frac{d^2\varphi''}{dy^2} + 2.3\varphi^{\text{IV}}\right)$$
$$+z^2\left(\frac{d^2\varphi'''}{dx^2} + \frac{d^2\varphi'''}{dy^2} + 3.4\varphi^{\text{V}}\right) + \cdots = 0$$

so that by equating separately to zero the expressions with different powers of z, one has

$$\varphi''' = -\frac{d^2\varphi'}{2\,dx^2} - \frac{d^2\varphi'}{2\,dy^2}$$
$$\varphi^{\text{IV}} = -\frac{d^2\varphi''}{2.3\,dx^2} - \frac{d^2\varphi''}{2.3\,dy^2}$$
$$\varphi^{\text{V}} = -\frac{d^2\varphi'''}{3.4\,dx^2} - \frac{d^2\varphi'''}{3.4\,dy^2} = \frac{d^4\varphi'}{2.3.4\,dx^4} + \frac{d^4\varphi'}{3.4\,dx^2\,dy^2} + \frac{d^4\varphi'}{2.3.4\,dy^4}$$

. . . .

Thus the expression for φ becomes

$$\varphi = \varphi' + z\varphi'' - \frac{z^2}{2}\left(\frac{d^2\varphi'}{dx^2} + \frac{d^2\varphi'}{dy^2}\right) - \frac{z^3}{2.3}\left(\frac{d^2\varphi''}{dx^2} + \frac{d^2\varphi''}{dy^2}\right)$$
$$+\frac{z^4}{2.3.4}\left(\frac{d^4\varphi'}{dx^4} + \frac{2d^4\varphi'}{dx^2\,dy^2} + \frac{d^4\varphi'}{dy^4}\right) + \cdots$$

in which the functions φ' and φ'' are indeterminate. This result shows that this expression is the complete integral of the proposed equation.

Having found the expression for φ, those for p, q, r will be found by differentiation as follows

$$p = \frac{d\varphi}{dx} = \frac{d\varphi'}{dx} + z\frac{d\varphi''}{dx} - \frac{z^2}{2}\left(\frac{d^3\varphi'}{dx^3} + \frac{d^3\varphi'}{dx\,dy^2}\right) - \frac{z^3}{2.3}\left(\frac{d^3\varphi''}{dx^3} + \frac{d^3\varphi''}{dx\,dy^2}\right) + \cdots$$
$$q = \frac{d\varphi}{dy} = \frac{d\varphi'}{dy} + z\frac{d\varphi''}{dy} - \frac{z^2}{2}\left(\frac{d^3\varphi'}{dx^2\,dy} + \frac{d^3\varphi'}{dy^3}\right) - \frac{z^3}{2.3}\left(\frac{d^3\varphi''}{dx^2\,dy} + \frac{d^3\varphi''}{dy^3}\right) + \cdots$$
$$r = \frac{d\varphi}{dz} = \varphi'' - z\left(\frac{d^2\varphi'}{dx^2} + \frac{d^2\varphi'}{dy^2}\right) - \frac{z^2}{2}\left(\frac{d^2\varphi''}{dx^2} + \frac{d^2\varphi''}{dy^2}\right)$$
$$+\frac{z^3}{2.3}\left(\frac{d^4\varphi'}{dx^4} + \frac{2\,d^4\varphi'}{dx^2\,dy^2} + \frac{d^4\varphi'}{dy^4}\right) + \cdots$$

After substituting these values in the expression for λ of Article 25, it will have the following form

$$\lambda = \lambda' + z\lambda'' + z^2\lambda''' + z^3\lambda^{\text{IV}} + \cdots$$

in which

$$\lambda' = -g(x\cos\xi + y\cos\eta) + \frac{d\varphi'}{dt} + \frac{1}{2}\left(\frac{d\varphi'}{dx}\right)^2 + \frac{1}{2}\left(\frac{d\varphi'}{dy}\right)^2 + \frac{1}{2}\varphi''^2$$
$$\lambda'' = -g\cos\zeta + \frac{d\varphi''}{dt} + \frac{d\varphi'}{dx}\frac{d\varphi''}{dx} + \frac{d\varphi'}{dy}\frac{d\varphi''}{dy} - \varphi''\left(\frac{d^2\varphi'}{dx^2} + \frac{d^2\varphi'}{dy^2}\right)$$
$$\lambda''' = -\frac{1}{2}\left(\frac{d^3\varphi'}{dt\,dx^2} + \frac{d^3\varphi'}{dt\,dy^2}\right) + \frac{1}{2}\left(\frac{d\varphi''}{dx}\right)^2 - \frac{1}{2}\frac{d\varphi'}{dx}\left(\frac{d^3\varphi'}{dx^3} + \frac{d^3\varphi'}{dx\,dy^2}\right) + \frac{1}{2}\left(\frac{d\varphi''}{dy}\right)^2$$
$$-\frac{1}{2}\frac{d\varphi'}{dy}\left(\frac{d^3\varphi'}{dx^2\,dy} + \frac{d^3\varphi'}{dy^3}\right) + \frac{1}{2}\left(\frac{d^2\varphi'}{dx^2} + \frac{d^2\varphi'}{dy^2}\right)^2 - \frac{1}{2}\varphi''\left(\frac{d^2\varphi''}{dx^2} + \frac{d^2\varphi''}{dy^2}\right)$$

and similarly for the others.

26. Now if $z = \alpha$ is the equation of the wall in which α is a very small function of x and y without z, the equation of condition for the same particles to be always contiguous to these walls (Article 24) becomes by the preceding substitutions

$$0 = \varphi'' - \frac{d\varphi'}{dx}\frac{d\alpha}{dx} - \frac{d\varphi'}{dy}\frac{d\alpha}{dy} - z\left(\frac{d^2\varphi'}{dx^2} + \frac{d^2\varphi'}{dy^2} + \frac{d\varphi''}{dx}\frac{d\alpha}{dx} + \frac{d\varphi''}{dy}\frac{d\alpha}{dy}\right)$$

$$
-\frac{1}{2}z^2\left[\frac{d^2\varphi''}{dx^2}+\frac{d^2\varphi''}{dy^2}-\left(\frac{d^3\varphi}{dx^3}+\frac{d^3\varphi'}{dx\,dy^2}\right)\frac{d\alpha}{dx}-\left(\frac{d^3\varphi'}{dx^2\,dy}+\frac{d^3\varphi'}{dy^3}\right)\frac{d\alpha}{dy}\right]
$$
$$+\cdots\cdots\cdots$$

which will hold when z is put equal to α, and which in this case, is reduced to this simple form

$$
\varphi''-\frac{d\left(\alpha\frac{d\varphi'}{dx}\right)}{dx}-\frac{d\left(\alpha\frac{d\varphi'}{dy}\right)}{dy}-\frac{d\left(\alpha^2\frac{d\varphi''}{dx}\right)}{2\,dx}-\frac{d\left(\alpha^2\frac{d\varphi''}{dy}\right)}{2\,dy}
$$
$$
+\frac{d\left(\alpha^3\left(\frac{d^3\varphi'}{dx^3}+\frac{d^3\varphi'}{dx\,dy^2}\right)\right)}{2.3\,dx}+\frac{d\left(\alpha^3\left(\frac{d^3\varphi'}{dx^2\,dy}+\frac{d^3\varphi'}{dy^3}\right)\right)}{2.3\,dy}
$$
$$+\cdots=0$$

and this equation should be valid for all the given walls.

27. Finally, the equation for the outer free surface of the fluid which is $\lambda=0$, is of the form

$$\lambda'+z\lambda''+z^2\lambda'''+z^3\lambda^{\text{IV}}+\cdots=0$$

The equation of condition, requiring that the same particles remain contiguous to the surface (Article 24), will be

$$
\frac{d\lambda'}{dt}+\frac{d\varphi'}{dx}\frac{d\lambda'}{dx}+\frac{d\varphi'}{dy}\frac{d\lambda'}{dy}+\varphi''\lambda''
$$
$$
+z\left\{\begin{array}{l}\frac{d\lambda''}{dt}+\frac{d\varphi''}{dx}\frac{d\lambda'}{dx}+\frac{d\varphi'}{dx}\frac{d\lambda''}{dx}+\frac{d\varphi''}{dy}\frac{d\lambda'}{dy}+\frac{d\varphi'}{dy}\frac{d\lambda''}{dy}\\ +2\varphi''\lambda'''-\left(\frac{d^2\varphi'}{dx^2}+\frac{d^2\varphi'}{dy^2}\right)\lambda''\end{array}\right\}
$$
$$
+z^2\left\{\begin{array}{l}\frac{d\lambda'''}{dt}+\frac{d\varphi''}{dx}\frac{d\lambda''}{dx}+\frac{d\varphi'}{dx}\frac{d\lambda'''}{dx}+\frac{d\varphi''}{dy}\frac{d\lambda''}{dy}+\frac{d\varphi'}{dy}\frac{d\lambda'''}{dy}\\ -\frac{1}{2}\left(\frac{d^3\varphi'}{dx^3}+\frac{d^3\varphi'}{dx\,dy^2}\right)\frac{d\lambda'}{dx}-\frac{1}{2}\left(\frac{d^3\varphi'}{dx^2\,dy}+\frac{d^3\varphi'}{dy^3}\right)\frac{d\lambda'}{dy}\\ -2\left(\frac{d^2\varphi'}{dx^2}+\frac{d^2\varphi'}{dy^2}\right)\lambda'''-\frac{1}{2}\left(\frac{d^2\varphi''}{dx^2}+\frac{d^2\varphi''}{dy^2}\right)\lambda''+3\varphi''\lambda^{\text{IV}}\end{array}\right\}
$$
$$+\cdots=0$$

If z is eliminated from these two equations, an equation is obtained which should hold for all points of the exterior surface.

Subsection III
Application Of These Formulas To The Motion Of A Fluid Which Is Flowing In A Narrow And Nearly Vertical Vessel

28. Let us now imagine that the fluid flows in a narrow and almost vertical vessel. Also, let us assume for more simplicity, that the abscissas x are vertical and oriented upward. Then there results (Article 23)

$$\xi = 0, \qquad \eta = 90^\circ, \qquad \zeta = 90^\circ$$

thus

$$\cos\xi = 1, \qquad \cos\eta = 0, \qquad \cos\zeta = 0$$

Let us also assume in order to simplify the problem as much as possible that the vessel is flat so that the y-coordinate is zero and the z-coordinate is very small.

Finally, let $z = \alpha$ and $z = \beta$ be the equations for the two walls of the vessel where α and β are known functions of x and very small. Thus one will have, relative to these walls, the two equations (Article 26)

$$\varphi'' - \frac{\mathrm{d}\left(\alpha\dfrac{\mathrm{d}\varphi'}{\mathrm{d}x}\right)}{\mathrm{d}x} - \frac{\mathrm{d}\left(\alpha^2\dfrac{\mathrm{d}\varphi''}{\mathrm{d}x}\right)}{2\,\mathrm{d}x} + \cdots = 0$$

$$\varphi'' - \frac{\mathrm{d}\left(\beta\dfrac{\mathrm{d}\varphi'}{\mathrm{d}x}\right)}{\mathrm{d}x} - \frac{\mathrm{d}\left(\beta^2\dfrac{\mathrm{d}\varphi''}{\mathrm{d}x}\right)}{2\,\mathrm{d}x} + \cdots = 0$$

which will be used to determine the functions φ' and φ''.

The quantities z, α, β will be considered to be very small of the first order and at least for a first approximation, the quantities of the second order and higher will be neglected. Thus the two preceding equations will be reduced to the following

$$\varphi'' - \frac{\mathrm{d}\left(\alpha\dfrac{\mathrm{d}\varphi'}{\mathrm{d}x}\right)}{\mathrm{d}x} = 0, \qquad \varphi'' - \frac{\mathrm{d}\left(\beta\dfrac{\mathrm{d}\varphi'}{\mathrm{d}x}\right)}{\mathrm{d}x} = 0$$

When the two equations are subtracted from one another, there results

$$\frac{\mathrm{d}\left((\alpha-\beta)\dfrac{\mathrm{d}\varphi'}{\mathrm{d}x}\right)}{\mathrm{d}x} = 0$$

an equation for which the integral is $(\alpha - \beta)(\mathrm{d}\varphi'/\mathrm{d}x) = \theta$, where θ is an arbitrary function of t, which should be very small and of the first order.

But it is clear that $\alpha - \beta$ is the horizontal dimension of the vessel which we will represent by γ. Therefore, we will have

$$\frac{\mathrm{d}\varphi'}{\mathrm{d}x} = \frac{\theta}{\gamma}$$

and integrating again with respect to x

$$\varphi' = \theta \int \frac{\mathrm{d}x}{\gamma} + \zeta^*$$

where ζ^* is a new arbitrary function of t.

If the same equations were added together and if one puts $(\alpha + \beta)/2 = \mu$, there will result

$$\varphi'' = \frac{\mathrm{d}\left(\mu \dfrac{\mathrm{d}\varphi'}{\mathrm{d}x}\right)}{\mathrm{d}x}$$

or, after substitution of the expression for $\mathrm{d}\varphi'/\mathrm{d}x$, the following equation is obtained

$$\varphi'' = \theta \frac{\mathrm{d}\left(\dfrac{\mu}{\gamma}\right)}{\mathrm{d}x}$$

from which it is seen that since γ, μ, θ are very small quantities of the first order, φ'' will also be very small of the same order.

Thus by always neglecting the quantities of the second order, one will have from the formulas of Article 25, the vertical velocity

$$p = \frac{\mathrm{d}\varphi'}{\mathrm{d}x} = \frac{\theta}{\gamma}$$

and the horizontal velocity

$$r = \varphi'' - z\frac{\mathrm{d}^2\varphi'}{\mathrm{d}x^2} = \theta\left(\frac{\mathrm{d}\left(\dfrac{\mu}{\gamma}\right)}{\mathrm{d}x} - z\frac{\mathrm{d}\left(\dfrac{1}{\gamma}\right)}{\mathrm{d}x}\right) = \frac{\theta}{\gamma}\left[\frac{\mathrm{d}\mu}{\mathrm{d}x} + (z - \mu)\frac{\mathrm{d}\gamma}{\gamma\,\mathrm{d}x}\right]$$

Since $\cos\zeta = 0$, the quantity λ'' will also be very small and of the first order. Consequently, the expression for λ will be reduced to (Article 25)

$$\lambda' = -gx + \frac{\mathrm{d}\theta}{\mathrm{d}t}\int \frac{\mathrm{d}x}{\gamma} + \frac{\mathrm{d}\zeta^*}{\mathrm{d}t} + \frac{\theta^2}{2\gamma^2}$$

Since this expression is equal to zero, it will give the shape of the surface of the fluid. Since it does not contain the z coordinate but only the abscissa x and the time t, it follows that the surface of the fluid should always be plane and horizontal.

Finally, the equation of condition which requires that the same particles are always contiguous to the surface will be reduced for the same reason to $\mathrm{d}\lambda'/\mathrm{d}t + (\mathrm{d}\varphi'/\mathrm{d}z)(\mathrm{d}\lambda'/\mathrm{d}x)$ (Article 27) knowing that $\mathrm{d}\lambda/\mathrm{d}t + (\theta/\gamma)(\mathrm{d}\lambda/\mathrm{d}x) = 0$ which also does not contain z but only x and t.

29. In order to distinguish the quantities related to the upper surface of the fluid from those related to the lower surface, the former will be denoted with one prime and the latter with two primes. Thus x', γ', etc. are the abscissa and the width of the vessel, etc., for the upper surface and x'', γ'', etc. are the same abscissa and width of the vessel, etc., for the lower surface.

Thus λ', λ'' will represent in what follows the values of λ for the two surfaces so that for the upper surface, one will have

$$\lambda' = -gx' + \frac{\mathrm{d}\theta}{\mathrm{d}t}\int\frac{\mathrm{d}x'}{\gamma'} + \frac{\mathrm{d}\zeta^*}{\mathrm{d}t} + \frac{\theta^2}{2\gamma'} = 0$$

and for the lower surface a similar equation

$$\lambda'' = -gx'' + \frac{\mathrm{d}\theta}{\mathrm{d}t}\int\frac{\mathrm{d}x''}{\gamma''} + \frac{\mathrm{d}\zeta^*}{\mathrm{d}t} + \frac{\theta^2}{2\gamma''} = 0$$

Finally, the expression $\mathrm{d}\lambda'/\mathrm{d}t + (\theta/\gamma')(\mathrm{d}\lambda'/\mathrm{d}x') = 0$ will be the equation of condition such that the particles which are at the surface always remain there and the expression $\mathrm{d}\lambda''/\mathrm{d}t + (\theta/\gamma'')(\mathrm{d}\lambda''/\mathrm{d}x'') = 0$ will be the equation of condition for the lower surface which also requires that this surface be composed of invariable particles of the fluid.

After stating these facts, one must distinguish between the four cases which describe how a fluid can flow in a vessel and each of these cases requires a particular solution.

30. The first case describes a given quantity of fluid flowing in an infinite vessel. In this case, it is obvious that both surfaces must always contain the same particles and thus the equations for the two surfaces are $\lambda' = 0$, $\lambda'' = 0$. Moreover

$$\frac{\mathrm{d}\lambda'}{\mathrm{d}t} + \frac{\theta}{\gamma'}\frac{\mathrm{d}\lambda'}{\mathrm{d}x'} = 0, \qquad \frac{\mathrm{d}\lambda''}{\mathrm{d}t} + \frac{\theta}{\gamma''}\frac{\mathrm{d}\lambda''}{\mathrm{d}x''} = 0$$

These four equations will be used to determine the variables x', x'', θ, ζ^* as functions of t.

The equation $\lambda' = 0$ gives after differentiation

$$\frac{\mathrm{d}\lambda'}{\mathrm{d}x'}\mathrm{d}x' + \frac{\mathrm{d}\lambda'}{\mathrm{d}t}\mathrm{d}t = 0$$

and thus

$$\frac{\mathrm{d}\lambda'}{\mathrm{d}t} = -\frac{\mathrm{d}\lambda'}{\mathrm{d}x'}\frac{\mathrm{d}x'}{\mathrm{d}t}$$

After substituting this expression in the following equation

$$\frac{\mathrm{d}\lambda'}{\mathrm{d}t} + \frac{\theta}{\gamma'}\frac{\mathrm{d}\lambda'}{\mathrm{d}x'} = 0$$

and after dividing by $\mathrm{d}\lambda'/\mathrm{d}x'$, one will obtain $\mathrm{d}x'/\mathrm{d}t = \theta/\gamma'$.

Similarly, after combining the equation $\lambda'' = 0$ with the equation $\mathrm{d}\lambda''/\mathrm{d}t = -(\mathrm{d}\lambda''/\mathrm{d}x'')$ $(\mathrm{d}x''/\mathrm{d}t)$, the following equation will be obtained

$$\frac{\mathrm{d}x''}{\mathrm{d}t} = \frac{\theta}{\gamma''}$$

Therefore, we have that

$$\theta\,\mathrm{d}t = \gamma'\,\mathrm{d}x' = \gamma''\,\mathrm{d}x''$$

The terms of these equations are independent and consequently, one has after integration

$$\int \gamma''\,\mathrm{d}x'' - \int \gamma'\,\mathrm{d}x' = m$$

in which m is a constant, which obviously expresses the given quantity of the fluid which flows in the vessel. This equation will thus give the expression for x'' as a function of x'.

Now, if in the equation $\lambda' = 0$ the quantity $\lambda'\,\mathrm{d}x'/\theta$ is substituted for $\mathrm{d}t$, it becomes

$$-gx' + \frac{\theta\,\mathrm{d}\theta}{\gamma'\,\mathrm{d}x'}\int\frac{\mathrm{d}x'}{\gamma'} + \frac{\theta\,\mathrm{d}\zeta^*}{\gamma'\,\mathrm{d}x'} + \frac{\theta^2}{2\gamma'^2} = 0$$

which, after multiplication by $-\gamma'\,\mathrm{d}x'$ gives the following equation

$$g\gamma' x'\,\mathrm{d}x' - \theta\,\mathrm{d}\theta\int\frac{\mathrm{d}x'}{\gamma'} - \theta\,\mathrm{d}\zeta^* - \frac{\theta^2\,\mathrm{d}x'}{2\gamma'} = 0$$

which is integrable and for which the integral is

$$g\int\gamma'\,x'\,\mathrm{d}x' - \frac{\theta^2}{2}\int\frac{\mathrm{d}x'}{\gamma'} - \int\theta\,\mathrm{d}\zeta^* = \text{const.}$$

In the same manner, after substituting $\gamma'' \,\mathrm{d}x''/\theta$ for $\mathrm{d}t$ in the equation $\lambda'' = 0$ and multiplying by $-\gamma'' \,\mathrm{d}x''$, a new integrable equation will be found for which the integral is

$$g\int \gamma'' \,\mathrm{d}x'' - \frac{\theta^2}{2}\int \frac{\mathrm{d}x''}{\gamma''} - \int \theta \,\mathrm{d}\zeta^* = \text{const.}$$

After subtracting the two equations from one another in order to eliminate the term $\int \theta \,\mathrm{d}\zeta^*$, the following equation is obtained

$$g(\int \gamma'' x'' \,\mathrm{d}x'' - \int \gamma' x' \,\mathrm{d}x') - \frac{\theta^2}{2}\left(\int \frac{\mathrm{d}x''}{\gamma''} - \int \frac{\mathrm{d}x'}{\gamma'}\right) = L$$

in which the quantities $\int \gamma'' x'' \,\mathrm{d}x'' - \int \gamma' x' \,\mathrm{d}x'$ and $\int \mathrm{d}x''/\gamma'' - \int \mathrm{d}x'/\gamma'$ are the integrals of $\gamma x \,\mathrm{d}x$ and $\mathrm{d}x/\gamma$, where the limits are from $x = x'$ to $x = x''$ and where L is a constant.

This equation gives θ as a function of x' since x'' is already known as a function of x' from the equation found above. Thus having θ as a function of x', the variable t will also be found as a function of x' from the equation $\mathrm{d}t = \gamma' \,\mathrm{d}x'/\theta$, for which the integral is $t = \int \gamma' \,\mathrm{d}x'/\theta + H$, where H is an arbitrary constant.

The two constants L and H, will be determined from the initial state of the fluid. Indeed, when $t = 0$, the value of x' is given by the initial state of the fluid in the vessel and if it is assumed that the initial velocities of the fluid are equal to zero, then θ should be equal to zero when t is equal to zero so that the expressions for p, q, r (Article 25) are equal to zero. But if the fluid were first put into motion by arbitrary impulses then the values of λ' and λ'' would have to be given when $t = 0$ because the quantity λ, calculated at the surface of the fluid, expresses the pressure imparted to the fluid which must be counterbalanced by the external pressure (Article 2). And from Article 29, we have the equation

$$\lambda'' - \lambda' = -g(x'' - x') + \frac{\mathrm{d}\theta}{\mathrm{d}t}\left(\int \frac{\mathrm{d}x''}{\gamma''} - \int \frac{\mathrm{d}x'}{\gamma'}\right) - \frac{\theta^2}{2}\left(\frac{1}{\gamma''^2} - \frac{1}{\gamma'^2}\right)$$

Thus by putting $t = 0$, an equation will be obtained with which to determine the initial value of θ.

Consequently, the problem is solved and the motion of the fluid is completely determined.

31. In the second case, the vessel is of finite length and the fluid is flowing from an orifice in its bottom. In this case, as in the preceding case, the two following equations are obtained for the upper surface

$$\lambda' = 0, \qquad \frac{\mathrm{d}\lambda'}{\mathrm{d}t} + \frac{\theta}{\gamma'}\frac{\mathrm{d}\lambda'}{\mathrm{d}x'} = 0$$

But for the lower surface, one will simply have the equation $\lambda'' = 0$. This is a result of the fact that since the fluid is flowing out of the vessel, there must be at each instant different particles at this surface. On the other hand, the abscissa x'' for this same surface is given and constant so that only three unknowns will have to be determined, x', θ and ζ^*.

The first two equations give at the outset, as in the preceding case

$$\mathrm{d}t = \frac{\gamma'\,\mathrm{d}x'}{\theta}, \qquad g\gamma' x'\,\mathrm{d}x' - \theta\,\mathrm{d}\theta \int \frac{\mathrm{d}x'}{\gamma'} - \theta\,\mathrm{d}\zeta^* - \frac{\theta^2\,\mathrm{d}x'}{2\gamma'} = 0$$

and then the equation $\lambda'' = 0$ gives

$$-gx'' + \frac{\mathrm{d}\theta}{\mathrm{d}t}\int \frac{\mathrm{d}x''}{\gamma''} + \frac{\mathrm{d}\zeta^*}{\mathrm{d}t} + \frac{\theta^2}{2\gamma''^2} = 0$$

in which x'', γ'' and $\int \mathrm{d}x''/\gamma''$ are constants which are denoted, for greater simplicity, by f, h and n. Thus, by substituting for $\mathrm{d}t$ the quantity $\gamma'\,\mathrm{d}x'/\theta$, and then multiplying by $-\gamma'\,\mathrm{d}x'$, the following equation is obtained

$$gf\gamma'\,\mathrm{d}x' - n\theta\,\mathrm{d}\theta - \theta\,\mathrm{d}\zeta^* - \frac{\theta^2\,\mathrm{d}x'}{2h} = 0$$

After subtracting this equation from the preceding equation to eliminate the term $\theta\,\mathrm{d}\zeta^*$, one has

$$g(f - x')\gamma'\,\mathrm{d}x' - \left(n - \int \frac{\mathrm{d}x'}{\gamma'}\right)\theta\,\mathrm{d}\theta - \left(\frac{1}{2h} - \frac{1}{2\gamma}\right)\theta^2\,\mathrm{d}x' = 0$$

an equation which only contains two variables x' and θ. From this equation, one of these variables can be determined as a function of the other.

Then the quantity t will be expressed by the same variable after integrating the following equation

$$\mathrm{d}t = \frac{\gamma'\,\mathrm{d}x'}{\theta}$$

The constant will be determined from the initial state of the fluid, as in the preceding problem.

32. In the third case, the fluid is flowing in a vessel of infinite length, but with the fluid level constant which continuously requires the addition of fluid. This case is the converse of the preceding one because now the lower surface is expressed by the two equations

$$\lambda'' = 0, \qquad \frac{\mathrm{d}\lambda''}{\mathrm{d}t} + \frac{\theta}{\gamma''}\frac{\mathrm{d}\lambda''}{\mathrm{d}x''} = 0$$

and the upper surface is simply defined by $\lambda' = 0$ because of the continual replacement of particles at this surface. Thus it will only be necessary to change the quantities x', γ' to

x'', γ'' in the equations of the preceding article and to take for f, h, n, the values of x', γ', $\int \mathrm{d}x'/\gamma'$.

Also, it is assumed that when fluid is added to the vessel each layer takes on the velocity of the next layer. Thus the increase or decrease in velocity of this layer, during the first instant, is the same as if the vessel were not filled to the same level during this instant.

33. Finally, in the fourth case, the fluid flows from a vessel of finite length in which the fluid level is constant. In this case, the particles of the upper surface are constantly replaced. Consequently, one will simply have for these two surfaces the following equations

$$\lambda' = 0, \qquad \lambda'' = 0$$

but at the same time the two abscissas x' and x'' are given and constant so that only two unknowns θ and Δ have to be determined as functions of t.

Now let

$$x' = f, \qquad \gamma' = h, \qquad \int \frac{\mathrm{d}x'}{\gamma'} = n, \qquad x'' = F, \qquad \gamma'' = H, \qquad \int \frac{\mathrm{d}x''}{\gamma''} = N$$

The two equations $\lambda' = 0$ and $\lambda'' = 0$ then become

$$-gf + \frac{\mathrm{d}\theta}{\mathrm{d}t} n + \frac{\mathrm{d}\zeta^*}{\mathrm{d}t} + \frac{\theta^2}{2h^2} = 0$$
$$-gF + \frac{\mathrm{d}\theta}{\mathrm{d}t} N + \frac{\mathrm{d}\zeta^*}{\mathrm{d}t} + \frac{\theta^2}{2H^2} = 0$$

from which, after eliminating $\mathrm{d}\zeta^*/\mathrm{d}t$, one will obtain

$$g(F - f) - (N - n)\frac{\mathrm{d}\theta}{\mathrm{d}t} - \left(\frac{1}{2H^2} - \frac{1}{2h^2}\right)\theta^2 = 0$$

from which after solving for t, the following equation is obtained

$$\mathrm{d}t = \frac{(N - n)\mathrm{d}\theta}{g(F - f) - \left(\frac{1}{2H^2} - \frac{2}{2h^2}\right)\theta^2}$$

This is an independent equation which is integrable using sines and cosines or logarithms.

34. The preceding solutions are consistent with those found by the original discoverers of the theories of motion of fluids in which it was assumed that the different layers of fluid kept exactly their parallelism during their descent in a vessel (see the *Hydrodynamique* of Daniel Bernoulli, the *Hydraulique*[54] of John Bernoulli and the *Traité des fluides* of

d'Alembert). Our analysis shows that this assumption is only valid when the width of the vessel is infinitesimal, but it can always be used as a first approximation in every problem and the solutions are exact to the second order when the dimensions of the vessel are of the first order.

But the great advantage of this analysis is that it is possible by this means to gradually obtain the true motion of fluids in vessels of arbitrary shape. Indeed, having found, as we just did, the first-order estimates of the unknowns and neglecting the second orders of the widths of the vessel, it is easy to increase the accuracy by considering the terms neglected. This detail presents a difficulty only in the length of the calculations and we will not treat it any further.

Subsection IV
Application Of The Same Formulas To The Motion Of A Fluid Contained In A Shallow And Nearly Horizontal Conduit And In Particular, To The Propagation Of Waves

35. Since it is assumed that the depth of the fluid is very small, the z coordinates should be taken vertical and oriented upward, the x abscissa and the other coordinate y should then be horizontal and one will have (Article 23)

$$\cos\xi = 0, \qquad \cos\eta = 0, \qquad \cos\zeta = 1$$

By taking the x- and y-axes in the horizontal plane formed by the upper surface of the fluid in the state of equilibrium, the equation $z = \alpha$, where α is a function of x and y, is the equation of the bottom of the conduit.

The quantities z and α will be treated as very small and of the first order and we will neglect quantities of the second order and higher, that is, those which contain squares and products of z and α.

The equation of condition relative to the bottom of the conduit gives (Article 26)

$$\varphi'' = \frac{\mathrm{d}\left(\alpha \dfrac{\mathrm{d}\varphi'}{\mathrm{d}x}\right)}{\mathrm{d}x} + \frac{\mathrm{d}\left(\alpha \dfrac{\mathrm{d}\varphi'}{\mathrm{d}y}\right)}{\mathrm{d}y}$$

from which it is seen that φ'' is a quantity of the first order.

Then the expression for the quantity λ is reduced to $\lambda' + \lambda'' z$ (Article 25) and the quantities of second order should be neglected in the expression for λ' and the quantities of the first order in the expression for λ''. Thus, because

$$\cos\xi = 0, \qquad \cos\eta = 0, \qquad \cos\zeta = 1$$

one will obtain from the formulas of the same article

$$\lambda' = \frac{\mathrm{d}\varphi'}{\mathrm{d}t} + \frac{1}{2}\left(\frac{\mathrm{d}\varphi'}{\mathrm{d}x}\right)^2 + \frac{1}{2}\left(\frac{\mathrm{d}\varphi'}{\mathrm{d}y}\right)^2, \qquad \lambda'' = -g$$

Thus the equation representing the upper surface of the fluid is obtained (Article 27) as

$$\lambda' - gz = 0$$

The equation of condition is

$$\frac{\mathrm{d}\lambda'}{\mathrm{d}t} + \frac{\mathrm{d}\varphi'}{\mathrm{d}x}\frac{\mathrm{d}\lambda'}{\mathrm{d}x} + \frac{\mathrm{d}\varphi'}{\mathrm{d}y}\frac{\mathrm{d}\lambda'}{\mathrm{d}y} - g\varphi'' + gz\left(\frac{\mathrm{d}^2\varphi'}{\mathrm{d}x^2} + \frac{\mathrm{d}^2\varphi'}{\mathrm{d}y^2}\right) = 0$$

The equation $\lambda' - gz = 0$ immediately gives $z = \lambda'/g$ for the shape of the upper surface of the fluid at each instant. Since the equation of condition should also hold relative to the same surface, it should be valid when substituting the quantity λ'/g for z. Using this result, this equation will become

$$\frac{\mathrm{d}\lambda'}{\mathrm{d}t} + \frac{\mathrm{d}\left(\lambda'\frac{\mathrm{d}\varphi'}{\mathrm{d}x}\right)}{\mathrm{d}x} + \frac{\mathrm{d}\left(\lambda'\frac{\mathrm{d}\varphi'}{\mathrm{d}y}\right)}{\mathrm{d}y} - g\varphi'' = 0$$

If the expression given earlier for φ'' is substituted into this equation, the preceding equation will be reduced to the following equation

$$\frac{\mathrm{d}\lambda'}{\mathrm{d}t} + \frac{\mathrm{d}(\lambda' - g\alpha)\frac{\mathrm{d}\varphi'}{\mathrm{d}x}}{\mathrm{d}x} + \frac{\mathrm{d}(\lambda' - g\alpha)\frac{\mathrm{d}\varphi'}{\mathrm{d}y}}{\mathrm{d}y} = 0$$

in which it will only remain to replace λ' by the expression

$$\frac{\mathrm{d}\varphi'}{\mathrm{d}t} + \frac{1}{2}\left(\frac{\mathrm{d}\varphi'}{\mathrm{d}x}\right)^2 + \frac{1}{2}\left(\frac{\mathrm{d}\varphi'}{\mathrm{d}y}\right)^2$$

in order to obtain an equation in partial differences of second order, which will be used to determine φ' as a function of x, y and t.

After this result is obtained, the shape of the upper surface is known from the equation

$$z = \frac{\mathrm{d}\varphi'}{g\,\mathrm{d}t} + \frac{1}{2g}\left(\frac{\mathrm{d}\varphi'}{\mathrm{d}x}\right)^2 + \frac{1}{2g}\left(\frac{\mathrm{d}\varphi'}{\mathrm{d}y}\right)^2$$

If it were required to know the horizontal velocities p and q of each particle, they would be found from the formulas (Article 25)

$$p = \frac{d\varphi'}{dx}, \qquad q = \frac{d\varphi'}{dy}$$

36. The integral calculus of partial differences has not been developed to a point where so complicated an equation can be easily integrated. Our only recourse is to simplify this equation using some limiting conditions.

At the outset, it will be assumed that the fluid in its motion neither falls below or ascends above a given level by more than an infinitesimal amount so that the z-coordinates for the upper surface are always very small. In addition, the horizontal velocities p and q are assumed to be infinitesimal.

Therefore, the quantities $d\varphi'/dt, d\varphi'/dx, d\varphi'/dy$ should be infinitesimal and consequently, the quantity φ' should also be infinitesimal.

Thus by neglecting in the equation proposed here, the infinitesimal quantities of second and higher order reduces it to this linear form

$$\frac{d^2\varphi'}{dt^2} - g\frac{d\left(\alpha\dfrac{d\varphi'}{dx}\right)}{dx} - g\frac{d\left(\alpha\dfrac{d\varphi'}{dy}\right)}{dy} = 0$$

and thus

$$z = \frac{d\varphi'}{gdt}, \qquad p = \frac{d\varphi'}{dx}, \qquad q = \frac{d\varphi'}{dy}$$

Thus this equation expresses the general theory of small perturbations of a shallow fluid and as a consequence, the true theory of waves created by successive infinitesimal ascents and descents of still water contained in a shallow basin or conduit. The theory of waves given by Newton in the forty-sixth proposition of the Second Book, is based on the precarious and unnatural assumption that the vertical oscillations of waves are analogous to those of water [moving] in a [sinusoidally] curved pipe. This assertion must be viewed as absolutely incapable of explaining this phenomenon.

37. If the basin or conduit is assumed to have a horizontal bottom then the quantity α is constant and equal to the depth of water and the equation for the motion of the waves becomes

$$\frac{d^2\varphi'}{dt^2} = g\alpha\left(\frac{d^2\varphi'}{dx^2} + \frac{d^2\varphi'}{dy^2}\right)$$

This equation is entirely similar to the one which determines the small oscillations of air in the creation of sound, considering only the motion of particles parallel to the horizon, as will be seen in Article 9 of the following section. The elevations z, above the level of water, correspond to the condensation of air and the depth α of water in the conduit corresponds to the height of the atmosphere, assumed to be homogeneous, which creates a perfect analogy between the waves created at the surface of still water from successive ascents and descents and the waves created in air from successive condensations and rarefactions of the air, an analogy that several authors had already assumed but which no one, until now, had rigorously demonstrated.

Since the velocity of the propagation of sound is equal to the velocity a heavy body would have by falling through half of the height of the atmosphere, assumed homogeneous, the velocity of wave propagation will similarly be the same as the velocity a heavy body would have falling from a height equal to half the depth of water in the conduit. Consequently, if this depth were one foot, the velocity of the waves would be 5,495 feet per second.[55] If the depth of the water is rather large, the velocity of the waves would vary as the square root of the depth, if the waves are not too large.

Also, whatever the depth of the water[56] and shape of its bottom, the preceding theory could always be applied if it is assumed that the water is perturbed only to a very shallow depth during the formation of waves. This assumption is very plausible because of the mutual interactions between the particles of water. In addition, this assumption is confirmed by experience, even for the large waves of the sea. Thus, in this manner, the velocity of the waves will determine the depth α to which the water is perturbed during their formation. Since this velocity is expressed as n feet per second, one has

$$\alpha = \frac{n^2}{30,196} \quad \text{(in feet)}$$

In Volume 10 of the Anciens Mémoires de l'Académie des Sciences of Paris test results will be found on the velocity of waves obtained by de la Hire and which give one and a half feet per second for this velocity or more exactly 1,412 feet per second. Thus with $n = 1,412$, the depth α is 66/1000 feet, that is, 8/10 inches or about 10 lines.

SECTION XII
THE MOTION OF COMPRESSIBLE AND ELASTIC FLUIDS

1. In order to apply the general equation of Article 2 of the preceding section to this type of fluid, it will be observed that the term $\mathrm{S}\,\lambda\,\delta L$ must be deleted because the condition of incompressibility to which this term refers does not exist in this case. But, on the other hand, the action of elasticity which resists compression and which tends to dilate the fluid must be accounted for.

Then let ϵ be the elasticity of an arbitrary particle $\mathrm{D}m$ of fluid. Since its effect is to increase the volume $\mathrm{D}x\,\mathrm{D}y\,\mathrm{D}z$ of this particle and as a consequence, to diminish the quantity $-\mathrm{D}x\,\mathrm{D}y\,\mathrm{D}z$, the moment $-\epsilon\,\delta(\mathrm{D}x\,\mathrm{D}y\,\mathrm{D}z)$ must be added to the first member of this equation. Therefore, for all the particles, the integral term $-\mathrm{S}\,\epsilon\,\delta(\mathrm{D}x\,\mathrm{D}y\,\mathrm{D}z)$ will have to be substituted for the term $\mathrm{S}\,\lambda\,\delta L$. But since δL is equal to $\delta(\mathrm{D}x\,\mathrm{D}y\,\mathrm{D}z)$, it is clear that the general equation will retain the same form by simply replacing λ with $-\epsilon$. Using the same methods, three equations similar to equations (A) will be obtained, that is

$$\text{(a)}\quad \begin{cases} \Delta\left(\dfrac{\mathrm{d}^2x}{\mathrm{d}t^2}+X\right)+\dfrac{\mathrm{D}\epsilon}{\mathrm{D}x}=0 \\ \Delta\left(\dfrac{\mathrm{d}^2y}{\mathrm{d}t^2}+Y\right)+\dfrac{\mathrm{D}\epsilon}{\mathrm{D}y}=0 \\ \Delta\left(\dfrac{\mathrm{d}^2z}{\mathrm{d}t^2}+Z\right)+\dfrac{\mathrm{D}\epsilon}{\mathrm{D}z}=0 \end{cases}$$

Similarly, the value of ϵ should be equal to zero at the surface of the fluid if that fluid surface is free. But where the fluid is contained by walls, the value of ϵ will be equal to the resistance exerted by the walls to contain the fluid, which is obvious since ϵ expresses the elastic force of its particles.

2. For the case of compressible fluids, the density Δ is always given by a known function of ϵ, x, y, z and t, which depend on the laws of elasticity and the thermal properties for the fluid which are assumed to hold at each instant at every point in space. Thus there are unknowns ϵ, x, y and z to determine as functions of t. Consequently, a fourth equation is still needed for the complete solution of the problem. For the case of incompressible fluids, the condition of volume invariability produced equation (B) of Article 3, and the conditions for the invariability of density from one instant to the next produced equation (H) of Article 11. For the case of compressible fluids, none of these conditions holds in particular because the volume and density both vary. But the mass, which is the product of these two quantities must remain invariable. Therefore $\mathrm{d}(\mathrm{D}m) = 0$ or more accurately $\mathrm{d}(\Delta\,\mathrm{D}x\,\mathrm{D}y\,\mathrm{D}z) = 0$. Thus, by differentiating logarithmically we obtain $\mathrm{d}\Delta/\Delta + \mathrm{d}(\mathrm{D}x\,\mathrm{D}y\,\mathrm{D}z)/\mathrm{D}x\,\mathrm{D}y\,\mathrm{D}z = 0$ and after substituting the value of $\mathrm{d}(\mathrm{D}x\,\mathrm{D}y\,\mathrm{D}z)$ (this expression is identical to $\delta(\mathrm{D}x\,\mathrm{D}y\,\mathrm{D}z)$ of Article 2 of the preceding section after replacing d by δ), the following equation is obtained

$$\text{(b)}\quad \frac{\mathrm{d}\Delta}{\Delta}+\frac{\mathrm{Dd}x}{\mathrm{D}x}+\frac{\mathrm{Dd}y}{\mathrm{D}y}+\frac{\mathrm{Dd}z}{\mathrm{D}z}=0$$

which corresponds to equation (B) of Article 3 of the preceding section. The difference is that the former is related to the invariability of volume and the latter to the invariability of the mass.

3. If the coordinates x, y, z are viewed as functions of the original coordinates a, b, c and the elapsed time t from the beginning of the motion, equations (a) will become, using methods similar to those of Article 5 of the preceding section, of the form

$$(c)\quad \begin{cases} \theta\Delta\left(\frac{d^2x}{dt^2}+X\right)+\alpha\frac{d\epsilon}{da}+\beta\frac{d\epsilon}{db}+\gamma\frac{d\epsilon}{dc}=0 \\ \theta\Delta\left(\frac{d^2y}{dt^2}+Y\right)+\alpha'\frac{d\epsilon}{da}+\beta'\frac{d\epsilon}{db}+\gamma'\frac{d\epsilon}{dc}=0 \\ \theta\Delta\left(\frac{d^2z}{dt^2}+Z\right)+\alpha''\frac{d\epsilon}{da}+\beta''\frac{d\epsilon}{db}+\gamma''\frac{d\epsilon}{dc}=0 \end{cases}$$

or in a simpler form

$$(d)\quad \begin{cases} \Delta\left[\left(\frac{d^2x}{dt^2}+X\right)\frac{dx}{da}+\left(\frac{d^2y}{dt^2}+Y\right)\frac{dy}{da}+\left(\frac{d^2z}{dt^2}+Z\right)\frac{dz}{da}\right]+\frac{d\epsilon}{da}=0 \\ \Delta\left[\left(\frac{d^2x}{dt^2}+X\right)\frac{dx}{db}+\left(\frac{d^2y}{dt^2}+Y\right)\frac{dy}{db}+\left(\frac{d^2z}{dt^2}+Z\right)\frac{dz}{db}\right]+\frac{d\epsilon}{db}=0 \\ \Delta\left[\left(\frac{d^2x}{dt^2}+X\right)\frac{dx}{dc}+\left(\frac{d^2y}{dt^2}+Y\right)\frac{dy}{dc}+\left(\frac{d^2z}{dt^2}+Z\right)\frac{dz}{dc}\right]+\frac{d\epsilon}{dc}=0 \end{cases}$$

These transformed equations are analogous to the transformed equations (C) and (D) of the cited article. With regard to equation (b), if it is transformed according to the procedure of Article 3 of the preceding section, it will reduce to an equation of the following form

$$\frac{d\Delta}{\Delta}+\frac{d\theta}{\theta}=0$$

in which the differentials $d\Delta$ and $d\theta$ are solely with respect to the variable t. Therefore, after integration, one has

$$\Delta\theta = f(a,b,c)$$

When $t = 0$, it was seen in the cited article that θ is equal to unity. Therefore, if it is assumed that H is then the value of Δ, one has

$$H = f(a,b,c)$$

and the equation becomes $\Delta\theta = H$ or equivalently $\theta = H/\Delta$ that is, after substituting for θ its equivalent

$$(e)\quad \begin{cases} \frac{dx}{da}\frac{dy}{db}\frac{dz}{dc}-\frac{dx}{db}\frac{dy}{da}\frac{dz}{dc}+\frac{dx}{db}\frac{dy}{dc}\frac{dz}{da} \\ -\frac{dx}{dc}\frac{dy}{db}\frac{dz}{da}+\frac{dx}{dc}\frac{dy}{da}\frac{dz}{db}-\frac{dx}{da}\frac{dy}{dc}\frac{dz}{db}=\frac{H}{\Delta} \end{cases}$$

a transformed equation which is analogous to the transformed equation (E) of the cited article.

Finally, what was said in Article 8 of the same section should be applied to these equations relative to the surface of the fluid.

4. If it is desired, which is simpler, to obtain the equations between the velocities p, q, r of the particles in the directions of the coordinates x, y, z by viewing these velocities, as well as the quantities Δ and ϵ, as functions of x, y, z and t, the transformations of Article 10 of the preceding section will be used and the equations (a) will immediately give these transformed equations which are analogous to the transformed equations (F) of this last article

$$(f)\quad \begin{cases} \Delta\left(\dfrac{dp}{dt}+p\dfrac{dp}{dx}+q\dfrac{dp}{dy}+r\dfrac{dp}{dz}+X\right)+\dfrac{d\epsilon}{dx}=0 \\ \Delta\left(\dfrac{dq}{dt}+p\dfrac{dq}{dx}+q\dfrac{dq}{dy}+r\dfrac{dq}{dz}+Y\right)+\dfrac{d\epsilon}{dy}=0 \\ \Delta\left(\dfrac{dr}{dt}+p\dfrac{dr}{dx}+q\dfrac{dr}{dy}+r\dfrac{dr}{dz}+Z\right)+\dfrac{d\epsilon}{dz}=0 \end{cases}$$

In equation (b), besides the substitution of $p\,dt$, $q\,dt$, $r\,dt$ for dx, dy, dz, respectively, and the replacement of D by d, one should also replace $d\Delta$ by its complete expression

$$\left(\frac{d\Delta}{dt}+\frac{d\Delta}{dx}p+\frac{d\Delta}{dy}q+\frac{d\Delta}{dz}r\right)dt$$

and after dividing by dt, the following transformed equation will be obtained

$$\frac{d\Delta}{\Delta dt}+\frac{d\Delta}{\Delta dx}p+\frac{d\Delta}{\Delta dy}q+\frac{d\Delta}{\Delta dz}r+\frac{dp}{dx}+\frac{dq}{dy}+\frac{dr}{dz}=0$$

which after multiplication by Δ reduces to this simpler form

$$(g)\quad \frac{d\Delta}{dt}+\frac{d(\Delta p)}{dx}+\frac{d(\Delta q)}{dy}+\frac{d(\Delta r)}{dz}=0$$

With regard to the condition relative to the motion of particles at the surface, it will also be represented by equation (I) of Article 12 of the preceding section, that is

$$(i)\quad \frac{dA}{dt}+p\frac{dA}{dx}+q\frac{dA}{dy}+r\frac{dA}{dz}=0$$

assuming that $A=0$ is the equation of the surface.

5. It is easy to satisfy equation (g) by assuming that

$$\Delta p = \frac{\mathrm{d}\alpha}{\mathrm{d}t}, \qquad \Delta q = \frac{\mathrm{d}\beta}{\mathrm{d}t}, \qquad \Delta r = \frac{\mathrm{d}\gamma}{\mathrm{d}t}$$

in which α, β, γ are functions of x, y, z and t. With these substitutions, the equation becomes

$$\frac{\mathrm{d}\Delta}{\mathrm{d}t} + \frac{\mathrm{d}^2\alpha}{\mathrm{d}t\,\mathrm{d}x} + \frac{\mathrm{d}^2\beta}{\mathrm{d}t\,\mathrm{d}y} + \frac{\mathrm{d}^2\gamma}{\mathrm{d}t\,\mathrm{d}z} = 0$$

which is integrable with respect to t and for which the integral is

$$\Delta = F - \frac{\mathrm{d}\alpha}{\mathrm{d}x} - \frac{\mathrm{d}\beta}{\mathrm{d}y} - \frac{\mathrm{d}\gamma}{\mathrm{d}z}$$

where F is a function of x, y, z without t and dependent on the law of the initial density of the fluid.

Therefore, one has

$$p = \frac{\dfrac{\mathrm{d}\alpha}{\mathrm{d}t}}{F - \dfrac{\mathrm{d}\alpha}{\mathrm{d}x} - \dfrac{\mathrm{d}\beta}{\mathrm{d}y} - \dfrac{\mathrm{d}\gamma}{\mathrm{d}z}}$$

$$q = \frac{\dfrac{\mathrm{d}\beta}{\mathrm{d}t}}{F - \dfrac{\mathrm{d}\alpha}{\mathrm{d}x} - \dfrac{\mathrm{d}\beta}{\mathrm{d}y} - \dfrac{\mathrm{d}\gamma}{\mathrm{d}z}}$$

$$r = \frac{\dfrac{\mathrm{d}\gamma}{\mathrm{d}t}}{F - \dfrac{\mathrm{d}\alpha}{\mathrm{d}x} - \dfrac{\mathrm{d}\beta}{\mathrm{d}y} - \dfrac{\mathrm{d}\gamma}{\mathrm{d}z}}$$

After substituting these expressions in equations (f) and replacing ϵ with its expression as a function of Δ, x, y, z, t (Article 2), three equations of partial differences will be obtained between the unknowns α, β, γ. Then the solution of the problem is only a matter of the integration of these equations. But this integration is beyond the capability of known analytical techniques.

6. By neglecting heat and other variables, which can cause the elasticity to vary independent of density, the value of the elasticity ϵ will be given by a function of the density Δ such that $\mathrm{d}\epsilon/\Delta$ will be a differential of one variable and as a consequence, integrable. The integral will be denoted by E.

Moreover, let $X\,\mathrm{d}x + Y\,\mathrm{d}y + Z\,\mathrm{d}z$ be an exact differential for which the integral is V, as in Article 15 of the preceding section. If equations (f) of Article 4 are multiplied by $\mathrm{d}x$, $\mathrm{d}y$,

$\mathrm{d}z$, respectively and then added together, they will give, after division by Δ, an equation of the form

$$(\ell)\qquad -\,\mathrm{d}E - \mathrm{d}V = \begin{cases} \left(\frac{\mathrm{d}p}{\mathrm{d}t} + p\frac{\mathrm{d}p}{\mathrm{d}x} + q\frac{\mathrm{d}p}{\mathrm{d}y} + r\frac{\mathrm{d}p}{\mathrm{d}z}\right)\mathrm{d}x \\ + \left(\frac{\mathrm{d}q}{\mathrm{d}t} + p\frac{\mathrm{d}q}{\mathrm{d}x} + q\frac{\mathrm{d}q}{\mathrm{d}y} + r\frac{\mathrm{d}q}{\mathrm{d}z}\right)\mathrm{d}y \\ + \left(\frac{\mathrm{d}r}{\mathrm{d}t} + p\frac{\mathrm{d}r}{\mathrm{d}x} + q\frac{\mathrm{d}r}{\mathrm{d}y} + r\frac{\mathrm{d}r}{\mathrm{d}z}\right)\mathrm{d}z \end{cases}$$

for which the first member is integrable. Thus the second member should also be integrable. Therefore, one will again have the case of equation (L) of Article 15 of the preceding section, and consequently, similar results will be obtained.

7. In general, if the quantity $p\,\mathrm{d}x + q\,\mathrm{d}y + r\,\mathrm{d}z$ is found to be an exact differential at an arbitrary instant, which always holds at the beginning of the motion when the fluid departs from rest or when it is put in motion by an impulse applied at the surface, then this quantity should always be an exact differential (Articles 17 and 18 of the preceding section).

With this assumption, one will have, as in Article 20 of the preceding section

$$p\,\mathrm{d}x + q\,\mathrm{d}y + r\,\mathrm{d}z = \mathrm{d}\varphi$$

which gives

$$p = \frac{\mathrm{d}\varphi}{\mathrm{d}x}, \qquad q = \frac{\mathrm{d}\varphi}{\mathrm{d}y}, \qquad r = \frac{\mathrm{d}\varphi}{\mathrm{d}z}$$

Since equation (ℓ) has been integrated, it will give after these substitutions

$$(\mathrm{m})\quad E = -V - \frac{\mathrm{d}\varphi}{\mathrm{d}t} - \frac{1}{2}\left(\frac{\mathrm{d}\varphi}{\mathrm{d}x}\right)^2 - \frac{1}{2}\left(\frac{\mathrm{d}\varphi}{\mathrm{d}y}\right)^2 - \frac{1}{2}\left(\frac{\mathrm{d}\varphi}{\mathrm{d}z}\right)^2$$

an expression which satisfies simultaneously the three equations (f) of Article 4.

Since $E = \int \mathrm{d}\epsilon/\Delta$ will be a function of Δ because ϵ is a known function of Δ, thus Δ will be a function of E. After substituting the value of Δ obtained from the preceding equation as well as those of p, q, r in equation (g) of Article 4, an equation in partial differences for φ will be obtained, which because it only contains this unknown, will be sufficient to determine it. All that remains is to integrate this equation.

8. For fluids which have been shown to be elastic, their elasticity is always proportional to density so that $\epsilon = i\Delta$ where i is a constant which is determined for a given fluid density for which the elasticity is known. Thus, for air, the elasticity is equal to the weight of a column of mercury in a barometer. If H is the height of the mercury in the barometer

for a given density of air which is taken for unity, n is the density of mercury, that is, the numerical ratio of the density of mercury to the density of air which is the same as the ratio of specific gravities, and g is the accelerating force of gravity, then when $\Delta = 1$ the following equation results

$$\epsilon = gnH$$

and as a consequence

$$i = gnH$$

Note that nH is the height of the atmosphere which is assumed to be homogeneous. Thus by designating this height by h, one simply has $i = gh$, and from there $\epsilon = gh\Delta$. Therefore, since $E = \int \mathrm{d}\epsilon/\Delta$, one has $E = gh \ln \Delta$.

Now equation (g) of Article 4 can be put in the following form

$$\frac{\mathrm{d}(\ln \Delta)}{\mathrm{d}t} + \frac{\mathrm{d}(\ln \Delta)}{\mathrm{d}x} p + \frac{\mathrm{d}(\ln \Delta)}{\mathrm{d}y} q + \frac{\mathrm{d}(\ln \Delta)}{\mathrm{d}z} r + \frac{\mathrm{d}p}{\mathrm{d}x} + \frac{\mathrm{d}q}{\mathrm{d}y} + \frac{\mathrm{d}r}{\mathrm{d}z} = 0$$

and after substituting $E/(gh)$, $\mathrm{d}\varphi/\mathrm{d}x$, $\mathrm{d}\varphi/\mathrm{d}y$, $\mathrm{d}\varphi/\mathrm{d}z$ for $\ln \Delta$, p, q, r, respectively, and then multiplying by gh, it becomes

$$gh\left(\frac{\mathrm{d}^2\varphi}{\mathrm{d}x^2} + \frac{\mathrm{d}^2\varphi}{\mathrm{d}y^2} + \frac{\mathrm{d}^2\varphi}{\mathrm{d}z^2}\right) + \frac{\mathrm{d}E}{\mathrm{d}t} + \frac{\mathrm{d}E}{\mathrm{d}x}\frac{\mathrm{d}\varphi}{\mathrm{d}x} + \frac{\mathrm{d}E}{\mathrm{d}y}\frac{\mathrm{d}\varphi}{\mathrm{d}y} + \frac{\mathrm{d}E}{\mathrm{d}z}\frac{\mathrm{d}\varphi}{\mathrm{d}z} = 0$$

It will only remain to substitute the expression for E found above, which will give the final equation in φ

$$(\mathrm{n}) \quad \left\{\begin{aligned}
&0 = gh\left(\frac{\mathrm{d}^2\varphi}{\mathrm{d}x^2} + \frac{\mathrm{d}^2\varphi}{\mathrm{d}y^2} + \frac{\mathrm{d}^2\varphi}{\mathrm{d}z^2}\right) - \frac{\mathrm{d}^2\varphi}{\mathrm{d}t^2} - \frac{\mathrm{d}V}{\mathrm{d}x}\frac{\mathrm{d}\varphi}{\mathrm{d}x} - \frac{\mathrm{d}V}{\mathrm{d}y}\frac{\mathrm{d}\varphi}{\mathrm{d}y} - \frac{\mathrm{d}V}{\mathrm{d}z}\frac{\mathrm{d}\varphi}{\mathrm{d}z}\\
&-2\frac{\mathrm{d}\varphi}{\mathrm{d}x}\frac{\mathrm{d}^2\varphi}{\mathrm{d}x\,\mathrm{d}t} - 2\frac{\mathrm{d}\varphi}{\mathrm{d}y}\frac{\mathrm{d}^2\varphi}{\mathrm{d}y\,\mathrm{d}t} - 2\frac{\mathrm{d}\varphi}{\mathrm{d}z}\frac{\mathrm{d}^2\varphi}{\mathrm{d}z\,\mathrm{d}t}\\
&-\left(\frac{\mathrm{d}\varphi}{\mathrm{d}x}\right)^2\frac{\mathrm{d}^2\varphi}{\mathrm{d}x^2} - \left(\frac{\mathrm{d}\varphi}{\mathrm{d}y}\right)^2\frac{\mathrm{d}^2\varphi}{\mathrm{d}y^2} - \left(\frac{\mathrm{d}\varphi}{\mathrm{d}z}\right)^2\frac{\mathrm{d}^2\varphi}{\mathrm{d}z^2}\\
&-2\frac{\mathrm{d}\varphi}{\mathrm{d}x}\frac{\mathrm{d}\varphi}{\mathrm{d}y}\frac{\mathrm{d}^2\varphi}{\mathrm{d}x\,\mathrm{d}y} - 2\frac{\mathrm{d}\varphi}{\mathrm{d}x}\frac{\mathrm{d}\varphi}{\mathrm{d}z}\frac{\mathrm{d}^2\varphi}{\mathrm{d}x\,\mathrm{d}z} - 2\frac{\mathrm{d}\varphi}{\mathrm{d}y}\frac{\mathrm{d}\varphi}{\mathrm{d}z}\frac{\mathrm{d}^2\varphi}{\mathrm{d}y\,\mathrm{d}z}
\end{aligned}\right.$$

which alone contains the theory of motion for elastic fluids under the assumptions adopted here.

9. When the velocity of the fluid is very small and the quantities of an order larger than the first are neglected, it is clear from Article 21 of the preceding section that the quantity

$p\,\mathrm{d}x + q\,\mathrm{d}y + r\,\mathrm{d}z$ must also be an exact differential. Hence, in this case, the preceding formulas will always hold however the motion was begun as long as it remains very small. Consequently, the function φ will be very small.

In the theory of sound, it is assumed that the motion of the particles of the air is very small. Thus, by viewing the quantity φ in equation (n) as very small and neglecting the terms which are of an order greater than the first, the following general equation for this theory is obtained

$$gh\left(\frac{\mathrm{d}^2\varphi}{\mathrm{d}x^2} + \frac{\mathrm{d}^2\varphi}{\mathrm{d}y^2} + \frac{\mathrm{d}^2\varphi}{\mathrm{d}z^2}\right) - \frac{\mathrm{d}^2\varphi}{\mathrm{d}t^2} - \frac{\mathrm{d}V}{\mathrm{d}x}\frac{\mathrm{d}\varphi}{\mathrm{d}x} - \frac{\mathrm{d}V}{\mathrm{d}y}\frac{\mathrm{d}\varphi}{\mathrm{d}y} - \frac{\mathrm{d}V}{\mathrm{d}z}\frac{\mathrm{d}\varphi}{\mathrm{d}z} = 0$$

Now by neglecting the second order quantities in φ in the expression for E of Article 7, one simply has (Article 8)

$$E = -V - \frac{\mathrm{d}\varphi}{\mathrm{d}t} = gh \ln \Delta$$

It can be assumed that the function φ is zero in the state of rest or of equilibrium. In this case, one has $\mathrm{d}\varphi/dt = 0$ and as a consequence $gh \ln \Delta = -V$, and $\Delta = \mathrm{e}^{-V/(gh)}$. When the air is vibrating, its natural density is increased in the ratio of $(1 + s)$ to 1, where s is a very small quantity. One has in general

$$\Delta = \mathrm{e}^{-V/(gh)}(1 + s)$$

and from this result, by neglecting the squared terms in s, one has $\ln \Delta = (-V/gh) - s$, then $s = \mathrm{d}\varphi/(gh\,\mathrm{d}t)$. With respect to the expression for V, which depends upon the accelerating forces and for which the fluid is assumed dense and with the z-ordinate taken for more simplicity vertical and downward, one has by the formula of Article 23 of the preceding section

$$V = -gz$$

where g is the force of gravity. Thus the equation for the propagation of sound is

$$gh\left(\frac{\mathrm{d}^2\varphi}{\mathrm{d}x^2} + \frac{\mathrm{d}^2\varphi}{\mathrm{d}y^2} + \frac{\mathrm{d}^2\varphi}{\mathrm{d}z^2}\right) + g\frac{\mathrm{d}\varphi}{\mathrm{d}z} = \frac{\mathrm{d}^2\varphi}{\mathrm{d}t^2}$$

After having determined φ from this equation, the velocities p, q, r of the air are obtained, as well as its condensation s from the following formulas

$$p = \frac{\mathrm{d}\varphi}{\mathrm{d}x}, \qquad q = \frac{\mathrm{d}\varphi}{\mathrm{d}y}, \qquad r = \frac{\mathrm{d}\varphi}{\mathrm{d}z}, \qquad s = \frac{\mathrm{d}\varphi}{gh\,\mathrm{d}t}$$

10. If only the horizontal motion of air is considered, the function φ is assumed not to contain z, but only x, y and t. Then, the equation for φ becomes

$$gh\left(\frac{d^2\varphi}{dx^2}+\frac{d^2\varphi}{dy^2}\right)=\frac{d^2\varphi}{dt^2}$$

But even with this simplification, this equation is still too complicated to be rigorously integrated.[57] Also, this equation is completely similar to the equation for the motion of waves in a horizontal shallow conduit. (See Article 37 of the preceding section.)

Until now, only the single case, where the mass of the air is considered to have one dimension, has been entirely resolved, that is, the case of a sonorous line for which the particles only move longitudinally. In this case, by taking this dimension to be the x-axis, the function φ does not contain y and the above equation is reduced to $gh(d^2\varphi/dx^2) = d^2\varphi/dt^2)$ which is similar to the equation for vibrating chords for which the complete solution is .

$$\varphi = F(x+t\sqrt{gh}) + f(x-t\sqrt{gh})$$

in which the functions F and f are arbitrary.

This formula expresses two important theories: the operation of a flute or organ pipe and the propagation of sound in free air. It is only a question of correctly determining the two arbitrary functions. In what follows, the principles which guide the determination of these two functions will be presented.

11. For the flute only the sonorous line is considered. It is assumed that the initial state of this line is given and that this state is a function of the vibrations impressed on the particles. The law of oscillation is needed in this problem.

At the outset, put the origin of the x-axis at one end of the line and let a be its length, that is, the length of the flute. The condensation s and the longitudinal velocities p are given when $t = 0$, from $x = 0$ to $x = a$. They will be denoted, S and P.

Now, since $s = d\varphi/(gh\,dt)$ and $p = d\varphi/dx$, if the general expression for φ of the preceding article is differentiated and if F' and f' are the differentials of the functions F and f, so that $F'(x) = dF(x)/dx$ and $f'(x) = df(x)/dx$ one has

$$p = F'(x+t\sqrt{gh}) + f'(x-t\sqrt{gh})$$
$$s\sqrt{gh} = F'(x+t\sqrt{gh}) - f'(x-t\sqrt{gh})$$

Now put $t = 0$ and replace p with P and s with S to obtain

$$P = F'(x) + f'(x), \qquad S\sqrt{gh} = F'(x) - f'(x)$$

Thus, since P and S are given for all values of x from $x = 0$ to $x = a$, one will have in this domain the values of $F'(x)$ and $f'(x)$. Consequently, the values of p and s for any value of x and for an arbitrary time as long as the quantity $x \pm t\sqrt{gh}$ lies between 0 and a will be known.

Since the time t is always increasing, the quantities $x + t\sqrt{gh}$ and $x - t\sqrt{gh}$ will soon be outside these limits and the determination of the functions

$$F'(x + t\sqrt{gh}), \qquad f'(x - t\sqrt{gh})$$

will then depend upon the conditions which hold at the ends of the sonorous line: that is, whether the flute is open or closed at its ends.

12. Let us assume at the outset that the flute is open at both ends so that the sonorous line is in contact with the surrounding air. It is clear that its elasticity at the two end points cannot be counterbalanced by the constant pressure of the atmosphere so the condensations should always be zero. Therefore, in this case, s should be equal to zero when $x = 0$ and $x = a$, whatever the value of t, which gives two conditions to be satisfied

$$F'(t\sqrt{gh}) - f'(-t\sqrt{gh}) = 0$$
$$F'(a + t\sqrt{gh}) - f'(a - t\sqrt{gh}) = 0$$

which should always exist for any positive value of t. Thus, in general, by taking z as an arbitrary positive value, one has

$$F'(a + z) = f'(a - z)$$

and

$$f'(-z) = F'(z)$$

Therefore, while z is less than a, the values of $F'(a+z)$ and $f'(-z)$ will be known because they are reduced to those of $f'(a - z)$ and of $F'(z)$, which are given.

Let us replace z by $a + z$ in these formulas so that

$$F'(2a + z) = f'(-z) = F'(z)$$
$$f'(-a - z) = F'(a + z) = f'(a - z)$$

Thus when z is less than a, the values of $F'(2a + z)$ and of $f'(-a - z)$ will also be known because they are reduced to those of $F'(z)$ and of $f'(a - z)$, which are given.

Again, let us replace z by $a + z$ in the last formulas and after having combined them with the prior formulas, because z is arbitrary, one has

$$F'(3a + z) = F'(a + z) = f'(a - z)$$
$$f'(-2a - z) = f'(-z) = F'(z)$$

Therefore, when z is less than a, the values of $F'(3a+z)$ and of $f'(-2a-z)$ will also be known because they are reduced to the given values of $F'(z)$ and of $f'(a-z)$.

In a similar manner, by replacing z by $a+z$

$$F'(4a+z) = f'(-z) = F'(z)$$
$$f'(-3a-z) = F'(a+z) = f'(a-z)$$

from which the values of $F'(4a+z)$ and of $f'(-3a-z)$ will be known when z is less than a and so on.

In this manner, the values of the functions $F'(x+t\sqrt{gh})$ and $f'(x-t\sqrt{gh})$ will be known for all values of t from the start of the motion of the sonorous line. Therefore, at each instant, the state of the sonorous line will be known. Thus the velocities p and the condensations s of each particle will be known.

It is clear from the preceding formulas that the values of these functions remain the same if the quantity $t\sqrt{gh}$ is increased by $2a$, $4a$, $6a$, etc. Consequently, the sonorous line will return to exactly the same state after each time interval given by the equation $t\sqrt{gh} = 2a$, which gives $2a/\sqrt{gh}$ for this interval of time.

Thus the duration of the oscillations of the sonorous line is independent of any initial vibrations and only depends upon the length a of this line and of the height h of the atmosphere.

By assuming that the force of gravity g is equal to unity, the unit of distance should be taken as double the distance a heavy body traverses freely in the unit of time (SECTION II, Article 2). Thus, if h is taken for the unit of space, which is permissible, the unit of time will be the time a heavy body takes to fall from the height $h/2$. The time for a complete oscillation of the sonorous line is $2a$ or, which is the same, the time of one oscillation will be equal to the time of fall of a body from the height $h/2$ in the ratio of $2a$ to h.

13. If the flute were closed at both ends, then the condensation s could be arbitrary because the elasticity of the particles would be counterbalanced by the resistance of the walls at the ends. But, for the same reason, the velocities p should be zero which would again give the conditions

$$F'(t\sqrt{gh}) + f'(-t\sqrt{gh}) = 0$$
$$F'(a+t\sqrt{gh}) + f'(a+t\sqrt{gh}) = 0$$

These formulas are those we have examined earlier, by assuming only the functions f' negative. Thus similar conclusions will result from these expressions and again we will obtain the same expression for the duration of the sonorous line.

The results would be different if the flute were open at one end and closed at the other. In this case s will always be zero at the open end and p will be equal to zero at the closed end.

Thus, by assuming the flute open at $x = 0$ and closed at $x = a$, the following conditions exist

$$F'(t\sqrt{gh}) - f'(-t\sqrt{gh}) = 0$$
$$F'(a + t\sqrt{gh}) + f'(a - t\sqrt{gh}) = 0$$

from which, by an analysis similar to the one described in Article 12, the following formulas are obtained

$$\begin{aligned} F'(a+z) &= -f'(a-z), & f'(-z) &= F'(z), \\ F'(2a+z) &= -F'(z), & f'(-a-z) &= -f'(a-z), \\ F'(3a+z) &= f'(a-z), & f'(-2a-z) &= -F'(z), \\ F'(4a+z) &= F'(z), & f'(-3a-z) &= f'(a-z) \end{aligned}$$

and so on.

Since z is less than a, the functions $F'(z)$ and $f'(a - z)$ are given by the initial state of the sonorous line and then by this means the values of the other functions are also known

$$F'(a+z), \qquad F'(2a+z), \qquad \ldots, \qquad f'(-z), \qquad f'(-a-z), \qquad \ldots$$

Consequently, the state of the line will be determined after an arbitrary time t. But, it is seen from the preceding formulas that this state will be the same after a time interval defined by the equation

$$t\sqrt{gh} = 4a$$

from which it follows that the duration of the vibrations are twice as long as in open flutes or flutes closed at both ends. This result is confirmed by experience, for the organs called *bourdons*, which are closed at the upper extremity opposite the mouth, give a note which is an octave lower than if they were open.

The theory of flutes is discussed in the first two volumes of Turin, the memoirs of Paris for 1762 and the Novi Commentarii of Petersburg, Volume XVI.

14. Let us now consider a sonorous line of infinite length vibrating at the outset only locally over a short length. This will be the case for the vibrations in air produced by sonorous bodies.

Let us assume that the initial vibrations extend from $x = 0$ to $x = a$, where a is a very small quantity. The initial velocities and condensations P and S will then be given at each positive or negative x-coordinate, but they will be real only from $x = 0$ to $x = a$. Outside these limits they will be zero. It will be the same for the functions $F'(x)$ and $f'(x)$ because by putting $t = 0$ the following equations result

$$P = F'(x) + f'(x), \qquad S\sqrt{gh} = F'(x) - f'(x)$$

and consequently

$$F'(x) = \frac{P + S\sqrt{gh}}{2}, \qquad f'(x) = \frac{P - S\sqrt{gh}}{2}$$

from which it results that by taking z as a positive quantity, less than a, the functions $F'(x + t\sqrt{gh})$ and $f'(x - t\sqrt{gh})$ will have real values for $x \pm t\sqrt{gh} = z$. Consequently, after an arbitrary time t, the velocities p and the condensations s will be zero for all points of the sonorous line with the exception of those for which the abscissas are $x = z \mp t\sqrt{gh}$.

These results explain how sound propagates and how it is created from both ends of a sonorous body and in equal times how sonorous lines are created for which the length is the initial length a.

The velocity of propagation of the sonorous lines is given by $\sqrt{gh}$. Consequently, it is constant and independent of any initial motion. This claim is confirmed by experience because all the low or high sounds seem to propagate with nearly equal velocity. With respect to the absolute value of this velocity it is equal to unity when $g = 1$ and $h = 1$, as in Article 12. But the unit of velocity is here the one that a heavy body must have after falling from half the height h, taken for unity (SECTION II, Article 2). Therefore, the velocity of sound will result from a descent from a height $h/2$.

15. If it is assumed, as most physicists have, that air is 850 times lighter than water and that water is 14 times lighter than mercury, then the density of mercury is 11900 times the density of air. But taking the mean height of the barometer equal to 28 French inches, the height h is 333200 inches or 27766 and 2/3 feet for a column of air of uniform density and in equilibrium with the column of mercury in the barometer. Therefore, the velocity of sound will result from a height of 13883 and 1/3 feet and it will, consequently be equal to 915 feet per second.

The results of experiment give about 1088, which is 1/6 larger than the calculated value. This difference can only be attributed to the uncertainty in the results given by experiment. For more information on this discrepancy refer to the memoir by the late Johann Lambert, among those of the Académie de Berlin for 1768.[58]

16. If the sonorous line were held at one end by an immobile obstacle, then the particle of air contiguous to this obstacle would have no motion. Consequently, if the quantity a were the value of the x-coordinate at this position, the velocity p should be zero when $x = a$, whatever the value of t, which gives the condition

$$F'(a + t\sqrt{gh}) + f'(a - t\sqrt{gh}) = 0$$

But it has been seen that the function $f'(a - t\sqrt{gh})$ has a real value when $a - t\sqrt{gh} = z$ (Article 14). Thus, since $F'(a + t\sqrt{gh}) = -f'(a - t\sqrt{gh})$ the function $F'(a + t\sqrt{gh})$ will also have real values when $a - t\sqrt{gh} = z$ that is, when $t\sqrt{gh} = a - z$. Consequently, the function $F'(x + t\sqrt{gh})$ will not only be real when $x + t\sqrt{gh} = z$ but also when

$x + t\sqrt{gh} = 2a - z$ from which it follows that, in this case, the velocities p and the condensations s will also be real for the abscissas

$$x = 2a - z - t\sqrt{gh}$$

Thus the sonorous lines, after having traversed the space a, will be reflected by the encountered obstacle and they will retrace their path with the same velocity, which explains ordinary echoes.

In the same manner, a composed echo can be explained by assuming the sonorous line ending at both ends with immobile obstacles which will reflect successively the sonorous lines and will make them execute some kind of continual oscillations. The reader should refer to the works cited above (Article 13) and the Mémoires of the Académie de Berlin for 1759 and 1765 for additional information.

NOTES

Translator's Introduction

[1] The title of the first edition of this work is *Méchanique analitique*. Both words are misspelled. The first word should be "Mécanique." The "h" is archaic and had earlier been dropped in French. As for the second word, it could be the influence of the Italian word "analisi," which caused the y to be replaced with an i. It is also strange that the publisher and the review of the work by the board appointed by the Académie des Sciences did not change the spelling of these two words. Cf. Endnote #4, pp. 474-475.

[2] *Oeuvres de Lagrange*, publiées par les soins de M. J.-A. Serret (1819-1885), sous les auspices de son Excellence Le Ministre de L'Instruction Publique, Paris, Gauthier-Villars, Imprimeur-Libraire. (Lagrange's works comprise fourteen volumes published between 1867 and 1892. The first ten volumes were edited by J.-A. Serret, volumes eleven and twelve by G. Darboux (1842-1917) and volumes thirteen and fourteen which deal with Lagrange's correspondence by L. Lalanne (1815-1898).)

[3] Taton, René, Inventaire Chronologique de l'oeuvre de Lagrange, Revue d'Histoire des Sciences et de Leurs Applications, 27, 1974, pp. 3-36.

[4] Sarton, George, Lagrange's Personality (1736–1813), Proceedings of the American Philosophical Society, 88, 6, December, 1944, pp. 457-496.

[5] Loria, Gino, Nel secondo centenario della nascita di G. L. Lagrange (1736-1936). Isis, 28, 1938, pp. 366-375.

[6] Sarton, George, René Taton et George Beaujouan, Documents nouveaux concernant Lagrange, Revue d'Histoire des Sciences et de Leurs Applications, 3, 1950, pp. 110-132.

[7] Burzio, Filippo, *Lagrange*, I Grandi Italiani, Collana Di Biografie, Diretta Da Luigi Federzoni, Unione Tipografico-Editrice Torinese, 1942.

[8] *Oeuvres de Lagrange*, Vol. XIV, p. 263.

[9] French troops under the generalship of Napoleon Bonaparte conquered Northern Italy and incorporated the entire region. Piedmont was annexed by France on December 16, 1798 and divided into six departments. Turin was the major city of the Département du Pô. Consequently, there was no need for this office.

[10] The Italian title reads, *Analytical Foundations for the Use of Italian Youth.* This work in two volumes presented a review of geometry and algebra as they were known in the early part of the 18th century. It stressed the power of analytical methods. It became very useful as a textbook and was used as such in Italy, France and England. The woman who wrote this book-Maria Gaetana Agnesi–abandoned mathematics soon after its publication and devoted the remainder of her life to charity.

[11] This is an abbreviation of the title of the proceedings of the Royal Academy of Turin. The complete title is *Miscellanea Philosophica Mathematica Societatis privatae Taurinensis* and is given in the first volume. With the second volume the title changed to *Mélanges de Philosophie et de Mathématique de la Societé Royale de Turin*, and beginning with the sixth volume the title became *Mémoires de l'Académie Royale des Sciences.*

[12] Lagrange, J. L., Recherches sur la Méthode de Maximis et Minimis, (*Miscellanea Taurinensia*, Vol. I, 1759) *Oeuvres de Lagrange*, Vol. I, pp. 3-20.

[13] Maclaurin, Colin, *Treatise of Fluxions*, 1742.

[14] Lagrange, J. L., Essai d'une Nouvelle Méthode pour Déterminer Les Maxima et Les Minima des Formulas Intégrales Indéfinies, (Miscellanea Taurinensia, Vol. II, 1760–1761). *Oeuvres de Lagrange*, Vol. I, pp. 335-362.

[15] Lagrange called his approach the Method of Variations but Euler renamed it the Calculus of Variations. Since the methodology had broad application, Euler's designation seemed more appropriate and consequently, it was generally adopted. Cf. Euler, Leonhard, *Elementa calculi variationum, Opera Omnia*, Series I, Vol. XXV, p. 145.

[16] *Oeuvres de Lagrange*, Vol. XIII, p. 154.

[17] Borgato, Maria Teresa and Luigi Pepe, *Lagrange a Torino (1750–1759) et le sue lezioni inedite nella Reale Scuole di Artiglieria*, Bollettino di Storia dell Scienze Matematiche, Vol. VII, numero 2, dicembre 1987, p. 32.

[18] There is a footnote in Fraser's memoir (Fraser, Craig, *J. L. Lagrange's Early Contributions to the Principles and Methods of Mechanics*, Archive for the History of Exact Sciences, Vol. 28, 1983, pp. 197-241) which bears upon what we are discussing here.

> ... At the end of his life Lagrange reportedly stated that d'Alembert and Euler were (in that order) the most important early influences on him. However, because of the difficulty of d'Alembert's treatises, he strongly advised anyone then taking up the study of mathematics to turn to Euler: "quand on voulait être géometre, il fallait êtudier Euler." (T.N. The French reads, when one wants to become a geometer, it is necessary to study Euler.) These remarks are attributed to Lagrange in a letter that appeared in the Moniteur Universalis in 1814, pp. 226-228. The letter, which is identified only by the initials L.B.M.D.G. (T.N. The initials refer to "le Baron Maurice de Génève") is a commentary on Delambre's eulogy of Lagrange. I have relied on excerpts from it reprinted in an Essay Review by I. Grattan-Guiness: "Recent researches in French mathematical physics of the early 19th century", Annals of Science 38 (1981) pp. 663-690 and p. 679.

[19] Hahn, Roger, Scientific Research as an Occupation in Eighteenth-Century Paris, Minerva, Vol. XIII, No. 4, Winter 1975 and The Anatomy of a Scientific Institution, The Paris Academy of Sciences, 1666–1803, University of California Press, Berkeley, 1971.

[20] Taton, René, Inventaire Chronologique de l'Oeuvre de Lagrange, Revue d'Histoire des Sciences et de Leurs Applications, 27, 1974, pp. 3-36.

[21] In 1788, Lagrange is living at rue Fromanteau (or Froidmanteau), No. 4, Section du Muséum. *Oeuvres de Lagrange*, Vol. XIV, pp. 286, 310, 314.

[22] Borgato, Maria Teresa and Luigi Pepe. L'inventaire des manuscrits de Lagrange et la Mécanique avec l'édition du manuscrit de Lagrange: ⟨⟨Differentes notes sur les ouvrages de mécanique⟩⟩, Supplemento al numero 124 (1990) degli Atti della Accademia delle Scienze di Torino, Classe di Scienze Fisiche, Matematiche e Naturali, Torino, 1990.

[23] Notice sur la vie et les ouvrages de M. Lagrange, *Oeuvres de Lagrange*, Vol. I, p. XLVII.

[24] These lectures have been translated into English with the title *Lectures on Elementary Mathematics* by Thomas J. McCormack, The Open Court Publishing Company, Chigago, 1901.

[25] At the time of his death, he lived at rue Faubourg S. Honorè, No. 128, quartier du Roule. This is a very fashionable and elegant quarter where Lagrange built his home in a style typical of the period.

[26] The Constituent Assembly decreed on April 4, 1791 that henceforth the Parisian church Sainte-Geneviève would be the site for the burial of outstanding men of the republic and that the name of the church would be changed to the Panthéon Français. Later, Napoleon decreed on February 20, 1806 that the Panthéon would again be a Catholic place of worship without losing its service as a burial site for notable men of the nation and for men who had rendered outstanding service to the nation. Lagrange, who had earlier become a French citizen, met both of these criteria.

[27] The eulogies delivered by Lacépède and Laplace can be found in the following memoir: Borgato, Maria Teresa and Luigi Pepe, Sulle lettere familiari di Giuseppe Luigi Lagrange, Bollettino di Storia delle Scienze Matematiche, Vol. IX, 1989, pp. 254-256.

[28] Notice sur la vie et les ouvrages de M. Lagrange, *Oeuvres de Lagrange*, Vol. I, p. xi.

[29] The French reads, A Dictionary of past and present Atheists. Maréchal, Sylvain, *Dictionaire des athées anciens et modernes,* Paris, Grabit, an VIII, p. 126.

[30] Sur la destructione des Jésuites en France, par un auteur désintéressé S.I, 1765. Lettre à M. . . . Conseiller au Parlement de . . . Pour servir de supplément . . . S.I. 1767. Seconde lettre, S.I.a., July 15, 1767.

[31] The French reads . . . a monk and Jesuit to burn.

[32] The French reads, Letters to a German Princess.

[33] *Oeuvres de Lagrange*, Vol. XIV, pp. 135-138. The given year is uncertain.

[34] *Oeuvres de Lagrange*, Vol. XIV, pp. 138-144.

[35] *Oeuvres de Lagrange*, Vol. XIV, pp. 144-146.

[36] *Oeuvres de Lagrange*, Vo.l XIV, pp. 154-156.

[37] Recherches sur la Méthode de Maximis et Minimis. (Miscellanea Taurinensia. Vol. I, 1759). *Oeuvres de Lagrange*, Vol. I. pp. 3-20. The quote is on page 15.

[38] Delsedime, Paolo, La disputa delle corde vibranti ed una lettera inedita di Lagrange a Daniel Bernoulli. Physis, 13(1971), pp. 117-146.

[39] Lagrange is referring to the following work by Euler: *Methodus Inveniendi Lineas Curvas Maximi Minimive Proprietate Gaudentes sive Solutio Problematis Isoperimentrici Latissimo Sensu Accepti*, Lausanne and Geneva, 1744.

[40] *Oeuvres de Lagrange*, Vol. XIV, pp. 170–174. The quote is at the end of the letter and begins on p. 173.

[41] This entire work has been published in the following reference: Borgato, Maria Teresa and Luigi Pepe, Lagrange a Torino (1750–1759) e le sue lezioni inedite nelle R. Scuole di Artiglieria, Bollettino di Storia delle Scienze Matematiche, Vol. VII, 2, 1987.

[42] The Italian reads, The Principles of Sublime Analysis written by La Grange at the Royal Artillery School. (The term "Sublime Analysis" refers to the infinitesimal calculus.)

[43] Application de la Méthode exposée dans le mémoire précédent a la solution de différents problémes de dynamique. (Miscellanea Taurinensia, Vol. II, 1760–1761.) *Oeuvres de Lagrange*, Vol. I, pp. 365-468.

[44] Recherches sur la libration de la Lune, (Prix de l'Académie Royale des Sciences de Paris, tome IX, 1764. This work was not published until 1777.) It can also be found in the *Oeuvres de Lagrange*, Vol. VI, pp. 5-61.

[45] *Oeuvres de Lagrange*, Vol. V, pp. 5-122.

[46] *Oeuvres de Lagrange*, Vol. XIV, pp. 115-117.

[47] This reference is to the memoir: Théorie de la libration de la Lune. Cf. Endnote #44. This letter can be found in: *Oeuvres de Lagrange*, Vol. XIII, p. 361.

[48] Credit for the re-discovery of the Principle of Virtual Work is given to Lagrange in the article on Varignon in the *Dictionary of Scientific Biography*, Charles Scribner's and Sons, New York, 1973. The same attribution is made by J. B. Fourier in Mémoire sur la Statique contenant la demonstration du Principe des Vitesses Virtuals et la Théorie des Moments, Journal de l'École Polytechnique, Ve Cahier, p. 20, 1798.

[49] Harmonie Entre Les Principes Generaux de Repos et de Mouvement de M. de Maupertuis, Leonhardi Euleri, Opera Omnia, Commentiones Mechanicae, 5, 2, pp. 152–176.

[50] D'Alembert, Jean le Rond, *Traité de Dynamique*, Chapter IV, (1743, pp. 182–183), (1758, p. 267).

[51] Euler, Leonhard, *Opera Omnia*, II 5 (1957), pp. 168–172.

[52] Jourdain, Philip E. B., *The Principle of Least Action*, The Open Court Publishing Co., Chicago, 1913.

[53] Ernst Mach, whose books on mechanics exercised a profound influence on the development of modern mechanics, states that his interest in the development of mechanics was stimulated "by the beautiful introductions of Lagrange to the various sections in the *Mécanique analytique* as well as by the lucid and spirited treatise of Jolly, *Principien der Mechanik*." Ernst Mach, *The Science of Mechanics*, A Critical and Historical Account of its Development, The Open Court Publishing Company, LaSalle, Illinois, Sixth Edition, 1960, Preface to the First German Edition, p. XXIV.

Excerpt

[1] A register of the minutes of the Academy's meetings was kept and excerpts from this register were printed later. This note is an excerpt from this register which has been taken from the first edition of the Méchanique analitique. It reveals a very important right which was associated with membership in the Academy and which was very vital to its membership. In France, a Royal board existed which censored all printed works. However, the Academy held an exclusive privilege from the King which permitted the works of its members to be printed without obtaining approval of the Royal board of censors. Lagrange's work was printed under this privilege.

Preface to the Second Edition

[1] T.N. Lagrange finds it necessary to claim that the calculus is as well-founded as geometry. In his day, the theoretical basis of the calculus was still tenuous and not until the time of Cauchy would this situation change. Although he applied the calculus to problems of mechanics forcefully, there is the intimation that either he expected his audience to entertain some doubt or that he himself may have entertained some doubt on its foundation.

[2] T.N. Lagrange designated the derivative of a function by the term "derived function". During his lifetime, the limit concept had not yet been accepted as the basis for the definition of the derivative. In fact, most definitions still employed geometric intuition and metaphysical notions. Lagrange tried to eliminate both by defining the derivatives of a function in terms of its infinite series. However, it was later shown that not every continuous function can be expanded in an infinite series and hence, his method was not general enough to serve as a basis for the calculus. In addition, a mathematical definition of continuity did not exist. Most mathematicians of the 18th century viewed continuity in terms of the motion of a particle which to them traversed an obviously continuous path. Hence, it was concluded that since motion is analogous to a continuum every continuum is infinitely divisible. Lagrange did not share this view and sought an algebraic definition. Yet, he used this analogy in his work. The controversy over the conceptual basis for the calculus is explained well by Carl B. Boyer, *The History of the Calculus and Its Conceptual Development*, Dover Publications, Inc. New York, 1959, pp. 251-255, and Judith V. Grabiner, *The Calculus as Algebra*, J. L. Lagrange, 1736-1813, Garland Publishing, Inc. New York, 1990.

[3] T.N. Lagrange is referring to the following memoirs:

> Sur la théorie des variations des élémens des planètes et en particulier des variations des grands axes de leurs orbites.
>
> Sur la théorie générale de la variation des constantes arbitraires, dans tous les problèmes de mécanique.
>
> Supplément au mémoire sur la théorie générale de la variation des constantes arbitraires, dans tous les problémes de la mécanique.

All three memoirs can be found in the *Oeuvres de Lagrange*, Volume VI.

[4] T.N. Lagrange's reference is to the "Recherches sur la nature et la propagation du son," *Oeuvres de Lagrange*, Volume I, pp. 39-148. (The French title reads, Research into the nature and propagation of sound.)

[5] T.N. This work is a series of eight volumes written by d'Alembert between 1761 and 1780. The full title of this work is *Opuscules mathématiques*, and in it, d'Alembert wrote on varied and diverse topics. (The Latin title reads, Minor Mathematical Works.)

Part I. Statics

[1] T.N. Force and moment equilibrium derive from the investigation of the equilibrium of the lever. Hence, it would have been more precise for Lagrange to list as the first principle, force and moment equilibrium.

[2] T.N. The Latin title reads, *On the Equilibrium of Planes*. This treatise deals with the foundations of statics: namely, the law of the lever and the determination of centers of gravity.

[3] T.N. The **Method of Exhaustion** is an approach to integration which is wholly geometrical and does not utilize the concept of limit. Archimedes summed a series of inscribed triangles to produce the integration but the approach always left a remainder. The remainder could be made as small as necessary by increasing the number of inscribed triangles.

[4] T.N. The Latin reads, *Demonstration of the Equilibrium of the Balance*. This work is accessible in the *Oeuvres Complètes de C. Huygens*, Vol. 19, Martinus Nijhoff, 1937.

[5] T.N. The French title reads, A Collection of Past Memoirs of the Academy of Sciences. The Académie des Sciences was founded in 1666 by Colbert under the auspices of Louis XIV. But prior to the reorganization of the academy in 1699 its memoirs were not published regularly.

[6] T.N. Archimedes assumes implicitly in his demonstration of the lever that moment is defined as force times perpendicular distance. Huygens attempts to demonstrate the principle of the lever without recourse to this definition of moment. However, his approach implicitly assumes the definition.

[7] T.N. A thorough discussion of this demonstration can be found in Ernst Mach, *The Science of Mechanics*, The Open Court Publishing Co., 6th Ed., 1960, pp. 20-32.

[8] I believe that d'Alembert was the first to try to demonstrate this proposition, but the demonstration which he gave in the Mémoires de l'Académie des Sciences of 1769 is not entirely satisfactory. The one given afterward by Fourier in the Fifth Volume of the Journal de l'Ecole Polytechnique is rigorous and very ingenious, but it is not drawn from the essence of the lever. (Lagrange) (T.N. The more interesting of these two memoirs is the one by Fourier. It is titled "Mémoire sur la statique contenant la demonstration du principe des vitesses virtuelles et la théorie des moments" and can be found in the *Oeuvres de Fourier*, Vol. II, pp. 477-521.)

[9] T.N. This 16th century Italian mechanician is more commonly known to English-speaking peoples as Guidobaldo del Monte.

[10] T.N. The Latin title reads, The Book on Mechanics. An abridged English translation of this work is available. Cf. *Mechanics in Sixteenth-Century Italy*, The University of Wisconsin Press, Madison, 1969.

[11] T.N. Guido Ubaldo may not have had a clear understanding of the concept of moment. Cf. Pierre Duhem, *The Origins of Statics*, Chapter X, Boston Studies in the Philosophy of Science, Vol. 123, Kluwer Academic Publishers, 1991.

[12] T.N. Lagrange overlooks the solution to the inclined plane in the *Liber Jordani de ratione ponderis* from the School of Jordanus de Nemore in the 13th century. It is difficult to explain why this solution was not given more credence by Lagrange or for that matter his contemporaries. He most likely was aware of the solution since it was published twice in the 16th century. Cf. Pierre Duhem, *The Origins of Statics*, Chapter VII, Boston Studies in the Philosophy of Science, Vol. 123, Kluwer Academic Publishers, 1991.

[13] T.N. Stevin called this wreath of spheres a "clootrans."

[14] T.N. The Latin title reads, *Mathematical Memoirs*. Lagrange is referring to the Latin translation by Willebrord Snel of the original work in Flemish by Stevin entitled *Wisconstige Gedachtenissen*. Volume 4 of this work deals with statics.

[15] T.N. In 1634, the works of Stevin were translated into French by Albert Girard under the title *Les Oeuvres mathématiques di Simon Stevin de Bruges* and printed by the Elseviers. It is also Volume 4 of this work which deals with statics.

[16] T.N. The Italian title reads, On Mechanics. This work enjoyed a wide circulation in manuscript among Galileo's students before it was translated into French by Father Mersenne and published. The Italian text was published after Galileo's death. An English translation of this work is available. Cf. On Motion and On Mechanics, The University of Wisconsin Press, Madison, 1960.

[17] T.N. The basic principle behind Galileo's theory of motion is that the velocity acquired by a heavy body descending from the same height on differently inclined planes depends solely on the elevation of the plane. This principle is formulated in the *Dialogo di Galileo Galilei delle due massimi Sistemi del Mondo, il Ptolemaico e il Copernicano* but without demonstration. It is demonstrated in the *Discorsi e dimostrazioni matematiche*

intorno a due nuove scienze attenenti alla Meccanica ed a movimenti locali, di Galileo Galilei. The *Dialogo* was published for the first time in 1632 and the *Discorsi* in 1638. (The Italian titles read respectively, Dialogue of Galileo Galilei Concerning the Two Principal World Systems: the Ptolemaic and Copernican, and Mathematical Discourses and Demonstrations on the Two New Sciences concerning Mechanics and Local Motion by Galileo Galilei.) Actually, the demonstration was not included in the first edition of this latter work but in subsequent editions.

[18] T.N. The full title of this book is: *Harmonie universelle, contenant la théorie et la pratique de la Musique, où est traité de la nature des sons, et des mouvemens, des consonances, des genres, des modes, de la composition, de la voix, des chants et de toutes sortes d'instruments harmoniques*: par F. Marin Mersenne du l'ordre des Minimes. A Paris chez Sèbastien Cramoisy, Imprimeur ordinaire du Roy, ruë S. Jacques, aux Cicognes, MDCXXXVI. (The French title reads, Universal Harmony, containing the theory and practice of music and treating the nature of sound, motion, consonances, categories, modes, composition, voice, songs and of all sorts of instruments of harmony, by F. Marin Mersenne, of the Order of the Minims. Paris, Sebastian Cramoisy, official printer to the King, ruë St. Jacques, at the sign of the Storks, 1636.)

[19] T.N. Roberval's treatise is included in the book *Harmonie universelle* by F. Marin Mersenne and is entitled: *Traité de Méchanique; des poids soustenus par des puissances sur les plans inclinez à l'horizon; des puissances qui soutiennent un poids suspendu a deux chordes*; par G. Pers. de Roberval, Professeur Royal des Mathematiques au Collège de Maistre Gervais, et en la chaire de Ramus au Collège Royal de France. (The French title reads, Treatise on Mechanics; on weights supported by forces on inclined planes; forces which support a weight held by two ropes by G. Persone de Roberval, Royal Professor of Mathematics at the College de Maistre Gervais and holding the Chair of Ramus at the Collège Royal de France.)

[20] T.N. Lagrange is referring to the treatise by Bernard Lamy (1640-1715) entitled *Traitez de méchanique, de l'equilibre des solides et des liqueurs*, Paris, 1679. The name of the author of this work is usually spelled "Lamy". (The French title reads, Treatise on mechanics and the equilibrium of solids and liquids.)

[21] T.N. The complete title of the work by Varignon is *Projet d'une nouvelle méchanique*, Paris, 1687.

[22] T.N. The Greek title reads, *Mechanical Problems.* This work is now commonly attributed to either Theophrastus (372?-287? B.C.) or Strato (?-271 B.C.) both Peripateticians. In this work it is recognized that if two orthogonal components of the velocity of a body in motion are in a constant ratio, the body traverses the diagonal of a parallelogram whose sides are the component velocities. If there is no fixed ratio, the trajectory is a curve.

[23] T.N. Lagrange is referring to the *Discorsi e dimostrazioni matematiche intorno a due nuove scienze attenenti alla Meccanica ed a movimenti locali.*

[24] T.N. The complete title of this treatise is *Della Scienza Meccanica e della utilitè che si traggano dagl'instrumenti di quella.* (The Italian reads, On the Science of Mechanics and on the utility to be derived from its instruments). This work contains in a more developed form everything which was published by Mersenne in the *Les Méchaniques de Galilée.*

[25] T.N. On the contrary, Duhem asserts that all mechanicians familiar with the Composition of Velocities from Aristotle's *Mechanical Problems* knew the Composition of Forces since Aristotle held that force is proportional to velocity. According to Duhem, mechanicians tried in the 16th and 17th century to avoid basing a principle of statics on Aristotelian dynamics which they knew was incorrect. Consequently, they sought to base statics on autonomous principles such as the Principles of the Lever and the Inclined Plane. Cf. Pierre Duhem, *The Origins of Statics*, Chapter XVII, The Parallelogram of Forces of Dynamics, Boston Studies in the Philosophy of Science, Vol. 123, Kluwer Academic Publishers, 1991.

[26] T.N. By combining the applied force at each end of the lever with the added horizontal force, it can be shown that the resultants at each end of the lever intersect at a point on a vertical line passing through the point of support. It follows that the moment is zero about the point of support.

[27] T.N. Lagrange is referring to the areas of the triangles constructed according to this rule.

[28] T.N. Using modern terminology, this law can be stated as follows: The moment of the resultant of two concurrent forces with respect to a point in their plane is equal to the algebraic sum of the moments of the components with respect to the same point.

[29] T.N. The "arbitrary point" is the moment center. If this point lies between the component forces the resultant moment will be found from the algebraic difference of the component moments. On the other hand, if this point lies outside of the component forces, the moments are added algebraically.

[30] T.N. Although Varignon claims to use the Composition of Velocities, Lagrange asserts that he is using the Composition of Forces. Hence, Lagrange claims that the work "lacks in accuracy."

[31] T.N. The French reads, A New and General Demonstration on the use of a Block and Tackle.

[32] T.N. Lagrange is referring to the following work by Ignace Gaston Pardies, *La Statique ou la science des forces mouvantes*, 1673. (The French reads, Statics or the science of moving forces.)

[33] T.N. Lagrange is referring to the following work by Claude François Milliet Dechales, *Cursus seu Mundus mathematicus*, 1674. (The Latin reads, The Course or the Mathematical Universe.)

[34] T.N. The French title reads, A New Method of Demonstrating the Principal Theorems of the Elements of Mechanics.

[35] T.N. The French title reads, A History of Scholarly Works. The author is Basnage who stated: "It appears that Father Lamy owes his discovery of the new principles of mechanics to Varignon."

[36] T.N. Although Duhem claims that mechanicians understood that forces and velocities could be combined by the parallelogram law, there is no early record of any effort to compose forces in the same manner as velocities. Lagrange has offered a plausible explanation for this fact. Cf. Endnote #25 above.

[37] T.N. The complete title of one series of memoirs from the Russian Academy of Sciences is: Novi commentarii academiae scientiarum imperialis Petropolitanae, Petropoli.

[38] T.N. Many investigators during the 18th century tried to distinguish between principles which were based on empirical results and those which had a rational basis. One of the principles which attracted attention is the Principle of the Parallelogram of Forces. This principle has an empirical origin which is accepted and carried on by Newton and Varignon. However, investigators such as Bernoulli tried to establish it rationally from geometric considerations but without much success. Cf. René Dugas, *A History of Mechanics*, Éditions Du Griffon, Neuchatel, Switzerland, 1957, pp. 233-234.

[39] T.N. This is the abbreviation for the title of the Memoirs of the Academy of Sciences of Turin. The full title is Miscellanea Philosophica Mathematica Societatis privatae Taurinensia, given in volume 1. With the second volume, the title was changed to Mélanges de Philosophie et de Mathématique de la Societé Royale de Turin. Beginning with the sixth volume the title became Mémoires de l'Académie Royal des Sciences.

[40] T.N. Credit for the formulation of the Principle of Virtual Velocities is generally given to Galileo although the truth and general applicability of the Principle had been recognized long before. The root of the principle can be traced to Aristotle's *Mechanical Problems*. Moreover, there were numerous investigators who preceded Galileo and contributed to its development, notably, the 13th century mathematician Jordanus de Nemore who made singular contributions. For 16th century mechanicians, the Aristotelian tradition lacked the rigor which they found in the Archimedean approach to mechanics and consequently, it was discarded and the latter accepted.

[41] T.N. Throughout this introduction, Lagrange is overly favorable to Guido Ubaldo. But his esteem is typical of 18th century mechanicians.

[42] T.N. Lagrange is referring to Father Mersenne's translation entitled *Les Méchaniques de Galilée* where the theory of the inclined plane is presented and to the demonstration supporting the theory inserted in the *Discorsi e dimostrazioni matematiche intorno a due nuove scienze attenenti alla Meccanica, ed ai movimenti locali.*

[43] T.N. Galileo used the word **momento** for moment. He understood this term in three ways:

1. To mean statical moment, i.e. the product of weight and perpendicular distance to an axis.
2. As the product of weight and velocity in the sense expressed by Aristotle in the *Mechanical Problems*.
3. As the measure of a body's positional gravity on an inclined plane.

[44] T.N. At one point, Wallis took a position on the definition of static equilibrium which was a compromise between the viewpoints of Galileo and Descartes. But, after some vacillation, he adopted the Cartesian viewpoint.

Cf. Pierre Duhem, *The Origins of Statics*, Chapter XVII, The Systematization of the Laws of Statics, pp. 395-399, Boston Studies in the Philosophy of Science, Vol. 123, Kluwer Academic Publishers, 1991.

[45] T.N. Lagrange views **moment** dynamically as force times initial velocity whereas it is commonly viewed statically as force times perpendicular distance to a point. The latter definition is not a special case of the former. Lagrange's view is similar to Galileo's and hence, it can be traced back to Aristotelian dynamics.

[46] T.N. Descartes used the word **force** where modern terminology would require **work** and Lagrange retains Descartes' terminology.

[47] T.N. *Oeuvres de Descartes*, published by Charles Adam and Paul Tannery: Correspondence, Vol. 11, 13 juillet 1638, p. 222. In this letter, Descartes correctly demonstrates that it is work, i.e. force times collinear distance, which must be considered to determine static equilibrium.

[48] T.N. Lagrange is referring to a short treatise entitled: *Explication des engins par l'ayde desquels on peut, avec une petite force, lever un fardeau fort pesant.* (The French reads, An explication of machines by means of which a very heavy load can be lifted with a small force) Cf. *Oeuvres de Descartes*, published by Victor Cousin, Vol. 5, 1824, pp. 431-442.

[49] T.N. The Latin reads, On the natural descent of the center of gravity. This work is part of a collection entitled *Opera geometrica Evangelistae Torricellii.*

[50] T.N. This letter appears in the *Nouvelle mécanique ou statique*, Paris, 1725. Lagrange is generally given credit for discovering and bringing this letter to the attention of mechanicians. Cf. The article on Varignon in the *Dictionary of Scientific Biography*, Charles Scribner's and Sons, New York, 1973.

[51] T.N. The complete French title is, Loi du Repos des Corps. (The French title reads, the Law of Rest of Bodies) This memoir was read to the Académie des Sciences in Paris on February 20, 1740. It is readily accessible in the *Oeuvres de Maupertuis*, Volume IV, pp. 45-63. In the **Addition** added later by Maupertuis at the end of this memoir, he states the Law of Rest in the following terms:

> Let a system of heavy bodies be pulled towards fixed centers by forces which are arbitrary functions of their distances from the centers. In order that all of the bodies of the system remain at rest, the sum of the products of each mass with the intensity of the applied force and with the integral of each distance function with respect to the differential element of the distance (which could be called the sum of the forces of rest) is a minimum.

The Loi du Repos des Corps is a law of statics and derives from the Principle of Virtual Work. In modern terms, it is identical to the Principle of Minimum Potential.

[52] T.N. Gaspard Courtivron, "Recherches de statique et de dynamique où l'on donne un nouveau principle général pour la considération des corps animés par des forces variables, suivant une lois quelconque." (The French reads, Research in statics and dynamics where a new general principle is presented for the consideration of bodies acted on by variable forces following an arbitrary law.)

[53] T.N. In a much later memoir entitled "Sur le Principe des Vitesses Virtuelle" (*Oeuvres de Lagrange*, Vol. 7, pp. 317-321) published in 1798, Lagrange added to the discussion presented here. However, the basic claim remains unchanged: to wit, the Principle of Pulleys provides a more obvious and direct approach to the demonstration of the Principle of Virtual Velocities than the Principle of the Composition of Forces.

[54] T.N. The concept of a "mass point" or "material point" assumes that all of a body's mass is concentrated at a single point such as the center of gravity for which only the motion of translation requires consideration. Euler is generally given credit for this concept.

[55] T.N. The Principle of Virtual Velocities has its origins in Aristotle's *Mechanical Problems*. It derives from a discredited Aristotelian dynamics which proposed that force is proportional to velocity. Lagrange retains the older terminology of this principle but will apply the principle as reformulated by Jordanus de Nemore and Descartes in the form known today as the Principle of Virtual Work.

[56] T.N. Cf. Endnote 43.

57 T.N. Lagrange is referring to a rigid body for which the number of unknown forces equals the number of equations. One independent equation is obtained for every independent coordinate. If there are more forces than equations, an exact solution based on statics alone is impossible.

58 T.N. It was unusual in Lagrange's day to use rectangular coordinates.

59 T.N. Lagrange is referring to Polar and Cylindrical Coordinates, respectively.

60 T.N. This statement is true for statics and dynamics but it would not be true generally for stress calculations.

61 T.N. A technical note by Louis Poinsot (1777-1859) entitled "Sur un point fondamental de la Mécanique analitique de Lagrange": is included in the third edition of this work edited by Joseph Bertrand. In this note, Poinsot shows that these equations of Lagrange are only applicable to transformations between orthogonal coordinate systems. If the transformation is from orthogonal coordinates to a set of coordinates ξ, ψ, φ which are not orthogonal but oblique to one another, then the equation given by Lagrange does not hold. The reason is that the expression on the right hand side of the equation does not represent the virtual work of the second system of forces since the displacements are not independent. Poinsot re-defines the forces so that the differentials $d\xi$, $d\psi$, $d\varphi$ represent a displacement pattern which multiplied by the re-defined forces are equal to the virtual work.

62 T.N. The global coordinates are with respect to the fixed axes.

63 T.N. The complete French title is: Recherches sur la précession des équinoxes et sur la nutation de la terre. (The French reads, Research on the precession of the equinoxes and the nutation of the Earth.) This work was published in 1749. In this work, d'Alembert followed closely Clairaut's procedure but employed more terms in the series expansion for the equation of motion of the Earth. Consequently, he obtained a more accurate solution.

64 T.N. A comparison of Lagrange's analytical demonstration with the cited geometrical demonstration is another example, as Lagrange probably intended, of the power of the analytical approach.

65 T.N. Cf. Endnote 51.

66 T.N. In order to have an invariant point called the center of gravity, the lines of force of the gravitational field must be considered parallel. This is a pure postulate justified solely by having its deductions agree with reality. The center of gravity was first defined in this fashion by Torricelli.

67 T.N. The French reads: Conservation of Live Forces. This law led to the modern principle of the Conservation of Energy but in the 18th Century it was understood only in a limited mechanical context.

68 T.N. In the following discussion, Lagrange considers only stable and unstable equilibrium. The question of neutral or indifferent equilibrium is not addressed.

69 T.N. Cf. Endnote 52.

70 T.N. A technical note by Gustav Lejeune Dirichlet (1805-1859) entitled "Sur la stabilité de l'équilibre" is included in the third edition of this work edited by Joseph Bertrand. Dirichlet demonstrates in this memoir that the variation of the potential function produces small oscillations by virtue of the equation of the Conservation des Forces Vives and that these oscillations are bounded when the potential function has a minimum value. Once it is known that the resulting velocities are bounded, it can be concluded that the equilibrium state is stable.

71 T.N. Lagrange begins with a Taylor Series expansion for the function II. Since the coefficients $B, C, D, \ldots$ represent the first partial derivatives with respect to $x, y, z, \ldots$ respectively, they must individually be equal to zero for the function II to be a minimum. This leads to an equation which is called today a Quadratic Form. From the Quadratic Form, Lagrange obtains the equation

$$x = f\xi^2 + g\eta^2 + h\zeta^2 + \cdots$$

This last equation is the general result of an analysis which can be found in modern terms in textbooks on linear algebra. Cf. Grossman, Stanley I., *Elementary Linear Algebra*, Saunders College Publishing, Fourth Edition, pp. 420-429.

72 T.N. The demonstration presented here is discussed further in two well-known treatises.

Mach, Ernst, *The Science of Mechanics*, The Open Court Publishing Company, Sixth Edition, pp. 77-84.

Routh, Edward J., *A Treatise on Analytical Statics*, Cambridge University Press, Second Edition, Vol. I, 1909, pp. 182-183.

73 T.N. In Lagrange's mechanics, the physical constraints between bodies are replaced by equations between the coordinates of the different bodies. This approach does away with the need to consider the physical nature of the forces between the bodies or by what means these forces are produced. By means of the undetermined coefficient, a fictitious force is introduced which maintains the constraints on the problem. Hence, this abstraction leads to a purely analytic procedure.

Although Lagrange is generally given credit for this discovery of the Method of Multipliers, Euler used the concept first in his research. But Euler unlike Lagrange did not appreciate the generality and broad applicability of the method. Cf. Goldstine, Herman H., *A History of the Calculus of Variations from the 17th through the 19th century*, Studies in the History of Mathematics and the Physical Sciences, Springer-Verlag, 1980, pp. 148-150.

74 T.N. These two symbols define two different mathematical operations and for this reason are generally called operators. The differential is denoted by the letter d and it is the rate of change of a function at a point while the symbol δ denotes the variation of a function which is obtained by varying the function by a small amount over its entire range. Thus the variation of a function is equivalent to a virtual displacement in mechanics. It is the recognition of this fact which helped Lagrange to develop a mechanics based on variations.

75 T.N. With the integral sign S, Lagrange takes into account the fact that a system may be composed of many bodies and that at given points, lines or surfaces, there may be discontinuities.

76 T.N. It is impossible to have even three equations because three equations will suffice to determine the location of each point uniquely without having to consider the forces applied to the system. (J. Bertrand, Third Edition)

77 T.N. Lagrange is referring to his work *Théorie des Fonctions Analytiques contenant les Principes du Calcul différential, dégagés de toute considération d'infiniment petits, d'evanouissans, de limites et de fluxions, et réduits a l'analyse algébrique des quantités finies*. (The French reads, The Theory of Analytical Functions containing the principles of differential calculus, free of all considerations of the infinitely small, of vanishing quantities, of limits and of fluxions, and reduced to the algebraic analysis of finite quantities.) Lagrange was the first to use this notation. It derives from his effort to avoid geometric arguments. His aim was to consider quantities as symbols and to operate on them algebraically.

78 T.N. The term "partial differentials" simply refers to the differentiation of a function of many variables. Lagrange has not used the usual notation for a partial derivative.

79 T.N. Lagrange prepared a series of lessons which were to serve as a commentary and supplement to Part I of the *Théorie des Fonctions*. Ultimately, twenty-two lessons were prepared and published in the XIV Cahier du Journal de l'Ecole Polytechnique (1808) as the *Leçons sur le Calcul des Fonctions* (The French reads, Lessons on the Calculus of Functions.). The last lesson dealt with the Calculus of Variations.

80 T.N. Cf. Endnote 78.

81 T.N. These are the equilibrium equations written with respect to the x-, y- and z-coordinate axes.

82 T.N. The forces are assumed to be proportional to their collinear directions. Consequently, the constant of proportionality is the same for all the forces. Hence, the constant does not have to be considered since it cancels out in the equations.

83 T.N. It is obvious that Lagrange means force. In his time, pressure and force were used interchangeably as evidenced by the writings of Lagrange's contemporaries notably Euler.

84 T.N. Lagrange writes fp for the function, where in modern notation it would be written $f(p)$.

85 T.N. In the Third Edition of this work, Bertrand gives this reference as *Mécanique céleste*, Vol. II, Book III, Chapter I. There exists an earlier work by Laplace which also deals with this subject. It is the Théorie des attractions des sphéroïdes et de la figure des planètes. (The French reads, Theory of the attractions of spheres and of the shape of the planets.) This memoir was read before the Académie des Sciences on Aug. 11, 1784 and it is readily available in the *Oeuvres complètes de Laplace*, Vol. X, pp. 209-291.

[86] T.N. Since df and dg are equal to zero, it is obvious that zero is being added to the equation resulting from the application of the Principle of Virtual Work.

[87] Each time a condition is placed on the problem, an additional equation of condition is also added. Hence the problem remains indeterminate. Cf. Endnote 86.

[88] T.N. Lagrange views the elastic restoring force as tending to restore the string to a horizontal and rectilinear line. This force would tend to increase the internal angle at the point of application of the second body and hence, decrease its complementary angle e external to the triangle.

[89] T.N. The quantity E has units of moment per angular measure. Since the quantity e has units of angular measure, the product of E and e will have units of work. It is clear that the virtual work of the moment E is given by $E\,de$. Euler had shown earlier how to calculate the work of a moment. Cf. Recherches sur les plus grandes et plus petits qui se trouvent dans les actions des forces, *Opera Omnia* II, 5, 1-37, (1748).

[90] T.N. The well-known theorem is the Law of Cosines.

[91] T.N. Contrary to Lagrange's statement, this problem can be easily solved by direct methods. It is clear that the resultant of the applied forces must be normal to the curved surface if the string is to be in equilibrium. This observation suffices to solve this problem.

[92] T.N. The investigations of James Bernoulli (1654-1705) and later, Euler (1707-1783) had demonstrated that the bending moment is proportional to curvature. Lagrange simply follows their lead on this point.

[93] T.N. Euler, Leonhard, "Découverte d'un nouveau principe de mécanique," *Opera omnia*, II, 5, 81-108, (1752).

Hydrostatics

[1] T.N. Lagrange is discussing the Archimedean treatise *De insidentibus acquae* or as it is known in English, On Floating Bodies. This work was published by Tartaglia in 1543 and later, corrected and a commentary added by Frederico Commandino and re-published in 1565.

[2] T.N. For this problem, Archimedes treats the Earth as if it were a spherical and fluid body. Furthermore, he conceives this body as composed of a number of solids each having the shape of a pyramid whose vertices lie at the center of the Earth and whose bases compose its surface. Since the pyramids are in equilibrium, they must all have the same weight and similarly positioned elements of the pyramid must suffer the same pressure. If a body is immersed in one of the pyramids, the center of the Earth will be subject to the developed buoyant force in addition to the weight of the pyramid.

[3] T.N. The Latin reads, *On those things which float on water.*

[4] T.N. The shape of a container filled with fluid has no effect on the pressure at its base. The pressure is determined by multiplying the specific weight of the fluid by its depth. Before the principles of hydrostatics were understood, this seemed to be a puzzling phenomenon and hence, it was called the "hydrostatic paradox."

[5] T.N. The Latin reads, *Mathematical Memoirs*. It was first translated from Flemish into Latin by Willebrord Snel or Snellius in 1605 and into French by Albert Girard in 1634.

[6] T.N. The Italian reads, Discourse on all bodies which float on water or which move in it. This work was published in Florence in 1612.

[7] T.N. Lagrange is referring to the work by Blaise Pascal (1623-1662) entitled *Traités de l'équilibre des liqueurs et de la pesanteur de la masse de l'air. Contenant l'explication des causes de divers effets de la nature qui n'avaient point été bien connus jusques ici et particulièrement de ceux que l'on avait attribués a l'horreur du vide*, Paris, 1663. (The French title reads, Treatises on the equilibrium of fluids and on the weight of air. Containing an explanation of the causes of diverse effects in nature which heretofore were not well understood and in particular, the effect which has been attributed to the horror of a vacuum.)

[8] T.N. Astronomical and physical data led Huygens and Newton to assume that the Earth is an oblate ellipsoid. Newton's approach to estimating the degree of oblateness was to consider the equilibrium of two orthogonal

columns of fluid, one from the Earth's center to a pole and the other from the Earth's center to the equator. Assuming that the same pressure is exerted by each column at the Earth's center, the presence of centrifugal force on the column extending to the equator would require a greater height. The ratio of the two heights of the columns is a measure of oblateness. Cf. *The Mathematical Principles of Natural Philosophy*, The Science Classics Library, Citadel Press, 1964. Book III, Proposition XIX, Problem III.

[9] T.N. Lagrange is referring to a work by Pierre Bouguer entitled, "Comparaison des deux lois que la Terre et les autres Planètes doivent observer dans la figure que la Pesanteur leur fait prendre", Mémoires de l'Académie des Sciences, 1734. (The French reads, Comparison of the two laws which the shape of the Earth and the other planets must follow due to the effect of gravity.)

[10] T.N. Lagrange is referring to a book by Alexis-Claude Clairaut entitled *Théorie de la figure de la Terre tirée des principes de l'hydrodynamique*, Durand, Paris, 1743, 1808. (The French title reads, Theory of the shape of the Earth deduced from the principles of hydrodynamics.)

[11] T.N. Lagrange is referring to a memoir by Leonhard Euler entitled, "Principes généraux du mouvement des fluides," Histoire de l'Académie de Berlin, 1755.

[12] T.N. In this derivation, Lagrange has assumed that the density of the fluid is equal to unity and consequently, he has omitted it from the derived equations. He mentions this fact later on page 143 of this work.

[13] T.N. In 18th century mechanics, the distinction between pressure and force is not observed punctiliously. Consequently, Lagrange is not consistent in his use of these two terms. Moreover, if the cross sectional area is very small, Lagrange will use force and pressure interchangeably.

[14] T.N. There is another way to view the equations which follow. They express the first variation of the quantities p, q, r, etc. For example, the first variation of the quantity p is by definition $\delta p = (\mathrm{d}p/\mathrm{d}s)\delta s$.

[15] T.N. These three expressions represent the components of what is called today the "curl" of a vector field, i.e. $\nabla \times F$ where $F(p, q, r)$. They express the fact that the function F represents a conservative field.

[16] T.N. If the phenomenon is conservative, that is, there is no energy dissipation, this equation will always be exact and integrable. All the results obtained by Lagrange in this section require the phenomenon to be conservative.

[17] T.N. The French reads, surfaces of equal potential.

[18] T.N. Lagrange is referring to Colin Maclaurin (1698-1746) a Scottish mathematician and physicist. In his *Treatise of Fluxions* (1742), he showed that a homogeneous fluid mass rotating uniformly about an axis under the action of gravity has the shape of an ellipsoid of revolution.

[19] T.N. The background to the theory of surfaces is discussed in nearly all textbooks on advanced calculus. Cf. John M. H. Olmsted, *Advanced Calculus*, The Appleton-Century Mathematics Series, Appleton-Century-Crofts, Inc. 1961, pp. 203-237.

[20] T.N. Lagrange is referring to the scientist Jean André Deluc (1727-1817).

[21] The toise is a unit of measure corresponding to the fathom or six feet.

Part II. Dynamics: Volume I

[1] T.N. The Italian title reads: Discourses and Mathematical Demonstrations Concerning Two New Sciences.

[2] T.N. The goal was to develop the governing differential equations for the problem. Once this equation had been found, its simple integration provided the solution to the problem.

[3] T.N. Lagrange is referring to the Dialogues Concerning the Two Principal World Systems. Galileo displays a vague understanding of centrifugal force in this work.

[4] T.N. The first complete account of centrifugal force was given by Christian Huygens (1629-1695). He used what is called today the Principle of Equivalence and Galileo's law of free fall for a body near the Earth's surface.

[5] T.N. A German translation of this work is available in Ostwald's Klassiker der Exakten Wissenschaften, No. 192, Leipzig, 1913. An English translation of this work also exists: CHRISTIAAN HUYGENS' THE PENDULUM CLOCK OR GEOMETRICAL DEMONSTRATIONS CONCERNING THE MOTION OF PENDULA AS APPLIED TO CLOCKS, translated by Richard J. Blackwell, Iowa State University Press, 1986.

[6] T.N. Lagrange is referring to using an orthogonal coordinate system which moves on the curve and for which one of the directions is tangent to the curve.

[7] T.N. It is generally thought that the first text on mechanics in a modern analytical format is by Jacob Hermann (1678-1733), *Phoronomia, sive de viribus et motibus corporum solidorum et fluidorum libri duo*, Amsterdam, 1716. (The Latin title reads, Phóronomy, or Two Books on the Forces and Motions of Solid and Fluid Bodies. (The translators wish to thank Professor Grant Leneaux of the University of Nevada for the translation of this Latin title).

[8] T.N. This result appears to be an immediate result of Newton's Laws. However, it was not used in this fashion by Newton and it cannot be found in this book by Maclaurin. It seems that it was first used by Euler. The following memoirs are examples

> "Découverte d'un nouveau principe de mécanique," *Opera omnia*, II, 5, 81-108, (1752).
>
> "Recherches sur le mouvement des corps céleste en général," *Opera omnia*, II, 25, 1-44, (1747-9).

[9] T.N. The complete title of this book by John Wallis (1616-1703) is *Mechanica, sive de Motu, Tractus Geometricus* published at London in 1671. The third part of this treatise contains a description of the phenomena of impact. In particular, Wallis' law of inelastic impact is derived there.

[10] T.N. Although the conceptualization of the two centers is quite different, the center of oscillation and the center of agitation are identical points in a body. The term center of agitation has been replaced by center of percussion.

[11] T.N. Descartes had incorrectly developed a procedure to locate the center of agitation. When Roberval pointed out the error, Descartes disputed his claim. But Descartes had erred in his procedure. He projected each point of the body on to a vertical line through the axis of rotation and the center of gravity but failed to attribute only the horizontal component of the velocity to this projected point. A concise discussion of this dispute is given in: René Dugas, *A History of Mechanics*, Editions Du Griffon, Neuchatel, Switzerland, pp. 163-165.

[12] T.N. Cf. Endnote 7.

[13] What makes these solutions still more complex is that the author wants to avoid making dt or the time increments constant as he cautions (Article 94). (Lagrange)

[14] T.N. Cf. Isaac Newton, *The Mathematical Principles of Natural Philosophy*, Axioms or Laws of Motion, Corollary IV. Lagrange gives Newton credit for the discovery of this fundamental principle but Huygens, Wallis and Wren had used it in their investigations on the impact of bodies carried out at the invitation of the Royal Society in 1668.

[15] T.N. The principle to which Lagrange refers is called today the Principle of Moment of Momentum. Briefly stated, it says that the rate of change of the angular momentum of a system about any fixed line is equal to the moment of the external forces about this line. For more information about this principle refer to: C. Truesdell, "Essays in the History of Mechanics," Springer-Verlag, New York, Inc. 1968.

[16] T.N. Lagrange considered theological and metaphysical speculations as foreign to science. Hence, his sarcasm, while directed here at d'Arcy, applies generally to those who would extrapolate from simple mechanical results to metaphysical principles.

[17] T.N. Lagrange is referring to Leonhard Euler's great treatise on isoperimetrical problems entitled: *Methodus Inveniendi Lineas Curvas Maximi Minimive Proprietate Gaudentes sive Solutio Problematis Isoperimetrici Latissimo Sensu Accepti*, Lausanne and Geneva, 1744. (The Latin title reads, A Method of Finding Plane Curves that Show Some Property of Maxima or Minima.)

[18] T.N. The choice of rectangular coordinates seems obvious today but it was far from obvious in Lagrange's time.

[19] T.N. The general formula of dynamics is simply the application of the Principle of Virtual Work to d'Alembert's Principle. Lagrange used this formulation for the first time in his memoir of 1764 entitled: Théorie de la Libration de la Lune, *Oeuvres de Lagrange*, Vol. 5, pp. 5–61.

[20] T.N. If v_1 and v_2 are the velocities of two spherical masses before direct central impact and v_1' and v_2' the velocities after impact then

$$v_1' - v_2' = -e(v_1 - v_2)$$

where e is the coefficient of restitution. If e is equal to one, the impact is elastic and thus there is no energy dissipation. If e is close or equal to zero, the impact is inelastic or as it was denoted in the 18th century a "hard" impact. The salient consideration in the 18th century is that the Principle of Vis Viva holds in the former case but not for the latter.

[21] T.N. Lagrange is referring to the memoir Application de la Méthode Exposée dans le Mémoire Précédent a la Solution de Différents Problèmes de Dynamique, (Misc. Taur., Vol. II, 1760-1761) *Oeuvres de Lagrange*, Vol. I, pp. 365-468.

[22] T.N. This integral is often called the action integral and it is usually written as

$$\delta \int 2T \, \mathrm{d}t = 0$$

[23] T.N. In a letter to Euler dated August 12, 1755, Lagrange described his formulation of the variational calculus which he called the Method of Variations. His method is the simple algebraic approach used today without the need of geometrical insight which earlier approaches required. Later, Euler changed the terminology to the Calculus of Variations. Cf. Elementa Calculi Variationum, *Opera Omnia*, (1), 25, p. 145.

[24] T.N. Lagrange did not vary time in this development because this operation does not simplify or further generalize the resulting differential equations. The nature of a variation is such that the operation requires that the variable be made a function of a parameter. Lagrange had demonstrated this point earlier. For further discussion of this topic, the reader should refer to Jourdain, Philip E. B., *The Principle of Least Action*, The Open Court Publishing Co., Chicago, 1913.

[25] T.N. This result is generally referred to as Euler's Theorem on Homogeneous Functions.

[26] T.N. The method discussed here by Lagrange is known today as the Method of the Variation of Parameters. It is still widely used to solve linear second order non-homogeneous equations.

Euler used this method first and applied it to treat perturbations in the orbits of the planets Jupiter and Saturn. Later, Lagrange developed the method more fully in two papers:

> Nouvelle Mémoire de l'Académie de Berlin 5, 1774, 201 ff. and 6, 1775, 190 ff. (These memoirs can also be found in: *Oeuvres de Lagrange*, Vol. 4, pp. 5-108 and 151-251.)

and more generally in

> Mémoire de l'Académie des Sciences, 1808, 267 ff. (This memoir can also be found in: *Oeuvres de Lagrange*, Vol. 6, pp. 713-768.)

[27] T.N. Astronomers classify the perturbations of planetary orbits as either long term or short term disturbances. The former are called <u>secular</u> and the latter are called <u>periodic</u>. A secular perturbation is due to the relative orientation of the planetary orbits in space. The period of these disturbances is greater than 50,000 years. The periodic perturbations arise from the relative positions of the planets in their orbits. The period of these disturbances is much smaller than those called secular.

[28] T.N. During the 18th century, mathematicians attempted to analyze music mathematically. Euler had even produced formulas which were to be used to write music. This research led to an understanding of the mechanics of vibration.

[29] T.N. Lagrange is referring to Ernst Florenz Friedrich Chladni (1756–1827) who devoted his research to the study of acoustics and the nature of vibration. His research in these areas was due to his interest in music. He deduced the velocity of sound in solids from the pitch of the sound that a long rod produces when made to vibrate longitudinally. His early experiments were described in the treatise *Entdeckungen über die Theorie des Klanges*, Leipzig, 1787 and his later discoveries were presented in *Die Akustik*, Leipzig, 1802.

[30] T.N. Lagrange is referring to dividing a string into an equal number of segments without any portion remaining.

[31] T.N. Joseph Sauveur (1653-1716) is sometimes called the "father of acoustics." He pursued research on the phenomenon of beats, the velocity of sound, sympathetic vibration, harmonics and as Lagrange points out, on the division of vibrating strings by nodes.

[32] T.N. The reference is to John Wallis (1616-1703). The text is his *Treatise of Algebra, Both Historical and Practical*, London, 1685.

[33] T.N. Jean-Philippe Rameau (1683-1764) was France's greatest composer in the 18th century. He tried to explain the consonance and dissonance of music mathematically. His basic principle is considered in the concept of a corps sonore. We have translated this term as a sonorous body.

[34] T.N. Lagrange knew that various functions could be represented by the superposition of sine and cosine functions. However, it is not until 1815 that J. B. Fourier introduces a general method for representing an arbitrary function by the superposition of trigonometric functions. Although Fourier first presented his results in 1807 and again in 1810, Lagrange held resolutely to the view that this operation could not be carried out generally.

Part II. Dynamics: Volume II

[1] T.N. These relations represent a spherical coordinate system except that the angle ψ is generally taken between the radius vector r and the z-axis.

[2] T.N. Lagrange uses the phrase "rectangular spherical triangle" to describe the triangle resulting from his choice of coordinates to define the position of the body in space. This phrase has been shortened to "triangle" since the single term is a sufficient description.

[3] If the formulas give a negative value for $2a$, the equation $b = a(1 - e^2)$ indicates that e is greater than unity because b is positive and equal to D^2/g. For this reason the trajectory becomes a hyperbola. (Bertrand)

[4] T.N. The system of Natural or Naperian logarithms have the irrational number 2.7182... for base. This system of logarithms is sometimes called the hyperbolic system of logarithms.

[5] The reader should refer to the Mémoires de Berlin for the years 1768-69; the *Théorie des fonctions*, Chapter XVI, PART I, and the *Traité de Résolution des équations*, note 11. (Lagrange)

[6] Cf. Endnote #4.

[7] There are several applications of this method in the Mémoires de l'Académie de Berlin for the year 1776. (Lagrange)

[8] T.N. Aries is a constellation in the northern hemisphere.

[9] T.N. Cf. Endnote #4.

[10] T.N. Henrich Wilhelm Olbers (1758-1840) discovered the asteroid Pallas on March 28, 1802. The orbit of Pallas about the Sun is between Mars and Jupiter.

[11] T.N. Cf. Endnote #8.

[12] T.N. Lagrange is referring to the properties of parabolic motion which he developed earlier.

[13] T.N. In modern notation, the equation would read: $(K/\rho^3 - R)/(K/\rho^3) = \rho^3/r^3$.

[14] T.N. Newton made the first attempt to calculate the orbit of a comet. His solution was not entirely satisfactory. A short time later, Edmund Halley calculated the closed elliptical orbit for the comet of 1680. The comet of 1744 induced Euler to attempt to calculate its orbit. In his memoir of 1744 entitled *Theoria motus Planetarum et Cometarum*, he presented an analytical solution to the problem of a comet's trajectory. However, his method contained several inaccuracies and besides, it required a fourth observation in order to obtain the solution. With regard to the solution offered by Johann Heinrich Lambert (1728-1777), Bertrand notes that Lagrange is referring to Lambert's second solution entitled *Insigniores orbitae cometarum properietates* of 1761.

[15] T.N. The perpendicular is the altitude which corresponds to the first face.

[16] T.N. Lagrange is referring to the circumference of the Earth when he says that a degree of circumferential arc subtends 25 leagues. This would mean that Lagrange assumes that a league is approximately equal to 2.65 English miles.

[17] T.N. Recent investigations into the creation of our solar system support a theory which claims that our solar system was created from the debris of an explosion of a large star. The debris was in the form of a large, rotating disk-like cloud. The material at its center congealed to become our Sun and other portions congealed to form the planets and comets which compose our solar system.

[18] Lagrange is referring to the formula which expresses the velocities and the directions of the motion. (Bertrand)

[19] The use of the function Ω in this case, where it does not exist, makes the development somewhat obscure but some reflection on the matter will force the obscurity to disappear. (Bertrand)

[20] The reader should refer to the First Class of the Institut for 1808. (Lagrange) T.N. Lagrange is referring to the memoir:

> Mémoire sur la théorie des variations des élémens des planètes et en particulier des variations des grands axes de leurs orbites. *Oeuvres de Lagrange*, Vol. VI, pp. 713-768.

[21] In this problem, there is only one variable independent of time. Consequently, all of the differential expressions can be integrated which makes Lagrange's remark appear useless. I would add that this remark may later confuse the reader who will see in Article 73 the letter χ adopted as one of the variables of the problem. (Bertrand)

[22] This subsequent application of the formulas of Article 64 necessitates some explanation because these formulas were assumed to be functions of the variables of the elliptical motion and the variable χ is not one of them. In fact, the variable χ is not even precisely defined since it is defined only by its differential. The reader should refer to the well-founded remarks of Binet, Journal de l'Ecole Polytechnique, Vol. XVII, p. 76 (28[e] cahier). (Bertrand)

[23] T.N. Astronomers classify the perturbations of planetary orbits as either long-term or short-term disturbances. The former are called secular and the latter are called periodic. A secular perturbation is due to the relative orientation of the planetary orbits in space. The period of these disturbances is greater than 50,000 years.

[24] T.N. Lagrange assumes that space is pervaded by an undetectable fluid. The retention of this medium was due to the fact that gravitational action at a distance without an intervening medium has always proved difficult to accept. The solution to this problem was the invention of an ether which served for the propagation of gravitational forces and light. The Michelson-Morley Experiment of 1881 could not detect any deviation in the velocity of light in the ether. Hence, the belief in an ether has been abandoned but during the 18th century it was believed to exist.

[25] The reader should refer to volume four of the Mémoires de Turin. (Lagrange)

[26] T.N. In a thesis presented to the Faculty of Sciences at Paris, Serret noted that this demonstration is not very rigorous. Indeed, further development is necessary in order to make it entirely satisfactory. (Bertrand includes in volume two of the third edition of this work a note by Serret which clarifies this claim.)

[27] T.N. Ossian Bonnet gave a very simple explanation for this fact in the Journal de Liouville, Vol. IX, p. 105. This explanation can easily be generalized. (Bertrand includes in volume two of the third edition of this work a note by Bonnet which gives this explanation.)

[28] Euler was first to present this theorem for the case of parabolic motion. Lagrange's demonstration is very indirect. Another demonstration by Lagrange is given in the Mémoires de l'Académie de Berlin for 1778. (Bertrand) T.N. Cf. Endnote 14.

[29] T.N. The introduction of this angle may not be necessary in view of the assumptions made earlier. Cf. M. Binet, Journal de l'Ecole Polytechnique, Vol. XVII, p. 76. (28e cahier)

[30] T.N. Laplace pointed out that Lagrange presented this theorem in an unsatisfactory form and in addition, that he had demonstrated the theorem much earlier. Cf. A memoir by Laplace in the Mémoires de l'Académie des Sciences de Paris of 1784, article 57 of the second book of the *Mécanique Celeste* and the second chapter of Book XV (volume V).

[31] T.N. Lagrange is referring to the memoir:

> Théorie des variations séculaires des élémens des Planetes.
> Oeuvres de Lagrange, Vol. V, pp. 125-207.

[32] T.N. Cf. Endnote 24.

[33] T.N. This equation appears to be in error. It is not clear how this approximation was made.

[34] T.N. This approximation is justified by the fact that the neglected terms are periodical and the secular variation is of such long period that they can be neglected.

[35] T.N. This notation is due to Lagrange and it was encountered earlier. It is usually written as $F'(p')$.

[36] T.N. Bertrand has reproduced in Note VII of the first volume of the third edition the demonstration by Poisson that the derivatives are independent of time.

[37] T.N. Bertrand objects to Lagrange's argument that the length of the trajectory of the body must be a maximum or a minimum. He points out that Lagrange omits the consideration of a point of inflection. A discussion of this objection is given in Note VI of the second volume of the third edition of this work.

[38] T.N. Bertrand objects to Lagrange's claims for the generality of these types of transformation. In volume one of this translation in the endnotes for the section on statics, namely, endnote #61, Bertrand's objection is discussed. Briefly stated, the transformation of coordinates must be orthogonal for Lagrange's claims to hold.

[39] This assertion is imprecise. The polynomial written in terms of $\cos\psi$ must never change sign. Consequently, the function $\cos\psi$ must always lie between the two roots. Thus no maximum or minimum is associated with one of the roots. (Bertrand)

[40] This formula is inexact. Bravais brought to my attention this oversight by Lagrange and gave me the correct calculation which I have reproduced at the end of this volume. (Bertrand) (T.N. Bertrand is referring to Note VII in volume two of the third edition of this work.)

[41] T.N. The translation of this sentence follows Bertrand's suggestion.

[42] It should be noted that Lagrange defines the quantities $\mathrm{d}P$, $\mathrm{d}Q$, $\mathrm{d}R$ without attempting to determine whether they are integrable. In reality, they are not integrable. This fact is essential in order to understand the substance of Article 15 which follows. (Bertrand)

[43] T.N. Lagrange has written "center of the system" where it should probably read "center of the body."

[44] The preceding equation defines the direction of the principal axes for surfaces of second order. The following derivation is the first direct demonstration to show that the roots of this equation have real values. (Bertrand)

[45] It is clear that the introduction of the quantities p'', p''', etc. is not necessary. It suffices to note that the last equation has for its first term, the sum of the products of three pairs of imaginary and conjugate expressions. (Bertrand)

[46] It is worthwhile to note that this problem was later resolved by Poisson without any reference to Lagrange. His solution is part of the sixteenth book of the **Journal de l'Ecole Polytechnique** which was published in 1815 and later, it was republished in the second edition of his *Mécanique*. (Bertrand)

[47] T.N. Cf. Endnote #4.

[48] T.N. The work cited by Lagrange probably refers to John Bernoulli's (1667-1748) *Hydraulica nunc primum detecta ac demonstrata directe ex fundamentis pure mechanicis*. It is in this work that the concept of internal hydraulic pressure appears for the first time.

[49] T.N. D'Alembert is generally passed over in favor of Euler as the seminal contributor to hydrodynamics. In the first edition, Lagrange made no reference to Euler at this point.

[50] T.N. The notation used by Lagrange for partial derivatives is the same as for the ordinary derivative. However, no agreed upon notation was in use in his time.

[51] The following development does not appear in the first edition of this work. (Bertrand)

[52] T.N. In view of the satisfaction of equation (L), the following relation is an exact differential.

[53] T.N. Since there are many solutions to this equation, which is known today as Laplace's Equation, the correct solution will be the one which meets the requirements of the preceding paragraph.

[54] T.N. Cf. Endnote #48.

[55] T.N. This result does not agree at all with modern theory.

[56] The assumption made here by Lagrange is inadmissible. The reader should refer to the note at the end of this volume. (Bertrand). (T.N. Bertrand is referring to Note VIII in volume two of the third edition of this work.)

[57] This equation and the more general equation in which the function φ is assumed to be a function of x, y, z has been integrated by Poisson. The reader should refer to the recent Mémoires de l'Académie des Sciences, volume III. (Bertrand)

[58] Laplace discussed the probable cause of this discrepancy between theoretical and experimental results. The reader should refer to volume five of the *Mécanique celeste*, Book XII, Chapter III. (Bertrand)

Boston Studies in the Philosophy of Science

147. L. Embree (ed.): *Metaarchaeology*. Reflections by Archaeologists and Philosophers. 1992 ISBN 0-7923-2023-9
148. S. French and H. Kamminga (eds.): *Correspondence, Invariance and Heuristics*. Essays in Honour of Heinz Post. 1993 ISBN 0-7923-2085-9
149. M. Bunzl: *The Context of Explanation*. 1993 ISBN 0-7923-2153-7
150. I.B. Cohen (ed.): *The Natural Sciences and the Social Sciences*. Some Critical and Historical Perspectives. 1994 ISBN 0-7923-2223-1
151. K. Gavroglu, Y. Christianidis and E. Nicolaidis (eds.): *Trends in the Historiography of Science*. 1994 ISBN 0-7923-2255-X
152. S. Poggi and M. Bossi (eds.): *Romanticism in Science*. Science in Europe, 1790–1840. 1994 ISBN 0-7923-2336-X
153. J. Faye and H.J. Folse (eds.): *Niels Bohr and Contemporary Philosophy*. 1994 ISBN 0-7923-2378-5
154. C.C. Gould and R.S. Cohen (eds.): *Artifacts, Representations, and Social Practice*. Essays for Marx W. Wartofsky. 1994 ISBN 0-7923-2481-1
155. R.E. Butts: *Historical Pragmatics*. Philosophical Essays. 1993 ISBN 0-7923-2498-6
156. R. Rashed: *The Development of Arabic Mathematics: Between Arithmetic and Algebra*. Translated from French by A.F.W. Armstrong. 1994 ISBN 0-7923-2565-6
157. I. Szumilewicz-Lachman (ed.): *Zygmunt Zawirski: His Life and Work*. With Selected Writings on Time, Logic and the Methodology of Science. Translations by Feliks Lachman. Ed. by R.S. Cohen, with the assistance of B. Bergo. 1994 ISBN 0-7923-2566-4
158. S.N. Haq: *Names, Natures and Things*. The Alchemist J̄abir ibn Ḥayyān and His *Kitāb al-Aḥjār* (Book of Stones). 1994 ISBN 0-7923-2587-7
159. P. Plaass: *Kant's Theory of Natural Science*. Translation, Analytic Introduction and Commentary by Alfred E. and Maria G. Miller. 1994 ISBN 0-7923-2750-0
160. J. Misiek (ed.): *The Problem of Rationality in Science and its Philosophy*. On Popper vs. Polanyi. The Polish Conferences 1988–89. 1995 ISBN 0-7923-2925-2
161. I.C. Jarvie and N. Laor (eds.): *Critical Rationalism, Metaphysics and Science*. Essays for Joseph Agassi, Volume I. 1995 ISBN 0-7923-2960-0
162. I.C. Jarvie and N. Laor (eds.): *Critical Rationalism, the Social Sciences and the Humanities*. Essays for Joseph Agassi, Volume II. 1995 ISBN 0-7923-2961-9
Set (161–162) ISBN 0-7923-2962-7
163. K. Gavroglu, J. Stachel and M.W. Wartofsky (eds.): *Physics, Philosophy, and the Scientific Community*. Essays in the Philosophy and History of the Natural Sciences and Mathematics. In Honor of Robert S. Cohen. 1995 ISBN 0-7923-2988-0
164. K. Gavroglu, J. Stachel and M.W. Wartofsky (eds.): *Science, Politics and Social Practice*. Essays on Marxism and Science, Philosophy of Culture and the Social Sciences. In Honor of Robert S. Cohen. 1995 ISBN 0-7923-2989-9
165. K. Gavroglu, J. Stachel and M.W. Wartofsky (eds.): *Science, Mind and Art*. Essays on Science and the Humanistic Understanding in Art, Epistemology, Religion and Ethics. Essays in Honor of Robert S. Cohen. 1995 ISBN 0-7923-2990-2
Set (163–165) ISBN 0-7923-2991-0
166. K.H. Wolff: *Transformation in the Writing*. A Case of Surrender-and-Catch. 1995 ISBN 0-7923-3178-8
167. A.J. Kox and D.M. Siegel (eds.): *No Truth Except in the Details*. Essays in Honor of Martin J. Klein. 1995 ISBN 0-7923-3195-8

Boston Studies in the Philosophy of Science

168. J. Blackmore: *Ludwig Boltzmann, His Later Life and Philosophy, 1900–1906*. Book One: A Documentary History. 1995 ISBN 0-7923-3231-8
169. R.S. Cohen, R. Hilpinen and R. Qiu (eds.): *Realism and Anti-Realism in the Philosophy of Science*. Beijing International Conference, 1992. 1996 ISBN 0-7923-3233-4
170. I. Kuçuradi and R.S. Cohen (eds.): *The Concept of Knowledge*. The Ankara Seminar. 1995 ISBN 0-7923-3241-5
171. M.A. Grodin (ed.): *Meta Medical Ethics*: The Philosophical Foundations of Bioethics. 1995 ISBN 0-7923-3344-6
172. S. Ramirez and R.S. Cohen (eds.): *Mexican Studies in the History and Philosophy of Science*. 1995 ISBN 0-7923-3462-0
173. C. Dilworth: *The Metaphysics of Science*. An Account of Modern Science in Terms of Principles, Laws and Theories. 1995 ISBN 0-7923-3693-3
174. J. Blackmore: *Ludwig Boltzmann, His Later Life and Philosophy, 1900–1906* Book Two: The Philosopher. 1995 ISBN 0-7923-3464-7
175. P. Damerow: *Abstraction and Representation*. Essays on the Cultural Evolution of Thinking. 1996 ISBN 0-7923-3816-2
176. G. Tarozzi (ed.): *Karl Popper, Philosopher of Science*. (in prep.)
177. M. Marion and R.S. Cohen (eds.): *Québec Studies in the Philosophy of Science*. Part I: Logic, Mathematics, Physics and History of Science. Essays in Honor of Hugues Leblanc. 1995 ISBN 0-7923-3559-7
178. M. Marion and R.S. Cohen (eds.): *Québec Studies in the Philosophy of Science*. Part II: Biology, Psychology, Cognitive Science and Economics. Essays in Honor of Hugues Leblanc. 1996 ISBN 0-7923-3560-0
    Set (177–178) ISBN 0-7923-3561-9
179. Fan Dainian and R.S. Cohen (eds.): *Chinese Studies in the History and Philosophy of Science and Technology*. 1996 ISBN 0-7923-3463-9
180. P. Forman and J.M. Sánchez-Ron (eds.): *National Military Establishments and the Advancement of Science and Technology*. Studies in 20th Century History. 1996 ISBN 0-7923-3541-4
181. E.J. Post: *Quantum Reprogramming*. Ensembles and Single Systems: A Two-Tier Approach to Quantum Mechanics. 1995 ISBN 0-7923-3565-1
182. A.I. Tauber (ed.): *The Elusive Synthesis: Aesthetics and Science*. 1996 ISBN 0-7923-3904-5
183. S. Sarkar (ed.): *The Philosophy and History of Molecular Biology: New Perspectives*. 1996 ISBN 0-7923-3947-9
184. J.T. Cushing, A. Fine and S. Goldstein (eds.): *Bohmian Mechanics and Quantum Theory: An Appraisal*. 1996 ISBN 0-7923-4028-0
185. K. Michalski: *Logic and Time*. An Essay on Husserl's Theory of Meaning. 1996 ISBN 0-7923-4082-5
186. G. Munévar (ed.): *Spanish Studies in the Philosophy of Science*. 1996 ISBN 0-7923-4147-3
187. G. Schubring (ed.): *Hermann Günther Graßmann (1809–1877): Visionary Mathematician, Scientist and Neohumanist Scholar*. Papers from a Sesquicentennial Conference. 1996 ISBN 0-7923-4261-5

GPSR Compliance
The European Union's (EU) General Product Safety Regulation (GPSR) is a set of rules that requires consumer products to be safe and our obligations to ensure this.

If you have any concerns about our products, you can contact us on

ProductSafety@springernature.com

In case Publisher is established outside the EU, the EU authorized representative is:

Springer Nature Customer Service Center GmbH
Europaplatz 3
69115 Heidelberg, Germany

www.ingramcontent.com/pod-product-compliance
Ingram Content Group UK Ltd.
Pitfield, Milton Keynes, MK11 3LW, UK
UKHW061655190726
13853UKWH00008B/2222
* 9 7 8 9 4 0 1 5 8 9 0 4 8 *